CHEMICAL TECHNICIANS' READY REFERENCE HANDBOOK

CHEMICAL TECHNICIANS' READY REFERENCE HANDBOOK

Jack T. Ballinger, B.S., M.S., Ed.D.
Professor (retired), St. Louis Community College, St. Louis, Missouri

Gershon J. Shugar, B.S., M.A., Ph.D.   (Deceased)
Professor, Essex County College, Newark, New Jersey

CONSULTING EDITORS

Barry A. Ballinger, B.S., M.S.
Certified Occupational Hygienist, Pfizer Inc., Canterbury, England

Clinton T. Ballinger, B.S., Ph.D.
Chief Executive Officer, Evident Technologies, Troy, New York

Fifth Edition

New York   Chicago   San Francisco   Lisbon   London   Madrid
Mexico City   Milan   New Delhi   San Juan   Seoul
Singapore   Sydney   Toronto

The **McGraw·Hill** Companies

Cataloging-in-Publication Data is on file with the Library of Congress

Copyright © 2011, 1996, 1990, 1981, 1973 by The McGraw-Hill Companies, Inc. All rights reserved. Printed in the United States of America. Except as permitted under the United States Copyright Act of 1976, no part of this publication may be reproduced or distributed in any form or by any means, or stored in a data base or retrieval system, without the prior written permission of the publisher.

1 2 3 4 5 6 7 8 9 0 DOC/DOC 1 7 6 5 4 3 2 1

ISBN 978-0-07-174592-5
MHID 0-07-174592-0

Sponsoring Editor
Larry S. Hager

Editing Supervisor
Stephen M. Smith

Production Supervisor
Richard C. Ruzycka

Acquisitions Coordinator
Michael Mulcahy

Project Manager
Ranjit Kaur, Cenveo Publisher Services

Copy Editor
Andy Saff

Proofreader
Manish Tiwari, Cenveo Publisher Services

Art Director, Cover
Jeff Weeks

Composition
Cenveo Publisher Services

Printed and bound by RR Donnelley.

McGraw-Hill books are available at special quantity discounts to use as premiums and sales promotions, or for use in corporate training programs. To contact a representative, please e-mail us at bulksales@mcgraw-hill.com.

This book is printed on acid-free paper.

Information contained in this work has been obtained by The McGraw-Hill Companies, Inc. ("McGraw-Hill") from sources believed to be reliable. However, neither McGraw-Hill nor its authors guarantee the accuracy or completeness of any information published herein, and neither McGraw-Hill nor its authors shall be responsible for any errors, omissions, or damages arising out of use of this information. This work is published with the understanding that McGraw-Hill and its authors are supplying information but are not attempting to render engineering or other professional services. If such services are required, the assistance of an appropriate professional should be sought.

ABOUT THE AUTHORS

Jack T. Ballinger, B.S., M.S., Ed.D., is an adjunct faculty member with the National Corn-to-Ethanol Research Center at Southern Illinois University, Edwardsville, Illinois. He taught chemistry and chemical technology at St. Louis Community College for 35 years as a full-time professor. Dr. Ballinger serves as the Education and Workforce Development Coordinator and presents both courses on ethanol process operations and training new process operators and laboratory technicians.

Gershon J. Shugar, B.S., M.A., Ph.D. (deceased), was a professor of engineering technologies at Essex County College, Newark, New Jersey. In 1947, he founded a chemical manufacturing business that became the largest exclusive pearlescent pigment manufacturing company in the United States. In 1968, Dr. Shugar was appointed assistant professor of chemistry at Rutgers University, where he taught until his appointment at Essex County College.

CONTENTS

Preface xxvii
Acknowledgments xxix

Chapter 1. Chemical Process Industry Workers and Government Regulations   1

The Role of the Chemical Process Industry Worker / 1
  A. Working in the Chemical Process Industries / 1
  B. Maintaining Safety, Health, and Environmental Standards in the Plant / 2
  C. Handling, Storing, and Transporting Chemical Materials / 3
  D. Operating, Monitoring, and Controlling Continuous and Batch Processes / 3
  E. Providing Routine Preventative Maintenance and Service to Processes, Equipment, and Instrumentation / 4
  F. Analyzing Plant Materials / 5
The Role of the Chemical Laboratory Technician / 6
  A. Working in the Chemical Process Industries / 6
  B. Maintaining a Safe and Clean Laboratory Adhering to Safety, Health, and Environmental Regulations / 7
  C. Sampling and Handling Chemical Materials / 7
  D. Measuring Physical Properties / 8
  E. Performing Chemical Analysis / 8
  F. Performing Instrumental Analysis / 9
  G. Planning, Designing, and Conducting Experiments / 9
  H. Synthesizing Compounds / 10
Government Regulations / 10
  A. Introduction / 10
  B. Occupational Safety and Health Administration / 11
      Chemical Hygiene Plan (CHP) / 12
  C. Department of Transportation / 13
  D. Environmental Protection Agency / 14
Governmental and Other Sources of Chemical Process Information / 14
References / 17

Chapter 2. Chemical Plant and Laboratory Safety   19

Introduction: Chemical Plant and Laboratory Safety / 19
Basic Safety Rules for Working With or Around Chemicals / 19
Toxic Chemicals / 21
  A. Introduction / 21
Chemical Exposure Hazards / 22
Guidelines to Minimize Toxic Exposure / 22
Risk Assessment / 23
Hierarchy of Control / 24
Personal Protective Equipment / 25
  A. Introduction / 25
  B. Eye Protection / 25
  C. Hearing Protection / 27

D. Hand Protection / 28
E. Protective Apparel / 28
F. Head Protection / 30
G. Respiratory Protection / 30
    Introduction / 30
    Dust Masks or Filtering Face Pieces / 31
    Cartridge Respirators / 31
    Respirator Fit Tests and Fit Checks / 33
    Supplied-Air Respirators / 34
    Self-Contained Breathing Apparatus / 34
    Care and Maintenance of Respirators / 35
Checklist for Wearing Personal Protective Equipment (PPE) / 36
Fire Safety / 37
  A. Introduction / 37
  B. Classifications of Fire / 37
    Class A Fires / 37
    Class B Fires / 37
    Class C Fires / 37
    Class D Fires / 37
  C. Types of Fire Extinguishers / 37
  D. How to Use a Fire Extinguisher / 38
Safety Equipment / 39
  A. Eyewash and Safety Showers / 39
  B. Fall Protection / 39
  C. Ladder Safety / 40
  D. Aerial Lift Safety / 40
  E. Lifting Safety / 42
Confined Spaces / 43
Lockout/Tagout / 44
References / 45

Chapter 3. Chemical Handling and Hazard Communication   47

Introduction / 47
Material Safety Data Sheet / 47
  A. MSDS on the Internet / 51
Container Labeling / 51
  A. U.S. Department of Transportation / 51
  B. National Fire Protection Association / 53
  C. American National Standards Institute / 53
  D. Hazardous Materials Information System / 53
  E. International Organization for Standardization / 54
  F. Globally Harmonized System / 60
  G. International Air Transportation Association / 60
Laboratory Information Management Systems / 60
Bulk Containers (United States Only) / 61
  A. Carboys / 61
  B. Safety Cans / 61
  C. Drums / 63
  D. Chemtotes / 63
  E. Rail Tank Cars / 64
  F. Highway Tankers / 64
Bonding and Grounding / 64
ATEX Directive (EU) / 65
Forklift Operations / 66
  A. General Forklift Safety Guidelines / 67
Crane Safety / 67
Transferring Liquids / 69
Controlling Spills / 69

Solid Chemical Spills / 70
Spill Pillows / 70
Acid Solution Spills / 70
Alkali Solution Spills / 71
Volatile and Flammable Solvent Spills / 71
Mercury Spills / 72
Hazardous Waste Disposal / 72
References / 73

Chapter 4. Handling Compressed Gases 75

Compressed Gases / 75
General Cylinder Handling Precautions / 75
Cylinder Markings / 76
Cylinder Sizes / 78
Common Compressed Gases and Their Properties / 79
   A. Inert Gases / 79
   B. Compressed Air / 80
   C. Chlorine Gas (Cl_2) / 80
   D. Acetylene (C_2H_2) Gas / 80
   E. Acid Gases / 81
   F. Alkaline Gases / 81
   G. Ammonia (NH_3) Gas / 81
   H. Hydrogen (H_2) Gas / 81
   I. Liquefied Petroleum (LP) Gases / 81
   J. Methyl Acetylene Propadiene (MAPP) / 82
   K. Oxygen (O_2) Gas / 82
Cryogenic Gases and Liquids / 82
   A. Introduction / 82
   B. Liquid Fluorine (F_2) / 83
   C. Liquid Nitrogen (N_2) / 83
   D. Liquid Oxygen (LOX) / 83
   E. Cryogenic Handling Precautions / 84
Control and Regulation of Gases / 84
   A. Gas Regulators / 84
   B. How a Gas Regulator Works / 85
   C. Pressure-Reduction Stages / 85
   D. Installation / 85
   E. Operation / 86
   F. Shutdown / 87
   G. Dismantling / 87
   H. Troubleshooting / 87
Gas Valves / 88
   A. Needle Valves / 88
   B. Metering Valves / 89
   C. Direction (Check) Valves / 89
   D. Relief Valves / 90
   E. Gauges / 91
   F. Manual Flow Controls / 91
   G. Safety Devices / 91
      Traps and Check Valves / 91
   H. Quick Couplers / 91
      Safety Relief Valves / 91
   I. Flowmeters / 92
   J. Gas-Leak Detectors / 93
   K. Gas-Washing Bottles / 93
      Procedure / 94
Collecting and Measuring Gases / 94
   A. Apparatus / 94

Drying and Humidifying Gases / 94
    A. Types of Gas Dispersers / 95
    B. Gas-Diffusing Stones / 96
Control of Gases Evolved in a Reaction / 96
    A. Precautions / 96
    B. Alternative Methods / 96
        Method 1 / 96
        Method 2 / 96
        Method 3 / 96
        Method 4 / 98
Compressors / 99
Relative Humidity / 99

Chapter 5. Pressure and Vacuum   105

Pressure / 105
Absolute, Gauge, Differential, and Standard Pressures / 105
Gas Laws / 105
    A. Boyle's Law / 105
    B. Charles' Law / 105
    C. Gay-Lussac's Law / 107
    D. Avogadro's Law / 107
    E. Dalton's Law / 107
    F. The Combined Gas Law / 107
    G. The Ideal Gas Law / 107
    H. Pascal's Law / 108
Vacuum / 108
Water Aspirator Guidelines / 109
Types of Vacuum Pumps / 109
Vacuum Pump Maintenance / 112
    A. When and How to Change the Oil / 112
    B. Vacuum Pump Oils / 112
    C. Filling Vacuum Pumps With Oil / 112
        Do Not Overfill the Pump / 113
    D. Vacuum Connections / 113
        Locating Leaks in Vacuum Systems / 113
    E. Gas Ballast / 113
    F. Belt Guards / 114
Troubleshooting Vacuum Pumps / 114
    A. Pump Runs Hot / 114
    B. Pump Is Noisy / 114
    C. Pump Does Not Achieve Expected Vacuum / 115
    D. Volatile Substances With Vacuum Pumps / 115
    E. Pump Spits Oil from Exhaust Port / 115
    F. Backstreaming / 115
Vacuum or Gas Flow Safety Traps / 116
Pressure and Vacuum Measuring Devices and Gauges / 116
    A. The Open-End Manometer / 116
    B. The Closed-End Manometer / 117
    C. Filling the Manometer / 118
    D. Bourdon-Type Gauges / 118
    E. McLeod Gauges / 119
    F. Pirani Gauges / 119
    G. Thermocouple Gauges / 120
References / 120

Chapter 6. Mathematics Review and Conversion Tables   121

Writing Numbers in Exponential Notation / 121
Grouping of Numbers and Decimal Points / 121

Scientific Calculators / 121
  A. Fixed Point Display / 122
  B. Scientific Notation Display / 122
  C. Engineering Notation Display / 122
The SI (Metric) System / 122
Significant Figures / 123
General Rules for Dealing With Significant Figures / 124
Rounding / 124
The Quadratic Equation / 125
Graphs / 125
Logarithms / 126
Dimensional Analysis / 127
Conversions between Engineering Units and SI / 130
Conversion of Millimeters to Decimal Inches / 133
Conversion of Standard Drill Bit Sizes / 136
References / 136

Chapter 7. Standard Operating Procedures   137

Standard Operating Procedures (SOPs) / 137
Standard Operating Conditions and Limits (SOCLs) / 137
Good Laboratory Practice (GLP) / 138
Current Good Manufacturing Practice (CGMP) / 138
  A. 21 CFR 211.180: General Requirements / 138
  B. 21 CRF 211.182: Equipment Cleaning and Use Log / 139
  C. 21 CFR 211.186: Master Production and Control Records / 139
  D. 21 CFR 211.188: Batch Production and Control Records / 139
General Guidelines of Log Books and Batch Records / 140
International Organization for Standardization (ISO) / 141
Statistics / 142
Statistical Terminology / 143
Method of Least Squares / 145
Statistical Process Control / 147
References / 148

Chapter 8. Laboratory Glassware   149

Introduction / 149
Ground-Glass Equipment / 150
  A. Joints and Clamps / 150
Care of Ground-Glass Surfaces / 151
  A. Lubrication / 151
  B. Choice of Lubricant / 152
  C. Lubricating Ground-Glass Joints / 152
  D. Components / 152
  E. Assemblies / 156
      Lubricating Stopcocks / 156
      Procedure / 157
Storage of Glassware / 158
  A. In Drawers / 158
  B. On Shelves and in Cabinets / 159
  C. Preparation for Long-Term Storage / 160
Assembly of Ground-Joint Glassware / 160
  A. Safety / 163
  B. Cutting Glass Tubing / 163
      Cutting Small-Diameter Tubing (25-mm OD or Less) / 163
  C. Dulling Sharp Glass Edges / 163
      End Finishing (Fire-Polishing) / 163

D. Common Glassblowing Techniques / 163
       Making Points / 163
       Procedure / 164
       Sealing the Ends of Capillary Tubes / 166
       Bending Glass Tubing / 166
Handling Stoppers and Stopcocks / 167
   A. Loosening Frozen Stoppers and Stopcocks / 167
References / 168

Chapter 9. pH Measurement   169

Acids and Bases / 169
   A. Strengths of Acids and Bases / 169
   B. The pH Scale / 170
pH Theory / 171
Acid/Base Indicators / 175
Methods for Determining pH / 176
   A. Colorimetric Determinations: Indicators / 176
       Principle / 176
       Procedure / 176
       Preparation of Indicator Solutions / 176
       The Importance of Selecting the Proper Indicator / 177
   B. pH Test Paper / 178
   C. pH Meter / 179
       Components / 179
       Procedure / 179
       Buffer Standardization of the pH Meter / 181
       Principle / 181
       Procedure / 181
       Electrode Maintenance / 182
       Hints for Precision Work / 182
pH Titration / 183
   A. Introduction / 183
   B. Strong Acid–Strong Base / 184
   C. Weak Acid–Strong Base / 186
   D. Strong Acid–Weak Base / 189
Buffer Solutions / 189
   A. Preparation of Buffer Solutions / 189
       Method 1 / 189
       Procedure / 189
       Method 2 / 191
       Procedure / 191
       Method 3 / 191
       Procedure / 191
Ion-Selective Electrodes / 192
   A. Selective Ion Calibration and Calculation Example / 193
References / 195

Chapter 10. Basic Electricity   197

Introduction / 197
Electrical Safety Practices / 197
Electricity Compared to Fluids / 197
Ohm's Law / 200
Conductors / 201
Conductivity of Various Metals / 202

How to Read a Resistor / 202
Alternating and Direct Current / 202
Fuses / 203
   A. Type of Fuses / 204
   B. Procedure for Replacing Fuses / 204
   C. Fuse-Holding Blocks / 204
   D. Determining the Cause for Blown Fuses / 205
Circuit Breakers / 205
Series and Parallel Circuits / 205
Electrical Power / 206
Servicing Inoperative Devices / 207
Electrical Testing Instruments: Voltmeter, Ammeter, Continuity Tester, and Multimeter / 207
Electric Motors / 208
   A. Starting Motors / 208
   B. Magnetic Starters / 209
   C. Using Motors Correctly / 209
   D. Types of Motors / 210
   E. Belts, Pulleys, and Power Transmission / 210
Making Electrical Connections / 210
   A. Electrical Terminal Connections / 211
   B. Splicing Wires Temporarily / 211
   C. Continuous Wire / 212
   D. Splicing Wires Permanently / 212
   E. Electrical Soldering / 212
   F. Taping / 213
Grounding Electrical Equipment / 213
   A. Summary of Safety Precautions / 214
   B. Importance of Grounding / 214
   C. Properly Grounded Items / 214
   D. Proper Grounding / 214
   E. How to Determine If Your Equipment Is Grounded / 215
   F. Ground Fault Circuit Interrupters / 216
Electric Circuit Elements / 217
References / 218

Chapter 11. Sampling

The Importance of Proper Sampling / 219
Heterogeneous versus Homogeneous Mixtures / 219
The Gross Sample / 219
Basic Sampling Rules / 220
Sampling Gases / 220
   A. Gas Sampling Techniques / 221
   B. Freeze-Out or Condensation / 221
Grab Sampling / 222
Sampling Liquids / 223
Sampling Homogeneous Solids / 225
Sampling Non-Homogeneous Solids / 226
   A. Coning and Quartering / 226
   B. Rolling and Quartering / 227
Sampling Metals / 227
Crushing and Grinding / 228
   A. Classifying Particulate Sizes / 228
   B. Crushing and Grinding Equipment / 230
      Ball Mill / 230
      Jaw Crusher / 230
      Pan Crusher / 230
      Hammer Mill / 230

Chapter 12. Laboratory Filtration — 233

Introduction / 233
Filtration Methods / 234
Filter Media / 235
   A. Paper / 235
   B. Membrane Filters / 236
   C. Fritted Glassware / 237
Filtering Accessories / 238
   A. Filter Supports / 238
   B. Filter Aids / 238
      Limitations of Filter Aids / 239
   C. Wash Bottles / 239
   D. Filter Pump and Accessories / 240
Manipulations Associated With the Filtration Process / 240
   A. Decantation / 241
   B. Washing / 241
      Transfer of the Precipitate / 242
Gravity Filtration / 242
      Procedure / 243
Vacuum Filtration / 246
      Procedure / 246
   A. Equipment and Filter Media / 247
      Büchner Funnels / 247
      Wire-Gauze Conical Funnel Liner / 248
      Crucibles / 248
      The Hirsch Funnel and Other Funnels / 250
   B. Large-Scale Vacuum Filtration / 250
Filtration for Gravimetric Analysis / 250
   A. Aging and Digestion of Precipitates / 251
   B. Filtration and Ignition / 252
      Preparation of Crucibles / 252
      Preparation of a Filter Paper / 252
      Transfer of Paper and Precipitate to Crucible / 252
      Ashing of a Filter Paper / 253
      Cooling the Crucible to Constant Mass / 254
      Summary / 254
References / 255

Chapter 13. Recrystallization — 257

Introduction / 257
Requirements of the Solvent / 257
   A. Solvency / 257
   B. Volatility / 258
   C. Solvent Pairs / 259
      Principle / 259
      Procedure / 259
Recrystallization of a Solid / 259
   A. Selecting the Funnel / 259
      Gravity Filtration / 259
      Vacuum Filtration / 259
   B. Heating the Funnel / 259
      Principle / 259
      Method 1 / 260
      Method 2 / 260
      Method 3 / 260
   C. Receiver Selection / 260

D. General Procedure / 260
       Recrystallizing / 260
       Completing Crystallization / 262
   E. Preventing Crystallization of a Solute During Extraction / 263
       Washing Crystals / 263
       Inducing Crystallization / 263
Decolorization / 263
       Principle / 263
   A. General Procedure / 263
   B. Additional Techniques and Hints / 264
   C. Batch Decolorization / 264
       Procedure / 265
Fractional Crystallization / 265
       Principle / 265
       Procedure / 265
Laboratory Use of Purification by Fractional Crystallization / 265
       Procedure / 265
       Procedure for Other Compounds / 266
Noncrystallization: "Oiling" of Compounds / 266
References / 266

Chapter 14. The Balance   267

Introduction / 267
Measurement / 267
Definitions of Terms / 268
   A. Factors Influencing Accuracy / 268
Errors in Determining Mass / 269
Types of Balances / 269
   A. Equal-Arm Balances / 269
       Principle of Operation / 269
       General Procedure / 270
   B. The Triple-Beam Balance / 271
       Procedure / 271
   C. The Dial-O-Gram Balance / 271
       Procedure / 272
   D. Two-Pan Equal-Arm Chainomatic Balances / 272
       Procedure / 272
       Using the Vernier on the Chainomatic Balance and Other Equipment / 272
       Principle / 272
   E. Unequal-Arm Balance—Substitution / 273
   F. Single-Pan Analytical Balance / 273
       General Procedure / 274
       Types of Single-Pan Analytical Balances / 274
   G. Electronic Balances / 275
       Procedure / 277
   H. Other Types of Balances / 277
Computerized Balances / 277
Choosing the Correct Balance / 278
   A. Criteria for Choosing a Balance / 278
   B. Model Calculation / 279
References / 279

Chapter 15. Gravimetric Analysis   281

Introduction / 281
Handling Samples in the Laboratory / 281
   A. Pretreatment of Samples / 281
   B. Crushing and Grinding / 282

C. Changes in Sample / 282
D. Grinding Surfaces / 283
E. Flame Tests for Identification of Elements / 283
Moisture in Samples / 284
Forms of Water in Solids / 284
Effects of Grinding on Moisture Content / 285
Drying Samples / 286
    Principle / 286
    Procedure / 286
Drying Collected Crystals / 286
    A. Gravity-Filtered Crystals / 286
    B. The Desiccator / 287
    C. Vacuum-Filtered Crystals / 288
    D. Drying Crystals in a Centrifuge Tube / 289
    E. Abderhalden Drying Pistol / 289
Drying Organic Solvents / 290
    Principle / 290
    Procedure / 290
    A. Efficiency of Drying Operations / 290
    B. Classification of Drying Agents / 291
    C. Determining If Organic Solvent Is "Dry" / 292
Freeze-Drying: Lyophilization / 292
    Procedure / 292
    A. Advantages of Freeze-Drying / 293
    B. Freeze-Drying Corrosive Materials / 293
Preparing the Sample for Final Analysis / 293
    A. Liquid Reagents Used for Dissolving or Decomposing Inorganic Samples / 293
        Hydrochloric Acid / 293
        Nitric Acid / 293
        Sulfuric Acid / 294
        Perchloric Acid / 294
        Oxidizing Mixtures / 294
        Hydrofluoric Acid / 294
    B. Decomposition of Samples by Fluxes / 294
        Method of Carrying out a Fusion / 295
        Types of Fluxes / 295
    C. Decomposition of Organic Compounds / 295
        Combustion-Tube Methods / 296
        Combustion With Oxygen in Sealed Containers / 297
        Peroxide Fusion / 299
        Wet-Ashing Procedures / 299
        Dry-Ashing Procedure / 299
Determining Mass of Samples / 300
    Procedure / 300
    A. Direct Mass Determination / 300
    B. Mass Determination by Difference / 301
    C. Gravimetric Calculations / 301
Microdetermination of Carbon and Hydrogen / 302
    A. Combustion of Organic Material / 303
    B. Removal of Interfering Substances / 303
        Removal of Sulfur / 303
        Removal of Halogens (Except Fluorine) / 303
        Removal of Nitrogen Oxides / 303
    C. Measurement of Carbon Dioxide and Water / 304
    D. Microprocedure for Determination of Carbon and Hydrogen in Organic Compounds / 304
        Principle / 304
        Scope / 304
    E. Total Oxidizable Carbon Analyzers / 304
Kjeldahl Analysis / 305
    A. Computerized Combustion Nitrogen Analyzers / 307
    B. Dumas Determination of Nitrogen / 308

Determination of Halogens / 309
  A. Decomposition of Organic Material / 309
  B. Removal of Interfering Ions / 309
    Ion Exchange / 310
  C. Other Methods of Determining Halogen / 310
Determination of Sulfur / 310
  A. Decomposition of Organic Material and Conversion of Sulfur to a Measurable Form / 310
      Oxidation / 310
      Tube Combustion (Dietert) / 311
      Parr Bomb / 311
      Schöninger Combustion / 311
References / 311

Chapter 16. Preparation of Solutions   313

Introduction / 313
Grades of Purity of Chemicals / 313
  A. Commercial or Technical Grade / 313
  B. Practical Grade / 313
  C. USP / 313
  D. CP / 313
  E. Spectroscopic Grade / 314
  F. Chromatography Grade / 314
  G. Reagent Analyzed (Reagent Grade) / 314
  H. Primary Standard / 314
Common Hazardous Chemicals / 314
  A. Incompatible Chemicals / 314
Water for Laboratory Use / 317
  A. Water-Purity Specifications / 317
  B. Softening Hard Water / 317
      Temporarily Hard Water / 317
      Permanently Hard Water / 317
      Ion-Exchange Resins / 317
  C. Purifying Water by Reverse Osmosis / 318
  D. Distilled Water / 318
  E. Demineralized Water / 319
  F. Reagent-Grade Water / 319
Solutions / 320
  A. The Dissolution Process / 320
  B. Solubility / 320
  C. Solubility Rules for Common Inorganic Compounds / 320
  D. Increasing the Rate of Solution of a Solute in a Liquid / 321
Colloidal Dispersions / 321
  A. Electrical Behavior / 323
  B. Distinguishing between Colloids and True Solutions / 323
      Method 1 / 323
      Method 2 / 323
  C. Distinguishing between Colloids and Ordinary Suspensions / 323
  D. Preparation of Standard Laboratory Solutions / 324
      Preparation and Storage of Carbonate-Free NaOH (0.1 N) / 324
      Procedure / 324
Reference / 331

Chapter 17. Process Analyzers   333

Introduction / 333
Process Sensors / 333
  A. Temperature Sensors / 333
  B. Resistance Temperature Detectors / 336

Conductivity Measurements / 336
Pressure Measurements / 338
Process pH Measurements / 338
Troubleshooting Process pH Electrodes / 340
   A. Potassium Chloride Soaking / 340
   B. Ammonia Soaking / 340
   C. Application of Vacuum / 340
Flow Rate Measurements / 340
Level Measurements / 342
Moisture Measurements / 342
References / 343

Chapter 18. Plumbing, Valves, and Pumps   345

Plumbing: Tubing versus Piping / 345
General Guidelines for Selecting Tubing / 345
Metal Tubing / 346
   A. Cutting Tubing / 346
   B. Removing Burs from Cut Tubing / 348
Bending Tubing / 348
   A. Bends Near Tube Fittings / 349
Connecting Tubing Mechanically / 350
   A. Flared Connections / 350
   B. Connecting Flared Tubing / 351
Compression Fittings / 351
Applying Pipe Thread Tape or Sealants / 354
Pipes and Fittings / 355
Tracing of Piping Systems / 356
Welding Safety / 356
   A. Hot Work Permits / 357
   B. Fire Prevention from Welding Activities / 358
Plumbing Soldering / 358
Fluid Flow / 359
Bernoulli's Principle / 360
Process Valves / 361
Flash Arrestors / 363
Pumps / 364
   A. Kinetic Pumps / 364
   B. Rotary Pumps / 366
   C. Reciprocating Pumps / 367
   D. Metering Pumps / 369
   E. Variable-Frequency Drive (VFD) Pumps / 369
   F. General Pump Maintenance / 369
      Bearing Lubrication / 369
      General Pump Maintenance Guidelines / 370
References / 371

Chapter 19. Physical Properties and Determinations   373

Physical Testing / 373
   A. Introduction / 373
   B. Physical and Chemical Changes / 373
Temperature Measurements / 373
   A. Introduction / 373
   B. Temperature Scale Conversions / 375
   C. Thermometer Corrections / 377
Melting-Point Determinations / 377
   A. Introduction / 377

  B. Capillary-Tube Method / 377
    Procedure / 377
  C. Thiele Tube Method / 378
    Procedure / 378
  D. Electric Melting-Point Apparatus / 380
    Procedure / 380
Boiling-Point Determinations / 381
  A. Introduction / 381
  B. Test Tube Method / 381
    Procedure / 381
  C. Capillary-Tube Method / 381
    Procedure / 381
Flash Point Determinations / 382
  A. Introduction / 382
  B. Open-Cup and Closed-Cup Method / 383
Density / 385
  A. Definition / 385
  B. Determining Solid Densities / 385
    Regularly Shaped Solids / 385
    Procedure / 385
    Irregularly Shaped Solids / 386
    Procedure / 386
  C. Determining Liquid Densities / 386
    Density-Bottle Method Procedure / 386
    Float Method / 387
    Procedure / 387
  D. Determining Gas Densities / 387
  E. Dumas Method / 387
    Procedure / 388
Specific Gravity / 388
  A. Introduction / 388
  B. Pyknometer Method / 388
    Procedure / 388
  C. Hydrometer Method / 388
  D. Specific Gravity Conversions / 389
Viscosity / 390
  A. Definition / 390
  B. Viscosity Standards / 390
  C. Small-Bore Tube Method / 391
  D. Saybolt Viscometer Method / 391
  E. The Falling-Piston Viscometer Method / 391
  F. Rotating Concentric Cylinder Viscometer Method / 391
  G. Falling-Ball Viscometer Method / 392
  H. Ostwald Viscometer Method / 393
    Procedure / 393
  I. Determining Kinematic Viscosities / 394
Surface Tension / 395
  A. Introduction / 395
  B. Fluid Surface-Tension Measurement / 396
Refractive Index / 396
  A. Introduction / 396
  B. Specific Refraction / 397
  C. Refractometers / 397
Polarimetry / 399
  A. Introduction / 399
  B. Direction of Rotation of Polarized Light / 400
  C. Specific Rotation / 401
  D. Causes of Inaccurate Measurements / 402
  E. How to Use a Polarimeter / 402
  F. Specific Rotation Calculations / 403

Rheology / 404
  A. Young's Modulus and Elasticity / 404
  B. Bulk Modulus and Compressibility / 405
  C. Hardness / 406
References / 406

Chapter 20. Extraction   407

Introduction / 407
Soxhlet Extraction / 407
    Procedure / 407
Extraction of a Solute Using Immiscible Solvents / 407
    Principle / 407
  A. Using Extraction Solvent of Higher Density / 408
    Cautions and Techniques / 409
Extraction Procedures in the Laboratory / 410
  A. Using Water / 410
  B. Using Dilute Aqueous Acid Solution / 410
  C. Using Dilute Aqueous Basic Solution / 410
  D. Selective Extraction / 410
Considerations in the Choice of a Solvent / 410
  A. Diethyl Ether / 411
Peroxides in Ether / 411
  A. Detection of Peroxides in Ethers / 412
    Test 1 / 412
    Test 2 / 412
    Test 3 / 412
  B. Removal of Peroxides from Ethers / 412
Recovery of the Desired Solute from the Extraction Solvent / 413
  A. Emulsions / 413
  B. Breaking Emulsions / 413
Extraction by Disposable Phase Separators / 414
    Procedure / 414
Continuous Liquid-Liquid Extractions / 415
  A. Higher-Density-Solvent Extraction / 415
  B. Lower-Density-Solvent Extraction / 415
Ultrafiltration / 416
Kuderna-Danish Sample Concentrator / 417
References / 418

Chapter 21. Distillation and Evaporation   419

Introduction / 419
  A. Vapor Pressure / 419
Simple Distillation / 419
    Procedure / 420
  B. Distillation of Pure Liquids / 421
  C. Distillation of a Solution / 422
  D. Distillation of a Mixture of Two Liquids / 422
    Principle / 422
  E. Rate of Distillation (No Fractionation) / 423
  F. Concentration of Large Volumes of Solutions / 423
Azeotropic Distillation / 423
Fractional Distillation / 426
    Principle / 426
    Procedure / 428
  A. Heating Fractionating Columns / 428
  B. Efficiency of Fractionating Columns / 428

C. Bubble-Cap Fractionating Columns / 428
D. Total-Reflux–Partial-Takeoff Distilling Heads / 429
Vacuum Distillation / 430
    Principle / 430
  A. Nomographs / 430
  B. General Requirements / 431
  C. Assemblies for Simple Vacuum Distillation and Fractionation / 434
    Simple Vacuum Distillation / 434
    Procedure / 434
    Vacuum Fractionation / 434
    Procedure / 435
  D. Modified Distillate Receiver / 435
Steam Distillation / 435
    Principle / 435
    Procedure / 436
  A. Steam Distillation With Superheated Steam / 439
Refluxing / 439
    Procedure / 442
Process Distillation / 442
Evaporation of Liquids / 443
  A. Small Volumes / 443
    Method 1 / 443
    Method 2 / 444
  B. Direct Heating of Evaporating Dishes / 444
  C. Transferring Residues from Watch Glasses or Evaporating Dishes / 444
  D. Large Volumes / 445
    Method 1 / 445
    Method 2 / 445
  E. Evaporation Under Reduced Pressure / 445
    Procedure / 445
  F. Evaporation Under Vacuum / 446
    Water Aspirator / 446
    Mechanical Vacuum Pump / 446
Rotary Evaporator / 447
Sublimation / 447
  A. Atmospheric and Vacuum Sublimation / 448
  B. Methods of Sublimation / 448
    Simple Laboratory Procedure at Atmospheric Pressure / 448
    Methods Useful at Atmospheric or Reduced Pressure / 448
References / 450

Chapter 22. Inorganic Chemistry Review   451

Matter / 451
The Periodic Table / 451
Chemical Bonding / 455
  A. Ionic Bonds / 455
  B. Covalent Bonds / 457
Molecular and Structural Formulas / 457
  A. Molecular Weights / 457
Writing Chemical Formulas / 458
Naming Inorganic Compounds / 458
Naming Binary Compounds / 460
  A. The "ous/ic" Nomenclature System / 460
Naming Compounds Containing Polyatomic Ions / 461
  A. Naming Covalent Compounds / 461
Naming Acids / 463
Hydrates / 463
Common or Trivial Names / 464
References / 465

Chapter 23. Organic Chemistry Review   467

Naming Organic Compounds / 467
Alkanes / 467
   A. IUPAC Rule 1 / 467
   B. IUPAC Rule 2 / 467
   C. IUPAC Rule 3 / 469
Alkenes / 469
Alkynes / 469
Cyclic Hydrocarbons / 470
Isomers / 471
Aromatic Compounds / 471
Organic Families and Their Functional Groups / 472
   A. Alkyl Halides / 472
Alcohols / 473
Amines / 475
Ethers / 475
Aldehydes / 476
Ketones / 477
Carboxylic Acids / 477
Amides / 478
Esters / 478
Amino Acids / 479
Polymers / 479
Summary of Organic Families and Their Functional Groups / 481
Organic Reactions / 481
   A. Cracking / 481
   B. Alkylation / 482
   C. Fermentation / 482
References / 482

Chapter 24. Chemical Calculations and Concentration Expressions   483

Material Balance / 483
Molecular Relationships from Equations / 483
Mass Relationships from Equations / 485
Mole Relationships from Chemical Equations / 486
Volume Relationships from Equations / 487
Percent Yield / 488
Solution Terminology / 488
Concentration Expressions / 489
   A. Percent / 489
      % (wt/wt) / 490
      % (wt/v) / 490
      % (v/v) / 490
   B. Parts per Million / 490
   C. Molarity / 491
   D. Normality / 492
Redox Reactions / 493
Dilutions / 493
Neutralization and Titration / 494
References / 495

Chapter 25. Volumetric Analysis   497

Introduction / 497
   A. Definitions of Terms / 497
      Units of Volume / 497

           Titration / 497
           Back Titration / 497
           Standard Solution / 497
           Equivalence Point / 497
           End Point / 497
           Titration Error / 497
           Meniscus / 497
       B. Typical Physical Changes During Volumetric Analysis / 498
           Reading the Meniscus / 498
           Procedure / 498
       C. Washing and Cleaning Laboratory Glassware / 498
           General Rules / 498
           Cleaning Volumetric Glassware / 500
           Cleaning Glassware Soiled With Stubborn Films and Residues / 501
           Metal Decontamination of Analytical Glassware / 502
           Ultrasonic Cleaning / 502
           Fluidized-Bath Method / 502
       D. Drying Laboratory Glassware / 503
           Drainboards and Drain Racks / 503
           Dryer Ovens / 503
           Quick Drying / 503
           Rinsing Wet Glassware With Acetone / 503
Tools of Volumetric Analysis / 504
   A. Volumetric Flasks / 504
       Directions for the Use of a Volumetric Flask / 504
       Introducing Standard Directly into a Volumetric Flask / 504
       Dilution to the Mark / 504
   B. Pipettes / 504
       Directions for the Use of a Pipette / 506
       Procedure / 506
   C. Other Pipettes / 508
   D. Burettes / 509
   E. Directions for the Use of a Burette / 509
   F. Filling / 509
   G. Holding the Stopcock / 510
Performing a Titration / 510
       Splitting a Drop of Titrant / 513
   A. Titration Curves / 513
Acid-Base Titrations and Calculations / 513
   A. Procedure for Standardization of a Base or Acid / 514
   B. Back Titrations / 515
Oxidation-Reduction Titrations and Calculations / 515
Chelates and Complexo-Metric Titrations / 517
Karl Fischer Method for Water Determination / 519
Electronic Pipettes and Dispensers / 520
References / 520

Chapter 26. Chromatography   521

Introduction / 521
Adsorption Chromatography / 521
Partition Chromatography / 521
Thin-Layer Chromatography / 522
   A. Introduction / 522
   B. Preparation of the Plate for Thin-Layer Chromatography / 522
   C. Thin-Layer Chromatography Procedure / 523
   D. Tips on Technique for Thin-Layer Chromatography / 525
       Precoated, Commercially Available Plates / 525

Activation of Plates / 526
Spotting / 526
Placing the Plate in the Tank / 526
Edge Effects / 526
Gas Chromatography / 526
Instrument Design and Components / 527
  A. Carrier Gas / 527
  B. Sample Injection / 527
  C. Columns / 528
  D. Column Oven / 529
  E. Detectors / 530
    Thermal-Conductivity Detectors (TCDs) / 530
    Flame-Ionization Detector / 531
  F. Recorder/Integrator / 531
General GC Operational Procedures / 532
  Instrument Start-Up Procedure / 533
  Instrument Shutdown Procedure / 534
  A. Microliter-Syringe-Handling Techniques / 534
  B. Gas Chromatographic Fittings and Valves / 535
Solid Supports / 535
Liquid Phases / 535
Column Efficiency / 536
Flowmeters / 538
Qualitative Analysis / 540
Integration Techniques / 541
  A. Cut and Weigh / 541
  B. Peak Height / 542
  C. Triangulation / 542
  D. Planimeter / 542
  E. Electronic Integrator/Computers / 543
Quantitative Analysis / 543
  A. Normalization / 543
  B. External Standard / 545
  C. Internal Standard / 546
  D. Standard Addition / 547
Troubleshooting Gas Chromatography / 548
Liquid Chromatography / 550
Column Chromatography / 552
  A. Liquid Column Techniques / 552
General Operating Procedure / 552
  A. General Precautions / 554
  B. Sorbents for Column Chromatography / 554
LC Stationary Phases / 554
  A. Alumina / 554
  B. Silica Gel / 555
  C. Bonded Stationary Phases / 555
LC Mobile Phases / 555
High-Performance Liquid Chromatography / 556
HPLC Instrument Design and Components / 556
  A. HPLC Solvent Delivery Systems / 557
  B. HPLC Solvent Mixing Systems / 557
  C. HPLC Sample Injection / 559
  D. HPLC Columns / 559
  E. HPLC Detectors / 561
  F. Recorder/Integrator / 561
Mobile-Phase Preparation and Storage / 562
Qualitative and Quantitative HPLC Techniques / 563
HPLC Pump Troubleshooting / 563
  A. No or Low Pump Pressure / 563
    Air or Gas Bubbles / 563

　　　　　Contaminated Check Valves / 563
　　　　　Broken Piston Plungers / 563
　　　　　Leaking Seals / 563
　　B. High Pump Pressure / 563
Automated HPLC Systems / 564
Ion-Exchange Chromatography / 564
　　A. Introduction / 564
　　B. Preparing an Ion-Exchange Column / 566
　　C. IC Detectors / 568
Supercritical-Fluid Chromatography / 569
　　A. Introduction / 569
Size-Exclusion Chromatography / 570
Process Chromatography / 570
References / 571

Chapter 27. Spectroscopy　　　　　　　　　　　　　　　　　　　　　573

The Electromagnetic Spectrum / 573
　　A. Introduction / 573
　　　　Types of Electromagnetic Radiation / 573
　　　　Conversion of Units / 573
　　　　Frequency and Wavelength Relationship / 573
　　B. The Visible Spectrum and Refraction / 573
　　C. Diffraction / 574
　　D. Sodium Vapor and Ultraviolet Lamps / 574
　　E. Use of the Electromagnetic Spectrum in the Study of Substances / 576
　　F. Colorimetry / 578
　　　　Duboscq Colorimeter / 578
　　　　Materials / 578
　　　　Procedure / 579
　　G. Nephelometry / 579
　　H. Spectroscopy and Spectrophotometry / 579
　　　　Definitions and Symbols / 581
　　I. Spectrophotometry Concentration Analysis / 581
　　　　Measurement of Absorption Spectra / 582
Visible and Ultraviolet Spectroscopy / 584
　　A. Ultraviolet Spectrophotometer Components / 585
　　B. Chromophores / 585
　　C. Solvents Used in Visible and Ultraviolet Spectroscopy / 586
　　D. Sample Preparation / 586
　　　　Matched Cuvettes / 587
　　E. Using the Spectrophotometer / 587
　　　　Bausch and Lomb Spectronic 20 / 587
　　　　Components / 587
　　　　Procedure / 587
　　F. Diode-Array Spectro-Photometers / 589
　　G. Computerized Spectra Analytical Systems and Fluorometers / 589
Infrared Spectroscopy / 590
　　A. Introduction / 590
　　B. Wavelength versus Wave Number / 590
　　C. Infrared Instrument Components and Design / 592
　　D. Infrared Cell Materials / 593
　　E. Infrared Liquid Cells / 593
　　F. Calculating Cell Path Length / 596
　　G. Solid Samples / 597
　　H. Fourier Transform Infrared Spectroscopy / 598
　　　　The Michelson Interferometer / 599
　　I. Infrared Absorption Spectra / 600

Near Infrared (NIR) Spectrometer / *601*
    A. Continuous Spectra Instruments / *602*
    B. Process Spectroscopy Analyzers / *604*
References / *605*

Chapter 28. Atomic Absorption Spectroscopy   607

Introduction / *607*
Atomic Absorption Spectrophotometer Components / *609*
Factors and Conditions That Effect AA Spectrophotometer Response / *611*
Atomic Absorption Safety Considerations / *612*
Atomic Absorption Calibration / *613*
    A. Calibration by Standard Additions Method / *615*
Atomic Absorption Interferences / *615*
Computerized Atomic Absorption Spectrophotometers / *616*
Graphite Furnace Technique / *617*
Mercury Analyzers / *618*
References / *618*

Appendix A. Commonly Used Abbreviations   619

Appendix B. Physical Properties of Water   627

Vapor Pressure and Density of Water at Various Temperatures / *627*

Appendix C. Glossary of Chemical Process Terms   629

Index   645

PREFACE

According to the Department of Labor, there are currently 750,000 individuals employed in the United States in the chemical process industries. This number could exceed 5 million workers if one considers employees in chemical-related industries throughout the world. In this country, 250,000 employees are classified as chemical technicians, with the remaining one-half million individuals employed as chemical process workers (manufacturing plants, food processing, petroleum refining, agro-chemical businesses, and so on) in some capacity. Several trends are apparent in these chemical process industries, including the following:

- As downsizing trends continue, chemical workers at all levels will need multiple skills and to be prepared to accept individual responsibilities.
- Various governmental agencies and international competition will require better employee training, quality control, and documentation.

Now in its fifth edition, the *Chemical Technicians' Ready Reference Handbook* is a proven, excellent resource for anyone working in the chemical process industries. These individuals can range from laboratory technicians and process control operators to chemists, engineers, and maintenance and management personnel. This Handbook has proven to be the definitive reference on chemical safety, laboratory procedures, chemical nomenclature, basic electricity, laboratory statistics, and instrumental techniques. For four previous editions, this Handbook has been hailed for its scope and depth of coverage and step-by-step directions for performing virtually every laboratory task. It remains the undisputed classic in the field.

The fourth edition of the Handbook was published in 1996, thus there was a need to modernize certain chapters and topics. For example, chemical plant and laboratory safety procedures were in need of revision. Specifically, Chapter 2, "Chemical Plant and Laboratory Safety," and Chapter 3, "Chemical Handling and Hazard Communication," have been revised to reflect current governmental guidelines, regulations, and recommended practices. Many users have requested that we expand this best-selling Handbook to include more information specifically for chemical process operations. Two brand-new chapters, 17, "Process Analyzers," and 18, "Plumbing, Valves, and Pumps," have been added. The Handbook still covers the classic wet methods in gravimetric and volumetric analysis, plus difficult analytical procedures such as Karl Fischer titrations and Kjeldahl nitrogen digestions.

This Handbook is designed to provide a practical guide to unskilled or novice employees as well as professionally trained chemical process operators, chemical laboratory technicians, chemists, biologists, and maintenance and management personnel. There are specific references listed at the end of each chapter for additional information. Current addresses, telephone numbers, and Internet addresses are also included. Twenty-eight chapters have been assembled dealing with topics such as:

- Chemical plant and laboratory safety and federal regulations
- Laboratory mathematics, chemical calculations, and unit conversions
- Basic electricity and techniques
- Standard operating procedures
- Inorganic chemistry

- Organic chemistry
- Proper sampling techniques and expanded physical testing procedures
- Plumbing, valves, and pumps used in process applications

In addition, conversion tables, easy-to-use professional references, and a chemical process glossary have been included. Many example problems have been worked out in detail, ranging from simple temperature conversions and concentration expressions to process unit parameters dealing with integrating the area of chromatographic peaks to understanding and calculating Reynolds numbers.

JACK T. BALLINGER

ACKNOWLEDGMENTS

A comprehensive chemical process workers' handbook dealing with both chemical laboratory technicians and chemical process operators cannot be produced by one individual. I am truly blessed to have two sons who "volunteered" to be consulting editors for this Handbook. My older son, Clint, is a nuclear engineer and my other son, Barry, is a certified industrial hygienist. Thus, safety and process engineering are well covered.

We have managed to cover a wide range of topics, including not only classical and instrumental chemical technology but current process technology as well. Topics range from plant and laboratory safety, federal regulations, process analyzers, chemical calculations, and conventional wet methods to both plant and laboratory pumps. We have many individuals to acknowledge and thank for their substantial contributions toward the completion of this Handbook. I am indebted to Larry Hager, senior editor at McGraw-Hill, for his encouragement that this fifth edition be written and eternally grateful to David Fogarty, editing manager at McGraw-Hill, for his tremendous technical assistance and patience. My two sons and I simply could not have completed this very ambitious manuscript without their guidance.

In addition, the following companies, publishers, and organizations have contributed to this fifth edition of the *Chemical Technicians' Ready Reference Handbook*:

ABB Instrumentation, Warminster, Pennsylvania, Robert Mapleston, Marketing Communications Manager, (215) 674-6580, Robert.Mapleston@us.abb.com

Ace Scientific Supply Company, 1420 East Linden Avenue, Linden, New Jersey 07036, Nate Schurer, Director of Marketing, (201) 925-3300

Alltech Associates, Inc., 2051 Waukegan Road, Deerfield, Illinois 60015-1899, Julia Poncher

American Chemical Society, 1155 16th Street, NW, Washington, D.C. 20036, Blake Aronson, Education Department, (202) 872-6108

Hewlett-Packard Company, 3000 Hanover Street, Palo Alto, California 94303-1185, Kevin A. O'Connor, Senior Community Relations Specialist

Hydraulic Institute, Inc., 6 Campus Drive, First Floor, North Parsippany, New Jersey 07054-4406, Gregg Romanyshyn, Technical Director, (973) 267-9700 x114, gromanyshyn@pumps.org

Matheson-Trigas, Secaucus, New Jersey, Mary Smickenbecker, Vice President, Marketing/Strategic Marketing Group, (215) 648-4020, msmickenbecker@matheson-trigas.com

McGraw-Hill, 1221 Avenue of the Americas, 45th Floor, New York, New York 10020-1095, Cynthia Aguilera, Permissions Department, cynthia_aguilera@mcgraw-hill.com

Mettler-Toledo, Inc., Public Relations Department, Box 71, Hightstown, New Jersey 08520

National Fire Protection Association, One Batterymarch Park, Quincy, Massachusetts 02169-7471, Dennis J. Berry, Secretary of the Corporation and Director of Licensing, (617) 770-3000, dberry@nfpa.org

Perkin-Elmer Corporation, 761 Main Avenue, Norwalk, Connecticut 06859, Carol Blaszczynski, Public Relations

Perten Instruments North America, Inc., 6444 South Sixth Street Road, Springfield, Illinois 62712, Wes Shadow, Business Development Manager, (801) 543-3346, wshadow@perten.com

Sargent-Welch, P.O. Box 4130, Buffalo, New York 14217, Noel France Vache, Communications Coordinator and Community Relations, (800) 727-4368, nvache@wardsci.com

Shimadzu Scientific Instruments, 7102 Riverdale Drive, Columbia, Maryland 21046, Paul J. Evans, Website Administrator, (410) 381-1227, pjevans@shimadzu.com

Swagelok Company, 29500 Solon Road, Solon, Ohio 44139, Richard Monreal, Operations Manager, Communications, (440) 248-4600, richard.monreal@swagelok.com

Thermo Fisher Scientific, 2000 Park Lane, Pittsburgh, Pennsylvania 15275, Janine Hochman, Manager, Design and Production Marketing and Product Services; Alan M. Doernberg, Senior Counsel, Intellectual Property, (412) 490-8451, alan.doernberg@thermofisher.com

Varian Associates, Inc., 3120 Hansen Way, Palo Alto, California 94304-1030, Judith Farrell, Product Promotion Manager

Waugh Controls Corporation, 200 East Randolph Street, Chicago, Illinois 60601, Roger May

We would especially like to thank all of the chemical process operators and chemical laboratory technicians involved daily in making our lives safer, healthier, more productive, and perhaps even easier. We invite comments, criticisms, and suggestions from these technicians, operators, and their employers as well. Good luck in your chemical processing and laboratory endeavors.

Most importantly, this expanded fifth edition of the *Chemical Technicians' Ready Reference Handbook* is dedicated to the memory of Gershon J. Shugar. Dr. Shugar and his family of experts (Ronald A. Shugar, Lawrence Bauman, and Rose Shugar Bauman) produced the first and second editions of this Handbook.

CHEMICAL TECHNICIANS' READY REFERENCE HANDBOOK

CHAPTER 1
CHEMICAL PROCESS INDUSTRY WORKERS AND GOVERNMENT REGULATIONS

THE ROLE OF THE CHEMICAL PROCESS INDUSTRY WORKER

The occupation of "chemical process industry worker" is difficult to define because the formal training, experience, and duties vary from company to company and region to region. Many countries have now developed occupational skill standards for most of these occupations, including chemical process operations employees and chemical laboratory technicians. The American Chemical Society, the U.S. Department of Labor, and the U.S. Department of Education have worked jointly with the U.S. chemical process industry (CPI) to develop "Voluntary Industry Skill Standards for CPI Technical Workers."

The U.S. Department of Labor studies show that nearly 750,000 CPI workers are employed by U.S. companies, government agencies, and other organizations. This number could exceed 5 million workers if one considers chemical-related employees throughout the world. According to the American Chemical Society, there are approximately 500,000 chemical plant operators and 240,000 chemical laboratory technicians currently employed in the United States. This group makes up the fourth largest U.S. manufacturing industry and it should be noted that these data do not include other chemical-related occupations (science technicians, clinical technicians, forensic technicians, waste water operators, etc.).

Six critical job functions specifically for process operators (POs) have been identified in the skill standards study and are presented in part in Table 1.1. These critical job functions, tasks, and technical workplace standards are published in detail in the American Chemical Society's "Foundations for Excellence in the Chemical Process Industries."

A. Working in the Chemical Process Industries

The following *core and recommended competencies* have been identified for process operators:

- Understand the importance of teamwork.
- Recognize that all members of a team have opinions that must be valued.
- Demonstrate skills in problem solving.
- Demonstrate responsibility for fellow workers' health and safety.
- Demonstrate critical thinking.
- Successfully coordinate several tasks simultaneously.
- Make decisions based on data and observations.
- Pay close attention to details and observe trends.
- Demonstrate high ethical standards in all aspects of work.
- Apply total quality management (TQM) principles to all aspects of work.

TABLE 1.1 Critical Job Functions for Chemical Process Operators

A. Working in the chemical process industries.
B. Maintaining safety/health and environmental standards in the plant.
C. Handling, storing, and transporting chemical materials.
D. Operating, monitoring, and controlling continuous and batch processes.
E. Providing routine and preventative maintenance and service to processes, equipment, and instrumentation.
F. Analyzing plant materials.

(With permission from the American Chemical Society's "Foundations for Excellence in the Chemical Process Industries.")

B. Maintaining Safety, Health, and Environmental Standards in the Plant

Safety, health, and environmental (S/H/E) protection must be a continuous, high-level concern to every technical worker in the chemical processing industry. Process operators working in this industry may handle hazardous materials and operate potentially dangerous equipment as a major part of their work. Improper handling can have serious effects on the environment, the PO, other individuals, and the physical facility. Process operators must be aware that compliance with safety, health, and environmental standards is the most important part of their work. Regulations specific to the CPI have been developed at the federal, state, and local levels. In addition, industry groups and individual companies have developed procedures to ensure the safety and health of employees, consumers, and the environment.

Regulations require continuous monitoring and extensive record keeping that address the environmental impact of chemicals, the processes used to make them, and their storage and disposal. Process operators are the employees who have the most direct contact with, and control of, the chemicals and related products. They must accept responsibility for ensuring that S/H/E considerations are addressed appropriately.

Process operators must be able to implement the regulations pertaining to the industry and to the processes in which they work. Knowing the regulations is not sufficient—the PO must also understand the reasons for them and the consequences of not adhering to them. Furthermore, POs must have significant knowledge about the materials and equipment they handle and the processes they control.

Tasks conducted by process operators related to this critical job function include:

- Accessing procedures, files, and other documentation for S/H/E guidelines before starting any job and responding appropriately.
- Inspecting all areas for hazards.
- Providing input to coworkers about unsafe conditions.
- Complying with regulatory standards, including procedures for red tag lockout, safe handling of all chemicals, and implementing confined-space entry procedures, among others.
- Investigating accidents and incidents as part of process safety management.
- Working closely with safety and regulatory contacts; participating in S/H/E audits.
- Reporting any hazardous conditions that may affect the safety and health of self or coworkers and those that could have a negative impact on the environment.
- Properly labeling all materials in the plant.
- Participating in public awareness activities.
- Participating in and conducting safety response drills.
- Verifying the completeness of appropriate paperwork.
- Participating in audits and safety reviews.

C. Handling, Storing, and Transporting Chemical Materials

Process operators working in the chemical processing industry are expected to handle, store, and transport chemical materials within the plant and are expected to arrange for transportation to and from the plant. To perform these jobs effectively and safely, POs must be familiar with chemical and physical properties of the materials they handle and with the regulations governing the safe use, storage, and handling of the materials.

Many workplace chemicals are considered hazardous materials and must be handled with respect, which requires knowledge and understanding of properties of the materials. The physical and chemical characteristics of the material being stored must be compatible with specifications for storage containers and vessels ranging from laboratory glassware to tank storage facilities.

Tasks conducted by process operators related to this critical job function include:

- Receiving, verifying, and identifying materials and products from ships, trucks, railroads, and other carriers.
- Completing the proper paperwork associated with receiving materials.
- Unloading, or arranging for someone to unload, materials and products from ships, trucks, railroads, and other carriers.
- Transferring, or arranging to transfer, materials to storage or processing unit.
- Properly labeling all materials.
- Ensuring that materials meet specifications by measuring specified properties.
- Maintaining material inventories.
- Maintaining storage facilities.
- Inspecting drums, barrels, and containers for proper use and storage.
- Cleaning, decontaminating, and disposing of storage drums, barrels, and containers in an appropriate manner.
- Inspecting the tank farm and verifying the identity and quantity of the contents of each container.
- Transferring materials from processes or units to containers; in the case of continuous processes, ensuring that the transfer follows the specifications of the plant and the customers.
- Verifying and preparing shipping papers or supplying the data to the appropriate shipping clerk.
- Loading materials onto ships, trucks, railroad cars, or other transportation vehicles.
- Handling all materials using environmental health and safety guidelines as outlined in the standard operating procedure (SOP) or material safety data sheet (MSDS).
- Cleaning up or arranging for cleanup of all spills.
- Responding to all emergencies.
- Developing and meeting shipment schedules.

D. Operating, Monitoring, and Controlling Continuous and Batch Processes

Process operators are responsible for operating, monitoring, and controlling the operations and processes used to make, blend, and separate chemical materials in a plant. Often, POs are assigned to a specific unit within a process operation for a long time. Process operators are the prime source of process information for engineers and managers seeking to ensure competitiveness and efficiency. To operate the process most effectively, the PO requires a fundamental understanding of how chemical processes work, the building blocks or unit operations of processes, the chemistry that takes place in the processes, and the technology used to control and optimize processes.

Tasks conducted by process operators related to this critical job function include:

- Reviewing and properly interpreting all checklists associated with a process.
- Completing all required reports to describe process activities, discrepancies, and maintenance.

- Adjusting control equipment as specified by procedures, setting operating parameters, and identifying abnormal conditions that require reporting.
- Checking equipment to ensure safety for electrical loading, physical stressing, and temperature variation.
- Starting up a process or processes according to specified procedures.
- Monitoring operating parameters by reading gauges, instruments, and meters, and logging or otherwise recording the information as necessary.
- Adjusting operating parameters to optimize conditions appropriate to the process.
- Recognizing and correcting any deviations.
- Responding to any alarms.
- Collecting appropriate samples.
- Conducting on-site inspections.
- Submitting samples for analysis.
- Recording and reporting data.
- Shutting down processes according to procedures.
- Shutting down processes in emergency situations.
- Maintaining piping systems.
- Participating in developing/revising standard operating procedures.
- Calculating and documenting production rates.
- Training new process operators.
- Maintaining quality control charts to determine whether the plant is under control.

E. Providing Routine Preventative Maintenance and Service to Processes, Equipment, and Instrumentation

Maintenance is essential to efficient operation and functioning in the chemical processing industry. If performed properly, routine and preventive maintenance enhances safety and improves the efficiency of the operation. In addition, emissions, leaks, and unplanned shutdowns are also reduced. In total, high costs and lost profitability associated with lost production are reduced. In large plants, maintenance is usually performed by specialists trained to maintain specific equipment and instruments. In smaller plants, the process operator is responsible not only for running but, frequently, for maintaining the overall operation. In both types of plants, the operator's role in requesting/performing maintenance and implementing maintenance schedules is critical to ensuring continuous operation.

Working knowledge of basic maintenance techniques is essential if the technician or operator is to participate in turnarounds or make necessary adjustments when processes deviate from normal conditions. Process operators are the first defense for recognizing the need for maintenance, proactively heading off minor problems before they become major upsets, correcting deteriorating situations at the onset, and, if not performing the maintenance tasks, arranging for them to take place.

Modern chemical processes use on-line controls for temperature, pressure, flow, level, and analytical targets. These on-line control instruments act as additional eyes and ears of process operators, aiding in the maintenance of a continuous, economical, safe, and stable mode of operation. Malfunctioning instruments can lead to upsets, equipment failure, hazardous conditions, and environmental incidents. Consequently, POs must have a working knowledge of instrumentation, particularly how to recognize abnormal indications and behavior and how to protect the operation from instrument-induced upset.

Tasks conducted by process operators related to this critical job function include:

- Reading and following all SOPs associated with the maintenance of processes, equipment, and instruments before starting any work.
- Observing, communicating, and recording any deviations from normal operations of processes, equipment, and instrumentation.
- Initiating maintenance requests.
- Implementing and developing, if necessary, a preventive maintenance schedule.
- Inspecting equipment.
- Preparing equipment for maintenance.
- Following maintenance schedules developed by plant personnel.
- Opening lines and equipment.
- Changing seals and valves on on-line equipment.
- Changing seals and packings on pumps and valves.
- Changing and replacing pipes as required.
- Checking fluid levels in process equipment.
- Performing vibration analysis.
- Performing steam tracing techniques.
- Testing and replacing pressure release valves as needed.
- Calibrating instruments and checking standards.
- Troubleshooting problems in processes and instrumentation as required.
- Inspecting and testing safety control equipment and devices.

F. Analyzing Plant Materials

On some occasions, chemical process operators might be required to perform work traditionally done by specialists (primarily chemical laboratory technicians, chemists, microbiologists, etc.). One such area is making chemical and physical measurements, where operators conduct standard procedures to measure product and process quality. To develop this skill, the PO will need to have basic training for laboratory operations in areas such as sampling, concepts of measurements, following procedures, and performing physical, chemical, and instrumental tests. The interface between a centralized laboratory and a unit laboratory will vary from company to company and location to location, and the jurisdictions of each will need to be determined at the local level.

Tasks conducted by process operators related to this critical job function include:

- Following specified procedures to collect appropriate samples for analysis from process streams or products (solids/liquids/gases).
- Visually inspecting samples to ensure adequate representation of the sampled materials and determining whether any immediate response is required.
- Labeling samples appropriately and according to any prescribed procedures.
- Delivering samples in a condition representative of the material to testing locations.
- Preparing necessary reagents and standards required to conduct tests.
- Calibrating instruments as required.
- Analyzing quality control standards to appropriate precision levels by using prescribed methods.
- Performing appropriate physical and chemical tests, according to standard procedures.
- Calculating results using calculators and computers.

- Determining whether reanalyzing is necessary.
- Reporting results using prescribed procedures or effective presentation techniques to appropriate personnel.
- Adjusting process parameters as necessary following standard practice or as appropriate in responding to analytical results.
- Entering data into computers and other appropriate logs.
- Reviewing trends of process variations and sampling analyses.
- Comparing sample analyses with control values and responding appropriately.
- Preparing proper paperwork to submit samples to the laboratory.

THE ROLE OF THE CHEMICAL LABORATORY TECHNICIAN

Eight critical job functions for chemical laboratory technicians have been identified in the previously mentioned skill standards study. These critical job functions, tasks, and technical workplace standards are published in detail in the American Chemical Society's "Foundations for Excellence in the Chemical Process Industries." Until the exact roles, formal training, and duties of a chemical laboratory technician can be identified and specified, the best description of this worker can be presented using a job task analysis as shown in Table 1.2.

TABLE 1.2 Critical Job Functions for Chemical Laboratory Technicians

A. Working in the chemical process industries.
B. Maintaining a safe and clean laboratory.
C. Sampling and handling chemical materials.
D. Measuring physical properties.
E. Performing chemical analysis.
F. Performing instrumental analysis.
G. Planning, designing, and conducting experiments.
H. Synthesizing compounds.

(With permission from the American Chemical Society's "Foundations for Excellence in the Chemical Process Industries.")

A tentative **job task analysis** covering the eight major critical job functions for a practicing chemical laboratory technician contains the following:

A. Working in the Chemical Process Industries

The following *core and recommended competencies* have been identified for chemical technicians:

- Understand the importance of teamwork and have experience working as a member of a team for planning, performing, analyzing, and reporting. Recognize that all members of a team have opinions that must be valued.
- Demonstrate skill in problem solving.
- Demonstrate responsibility for fellow workers' health and safety.
- Demonstrate critical thinking skills.
- Coordinate several tasks simultaneously.
- Make decisions based on data and observations.

- Pay close attention to details and observe trends.
- Demonstrate high ethical standards in all aspects of work.
- Apply quality principles to all aspects of work.

B. Maintaining a Safe and Clean Laboratory Adhering to Safety, Health, and Environmental Regulations

Tasks conducted by chemical technicians related to this critical job function include:

- Performing safety inspections.
- Participating in safety audits.
- Participating in S/H/E training.
- Conducting and participating in safety demonstrations, drills, and meetings.
- Using safety monitoring equipment.
- Labeling all chemicals, materials, tools, and equipment with appropriate S/H/E details.
- Organizing and storing chemicals, glassware, and other equipment properly.
- Keeping laboratories clean and orderly.
- Selecting appropriate safety equipment, including hoods, and using it correctly when conducting laboratory tasks.
- Ensuring that warning labels are displayed appropriately.
- Using ergonomic procedures.
- Reporting and taking action on all unsafe or potentially unsafe conditions and acts.
- Identifying and responding to emergencies, alarms, and any abnormal situations.
- Reporting and taking action on any potential environmental noncompliance.
- Reporting and taking action on any potential health or industrial hygiene problem.
- Encouraging others to act in accordance with good S/H/E standards (providing peer support).
- Disposing of waste chemicals and materials.
- Interpreting MSDSs.

C. Sampling and Handling Chemical Materials

Tasks conducted by chemical technicians related to this critical job function include:

- Selecting containers and preparing and storing samples and materials in compliance with both regulations and compatibility.
- Labeling all samples and chemical materials with information containing chemical name, formula, toxicity, date stored, expiration date, appropriate symbols, and other pertinent information.
- Storing samples and chemical materials.
- Preserving materials as recommended, in accordance with MSDSs.
- Maintaining inventory, with information regarding expiration, toxicity, etc.
- Developing materials and sampling inventory database and schedule.
- Maintaining inventories within prescribed ranges of quantity and as required to maintain safety and environmental standards.
- Ordering materials as required and evaluating new materials and suppliers when necessary.

8 CHAPTER ONE

- Disposing of materials and samples in compliance with all federal, state, local, and employer regulations.
- Handling hazardous wastes in compliance with all federal, state, local, and employer regulations.
- Preparing samples for shipment in compliance with employer's shipping and receiving rules and regulations.
- Responding appropriately to chemicals spilled in the laboratory.
- Ensuring that appropriate heating, ventilation, and electrical services are used in chemical storage areas.
- Preparing materials for testing and analysis.
- Recognizing hazards associated with handling radioactive materials.

D. Measuring Physical Properties

Tasks conducted by chemical technicians related to this critical job function include:

- Obtaining representative samples.
- Preparing samples appropriately for tests or analyses.
- Choosing appropriate test equipment to make a required measurement.
- Checking instruments for correct operation.
- Preparing or acquiring calibration standards.
- Calibrating equipment.
- Testing or analyzing control "standard" samples; calculating results and comparing with control values.
- If within statistical range, analyzing samples; if not, troubleshooting causes of error by repeating calibration, instrument check, and maintenance.
- Calculating results.
- Recording and reporting data.
- Cleaning and maintaining apparatus.

E. Performing Chemical Analysis

Tasks conducted by chemical technicians related to this critical job function include:

- Obtaining representative samples.
- Making observations regarding condition of sample and recording any notable characteristics.
- Responding to problems by reading test documents or procedures and implementing appropriate information.
- Identifying appropriate equipment for the analysis to be conducted.
- Gathering, cleaning, and calibrating all necessary glassware, reagents, chemicals, electrodes, and other equipment required to complete the specified analysis.
- Preparing and standardizing any required reagents.
- Preparing samples for analysis (dissolve, digest, combust, ash, separate interfering material, etc.).
- Analyzing standards or control samples using specified techniques.
- Recording data and, if within limits, proceeding to analyze sample.
- Calculating results to appropriate significant figures.

- Recording data and presenting results, as appropriate, for single samples and for multiple samples to display trends.
- Evaluating analytical results and responding appropriately.
- Identifying conditions that indicate need for an analysis to be repeated.
- Recording and reporting data.
- Maintaining laboratory work areas and returning all equipment and materials to original storage locations.
- Modifying or developing analytical methods to be appropriate to necessary test methods, required analyses, the implementation of personnel qualifications, and working environment where methods are to be used.

F. Performing Instrumental Analysis

Tasks conducted by chemical technicians related to this critical job function include:

- Obtaining representative samples.
- Determining appropriate treatment of the sample before conducting an analysis.
- Preparing a sample for analysis according to specifications.
- Selecting the analytical instrument to be used as appropriate to the results needed and other constraints.
- Starting up instrument by checking all connections and gas cylinders and implementing procedures to ensure reliable results.
- Setting all the instrumental parameters properly using manual and/or program microprocessor settings.
- Calibrating and standardizing equipment and materials as appropriate.
- Developing necessary calibration charts.
- Analyzing standards and control materials.
- Evaluating results of testing or analyzing standards and control materials.
- Adjusting operating parameters as necessary.
- Conducting analyses.
- Reviewing and interpreting results.
- Recording results with appropriate detail.
- Reporting results as appropriate.
- Identifying the need for and performing routine maintenance as required.
- Shutting down instrument and cleaning up work area.
- Maintaining and/or ordering spare parts necessary to ensure consistent operation.

G. Planning, Designing, and Conducting Experiments

Tasks conducted by chemical technicians related to this critical job function include:

- Working with team members to set goals and divide the work to be done.
- Conducting literature searches.
- Identifying resources (e.g., people, equipment, chemicals, and methods).
- Gathering chemicals and obtaining resources.
- Creating a statistical design for the experiment using a quality model.

- Designing control ranges as appropriate for the defined needs and the experimental conditions.
- Describing procedures in writing as appropriate for the intended audience.
- If appropriate, designing and running computer simulations.
- Performing experiments and procedures in the laboratory or field.
- Initializing and monitoring automated experiments.
- Evaluating and presenting results.
- Assessing and redesigning experiments as necessary.
- Reporting results.
- Working with team members to determine required follow-up activities.
- Implementing results, as appropriate.

H. Synthesizing Compounds

Tasks conducted by chemical technicians related to this critical job function include:

- Familiarizing oneself with reaction characteristics before performing syntheses, including the nature of the reaction (kinetics, equilibrium, exothermic/endothermic, etc.).
- Determining all safety and handling aspects of the work to be conducted (e.g., consider side reactions).
- Identifying and obtaining information relevant to specific laboratory activities.
- Obtaining starting materials and solvents in appropriate quantities and conditions.
- Determining the purity of the starting materials, if appropriate.
- Obtaining and calibrating all the monitoring instruments required.
- Assembling necessary glassware, instruments, and equipment for a synthesis procedure.
- Establishing the proper starting conditions and setting the required instrument parameters as appropriate for the experiment to be conducted.
- Initiating the reaction.
- Monitoring the reaction.
- Collecting and purifying the product of the reaction (e.g., filter, evaporate, crystallize).
- Verifying the identity and characterizing the purity of the product, as appropriate.
- Analyzing product sample (or submitting for analysis) to determine the purity and for confirmation of structure.
- Disassembling and cleaning equipment, glassware, and other materials used, and returning them to properly stored conditions.
- Documenting all findings in a laboratory notebook, obtaining witness signatures, and discussing results with team members.
- Participating in planning next steps.

GOVERNMENT REGULATIONS

A. Introduction

There are hundreds of laws, regulations, and codes regulating the chemical process industries and process operators. Four federal agencies have been selected and some of the major laws and regulations, but certainly not all, have been listed in this chapter. These regulation titles and numbers are listed as public laws, the Code of Federal Regulations (CFR), and laws of the United States

Congress (USC). If more details are needed, please see the references at the end of this Handbook. The CFR falls into 50 title numbers and subject areas, for example, 29 = Labor—OSHA, 40 = Protection of the Environment—EPA, and 49 = Department of Transportation—DOT. The Food and Drug Administration has a CFR number of 21 and is discussed in Chapter 7 "Standard Operating Procedures" of this Handbook.

B. Occupational Safety and Health Administration

The U.S. Department of Labor's Occupational Safety and Health Administration (OSHA) is responsible for enforcing various federal laws dealing with health and safety hazards in the workplace. OSHA became part of the National Labor Law in April 1971 and its primary mission is "to assure so far as possible every working man and woman in the nation safe and healthful working conditions and to preserve our human resources." These regulations are published as part of the CFR, Title 29, Section 1910, U.S. Department of Labor.

Specifically, in 1983 federal regulation 29 CFR 1910.1200 was promulgated requiring a "Hazard Communication Program" to ensure that the hazards of all chemicals produced or imported are evaluated, and that information concerning their hazards is transmitted to employers and employees. This transmittal of information must be comprehensive and include container labeling, MSDSs, and employee training. This law affects employees who handle hazardous chemicals in the workplace: "(a) Employers shall ensure that labels on incoming containers of hazardous chemicals are not removed or defaced; (b) Employers shall maintain all material safety data sheets that are received with incoming shipments of hazardous chemicals, and ensure that they are readily accessible to all employees; (c) Employers shall ensure that employees are appraised of the hazards of the chemicals in their workplaces." This particular OSHA standard is sometimes referred to as the "Right to Know" law. Table 1.3 lists some of the more common OSHA regulations that affect the chemical process industries.

TABLE 1.3 Some Major Occupational Safety and Health Administration Regulations Affecting the Chemical Process Industries

Law/regulation	Regulation number	General description
Hazardous Materials Transportation Act (HMTA)	49 USC 1801	Control of movement of hazardous materials.
Hazardous Material Regulations	49 CFR 100-180	Regulation of packaging, labeling, placarding, and transporting.
Hazardous Materials Training (DOT Docket HM-126F)	49 CFR 172.700-704	Assurance of training for all persons involved in handling, storing, or transportation of hazardous materials.
DOT Hazard Communication Proper Names	49 CFR 172.101	Gives proper shipping for hazardous materials.
DOT Hazard Communication Placards	49 CFR 172.504	Gives proper placard for hazardous materials.
DOT Hazardous Materials	49 CFR 173.1	General requirements for shipments and packaging.
Segregation Table for Hazardous Materials	49 CFR 177.848	Lists which classes of hazardous chemicals must be segregated.

The U.S. Environmental Protection Agency (see the section "Environmental Protection Agency" later in this chapter for more details) has a similar rule designed to protect the entire community surrounding the facilities that handle specifically listed hazardous materials in excess of certain quantities. Since 1986, the EPA's Superfund Amendments and Reauthorization Act (SARA) has required certain employers to develop emergency plans and guidelines for informing community members in the event of a hazardous chemical release or other crisis situation.

In 1987, OSHA promulgated the Hazard Communication Standard (perhaps better known as the "Right to Know" Law). This standard provides for the complete disclosure of the presence and properties of any hazardous chemical with which employees may have contact in the workplace. Originally this standard was directed for the protection of all workers in general industry (e.g., health care workers, transportation workers, college employees, manufacturing workers, etc.).

In 1991, OSHA promulgated the Occupational Exposure to Hazardous Chemicals in Laboratories Standard, more commonly referred to as the "Lab Standard." This standard was specifically written to address laboratory-type operations that were found not to be adequately covered under the hazard communication standard. The hazard communication stand still applies to all workers handling hazardous chemicals. However, laboratory workers must comply with the lab standards requirements as well. The lab standard is deemed a performance standard, as such employees are free to determine the best manner to comply with its mandated requirements. The main requirement of the standard is the development of a **chemical hygiene plan** (CHP).

Chemical Hygiene Plan (CHP). The following outline was extracted from "Prudent Practices for Handling Hazardous Chemicals in Laboratories," which was published in 1981 by the National Research Council and is available from the National Academy Press. This publication is cited here because of its wide acceptance in the laboratory community and the sponsorship by the National Research Council.

Another valuable reference for dealing with the complex legislation and regulations governing hazardous chemicals is "Prudent Practices for Disposal of Chemicals from Laboratories."

General Principles for Work with Laboratory Chemicals

1. It is prudent to minimize all chemical exposures.
2. Avoid underestimation of risk.
3. Provide adequate ventilation.
4. Institute a chemical hygiene program.
5. Observe the PELs, TLVs.

Chemical Hygiene Responsibilities

1. Chief executive officer.
2. Supervisor of the department.
3. Chemical hygiene officer(s).
4. Laboratory supervisor.
5. Project director.
6. Laboratory worker.

The Laboratory Facility

1. Design.
2. Maintenance.
3. Usage.
4. Ventilation.

Components of the CHP

1. Basic Rules and Procedures.
2. Chemical Procurement, Distribution, and Storage.
3. Environmental Monitoring.

4. Housekeeping, Maintenance, and Inspections.
5. Medical Program.
6. Protective Apparel and Equipment.
7. Records.
8. Signs and Labels.
9. Spills and Accidents.
10. Information and Training Program.
11. Waste Disposal Program.

None of the recommendations given here will modify any requirements of OSHA's Laboratory Standard. In general, "Prudent Practices for Handling Hazardous Chemicals in Laboratories" deals with both safety and chemical hazards while the OSHA Laboratory Standard is concerned primarily with chemical hazards. These CHPs must be customized to meet the requirements of the federal, state, and local compliance guides. Updating services are now commercially available to provide chemical users with the latest laws and regulations on a state-by-state basis. The topics can include asbestos removal, PCB disposal, small-quantity generators, permits, etc.

Chemical hygiene plans must have provisions for the determination of employee exposure to hazardous chemicals, and must provide monitoring of hazardous material levels, provisions for medical attention, employee training, employee medical and exposure record keeping, and hazard materials identification. This hazard identification and information is usually provided in the form of **MSDSs,** as described in more detail in Chapter 3, "Chemical Handling and Hazard Communication."

C. Department of Transportation

The federal Department of Transportation (DOT) is an agency that exercises control of the transportation of hazardous materials in commerce. This agency sets rules to protect life and property from releases of hazardous materials during handling and transport. These rules and regulations dictate specific requirements for labeling of packages, required paperwork to accompany hazardous material shipments, packaging specifications, and even training of individuals who handle or transport hazardous materials. Table 1.4 lists some of the major DOT regulations that affect the chemical process industries. Approximately one and a half billion tons of hazardous materials are transported annually in the United States. This amounts to over half a million shipments per day. The DOT estimates that a full two-thirds of the total volume travels over public roadways and that 64 percent of all incidents involving transporting hazardous waste are due to human error.

TABLE 1.4 Some Major DOT Regulations Affecting the Chemical Process Industries

Law/regulation	Regulation number	General description
Atomic Energy Act (AEA)	42 USC 2073	Establish standards for protection against radiation hazards.
Energy Reorgaization Act (ERA)	42 YSC 5841	
Standards for Protection Radiation; Licenses	10 CFR 20	Establish exposure 10 CFR 30-35 limits and license.

Hazardous materials are categorized by DOT according to their particular hazard and are listed in Part 172.101 of Title 49 of the code of Federal Regulations, called the Hazardous Materials (HM) Table. For example, HM-126F was issued by DOT to prevent hazardous materials transportation incidents from occurring.

Another DOT regulation, HM-126F, requires employees involved in the handling, storing, or transporting of hazardous materials to receive specialized training in safe work practices, general awareness, function specific responsibilities, and professional driving skills.

HM-181 revisions have been recently issued by DOT to ensure that the standards of the United States and the United Nations are compatible. It requires that all employees involved in the handling, storing, or transporting of hazardous materials receive the proper training, and sets provisions for minimum performance requirements for packaging materials. DOT also provides a Segregation Table for Hazardous Materials, found in Part 177.848 of Title 49, CFR, indicating which classes of hazardous materials must be segregated during shipment. This ensures that incompatible materials are not mixed in the event of an accident. Chapter 3, "Chemical Handling and Hazard Communication," has detailed information on DOT regulations, placards, and labeling.

D. Environmental Protection Agency

Environmental regulations in the United States date back to the turn of the century with the Rivers and Harbors Act of 1899. The Clean Air Act of 1955 was one of the first air pollutant regulations, set out to address air pollution, which was a common problem in large cities. The Resources Conservation Recovery Act (RCRA) of 1976 set the stage for regulating and defining hazardous wastes. It also created a waste-tracking system using manifests so waste could be tracked from "cradle to grave." This manifest system made it easier for regulators to track hazardous waste from the point of generation to the ultimate disposal site.

More recently, the EPA has created new regulations or expanded existing regulations to promote a cleaner and safer community. The regulations now include the regulation of pesticides, limits on pollutants in drinking water, mandatory requirements for cleaning up spilled materials, registering of toxic materials with the EPA, and emergency response planning requirements to name a few. Table 1.5 lists some of the major EPA regulations that affect the chemical process industries.

GOVERNMENTAL AND OTHER SOURCES OF CHEMICAL PROCESS INFORMATION

American Board of Industrial Hygiene (ABIH)

(517) 321-2638; www.abih@abih.org

American Chemical Society (ACS)

(800) 227-5558; www.help@acs.org

American Conference of Governmental Industrial Hygienists (ACGIH)

(513) 742-2020; www.mail@acgih.org

American Industrial Hygiene Association (AIHA)

(703) 849-8888; www.infonet@aiha.org

American National Standards Institute (ANSI)

(202) 293-8020; www.ansi.org

American Organization of Analytical Chemists International (AOAC)

(800) 379-2622; www.aoac@aoac.org

American Petroleum Institute (API)

(202) 682-8000; www.api.org

American Red Cross

(202) 303-5000; www.redcross.org

TABLE 1.5 Some Major Environmental Protection Agency Regulations Affecting the Chemical Process Industries

Law/regulation	Regulation number	General description
Clean Air Act (CAA)	42 USC 7401	Congress first passed CAA in 1963 to protect air quality and human health.
Clean Air Act Amendments of 1990 (CAAA)	42 USC 7409	Expansion of air quality protection. This regulation sets restrictions for Hazardous Air Pollutants (HAPs).
National Emission Standards for Hazardous Air Pollutants (NESHAP)	40 CFR 70	Control of air pollutant emissions.
Resource Conservation and Recovery Act (RCRA)	42 USC 6901	Protection of health and the environment.
Hazardous Waste Management	40 CFR 260-272	"Cradle-to-grave" control of chemical waste.
Underground Storage Tanks	40 CFR 280	Protection against groundwater and soil contamination.
Comprehensive Environmental Response, Compensation, and Liability Act (CERCLA)	42 USC 9601	Remediation of past and disposal sites and of liability "Superfund Law."
Superfund Amendments and Reauthorization (SARA)	42 USC 9601 42 USC 11000	Planning for emergencies and reporting of hazardous material releases into the environment.
National Contingency Plan	40 CFR 300-302	Cleanup requirements for spills and disposal sites.
Emergency Planning and Notification	40 CFR 355	Requirements for reporting of extremely hazardous materials and unplanned releases.
Hazardous Chemical Reporting: Community Right-to-Know Act	40 CFR 370	Requirements for reporting hazardous chemicals in use.
Toxic Chemical Release Reporting (SARA 313)	40 CFR 372	Requirements for reporting of chemical releases.
Toxic Substances Control Act (TSCA)	15 USC 2601	Protection of humans and the environment by requiring testing and necessary restrictions on use of certain chemical substances.
Reporting and Record keeping	40 CFR 704	One provision exempts use of small quantities solely for research and development.
Federal Water Pollution Act (FWPCA)	33 USC 1251	Improvement and protection of water quality.
Criteria and Standards for the National Pollutant Discharge Elimination System (NPDES)	40 CFR 12	Control of discharge to public waters.
General Pretreatment Regulations for Existing and New Sources of Pollution	40 CFR 40	Control of discharge of pollutants to public treatment works.

American Society for Testing and Materials (ASTM)

(610) 832-9500; www.astm.org

American Society of Safety Engineers (ASSE)

(847) 699-2929; www.asse.org

Canadian Centre for Occupational Health and Safety (CCOHS)
(800) 668-4284; www.ccohs.ca

Canadian Society of Safety Engineering
(416) 646-1600; www.csse.org

Canadian Standards Association (CSA)
(800) 463-6727; www.csa.ca

Centers for Disease Control (CDC)
(800) 232-4636; www.cdc.gov

Chemical Abstracts Service
(800) 848-6538; www.help@cas.org

Chemical Transportation Emergency Center (CHEMTREC)
(800) 262-8200; www.chemtrec.com

U.S. Coast Guard
(617) 223-8480; www.uscg.mil

Compressed Gas Association, Inc. (CGA)
(703) 412-0900; www.cganet.com

Department of Transportation (DOT)
www.dot.gov

Environmental Protection Agency (EPA)
(202) 272-0167; www.epa.gov

EPA RCRA, Superfund, Hazardous Waste Hotline
(800) 424-9346; www.epa.gov

Factory Mutual (FM)
(800) 320-6808; www.fmglobal.com

Food and Drug Administration (FDA)
(800) 216-7331; www.fda.gov

International Organization for Standardization (ISO)
+ 41 22 749 01 11; www.iso.ch

Mine Safety and Health Administration (MSHA)
(202) 693-9400; www.msha.gov

National Fire Protection Association (NFPA)
(617) 770-3000; www.NFPA.org

National Institute of Standards and Technology (NIST)
(800) 321-6742; www.osha.gov

National Response Center (NRC)
(202) 321-6742; www.epa.gov/oem

National Safety Council (NSC)
(800) 621-7615; www.info@nsc.org

Nuclear Regulatory Commission (NRC)

(800) 368-5642; www.nrc.gov

Occupational Safety & Health Administration (OSHA)

(800) 321-6742; www.osha.gov

OSHA Hotline (24 hours)

(800) 321-6742 or (800) 321-OSHA

Underwriters Laboratories Inc. (UL)

(877) 854-3577; www.ul.com

REFERENCES

Much of the information that a process operator needs can be found in various reference handbooks, indexes, dictionaries, government publications, abstracting services, the Internet, etc., including the following:

1. "Foundations for Excellence in the Chemical Process Industries, Voluntary Industry Standards for Chemical Process Industries Technical Workers." Robert Hofstader and Kenneth Chapman, ISBN: 1-8412-3492-2, American Chemical Society, 1155 Sixteenth Street NW, Washington, DC 20036 (1997).
2. Code of Federal Regulations 29, Labor/OSHA. Parts 1900–1910.999. General Industrial Standards, OSHA Subpart A-T. Revised every July. U.S. Government Printing Office, Washington, DC 20402. This reference contains the OSHA Standards for the workplace.
3. Code of Federal Regulations 29, Labor/OSHA. Parts 1910.1000–end. General Industrial Standards, OSHA Subpart Z (Z tables) and Haz-Com. U.S. Government Printing Office, Washington, DC 20402. Hazard Communication regulations with the revised OSHA Z list of hazardous chemicals, new PELs, and respiratory regulations.
4. Code of Federal Regulations 49, Department of Transportation. Parts 100–185, Hazardous Materials Regulations. Revised every October. U.S. Government Printing Office, Washington, DC 20402. This reference contains DOT rules for packaging, classification, labeling, placarding, and shipping of hazardous materials and waste.
5. Code of Federal Regulations 49, Department of Transportation. Parts 186–199, Hazardous Materials Regulations, Revised every October. Includes Performance-Oriented Packaging Standards (HM-181). U.S. Government Printing Office, Washington, DC 20402. This reference contains DOT rules for packaging, classification, labeling, placarding, and shipping of hazardous materials and waste.
6. Code of Federal Regulations 49, Department of Transportation. Parts 400–999, Coast Guard, National Highway Traffic Safety Administration, Urban Mass Transportation Administration, etc. Revised every October. U.S. Government Printing Office, Washington, DC 20402.
7. Code of Federal Regulations 40, Environmental/EPA. Parts 1–49, General Regulations. Revised every July. U.S. Government Printing Office, Washington, DC 20402.
8. Code of Federal Regulations 10, Department of Energy. Radiation. U.S. Government Printing Office, Washington, DC 20402.
9. Prudent Practices in the Laboratory: Handling and Disposal of Chemicals. ISBN 0-309-05229-7. National Research Council, National Academy Press, Washington, DC (1995).

CHAPTER 2
CHEMICAL PLANT AND LABORATORY SAFETY

INTRODUCTION: CHEMICAL PLANT AND LABORATORY SAFETY

In the United States, occupational exposure to hazardous chemicals in the workplace is covered by many Occupational Safety and Health Administration (OSHA) regulations, specifically the "Hazard Communication Standard" [29 CFR 1910.1200], the "Laboratory Standard" [29 CFR 1910.1450], and other federal, state, and local regulations. Internationally, occupational exposure to hazardous chemicals is regulated by a wide variety of risk-based approached sets of legislation. In the United Kingdom, legislation is driven by the Control of Substances Hazardous to Health (commonly known as CSHH). In most countries the prevention of accidents and providing a safe working environment for all employees is generally a high priority. However, according to the Bureau of Labor Statistics, workplace injuries and illnesses among private industry in 2008 occurred at a rate of 3.9 cases per 100 equivalent full-time workers. Clearly there is room for improvement. Chemical laboratories and chemical handling facilities typically possess a greater number of unique hazards than most other general worksites and as a result present some of the greatest challenges in the prevention of injury and illness. The guiding principles in any robust health and safety program are based on identification and characterization of risk, with the implementation of the hierarchy of control principle which is explained in more detail within this chapter.

BASIC SAFETY RULES FOR WORKING WITH OR AROUND CHEMICALS

1. Never attempt to perform any procedure without being adequately trained and authorized in the safe operation of equipment and also in knowledge of the job-specific details through the reading of standard operating procedures. Fully understand the hazards and risks of the job before commencing any task.
2. Follow the handling guidelines and control measures as outlined in the procedure or in the Material Safety Data Sheet (MSDS, described later in this chapter) for the chemical agent or compounds being handled.
3. Wear appropriate eye protection at all times. Ordinary prescription glasses are not adequate protection; glasses equipped with hardened glass or plastic safety lenses with side shields are recommended. Most plant and laboratory areas have mandatory safety-glasses use policies in place to protect against the hazards in the area. (See "Eye Protection" section later in this chapter.) Depending on the hazard, goggles or a full-face shield may be required.
4. Wear appropriate protective gloves when handling chemicals. Dermal contact through the skin is a potential source of exposure to toxic, sensitizers, and/or corrosive materials. (See "Hand Protection" section later in this chapter.)
5. Wear appropriate protective apparel. Steel-toed safety shoes are normally required in plant areas; sandals, high-heeled shoes, and open-toe shoes should never be allowed. General plant personal protective equipment can include hearing protection, helmets, safety glasses, dedicated

work uniforms, and may also include respiratory protection. (See "Protective Apparel" section later in this chapter.)

6. Never eat, drink, apply cosmetics, lip balm, or insert or remove contact lenses in a chemical environment.
7. Never work without knowing the location and operation of all emergency equipment (eyewash station, fire extinguishers, safety showers, etc.). All employees should know how to obtain emergency assistance (Fire Department, Plant Security, Medical Department, etc.). All safety and emergency response equipment should be regularly inspected and tested to ensure its functionality.
8. Never wear loose-fitting clothing or unrestrained hair in the laboratory, plant room, machine shop, or on the plant floor, where mechanical moving parts (belts, pulley, gears, etc.) may present an entanglement hazard.
9. Never leave the laboratory or workplace after using chemicals without washing hands thoroughly. Always wash hands after removal of your gloves. Do not wear contaminated laboratory coats or other clothing into nonplant or nonlaboratory areas.
10. Never purposely inhale a chemical to facilitate its identification. Although chemicals can be detected at very small concentrations, their toxicity even at such low levels can present a significant danger and should never purposely be inhaled. Always use a laboratory or identification kit to identify chemical container contents.
11. Never pipette liquids by mouth; *always* use a pipette bulb, vacuum system, autopipette, or a water aspirator to provide the vacuum.
12. Never pour water into concentrated acid, especially H_2SO_4 (sulfuric acid), because the excessive heat generated may cause it to spatter or break the glass container. It is best to add acid into water slowly to allow heat to dissipate without spattering.
13. Never use an open flame or nonintrinsically safe equipment to heat flammable materials. In general, open-flame heating devices are not as safe as nonsparking electrical equipment. It is recommended to turn off heaters, burners, and any other sources of ignition when they are unattended.
14. Never bring chemicals into contact with the skin. Use a scoop or other sampling device for solid chemicals as many chemicals can be corrosive or toxic, or they may be strong sensitizing agents that can cause significant ill-health effects.
15. Never perform experiments or reactions that could produce objectionable or unknown gases without using a fully functional fume hood or local exhaust ventilation system.
16. Never heat "soft" glass containers (most bottles, funnels, graduated cylinders, thick-walled glassware, etc.) in an open flame. Unlike Kimax® or Pyrex®, soft glassware is not designed and treated to withstand high temperatures or thermal shock. All broken or cracked glassware should be discarded immediately in special broken glass containers. Broken glass should never be discarded in normal trash containers as this might present an injury risk to janitorial employees.
17. Never attempt to insert a glass tube or thermometer into a cork or rubber stopper without a lubricant (water or glycerin). Glass tubing should be "fire polished" and held in a cloth or special insertion device to minimize the hazard due to breakage. Never use force when working with glassware.
18. Never attempt to carry out a reaction without first having read the approved standard operating procedure (SOP) and knowing the all control measures to be implemented.
19. Never engage in horseplay or distract other employees.
20. Never use any chemicals found in unlabeled containers and, conversely, never store chemicals in a container without labeling it.
21. Never work alone or leave chemical reactions unattended without arranging appropriate safeguards.

22. Never return excess chemicals to their original container. The problem of disposing of excess chemicals can be minimized by taking only the amount of material required for the reaction.

23. Always report a chemical spill immediately to your supervisor and other employees at risk. Follow your emergency spill procedures(see Table 2.1 for chemical spill types and corrective action summary). More details about generic spill procedures are given in "Controlling Spills," in Chapter 3).

TABLE 2.1 Chemical Spill Types and Corrective Action Summary

Spill type	Corrective action
Minor Liquid Chemical Spills	Absorb material, place in compatible container and make ready for appropriate disposal and report the event.
Strong Acid Spills	Follow instructions in MSDS, or when not available, dust area with solid sodium bicarbonate (baking soda), sweep up, place in a suitable container and report the event.
Strong Base Spills	Follow instructions in MSDS, or when not available, dust area with solid boric acid, place in a suitable container and report the event.
Large Chemical Spills	Contain spill immediately only if it is safe to do so, evacuate the immediate area, keep out of the area, warn others, notify emergency services.

TOXIC CHEMICALS

A. Introduction

There are an estimated 100,000 chemicals used in the workplace today of which 60,000 are considered hazardous. A toxic chemical is any substance that has the ability to damage, alter, or interfere with human metabolic systems. There are four primary routes by which substances can enter the body: (1) inhalation, (2) absorption through the skin, (3) ingestion, and (4) injection. In the following discussion, the words *toxic* and *poison* are synonymous. Where legislation exists, such as the OSHA hazard communication standard, all employees are to be made aware of the toxicity of any chemicals that they might be exposed to in their employment.

Did you know that 1 g of table salt (sodium chloride) will kill a rat? Toxicologists acquired data using test animals to determine lethal dosages per body mass and means of exposure. A person may receive the **lethal dose (LD)** or **lethal concentration (LC)** of a toxin through inhalation, ingestion, injection, or skin contact. A standardized expression of toxicity is the **LD_{50}** which represents the quantity, or dose, of toxic material necessary to cause death in 50 percent of the test animals. This term is normalized by including the animal's body weight. For example, the toxicity (LD_{50}) of mercury is expressed as 50 mg/kg, thus mercury would be classified as highly toxic. **Acute toxicity** is defined as the ability of a chemical to cause a harmful effect after a single exposure (Table 2.2). This terminology is present in many MSDSs and is presented here for informational purposes only, as it is but one method used to characterize the degree of toxicity of a chemical compound.

TABLE 2.2 Acute Toxicity Hazard Levels

Hazard level	Toxicity rating	Oral LD_{50} (rat, per kg)
Low	Slightly toxic	500 mg to 5 g
Medium	Moderately toxic	50–500 mg
High	Highly toxic	<50 mg

CHEMICAL EXPOSURE HAZARDS

The Occupational Safety and Health Administration (OSHA) and **American Conference of Governmental Industrial Hygienist (ACGIH)** have established guidelines to protect workers from overexposure to potentially hazardous chemicals. The two systems are similar and both refer to an airborne concentration to which an individual may be repeatedly exposed, day after day, without suffering adverse effects. Mandatory or even recommended **occupational exposure limits (OELs)** are common in most industrialized nations for providing limits against exposure to gases, vapors, and particulates. Internationally, the most commonly used limits are the **threshold limit values (TLVs)** as established by the American Conference of Governmental Industrial Hygienists (ACGIH).

In the United States, OSHA has permissible exposure limits (PEL) which are written into enforceable standards for exposure to toxic and hazardous air contaminants. The TLVs are considered consensus standards and typically not enforceable under law; however, they are routinely used in industry as a guide for protecting employee health. Unlike PELs, TLVs are updated on an annual basis after considerable review of current information regarding the health effects of certain chemicals.

For airborne exposures, there are five types of limits as defined here.

1. **Time-Weighted Average** (Can be noted as a PEL, WEL, REL, MAK, etc.) depending on the country legislation) is defined as "the average concentration for a normal 8-hour workday or 40-hour workweek, to which nearly all workers may be repeatedly exposed, day after day, without adverse effects."
2. **Short-Term Exposure Limit (STEL)** is defined as "the maximum concentration to which workers can be exposed for a period of up to 15 minutes continuously without suffering from (1) irritation, (2) chronic or irreversible tissue change, (3) narcosis of sufficient degree to increase the likelihood of accidental injury, impair self-rescue, or reduce work efficiency."
3. **Ceiling (C)** is defined as "the concentration that should not be exceeded even instantaneously." This value of C cannot be exceeded without proper protective actions or equipment.
4. **Immediately Dangerous to Life or Health (IDLH)** constitutes an atmosphere that poses an immediate hazard to life or produces immediate irreversible health effects. The National Institute for Occupational Safety and Health (NIOSH) Pocket Guide to Chemical Hazards defines IDLH concentration as the "maximum concentration level from which one could escape within 30 minutes without any escape-impaired symptoms or irreversible health effects." The American National Standards Institute (ANSI) defines IDLH as "any atmosphere that poses an immediate irreversible debilitating effects on health."
5. **Oxygen-Deficient Atmosphere** is an atmosphere of less than 19.5 percent oxygen by volume measured at sea level.

Although this section refers to airborne contaminants, many standards also make a note of certain chemicals when there is a dermal adsorption hazard. These chemicals are typically given a "skin notation" to indicate this hazard. The "skin notation" indicates that precautions should be taken to prevent exposure through the skin. A complete listing and specific details about these various exposure limits is too extensive for this handbook, but TLV booklet copies are available from the American Conference of Governmental Industrial Hygienists, Inc., or can be found through most government health and safety websites.

GUIDELINES TO MINIMIZE TOXIC EXPOSURE

A worker handling chemicals can minimize exposure to toxic substances by following some general guidelines:

CHEMICAL PLANT AND LABORATORY SAFETY

1. Know the chemical and toxic properties of all materials involved by reviewing the material safety data sheets.
2. Know that toxic exposure can occur through inhalation, ingestion, or through the skin. Always attempt to minimize exposure through the use of engineering controls (ventilation, enclosures, etc.) and as a last resort where suitable controls are not available, use personal protective equipment (gloves, respirators etc.).
3. Always use a fume hood or local exhaust ventilation when there is any question about potential toxicity.

Fire and chemical process safety risks should be evaluated first to determine if the general ventilation system is an integral part of the assessment of controlling the risks involved, otherwise known as a basis of safety. Globally, regulatory agencies have established required minimum air exchange rates, and these criteria must be met before evaluating any exposure concerns.

Commonly accepted ventilation rates where chemicals are handled are as follows: processing areas capable of 10 to 15 **air changes per hour** (**ACH**), laboratories 8 to 12 ACH, mechanical/service and utility rooms 6 to 10 ACH. This is, however, dependent on many factors, including the work activities and specific chemicals being handled. Airflow patterns and exchange rates should be periodically evaluated to verify their effectiveness. Aerosol generators and commercial smoke sources are available for these studies. Toxic and flammable substances should be stored in cabinets fitted with auxiliary ventilation.

In addition to proper room ventilation, **fume hoods**, **fume cupboards**, and **local ventilation systems** should be readily available for the effective containment and removal of flammable, toxic vapors, and particulates. Again, there are globally recommended guidelines to be followed. For fume hoods/cupboards face velocities around 100 ft/min or 0.5 m/s is considered the industry norm. With energy costs in consideration, newer containment technologies are now available with lower face velocities, which as a result save considerable money on heating and cooling costs. As such, some manufacturers have face velocities in the range of 0.4 m/s without compromising on containment. The hood sash (sliding glass door) has a tremendous effect on hood efficiency; the hood is least efficient when the sash is fully opened. Hoods are not intended for the storage of chemicals or equipment, these materials can adversely affect the airflow and performance of a hood. Local exhaust ventilation systems are subject to minimum air flow standards as well. The site safety specialist or industrial hygienist should be knowledgeable in these requirements. All ventilation systems should be checked periodically to ensure continued adequate performance over time and in some cases this verification is required by law at a specific timeframe.

Exposure concerns do not end by simply exhausting contaminated air away from the worker. Externally, most chemical and industrial plants have **wind socks** to indicate the prevailing wind direction. Where workers must be in the vicinity of an exhaust stack, such as on a rooftop, caution must be taken in order to minimize chances of exposure to any potentially contaminated air. There are stringent rules governing air emissions. In no case should the exhausted air be allowed to be in excess of the permitted concentrations as stated by the local environmental agency. In case of an emergency evacuation involving fire or potentially hazardous airborne contaminants, always attempt to depart from an area by using a path upwind from any contamination sources.

RISK ASSESSMENT

Modern occupational health and safety legislation typically requires that a risk assessment be carried out prior to initiation of an activity. One should bear in mind that risk management legislation in many countries requires the risk to be managed to a level that is termed "as low as is reasonably practical."

Risk assessment should follow these steps:

- Identify the hazards.
- Identify who may be affected by the hazard and how.

- Evaluate the risk (or chance of something happening).
- Identify and prioritize appropriate control measures.
- Implement the controls.
- Periodically review the assessment to verify continuing efficacy.

The calculation of risk is based on the likelihood of the event happening and the severity of the event consequences. Many companies use risk assessment matrix charts, which have numerical values associated with both the severity and likelihood. The output of such a matrix results in a semiquantitative degree of risk of the task being risk assessed. Generally, the higher the number, the greater the risk and consequently more controls are required before allowing the task to commence.

As with any risk assessment, it should be recorded and reviewed for new operations and also periodically when there are any significant changes to existing work practices. This assessment should include recommendations to minimize the risks (e.g. control measures). After these additional controls have been identified, the risk ranking process can be performed again to measure the level of residual risk with the full control measures in place. Generally with the extra controls in place the risk is deemed acceptable.

HIERARCHY OF CONTROL

From a chemical exposure perspective, a key part of a good risk assessment lies within the identification and management of potential exposures to occupational hazards. Protecting workers health and safety requires the implementation of adequate control measures. The **hierarchy of control principle** has been used as a means of determining how to implement practical and feasible controls measures. The most commonly stated summary of this hierarchy can be summarized as follows:

- Elimination.
- Substitution.
- Engineering controls.
- Administrative controls.
- Personal protective equipment (PPE).

The core thought behind this hierarchy is the control methods listed first are, generally speaking, more effective and consistently protective than those at the bottom of the list. Where elimination removes the hazard entirely, PPE can only control against it to a point. Typically, following the hierarchy principle leads to the implementation of safer work systems and can thereby reduce the risk of illness or injury.

Attempting to follow this hierarchy can be challenging even for the most seasoned professional. Many times elimination or substitution cannot be done for existing processes. For example, elimination of a hazardous chemical that is manufactured to be sold is not a viable or sensible action. For most new processes, it may be early enough in the project design/implementation phase to find suitable substitutes that are not detrimental to the overall success of the process and can be changed at a reasonable cost. For existing processes, significant changes in equipment, raw materials, and procedures may be required to eliminate or substitute for a hazard and this can be quite costly to the point where it is not a viable option. At this point in the hierarchy, other options must be considered.

Engineering controls such as barriers, shields, local exhaust ventilation systems, and even isolators are quite commonly employed with great success. These hard systems can cost significant money up front but over time tend to be better options than PPE or administrative controls, which aren't always effective as they tend to be dependent on human behavior. If an individual chooses not to wear his or her PPE, chooses not to follow the appropriate procedure, or chooses the wrong PPE or wears it improperly, the control measure doesn't work and the worker can be injured. This is not to say that PPE does not work. It can work very well but it must be selected and used properly for it to be effective and this is why it is considered a last choice in the overall hierarchy of control.

Keep in mind PPE may be a seemingly cheap and easy option, but it can cost a significant amount of money if you multiply the costs by the number of employees over a long period of time.

PERSONAL PROTECTIVE EQUIPMENT

A. Introduction

Personal protective equipment refers to clothing, eyewear, gloves, helmets, footwear, or other garments that provide protection to the wearer from bodily injury or from exposure to chemical, biological, or some radiological agents. There are various degrees of protection afforded by PPE. Occupational Safety and Health Administration and the EPA use a system of four levels of protective clothing ensembles for workers involved in hazardous waste operations or responding to hazardous chemical releases.

- **Level A** is the highest level of respiratory, skin, and eye protection available. This fully encapsulated suit must be compatible with the chemicals involved and provides the highest level of potential splash or vapor exposure. Pressure-demand, full-face self-contained breathing apparatus (SCBA) supplied-air respirator is required. This is the level recommended for site entries if operations involve high potential for splash or exposure to vapors, gases, or particulates that may have a high degree of toxicity, or are of unknown nature.
- **Level B** provides the highest level of respiratory protection but limited skin protection from airborne (gases, vapors, dusts, and mists) hazards. This level of protection should be used only when the vapors or gases are not suspected of containing high concentrations of chemical that are harmful to the skin or capable of being absorbed through the skin.
- **Level C** provides a limited level of respiratory protection through the use of an air-purifying respirator and skin protection from airborne (gases, vapors, dusts, and mists) hazards. Atmospheric concentrations of chemicals must not exceed IDLH levels and the atmosphere must contain at least 19.5 percent oxygen.
- **Level D** requires no respiratory protection and limited skin protection from airborne hazards. Atmospheric concentrations of chemicals must not exceed PEL levels and the atmosphere must contain at least 19.5 percent oxygen.

B. Eye Protection

The National Safety Council estimates that job-related eye trauma costs amount to $300 million annually in the United States. Additional statistics claim that nearly 90 percent of all eye injuries are preventable. With these statistics in mind, it is important to discuss the importance of eye protection. Eye safety is a major concern for the process industry. Proper eye protection is generally a requirement in plant operations where there are chemical or physical hazards present that may be harmful to the eyes. As with other personal protective equipment, there are a wide variety of styles available, which offer varying degrees of protection. Common prescription glasses are generally not approved for safety. The only approved eye protection are those that have passed the rigorous safety standard testing. Most countries have their own test criteria; in the United States, American National Standards Institute establishes basic safety eyewear performance criteria (Z87.1) and in the European Union, basic safety glasses standards must meet the criteria established by EN166.

Everyday safety glasses with side shields are worn for general impact resistance against flying particles (see Fig. 2.1). This form of eye protection is the most common in general industry as it is usually a requirement for all persons entering a plant and laboratory to wear them at all times. Safety glasses do not provide adequate protection against splashes of hazardous chemicals.

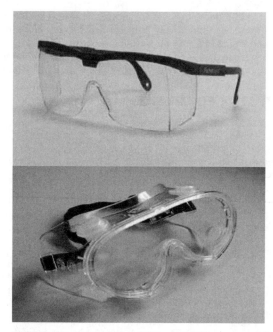

FIGURE 2.1 Safety glasses and goggles. (*Photo courtesy of Thermo Fisher Scientific.*)

For protection against harmful chemicals where there may be a splash or spray hazard, chemical splash goggles or shields should be worn (see Fig. 2.2). Splash goggles are tight fitting to the face and thus form a seal to prevent liquid from entering in the event of a splash to the face. Splash goggles are also effective for protection against airborne dust which may be harmful to the eyes, and are effective for impact protection from such operations where quantities of fast moving particles may be generated (e.g. grinding, cutting wheels, etc.).

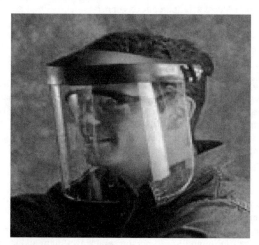

FIGURE 2.2 Face shield. (*Photo courtesy of Thermo Fisher Scientific.*)

In situations where there are chemical or physical hazards of great magnitude, it may be necessary to protect the entire face from potential injury. Full-face shields offer protection for the entire face of the wearer. Face shields are considered to be a secondary form of protection and safety glasses or goggles MUST always be worn with a face shield. Full-face shields are ideal for protection against chemical splashes and also for impact protection.

Another form of full-face protection is that of a welding helmet. Welding helmets not only provide the wearers eyes protection from the hot flying particles but also from ultraviolet light and radiant heat generated in welding operations. Welding helmet lenses are manufactured with varying degrees of darkness depending on the intensity of the welding application for which it is to be used. Darker lenses are needed for adequate protection against the higher-intensity welding operations such as plasma arc welding. In gas welding or oxygen cutting where the torch produces a high yellow light, it is desirable to use a filter or lens that absorbs the yellow light (Table 2.3). (See "Welding Safety" in Chapter 18 for more details.)

TABLE 2.3 Welding Filter Lenses for All Exposed Personnel

Welding operation	Shade number
Soldering	2
Torch brazing	3 or 4
Light cutting, up to 1 in	3 or 4
Medium cutting, 1 to 6 in	4 or 5
Heavy cutting, 6 in and over	5 or 6
Light gas welding, up to 1/8 in	4 or 5
Medium gas welding, 1/8 to ½ in	5 or 6
Heavy gas welding, ½ in and over	6 or 8
Shielded metal-arc welding, 1/16- through 5/32-in electrode	10
Gas shielded arc welding (nonferrous) 1/16 through 5/32 in	11
Gas shielded arc welding (ferrous) 1/16 through 5/32 in	12
Shielded metal-arc welding, 3/16- through ¼-in electrodes	12
Shielded metal-arc welding, 5/16- through 3/8-in electrodes	14
Atomic hydrogen welding	10–14
Carbon arc welding	14

C. Hearing Protection

There are several reasons individuals lose their ability to hear; the normal aging process, disease, some medications, genetics, and exposure to high noise can all cause hearing loss. With some noise exposures an individual's hearing may be affected for a short duration and hearing levels eventually return to their normal state. This is known as a **temporary threshold shift** or TTS. However, for some high noise exposures, a **permanent threshold shift** or PTS can occur where hearing levels do not return to normal. Even where temporary threshold shifts occur, these can become permanent if the noise exposures are repeated on a regular basis.

Most countries have established noise exposure limits (European Union: 87 dBA TWA, U.S.: 90 dBA TWA) to which it is believed the general working population can be routinely exposed before suffering a noise-induced hearing loss. The importance of the use of hearing protection is clearly evident in the medical literature for the prevention of noise-induced hearing loss. Controlling exposure to high-noise environments is relatively straightforward through the use engineering controls. However, where the noise source cannot be eliminated or isolated, the use of hearing protective devices (HPDs) is necessary.

There are two main types of hearing protective devices—ear plugs and ear muffs. Before putting in ear plugs, roll them tightly between the thumb and finger. Pull the ear back with one hand and insert the compressed ear plug in with the other hand. Pulling on the ear lessens the natural bend of the ear canal and causes the plug to fit more tightly. Reusable ear plugs can be cleaned with soap and water. If ear muffs are worn for sound protection, do not wear earrings as they can reduce the effectiveness of the muffs or cause injury if the muff gets caught on the earring. In addition, some safety glasses can interfere with the performance of hearing protective devices just by the nature of interaction with the safety glasses temple bars sitting under the ear muff. Ear muffs and ear plugs each have their own advantages and disadvantages, so it is important to determine what works best for a particular application.

D. Hand Protection

Gloves are very important for providing a barrier between the worker and possible exposure to a harmful physical or chemical hazard. There are a variety of gloves that provide protection from virtually all types of physical and chemical hazards, including extreme heat, cold, abrasions, cuts, and chemical exposure. Some newer gloves are constructed of multilayered materials and thus can provide protection against multiple chemical and physical hazards. When confronted with a situation of two or more hazards each of which requires different glove materials, gloves can be layered to achieve adequate protection. Even though dexterity of the hand is usually encumbered somewhat, this is an acceptable method to provide protection against multiple chemical/physical hazards where the exposure itself cannot be eliminated. Some of the most commonly encountered glove materials and their applications are discussed in the following paragraphs.

Leather gloves are a common multipurpose glove material for protection against most physical hazards. Leather can provide protection against heat, cold, cuts, and abrasions, and can also provide for a better grip in some situations. Leather gloves offer little protection against chemical exposures.

Steel mesh gloves made of small interlinked rings can provide excellent protection against cuts and exposure to heat. These gloves are used in the meat packing industry for protection against sharp cutting utensils. The interlinked ring design provides an air space that allows for the heat to dissipate before causing a thermal burn to the user. These steel mesh gloves are usually attached to a leather glove for added protection to the wearer.

An increasingly popular glove material for protection against physical hazards is Kevlar®. Its resistance to cuts and abrasions along with a longer lasting life than that of leather makes it a desirable glove material. Chemically resistant gloves are commercially available, but no one type of glove is resistant to all chemicals; therefore care should be exercised in selecting the proper gloves. Table 2.4 is an abbreviated listing of gloves and their recommended applications for use against exposure to chemicals. Always confirm with the glove manufacturer what the limitations of the glove material may be.

E. Protective Apparel

Sandals, cloth sneakers, and open-toed shoes are not recommended in chemical or manufacturing facilities and may be banned in some areas for safety reasons. Safety shoes equipped with protective, steel-toe reinforcement are suggested for all industrial workers. Some common examples of foot injuries come from crushed or broken feet, punctures of the sole, cuts, burns, electric shocks, and sprained or twisted ankles, to name a few. Over 100,000 disabling work injuries to feet and toes occur every year in the United States. The greatest majority of workers who sustained injury were found not to be wearing appropriate foot protection or were working in an area where the risks were not adequately controlled. In some cases specialized antistatic (or static dissipative) shoes are required where static electricity can present a fire and explosion risk where flammable liquids, powders, or gases may be present. It is important to note what the footwear requirements are before entering a work area.

Protective clothing ranging from rubber aprons, Tyvek™ coveralls, high visibility jackets, and laboratory coats to entire outer garments is commercially available. The use of garments to guard against chemical exposures can prove to be very effective. Guarding against exposure does not end

TABLE 2.4 Glove Materials and Compatibilities*

	Neoprene	Nitrile	Rubber	PVC	Butyl
Acetaldehyde	F	F	P	P	
Acetic acid	G-E	E	G		
Acetone	E	F			E
Acetonitrile	F-G	P	P	NR	G
Alkali hydroxides (Na, K, Ca)	E	E	E-G	E-G	E
Ammonium hydroxide	E				
Amyl alcohol	E	E	G		
Aniline	P-NR	P	NR	NR	
Animal or vegetable fats and oils	E		F-P	G	
Benzaldehyde	P	P	P	P	
Benzene	NR	NR	NR	NR	P
Benzyl alcohol	G	G	F	G	
Benzyl chloride	P	P	NR	P	
Butyl acetate	F	F	P	P	G
Carbon tetrachloride	F-P	F	NR	F	P
Chlorobenzene	NR	P	NR	P	P
Chloroform	F-P	F	NR	F	P
Cresols	F-P	F	NR	F	
Dibutyl phthalate	E		F	P	
Diethylamine	F-P	F	P	P	P
Diethyl ether	P	F	P	P	
Dimethyl formamide	F-G	G	P	P	E
Dimethyl sulfoxide (DMSO)	G			F	E
Dioctyl phthalate	G		F	P	
Dioxane	F	F	P	F	
Ethyl acetate	F-G	F	F	P	G
Ethyl benzene	P-NR	P	NR	P	
Ethylene glycol	F-G	G	F	G	
Formalin, 37%				E	E
Formic acid	G	F	G	G	
Gasoline	G		P	F	
Glycerine (glycerol)			G	E	
n-Heptane	G		P	G	
n-Hexane	G		P	G	P
Hydraulic fluid, cutting oil,	G		P	G	
Hydraulic oil	G	G	NR	F	
Hydrochloric acid	F-P	F	P	P	E
Isooctane	G		P	F	
Isopropyl alcohol	G	G	G	G	
Kerosene	G		P	G	
Methyl alcohol	G	G	G	F	
Methylene chloride	NR	NR	NR	NR	P
Methyl ethyl ketone	P	NR	F	NR	E
Methyl isobutyl ketone	P	NR	P	NR	
Mineral Oils	E-G	E	P	E	
Nitric acid	P	P	NR	NR	
Nitrobenzene	P	P	NR	P	
Octane	E-G	E	P	F	
Perchloroethylene	F-P	F-G	NR	P	
Sulfuric acid 20%	F	F	F	F	G
Toluene	F-P	F-G	NR	F	F
Xylene	F-P	F-G	NR	F	P

*E=excellent, G=Good, F=Fair, P=Poor, NR=Not Recommended, Blank=No Data

simply when the task stops. Chemical workers must know the appropriate technique for the safe removal, donning, cleaning, inspecting, and disposal of contaminated protective apparel. The act of removing a contaminated garment can expose the worker to the very chemical it was designed to guard against. Always use good hygienic methods, follow the manufacturer suggestions, and where available use a misting shower to damp down dusts present on the PPE or a full shower decontamination to remove other hazardous contaminants.

F. Head Protection

The National Safety Council reports there are in excess of 300 head injuries sustained each day on the job and over 120,000 every year. More than 50 percent of these cases required days away from work, job restrictions, or a transfer to a new role. It is suggested employers provide head protection where there is a possibility of falling or flying objects that could strike the head, where workers could bump into fixed objects, be exposed to electric shock, or where loose hair could become entangled in machinery. Before using a hard hat, check the label to determine the proper type and classification of head protection for the job. In addition, verify the hard hat has not exceeded the manufacturer expiration date, as plastics degrade over time. Hard hats provide protection from falling objects, bumps, burns, and electric shock. Never wear a hard hat backward; do not make modifications to them nor drill holes or apply stickers to the shell as this can compromise the performance of the hard hat. The suspension system inside the shell of the hat is designed to distribute an impact throughout the shell. It is imperative to the performance of the hard hat that the suspension is in good condition and appropriately installed inside the hard hat. Check periodically that at least a 1-in gap exists between the crown strap system and the shell. A common method of cleaning shells is dipping them in warm water containing a mild detergent for at least one minute, but it is strongly suggested that all care and use recommendations from the manufacturer be followed.

Under the U.S. classifications system, there are two types of hard hats and three classifications as follow. Notice that only type, Class B, protects workers from electric shock. Other countries have similar class systems as well.

- **Type 1**. Helmets with a full brim (minimum 1-¼ in wide).
- **Type 2**. Brimless hard hats with peak extending forward from the crown.
- **Class A**. General service with limited voltage protection.
- **Class B**. Utility service, high-voltage protection.
- **Class C**. Special service, no voltage protection.

G. Respiratory Protection

Introduction. Respiratory-related illnesses caused by inhalation of toxic materials have been noted for hundreds of years. Classic examples of such diseases are black lung disease suffered by coal miners or even lung cancers from exposure to asbestos fibers. Respiratory protection is a vital last line of defense for the minimization of exposure to hazardous substances. There are many regulations governing the allowable concentrations of chemicals one can be exposed to, and one way to attain safe exposures for workers after considering other options is through the use of respiratory protective equipment (RPE). In the United States, OSHA 29 CFR 1910.134 Respiratory Protection Standard outlines the requirements around the use of RPE. There are three types of respiratory hazards: oxygen-deficient atmospheres, gases and vapors, and airborne particulate contaminants. Normal air contains approximately 21 percent oxygen, 78 percent nitrogen, and 1 percent other gases, mostly argon. Air is commonly classified as oxygen deficient if it contains less than 19.5 percent oxygen. An oxygen-deficient atmosphere is most commonly found in confined spaces, like tanks and sewers. Workers are not always aware that they are in an oxygen-deficient atmosphere. Low oxygen levels below 19 percent can impair judgment and coordination and the effects become more pronounced and serious at lower oxygen concentrations. Continued exposure to an oxygen-deficient atmosphere can lead to unconsciousness and even death. The various types of respiratory protective equipment as well as their applications and limitations are described in detail in the following sections.

Dust Masks or Filtering Face Pieces. The dust mask is generally a disposable mask of varying filtration efficiencies, consisting of a fibrous pad that is contoured for the face and is secured to the wearer with two elastic bands. This mask is used to protect the worker from nuisance dusts, fumes, and some aerosols/mist, but it does not provide protection against vapors or gases. Dusts are composed of solid airborne particles (e.g., from processes like crushing, grinding, milling, solids mixing, and sanding). Dust particles tend to be larger than fumes. Metal fumes are actually made from welding operations where metals are vaporized and the gaseous metals condense and solidify back into small solid metal particulate. Fumes are generated from processes like welding, smelting, soldering, and general furnace work. Mists are small droplets of airborne liquids. Mists can be created by processes like spray painting and machining. A mist that is concentrated enough to block light or visibility is referred to as a fog. Smoke is also concentrated enough to block vision and is composed of vapors, gases, particles, and mists. Gases are defined as substances that exist in a gaseous form at room temperature. Gases tend to be colorless, odorless, and tasteless. Vapors are the gaseous form of a material that is typically in a solid or liquid state at room temperature. Dust masks provide the lowest level of protection compared with all types of respirators, but they do serve a function in many industries where exposures are relatively low, for example, in odd jobs where there may be dust generated from sweeping a floor, or in grinding operations where particles are generated. Dust mask must not be confused with the surgical masks seen in use in medical facilities, as this type of mask is not actually a respirator but rather a barrier to minimize exposure to biological hazards from blood droplets, saliva, sneezing, or cough sputum.

Dust and particulate respirators (Fig. 2.3) use fiber filters to trap the contaminants as recommended by the manufacturer. This type of respirator is usually disposable and afford a modest level of protection against toxic particulates, aerosols, and mists. These masks are available with variable levels of filtration efficiency, anywhere from 95 percent up to 99.9+ percent efficient.

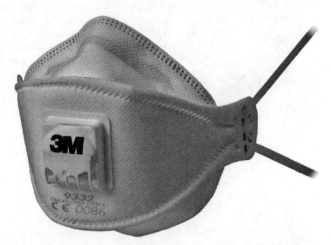

FIGURE 2.3 Dust and particulate (mask) respirator. (*Photo courtesy of Thermo Fisher Scientific.*)

Cartridge Respirators. Cartridge respirators are another type of air purifying respirators and are effective against gases, vapors, and particulates at concentration levels specified by the manufacturer. Before selecting a respirator, always consult the appropriate MSDS for the chemicals being used and the allowable exposure limits. Chemical cartridge respirators have a replaceable color-coded canister (cartridge)

TABLE 2.5 Respirator Cartridge Color Coding and Use

Atmospheric contaminates	Color of cartridge
Acid gases	white
Hydrocyanic acid gas	white with ½-in green stripe
Chlorine gas	white with ½-in yellow stripe
Organic vapors	black
Ammonia gas	green
Acid gas and ammonia gas	green with ½-in white stripe
Carbon monoxide	blue
Acid gases and organic vapor	yellow
Hydrocyanic acid gas and chlorpicrin vapor	yellow with ½-in blue stripe
Acid gases, organic vapors, and Ammonia gases	brown
Radioactive materials (except Tritium and Noble gases)	purple
Particulates (dust, fume, mist, fogs, smoke) in combination with any of the above gases and vapors	canister color for contaminate, as designated above, with ½-in gray stripe
All of the above atmospheric contaminates	red with ½-in gray stripe

which is filled with an adsorptive material (see Table 2.5 for a listing of cartridge types). This material purifies the air by attracting and binding specific airborne contaminants. These cartridges have a limited service life and become ineffective when they become saturated to the point that breathing becomes labored or chemical odors are detected by the wearer. At this point the cartridges must be replaced with a new set.

Chemical cartridge respirators (Fig. 2.4) are primarily effective only against gases and vapors at concentration levels specified by the manufacturer. The contaminant is usually adsorbed in a sorbent material, like activated charcoal; thus the cartridge must be replaced periodically. Saturation is usually indicated by difficulty in breathing or the odor of the contaminant passing through the cartridge. Operators should test their individual mask for proper fit before entering contaminated areas. This type of respirator is not to be used in atmospheres containing less than 19.5 percent oxygen.

There are two main styles of cartridge respirators, the full-face cartridge respirator and the half-face cartridge respirator. Full-face respirators provide the wearer better protection because the entire face is enclosed and toxins are less likely to leak around the edges. Full-face respirators have a transparent lens which gives the wearer eye protection in addition to the respiratory protection provided. Half-mask respirators operate on the same principle of air purification as the full-face respirators; however, half masks only cover the mouth and the nose and are slightly less effective in sealing out air contaminants than a full-face respirators.

In addition to the organic vapor cartridges, there are cartridges available that can purify harmful or toxic particulates such as toxic dusts,

FIGURE 2.4 Chemical cartridge respirator and color-coded disposable cartridge. (*Photo courtesy of Thermo Fisher Scientific.*)

welding fumes, and asbestos fibers. High-efficiency particulate air (HEPA) respirators are filters capable of a filtering efficiency of 99.97 percent at 0.3-μm-diameter particulate (see Table 2.6 for HEPA filter classifications).

TABLE 2.6 HEPA Filter Classifications for OSHA and EU

	N-Series	R-Series	P-Series
99.97% efficient	N100	R100	P100
99% efficient	N99	R99	P99
95% efficient	N95	R95	P95
European Union HEPA Specifications—EN 1822-1			
HEPA class	Retention (total)		
H10	>85%		
H11	>95%		
H12	>99.5%		
H13	>99.95%		
H14	>99.995%		

Respirator Fit Tests and Fit Checks. Each employee should be medically examined and must be fit-tested before being assigned a respirator. The respirator face piece must form a complete seal with the face. Employees should be cautioned that facial hair can prevent a proper respirator fit or seal. Fit testing is used to ensure that a proper seal is formed and contaminated air cannot enter. A **qualitative fit test** (see Fig. 2.5) exposes the wearer to an odorous chemical or an agent that is sweet or bitter tasting.

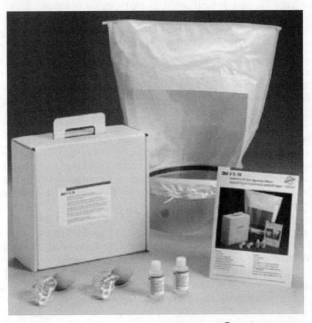

FIGURE 2.5 Qualitative fit test kit. Includes Bitrex® sensitive solution, fit test solution, testing hood, collar assembly, and spray nebulizers. (*Photo courtesy of Thermo Fisher Scientific.*)

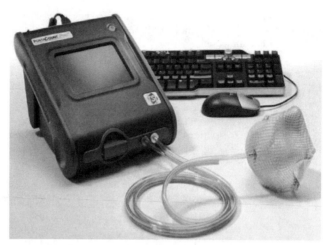

FIGURE 2.6 Quantitative fit test instrument. TSI Portacount (Model 8038) Respirator Face Testing Equipment. (*Photo courtesy Thermo Fisher Scientific.*)

If any odor or taste is detected, then the respirator fails the test. This type of testing is very subjective to the wearer and as such is not always a reliable method for determining fit.

A **quantitative fit check** involves an electronic instrument that samples air inside the wearer's respirator. The instrument (see Fig. 2.6) actually compares the air quality inside and outside the mask to determine if the respirator is performing adequately. Quantitative fit testing is the gold standard in fit-testing processes; however, it is more expensive to perform, is more time consuming, and requires a trained operator as well as a dedicated area for fit testing.

Both a positive and negative pressure user fit checks should be performed each time a respirator is to be used to determine if the respirator is fitting properly. A **negative pressure fit check** requires the wearer to adjust the respirator for a comfortable fit. The wearer places his or her palm over the inhalation inlets to prevent air from entering. The tester inhales slowly for about 10 seconds. A properly fitted face piece should collapse inwardly from the vacuum created. An employee can perform a **positive pressure fit check** by blocking the exhalation valve with the palm of his or her hand and blowing air slowly into the face piece. The face piece should bulge out slightly without significant leakage around the face piece.

Supplied-Air Respirators. **Supplied-air respirators** provide effective protection against a wide range of gases, chemicals, and particulate contaminants and can be used in an oxygen-deficient atmosphere. Breathing air is supplied to the wearer's face mask under positive pressure from a compressor which reduces the likelihood of contaminants entering the face mask. It is critical that the compressed air be of suitable quality for breathing and be free of toxic contaminants and also provide sufficient volumes for the wearer(s) at the end of the airline. Breathing air can be supplied via a manifold of compressed air cylinders, with diesel-powered or electric compressors. Air pumps are available that are electrical powered and produce no carbon monoxide, oil vapor, or mist, which is a concern when it comes to supplying a clean breathable air supply. Some commercial units come equipped with a range of filters and in-line sensors (commonly testing for carbon monoxide, carbon dioxide, oxygen level, temperature, and humidity) to ensure the appropriate air quality is maintained. Some permanent installations of industrial units can supply enough breathing air for up to several workers wearing continuous-flow half-mask or full-face piece respirators or for workers wearing continuous-flow air hoods.

Self-Contained Breathing Apparatus. **Self-contained breathing apparatus (SCBA)** is the ultimate in protective respirators for emergency work. This equipment consists of a full-face mask

FIGURE 2.7 Self-contained breathing apparatus (SCBA). (*Photo courtesy of Thermo Fisher Scientific.*)

(see Fig. 2.7) connected to a cylinder of compressed air. In general, there are no restrictions as to contaminants, concentrations, or atmospheric oxygen concentrations as the wearer has his or her own supply of air. However, the tank does have a limited capacity (usually 30 minutes for a standard 200 Bar or 3000 psi air cylinder). The SCBA units are bulky, heavy, and additional safety gear may be required such as gas tight suits if certain toxic contaminants are present. This type of respiratory protection is typically limited to specialized work activities such as firefighting, hazardous materials response teams, and confined-space entry.

One brand of SCBA equipment that is commonly used is the Scott Air Pack. The Scott Air Pack comes in a variety of styles: lightweight composite cylinders, steel cylinders, models designed for firefighting applications, and even a recirculating model which reuses the exhaled air of the wearer to allow for wearing times of up to an hour or more.

All self-contained breathing apparatus wearers must have the necessary training to apprise them of the physical limitations of such equipment, the applications they may be used under, and the proper donning and doffing procedures before entering an atmosphere dangerous to life and health. It is generally accepted that high-risk operations are subcontracted to specialist firms rather than companies providing their own suitably equipped and competently trained individuals to perform the task.

Care and Maintenance of Respirators. Where respirators are not disposable, it is required to maintain your RPE in a clean and functional state. Cleaning of personal respirators can be done by following the manufacturer's recommendations. Generally, the following guidelines apply for cleaning:

- Following the manufacturer's guidelines, disassemble the face piece, remove the cartridges, and wash with warm water containing a mild detergent. Do not wash the cartridges.
- Rinse face piece with clean water and repeat step one again and allow to dry. Do not attempt to dry the plastic lens as this may scratch the lens.
- Reassemble, attach new cartridges, and store clean respirator in a dust-free, clean environment in a sealable plastic bag.

CHECKLIST FOR WEARING PERSONAL PROTECTIVE EQUIPMENT (PPE)

All personnel must have specific training in the proper care and use of PPE. The checklists shown in Tables 2.7 and 2.8 are not necessarily complete and are presented only as a general guideline for wearing protective clothing and gear. It should be noted that heat stress can become a major factor in wearing PPE. Personal protective equipment can add significant weight and bulk to the worker and severely reduces the body's access to normal heat exchange mechanisms. In hot work environments (hot climates, foundries, etc.) a proper work/rest regime must be established, with frequent hydration breaks, to minimize possible heat stress.

TABLE 2.7 General PPE Pre-use Checklist

1. Ensure the appropriate PPE has been chosen for the job as stated in the work procedure.
2. Check that the clothing, gloves, and respiratory equipment are appropriate (compatible) for specific hazard(s).
3. Inspect PPE for tears, pinholes, cracks, malfunctioning closures, etc. Check interiors for discoloration, stiffness, and/or swelling.
4. Check gloves for holes, discoloration, cracks, or tears.
5. Adjust headpiece of hard hat to fit user's head.
6. Inspect and clean all eye protection (glasses, goggles, face shields, etc.) equipment so that clear vision is possible.
7. Ensure any expiration dates on the PPE have not been exceeded.

TABLE 2.8 Complete Suiting Procedure and Checklist (Only by Trained Personnel)

1. Step into the legs of suit from a standing or sitting position; ensure proper placement of feet in suit. The suit should be donned to the waist only at this point.
2. Put chemical-resistant boots on over the feet of the suit. Tape the suit's leg cuff over the top of each boot. Repeat this taping procedure if over boots are to be worn.
3. Put on the SCBA harness and air tanks assembly. Don the face piece and adjust for comfortable but secure fit.
4. Without connecting the breathing hose, open the valve on the air tank.
5. Perform positive and negative respirator face piece seal procedures.
6. Put on inner (surgical) gloves.
7. Put on basic gloves if not attached to the suit.
8. With the aid of an assistant, insert arms into suit sleeves and pull suit over the SCBA assembly. Adjust the suit for unrestricted motion.
9. Put on hard hat if required.
10. Put on outer gloves and tape suit sleeves to gloves.
11. Carefully raise hood over head so as not to disrupt SCBA mask face seal. With the aid of an assistant, adjust hood for comfort.
12. Secure all suit closures, except area for breathing hose connection. Secure all belts and clothing adjustments.
13. Connect the breathing hose while opening the main valve.
14. Assistant should check the wearer for proper breathing and make final closure on the suit.
15. Assistant should observe the wearer to ensure comfort, stability, and proper equipment operation.
16. At least one assistant should continuously monitor this PPE-equipped worker for heat stress and other problems.

FIRE SAFETY

A. Introduction

Portable fire extinguishers are covered in a comprehensive format in OSHA 29 CFR 1910.157. It is estimated that 28,000 industrial fires occur each year in the United States and result in some 596 million dollars in property damage. Ninety percent of these industrial fires are caused by human error. This section of fire safety will provide an overview and some of the equipment involved in fire safety. *Fire* can be defined as the rapid combination of a combustible substance with oxygen, accompanied by the rapid evolution of heat and light. When most substances burn, the actual combustion takes place only after the solid or liquid material has been vaporized or has been decomposed by the heat to produce a gas. The visible flame is the burning gas or vapor. All burners and other sources of ignition should be placed safely away from all flammable materials. For combustion to take place, three components must be present (see Fig. 2.8):

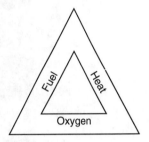

FIGURE 2.8 The fire triangle (oxygen/fuel/heat).

- Fuel.
- Oxidizer (oxygen, in the case of air).
- Heat (an ignition source to bring the temperature up to the ignition point of the fuel).

A fire will not start if any one of these three components is missing. However, once the fire has started, the heat evolved by the fire itself becomes the ignition source.

B. Classifications of Fire

The National Fire Protection Association (NFPA) has established four categories or "classes of fires." (see Chapter 3 for details). Other country fire classifications are similar in principle, but may contain additional classifications in addition to those following.

Class A Fires. Involving routine combustibles such as wood, cloth, paper, rubber, plastics, and so on.

Class B Fires. All flammable liquids (Classes I, II, and III liquids, greases, solvents, paints, etc.). Details of these classifications are shown in Table 3.2, "NFPA Classifications for Combustible Chemicals."

Class C Fires. All fires involving energized electrical equipment and apparatus (computers, motors, hot plates, ovens, instruments, etc.).

Class D Fires. All fires involving combustible metals (magnesium, potassium, sodium, lithium aluminum hydride, etc.).

Underwriters Laboratories (UL approved) tests and assigns numerical value to each size and type of fire extinguishers indicating the approximate square-foot area of fire-extinguishing capacity. For example, a 5-B extinguisher should be able to extinguish a 5-ft^2 area of Class B combustibles.

C. Types of Fire Extinguishers

There are five different types of fire extinguishers and they are classified by the type of fire for which they are suited:

1. Water extinguishers are effective against Class A type fires. Water extinguishers should *never* be used for extinguishing organic liquid (Class B), electrical (Class C), or metal (Class D) fires. Obviously, this type of fire extinguisher has very limited uses and could possibly be hazardous in a chemical process environment.

2. Carbon dioxide extinguishers are effective against burning liquids (Class B) and electrical (Class C) fires. They are especially recommended around chemical instrumentation containing delicate electronic systems or optics. Caution should be used with this type of extinguisher because the tremendous force of the escaping gas can break expensive glassware. Operator asphyxiation can also be a problem in closed areas. This type of extinguisher is not very effective against paper or trash fires and should not be used on some combustible metal and metal hydride fires (sodium, potassium, lithium aluminum hydride, etc.) because it produces water from atmospheric condensation.

3. Dry chemical extinguishers are effective against Classes A, B, and C fires and are especially useful when large volumes of burning liquids are involved (see Fig. 2.9). The solid powder contained in this type of fire extinguisher tends to smother the oxygen from the fuel. These extinguishers are usually filled with an inorganic compound like ammonium phosphate under nitrogen pressure; shaker-type canisters are also available for small fires. This type of extinguisher is not very effective against metal fires. Even though effective against electrical fires, it is not recommended for instrumentation fires because of the difficulty in cleaning delicate electronic and optical systems.

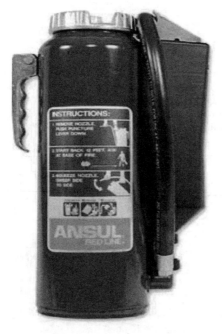

FIGURE 2.9 Typical multipurpose, dry chemical fire extinguisher. (*Photo courtesy of Thermo Fisher Scientific.*)

Another type of dry chemical extinguisher is the Class BC-rated extinguisher. These units are typically filled with either potassium or sodium bicarbonate. These chemical agents are nontoxic, nonconductive, and relatively easy to clean up.

Met-L-X extinguishers are specialized for burning metal (Class D) type fires. They contain a granulated sodium chloride formulation which tends to be very effective against burning metals, metal hydrides, and difficult organometallic compound fires. However, these extinguishers tend not to be very effective against the Class A, B, or C types of fire. Met-L-X extinguishers come in two varieties: a conventional gas cartridge-operated canister and a simple shaker-type container. Obviously, the shaker type is intended for small fires and the powder should be permitted to fall gently on the burning metal surface. Dry sand is also very effective for this type of fire.

4. Halogenated hydrocarbon extinguishers represent a type of fire extinguisher that has very limited usage. Once deemed a clean extinguisher, it became evident that the fire-extinguishing agent is a powerful ozone-depleting substance and as such is being phased out rapidly in many countries. This type of extinguishing system is not in general use but is still in use in specialized applications such as police, armed services, and in some aircraft.

D. How to Use a Fire Extinguisher

Remember PASS:

Pull the pin.

Aim at the base of the fire.

Squeeze the handle slowly.

Sweep the extinguisher nozzle side to side.

SAFETY EQUIPMENT

A. Eyewash and Safety Showers

Eyewash stations and safety showers should be readily available (see Fig. 2.10). Each worker should know and understand emergency procedures and how to obtain assistance, evacuation procedures, alarm system, and shutdown and personnel return to work procedures, and so on. Safety drills should be a routine practice for all employees. Regular functional checks and maintenance of this safety equipment are strongly recommended to ensure it remains functional.

B. Fall Protection

According to Department of Labor statistics, nearly 14 percent of all falls from height are fatal. In 2006 alone there were 809 fatalities in the United States, from activities such as falling from roofs, unprotected walkways, ladders, and scaffolds. Where falls are not fatal, over 300,000 workers suffer disabling injuries from falls each year. Nearly every industrialized country has regulations dealing with personal fall protection programs, including the implementation and use of body harnesses, positioning devices, scaffolds, safety nets, ladders, safety belts, lifelines, and lanyards. The best protection against falls is not to work from heights to begin with wherever possible. Wherever the height work must be performed, the use of suitable controls such as railings and barriers must be implemented to prevent a fall. There are some tasks where PPE must be used in order to minimize the significance of a fall and this is where the use of harnesses, lanyards, and fall-arrest systems come into use. Lanyards (Fig. 2.11) must be anchored to a fixed point, lifeline, or drop line capable of supporting a dead weight of 5400 lb per worker.

The American National Standards Institute (ANSI) Standard A10.14-1975 covers safety belts, harnesses, lanyards, and lifelines for industrial use. The ANSI standard specifies four classes of equipment:

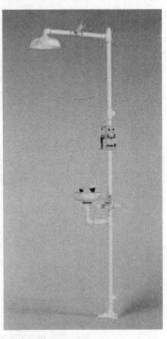

FIGURE 2.10 Eyewash fountain and safety shower. (*Photo courtesy of Thermo Fisher Scientific.*)

- Class I covers body belts used to restrain a worker performing tasks in a hazardous work position and to reduce the probability of falls. It should be noted that body belts (Fig. 2.12) are not recommended for fall protection.
- Class II covers chest harnesses which are to be used for limited fall hazards or for retrieval of a person from a tank or bin.

FIGURE 2.11 Shock-absorbing lanyard. (*Photo courtesy of Thermo Fisher Scientific.*)

FIGURE 2.12 Body belts are not recommended for fall protection use. (*Photo courtesy of Thermo Fisher Scientific.*)

- Class III covers body harnesses (Fig. 2.13) which are to be used to arrest the most severe free falls.
- Class IV covers suspension belts which are to be used to provide independent work support, as a tree trimmer belt or raising/lowering harness.

C. Ladder Safety

OSHA 29 CFR 1926.1053 and OSHA 29 CFR 1910.25-1910.29 cover ladder safety in the process industry. Over 300 deaths and 65,000 disabling injuries occur each year because of improper ladder use. Almost 10 percent of all fatal occupational injuries are due to falls from ladders and other elevated worksites (see Table 2.9). When appropriate, every process operator should receive specialized training in the use of full-body harnesses, rope grabs, lifelines, anchor points, and other fall-arrest equipment. All ladders sold in the United States must have a posted duty rating based on American National Standards Institute (ANSI) Standards A14.1 and A14.2. Table 2.10 lists the standard duty rating of these ladders; since OSHA generally considers the nominal weight of an industrial worker to be 250 lb (with clothing, PPE, and tools), only Types I and IA are permitted for employees in industrial settings.

FIGURE 2.13 Full-body harness. (*Photo courtesy of Thermo Fisher Scientific.*)

D. Aerial Lift Safety

The OSHA 29 CFR 1926.556 and OSHA 29 CFR 1910.67 standards deal primarily with aerial lift safety. Federal and local regulations require that all aerial lift operators be trained. These regulations require that the aerial lift's manual be stored at all times on the unit and in a designated compartment. All aerial and ground hazards (power lines, traffic, and ground surface conditions) must be eliminated or identified before an aerial lift is operated. An aerial lift should never be used as a crane. Never exceed the manufacturer's load capacity guidelines. It is recommended that a block at least

CHEMICAL PLANT AND LABORATORY SAFETY

TABLE 2.9 Fall Protection Guidelines

- Only trained workers should attempt to use fall protection equipment.
- Always evaluate fall protection equipment as a total system. Remember that a chain is only as strong as its weakest link.
- Inspect fall protection equipment before each use and inspect at least twice a year according to the manufacturer's instructions.
- Destroy safety belts and lanyards that have been subjected to any fall.
- Know your fall protection equipment and its limitations. This includes: lanyard, deceleration devices, lifelines, pole strap, window cleaner's belt, lineman's body belt, full-body harness, D-rings, snap-hooks, rope grabs, etc.
- Know proper anchoring and tie-off techniques.
- Know how to inspect, maintain, and store personal fall protective equipment.
- Always keep three points of contact when climbing a ladder.
- Do not carry heavy objects while climbing; always wear a tool belt and/or use a towline.
- Never allow free fall to exceed 6 ft before the fall arrest system activates.
- Never allow the stopping distance (deceleration distance) to exceed 3.5 ft.
- Only full-body harnesses are recommended for personal fall protection, not body belts.
- Lanyards and lifelines must have a breaking strength of 5000 lb or greater.
- The anchorage point should be able to support a weight of at least 5000 lb for each worker.
- The anchorage location should not allow a worker to collide with a lower-level hazard in the event of a fall.
- Always determine that the locking snap-hook is in proper working condition. The spring-loaded keeper should not open under pressure and only open by releasing the mechanism.
- Keep the tie-off point to the lifeline or anchorage at or above the D-ring on the worker's full-body harness. This will keep the free fall distance below the required 6 ft.

TABLE 2.10 Standard Duty Ratings of Manufactured Portable Ladders

Type IA	Extra Heavy Duty, Industrial	Rating: 300 lb
Type I	Heavy Duty, Industrial	Rating: 250 lb
Type II	Medium Duty, Commercial	Rating: 225 lb
Type III	Light Duty, Household	Rating: 200 lb

three times the size of the outrigger floats be used to distribute the aerial lift's weight. Always cordon off the aerial lift's work area to prevent accidents with other moving vehicles. Never climb from the platform or aerial basket to an adjacent structure. Never use a ladder or other device to extend your reach in an aerial lift. Never tamper with or block the dead man's switch since this safety device is designed to prevent the movement of the aerial lift. Before use or at the beginning of each shift, all aerial lifts should be inspected.

Most self-propelled aerial lifts do not have insulated platforms; thus electrically energized systems must be avoided. Table 10.1 shows minimum safe guidelines for safe distance approach to energized lines by unauthorized employees. If work must be done by qualified employees in violation of these distance guidelines, the utility company must be notified and additional safety precautions taken. Table 2.11 gives a typical aerial lift inspection checklist.

TABLE 2.11 Routine Inspection Checklist of Aerial Lift

1. Park vehicle on flat, level surface.
2. Inspect tires for damage or excessive wear.
3. Check air pressure in tires.
4. Check hydraulic fluid levels and pressure.
5. Check oil, coolant, and fuel levels.
6. Check battery fluid level and charge status.
7. Check all hoses for damage or excessive wear.
8. Check all guardrails, platforms, and welded parts for security.
9. Check outriggers and stabilizers for proper size and condition.
10. Check pivot pins for security.
11. Check all safety limit switches.
12. Operate the boom systems and check for unusual noises, vibrations, hydraulic leaks, and uneven operations.
13. Check all ground and platform control functions.

E. Lifting Safety

According to the Bureau of Labor Statistics (BLS) more that one million workers suffer back injuries each year in the United States. This BLS study also found that approximately 75 percent of these injuries occurred while lifting. Lifting in itself is not inherently hazardous; it depends on the presence of certain risk factors that make a lifting operation more likely to cause injuries. Risk factors such as lifting loads that are too heavy, lifting unstable loads, lifting unwieldy loads, lifting and twisting one's back, lifting from an awkward body position, frequent lifting activities, and using poor lifting technique can all be significant for back injuries. Where these risks cannot be eliminated from the job, implementation of other controls involving the design of workstations, using lifting aids or tools to minimize manual lifting, and the implementation of administrative controls involving the training to minimize this problem can be effective. A current good reference available is Fact Sheet No. OSHA 89-Department of Labor, entitled "Back Injuries—Nation's #1 Workplace Safety Problem."

Attempt always to keep your back vertical and bend at the knees (Fig. 2.14).

FIGURE 2.14 Proper lifting technique: Always bend at the knees and keep the back as vertical as possible. (*Photo courtesy of Thermo Fisher Scientific*.)

The use of back belts will not increase your maximum lifting potential nor will it prevent an injury. Data have shown these to be ineffective and resources are better spent on proper solutions. Table 2.12 gives six suggested steps on how to lift properly.

TABLE 2.12 Checklist on How to Lift a Load Properly

1. Before lifting, remove any potential tripping or obstruction hazards. Know your lifting limits and the weight of the object.
2. Stand as close to the load as possible with your feet apart and well balanced.
3. Bending at the knees (not the waist), squat down while keeping your body as vertical as possible.
4. Secure a firm grasp of the load before attempting the lift.
5. Slowly begin the lifting using your legs, not your back.
6. Once lifted, keep the load as close to your body as possible. The back and arms are placed under more stress as the load's center of gravity moves away from the body.

CONFINED SPACES

A **confined space** has the following characteristics:

- Any space not designed for continuous employee occupancy.
- Any space that is large enough for an employee(s) to enter and perform work.
- Any space that has restricted or limited access to entry or exit. (e.g., reaction vessel, boiler, storage tank, etc.).

Confined-space activities in the workplace are covered by OSHA 29 CFR 1910.146 standard. An employer is required by this standard to identify and label all confined spaces in the workplace. This determination is to include all locations, number, and size of entry/exit ports, specific hazards involved, number of rescue personnel required, specific training needed, actual rescue procedures, and personal protective equipment needed. Each rescue member must be trained in the use of PPE, use of rescue equipment, and in basic first aid and cardiopulmonary resuscitation.

A **permit-required confined space** has the following characteristics:

- Any confined space that could possibly contain a hazardous atmosphere or other serious safety or health hazards to workers.
- Any confined space that contains materials that could engulf a worker and cause suffocation, for example, grains, sand, chemicals, and so on.

If an employee is to enter a confined space, specifically written procedures must be followed. Typically, a permit is issued for the worker to enter a confined space. Before a permit is issued, the confined space must be monitored for hazardous atmospheres, checked for dangerous conditions, and, maybe, even ventilated for several hours prior to entry.

Most confined space fatalities are caused by atmospheric hazards (carbon monoxide, carbon dioxide, hydrogen sulfide, flammable gases, and oxygen deficiency). An atmosphere is considered oxygen deficient when the oxygen level falls below 19.5 percent. Remember that a simple particulate or cartridge respirator is not appropriate protection for working in an oxygen-deficient atmosphere. All potentially hazardous confined space activities require an authorized **confined space entry permit**. This permit should include:

- Location
- Purpose of entry
- Time and date
- Description of work
- Authorized tools and heat-producing devices
- MSDS available
- All special requirements (lockout/tagout, PPE, etc.) for structural stability, material release, engulfment, falling hazards, electrical/mechanical hazards, etc.)
- Rescue equipment and preparation
- Environmental tests results (oxygen level, flammability levels, toxicity levels, noise, temperature, visibility, etc.)
- Special entry instructions
- Authorized entrants
- Authorization signature(s)

Protection for workers against electrocution and mechanical moving parts in confined spaces, lockout/tagout procedures should be strictly followed (see the following section). Falls can be fatal in confined-space activities; follow the appropriate OSHA guidelines for fall protection. Welding activities can be extremely hazardous in confined-space activities. Determine that all welding safety requirements are being followed before entering a confined space.

LOCKOUT/TAGOUT

The OSHA 29 CFR 1910.147 standard covers the control of hazardous energy in the workplace and is routinely referred to as lockout/tagout. Two other OSHA standards involve process industry lockout/tagout procedures: 29 CFR 1910.212 entitled "General Requirements for All Machines" and OSHA's 29 CFR 1910.219 entitled "Mechanical Power Transmission Apparatus." In addition, the American National Standards Institute's Z244.1-1982 provides personal protection lockout/tagout of energy sources guidelines.

Many different forms of energy (electrical, mechanical, pneumatic, chemical, thermal, hydraulic, and even gravity) are used to power process equipment. All these energy sources must be isolated when equipment is to be serviced or repaired (Fig. 2.15). This lockout/tagout procedure is not required for simple, cord-and-plug-connected electrical equipment exclusively controlled by the employee performing the servicing.

FIGURE 2.15 Lockout/tagout multiple lock hasp. (*Photo courtesy of Thermo Fisher Scientific.*)

All hydraulic/pneumatic/process material lines must be blocked during a lockout/tagout procedure. One technique called **blanking** is shown in Fig. 2.16.

This OSHA standard requires that **authorized employee**(s) be trained to perform lockout and tagout procedures on equipment to be serviced or repaired (Table 2.13).

OSHA defines an **affected employee** as one who operates or works in the area of the process equipment being locked or tagged. An affected employee is not authorized to perform lockout/tagout procedures. If more than one authorized employee is involved in the lockout/tagout procedure, each person will place a personal lock on the isolation hasp. This one person-one lock system is necessary to ensure accountability when the equipment is reenergized.

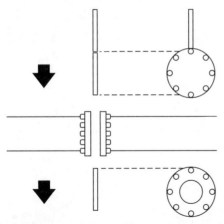

FIGURE 2.16 Blanking of process fluid lines.

TABLE 2.13 Lockout/Tagout Procedure

1 Plan: identify hazards, PPE, energy sources, locks necessary
2 Notify all affected employees
3 Shut down equipment at Operating controls
4 Isolate ALL potential energy sources
5 Lock and tag all energy sources
6 Release all residual and stored energy
7 Verify that energy isolation procedure is complete

REFERENCES

1. *Code of Federal Regulations 29, Labor/OSHA,* Parts 1900 1910.999. General Industrial Standards, OSHA Subpart A-T. Revised every July. U.S. Government Printing Office, Washington, DC 20402. This reference contains the OSHA Standards for the workplace.

2. *Code of Federal Regulations 29, Labor/OSHA,* Parts 1910.1000–end. General Industrial Standards, OSHA Subpart Z (Z tables) and Haz-Com. U.S. Government Printing Office, Washington, DC 20402. Hazard Communication regulations with the revised OSHA Z list of hazardous chemicals, new PEL's, and respiratory regulations.

3. *Improving Safety in the Chemical Laboratory A Practical Guide.* Edited by Jay A. Young, 368 pages, ISBN 0-471-84693-7, Wiley and Sons (1987). Organization for Safety in Laboratories; Precautionary Labels and Material Safety Data Sheets. U.S. Government Printing Office, Washington, DC 20402.

4. *Handbook of Ventilation for Contaminant Control,* 2d edition. Henry J. McDermott, 416 pages, ISBN 0-250-40641-1, Butterworth Publishers (1985).

5. *Rapid Guide to Hazardous Chemicals in the Workplace.* N. Irving Sax and Richard J. Lewis, 256 pages, ISBN 0-442-28220-6, Van Nostrand Reinhold (1987). Gives rapid access to information on 700 chemicals frequently used in the laboratory.

6. *Dangerous Properties of Industrial Materials,* 2d edition. N. Irving Sax and Richard J. Lewis 4368 pages, ISBN 0-442-28020-3, Van Nostrand Reinhold (1995). A three-volume set containing over 20,000 entries.

7. "Threshold Limit Values for Chemical Substances and Physical Agents in the Workroom Environment with Intended Changes." American Conference of Governmental Industrial Hygienists, PO. Box 1937 Cincinnati, OH 45201 (2009, or latest edition).

8. "NIOSH/OSHA Pocket Guide to Chemical Hazards." NIOSH Publication No. 85-114, U.S. Government Printing Office, Washington, DC (1994, or latest edition).

9. *Prudent Practices in the Laboratory: Handling and Disposal of Chemicals.* ISBN 0-309-05229-7. National Research Council, National Academy Press Washington, DC (1995).

10. *NFPA Hazardous Materials Response Handbook,* 2d edition. Edited by G. Tokle, 700 pages. National Fire Protection Association, Quincy, MA 02269-9101 (1992). Contains the complete text of NFPA 471 "Recommended Practices for Responding to Hazardous Materials Incidents" and NFPA 472 "Standards for Professional Competence of Responders to Hazardous Materials Incidents."

11. *NFPA Fire Protection Guide to Hazardous Materials,* 11th edition. 544 pages. National Fire Protection Association, Quincy, MA 02269-9101 (1994). Contains the NFPA 704 "Hazards Identification System" and guide for taking proper steps to prevent fires while handling chemicals and emergency situations.

CHAPTER 3
CHEMICAL HANDLING AND HAZARD COMMUNICATION

INTRODUCTION

The lifecycle of chemicals from production through to the disposal presents a hazard to human health and the environment. Global trade has a significant impact on the creation of difficulties in the communication of hazard information. People of all ages, social status, and literacy levels around the world use and are exposed to these hazards all while using different languages and alphabets. This fact illustrates the challenges presented to effective hazard communication systems. Of the industrialized countries that participate in the trade and transport of hazardous substances there are many regulations that apply and sifting through them is not easy even with professional help. Standards organizations in the Europe and the United States are continually developing a greater understanding for one another's hazard and safety label systems, which is slowly resulting in continued alignment. Until the last decade there was no successful effort to bring together a common chemical safety terminology or handling practice. With the recent phasing in of the **Globally Harmonized System (GHS)**, there is significant progress to harmonize hazard communication strategies. Each country continues to have its own set of regulations governing the storage, use, transportation, classification, and labeling of hazardous substances; however; these regulations are adopting the code of practice outlined in GHS. The emphasis of this chapter is on the importance of communication of hazard information during handling of hazardous substances, specifically chemicals and chemical mixtures. Chapter 2 discussed the details of Occupational Safety and Health Administration (OSHA)'s Hazard Communication Standard with the discussion of hazard, risk, risk control, and safe work practices. This chapter goes into further detail regarding safety data sheets, the specifics of labeling practices, general transportation, and handling practices from point of generation to disposal as they are outlined in various global laws.

MATERIAL SAFETY DATA SHEET

Material safety data sheets (MSDS), or safety data sheets (SDS) as they are referred to under GHS rules, are a globally recognized source of information for the inherent properties, safe handling guidelines and hazards of a chemical. SDSs are an integral part of workplace safety as they provide the end users and emergency response personnel with manufacturer information for safe handling. They include information such as physical data, toxicity, health effects, first aid, reactivity, storage, disposal, personal protective equipment, environmental risks, and spill-handling procedures. SDS formats and content vary from country to country but increasingly follow a common structure under the GHS guidelines. The GHS guidelines for each section of an SDS are as follows:

1. Identification.
2. Hazard(s) identification.

3. Composition/information on ingredients.
4. First-aid measures.
5. Fire-fighting measures.
6. Accidental release measures.
7. Handling and storage.
8. Exposure control/personal protection.
9. Physical and chemical properties.
10. Stability and reactivity.
11. Toxicological information.
12. Ecological information.
13. Disposal considerations.
14. Transport information.
15. Regulatory information.
16. Other information.

Some examples of country-specific regulatory requirements for communication of hazards information are as follows:

The European Union (EU) currently requires that Risk and Safety Statements (R and S phrases) and a symbol appear on each label and safety data sheet for hazardous chemicals. However, beginning in 2010 these R and S phrases are being phased out and replaced with hazard and precautionary statements.

- **R phrases** consist of the letter R followed by a unique number. The number correlates to a specific type of risk.
- **S phrases** consist of the letter S followed by a unique number. The number correlates to a specific safety requirement.
- **Hazard statements:** Standard phrases assigned to a hazard class and category that describe the nature of the hazard.
- **Precautionary statements:** These are measures to minimize or prevent adverse effects.

In Canada, Workplace Hazardous Materials Information System (WHMIS) establishes the requirements for MSDSs in workplaces and is administered at the federal level by Health Canada, provincial ministries, and also the Department of Labor.

In the United Kingdom, the **Control of Substances Hazardous to Health (CSHH)** regulations mandate that employers provide adequate information to workers who may have exposure to any hazardous substances. In addition, the CHIP (Chemicals Hazard Information and Packaging for Supply) regulations outline the requirements for labeling and transportation. These too are changing with the implementation of new legislation such as GHS and Registration, Evaluation, Authorization and Restriction of Chemicals (REACH) regulations.

In the United States, OSHA Hazard Communication Standard requires that MSDS be available to employees for all potentially harmful substances handled in the workplace (see Fig. 3.1). These MSDSs are also sometimes required by other state or local agencies such as fire departments where high-risk operations may take place.

Material Safety Data Sheet

May be used to comply with
OSHA's Hazard Communication Standard,
29 CFR 1910.1200. Standard must be
consulted for specific requirements.

U.S. Department of Labor
Occupational Safety and Health Administration
(Non-Mandatory Form)
Form Approved
OMB No. 1218-0072

IDENTITY (*As used on Label and List*)	Note: Blank spaces are not permitted. If any item is not applicable, or no information is available, the space must be marked to indicate that.

Section I

Manufacturer's Name	Emergency Telephone Number
Address (Number, Street, City, State, and ZIP Code)	Telephone Number for Information
	Data Prepared
	Signature of Preparer (*Optional*)

Section II—Hazardous Ingredients/Identity Information

Hazardous Components (Specific Chemical Identity: Common Name(s))	OSHA PEL	ACGIH TLV	Other Limits Recommended	% (*optional*)

Section III—Physical/Chemical Characteristics

Boiling Point		Specific Gravity (H_2O = 1)	
Vapor Pressure (mm Hg)		Melting Point	
Vapor Density (AIR = 1)		Evaporation Rate (Butyl Acetate = 1)	
Solubility in Water			
Appearance and Odor			

Section IV—Fire and Explosion Hazard Data

Flash Point (Method Used)	Flammable Limits		LEL	UEL
Extinguishing Media				
Special Fire-Fighting Procedures				
Unusual Fire and Explosion Hazards				

FIGURE 3.1 Material safety data sheet. (*U.S. Department of Labor*, OSHA's Material Safety Data Sheet, OMB # 1218-0072.)

Section V—Reactivity Data

Stability	Unstable		Conditions to Avoid	
	Stable			
Incompatibility (*Materials to Avoid*)				
Hazardous Decomposition or Byproducts				
Hazardous Polymerization	May Occur		Conditions to Avoid	
	Will not Occur			

Section VI—Health Hazard Data

Route(s) of Entry:	Inhalation?	Skin?	Ingestion?
Health Hazards (*Acute and Chronic*)			
Carcinogenicity:	NTP?	IARC Monographs?	OSHA regulated?
Signs and Symptoms of Exposure			
Medical Conditions Generally Aggravated by Exposure			
Emergency and First-Aid Procedures			

Section VII—Precautions for Safe Handling and Use

Steps to Be Taken in Case Material is Released or Spilled
Waste Disposal Method
Precautions to Be Taken in Handling and Storing
Other Precautions

Section VIII—Control Measures

Respiratory Protection (*Specify Type*)			
Ventilation	Local Exhaust		Special
	Mechanical (*General*)		Other
Protective Gloves			Eye Protection
Other Protective Clothing or Equipment			
Work/Hygienic Practices			

FIGURE 3.1 (*Continued*)

CHEMICAL HANDLING AND HAZARD COMMUNICATION 51

A. MSDS on the Internet

Several excellent MSDS archives sites are available on the Internet.

- **Aldrich Chemical** http://www.msdsonline.com.
- **Cornell University** www.cnfusers.cornell.edu/cnf5_msds.taf.
- **Canadian Center for Occupational Health and Safety.** http://ccinfoweb.ccohs.ca/msds/search.htm.
- **Fisher Scientific** http://fishersci.com.
- **Oxford University** http://msds.chem.ox.ac.uk.
- **Vermont SIRI MSDS Collection (University of Vermont).** http://hazard.com/msds/index.html.

CONTAINER LABELING

There are seven major types of container labeling systems currently in use between the European Union and the United States:

A. Department of Transportation (DOT).
B. National Fire Protection Association (NFPA).
C. American National Standards Institute (ANSI).
D. Hazardous Materials Information System (HMIS).
E. International Standards Organization (ISO).
F. Globally Harmonized System (GHS).
G. International Air Transportation Association (IATA).

As good practice, all labels must contain a minimum of three types of information: (1) the identity or name of the chemical; (2) the name and address of the manufacturer or importer; and (3) the physical and health hazard warning, including potential target organs. In addition, recommended handling and storage instructions, first-aid information, and personal protective equipment should be included.

A. U.S. Department of Transportation

Department of Transportation labels and placards are diamond shaped (Fig. 3.2). Guidelines for using these DOT labels and placards are described in 49 CFR Part 172, Subpart E and 49 CFR Part 172, Subpart F, respectively. Department of Transportation uses colors, symbols, and numbers

FIGURE 3.2 DOT signs and labels.

TABLE 3.1 DOT Hazard Labels Color Coding

Hazard type	Color code
Corrosive	Black and White
Explosive	Orange
Flammable Gas or Liquid	Red
Non-flammable Gas	Green
Oxidizer	Yellow
Dangerous When Wet	Blue and White
Radioactive	Yellow and White
Solid Flammable	Red Striped
Spontaneously Combustible	Red
Toxic or Poison Gas	White

to represent and identify chemicals and potential hazards (Table 3.1). The number at the bottom of the label represents the hazard classification (total of nine) or division, not the degree of hazard. Table 3.2 lists some of these DOT hazard class or division numbers and shows materials that require placards if in a bulk packaging or more than 1000 lb. Remember that NFPA and other organizations use numbers to represent the degree of hazard. DOT uses diamond-shaped placards (49 CFR, Part 172, Subpart F); warning signs attached to railroad tank cars, trucks, cargo tanks, portable tanks; and freight containers. In general, placards must be visible on all four sides of the container, vehicle, or tank.

The Department of Transportation uses four-digit identification numbers (**ID Numbers**) to specifically identify materials, for example: 1256, naphtha solvent; 3077, environmentally hazardous substance (solid); and 1671, phenol (carbolic acid) gasoline. This identification system is used to

TABLE 3.2 DOT Hazard Class or Division Number for Materials

Hazard class or division number	Placard name	Placard design section reference
1.1	Explosive 1.1	172.411
1.2	Explosive 1.2	172.411
1.3	Explosive 1.3	172.411
1.4	Explosive 1.4	172.411
1.5	Explosive 1.5	172.411
1.6	Explosive 1.6	172.411
2.1	Flammable Gas	172.417
2.2	Non-flammable Gas	172.415
2.3	Poison Gas	172.416
3	Flammable Liquid	172.419
4.1	Flammable Solid	172.420
4.2	Spontaneous Combustion	172.422
4.3	Dangerous When Wet	172.423
5.1	Oxidizer	172.426
5.2	Organic Peroxide	172.427
6.1	Poison	172.430
6.2	Keep Away from Food	172.431
6.3	Infectious Substance	172.432
7	Radioactive, White I	172.436
7	Radioactive, Yellow II	172.438
7	Radioactive, Yellow III	172.440
7	Empty	172.450
8	Corrosive	172.442
9	Class 9	172.446

identify hazardous materials in tanker trucks and trailers, barges, and rail cars as shown on the placards in Fig. 3.2. In addition, DOT uses two-digit numbers called **Guide Numbers** to be used if the ID number is not available. These guide numbers provide only general guidelines to deal with the potential hazards and emergency action(s) to be taken.

The DOT uses labels similar to the placards on smaller packages shipped through the mail or courier services. This classification quickly identifies the hazardous characteristics of a material and what forms of personal protective equipment may subsequently be needed to provide adequate protection for an emergency responder in the event of an accidental release. See Chapter 1 section on "Department of Transportation" for additional details.

B. National Fire Protection Association

The **National Fire Protection Association** (NFPA) has been the most recognized organization in the world dedicated to fire protection and fire safety. The OSHA Hazard Communication Standard requires the employer to ensure that each container of hazardous chemicals in the workplace be labeled, tagged, or marked with the identity of the hazardous chemical(s) and the appropriate hazard warnings. These labeling requirements are commonly met through the use of the NFPA hazard identification system. This system consists of a diamond-shaped label which is divided into four smaller, colored diamonds, as shown in Fig. 3.3.

Each colored section represents a specific hazard with each hazard being rated on a 0 to 4 scale. Zero indicates a very low degree of hazard and a level 4 would indicate a very high degree of hazard. The specific hazard sections and numerical codes are shown in Table 3.3 and are described in detail following. A numbering scale from 0 to 4 is used to represent the level of hazard within each NFPA category. The following three categories are copied verbatim from the NFPA coding system for **health (blue)**, **flammability (red)**, and **instability hazard (yellow)**. A fourth category covering **special hazards (white)** is also detailed. Also listed are special considerations or properties like radioactive materials, water reactive materials, strong oxidizers, strong alkali, corrosive materials, and so on. The various levels of 0 through 4 are not attempted in this descriptive-only, special hazards category.

C. American National Standards Institute

The ANSI is a nonprofit organization funded by private membership made up of over 1300 companies, 250 professional organizations, and 30 different government agencies. ANSI is the representative agency with membership to the ISO. So most ISO rules and regulations will be applied through ANSI in the United States. The primary goal of ANSI is to make U.S. businesses competitive in the global market by developing and providing written standards covering topics on safety, health, and specific industrial topics. The ANSI relies more on word descriptions than the pictures, icons, numbers, and colors that the DOT and NFPA use for labeling (Fig. 3.4). The ANSI has three levels of hazard, normally printed on the upper left-hand corner of their labels. The ANSI label also contains the DOT diamond and related information (first-aid measures, storage requirements, spill or fire procedures, etc.).

CAUTION: Moderate hazard but still of concern.

WARNING: Between "caution" and "serious" levels.

DANGER: Serious hazard.

D. Hazardous Materials Information System

The HMIS is very similar to the NFPA system. However, HMIS uses a vertical order and includes the chemical name and a personal protective equipment entry, instead of the special hazards category used by NFPA. Organizations like the National Paint and Coating Association (NPCA) use this system. The numerical codes of 0 (least) to 4 (greatest) hazard is identical.

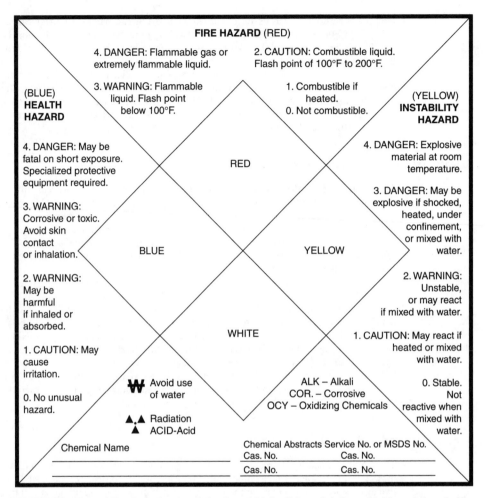

FIGURE 3.3 National Fire Protection Association hazard labels. Color-coded diamond with symbols and abbreviations. (*Reprinted with permission* from NFPA 704-2007, "Standard System for the Identification of the Hazards of Materials for Emergency Response." Copyright © 2007, National Fire Protection Association.)

All HMIS labels list in vertical order: the name of the chemical, health hazard, flammability hazard, reactivity hazard, and personal protective equipment needed (Fig. 3.5). The HMIS and the NPCA use the same signs. The color code (blue/health, red/flammability, and yellow/instability) and numerical code (0/minimal to 4/severe hazard) used by HMIS are identical to the NFPA codes.

Letters and icons are used by HMIS to indicate the recommended personal protective equipment as shown in Fig. 3.6.

E. International Organization for Standardization

The ISO has established a system to communicate safety information globally without the use of words. These ISO3864 defines four basic formats for safety labels using labels, shapes, colors, and symbols that are accepted and recognized worldwide. They are available as warning labels (such as flammable, radioactive, toxic, etc.), prohibition labels (such as no smoking or open flames), mandatory action labels (such as electrical warning), and safety information symbols.

TABLE 3.3 Identification of NFPA Numerical Codes* (Courtesy of National Fire Protection Association.)

\multicolumn{2}{	c	}{Identification of health hazard color code: Blue}	\multicolumn{2}{c	}{Identification of flammability color code: Red}	\multicolumn{2}{c	}{Identification of reactivity (stability) color code: Yellow}
Signal	Type of possible injury	Signal	Susceptibility of materials to burning	Signal	Susceptibility to release of energy
4	Materials which on very short exposure could cause death or major residual injury even though prompt medical treatment were given.	4	Materials which will rapidly or completely vaporize at atmospheric pressure and normal ambient temperature, or which are readily dispersed in air and which will burn readily.	4	Materials which in themselves are readily capable of detonation or of explosive decomposition or reaction at normal temperatures and pressures.
3	Materials which on short exposure could cause serious temporary or residual injury even though prompt medical treatment were given.	3	Liquids and solids that can be ignited under almost all ambient temperature conditions.	3	Materials which in themselves are capable of detonation or of explosive reaction but require a strong initiating source or which must be heated under confinement before initiation or which react explosively with water.
2	Materials which on intense or continued exposure could cause temporary incapacitation or possible residual injury unless prompt medical treatment is given.	2	Materials that must be moderately heated or exposed to relatively high ambient temperatures before ignition can occur.	2	Materials which in themselves are normally unstable and readily undergo violent chemical change but do not detonate. Also materials which may react violently with water or which may form potentially explosive mixtures with water.
1	Materials which on exposure would cause irritation but only minor residual injury even if no treatment is given.	1	Materials that must be preheated before ignition can occur.	1	Materials which in themselves are normally stable, but which can become unstable at elevated temperatures and pressures or which may react with water with some release of energy but not violently.
0	Materials which on exposure under fire conditions would offer no hazard beyond that of ordinary combustible material.	0	Materials that will not burn.	0	Materials which in themselves are normally stable, even under fire exposure conditions, and which are not reactive with water.

*Reprinted with permission from NFPA 704-2007, "Standard System for the Identification of the Hazards of Materials for Emergency Response." Copyright © 2007, National Fire Protection Association. This reprint material is not complete and official position of the NFPA on this referenced subject, which is represented solely by the standard in its entirety. The classification of any particular material within this system is the sole responsibility of the user and not the NFPA. NFPA bears no responsibility for any determinations of any values for any particular material classified or represented using this system.

TABLE 3.3(a) NFPA Health Hazards

Degree of hazard*	Criteria†
4—Materials that, under emergency conditions, can be lethal.	Gases whose LC_{50} for acute inhalation toxicity is less than or equal to 1000 parts per million (ppm) Any liquid whose saturated vapor concentration at 20°C (68°F) is equal to or greater than 10 times its LC_{50} for acute inhalation toxicity, if its LC_{50} is less than or equal to 1000 ppm Dusts and mists whose LC_{50} for acute inhalation toxicity is less than or equal to 0.5 milligram per liter (mg/L) Materials whose LD_{50} for acute dermal toxicity is less than or equal to 40 milligrams per kilogram (mg/kg) Material whose LD_{50} for acute oral toxicity is less than or equal to 5 mg/kg
3—Materials that, under emergency conditions, can cause serious or permanent injury.	Gases whose LC_{50} for acute inhalation toxicity is greater than 1000 ppm but less than or equal to 3000 ppm Any liquid whose saturated vapor concentration at 20°C (68°F) is equal to or greater than its LC_{50} for acute inhalation toxicity, if its LC_{50} is less than or equal to 3000 ppm, and that does not meet the criteria for degree of hazard 4 Dusts and mists whose LC_{50} for acute inhalation toxicity is greater than 0.5 mg/L but less than or equal to 2 mg/L Materials whose LD_{50} for acute dermal toxicity is greater than 40 mg/kg but less than or equal to 200 mg/kg Materials that are corrosive to the respiratory tract Materials that are corrosive to the eye or cause irreversible corneal opacity Materials that are corrosive to skin Cryogenic gases that cause frostbite and irreversible tissue damage Compressed liquefied gases with boiling points at or below −55°C (−66.5°F) that cause frostbite and irreversible tissue damage Materials whose LD_{50} for acute oral toxicity is greater than 5 mg/kg but less than or equal to 50 mg/kg
2—Materials that, under emergency conditions, can cause temporary incapacitation or residual injury.	Gases whose LC_{50} for acute inhalation toxicity is greater than 3000 ppm but less than or equal to 5000 ppm Any liquid whose saturated vapor concentration at 20°C (68°F) is equal to or greater than one-fifth its LC_{50} for acute inhalation toxicity, if its LC_{50} is less than or equal to 5000 ppm, and that does not meet the criteria for either degree of hazard 3 or degree of hazard 4 Dusts and mists whose LC_{50} for acute inhalation toxicity is greater than 2 mg/L but less than or equal to 10 mg/L Materials whose LD_{50} for acute dermal toxicity is greater than 200 mg/kg but less than or equal to 1000 mg/kg Compressed liquefied gases with boiling points between −30°C (−22°F) and −55°C (−66.5°F) that can cause severe tissue damage, depending on duration of exposure Materials that are respiratory irritants Materials that cause severe but reversible irritation to the eyes or lacrimators Materials that are primary skin irritants or sensitizers Materials whose LD_{50} for acute oral toxicity is greater than 50 mg/kg but less than or equal to 500 mg/kg
1—Materials that, under emergency conditions, can cause significant irritation.	Gases and vapors whose LC_{50} for acute inhalation toxicity is greater than 5000 ppm but less than or equal to 10,000 ppm Dusts and mists whose LC_{50} for acute inhalation toxicity is greater than 10 mg/L but less than or equal to 200 mg/L Materials whose LD_{50} for acute dermal toxicity is greater than 1000 mg/kg but less than or equal to 2000 mg/kg Materials that cause slight to moderate irritation to the respiratory tract, eyes, and skin Materials whose LD_{50} for acute oral toxicity is greater than 500 mg/kg but less than or equal to 2000 mg/kg

(Continued)

TABLE 3.3(a) NFPA Health Hazards (*Continued*)

Degree of hazard*	Criteria†
0—Materials that, under emergency conditions, would offer no hazard beyond that of ordinary combustible materials.	Gases and vapors whose LC_{50} for acute inhalation toxicity is greater than 10,000 ppm Dusts and mists whose LC_{50} for acute inhalation toxicity is greater than 200 mg/L Materials whose LD_{50} for acute dermal toxicity is greater than 2000 mg/kg Materials whose LD_{50} for acute oral toxicity is greater than 2000 mg/kg Materials that are essentially nonirritating to the respiratory tract, eyes, and skin

*For each degree of hazard, the criteria are listed in a priority order based on the likelihood of exposure.
†See Section B.3 for definitions of LC_{50} and LD_{50}.

TABLE 3.3(b) NFPA Flammability Hazards

Degree of hazard	Criteria
4—Materials that in themselves are readily capable of detonation or explosive decomposition or explosive reaction at normal temperatures and pressures.	Materials that are sensitive to localized thermal or mechanical shock at normal temperatures and pressures Materials that have an instantaneous power density (product of heat of reaction and reaction rate) at 250°C (482°F) of 1000 watts per milliliter (W/mL) or greater
3—Materials that in themselves are capable of detonation or explosive decomposition or explosive reaction but that require a strong initiating source or must be heated under confinement before initiation.	Materials that have an instantaneous power density (product of heat of reaction and reaction rate) at 250°C (482°F) at or above 100 W/mL and below 1000 W/mL Materials that are sensitive to thermal or mechanical shock at elevated temperatures and pressures
2—Materials that readily undergo violent chemical change at elevated temperatures and pressures.	Materials that have an instantaneous power density (product of heat of reaction and reaction rate) at 250°C (482°F) at or above 10 W/mL and below 100 W/mL
1—Materials that in themselves are normally stable but that can become unstable at elevated temperatures and pressures.	Materials that have an instantaneous power density (product of heat of reaction and reaction rate) at 250°C (482°F) at or above 0.01 W/mL and below 10 W/mL
0—Materials that in themselves are normally stable, even under fire conditions.	Materials that have an instantaneous power density (product of heat of reaction and reaction rate) at 250°C (482°F) below 0.01 W/mL Materials that do not exhibit an exotherm at temperatures less than or equal to 500°C (932°F) when tested by differential scanning calorimetry

TABLE 3.3(c) NFPA Instability Hazards

Degree of hazard	Criteria
4—Materials that rapidly or completely vaporize at atmospheric pressure and normal ambient temperature or that are readily dispersed in air and burn readily.	Flammable gases Flammable cryogenic materials Any liquid or gaseous material that is liquid while under pressure and has a flash point below 22.8°C (73°F) and a boiling point below 37.8°C (100°F) (i.e., Class IA liquids) Materials that ignite spontaneously when exposed to air Solids containing greater than 0.5% by weight of a flammable or combustible solvent are rated by the closed cup flash point of the solvent

(*Continued*)

TABLE 3.3(c) NFPA Instability Hazards (*Continued*)

Degree of hazard	Criteria
3—Liquids and solids that can be ignited under almost all ambient temperature conditions. Materials in this degree produce hazardous atmospheres with air under almost all ambient temperatures or, though unaffected by ambient temperatures, are readily ignited under almost all conditions.	Liquids having a flash point below 22.8°C (73°F) and a boiling point at or above 37.8°C (100°F) and those liquids having a flash point at or above 22.8°C (73°F) and below 37.8°C (100°F) (i.e., Class IB and Class IC liquids) Finely divided solids, typically less than 75 micrometers (μm) (200 mesh), that present an elevated risk of forming an ignitible dust cloud, such as finely divided sulfur, *National Electrical Code* Group E dusts (e.g., aluminum, zirconium, and titanium), and bis-phenol A Materials that burn with extreme rapidity, usually by reason of self-contained oxygen (e.g., dry nitrocellulose and many organic peroxides) Solids containing greater than 0.5% by weight of a flammable or combustible solvent are rated by the closed cup flash point of the solvent
2—Materials that must be moderately heated or exposed to relatively high ambient temperatures before ignition can occur. Materials in this degree would not, under normal conditions, form hazardous atmospheres with air, but, under high ambient temperatures or under moderate heating, could release vapor in sufficient quantities to produce hazardous atmospheres with air.	Liquids having a flash point at or above 37.8°C (100°F) and below 93.4°C (200°F) (i.e., Class II and Class IIIA liquids) Finely divided solids less than 420 μm (40 mesh) that present an ordinary risk of forming an ignitible dust cloud Solid materials in a flake, fibrous, or shredded form that burn rapidly and create flash fire hazards, such as cotton, sisal, and hemp Solids and semisolids that readily give off flammable vapors Solids containing greater than 0.5% by weight of a flammable or combustible solvent are rated by the closed cup flash point of the solvent
1—Materials that must be preheated before ignition can occur. Materials in this degree require considerable preheating, under all ambient temperature conditions, before ignition and combustion can occur.	Materials that will burn in air when exposed to a temperature of 815.5°C (1500°F) for a period of 5 minutes in accordance with ASTM D 6668, *Standard Test Method for the Discrimination between Flammability Ratings of F = 0 and F = 1* Liquids, solids, and semisolids having a flash point at or above 93.4°C (200°F) (i.e., Class IIIB liquids) Liquids with a flash point greater than 35°C (95°F) that do not sustain combustion when tested using the "Method of Testing for Sustained Combustibility," per 49 CFR 173, Appendix H, or the UN publications *Recommendations on the Transport of Dangerous Goods, Model Regulations and Manual of Tests and Criteria* Liquids with a flash point greater than 35°C (95°F) in a water-miscible solution or dispersion with a water noncombustible liquid/solid content of more than 85% by weight Liquids that have no fire point when tested by ASTM D 92, *Standard Test Method for Flash and Fire Points by Cleveland Open Cup*, up to the boiling point of the liquid or up to a temperature at which the sample being tested shows an obvious physical change Combustible pellets, powders, or granules greater than 420 μm (40 mesh) Finely divided solids less than 420 μm that are nonexplosible in air at ambient conditions, such as low volatile carbon black and polyvinylchloride (PVC) Most ordinary combustible materials Solids containing greater than 0.5% by weight of a flammable or combustible solvent are rated by the closed cup flash point of the solvent
0—Materials that will not burn under typical fire conditions, including intrinsically noncombustible materials such as concrete, stone, and sand.	Materials that will not burn in air when exposed to a temperature of 816°C (1500°F) for a period of 5 minutes in accordance with ASTM D 6668, *Standard Test Method for the Discrimination between Flammability Ratings of F = 0 and F = 1*

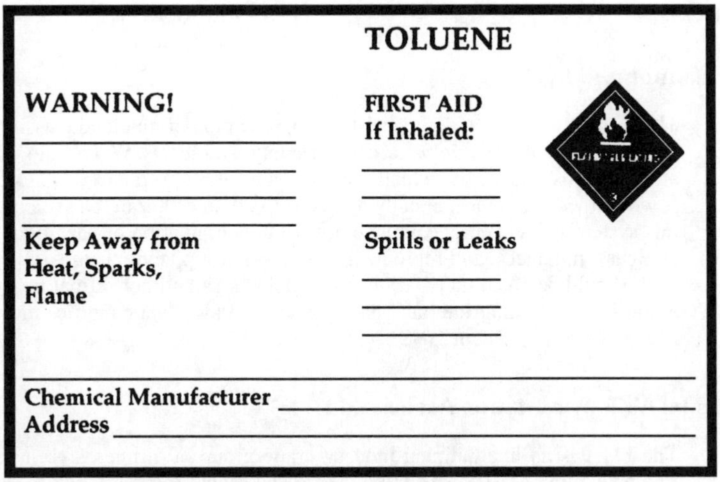

FIGURE 3.4 American National Standards Institute sign.

* See Figure 3.6 for Personal Protective Index

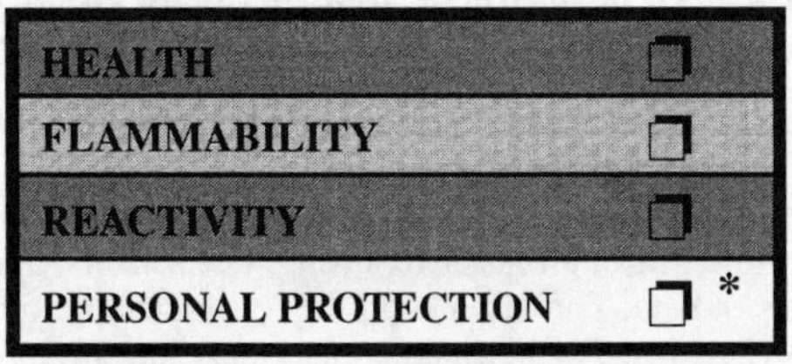

FIGURE 3.5 Hazardous materials identification system labels.

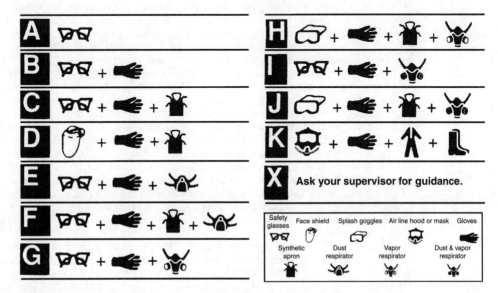

FIGURE 3.6 Hazardous materials identification system personal protective index.

F. Globally Harmonized System

The GHS is a common and consistent approach to defining and classifying hazards, and communicating hazard information on labels and safety data sheets. When fully implemented, GHS will cover all hazardous chemicals, which includes substances, products, mixtures, preparations, and so on. It will provide an internationally agreed-on system of hazard classification and labeling. Ultimately this system will encourage a move toward the elimination of hazardous chemicals—especially carcinogens, mutagens, and reproductive toxins—or replacing them with less hazardous ones.

It should be noted that GHS is not a global standard nor formal treaty nor legislation; rather it is a non-legally binding international agreement. Individual countries must therefore create their own legislation to implement GHS.

G. International Air Transportation Association

The IATA is an international industry trade group of airlines present in more than 150 countries. The main aim of IATA is to provide safe and secure transportation to its passengers. IATA labeling requirements are markedly similar to DOT with nine separate hazard classification labels.

LABORATORY INFORMATION MANAGEMENT SYSTEMS

A **Laboratory Information Management System (LIMS)** can be described as a computer system that processes and dispenses production and laboratory data. A LIMS is exceptionally helpful in process environments because it can increase performance, help make decisions, and improve production work flow—through faster reporting, test sensors, and collection programming, and the ability to view sets of data. The faster reporting characteristic of a LIMS can make the laboratory or plant run more efficiently. For instance, a plant may have to halt production until the results of particular tests are known. Using a LIMS in the control lab, the results can be immediately transmitted, faxed, or e-mailed directly to all interested and responsible individuals.

A LIMS can also help improve the quality of a production. Collection programs and testing sensors in the processes unit can immediately identify any variables in the process operation. It can provide the operator with up-to-the-minute information, such as purity, pH (acidity or basicity), temperature, and so on. The data are then transferred to permanent records and the LIMS correlates the results of numerous sensor testing sites with the values of information generated by the test. A LIMS is able to compare and evaluate entire sets of data instantly. The LIMS can put all data points together graphically and display the information for easier interpretation.

Tracking samples and chemicals are tasks that can be simplified with a LIMS. A supervisor or process operator can look up the sample on the LIMS and know who has finished their tests on the sample, which tests still need to be done, who is scheduled to perform those tests, and approximately how long. Without a computer system, it could take hours to determine the status of a sample. Using a LIMS can result in fewer mistakes, print data in standardized formats, and reduce paper work (see Fig. 3.7). Computers and **bar coding** are very effective in tasks ranging from tool/equipment management to personnel timekeeping and plant access control.

Typing hundreds of determinations into a computer can result in typing errors. A LIMS combats this problem with a "verifying step." When the data are entered, they are compared with past values of similar samples. If the value is outside the accepted range it will ask you to verify the value. Government regulatory agencies many times require a very standardized format for receiving data. These formats can easily be programmed into the LIMS and printed over and over, substituting data, dates, and lot numbers.

Efficiency can be greatly improved using a LIMS by reducing paper work. In the control lab, there can be hundreds of samples that need several tests done. Regulating agencies such as the Food and Drug Administration (FDA), OSHA, the Environmental Protection Agency (EPA), and the U.S. Department of Agriculture (USDA) frequently require large quantities of documentation. The FDA requires that documentation for each solution used in a test include who prepared the solution, an expiration date, and the lot number of chemicals/batches used.

FIGURE 3.7 Typical LIMS chemical bar code. (*Courtesy Mallinckrodt Speciality Chemicals Company*)

A chemical tracking and inventory system that includes a program of comprehensive chemical waste minimization is absolutely necessary in modern chemical processing industries. The system can include information about the kinds, quantities, location, and status of chemicals within a particular facility. Tracking a chemical from purchase to disposal can benefit any organization by reducing duplicate purchases and possibly reusing waste chemicals.

There are a number of standard systems that use a bar coding system for chemical tracking and inventory control. For example, Chemical Abstract Service (CAS) registry numbers are universally accepted for identifying specific chemicals and can be used in such a system. Costs can be further controlled by linking chemical purchasing requests into the computerized inventory system, so that orders can sometimes be filled with surplus chemicals. Some major chemical suppliers provide electronic access to their catalogs. Safety information can be added to the inventory system to promote proper handling of chemicals and to aid in compliance with hazard communication and waste management regulations.

BULK CONTAINERS (UNITED STATES ONLY)

In the United States, bulk containers are defined as any packaging (tank trucks, railroad cars, portable tanks, chemtotes, drums, etc.) that (a) has a maximum capacity greater than 119 gal, (b) has a maximum net mass greater than 882 lb, and (c) has a water capacity greater than 1000 lb.

A. Carboys

Carboys and other large containers of acids, caustics, flammable liquids, and corrosive materials are dangerous to handle and to transport. They tend to be fragile, and an even slight mechanical shock can cause breakage. It is also very difficult to pour directly from these vessels without spillage. In dealing with carboys it is recommended that a safety carboy tilter and an appropriate pouring spout always be used. Specially designed pouring spouts are available that permit air to enter the emptying container thereby preventing spurts and splashing. The transporting of large carboys should only be performed with safety carts and safety bottle carriers. These carriers securely hold the bottle or carboy in place and cushion the bottom from shock.

B. Safety Cans

OSHA covers safety cans in 29 CFR 1910.106(a)(29) and defines a safety can as "an approved container, of not more than 5 gal of capacity, having a spring-closing lid and spout cover and so designed that it will relieve internal pressure when subject to fire exposure." In addition, many local insurance carriers require Factory Mutual (FM) and/or Underwriter Laboratories (UL) approval.

TABLE 3.4 Maximum Allowable Container and Safety Storage Can Sizes Based on Flammability Classification

Container type	Class IA	Class IE	Class IC	Class II
Glass or approved plastic	1 pt	1 qt	1 gal	1 gal
Metal (other than DOT drums)	1 gal	5 gal	5 gal	5 gal
Safety cans	2 gal	5 gal	5 gal	5 gal
Metal Drum DOT specification	60 gal	60 gal	60 gal	60 gal
Metal Portable Tanks, Approved	660 gal	660 gal	660 gal	660 gal

Small quantities of liquids can safely be stored and transported in 5-gal or smaller safety cans. A safety can is a durable metal or plastic can that is hand-carried and used for smaller jobs. Many safety cans have a spring-loaded closable lid, which ensures a positive seal in the event it should get turned over and also allows for bleeding of pressure without chance of the vessel rupturing under extreme temperature conditions. The primary feature of a safety can is the flame arrestor. The **flame arrestor** is generally constructed of a metal wire mesh or perforated metal device situated inside the safety can. This device allows for liquids to pass through freely during pouring; however, it will stop sources of ignition from igniting the flammable contents inside the can. The flame arrestor works by dissipating any heat that may be introduced into the safety can and thus does not allow the flammable vapors to be ignited. Hot air is cooled as it enters the can and thus keeps the vapor temperature below its flash point. Safety cans provide a safe means of storing, dispensing, and transporting liquids. A **Type I** safety can has a single spout used for both pouring and filling. The spout includes a flame arrestor and a self-closing cap with pressure relief vent. A **Type II** safety cans come equipped with two openings, both equipped with flame arrestors to prevent flashbacks. Type II cans have a valve-equipped pouring spout with an attached flexible metal hose and a separate fill opening with a spring-loaded, self-closing cap. The fill opening also serves as a vent to allow air to displace liquid during pouring operations.

Metal safety cans which contain flammable liquids should NOT be poured into another metal container without grounding and bonding the cans. Bonding and grounding must be done to prevent the generation of static electricity which can ignite the flammable liquid (see the following section). Table 3.4 lists the maximum allowable container sizes (OSHA 29 CFR 1910.106) based on container type and flammability of solvent. This table does not apply to solvent mixtures, only individual solvents.

In addition to storage of individual container quantities, the total storage capacities of individual containers in cabinets (Table 3.5) are regulated by OSHA 29 CFR 1910.106(d)(3) and NFPA 30 Section 4-3.1.

For example, not more than 60 gal of Class I and/or Class II liquids, nor more than 120 gal of Class III liquids, may be stored in a single storage cabinet and not more than three such cabinets may be located in a single fire area.

TABLE 3.5 Maximum Storage Quantities for Cabinets

Liquid class	Maximum storage capacity
Flammable/Class I	60 gal
Combustible/Class II	60 gal
Combustible/Class III	120 gal
Combination of Classes	120 gal

C. Drums

Large drums, for example, 55-gal capacity, can be moved in a variety of manners. These filled drums can weigh in excess of 800 lb. Drums can be moved manually for relatively short distances by means of a drum dolly, which uses a levering arm to lift the weight of the drum. Some drum dollies can be set down horizontally so liquids may be dispensed from the drum through a spigot on the bung. Drum dollies are ideally suited for moving drums around the immediate work area, but to move drums long distances or to lift them onto platforms or pallets, an industrial truck or forklift must be used.

The most commonly used metal drums are made of steel with 18-gauge sides with 20-gauge tops and bottoms. Some drums are made of high-strength plastics and work well for holding corrosive or other materials that may be incompatible with steel. All drums are very durable and in order to be shipped off-site by rail, car, truck, or plane must meet strict performance packaging standards set by the DOT. All drums are designed and rated for holding specific materials and these ratings must not be exceeded. Drums must be compatible for the material they are intended to hold. Drums are rated by their maximum volume, the maximum density of material it may hold, and the maximum pressure it may withstand (Fig. 3.8). A code number stamped on each drum indicates these ratings as well as when and where the drum was manufactured. Identification numbers begin with the letters "NA" or "UN." NA means the material may only be transported within North America. UN means that the material is recognized for domestic or international transportation.

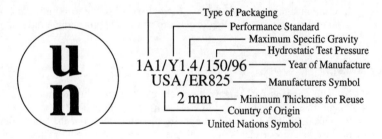

FIGURE 3.8 Container specification code.

All drums must be treated with respect because they contain large volumes of material and are very heavy. Drums should not be rolled or dragged to maneuver them into position. Using the right tool for the job will lessen the chance of an injury or spillage of a drums contents.

D. Chemtotes

Chemtotes (totes) are a common means for storage of both liquid and dry manufactured products. Liquid totes are generally made of metal alloys, aluminum, stainless, or Teflon® lined totes depending on their application. Liquid totes generally range in size from 200 to several hundred gallons. Liquid totes have a valve mechanism on the bottom side to allow a pipe or transfer line to be attached to remove the contents. Liquid totes are generally accessible by forklift from only one side because of the valving configuration located under the tote. Caution must be used when transporting totes as they are very heavy and can contain large volumes of hazardous materials.

Dry material totes are constructed of the same type materials as liquid totes. The dry material totes have a large opening on the top with a closeable lid for loading of material into the container.

Unloading can usually be accomplished by opening a panel at the base of the tote. Some totes can be unloaded from the top by tilting and pouring the material out, using a vacuum type mechanism to aspirate the material or an auger can be inserted to remove the contents.

E. Rail Tank Cars

Gases and bulk liquids can be shipped by rail in **TMU (ton multi unit or ton containers)** cars or tank cars, respectively. **Rail tank cars** tend to be single units with capacities ranging from 10,000 to 60,000 gal and a maximum pressure allowed of about 600 psig. The largest TMU cars are used primarily for liquefied petroleum products. These tanks cars can be insulated or noninsulated. The rail car consists of 15 cylinder tanks on a flat bed. Each cylinder normally has a capacity of between 100 and 200 gal depending on the gas being transported and pressures ranging from 500 to 1000 psig. These TMUs are sometimes equipped with two valves, one for vapor and another for liquids. These valves are normally protected with a hood and contain fusible plugs which melt between 155°F and 165°F.

While unloading tank cars, these rules must be followed:

1. All TMUs and tank cars must be unloaded by a person properly instructed in unloading hazardous materials.
2. All transfer lines must be properly bonded and grounded.
3. Brakes must be set and wheels blocked on all cars being unloaded.
4. Caution signs indicating "Stop—Men at Work" or "Stop—Tank Car Connected" must be placed to ensure adequate warning is given to workers approaching a rail car.

F. Highway Tankers

As with rail tank cars, highway tankers can transport liquids and gases as well. Highway-driven tanker trucks generally range in size from 1000- to 6000-gal capacity. Bulk liquids are loaded into the tanker truck from manhole openings located on top of the truck. Tanker trucks can have one or several compartments for multiple chemical shipments. These tanker trucks are generally unloaded through valve mechanisms located underneath. The same safety rules apply (trained personnel, bonding, grounding, locking the vehicle in place) to unloading tank trucks as are required with rail cars.

There are three types of vehicles used to transport compressed gases in industry: cargo tanks, portable tanks, and tube trailers. The **cargo tanks** are single units mounted directly on small truck bodies or on large beds to form semitrailers. These cargo tanks can be noninsulated, insulated, or equipped with cooling/heating units. **Portable tanks** are cylindrical tanks with flat legs on either end. They range in capacity from 120 to 2000 gal or more, with pressures ranging from 100 to approximately 500 psig. **Tube trailers** are a series of cylinders joined with a common head and mounted on a flat bed. Tube trailers can routinely be pressurized up to 2000 psig with gases like oxygen, nitrogen, and helium, and left at the user's location.

BONDING AND GROUNDING

Three major regulations cover the "bonding and grounding" of flammable liquids containers: OSHA 29 CFR 1910.106(e)(6)(ii), ATmosphères EXplosibles (ATEX) directive (from the French words ATmospheres Explosibles applicable in the European Union.), and the NFPA's Uniform Fire Code, UFC Div. VIII, Sec. 79.803(a). "Class I liquids shall not be run or dispensed into a container unless the nozzle and container are electrically interconnected." Bonding and grounding can be defined as having no voltage between two conductive objects (e.g., drum and solvent safety can) that are connected together (Fig. 3.9). Whenever transferring flammable liquids between metal containers, it is important to protect against accidental ignition caused by static electricity. Friction generated by the moving liquids creates static electricity. This static electricity is capable of causing a spark of great enough potential to ignite flammable vapors. Therefore, it is imperative to provide a means for static charge buildup to dissipate by bonding and grounding all metal containers during dispersing and storing of flammable liquids. Permanent connections are made by using solid or braided wires equipped with screw-type clamps or welded connections. Temporary connections should use only braided wire in conjunction with spring-loaded, magnetic clamps to maintain the metal to metal contact. Solid wire is not recommended in making temporary connections since it is not as durable with frequent flexing as braided wire. Both insulated and uninsulated type wires can be used for bonding and grounding connections; the uninsulated type wire can be inspected easier.

FIGURE 3.9 Always use proper bonding and grounding of a drum, solvent can, and containers. (*Photo courtesy of Thermo Fisher Scientific.*)

To dispense a flammable liquid safely from a metal drum, securely connect the drum to a "common ground line" such as a water pipe, building static grounding bus, or copper wire buried underground. Second, connect the drum to the receiving container by means of a bonding wire before dispensing the flammable liquid. When finished dispensing the flammable liquid, close the lid on the receiving container and then disconnect the bonding wire.

ATEX DIRECTIVE (EU)

The ATEX Directive, implemented as DSEAR (Dangerous Substances and Explosive Atmospheres Regulations) in the United Kingdom are sets of regulations introduced to protect employees from fires and similar events arising from potentially explosive atmospheres in the workplace while handling flammable liquids, dusts, and gases.

The ATEX directive is enforced by the Health and Safety Executive in the United Kingdom as well as other country-specific enforcement agencies. Under the ATEX rules, employers must classify into zones the areas where hazardous explosive atmospheres may occur (vapors—0, 1, 2; dusts—20, 21, 22). The classification given to a particular zone, and its size and location, depends on the chances of an explosive atmosphere being present and its persistence if it does. ATEX is most applicable to industries, including chemical manufacture, petrochemical industry, manufacturing operations where there are large quantities bulk of powders, flammable liquids, and other dangerous substances. Mining industries also must comply with these regulations. An additional broad spanning requirement in the European Union is for items of equipment (including forklifts as well as any chemical processing equipment) for use within the European Union must have a mandatory conformance marking, similar to a UL (underwriters laboratories) rating in the United States. Any product placed on the market in the European Economic Area must have a *Conformité Européenne* (CE) marking (see Fig. 3.10) The CE mark certifies it has met EU consumer safety, health, or environmental requirements.

The **CE marking** (also known as **CE mark**) is a mandatory conformance mark on many products placed on the single market in the European Economic Area (EEA).

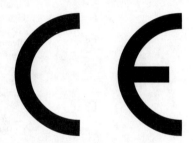

FIGURE 3.10 Conformance mark, CE mark.

The **CE marking** certifies that a product has met EU consumer safety, health, or environmental requirements. *Conformité Européenne* is French for "European conformity."

FORKLIFT OPERATIONS

OSHA 29 CFR 1910.178 standard for powered industrial trucks requires that only trained and authorized employees be permitted to operate a powered industrial truck. The American Society of Mechanical Engineers (ASME) defines a **powered industrial truck** as a mobile, power-propelled truck used to carry, push, pull, lift, stack, or tier material. Current National Institute for Occupational Safety and Health (NIOSH) statistics indicate that an estimated 35,000 forklift-related injuries occur annually in this country. The forklift is the most common powered industrial truck used to lift heavy objects and maneuver them into new positions. There are many types of forklifts, all of which have specific applications. Forklifts are powered in a variety of manners. Some forklifts operate electrically; these units operate with several sets of lead/acid-type batteries which require frequent recharging. However, there are no hazardous exhaust pollutants associated with this type. Other, older types run off of gasoline or diesel engines. These machines are being phased out because of their inefficiency and the fact that they produce high volumes of carbon monoxide. These units become troublesome if they are used in an area with poor ventilation. The more common forklifts today are powered by compressed natural gas or propane. Both of these generate significantly less air pollutants because they burn much cleaner and are more efficient.

There are 11 different categories of powered industrial trucks (Table 3.6). Proper selection depends on the power type and potential hazards in the work area. Always consult standard operating procedures (SOPs) or a supervisor before using any powered industrial truck in a potentially hazardous area.

TABLE 3.6 Powered Industrial Trucks Designations

Designation	Description
G	Gasoline-powered equipped with minimum acceptable safeguards against inherent fire hazards.
GS	Gasoline-powered equipped with additional safeguards to the fuel, exhaust, and electrical systems.
D	Diesel-powered equipped with minimum acceptable safeguards against inherent fire hazards.
DS	Diesel-powered equipped with additional safeguards to the fuel, exhaust, and electrical systems.
DY	Diesel-powered equipped with additional safeguards to the fuel, exhaust, and electrical systems. In addition, this model has no electrical ignition system or electrical equipment and has special temperature-limiting features.
E	Electric-powered equipped with minimum acceptable safeguards against inherent fire hazards.
ES	Electric-powered similar to E designation, but with additional safeguards to the electrical system to limit surface temperatures and prevent emission of hazardous sparks.
EX	Electric-powered but different from E, ES, and EE models. All electrical fittings and equipment are designed to be used in certain atmospheres containing flammable vapors and dusts.
LP	Liquefied petroleum-powered with minimum acceptable safeguards against inherent fire hazards.
LPS	Liquefied petroleum-powered with additional safeguards to fuel, exhaust, and electrical systems.

Powered industrial truck operators must be properly trained in the safe and proper use of heavy equipment. Many industrial accidents occur from inappropriately trained individuals who attempt to undertake a task for which they haven't had training. Some operators run into trouble when the **load capacity** of the forklift is exceeded. This can result in breakage of the equipment, dropping loads, and possibly injuries. This load capacity should be listed by the manufacturer's data plate on the body of the powered industrial truck. This data plate should also give the load center. The **load center** can be defined as one-half the length of the load when the weight is evenly distributed. For example, a load that is 50 in long has a load center of 25 in. Note: The farther the load center is from the fulcrum of the forklift, the less stable the load. The **fulcrum** on most forklifts is the front wheels. All forklifts have counterweights located at the rear of the unit. This counterweight keeps the center of gravity inside the stability triangle of the forklift. Forklifts have three points of suspension called the **stability triangle** (Fig. 3.11). Two of the points are the front wheels and the third point is at the center point of the rear. It should be noted that the closer the load center is to the fulcrum (front wheels), the greater the load is that can be lifted. All forklifts have a **maximum fork height**. This is the distance from the floor to the forks in their highest position. Note: The higher the forks are raised with a load, the less stable the forklift unit becomes.

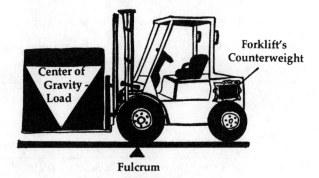

FIGURE 3.11   Forklift design and load balance.

A. General Forklift Safety Guidelines

There are many applications where fork trucks are restricted from certain areas. Forklifts which have not been certified as intrinsically safe may not enter an area where there is a potential for an explosive atmosphere. Intrinsically safe equipment is the equipment that has any source of ignition removed or shielded in such a way as to not provide a means for igniting a flammable atmosphere (Table 3.7).

CRANE SAFETY

OSHA 29 CFR 1910.179 deals with the operation and safety compliance of overhead and gantry cranes. Process operators should become familiar with standard crane signals. OSHA does not specify standard hand signals in its Overhead and Gantry Crane regulations; however, ANSI provides standard signs in ANSI B30.2-1983. The more common hand signals for controlling overhead and gantry cranes can be found in ASME publication USAS B30.5 (1968).

Cranes are equipped with a series of lights (Table 3.9) to indicate their operating status to all workers. Process operators rarely operate these cranes, but should be familiar with this crane status light system for safety. It is the crane operator's responsibility to see that the proper status lights are illuminated.

68 CHAPTER THREE

TABLE 3.7 General Forklift Safety Guidelines

Only trained and authorized personnel should operate a powered industrial truck.

- Know the specific designation of your powered industrial truck(s) and their limitations (capacities, etc.) and restrictions (environmental, etc.).
- Always perform a pre-operation inspection (written checklist recommended) of any powered industrial truck before using.
- Forklifts are less stable than most other vehicles because they have only three points of suspension.
- The back wheels provide the steering which means that the rear of the unit swings out in a turn. Many foot injuries have occurred with pedestrians because of this unexpected steering feature.
- A fully loaded forklift is physically difficult to maneuver, especially in tight spaces and more difficult to bring to a complete stop.
- Never modify a powered industrial truck without consulting both supervisor and the manufacturer.
- Always sound horn and use any convex mirrors whenever visibility is blocked or questionable. Pedestrians must be given the right-of-way.
- Always be aware of potential atmospheric hazards in the workplace environment.
- Do not allow riders on the forklift.
- Know the physical and chemical characteristics of substances being transported, including appropriate spill response.
- Never travel with the load raised. Travel with load just high enough to clear ground hazards, usually 6 to 10 in.
- Never allow anyone to get in front of, stand on, or walk underneath raised forks.
- Follow the specific instruction for refueling a powered industrial truck.
- Drive only in the designated operating areas.

TABLE 3.8 Standards for Various Types of Cranes

Crawler Locomotive and Truck Cranes: OSHA 29 CFR 1910.180
Derricks: OSHA 29 CFR 1910.181
Helicopter Cranes: OSHA 29 CFR 1910.183
Powered Industrial Trucks: OSHA 29 CFR 1910.178
Portable Pillar Cranes: ANSI B30.2
Mobile and Locomotive Cranes: ANSI B30.5
Monorails and Underhung Cranes: ANSI B30

TABLE 3.9 Crane Safety Lights

Green Light—The crane is energized and under control of an operator. Boarding, bumping, or pushing the crane is strictly prohibited.
Red Light—The main power disconnect is open. Red indicates that the crane can be boarded or deboarded. Bumping or pushing the crane is prohibited, except for troubleshooting.
Yellow Light—Boarding and deboarding the crane is prohibited. The crane is not under the control of an operator. The crane can be pushed with caution.
No Light—The crane should be considered not available for operation. The crane shall not be pushed or boarded without permission from the authorized person.
Blue Light—If light is ON, the magnet is not energized and may be connected or not connected. If light is OFF, the magnet is energized and should not be touched. The light being on to indicate that the magnet is off may seem odd. This sequence will prevent any worker from touching an energized magnet in case the light bulb were to burn out.

TRANSFERRING LIQUIDS

In general, the centrifugal pump has the widest use for recirculating liquids and general transfer work. When pumps are used to transfer reaction solutions, it is important to avoid contamination, to clean the pump after use and to get the centrifugal pump primed. Before centrifugal pumps will work, they must be primed and the inlet pipe and impeller must be completely filled with liquid. This can present a problem when working with corrosive solutions.

CONTROLLING SPILLS

Controlling spills in the workplace (Table 3.10) is covered by OSHA 29 CFR 1910.120, Hazardous Waste Operations and Emergency Response (HAZWOPER). Spills of chemicals anywhere are dangerous regardless of their nature because they cause conditions which can result in fire, personal injury, property damage, environmental contamination, or release of toxic fumes.

If a spill occurs in transporting bulk chemicals and the shipping paper or emergency response telephone number is not available, call the **Chemical Transportation Emergency Center (CHEMTEC) number at 1-800-424-9300.** CHEMTREC is a special service provided by the chemical industry to give shipper information and advice on dealing with spill situations. The federal government through the U.S. Coast Guard maintains a **National Response Center (NRC)** to deal with hazardous substance spills during transport. **Call NRC at 1-800-424-88021.** Your company or organization should have an established **Emergency Response Plan (ERP)** for dealing with chemical spills. Note: Only properly trained individuals may perform containment and cleanup of hazardous chemical spills as required by the HAZWOPER standard. Table 3.11 lists the emergency response plan that all process facilities must identify. In short, the HAZWOPER spill containment procedure says:

> IDENTIFICATION/EVACUATION/NOTIFICATION

TABLE 3.10 General Spill Containment Procedure

Step 1—Acknowledge that a spill has occurred.
Step 2—Identify the substances involved and their potential hazards.
Step 3—Shut off the source of the spill if it can be done without posing a risk to the employee.
Step 4—Determine and eliminate any sources of ignition.
Step 5—Secure the contaminated area and notify others according to company's emergency response plan.
Step 6—Report spill according to emergency response plan. This report should include reporting employee's name, location of spill, quantity of spilled substance, and any other ERP information.
Step 7—Evacuate all personnel in contaminated area not involved in controlling the spill. This action will minimize potential exposure to hazardous substances.
Step 8—Contain the spill. Prevent the spilled substance from entering drains, manholes, storm drains, air vents, and other environmental systems.
Step 9—Clean up spill, only if authorized and properly trained, using ERP procedure. Absorbents, described below, have proven to be very effective.
Step 10—Decontaminate all surfaces, reusable personal protective equipment, and process equipment. OSHA requirements for decontamination procedures are specified in the Hazardous Waste Operations and Emergency Response Standard in 29 Code of Federal Regulation 1910.120.
Step 11—Disposal of all spilled substances, spent absorbents, contaminated equipment, etc. in accordance with ERP. Used sorbent and neutralized chemical spills must be disposed of in a way that meets all local, state, and/or federal regulations.

TABLE 3.11 HAZWOPER Regulation Requirements

- The different levels of response and the specific chain of command.
- Emergency evacuation and medical treatment procedures.
- Procedure to be followed for notification, containment, assessment, and spill cleanup.
- Health and safety information regarding the specific risks and protective measures.
- Interfacing with outside sources.

SOLID CHEMICAL SPILLS

Every company that handles large quantities of chemicals should have a written procedure for its disposal according to ERP. These procedures should outline the custody of chemical waste and the individuals or departments responsible for proper disposal. Solid chemical wastes should always be identified, labeled, and segregated to prevent reactions with incompatible materials. Small solid waste spills can be swept together, brushed into a dustpan or shovel, and then deposited in the proper waste container. Separate broken glassware from other waste whenever possible to prevent cuts and other serious injuries. Broken glass should be disposed of in specially marked glass disposal containers. Biohazard waste should also be segregated and stored only in readily identified containers with plastic liners that are incinerable. For large and potentially toxic solid spills, a special vacuum cleaner equipped with a **high efficiency particulate air (HEPA)** filter should be used.

SPILL PILLOWS

The most common absorbents used on chemical spills include diatomaceous earth, vermiculite, expanded clay, sand, polypropylene fibers, sawdust, activated charcoal, and even ground corncobs. Many of the absorbents can be purchased packaged in spill pillows specifically labeled for acids, caustics, and organic solvents. These spill pillows should not be used with hydrofluoric acid spills unless specifically labeled. These pillows generally are packaged in four different sizes with the capacity to absorb a pint, quart, gallon, or 2 gal. Most of these spill pillows have unlimited shelf life so they can last indefinitely. Spill pillows are sometimes referred to as control pigs, dikes, or socks. Figure 3.12 shows a typical chemical spill control cabinet.

ACID SOLUTION SPILLS

Sodium carbonate (Na_2CO_3), commonly referred to as soda ash, is one of the most common and effective reagents used to neutralize acid spills. Large 50-lb bags can be purchased and stored until needed for neutralization of large acid spills. Smaller acid spills can be effectively neutralized with sodium bicarbonate ($NaHCO_3$), also known as baking soda. Acids are neutralized by sprinkling the neutralizing powder directly on the spilled liquid. Bubbles of carbon dioxide gas will be generated during the neutralization process. The acid is considered to be neutralized when the formation of bubbles stops or the pH is measured and determined to be neutral.

Many acid spills, once neutralized, are safe to be washed down the floor drain or sewer, provided the pH is between the regulatory limits (generally 5 to 11—always consult the company's SOPs). Some acid spills are to be treated as hazardous waste if they contain a regulated hazardous material. Check with a hazardous waste specialist to see how the spilled material is to be disposed of properly. Proper protective personal equipment (PPE) and caution must be used when neutralizing acids, particularly sulfuric acid (H_2SO_4) because of the generation of excess heat and possible spattering of the spilled material.

FIGURE 3.12 Liquid chemical spill control cabinet. (*Photo courtesy of Thermo Fisher Scientific*)

ALKALI SOLUTION SPILLS

Large base spills (alkali, alkaline, or caustic) can be neutralized with liquids like citric acid or dilute hydrochloric acid. Small caustic spills can be treated directly with solid boric acid (H_3BO_3) or dilute acetic acid solutions. A mop and bucket can be used, but avoid any splattering when squeezing out the mop. Flush the mop and bucket often and replace water frequently. As with acid spills, bases may be able to be sewered. However, check with a hazardous waste specialist before disposal.

CAUTION: Alkali solution can make the floor slippery. Alkali (caustic) materials are particularly injurious to the eyes. Proper eye protection must be worn to prevent eye injury.

VOLATILE AND FLAMMABLE SOLVENT SPILLS

Volatile and flammable materials tend to be organic chemicals and insoluble in water. Several manufacturers now supply universal sorbent kits for almost any kind or size of liquid spill. These sorbent materials come packaged in booms, socks, pillows, and as a sweeping compound. The booms for example are designed to float on water and selectively remove organic compounds. These sorbent materials are usually an amorphous silicate, polypropylene, or surfactant-treated polypropylene that can absorb an average of 14 times its weight in liquids and can fully saturate in as little as 20 seconds. These sorbents come packaged in casings made of chemical-resistant polyester blend. The bulk of the sorbent material should be disposed of in a waste-disposal safety container. Discard all volatile solvents in a waste-solvent receptacle that will contain the vapors and will not constitute a fire hazard. Volatile solvents are those solvents which vaporize readily at relatively low temperatures. The vapors that result can be toxic, nauseating, irritating, flammable, or other unpleasant side effects.

Volatile solvent spills pose special problems since they can evaporate very rapidly and occupy a rather large volume of space. This kind of spill can create a fire hazard if the solvent is flammable and it will invariably cause a high level of vapors that can be toxic as well. The vapors from a hazardous organic solvent spill can be significantly reduced by the application of a blanket of foam. Foam is sprayed directly onto the spilled solvent by means of a foam fire extinguisher. Aqueous film-forming foam (AFFF) fire extinguishers are designed specifically for this purpose. Once the spilled solvent is covered with foam, it is safer for responders to treat the spill as the vapor concentration will be significantly lessened. Discard the contaminated material in the company's designated waste chemical receptacle after cleanup.

MERCURY SPILLS

Mercury exposure in the workplace is covered by OSHA 29 CFR 1910.1000. The American Conference of Governmental Industrial Hygienists (ACGIH) is currently considering lowering the time weighted average (TWA) for mercury to 0.025 mg/m^3 and no allowable short-term exposure limit (STEL). Spills from broken thermometers, pressure measuring equipment, process catalyst, and electrical equipment are the most common sources of mercury spills. As a result of a spill, mercury can be distributed over a wide area, exposing a large surface area of the metal. In any mercury spill, unseen droplets are trapped in crevices in tables and floor surfaces. Unless the area has adequate ventilation, the combined mercury vapor concentrations may exceed the recommended limit. Surfaces which are apparently free of mercury will still harbor microscopic droplets. Vibrations in the work environment can rapidly increase the vaporization and distribution of mercury vapors. Sulfur dust can be used in an emergency for fixing the elementary mercury.

Effective mercury spill cleanup kits are now commercially available. The spill powder contains iron, copper, or zinc dust, which immediately amalgamates with the mercury. The amalgam can then quickly be picked up with a magnet or swept up with a whiskbroom and dust pan to be placed in disposable bags. Seal the mercury in an appropriate container. The sealed mercury amalgam can now be shipped to a designated hazardous waste treatment facility.

HAZARDOUS WASTE DISPOSAL

Hazardous waste disposal can be a difficult task to accomplish without an in-depth knowledge of the rules and regulations in the area where you work. Even with a broad-base knowledge, it can be difficult at times to keep up with and understand all the newly issued regulations.

The days of dumping drums of waste off into a ditch or pouring liquid waste onto the ground are gone. However, it does happen occasionally and many of those responsible for such grossly negligent acts are prosecuted and sentenced to prison terms or given heavy fines. The U.S. government has introduced legislation for the proper treatment, storage, and disposal of hazardous waste. The Congress passed legislation in the form of the **Resource Conservation and Recovery Act (RCRA)** to protect human health and the environment. In 1980 the EPA issued regulations which became enforceable under the act. It is these rules that define what constitutes a hazardous waste and how it must be handled. Wastes become classified as a hazardous waste under the RCRA rules if they meet any of the definitions of a "characteristic waste" such as ignitability, corrosivity, reactivity, and toxicity. Wastes can also become a regulated RCRA hazardous waste if the chemical is "listed." The "listed" waste sections consist of listings of solvents, unused or discarded commercial chemical products, wastes generated from specific industrial processes, and a listing of acutely toxic substances.

The amount of hazardous waste that is generated in a calendar month is used as a guide to indicate the generator status of an operation (Table 3.12). Those operations which generate waste in quantities greater than 1000 kg in a given month are deemed "**Large Quantity Generators (LQG)**." Operations which generate between 100 and 1000 kg in a month are deemed "**Small Quantity Generators (SQG)**." Those with 100 kg or less in a month are **called "Very Small Quantity Generators (VSQG)**." All three classifications of hazardous waste generators must abide by the rules set forth by the EPA. However, the more hazardous waste that is generated, the more stringent the requirements become. The following is a table of some of the requirements for the three levels of waste generators.

TABLE 3.12 EPA Classification of Hazardous Waste Generators

Generator requirements	Very small quantity generator	Small quantity generator	Large quantity generator
Quantity Limits	<100 kg	100–1000 kg	>1000 kg
EPA ID Number	Not Required	Required	Required
Storage Limits	Up to 999 kg	6000 kg/180 days	90 days
Waste Manifest	Not Required	Required	Required
Personnel Training	Not Required	Basic Training	Full Training Required
Spill Contingency Plan	Not Required	Basic Plan	Full Plan

Storage limits vary according to the generator status. VSQGs can store waste indefinitely without having to ship the waste off-site provided that they accumulate no more than 999 kg of waste. Small quantity generators are required to ship hazardous wastes off-site for treatment or disposal within 180 days and can accumulate no more than 6000 kg during that time period. There is a provision that SQGs may store hazardous wastes for up to 270 days if the waste is to be shipped over 200 miles for disposal. LQGs must store wastes for no more than 90 days prior to shipment off-site for disposal.

Both SQGs and LQGs are required to use a waste manifest with each waste shipment. The waste manifest is a document which assists the generator and disposal facility to keep track of the waste that is shipped. This document also assists governmental regulators in tracking waste from the point of generation to the ultimate disposal facility. The waste manifest was developed as part of the "cradle to grave" ideology that was introduced when RCRA became enforceable. The documentation helps to ensure that waste is disposed of in an environmentally safe manner as stipulated by the RCRA regulations.

An EPA identification number is required for small and large quantity generators of hazardous waste. The ID number is a unique number given to a generator of hazardous waste after submitting a request to the EPA. This number is used on waste manifests, reports and is used by the EPA to monitor the waste activities for individual waste generators.

REFERENCES

1. *Improving Safety in the Chemical Laboratory A Practical Guide*. Edited by Jay A. Young, 368 pages, ISBN 0-471-84693-7. Organization for Safety in Laboratories; Precautionary labels and Material Safety Data Sheets; Doing it Right; and so on Wiley and Sons (1987).

2. *Rapid Guide to Hazardous Chemicals in the Workplace*. N. Irving Sax and Richard J. Lewis, 256 pages, ISBN 0-442-28220-6, Van Nostrand Reinhold (1987). Gives rapid access to information on 700 chemicals frequently used in the Laboratory.

3. *Dangerous Properties of Industrial Materials*, 9th edition. By N. Irving Sax and Richard J. Lewis, 4368 pages, ISBN 0-442-28020-3, Van Nostrand Reinhold (1995). A three-volume set containing over 20,000 entries.

4. *The Merck Index*, 12th edition. Edited by S. Budavari, 2303 pages, ISBN 911910-27-1, Merck & Co. (1996). Covers over 10,000 chemicals, drugs, and biological substances. A partial table of contents includes abbreviations, monographs on individual compounds, organic reactions names, chemical abstract names and registry numbers, formula indexes, atomic weights, and so on.

5. *Chemical Abstracts.* American Chemical Society, Office of Chemical Abstracts, Columbus, Ohio (1-800-753-4227), http://www.cas.org. The ACS has abstracted continuously since 1907. Chemical Abstracts are listed by Subject, Author, Formula, and Patent Indexes and constitute the most complete reporting of all chemistry and chemically related publications in the United States and other countries.

6. *Chemical Data Guide for Bulk Shipment by Water.* U.S. Coast Guard, Department of Transportation, 400 Seventh Street, SW, Washington, DC 20520. The coverage of hazardous materials and bulk chemicals to be shipped by water.

7. "NIOSH/OSHA Pocket Guide to Chemical Hazards," NIOSH Pub. No. 85-114, U.S. Government Printing Office, Washington, DC (2008 or latest edition).

8. "Occupational Health Guidelines," NIOSH/OSHA NIOSH Pub. No. 81-123, U.S. Government Printing Office, Washington, DC (1981).

9. *Patty's Industrial Hygiene and Toxicology,* 4th edition, Wiley (1991). Comprehensive series of industrial hygiene practices and an updated rationale for setting and interpreting industrial standards. Ten volumes and parts set. Also available in CD-ROM version.

CHAPTER 4
HANDLING COMPRESSED GASES

COMPRESSED GASES

Most manufacturing facilities use compressed gases and liquefied compressed gases, which are contained under very high pressures (typically 100 to 2500 psi) in metal cylinders. A **compressed gas** is defined by the U.S. Department of Transportation (DOT) as "any material or mixture having in the container an absolute pressure exceeding 40 psi at 70°F (21°C) or regardless of the pressure at 70°F, having an absolute pressure exceeding 104 psi at 130°F (54°C); or any flammable liquid material having a vapor pressure exceeding 40 psi absolute at 100°F (38°C) as determined by ASTM Test D-323." These compressed substances are potentially dangerous because they are pressurized, flammable, corrosive, toxic, and/or extremely cold. When using compressed gases, wear appropriate protective equipment (review Chapter 2 "Chemical Plant and Laboratory Safety"). Respiratory protection should be available for immediate use when working with potentially toxic gases. Masks should be placed in a convenient location where they are not likely to become contaminated, and should be approved by the National Institute for Occupational Safety and Health (NIOSH) for the service intended. Those involved in the handling of compressed gases should become familiar with the proper application and limitations of the various types of respirators.

Many industrial accidents have occurred from the mishandling of these cylinders and their contents. Table 4.1 lists the most common Occupational Health and Safety Administration (OSHA) violations involving compressed gases.

GENERAL CYLINDER HANDLING PRECAUTIONS

1. Before using any compressed gases, read all cylinder label information and Material Safety Data Sheets (MSDS) associated with the compressed gas. Do not accept cylinders that do not identify the contents by name; do not rely on the cylinder's color code for content identification. Figure 4.1 gives a pictorial overview for the general handling of gas cylinders.
2. The cylinder cap should be left on each cylinder until it has been secured against a wall or bench, or placed in a cylinder stand, and is ready to be used.
3. Cylinders may be stored in the open, but should be protected from the ground beneath to prevent rusting. Cylinders may be stored in the sun, except in localities where extreme temperatures prevail. In the case of certain gases, the supplier's recommendation for shading should be observed. If ice or snow accumulates on a cylinder, thaw at room temperature or with water at a temperature not to exceed 125°F (52°C).
4. Avoid dragging, rolling, or sliding cylinders, even for short distances. They should be moved by a suitable hand or industrial truck.
5. Never tamper with safety devices in valves or cylinders.
6. Do not store filled and empty cylinders together. In addition, serious suckback can occur when an empty cylinder is attached to a pressurized system.
7. No part of a cylinder should be subjected to a temperature higher than 125°F (52°C). A flame should never be permitted to come in contact with any part of a compressed gas cylinder.

TABLE 4.1 Common OSHA Violations Involving Compressed Gases

1. Unsecured cylinders.
2. Cylinders stored without protective caps in place.
3. Noncompatible gases (such as hydrogen and oxygen) stored together.
4. Cylinder valves open when the cylinder is not in use (an attached regulator with a closed discharge valve is not sufficient).
5. Fire extinguishers not present during welding, burning, cutting, or brazing operations.
6. No safety showers and eyewash fountains where corrosive gases are used.
7. No gas masks and/or self-contained breathing apparatus conveniently located near areas where toxic gases are used or stored.

8. Cylinders should not be subjected to artificially created low temperatures −22°F (−30°C or lower), since many types of steel will lose their ductility and impact strength at lower temperatures. Special stainless steel cylinders are available for low-temperature use.
9. Do not place cylinders where they may become part of an electric circuit.
10. Bond and ground all cylinders, lines, and equipment used with flammable compressed gases.
11. Use compressed gases only in well-ventilated areas. Table 4.2 contains a partial listing of compressed gas cylinder sizes.
12. Cylinders should be used in rotation as received from the supplier. Storage areas should be set up to permit proper inventory rotation.
13. When discharging gas into a liquid, a trap or suitable check valve should be used to prevent liquid from coming back into the cylinder or regulator.
14. When returning empty cylinders, close the valve before shipment, leaving some positive pressure in the cylinder. Replace any valve outlet and protective caps originally shipped with the cylinder. Mark or label the cylinder "EMPTY" or "MT" (or utilize standard DOT "empty" labels) and store in a designated area for return to the supplier.
15. Always leak-check installed regulators, manifolds, connections, etc., with appropriate leak-detection solutions or devices. Never use an open flame for leak detection.
16. Never drop cylinders or permit them to strike each other violently.
17. In preparation for a possible large release of gas, eyewash fountains, safety showers, SCBAs, respirators, and/or resuscitators should be located nearby, but out of the immediate area that is likely to become contaminated in the event of a large release of gas.
18. Fire extinguishers, preferably of the dry-chemical type, should be kept close at hand and should be checked periodically to ensure their proper operation.
19. Low-boiling-point liquids or cryogenic materials (dry ice, fluorocarbons, liquid oxygen, liquid nitrogen, etc.) can cause frostbite on contact with living tissue and should be handled with care.
20. Use the proper wrench or tool for attaching a cylinder regulator or valve. A pipe wrench should never be used, and excessive force to tighten a regulator should be avoided.

CYLINDER MARKINGS

The DOT has developed regulations and markings to assure compressed-gas cylinder safety. Figure 4.2 shows a typical gas cylinder's parts and markings. The cylinder cap (1) protects the main cylinder valve. The valve handwheel (2) is used to open and close the cylinder valve. Certain valves are not equipped with handwheels (for example, those on acetylene tanks) and require special wrenches

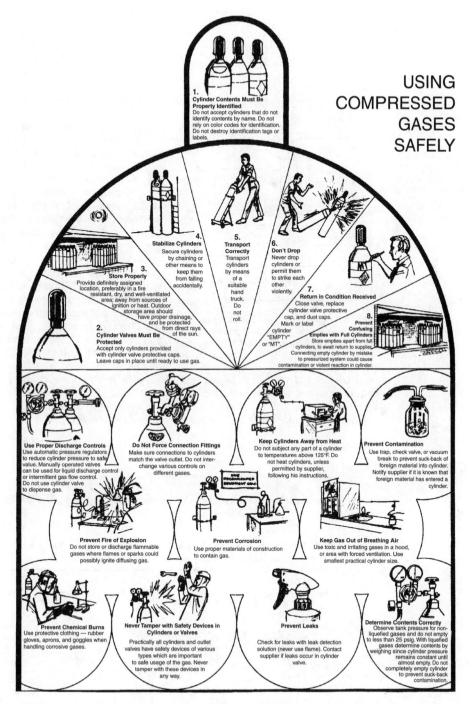

FIGURE 4.1 A pictorial overview of general handling procedures for gas cylinders, and the most common general precautions. (*Courtesy Matheson-Trigas, Inc., Guide to Safe Handing of Compressed Gases.*)

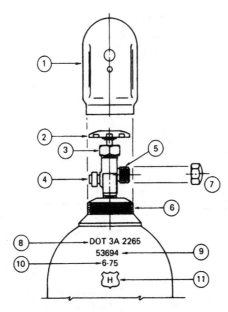

FIGURE 4.2 Compressed gas cylinder parts and DOT marking. (*Courtesy Matheson-Trigas, Inc., Guide to Safe Handling of Compressed Gases.*)

for operation. The valve packing nut (3) contains a packing gland and packing around the stem. It is adjusted only occasionally and is usually tightened if leakage is observed around the valve stem. It should not be tampered with when used in conjunction with diaphragm type valves. A safety device (4) permits gas to escape if the temperature becomes high enough to possibly rapture the cylinder by increased unsafe pressures. The valve outlet (5) connection to pressure and/or flow regulating equipment. Various types of connections are provided to prevent the interchanging of equipment for incompatible gases, usually identified by the **Compressed Gas Association (CGA)**. For example No 350 is used for hydrogen service. A cylinder collar (6) holds the cylinder cap at all times, except when regulating equipment is attached to the cylinder valve. The valve outlet cap (7) protects valve threads from damage and keeps the outlet clean; it is not used universally. Specification number (8) signifies that the cylinder conforms to DOT specification DOT-3A, governing material of construction, capacities, testprocedures, and also that the service pressure for which the cylinder is designated is 2265 psig at 70°F (21°C). The cylinder serial number is indicated by (9), and (10) indicates the date (month and year; in this case, June 1975) of initial hydrostatic testing. For most gases, hydrostatic pressure tests are performed on cylinders every 5 years to determine their fitness for further use. The original inspector's insignia for conducting hydrostatic and other required tests to approve the cylinder under DOT specifications is shown by (11).

CYLINDER SIZES

Normally, compressed gas cylinders range in capacity from approximately 0.04 m^3 (L.B.—lecture bottles) to 15 m^3, as shown in Table 4.2.

TABLE 4.2 Compressed Gas Cylinder Sizes. A Variety of Cylinders Is Available from Different Vendors. Actual Mass of Content Will Vary with Cylinder Pressure

Approximate dimensions (inches)	Air products	BOC (Airco)	Matheson	Scott specialty gases
High pressure				
9 × 55	A	300	1L	K
9 × 51	B	200	1A	A
7 × 33	C	80	2	B
7 × 19	D-1	30	3	C
4 × 17	D	12	4	D
2 × 12	L.B.	L.B.	L.B.	L.B.
4 × 26	E	E	3L	ER
10 × 51	BX	500	1U	—
9 × 51	BY	—	1H	—

COMMON COMPRESSED GASES AND THEIR PROPERTIES

Table 4.3 gives the physical properties of several compressed gases. Detailed physical and chemical properties information is provided for some of the more common process gases.

A. Inert Gases

Leaking cylinders of inert gases, such as argon, helium, nitrogen, etc., do not represent a hazard unless they are situated in confined places where an oxygen-deficient atmosphere may be created.

TABLE 4.3 Properties of Some Compressed Gases

Substance	Density (@ 0°C) g/L	Density (@ 0°C) lb/ft^3	Melting point °C	Boiling point °C	Specific heat cal/g°C
Acetylene	1.173	0.07323	−81.3	−83.6	0.3832
Air	1.2929	0.08071	—	—	0.2377
Ammonia	0.7710	0.04813	−75	−33.5	0.5202
Argon	1.7837	0.11135	−189.2	−185.7	0.1233
Arsine	3.48	0.217	−113.5	−54.8	—
Carbon dioxide	1.9769	0.12341	−57	−80 Subl.	0.2025
Carbon monoxide	1.2504	0.07807	−207	−191.5	0.2425
Chlorine	3.214	0.2006	−101.6	−34.7	0.1125
Ethane	1.3566	0.08469	−172.0	−88.3	0.3861
Ethyl chloride	2.870	0.1793	−138.7	12.2	0.2750
Ethylene	1.2604	0.07868	−169.4	−103.8	—
Fluorine	1.696	0.1059	−223	−187	0.182
Helium	0.1784	0.01114	−272	−268.94	1.25
Hydrogen	0.08988	0.005611	−259.14	−252.8	3.409
Hydrogen bromide	3.6445	0.22752	−86.7	−68.7	0.082
Hydrogen chloride	1.6392	0.10233	−111.3	−83.1	0.194
Hydrogen fluoride	0.922	0.0576	−92.3	−36.7	0.343
Hydrogen iodide	5.789	0.3614	−51.3	−35.7	0.06
Hydrogen selenide	3.670	0.229	−64	−42	—
Hydrogen sulfide	1.539	0.09608	−86	−62	0.2451
Hydrogen telluride	5.803	0.363	−48	−1.8	—
Isobutane	2.637	0.1669	−145.0	−10.2	—
Krypton	3.708	0.2315	−169	−151.8	—
Methane	0.717	0.0448	−182.5	−161.4	0.5929
Methyl ether	2.1098	0.13171	−138	−24.9	—
Neon	0.9004	0.05621	−248.67	−245.9	—
Nitric oxide	1.3402	0.08367	−167	−153	0.232
Nitrogen	1.2506	0.07807	−209.86	−195.8	0.2438
Nitrous oxide	1.978	0.1235	−102.4	−89.8	0.2126
Oxygen	1.42904	0.089212	−218.4	−183.0	0.2175
Phosgene	4.531	0.283	−118	8.3	—
Phosphine	1.529	0.0955	−113.5	−87.4	—
Propane	2.020	0.1261	−189.9	−44.5	—
Silicon tetrafluoride	4.68	0.292	—	−68	—
Sulfur dioxide	2.9269	0.18272	−76	−10	0.1544

B. Compressed Air

Compressed, clean, oil-free air is a common utility to have in a process operation. Compressed air in the plant is usually mechanically compressed in an air compressor. This means that the air is compressed by a piston which must be lubricated, and probably contains water vapor, oil, and possibly some debris from the intake air. Compressed air can be filtered through glass-wool plugs or dried by passing it through gas-drying towers.

C. Chlorine Gas (Cl_2)

Chlorine is highly irritating in low concentrations and can be fatal at higher concentrations. Chlorine is a greenish-yellow gas and has a disagreeable, pungent odor. Chlorine is 2.5 times more dense than air and therefore tends to settle in lower areas. Even though chlorine is non-flammable, it can support combustion and reacts violently with certain inorganic (hot iron, ammonia, etc.) and organic compounds (hydrocarbons, turpentine, etc.). Chlorine can react with the gas cylinder and water at high heat 149°C (above 300°F). Therefore, water is not usually suggested for use on leaking chlorine tanks; water and chlorine can combine to produce hydrochloric acid which is very corrosive. In general, if a large chlorine leak is involved, the immediate area should be evacuated.

D. Acetylene (C_2H_2) Gas

Acetylene (ethine, ethyne) C_2H_2, molecular weight 26.02, is a colorless, toxic gas. Acetylene has an ethereal odor when pure and a disagreeable odor when impure due normally to phosphine and/or hydrogen sulfide contamination. Acetylene is highly flammable and burns brilliantly in air with a very sooty flame. It is not explosive at ordinary pressure, but will form explosive mixtures when compressed with air to 2 atmospheres or greater. A major hazard that exists with acetylene gas is that of its explosive limits (LEL = 2.5 percent and UEL = 82 percent). Review Chapter 3, "Flash Point Determinations" section, for more about explosion limits. Acetylene forms insoluble explosive compounds with copperand silver, hence copper and brass containers, tubing, regulators, and so forth should be avoided. Specific gravity is 0.91 (air = 1); density = 1.165 g/L @ STP; melting point = −81.8°C (−115°F) (@ 890 Torr); boiling point = −84°C (−119°F) and it is soluble in water, alcohol, and acetone.

Acetylene gas cylinders are normally shipped with no more than 250 psig. Acetylene is unstable as a compressed gas and therefore must be supplied with a "filler" for stabilization. Acetylene gas cylinders are filled, with fuller's earth, lime, silica, and calcium sulfate (a solid) which is very porous and occupies approximately 8 percent of the cylinder's volume. This porous filler prevents flashbacks into the cylinder in the event of a fire. Since pure acetylene gas cannot be safely shipped at greater than 15 psig, it is necessary to dissolve the acetylene in an organic solvent like acetone. Acetone dissolves about 25 volumes of acetylene at 15°C (59°F) and 1.0 atmospheres and 300 volumes at 12 atmospheres. Acetylene cylinders are available in capacities ranging from 10 to 1400 ft^3 and protected only by fusible plugs. These fusible plugs which melt at 212°F (100°C) can be found on the top and bottom of the cylinder, and may number as high as four depending on the gas cylinder capacity. As with all flammable gases, acetylene gas must be stored separately from oxygen gas cylinders. The only allowable deviation to this rule is when oxygen and acetylene are stored on a gas carrier cart used for welding purposes.

Two types of valves are normally found on acetylene cylinders. A simple handwheel is attached to the valve. The second type has a valve equipped with a 3/8 in. square shank. It is operated by means of a square box socket wrench. It is recommended that the cylinder valve only be opened from ¼ to 1½ turns. The wrench should be left on the valve at all times in order to quickly close the valve in an emergency.

E. Acid Gases

Acid gases are corrosive and toxic. Therefore, put on appropriate personal protective equipment (faceshield, rubber gloves, breathing equipment) before transporting the leaking cylinder to a safe out-of-doors area or use a hood with forced ventilation. Typical acidic gases are hydrogen chloride, hydrogen sulfide, and sulfur dioxide.

F. Alkaline Gases

The alkaline gases are corrosive, flammable, and toxic. Put on appropriate personal protective equipment (face mask, rubber gloves, breathing equipment) before transporting the cylinder to a hood with forced ventilation or to a safe out-of-doors area.

G. Ammonia (NH_3) Gas

This very alkaline gas has a boiling point of −28°F (−33°C) at normal pressure. Ammonia has a vapor density of 0.6 compared to air, thus it tends to rise upon escaping. It is flammable, toxic, and very irritating to the eyes, skin, and mucous membranes. There are two types of cylinders for the liquid-vapor ammonia composition. The first is a normal cylinder which releases gaseous ammonia only. The second type of cylinder contains a **eductor tube** or **dip tube**. A full length eductor tube allows only liquid to be withdrawn, not gases. It allows liquid ammonia to be drawn from the cylinder (Fig. 4.3). A second type of liquefied compressed gas cylinder comes equipped with a **gooseneck eductor tube** which requires that the cylinder be placed on its side or inverted to withdraw liquid from the cylinder.

For ammonia leaks, the exact location can be determined with moist, blue litmus paper (which turns red when exposed). An open bottle of hydrochloric acid can also be used for leak detection because it produces a white, dense cloud as ammonium chloride is formed.

FIGURE 4.3 Liquefied compressed gas cylinder equipped with a full-length eductor tube.

H. Hydrogen (H_2) Gas

Hydrogen is a colorless, odorless, tasteless, and highly flammable gas. Hydrogen molecules contain two atoms of hydrogen, thus has a molecular weight of 2.01594 g/mole. Hydrogen forms explosive mixtures with air, oxygen, chlorine, and many other substances. Hydrogen has a specific gravity of 0.069 (air = 1) and a density of 0.08987 g/L @ 0°C. Hydrogen liquefies at −220°C (−364°F), solidifies at −257°C (−431°F), and has a boiling point of −252.7°C (−422.9°F). It has a critical temperature and pressure of −239.9°C (−399.8°F) and 12.8 atmospheres. One volume of hydrogen gas is soluble in approximately 50 volumes of water at 0°C. Extreme caution should be exercised with cylinders of hydrogen gas: Hydrogen is so flammable that it can self-ignite by the friction created by the escaping molecules from a leak.

I. Liquefied Petroleum (LP) Gases

Propane (C_3H_8) liquid boils at −44°F (−42°C) while butane (C_4H_{10}) boils at 31°F (−1°C). Thus, butane finds limited use in the northern areas of the United States where the temperature frequently falls below freezing. Many suppliers use a mixture of propane and butane. Propane and butane leaks would be difficult to smell, thus impurities, like mercaptans (sulfur-containing organic compound), are added. Leaking LPG tanks should not be approached unless done so by qualified hazardous materials emergency responders. Even small LPG tanks should be handled with care since the escaping liquid or gas can be below the freezing point of water. A **BLEVE** (boiling liquid expanding vapor explosion) is always a danger with LP gases. A BLEVE condition can occur when heat builds up around the vapor area of the tank from the direct impingement of flames and the structural strength is weakened. The boiling liquid vapors exert enough pressure on the weakened structure to cause rupture. This rapid deterioration of the tank and tremendous release of energy can propel the tank and equipment for great distances.

J. Methyl Acetylene Propadiene (MAPP)

MAPP is a mixture of methyl acetylene and propadiene with a specific gravity of 1.48. MAPP is currently being substituted for acetylene in many industrial applications. MAPP has several advantages over acetylene: (a) it is not shock sensitive and can be stored in normal cylinders; (b) its explosion limits range only from 3.4 percent to 10.8 percent; (c) MAPP cylinders can be pressurized to as high as 250 psig at 130°F (54°C) while pure acetylene is limited to less than 15 psig; and (d) gas cylinders containing MAPP can be protected with safety relief valves which open at 375 psig at approximately 175°F (79°C). MAPP burns with a temperature of 5300°F (2927°C) while acetylene can reach a temperature of 5700°F (3149°C) with oxygen.

K. Oxygen (O_2) Gas

Oxygen is a colorless, odorless, tasteless, neutral gas and has a molecular weight of 32.0 g/mole. Normal air is approximately 20 percent by volume oxygen with the remaining gases being nitrogen (79 percent) and argon (1 percent). Elemental oxygen is, by definition, a nonflammable gas, but it readily supports combustion. Oxygen gas condenses to a liquid at −183°C and 1 atmosphere pressure. Liquefied oxygen is abbreviated LOX. Oxygen has a melting point of −361.1°F (−218.4°C), boiling point −297.33°F (−182.96°C), and a critical temperature and pressure of −181.8°F (−118.8°C) and 49.7 atmospheres. One volume of oxygen is soluble in 32 volumes of water at 20°C. Caution: smoking, flames, and sparks must be avoided with oxygen gas because of the explosive hazard. Only clean, oxygen-approved regulators should be used with oxygen cylinders. Even trace amounts of oil, grease, or other organic contaminants can explode on contact with oxygen gas. Never use oil or other lubricants with oxygen. Never substitute pure oxygen for compressed air. Do not blow off clothing with oxygen gas. Never store oxygen cylinders in the same location as combustible gases.

CRYOGENIC GASES AND LIQUIDS

A. Introduction

Rapid removal of the gas phase from a **liquefied gas** may cause the liquid to cool too rapidly, causing the pressure and flow to drop below the required level. In such cases, cylinders may be heated in a water bath with the temperature controlled to go no higher than 125°F (52°C). Safety relief devices should be installed in all liquid transfer lines to relieve sudden, dangerous hydrostatic or vapor-pressure buildups.

The most common device used to reduce the pressure of nonliquefied gases to a safe level for gas removal is a pressure regulator. Delivery pressure will exactly balance the delivery pressure spring to give a relatively constant delivery pressure.

All gases can be eventually reduced to liquids and the liquids reduced to solids by an appropriate decrease in temperature and/or an increase in the applied pressure. However, there is a temperature for all gases above which pressure alone cannot condense them into liquids. This temperature is known as the **critical temperature**. For example, carbon dioxide easily liquefies if sufficient pressure is applied while the gas is below 87.87°F (31.04°C); however, no amount of applied pressure can cause liquefaction of the gas above this temperature.

The pressure required to liquefy a gas at its critical temperature is called the **critical pressure**. The critical pressure for carbon dioxide is 72.85 atm (7.358 × 10^6 Pa). The density of a substance at its critical temperature and critical pressure is called **its critical density**. The critical density of carbon dioxide is 0.468 g/cm^3. The critical temperature, critical pressure, and density for several gases are given in Table 4.4.

Gases that are liquefied by applying pressure and stored in insulated containers are referred to as **cryogenic liquids**. This name is derived from the Greek word "kryos" meaning ice cold. These very cold liquids (−150°F and below) require specially designed vessels for storage. A common device is the **Dewar flask**, which is a double-walled glass vessel with a silvered interior, similar to the lining found in a Thermos® jug. The extremely cold temperatures and potentially high pressures of these cryogenic liquids (gases) present many hazards.

TABLE 4.4 Cryogenic Properties of Some Common Gases

Substance	Density @ 1 atm, 0°C g/L	Boiling point 1 atm °C	Melting point 1 atm °C	Vapor density @ BP g/L	Critical temp °C	Critical pressure atm
Hydrogen H$_2$	0.0899	−252.5	−259.1	1.335	−239.9	12.8
Methane CH$_4$	0.5547	−164	−182.5	1.800	−82.1	45.8
Ammonia NH$_3$	0.771 g/mL	−33.35	−77.7	0.891	132.5	112.5
Nitrogen N$_2$	1.2506	−195.8	−209.9	4.610	−147	33.5
Oxygen O$_2$	1.429	−183.0	−218.4	4.741	−118.4	50.1
Helium He	0.1785	−268.6	−272.2 @ 26 atm	16.00	−267.9	2.26
Neon Ne	0.9002	−245.9	−248.7	9.499	−228.7	26.8
Argon Ar	1.784	−185.7	−189.2	5.895	−122.3	48.0
Carbon dioxide CO$_2$	1.98	−78.5	−56.6 @ 5.2 atm	0.518	31.04	72.85

B. Liquid Fluorine (F$_2$)

Fluorine is extremely hazardous (corrosive and poisonous) since it is the most reactive of all the nonmetals. Fluorine can be liquefied at −188°C (−306°F) and is slightly more dense than air with a specific gravity of 1.1. It can cause severe burns to the eyes and skin. Fluorine is a strong oxidizing agent, like oxygen, and will chemically react with almost any other substance. Fluorine will react with concrete and steel. As with chlorine leaks, it is always advisable to evacuate the immediate area. Water is normally not used in fighting a fluorine fire as it tends to intensify the fire.

C. Liquid Nitrogen (N$_2$)

Liquid nitrogen is colorless and odorless and has a specific gravity of approximately 1.0. Liquid nitrogen is inert and does not support combustion. Liquid nitrogen is a very cold liquid −196°C (−320°F). The high gas volume to liquid volume ratio causes liquid nitrogen to be considered hazardous. A cubic foot of liquid nitrogen will expand to 681 ft^3 at its boiling point.

D. Liquid Oxygen (LOX)

Liquid oxygen is a pale blue color, slightly more dense than water, and has a boiling point of −315°F. Liquid oxygen is extremely hazardous, although it is classified as a nonflammable gas. Liquid oxygen is transported in insulated trailers with capacities of up to 10,000 gal of LOX. Storage tanks vary in capacity of 500 to 11,000 gal and are equipped with frangible discs and safety relief valves. LOX supports combustion with almost all combustible materials. Valves or fittings on oxygen cylinders should never be oiled. Leaking liquid oxygen will even cause combustion to occur with asphalt (organic) pavement.

E. Cryogenic Handling Precautions

Always wear protective gloves, clothing, and eye shields when working with cryogenic fluids. The following are some hazards associated with cryogenic fluids.

- *Damage to living tissue*. Contact with cryogenic fluids can cause a localized solidification of tissue and produce a burn as painful as that received from a heat source. The extreme coldness causes a local arrest in the circulation of blood, and any exposed skin tissue should be restored to normal temperature as quickly as possible. This is most easily accomplished by simply immersing the damaged skin in warm water at approximately 45°C and then seeking medical attention for treatment of potential frostbite.
- *A high expansion rate upon vaporization*. Cryogenic liquids can expand to great volumes upon evaporation; for example, liquid methane expands to approximately 630 times its initial liquid volume when converted to its gaseous state. Therefore, if the mechanism of cooling this liquefied methane fails or is inadequate, the internal pressure within the storage container can increase rapidly. If proper ventilation is not provided, the container can explode.
- *An ability to liquefy other gases*. Cryogenic fluids are capable of condensing and even solidifying other gases. Gaseous air, for example, solidifies upon exposure to a number of cryogenic liquids. This solidification of air could present a hazard by blocking vent tubes on the cryogenic storage containers and preventing the release of pressure buildup.

CONTROL AND REGULATION OF GASES

A. Gas Regulators

A gas-pressure regulator is a precision instrument designed to reduce high-source pressures (cylinders or compression systems) to a safe value, one consistent with a system's design. Each regulator will control a chosen delivery pressure within the bounds of the regulator's delivery-pressure range. This constant delivery pressure prevents the over-pressurization of any apparatus downstream of the regulator and permits stable flow rates to be established.

CAUTION: Pressure regulators are not flow regulators. They do maintain a constant pressure, which in turn will maintain a constant flow providing the pressure downstream does not vary. Use flow controllers to control flow if the downstream pressures are subject to variations.

Three criteria are used to measure the performance of a regulator:

1. The regulator's ability to maintain a constant delivery pressure, regardless of the rate of gas discharge. All regulators will show a drop in delivery pressure with increased flow. The smaller the drop, the better the regulator performance.
2. The regulator's ability to maintain a constant delivery pressure as source pressure varies. This is very important.
3. The **lockup** of the regulator. This is defined as the final pressure attained by a system when all flow is stopped. It is usually slightly above the delivery pressure when set at flowing conditions. All regulators are chosen to give the best possible lockup performance, with slight deviation from delivery pressure.

B. How a Gas Regulator Works

A regulator reduces gas pressure by the counteraction of gas pressure on a diaphragm against the compression of a spring which can be adjusted externally with the pressure-adjusting screw (Fig. 4.4).

In operation, the pressure-adjusting screw is turned to exert force on the spring and diaphragm. This force is transmitted to the valve assembly, pushing the valve away from the seat. The high-pressure gas will flow past the valve into the low-pressure chamber. When the force of gas pressure on the diaphragm equals the force of the spring, the valve and seat assemblies close, preventing the flow of additional gas into the low-pressure chamber.

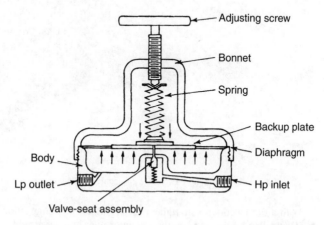

FIGURE 4.4 Schematic of a single-stage regulator (typical construction).

C. Pressure-Reduction Stages

Removal of gas from the low-pressure chamber will permit downward deflection of the diaphragm, opening the valve assembly, and thereby permitting a pressure increase in the low-pressure chamber. This constant throttling action permits a pressure balance in the regulator's low-pressure chamber, thus yielding a steady delivery pressure relatively independent of normal flow fluctuations and falling cylinder pressure. Every gas cylinder has a matching connector on the regulator. If the inlet of the regulator does not fit the cylinder outlet, do not force the fitting. It may not be the correct one for the situation.

Controlled pressure reduction requires the use of two-stage pressure reduction. Two stages of reduction constitute the same action in series, with the delivery pressure from one stage becoming the source pressure for the second stage. Most gas regulators employed for use on high-pressure cylinders are of either the single- or two-stage variety. Generally, the reduction of pressure in two stages (Fig. 4.5) permits a closer control of the delivery pressure over a wider range of inlet pressures.

D. Installation

1. Before connecting the regulator to the cylinder-valve outlet, be sure the regulator has the proper connection to fit the cylinder valve. If there is some doubt about the connections being correct, check the manufacturer's catalog for valve-outlet designation and description. Inspect the regulator inlet and cylinder-valve outlet for foreign matter. Remove foreign matter with a clean cloth *except in the case of oxygen.* In the case of oxygen, open the cylinder valve slightly to blow any dirt out of the outlet. A dirty oxygen-regulator inlet can be rinsed clean in fresh tetrachloromethane, CCl_4 and blown dry with oil-free nitrogen.

86 CHAPTER FOUR

FIGURE 4.5 Two-stage gas regulator. (*Photo courtesy Matheson-Trigas, Inc.*)

2. With a flat-faced wrench, tighten the regulator-inlet connection nut to the cylinder-valve outlet. (Depending on gas service, the regulator inlet may be a right-hand thread or a left-hand thread. Make sure that proper identification of the mating connections has been made.) *Do not force the threads.* Some regulator connections require the use of a flat gasket to provide a leaktight seal between the regulator and valve outlet. In this instance, gaskets are supplied with the regulator and should be replaced when they become worn. When utilizing Teflon® gaskets, do not exert excessive force in tightening the connection or the gasket may force its way into the valve opening and impede the discharge of gas.
3. Close the regulator by releasing the pressure-adjusting screw. Turn counterclockwise until the screw turns freely without tension.
4. Check to see that the needle valve on the regulator outlet is closed.
5. Attach tubing or piping to the regulator-valve outlet. Except for high-pressure regulators, a hose end is provided with the regulator.

CAUTION: Regulators and valves used with oxygen must not come into contact with oil and grease. In case of such contamination, do not connect the regulator. This problem must be referred to personnel trained in handling this situation.

E. Operation

1. Slowly open the cylinder valve until full cylinder pressure is registered on the tank gauge. (In the case of liquefied gases, a tank gauge is not usually provided.) It is recommended that the cylinder valve be fully opened to prevent limiting the flow to the regulator which would result in the failure of the regulator to maintain required delivery pressure. Never stand in front of the regulator when opening the cylinder valve.
2. Adjust the delivery pressure to the desired pressure setting by turning the pressure-adjusting screw clockwise and noting the delivery pressure which is registered on the delivery-pressure gauge.
3. The flow may now be regulated by proper adjustment of the needle valve.

F. Shutdown

1. Close the main cylinder valve.
2. Relieve all the pressure from the regulator through the needle valve until both gauges register zero.
3. Turn the adjusting screw counterclockwise until the screw turns freely without tension.
4. Close the regulator-outlet needle valve.

G. Dismantling

1. If the regulator will not be used for a while, store it in a clean, dry location, free of corrosive vapors.
2. If the regulator has been used with corrosive or flammable gases, flush it with dry nitrogen. This can be done by screwing in the pressure-adjusting screw (clockwise), opening the outlet valve, and directing a stream of dry nitrogen into the regulator inlet by means of a flexible tube or rubber hose. After flushing, turn out the adjusting screw and close the outlet valve.
3. Cap or seal the regulator inlet or simply store it in its original plastic bag. This will prevent dirt from clogging the regulator inlet and will extend the life of the regulator.

H. Troubleshooting

Regulators should be checked periodically to ensure proper and safe operation. This periodic check will vary, depending on gas service and usage.

Regulators in noncorrosive gas service such as nitrogen, hydrogen, and helium require relatively little maintenance and a quick check on a monthly basis is usually adequate. Regulators in corrosive-gas service such as hydrogen chloride, chlorine, and hydrogen sulfide require considerably more checking. At least once a week is recommended.

The procedure for inspecting any regulator is as follows:

1. Be sure gauges read zero when all pressure is drained from the system.
2. Open the cylinder valve and turn the adjusting screw counterclockwise; the high-pressure gauge should read the cylinder pressure.
3. Close the regulator-outlet needle valve and wait 5 to 10 minutes; the delivery pressure gauge should not indicate a pressure increase. A pressure increase would indicate leakage across the internal valve system.
4. Next, turn the adjusting screw clockwise until a nominal delivery pressure is indicated. Inability to attain a proper delivery-pressure setting or abnormal adjustment of the screw indicates improper operation, which may be attributed to blockage of the gas passage or a leak in the low-pressure side of the regulator. Continued wear on a regulator valve-and-seat assembly will cause a rise above a set delivery pressure, which is termed **crawl**. A regulator exhibiting crawl should not be used.
5. Close the cylinder valve and observe the pressure on both the inlet and delivery side of the regulator after 5 or 10 minutes. A drop in the pressure reading after this period of time may indicate a leak in. the system, possibly at the inlet or through the needle valve, safety devices, or diaphragm.
6. An excessive fall in delivery pressure under normal operating conditions and flow indicates an internal blockage. Any deviation from normal in the preceding checkout will require servicing by qualified repair personnel.

WARNING: A regulator, valve, or other equipment that has been used with another gas should never be used with oxygen. A regulator or control should never be used on more than one gas unless the user is fully familiar with the properties of the gases involved or has obtained assurance from the gas supplier that the interchange is permissible and there is no safety hazard. When a regulator shows signs of wear, it should be serviced only by reputable repair personnel.

GAS VALVES

Gas valves perform one of four basic functions: (a) isolation; (b) regulation; (c) direction; and (d) protection. Chapter 18 "Plumbing, Valves, and Pumps" contains additional information on process valves, plumbing, fittings, etc.

A. Needle Valves

(Figure 4.6). Used for isolation. Screw actuation allows the valve to be opened gradually. The threaded stem requires three to four turns of the handle to open or close the valve. At exceptionally high pressures (above 10,000 psig), a special variation of needle valve called ball and plug valve is recommended. A quarter turn of the handle quickly opens or closes the valve. However, because of their quick open and close design these valves do not allow a system to be turned on slowly and should not be used at the partially open position for throttling.

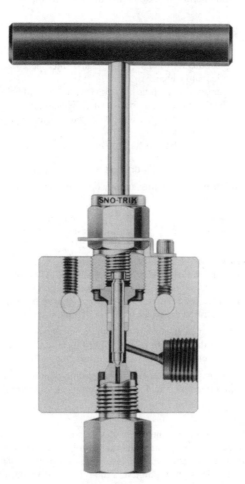

FIGURE 4.6 Needle valve. (*Reproduced with the permission of Swagelok Marketing Company.*)

B. Metering Valves

Designed to regulate (Fig. 4.7) or adjust the rate of flow. They usually have a long, finely tapered stem tip and a fine-pitch stem thread and open in five to ten turns. Valves of this type are not recommended for isolation service and are generally designed so they cannot be shut off.

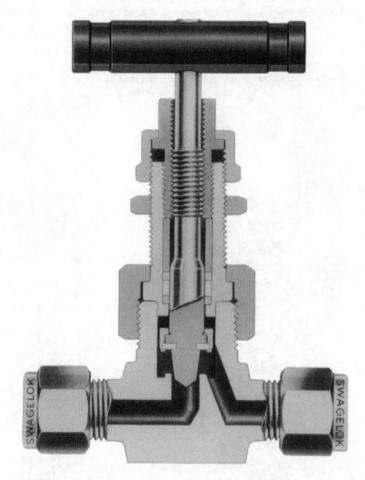

FIGURE 4.7 Regulating valve. (*Reproduced with the permission of Swagelok Marketing Company.*)

C. Direction (Check) Valves

Designed to have three or more connections, allowing the flow to be directed toward two or more systems. **Check valves** (Fig. 4.8) are a type of direction valve because they are held closed by a spring. These valves open automatically when the inlet pressure exceeds the outlet pressure by enough to overcome the spring force. When the flow stops, the check valve closes to prevent flow in the reverse direction.

D. Relief Valves

Protect a fluid system from excessive pressure. Pressure increase due to uncontrolled reactions or unexpected surges of pressure can be relieved by means of a **safety relief valve** (Fig. 4.9) installed in the gas line or cylinder. The valve is held closed by a spring as the system operates at its normal (designated) pressure at 150°F (66°C). When the pressure increases to a set point of the valve, it opens automatically and remains open until the system pressure decreases to below the set point. A **frangible disc** also serves the purpose of a safety relief valve, but cannot be reset. A metal disc is installed to burst at a predetermined pressure and to release the entire contents of the cylinder. A third type of relief assembly is the fusible plug. **Fusible plugs** do not operate on pressure but on temperature changes. These plugs are normally designed to melt at 100°C (212°F) for acetylene tanks and generally around 74°C (165°F) for other gases. Not all gas cylinders are equipped with safety valves. Certain compressed gases like fluorine and the Class A poisons are examples.

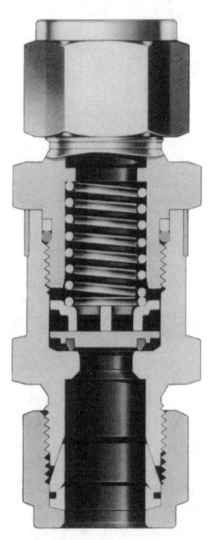

FIGURE 4.8 Check valve. (*Reproduced with the permission of Swagelok Marketing Company.*)

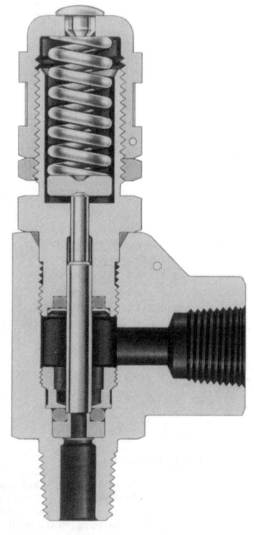

FIGURE 4.9 Safety relief valve. (*Reproduced with the permission of Swagelok Marketing Company.*)

E. Gauges

Since the performance of pumps will be measured by gauges, a brief comment on a few of the most popular types of gauges is needed. The gauge must be appropriately placed. If it is placed near the pump, it measures the vacuum at this location. If the system is large or has restrictions, the gauge reading may not be an accurate measure of the test area. If the gauge is placed in the test area (as it should be) and there are restrictions in the system, the pump performance may be unjustly judged as too slow. (Refer to Chapter 5, "Pressure and Vacuum.")

All the gauges except the McLeod measure the total pressure exerted by both gases and vapors, and different gases give different readings at the same pressure. Since all mechanical vacuum pumps are tested with the McLeod gauge, because it is a primary standard, any other type of gauge will indicate a higher pressure because it registers both gas and vapor. The exact reading of the gauge will depend, therefore, upon the gases and vapors present and the calibration of the gauge.

F. Manual Flow Controls

Where intermittent flow control is needed and an operator will be present at all times, a manual type of flow control may be used. This typeof control is simply a valve that is operated manually to deliver the proper amount of gas. Fine flow control can be obtained, but it must be remembered that dangerous pressures can build up in a closed system or in one that becomes plugged, since there is no provision for automatic prevention of excessive pressures.

G. Safety Devices

It is necessary to provide further supplementary safety devices to prevent over-pressurizing of lines and to prevent suck-back of materials into the cylinder controls—possibly into the cylinder itself. Aside from the possibility of causing rapid corrosion, the reaction of a gas with material that has been sucked back may be violent enough to cause extensive equipment and cylinder damage.

Traps and Check Valves. The danger of suck-back can be eliminated by providing a trap (Fig. 4.10) that will hold all material that can possibly be sucked back, or by using a check valve or suitable vacuum break. Check valves prevent the return flow of gas and thus keep foreign matter out of gas lines, regulators, and cylinders ahead of the valve. The valves are spring-loaded.

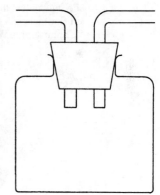

FIGURE 4.10 Safety bottle used to trap materials that are sucked back. Absolutely necessary in protecting vacuum systems and filtrates from contamination.

H. Quick Couplers

Quick couplers permit regulators, needle valves, and other components of a gas system to be connected and interconnected to cylinder outlets quickly and safely without the use of wrenches.

Safety Relief Valves. Pressure increases due to uncontrolled reactions or unexpected surges of pressure can be relieved by means of a safety relief device installed in the gas line.

For experiments conducted in glassware, such a pressure-relief device can be improvised by using a U tube filled with mercury (or other inert liquid), with one end attached by means of a T tube to the gas line, and with the other end free to exhaust into an open flask that will contain the mercury in case of overpressure (Fig. 4.11).

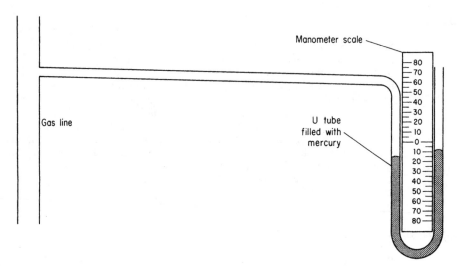

FIGURE 4.11 Pressure-relief U tube.

I. Flowmeters

Flowmeters (Fig. 4.12) are used in fluid systems to indicate the rate of flow of the fluid by registering the scale graduation at the center of the spherical float. Flowmeters do not control the rate of flow of the fluid unless they are specifically equipped with control valves or flow controllers. See the section on flowmeters in Chapter 17 "Process Analyzers" section on variable area flowmeters and Chapter 26 "Chromatography" section on flowmeters.

FIGURE 4.12 Gas or liquid flowmeter.

J. Gas-Leak Detectors

Gas leaks in systems can cause problems in flow control and can totally invalidate analytical results, especially in gas chromatography. The gas-leak detector enables you to pinpoint gas leaks easily and quickly. Some types feature meter readouts; others emit an audio signal when the leak is located. The gas-leak detector can detect helium that is leaking at the rate of 1×10^{-6} mL/s, a leak too small to bubble, yet one which could be really critical.

K. Gas-Washing Bottles

The flow of gases can be monitored by means of a gas-washing bottle, which contains a liquid through which the gas is passed (Fig. 4.13). The liquid must be inert to the gas, and the gas must not be soluble in the liquid. Solubility of common laboratory gases in water is shown in Table 4.5.

Flowmeters are designed for specific gases and for varying amounts of flow. The center of the float is the register of the fluid flow. The higher the float rises, the greater the flow rate. Parallax can affect the accuracy and reproducibility of measurements.

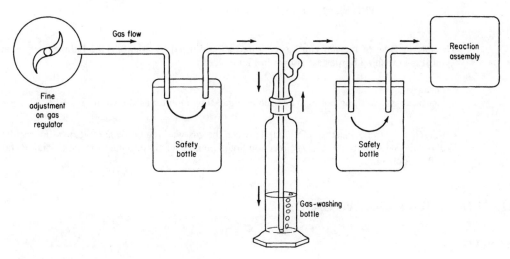

FIGURE 4.13 Typical gas-washer or -scrubber setup.

TABLE 4.5 Solubility of Common Laboratory Gases in Water

Class of solubility at 25°C	Formula of gas	Approximate solubility, L/100 L H_2O
Extremely soluble	HCL	50,000
	NH_3	75,000
	SO_2	4000
Moderately soluble	Cl_2	250
	CO_2	100
	H_2S	250
Very slightly soluble	H_2	2
	O_2	3
	N_2	1+
	CO	2

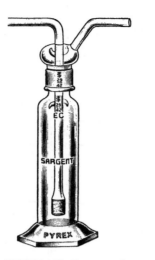 

FIGURE 4.14 Two types of gas-washing bottles.

The rate of flow of the gas can be measured and adjusted by counting the bubbles per minute. Typical gas-washing bottles are shown in Figure 4.14.

Procedure
Connect the gas-washing bottle to be used as a gas-flow monitor as shown in Fig. 4.13.

CAUTION: Always insert a safety trap between the gas-washing bottle and the pressure regulator. Without the safety bottle or trap the reaction liquid could back up into the wash bottle and contaminate the regulator.

COLLECTING AND MEASURING GASES

A. Apparatus

Gases evolved in a reaction can be collected in an inverted burette, previously filled with an inert liquid (see Fig. 4.15). The evolved gas cannot be soluble or react with the liquid. As the gas displaces the liquid, the rate of evolution can be calculated from the volume of gas and the time.

Gases have greater compressibility and greater thermal expansion than liquids and solids. Gas volumes are sensitive to the pressure and temperature.

DRYING AND HUMIDIFYING GASES

If a dry gas is needed for the reaction assembly, concentrated sulfuric acid can be used as the liquid in the gas bubbler. The sulfuric acid will dry the gas being bubbled through it.

Gases can be dried by passing them through a drying tower or tube packed with an absorbent, such as Drierite®, $CaCl_2$, and so on. (See Fig. 4.16.)

If the gas is to be humidified (moistened), water is used in the gas bubbler. The amount of moisture absorbed by the gas depends on the rate or flow, temperature of the gas and liquid, and size of the bubbles.

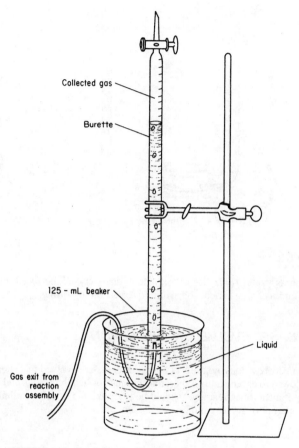

FIGURE 4.15 Setup for gas collection and measurement.

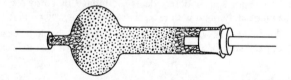

FIGURE 4.16 Gas-drying tube packed with an absorbent and glass-wool plugs at each end.

A. Types of Gas Dispersers

Gas dispersers range from a simple piece of glass tubing immersed beneath the liquid to more sophisticated and efficient types. The choice of gas disperser depends upon the reaction conditions, the reactants, the gas, the rate of flow of the gas, and the rate of reaction desired. The smaller the size of the gas bubbles, the greater the area of contact between the gas and the liquid.

A *fritted disk* (Fig. 4.17) gives very small gas bubbles, depending on its porosity. It affords the greatest gas contact with the liquid but also has the greatest tendency to clog.

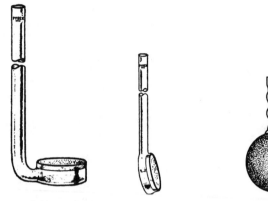

FIGURE 4.17 Fritted-glass gas dispersers.   **FIGURE 4.18** Gas-diffusing stone.

B. Gas-Diffusing Stones

Porous stones (Fig. 4.18) are made of inert crystalline-fused aluminum oxide, and they come in cylindrical or spherical shapes fitted with a tubing fitting.

CONTROL OF GASES EVOLVED IN A REACTION

In many reactions, noxious or corrosive gases are evolved in a chemical reaction. They should not be allowed to escape in the laboratory because of health reasons and because the gases will corrode the equipment. Such reactions should be performed in a well-ventilated hood. Table 4.3 shows physical properties of some common gases.

A. Precautions

1. Always use a hood when working with toxic or irritating chemicals.
2. The major source of accidents is spillage of corrosive chemicals on the clothing and skin. Immediately flood with excessive amounts of water and then consult the medical service.
3. Anything or any operation that must be forced should be examined very carefully. The application of excessive force to make something work can lead to accidents and broken equipment.

B. Alternative Methods

Method 1.   When a hood is not available and noxious or corrosive fumes are emitted from a reaction flask or from a concentration-solution evaporation, an *emergency hood* composed of a glass funnel, rubber tubing, an aspirator, and a water pump should be assembled as shown in Fig. 4.19.

Method 2.   Pass the gases through a gas-washing bottle that contains an absorbent for the gases. Use a basic absorbent (NaOH or NH_4OH solution) for acid gases; use an acid absorbent (dilute HCl or H_2SO_4) for basic gases.

Method 3.   Aspirate the opening from which the gases are exiting the reaction assembly with a T or Y tube which is attached to a water aspirator (Fig. 4.20). One end should be open to the atmosphere to maintain atmospheric pressure in the assembly.

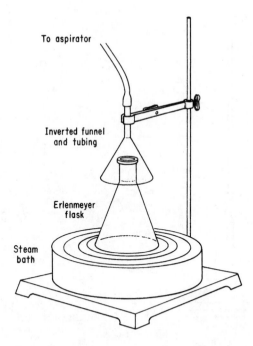

FIGURE 4.19 Temporary hood (not recommended for potentially hazardous fumes of vapors).

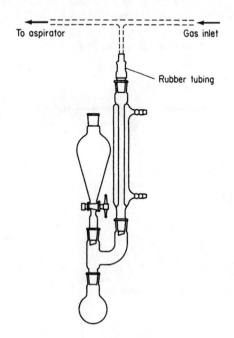

FIGURE 4.20 Aspirating gas from a reaction assembly.

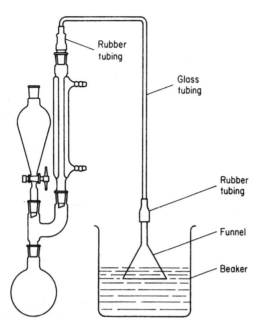

FIGURE 4.21 Trapping evolved gases.

Method 4. Pass the gas over an absorbing solution (Fig. 4.21). Gaseous HCl is very soluble in water. Suspend a funnel just over the surface of a container of water. When absorbing gases that are very soluble in water, do not immerse the funnel because there may be so much gas absorbed or dissolved that the water may be drawn back into the reaction assembly.

Alternative setups for absorbing gases are shown in Fig. 4.22.

To absorb a gas which is not vigorously soluble in water, the funnel may be immersed beneath the liquid surface (Fig. 4.22a).

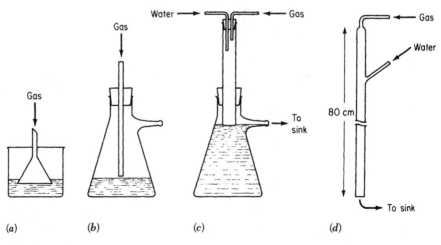

FIGURE 4.22 Alternate setups to absorb gases.

To absorb a moderately soluble gas, the setup shown in Fig. 4.22b may be used, with the tube ending *above* the liquid.

To absorb very large volumes of gas, or rapidly evolved gas, the setups shown in Fig. 4.22c and Fig. 4.22d may be used. In the fonner, the constant input of water overflows at a constant level from the outlet of the suction flask. Water level is above the lower end of a large-bore tube to prevent escape of the gas into the atmosphere. In the latter, the flowing water absorbs the gas, exiting to the drain. A tube about 80 cm long and about 25 mm in diameter is generally a convenient size to use in this setup.

COMPRESSORS

Compressors are used extensively in chemical processes: (1) to provide dry, clean air for control devices, pneumatic operations, and instruments; (2) to compress other gases, including chlorine and hydrogen, which in some processes are used as reactant gases. Compress inert gases like nitrogen or carbon dioxide to provide inert, process gas atmospheres; and (3) to assist in transferring granulated powders or particulates from one location to another. Process compressors are normally of two types: positive displacement or centrifugal (see pumps in Chapter 18 for details). All compressors require some type of cooling capability (intercooler and/or aftercooler heat exchangers), since the compression of gases generates heat. Excessive temperatures can create high pressures and dangerous conditions, thus all compressors must have specifically designed and sized safety relief valves. These **safety relief valves** should never be removed, by-passed, or modified without ensuring manufacturer's specifications. Compressors tend to be very noisy and routinely exceed OSHA's rules on noise exposure even when equipped with **silencers.** Do not work near an operating compressor without first determining the appropriate noise protection required. All compressed air contains residual moisture. Moisture can be minimized by passing the compressed air through a demister and a series of dryer tubes. A demister is any device that physically separates water from gases (see Chapter 5, "Pressure and Vacuum" for knockout drum type). The water can be removed by a centrifugal motion inside the demister which causes the heavy water molecules to fall to the bottom and the gases to rise to the top and escape. **Dryer tubes** normally contain moisture-absorbing chemicals: molecular sieves, alumina, or silical gel. Most of these dryer tubes can be regenerated by heating them and purging with dry gas flowing in the reverse direction.

RELATIVE HUMIDITY

The usual way of measuring the humidity of a gas is by wet- and dry-bulb thermometry, using an apparatus called a **psychrometer**. A psychrometer consists of two thermometers, one of which has its temperature-sensitive bulb covered with a wetted wicking material. When air is passed over both thermometers, the evaporation of water from the wick covering one of the thermometer bulbs will cause the temperature of that bulb to be reduced in proportion to the rate of evaporation of water. In turn, the rate of evaporation of the water will be a function of the humidity of the air passing over the wick. The difference between the wet-bulb and dry-bulb temperatures can then be used to determine the relative humidity by means of a humidity chart, or from a psychrometric chart, that is, a chart relating the humidity to the wet-bulb and dry-bulb temperatures. Such a chart is given in Table 4.6. The **percentage relative humidity** (RH) is defined as the partial pressure of the water vapor in air times 100 and divided by the saturated vapor pressure of pure water at the same temperature. The **dewpoint** can be defined as the temperature at which the water vapor in the air begins to condense. Dewpoint can also be defined as the temperature at which air is saturated with moisture.

EXAMPLE 4.1 What is the percent RH of air for which the dry-bulb temperature is 80°F (27°C) and the wet-bulb temperature is 70°F (21°C)? From Table 4.6, the percent RH is between 61 percent.

TABLE 4.6 Relative Humidity from Wet and Dry-Bulb Temperatures

Wet-bulb temp (°F)	21	22	23	24	25	26	27	28	29	30	31	32	33	34	35	36	37	38	39	40
14	1																			
15	15	4																		
16	28	17	7																	
17	42	31	20	10	1															
18	56	44	33	22	13	4														
19	71	58	46	35	25	16	7													
20	85	71	59	47	37	27	18	10	3											
21	100	86	72	60	49	39	29	21	13	6										
22		100	86	73	62	51	41	32	23	16	8	2								
23			100	87	74	63	52	43	34	36	18	11	5							
24				100	87	75	64	54	44	36	28	20	14	8	2					
25					100	87	76	65	55	46	37	30	23	16	10	5				
26						100	88	76	66	56	47	39	32	25	19	13	7	2		
27							100	88	77	67	58	49	41	34	27	21	15	10	5	
28								100	88	78	68	59	51	43	36	29	23	17	12	7
29									100	89	78	69	60	52	45	38	31	25	20	15
30										100	89	76	70	62	54	46	40	33	27	22
31											100	89	80	71	63	55	48	42	35	29
32												100	90	81	72	64	57	50	43	37
33													100	90	81	73	65	58	51	45
34														100	91	82	74	66	59	52
35															100	91	83	75	67	60
36																100	91	83	75	68
37																	100	91	83	75
38																		100	92	83
39																			100	92
40																				100
41																				
42																				
43																				
44																				
45																				
46																				
47																				
48																				
49																				
50																				
51																				
52																				
53																				
54																				

Dry-bulb temp (°F)

From Norbert A. Lange (ed.), "Handbook of Chemistry," 10th ed., McGraw-Hill Book Company, New York, 1967.

	41	42	43	44	45	46	47	48	49	50	51	52	53	54	55	56	57	58	59	60	
																					14
																					15
																					16
																					17
																					18
																					19
																					20
																					21
																					22
																					23
																					24
																					25
																					26
																					27
	3																				28
	10	5																			29
	17	12	8	4																	30
	24	19	14	10	6																31
	31	26	21	16	12	8	5	1													32
	39	33	28	23	18	14	10	7	3												33
	46	40	35	30	25	20	16	12	9	5											34
	54	47	42	36	31	26	22	18	14	10	7	4	1								35
	61	55	48	43	38	32	28	23	19	16	12	9	6	3							36
	69	62	55	49	44	39	34	29	25	21	17	14	10	8	5	2					37
	76	69	63	56	51	45	40	35	31	27	23	19	16	12	9	7	4	1			38
	84	77	70	63	57	52	46	41	36	32	28	24	20	17	14	11	8	6	3	1	39
	92	85	77	71	64	58	52	47	42	48	34	29	26	22	19	16	13	10	7	5	40
	100	92	85	78	71	65	59	54	48	43	35	35	31	27	23	20	17	14	11	9	41
		100	92	85	78	72	66	60	54	49	45	40	36	32	28	25	22	18	16	13	42
			100	93	86	79	72	66	61	55	50	46	41	37	33	30	26	23	20	17	43
				100	93	86	79	73	67	61	56	51	47	42	38	34	31	27	24	21	44
					100	93	86	79	73	67	62	57	52	48	43	39	35	32	29	26	45
						100	93	86	80	74	68	63	58	53	49	44	40	37	33	30	46
							100	93	86	80	75	69	63	59	54	50	45	41	38	34	47
								100	93	87	81	75	69	64	59	55	50	46	42	39	48
									100	93	87	81	75	70	65	60	55	51	47	43	49
										100	94	87	81	76	70	65	61	56	52	48	50
											100	94	87	82	76	71	66	61	57	53	51
												100	94	88	82	76	71	66	62	58	52
													100	94	88	82	77	72	67	63	53
														100	94	88	82	77	72	68	54
	41	42	43	44	45	46	47	48	49	50	51	52	53	54	55	56	57	58	59	60	

(*Continued*)

TABLE 4.6 Relative Humidity from Wet and Dry-Bulb Temperatures (*Continued*)

Wet-bulb temp (°F)	41	42	43	44	45	46	47	48	49	50	51	52	53	54	55	56	57	58	59	60
41	100	92	85	78	71	65	59	54	48	43	39	35	31	27	23	20	17	14	11	9
42		100	92	85	78	72	66	60	54	49	45	40	36	32	28	25	22	18	16	13
43			100	93	86	79	72	66	61	55	50	46	41	37	33	30	26	23	20	17
44				100	93	86	79	73	67	61	56	51	47	42	38	34	31	27	24	21
45					100	93	86	79	73	67	62	57	52	48	43	39	35	32	29	26
46						100	93	86	80	74	68	63	58	53	49	44	40	37	33	30
47							100	93	86	80	75	69	63	59	54	50	45	41	38	34
48								100	93	87	81	75	69	64	59	55	50	46	42	39
49									100	93	87	81	75	70	65	60	55	51	47	43
50										100	94	87	81	76	70	65	61	56	52	48
51											100	94	87	82	76	71	66	61	57	53
52												100	94	88	82	76	71	66	62	58
53													100	94	88	82	77	72	67	63
54														100	94	88	82	77	72	68
55															100	94	88	83	78	73
56																100	94	88	83	78
57																	100	94	89	83
58																		100	94	89
59																			100	94
60																				100
61																				
62																				
63																				
64																				
65																				
66																				
67																				
68																				
69																				
70																				
71																				
72																				
73																				
74																				
75																				
76																				
77																				
78																				
79																				
80																				
	41	42	43	44	45	46	47	48	49	50	51	52	53	54	55	56	57	58	59	60

Dry-bulb temp (°F)

HANDLING COMPRESSED GASES

	61	62	63	64	65	66	67	68	69	70	71	72	73	74	75	76	77	78	79	80	
	7	4	2																		41
	10	8	6	4	2																42
	14	12	10	7	5	3	2														43
	18	16	13	11	9	7	5	3	1												44
	22	20	17	15	12	10	8	6	5	3	1										45
	27	24	21	18	16	14	12	10	8	6	4	3	1								46
	31	28	25	22	20	17	15	13	11	9	7	6	4	3	1						47
	35	32	29	26	24	21	19	16	14	12	10	9	7	5	4	3	1				48
	40	36	33	30	27	25	22	20	18	15	13	12	10	8	7	5	4	3	1		49
	44	41	37	34	31	29	26	23	21	19	17	15	13	11	9	8	6	5	4	3	50
	49	45	42	38	35	32	30	25	24	22	20	18	16	14	12	11	9	8	6	5	51
	54	50	46	43	39	36	33	31	28	25	23	21	19	17	15	13	12	10	9	7	52
	58	54	50	47	44	40	37	34	32	29	27	24	22	20	18	16	14	13	11	10	53
	63	59	55	51	48	44	41	38	35	33	30	28	25	23	21	19	17	16	14	12	54
	68	64	60	56	52	48	45	42	39	36	33	31	29	26	24	22	20	18	17	15	55
	73	69	64	60	56	53	49	46	43	40	37	34	32	29	27	25	23	21	19	18	56
	78	74	69	65	61	57	53	50	47	44	41	38	35	33	30	28	26	24	22	20	57
	84	79	74	70	66	61	58	54	51	48	45	42	39	36	34	31	29	27	25	23	58
	89	84	79	74	70	66	62	58	55	51	48	45	42	39	37	34	32	30	28	26	59
	94	89	84	79	75	71	66	62	59	55	52	49	46	43	40	38	35	33	31	29	60
	100	94	89	84	80	75	71	67	63	59	56	53	50	47	44	41	39	36	34	32	61
		100	95	90	85	80	75	71	67	64	60	57	53	50	47	44	42	39	37	35	62
			100	95	90	85	80	76	72	68	64	61	57	54	51	48	45	43	40	38	63
				100	95	90	85	80	76	72	68	65	61	58	54	51	48	46	43	41	64
					100	95	90	85	81	77	72	69	65	61	58	55	52	49	46	44	65
						100	95	90	85	81	77	73	69	65	62	59	56	53	50	47	66
							100	95	90	86	81	77	73	69	66	62	59	56	53	50	67
								100	95	90	86	82	78	74	70	66	63	60	57	54	68
									100	95	90	86	82	78	74	70	67	63	60	57	69
										100	95	91	86	82	78	74	71	67	64	61	70
											100	95	91	86	82	78	74	71	68	64	71
												100	95	91	86	82	79	75	71	68	72
													100	95	91	87	83	79	75	72	73
														100	96	91	87	83	79	75	74
															100	96	91	87	83	79	75
																100	96	91	87	83	76
																	100	96	91	87	77
																		100	96	91	78
																			100	96	79
																				100	80
	61	62	63	64	65	66	67	68	69	70	71	72	73	74	75	76	77	78	79	80	

CHAPTER 5
PRESSURE AND VACUUM

PRESSURE

Pressure is defined as force per unit area. Thus two appropriate units would be pounds per square inch (lb/in^2 or psi) and, in the SI system, the **Pascal** (Pa) or Newtons per square meter (N/m^2). The SI unit of force equals the SI unit of mass times the SI unit of acceleration (kg m/s^2) and is called the **Newton**. Pressure therefore would be expressed in the SI system as Newtons per unit of area (m^2). Another relative expression of pressure is the **atmosphere** (atm) which, at sea level, will cause a column of mercury to rise 760 mm in an evacuated tube. Both 760 mm Hg and the atmosphere are commonly used pressure units and are equivalent to 14.696 psi or 101,325 Pa. These and other atmosphere equivalents and conversions are given in Tables 5.1 and 5.2.

ABSOLUTE, GAUGE, DIFFERENTIAL, AND STANDARD PRESSURES

Absolute pressure intensities are related to absolute zero pressure. Absolute zero pressure means the complete absence of any and all molecules, and therefore corresponds to a complete absence of pressure. It is a purely theoretical concept.

Gauge pressure, which is normally used to designate pressures *above* atmospheric pressure, does not include atmospheric pressure. At atmospheric pressure, the gauge reads zero. Thus, gauge pressures and absolute pressures differ from each other by the zero-point location. To convert a gauge pressure to the equivalent absolute pressure, add the atmospheric pressure.

$$\text{Absolute pressure} = \text{gauge pressure} + \text{local atmospheric pressure}$$

GAS LAWS

A. Boyle's Law

States that the pressure (P) exerted by a given mass of any ideal gas times its volume (V) will give a constant value if the temperature (T) is held constant.

$$PV = \text{constant} \quad (T \text{ and mass held constant})$$

B. Charles' Law

States that the volume (V) of a given mass of gas will be proportional to its temperature (T) expressed in Kelvin.

$$V \propto T \quad (T \text{ in Kelvin and mass held constant})$$

TABLE 5.1 Atmosphere Equivalents

1 atm = 14.696 lb/in² (psi)
= 29.921 in Hg
= 76 cm Hg
= 760 mm Hg (exact definition)
= 760 torr
= 33.899 ft H₂O
= 10.33 m
= 1.01325 × 10⁵ Pa
= 101.325 kPa (exact definition)
= 1.0133 bars
= 1.01325 × 10⁶ dyn/cm²
Other pressure equivalents are
1 torr = 1 mm Hg
= 0.03937 in Hg
= 0.53524 in H₂O
= 0.0013 bar
= 1000 μm Hg
= 1.000 × 10³ μm Hg

See Chapter 6, "Mathematical Review and Conversion Tables," for additional pressure unit conversion factors.

TABLE 5.2 Absolute, Gauge, Differential, and Standard Pressures

PSIG = pounds/square inch gauge (pressure referenced to vacuum)
PSIG = pounds/square inch gauge (pressure referenced to ambient air)
PSID = pounds/square inch differential (pressure difference between two points)
PSIS = pounds/square inch standard (pressure referenced to standard atmosphere)

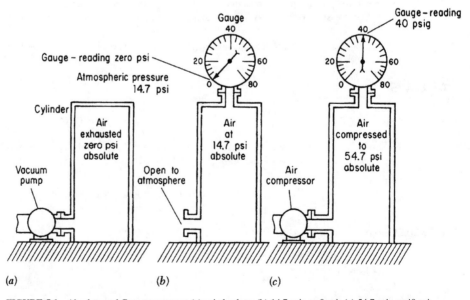

FIGURE 5.1 Absolute and Gauge pressures: (a) psi absolute; (b) 14.7 psia or 0 psi; (c) 54.7 psia or 40 psi.

C. Gay-Lussac's Law

States that the pressure (P) of any gas will change directly as the temperature (T) changes if the volume is held constant.

$$P \propto T \quad (V \text{ and mass held constant})$$

D. Avogadro's Law

States that equal volumes (V) of gases at the same temperature (T) and pressure (P) contain equal numbers of molecules. If *standard temperature and pressure conditions (STP)* of 0°C (273 K) and 1 atm are used, then Avogadro's law can be restated as *1 mole of any gas at STP occupies 22.4 liters.*

E. Dalton's Law

States that the total pressure of a gas mixture is the sum of the pressures from the individual gases under the same conditions.

$$P_{total} = P_1 + P_2 + P_3$$

F. The Combined Gas Law

Combines Boyle's, Charles', and Gay-Lussac's laws into a useful mathematical relationship for calculating changes in a gaseous substance at some starting conditions (1) to different conditions (2).

$$\frac{P_1 V_1}{T_1} = \frac{P_2 V_2}{T_2}$$

G. The Ideal Gas Law

A mathematical application of Avogadro's law, which provides a means for calculating the pressure, temperature, volume, or moles of any ideal gas.

$$PV = nRT$$

where P = pressure of the gas in atmosphere units
V = volume of the gas expressed in liters
n = number of moles of the gas
T = temperature of the gas in Kelvin
R = ideal gas constant: 0.08206 L atm/mol K*

---

*Other ideal gas constants are available but require different units for pressure and volume in the calculation. For example, 8.3145 kPa dm³/mol K, 8.3145 J/mol K, and 1.987 cal/mol K.

H. Pascal's Law

States that in a confined fluid, any externally applied pressure is transmitted equally in all directions. This principle is utilized in all hydraulic systems to move pistons to do work as well as to measure the pressure applied (Fig. 5.2). Since the applied pressure is transmitted equally in all directions throughout the confined volume, the location of the pressure-indicating device (or vacuum-measuring device) is a matter of convenience and choice.

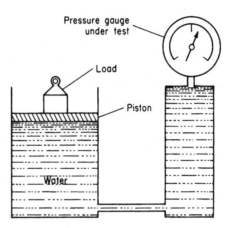

FIGURE 5.2 Pascal's law: transmission of pressure in a closed system.

VACUUM

Vacuum can be considered as space in which there are relatively few molecules. We say "relatively few" because there is no such thing as an absolute vacuum; every substance does exert a definite vapor pressure. Generally speaking, a vacuum is a state of reduced atmospheric pressure, i.e., some point below normal atmospheric pressure. A shorthand method to indicate the interconnections and components of a vacuum in a flow diagram uses graphic symbols (Fig. 5.3). These symbols provide a schematic flow diagram of the system, but do not specify the size, shape, or actual physical location of the components.

There are three general kinds of equipment used to produce a vacuum: (a) a water aspirator; (b) a mechanical vacuum pump or a rotary pump; and (c) a vapor-diffusion pump, filled with mercury, silicone oils, or specialized fluid. The water aspirator is the most commonly used source of moderate vacuum in the laboratory and small pilot plant applications. Water aspirators are inexpensive, almost infallible, and do not require complicated traps for protection. In fact, water aspirators provide an easy method of disposing of some acidic gases and vapors, corrosive fumes, and irritating or nauseating gases because they are dissolved in the water flow and are eliminated in the drain. The aspirator works on **Bernoulli's principle** (see Chapter 18, "Bernoulli's Principal" section for additional information), which states that as the velocity of the fluid is increased, the lateral pressure to the flow is decreased, resulting in the formation of a partial vacuum by the rapidly moving water. The aspirator, however, cannot provide a pressure lower than the vapor pressure of the flowing water, and that pressure is dependent on the temperature of the water. The vapor pressure of water at various temperatures is given in Appendix B. For example, on a cold day (H_2O at 10–15°C), the lowest pressure attainable is about 10 torr.

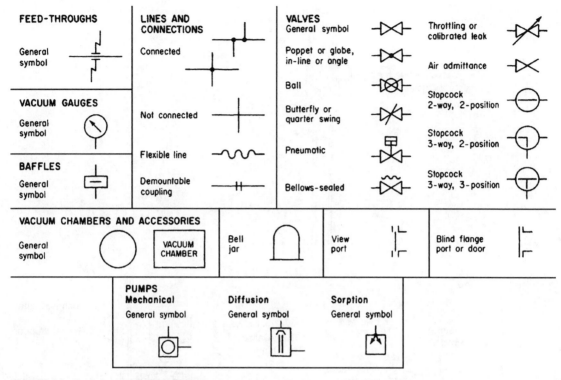

FIGURE 5.3 Symbols used in vacuum technology.

WATER ASPIRATOR GUIDELINES

1. Always insert a bottle trap between the aspirator and the apparatus under vacuum (Fig. 5.4(b)). The water pressure may drop suddenly, and when it does, the pressure in the apparatus may become less than that of the aspirator. Water would be drawn back from the aspirator into the apparatus, causing contamination.
2. Always use vacuum rated glassware as implosion hazards exist under vacuum conditions.
3. Always disconnect the aspirator from the apparatus before turning off the water; otherwise the water will be drawn back into the apparatus as described above.
4. If water does back up for any reason, immediately disconnect the tubing from the aspirator.

*TYPES OF VACUUM PUMPS**

A **mechanical pump** uses a rotary vane to produce a rough vacuum (approximately 10 torr) which is slightly better than that of a water aspirator. This device relies on the sweeping action of multiple vanes turning within a cylindrical housing. An electric motor usually provides the driving force. Pumps of this type may include oiling devices for lubricating or may use vanes of self-lubricating material. Large pumps of this type, together with ballast tanks, are used as the basis for central vacuum systems. A cross-sectional diagram of a typical rotary-vane, oil-sealed mechanical pump is shown in Fig. 5.5a.

*courtesy of SARGENT-WELCH

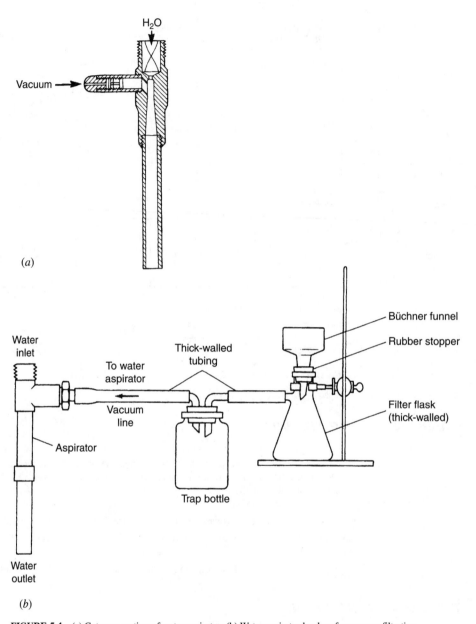

FIGURE 5.4 (a) Cutaway section of water aspirator; (b) Water-aspirator hookup for vacuum filtration.

Rotary-vane pumps are composed of either one or two stages, the number of stages being determined by the degree of vacuum desired. The stages of a two-stage pump are connected in series and mounted on a common shaft in such a manner that one stage backs up the other to obtain the best possible vacuum. Each stage consists of a rotor mounted concentrically on a shaft and located eccentrically in a generally cylindrical stator or ring, together with two movable vanes located diametrically opposite each other in slots of the rotor.

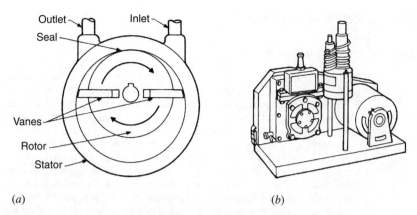

FIGURE 5.5 (a) Cross-sectional diagram of a typical mechanical (oil-sealed rotary-vane) vacuum pump; (b) combined mechanical oil-diffusion vacuum pump.

A **vapor-diffusion pump** is similar in principle to a water aspirator; the steam which entrains the undesired gases consists of a heavy vapor generated by evaporation of a pump oil. The pump-oil vapor is condensed after serving this purpose and is returned to the boiling pot, thus comprising a closed system. A combined mechanical-oil-diffusion pump is shown in Fig. 5.5b.

This two-stage mechanical pump is coupled to an all-metal, water-cooled, oil-diffusion pump. This type of pump is capable of producing a vacuum in the 1×10^{-6} torr range without using a liquid nitrogen or other cold trap (see Fig. 5.4 for conventional vacuum trap). Originally, most diffusion pumps used mercury, but this process is seldom used now because of possible toxicity hazards. Organic oils for diffusion pumps are superior to mercury, which has a vapor pressure of 10^{-4} torr at 0°C, 10^{-2} torr at 50°C, and 5 torr at 150°C. These diffusion pump oils are esters (E), hydrocarbons (H), polyphenyl ethers (P), or silicones (S), having excellent stability toward heat and extremely low vapor pressures, as shown in Table 5.3. The molecules of these oils, while passing through the space where the gas molecules of the system are being evacuated, strike those gas molecules and push them toward the outlet. The gas molecules are then replaced by the vapor molecules from the pump until practically all have been exhausted.

TABLE 5.3 Characteristics of Diffusion Pump Oils

Type	Chemical structure	Vapor pressure (@20°C), torr
Apiezon A	H	1×10^{-6}
Octoil	E	2×10^{-7}
Apiezon B	H	5×10^{-8}
DC 702	S	5×10^{-8}
DC 704	S	1×10^{-8}
Apiezon C	H	5×10^{-9}
Convalex	P	Very low
Santovac 5	P	Very low

VACUUM PUMP MAINTENANCE*

A. When and How to Change the Oil

The most common cause of pump failure and unsatisfactory performance is contaminated oil. If water or acid vapors have been passed through the pump and the oil is allowed to stand for any length of time, severe corrosion and extensive damage to any pump may occur. The simplest way to determine if an oil change is needed is to connect a gauge directly to the pump and ascertain if the rated ultimate vacuum can be attained.

An odor indicating the presence of a solvent indicates the need for an oil change. Light-brown, cloudy oil may indicate the presence of water. When in doubt, change the oil. A blackish color does not necessarily indicate bad oil in smaller models. The longer the vanes are allowed to wear in, the more perfect the internal seat of the moving parts. The carbon that is thus added to the oil will discolor it but will not reduce its vapor pressure and will even add to its lubricity. If the pump is used only in occasional service or is loaned to another operator, it is a good policy to drain and refill it with fresh oil before it is set aside for temporary storage. Many users find it convenient to attach a tag upon which the dates of oil changes may be recorded.

In most cases, draining a pump is quite simple. The pump should be operated so the oil will be warmed and become as thin as possible. Open the vacuum intake to the atmosphere, shut off the power, and open the drain valve. If possible, tip up the end opposite the valve to ensure complete drainage. When the flow has stopped, turn the power on for a few seconds or turn the pulley by hand to clear any oil that may still be in the stator cavities. Sufficient oil film is present on the moving parts to prevent damage if the pump is operated under power for a few moments while draining. Remember, any contaminated oil left in the pump will only serve to contaminate the fresh oil when it is added. The use of detergent oil is not recommended by most manufacturers.

B. Vacuum Pump Oils

Vacuum pump oil serves three vital functions: (1) it seals the internal portion of the pump from the atmosphere; (2) it lubricates the pump; and (3) it cools the pump by conducting heat to the outer housing. Use only approved high-vacuum oil for guaranteed results, as this is a pure mineral oil with an extremely low vapor pressure, possessing the proper flow characteristics and lubricating properties to ensure the very best pump service. Proper care must be taken to prevent foreign materials from entering the pump. A small particle of glass, a metal filing, a globule of mercury, or even certain condensable gas vapors will impair the normal life expectancy of any pump. All gases being evacuated will pass through the pump and its oil, unless adequate traps are employed. If gases are allowed to condense in the pump oil, not only will the rated ultimate vacuum not be reached, but corrosion of the finely machined parts is likely to occur.

A properly filled pump will have sufficient oil above the internal exhaust valve to seal the pump from the atmosphere and also to damp the sound of the gas as it passes through the valve. Since an overfilled pump may spit oil during operation and back up the oil beyond the intake during off cycles, it is better to start with less and add oil gradually to the desired level. A sight gauge is provided to indicate the amount of oil present.

C. Filling Vacuum Pumps with Oil

When filling the pump for the first time or refilling it with fresh oil, it is important to clear out any oil that may be in the stators. For specific draining information, see the vacuum pump manual. In general, turn the pump pulley at least two revolutions clock-wise (facing the pulley), or turn the pump on and allow it to operate under motor power for a few seconds. There need be no fear of damage since there is always sufficient oil present for lubricating purposes, and this action will clear any oil from the stators into the housing.

---

*courtesy of SARGENT-WELCH

If oil is added too fast at this point, spitting may occur. When the pump is quiet, alternately open and close the intake to the atmosphere. Each time it is opened, popping may occur. This noise is caused by the larger inrush and exhaust of air. When the pump has been closed for a few seconds, the noise should disappear. If it doesn't, add a little more oil and repeat the above procedure.

Do Not Overfill the Pump. If any pump continually spits oil, it is overfilled. When the pump is shut off, that amount of oil above the exhaust valve will drain into the pump and open the pump to the atmosphere. This feature prevents oil from backing up into a system that is not separated from the pump by a valve. A pump cannot be expected to act as a valve and "hold" vacuum. Pumps are designed with sufficient volume in the stators and trap to accommodate the correct amount of oil above the exhaust valve. A properly filled pump will not have oil rising above the intake connection when it is shut off. Since oil is continually being consumed, either by entrainment in the gases leaving the pump or by backstreaming, it will be necessary to add oil from time to time. The best gauge for this is the sound of the pump with regard to the pumping noise discussed above.

D. Vacuum Connections

The simplest system consists of a pump connected directly to the vessel to be evacuated. The speed of evacuation will be a function of: (1) the pump-down curve of the pump; (2) the vessel size and type; and (3) the length, diameter, and bends in the connecting tubing. Basic rules that apply, whenever practical, are:

1. Choose the correct size of pump for the time cycle desired.
2. The vessel should be nonporous and have a smooth surface.
3. Eliminate condensable vapors with a trap.
4. Keep vacuum lines short in length, large in diameter, and as direct as possible.
5. Always use thick-walled tubing to prevent or minimize collapsing.

Locating Leaks in Vacuum Systems. Leaks and cracks in vacuum systems destroy their effectiveness and can be easily detected and located by the use of a high-frequency **Tesla coil**. When the electrode is passed over the surface of evacuated vacuum-glassware assemblies, any leakage that is caused by a poor seal, a flaw in the glass, or an imperfect joint is pin-pointed by a yellow glow at the point where the electrical discharge enters the system.

E. Gas Ballast

This is a device used to help prevent the condensation of contaminant vapors within the pump, thereby protecting the pump from the corrosive action of these condensed vapors. Air is bled into the pump via the adjustable valve on top of the pump just before the gas is exhausted through the oil. Because of the pressure increase, the exhaust valve is forced open without subjecting the contaminant vapor molecules to as great a compression. Compression of the vapor would cause condensation. With the gas ballast open, the pump will not reach its rated vacuum because gas ballast is actually a controllable leak. The pump will also run warmer with the gas ballast valve open because of the greater amount of gas (air) it is handling. When all traces of the contaminant vapors have disappeared from the system and the oil, the gas-ballast valve may be closed to permit the pump to attain its ultimate vacuum. The amount of moisture that can successfully be handled by a gas ballast will vary with the size of the pump. *Caution:* If the contaminant vapor has reacted with the oil, it cannot be separated from the oil by the gas ballast. When this happens, the oil must be changed immediately.

F. Belt Guards

Most vacuum pumps are designed with a belt-guard assembly in place to protect workers from accidental entanglement in the belt and/or pulley mechanism, If a vacuum pump to be used does not have a belt guard, one should be installed prior to use. All belt guards should be easily opened for access to the belt and pulleys. Belt guards are required by OSHA machine guarding regulations, and their removal can result in the potential for personal injuries and fines.

Check the V belt for tightness, cleanliness, and wear. A loose belt will slip, and a tight belt will cause excess bearing wear. Belt tightness is adjusted by moving the motor toward or away from the pump. Check, by observation, for oil leaks at the drain valve, housing gasket, and shaft seal. If any leakage is observed, replace that component. If the pump is equipped with a smoke eliminator, check to see that the element is not clogged.

*TROUBLESHOOTING VACUUM PUMPS**

A pump that is allowed to run continuously will last longer and remain cleaner. Never cause a pump to cycle on and off. If the desired vacuum has been reached in a system, block off the pump with a valve and allow it to continue operating at low pressure. This will actually bring about a better "wear-in" of the moving parts and reduce the possibility of corrosion, which is more likely to occur while the pump is standing idle. When a task is complete and the pump is to be removed from the system, bleed air into the system and allow the pump to come to atmospheric pressure. If the pump is to be stored before it is used again, drain, flush, and refill it with fresh high-vacuum oil. Stopper the intake and exhaust openings. *Caution:* Refer to the operating manual of your particular pump for specific directions.

A. Pump Runs Hot

Operating temperatures will be related to surrounding ambient conditions. Under normal conditions when operating at low pressures and depending on the model, pump-oil temperatures may be expected to be approximately $65 \pm 10°C$. While small models may run slightly cooler, the larger single-stage pumps may run slightly warmer. Operation is satisfactory if the oil temperature does not exceed 80°C. The most frequent cause of overheating is handling too large a volume of air for prolonged periods or operating with contaminated oil. A pump should never be used as a control to regulate a specific vacuum. If it is allowed to cycle on and off frequently, it will probably overheat and fail to start, since the motor's "thermoprotector" will prevent the motor from operating. Check the following:

1. Oil level is low. Add oil.
2. Oil is gummy. Drain, flush, and refill with fresh oil.
3. Gas ballast is open. Close ballast valve.
4. The V belt is too tight. Loosen.
5. Abrasive particles have entered the pump. Disassemble and clean.
6. Pump is binding mechanically because of misalignment of parts damaged during shipping. Return for replacement.

B. Pump Is Noisy

Noise, of course, is a relative thing. When the noise level of a new pump is evaluated, it must be compared with another pump of comparable performance with regard to its free-air displacement, and it must not be on a platform that will amplify the normal operating sound. The sound should be analyzed as to its probable origin with respect to the following points:

---

*courtesy of SARGENT-WELCH

1. Oil level is low. Refer to manufacturer's manual on proper filling procedure, and add oil.
2. System is too large for the pump, causing prolonged operation at intermediate pressures with the resultant normal pumping sound. Add a smoke eliminator or select a larger pump.
3. System has pronounced leaks, causing prolonged operation at intermediate pressures with the resultant normal pumping sound. Locate and seal the leaks.
4. Exhaust valve is damaged or corroded. Return to the factory.
5. Vane springs are malfunctioning as a result of damage in shipment or the introduction of some foreign material. Return to the factory.

C. Pump Does Not Achieve Expected Vacuum

When a pump is connected to a system, the rated ultimate vacuum in most cases will not be achieved because of the configuration of the system. Always check to see that the oil in both the mechanical and diffusion pumps is at the proper level. There is always a possibility of leaks, and a complete check with a leak detector, if available, should be made. Quite frequently, material in the system (volatiles) will be releasing vapor (offgassing) at such a rate that the vacuum obtained will seem higher than expected. Water vapor in the air is a prime example of such foreign material. If the pump's performance must be substantiated, separate it from the system and gauge it directly. The following points may also be investigated:

1. Gas ballast is open. Close completely.
2. Plain grease instead of high-vacuum silicone grease may have been used at slip joints or seals.
3. Remove any excess vacuum grease at joints. Only a very thin film should be necessary.
4. Check seal around threaded hose connection at the pump.
5. Oil is contaminated or improper oil was used. Drain, flush, and refill.
6. Check gauge calibration.

D. Volatile Substances with Vacuum Pumps

Under vacuum most liquids turn to vapor. There are many tables showing the vapor pressure of a liquid relative to temperature. For vacuum work, this means that as soon as that pressure is reached, the equilibrium shifts to the vapor phase. Water, the most common liquid, has a vapor pressure of about 17 torr at room temperature. This means that as long as there is any liquid water present, no vacuum pump can achieve a vacuum greater than 17 torr. The same phenomenon occurs with all other liquids. If the pump is permitted to continue to run with the gas ballast open, the water will be separated from the oil and pass from the pump as a vapor. Until the water (or any other volatile) is out of the oil, the pump cannot reach its ultimate pressure. A suitable trap would prevent these vapors from reaching the pump.

E. Pump Spits Oil from Exhaust Port

Oil level is too high. Drain oil until the proper level is reached. Oil expands with heat after the Dump is started. If the pump was overfilled slightly while cold, it may spit some oil at operating temperature. Other liquids or condensed vapor may also have entered the pump from other components in the system to raise the oil level.

F. Backstreaming

Most vacuum pumps are designed so that when they are not overfilled, the oil will not back up beyond the intake connection. Backstreaming, however, occurs with all high-vacuum pumps. As the system is being evacuated, fewer gas molecules are passing through the lines, and oil vapor will pass

VACUUM OR GAS FLOW SAFETY TRAPS

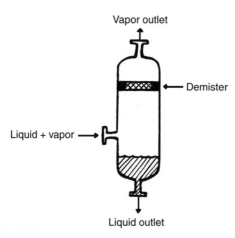

FIGURE 5.6 Knock-out drum and demister.

Risk of pump damage can be reduced and money will be saved in the long run if routine maintenance is performed on the pump. While operating, the pump should be protected from destructive vapors by a cold trap or a sorption trap before the pump intake for heavy vapor loads and by use of the gas ballast for lighter vapor loads. If these vapors contaminate or react with the oil, it should be changed promptly. During normal operation, some oil will be lost from the pump. Periodically check and refill to the proper level when necessary.

Knock-out (KO) drums are used on process units to remove entrained liquids and solids in gas streams. Droplets of liquids entrained in the vapor impinge and collect on a **demister**, usually just a synthetic or metal mess insert (Fig. 5.6). The vapor passes through the demister, usually with very little pressure drop. The liquids, and sometimes solids, that are "knocked out" in this process escape through a drain in the bottom.

PRESSURE AND VACUUM MEASURING DEVICES AND GAUGES

A. The Open-End Manometer

An **open-end** or **U-tube manometer** (Fig. 5.7) is a pressure-measuring device consisting of a U-tube filled with mercury, water, or even process fluid and which has one end open to the atmosphere and the other end attached to the system being measured. The atmosphere exerts pressure on the open end and the system exerts pressure from the other end. When the pressure in the system is equal to the atmospheric pressure, the heights of the liquids are the same; there is no difference in the levels.

When there is a difference between the pressure of the atmosphere and that of the system, there is a difference in the levels of the liquid in the manometer, and the difference in the levels is a measure of the difference in pressure. When the pressure of the system is higher than atmospheric pressure, the end attached to the system has a lower level because of the higher pressure.

The open-end manometer is limited to the measurement of moderately higher-than-atmospheric pressures. Too high a system pressure will exceed the capacity of the manometer and will blow the liquid out of the open tube. The difference in the heights of the mercury column is the pressure on the system, stated in millimeters of mercury or torr. The following corrections must be considered with all open-end manometers:

1. The vapor pressure of the liquid used in the manometer must also be taken into consideration because it exerts pressure.
2. The ambient temperature and hence the temperature of the liquid in the manometer must be carefully controlled because changes in temperature change the density of the liquid.
3. In a vertical manometer, the viscosity of the liquid used is relatively unimportant, but it is very significant when used in an inclined manometer. Furthermore, when very viscous liquids are used, sufficient time must elapse to allow the true level to be attained before taking the reading.

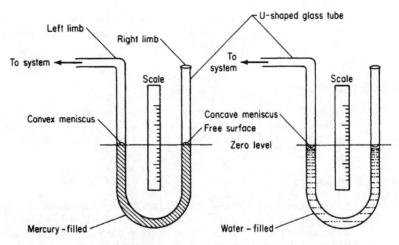

FIGURE 5.7 The U-tube manometer. When no pressure is applied to either limb, the manometer liquid remains at zero reading on the scale; the level of the meniscus in the left limb is the same as that in the right limb.

B. The Closed-End Manometer

The closed-end manometer (Fig. 5.8) is normally filled with mercury and is normally used to measure pressures less than atmospheric pressure. When a closed-end manometer is used to measure these low pressures, it is usually connected to the system with a Y connector. As the pressure is

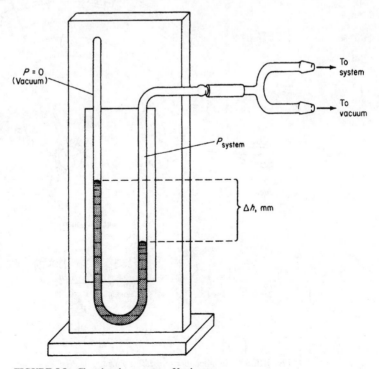

FIGURE 5.8 Closed-end manometer, U-tube type.

reduced, the mercury will rise in the tube that is connected to the vacuum system. When the mercury column stops rising, the vacuum in the system is determined by reading the difference in the heights of the mercury columns in the two tubes.

C. Filling the Manometer

Caution: Untreated or inadequately treated spills of mercury can release toxic levels of mercury into the air. This release could result in neurological damage to personnel who may be exposed. All mercury spills should be immediately reported to your supervisor and/or spill-response personnel.

D. Bourdon-Type Gauges

A bellows pressure element is a corrugated thin-walled metal tube that expands and contracts with applied pressure changes in a closed system. The Bourdon-type gauges are the simple mechanical types used to measure pressure in process operations using a bellows-type pressure element. **A Bourdon gauge** consists of a length of thin-walled metal tubing, flattened into an elliptical cross section, and then rolled into a C shape. When pressure is applied, the tube tends to straighten out, and that motion is transmitted by levers and gears to the pointer, indicating the pressure (Fig. 5.9).

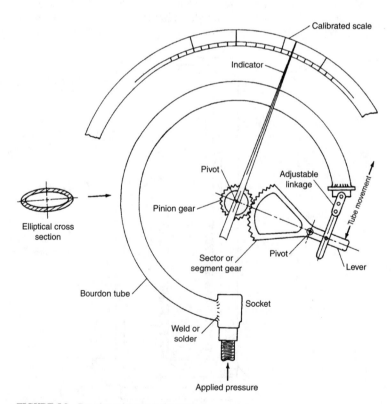

FIGURE 5.9 Bourdon tube or pressure gauge.

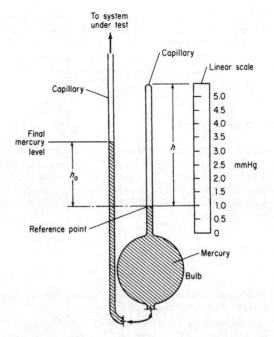

FIGURE 5.10 Uniform-scale method for reading McLeod gauges.

As a vacuum gauge, it is usually used only to indicate the condition of the system. Bourdon gauges (Fig. 5.10) are not suitable for accurate high-vacuum measurements. A U-tube manometer or McLeod gauge would be more accurate than a Bourdon tube for low pressure work.

E. McLeod Gauges

The McLeod gauge is the primary standard for the absolute measurement of pressure. A chamber (part of the gauge) of known volume is evacuated and then filled with mercury. This chamber terminates in a sealed capillary which is calibrated in micrometers of mercury, absolute pressure. When the mercury fills the chamber, the gas therein is compressed to approximately atmospheric pressure, and any trapped vapors are liquefied and have no significant volume. In other words, this gauge does not measure the pressure caused by any vapors present. Its reading represents only the total pressure of gases.

A cold trap is sometimes employed in conjunction with this gauge to prevent the transfer of vapors from the pump to the gauge or vice versa. These gauges, while accurate to approximately 10^{-5} torr, are not considered suitable for ordinary use because they do not read continuously and are usually fragile. The uniform scale (linear) McLeod gauge is shown in Fig. 5.10.

F. Pirani Gauges

A Pirani gauge measures the pressure of the gas by indicating the ability of the gas to conduct heat away from a hot filament. The greater the density (or pressure) of a gas, the greater the conduction of heat from the filament. As the temperature of the filament varies, so does its ability to carry current. Usually a Wheatstone bridge circuit is employed with a microammeter calibrated to read in micrometers of mercury.

Pressure is read on the basis of the current flow through the filament, which is a function of the filament temperature. Since heat is conducted away from the filament by vapors as well as by gases, these gauges measure the presence of the vapors. Also, different gases have different thermal conductivities. Hence, in the same system, they will yield a higher reading than a McLeod gauge, which measures only the pressure caused by the gases. They are usually used in the pressure range of 1 μm Hg to 1 mm Hg.

G. Thermocouple Gauges

A **thermocouple gauge** is similar to a Pirani gauge in many ways. The difference, basically, is that a thermocouple measures the filament temperature and the thermocouple output is shown on a meter calibrated in micrometers of mercury. Just as a Pirani gauge will give a different reading for the same pressure of different gases and vapors, so will a thermocouple gauge. The thermocouple gauge is more rugged, smaller, and slightly less sensitive than the Pirani. Its range is approximately the same, 1 μm Hg to 1 mm Hg.

REFERENCES

1. Flow Measurements for Engineers and Scientists. Nichkolas P. Cheremisinoff and Paul Cheremisinoff, ISBN 0-8247-7831-6. CRC Press.
2. Flow Measurement. Bela G. Liptak, ISBN 0-8019-8386-X. CRC Press.
3. Piping Calculations Manual. Shashi Menon, ISBN 0-0714-4090-9. McGraw-Hill.

CHAPTER 6
MATHEMATICS REVIEW AND CONVERSION TABLES

WRITING NUMBERS IN EXPONENTIAL NOTATION

In most scientific work, numbers can range from extremely small, like the mass of a carbon atom (0.00,000,000,000,000,000,000,002 g), to very large, like the mass of a train load of coal (9,000,000,000,000,000 g). These masses, written out in ordinary form, cannot be entered into most calculators which have perhaps 8 to 12 windows for displaying the digits. These very small and very large numbers must be written in exponential form. A number written in exponential notation consists of two parts. There is a **coefficient,** the numerical part, followed by 10 raised to some **power**. A carbon atom weighs 2×10^{-23} g and the train load of coal could be expressed as 9×10^{15} g. A scientific calculator is defined as one that has exponential capabilities. This type of calculator is absolutely necessary in a process operation when very large and very small numbers can be encountered. All scientific calculators have an exponential notation key labeled either EE or EXP. By convention, in scientific notation, the numerical part is always written with one nonzero digit before the decimal point.

EXAMPLE 6.1   The number 5678 should be expressed as

$$5.678 \times 10^3$$

(not 56.78×10^2 or 0.5678×10^4).

Using a scientific calculator, enter the number 5.678 (the coefficient) followed by the "EE" or "EXP" key, then type in 3 (the power).

After a calculation is completed, always ask two questions:

- Does the answer seem reasonable and make sense?
- Does the answer have the correct number of significant figures?

GROUPING OF NUMBERS AND DECIMAL POINTS

The use of commas to separate multiples of 1000 is not recommended in the SI system. Spaces are to be used instead of commas. Some European countries use commas as their decimal points. When writing numbers of less than one, a zero should be placed ahead of the decimal.

EXAMPLE 6.2   34,908,788.762,433 = 34 908 788.762 433
.3456567 = 0.345 656 7

SCIENTIFIC CALCULATORS

All scientific calculators use a display format with exponential numbers. Software and electronic devices are available which allow sophisticated calculators to be connected directly to computers and printers for additional display capabilities.

A. Fixed Point Display

The fixed point display mode, which is common to most advanced calculators, permits you to select the number of digits to be displayed *after* the decimal point. If a numerical entry or result exceeds the number of digits selected, the calculator automatically rounds the least significant figures shown in the display while continuing to perform all calculations at the machine's maximum accuracy.

B. Scientific Notation Display

Virtually all serious applications for advanced hand-held calculators involve very small or very large numbers. Since a calculator display has a limited number of digits, most advanced models express numbers which exceed the capacity of the display in scientific notation.

Scientific notation is a simple shorthand method of expressing a number as a multiple of a power of 10. For example, 10,000 in scientific notation is 1×10^4. Numbers smaller than one require a negative exponent: for example, $1 \times 10^{-4} = 0.001$. In each case, the exponent defines the number of digits in which the number is from the decimal point.

EXAMPLE 6.3 Light travels at approximately 30,000,000,000 cm/s. How much time is required for a ray of light to travel a distance of 0.00001 cm?

To get the travel time you would divide the distance by the velocity. The calculator will give you the answer: 3.3333333×10^{-6}. Without scientific notation, a 17-digit display would be required just to show the first 3 in fixed notation (0.000 000 000 000 000 333) which is not possible for most calculators.

In addition to permitting the processing of numbers far smaller or larger than the digit capacity of its display, a calculator with scientific notation simplifies many types of computations.

C. Engineering Notation Display

Engineers, chemists, operators, and technicians who frequently encounter numbers expressed in exponential increments of 3 (e.g., milli-, mega-, tera-, etc.) will find engineering notation a handy feature. This capability is found on all scientific calculators and, when selected, automatically converts entries and results to a modified form of scientific notation wherein the exponent of 10 is a multiple of 3 (e.g., 10^3, 10^9, 10^{-12}, etc.). See Figure 6.1 for various display examples.

| Fixed: 0.016666667 |

| Scientific: 1.666666 -02 |

| Engineering 16.66666 -03 |

FIGURE 6.1 Calculator display of same number with a 10 window calculator by fixed, scientific, and engineering notation.

THE SI (METRIC) SYSTEM

All scientific measurements consist of two parts: a number and a unit of measure. By international agreement in 1960, scientists throughout the world established a system of standardized units called the International System (French abbreviated SI, Le Systém International d'Unités).

This system is commonly referred to as the **metric system** and has seven base units as indicated in Table 6.1.

TABLE 6.1 SI Base Units

Quantity	Name	Symbol
Length	meter	m
Mass	kilogram	kg
Time	second	s
Electric current	ampere	A
Temperature	Kelvin	K
Amount of substance	mole	mol
Luminous intensity	candela	cd

The scientific community in the United States and most of the world uses the SI system for expressing measured units. The SI system consists of two parts: base units as shown in Table 6.1 and a system of numerical prefix as shown in Table 6.2.

TABLE 6.2 SI (Metric) Prefixes

Prefix	Symbol	Decimal equivalent	Exponential equivalent
tera	T	1 000 000 000 000	1×10^{12}
giga	G	1 000 000 000	1×10^{9}
mega	M	1 000 000	1×10^{6}
kilo	k	1 000	1×10^{3}
hecto	h	100	1×10^{2}
deka	da	10	1×10^{1}
SI units: meters, liters, and grams			
deci	d	0.1	1×10^{-2}
centi	c	0.01	1×10^{-2}
milli	m	0.001	1×10^{-3}
micro	μ	0.000 001	1×10^{-6}
nano	n	0.000 000 001	1×10^{-9}
pico	p	0.000 000 000 001	1×10^{-12}

SIGNIFICANT FIGURES

The proper handling of data is a prerequisite for any accurate statistical process analysis (see Chapter 7 "Standard Operating Procedures") and reporting. Different laboratory devices and procedures produce different degrees of accuracy. For example, microbalances are capable of measurements to 8 figures (e.g., 13.023456 g) while the same sample would be reported to only 4 figures (13.02 g) on a less accurate triple-beam balance. The number of **significant figures** in a measurement is the number of digits that is accurately known, plus one that is uncertain or doubtful. In the microbalance result above, the digits in the first seven places are known accurately but the last number (0.000006) is doubtful. The number of significant figures in a measurement are counted beginning with the first nonzero digit and stopping after including the first doubtful digit. The microbalance measurement has 8 significant figures while the triple beam balance has only 4 significant figures. Note that the zero is a significant figure in both cases. Calculations should never "improve" the precision or number of significant figures in an answer. Thus, a laboratory report should not contain any more significant figures in the final answer than found in the least precise data to obtain that answer.

The cardinal rule of significant figures: *A final result should never contain any more significant figures than the least precise data used to calculate it.*

GENERAL RULES FOR DEALING WITH SIGNIFICANT FIGURES

1. The number of significant figures in a measured quantity is the number of digits that are known accurately, plus one that is in doubt. For example, if the time is measured as 15.8 ± 0.1 seconds then it should be reported as 15.8 seconds.

2. The concept of significant figures only applies to measured quantities and results calculated from measured quantities. It is never applicable to exact numbers or definitions.

EXAMPLE 6.4 In converting 15.8 seconds to minutes:

$$15.8 \text{ seconds} \times 60 \text{ s/min} = 0.263333333 \text{ minute}$$

should be reported as **0.263** since the measured time had 3 significant figures and the conversion factor (60 s/min) has an unlimited number of significant figures since it is a definition. Again, the answer in a multiplication or division problem must be rounded off to contain the same number of significant figures as the smallest number of significant figures in any measurement or factor.

3. All significant figures are counted from the first nonzero digit. Note that leading zeros (zeros that occur at the start of a number) are not counted as significant figures.

$$0.0049 \quad (2 \text{ significant figures})$$
$$0.00000011366 \quad (5 \text{ significant figures})$$
$$90009 \quad (5 \text{ significant figures})$$

4. All confined zeros (zeros found between nonzero numbers) in a number are significant. For example:

$$5004 \quad (4 \text{ significant figures})$$
$$0.0400008 \quad (6 \text{ significant figures})$$
$$2.0981 \quad (5 \text{ significant figures})$$

5. The answer in an addition or subtraction problem must be rounded off to the first column that has a doubtful digit. For example, in this addition problem:

$$2.3340 + 0.2678 + 345.8 = 348.4018$$

The final answer should be rounded off to **348.4** since the measurement 345.8 only contains a value to the nearest 0.1. Again, in addition and subtraction, the last digit retained in the sum or difference should correspond to the first doubtful decimal position in the measurement numbers.

Since different final answers can be obtained by rounding either at each step or at the end of the total calculation, many companies and agencies have very specific policies called Standard Operating Procedures (SOP) on calculations (see Chapter 7). Consult your supervisor when in doubt.

ROUNDING

If a calculation yields a result that would suggest more precision than is justified, rounding off to the proper number of significant figures is required. Since the rounding rules have changed recently, it would be advisable to check with a supervisor or consult your company's SOPs dealing with calculation procedures. Final results can be effected if you round at the end of each step versus rounding only the final answer.

There are three basic rules of rounding:

1. If the digit following the last significant figure is *greater than 5,* the number is rounded up to the next higher digit. *NOTE:* This is a slight departure from the old 5 or greater rounding rule.

EXAMPLE 6.5 A mass of 51.226 lb expressed to 4 significant figures would be rounded to 51.23 lb.

2. If the digit following the last significant figure is *less than 5,* the number is rounded off to the present value of the last significant figure.

EXAMPLE 6.6 A mass of 51.224 lb expressed to 4 significant figures would be rounded to 51.22 lb.

3. If the digit following the last significant figure is *exactly 5,* the number is rounded off to the nearest even number.

EXAMPLE 6.7 A mass of 51.225 lb expressed to 4 significant figures would be rounded to 51.22 lb. *NOTE:* A mass of 51.22512 lb expressed to 4 significant figures would be rounded to 51.23 lb (rule 1) since it is *greater than 5.*

THE QUADRATIC EQUATION

A **quadratic equation** is one in which the unknown is raised to the second power. Sometimes you find a problem in which the unknown is squared in one term, appears to the first power (exponent 1, or no exponent) in another term, and is absent in a third term that is a real number. There are several methods by which quadratic equations can be solved, but only one method that can solve them all. It is necessary to rearrange the equation so it has the form $ax^2 + bx + c = 0$. The solution is then given by the **quadratic formula:**

$$x = \frac{-b \pm \sqrt{b^2 - 4ac}}{2a}$$

The ± in the formula indicates that there are two values of "x" that can satisfy the equation. There are two roots or answers for most quadratic equations. However, in most chemical calculations, one solution will seem reasonable and the other solution will appear to be impractical. For example, a negative ("−") value for a mass or pH would obviously be eliminated as one of the roots. The following quadratic equation would be solved using the quadratic formula as shown below:

quadratic equation $2x^2 - 3x - 9 = 0$

quadratic formula $x = \dfrac{-(-3) \pm \sqrt{-3^2 - 4(2)(-9)}}{2(2)}$

$x = \dfrac{3 \pm 9}{4}$

first possible solution $x = 3$

second possible solution $x = -1.5$

GRAPHS

A **graph** is a drawing that represents specific numerical values and their relationship (general pattern) to each other. Graphs are used in the process industries to display trends, to determine values of quantities, and even to display the history of physical and chemical data. The most common graphs use the Cartesian coordinates, which use two axes at right angles to each other. The horizontal axis is called the **abscissa** and the vertical axis is called the **ordinate**. If the numbers on the axes start at zero, they are said to be at the origin. However, many graphs show only the range of numerical values of interest to save space. Plotting the data points of the two variables on a grid produces a useful curve called the **slope**. If the shape of this slope is a straight line, then the relationship between the two variables is said to be **linear**.

The real advantage to graphing data comes from the fact that the value of one quantity (plotted on the abscissa) can be determined by its association or relationship to another value (plotted on the ordinate). If the plot's slope is linear, then data that was never actually determined can be read directly from the graph. Determining values between actual data points is called **interpolation**. Graphs can even be used to estimate a value beyond those measured data points by extrapolation. **Extrapolation** means to add extra length to a graph's plotted curve. Many process control charts are expressed in graphical form as discussed in Chapter 7 "Standard Operating Procedures."

LOGARITHMS

There are two kinds of logarithms: the **common logarithm** of a number is the power or exponent to which 10 must be raised to be equal to the number ($n = 10^x$). The second type of logarithm, called a **natural logarithm,** has a natural base of 2.718, sometimes called base $e (n = e^x)$. To distinguish between the two bases, the common logarithm of x is written log $10x$ or simply log x and the natural logarithm is written log $e\ x$ or ln x. A simple mathematical relationship exists between the two logarithmic forms:

$$\ln x = 2.303 \log x$$

The relationship between ordinary numbers and their common logarithmic expression is shown in Table 6.3. It should be noted from the table that the logarithm of 1 is 0, the logarithm of any number greater than 1 is a positive number, and the logarithm of a number less than 1 is a negative number. Table 6.3 indicates that whole-number multiples of 10 have logarithms that are whole numbers. However, most numbers are not whole-number multiples of 10 and require two sets of numbers to define the logarithmic expression.

TABLE 6.3 Numbers and Their Logarithms

Number	Fraction	Exponential form	Logarithm
1000000	1000000/1	10^6	6
100000	100000/1	10^5	5
10000	10000/1	10^4	4
1000	1000/1	10^3	3
100	100/1	10^2	2
10	10/1	10^1	1
1	1/1	10^0	0
0.1	1/10	10^{-1}	−1
0.01	1/100	10^{-2}	−2
0.001	1/1000	10^{-3}	−3
0.0001	1/10000	10^{-4}	−4
0.00001	1/100000	10^{-5}	−5
0.000001	1/1000000	10^{-6}	−6

EXAMPLE 6.8   Express the number 200 in logarithmic form.

SOLUTION

1. First write the number in scientific form:

$$\log 200 = \log(2 \times 10^2)$$

2. Second, determine the logarithm of the numbers individually from a common logarithm table or calculator and add them together.

$$\log 200 = \log 2 + \log 10^2 = 0.3010 + 2 = 2.3010$$

The number on the left of the decimal point in a logarithmic expression is called the **characteristic**, and the digits on the right of the decimal point are called the **mantissa**. The characteristic determines the order of magnitude (power) of the number, and the mantissa represents the number itself expressed as a value between 1 and 10. Most scientific calculators have both common and natural logarithmic functions available. Logarithms are used primarily in pH calculation (see Chapter 9 "pH Measurement") and nonlinear graphing.

DIMENSIONAL ANALYSIS

It is often necessary to convert one system of units to another: for example, English to SI or even one SI expression to another SI form. This mathematical technique is often called **dimensional analysis,** the **factor-label method,** or the **unit-factor method.**

Notice how the undesired units are canceled out with this mathematical method. Dimensional analysis is a method that facilitates the conversion of units from one system to another or within the same system of measurements. Table 6.5 will provide some of the common conversion factors.

TABLE 6.4   Commonly Used Mathematical Signs and Symbols

Symbol	Meaning
$+$	plus; add; positive
$-$	minus; subtract; negative
$\pm$	plus or minus; positive or negative
$\times$ or $\cdot$	times; multiplied by
$\div$ or $/$	is divided by
$=$	is equals; as
$\equiv$	is identical to; congruent with
$\neq$	is not identical
$\approx$	approximately
$<$	is less than
$>$	is greater than
$\leq$	is equal to or less than
$\geq$	is equal to or greater than
$:$	is ratio of
$\propto$	varies as (is proportional to)
a^2	a squared, the second power of a; $a \times a$
a^3	a cubed, the third power of a; $a \times a \times a$
a^{-1}	$1/a$ (reciprocal of a)
a^{-2}	$1/a^2$ (reciprocal of a^2)
Σ	summation of
Δ	difference in
k	a mathematical constant
π	"pi," a mathematical constant, 3.141592+

TABLE 6.5 Unit Conversion Table

Units of Area

Units	Square inches	Square feet	Square yards	Square miles	Square centimeters	Square meters
1 square inch	1	0.006 944 44	0.000 771 605	0.000 000 000 249 1	6.451 626	0.000 645 162 6
1 square foot	144	1	0.111 111 1	0.000 000 035 870 1	929.0341	0.092 903 41
1 square yard	1 296	9	1	0.000 000 322 831	8361.307	0.836 130 7
1 square mile	4 014 489 600	27 878 400	3 097 600	1	25 899 984 703	2.589 998
1 square centimeter	0.154 996 9	0.001 076 387	0.000 119 598 5	0.000 000 000 038 610 06	1	0.000 1
1 square meter	1549.9969	10.763 87	1.195 985	0.000 000 386 100 6	10 000	1

Units of Length

Units	Inches	Feet	Yards	Miles	Centimeters	Meters
1 inch	1	0.083 333 3	0.027 777 8	0.000 015 782 8	2.540 005	0.025 400 05
1 foot	12	1	0.333 333	0.000 189 393 9	30.480 06	0.304 800 6
1 yard	36	3	1	0.000 568 182	91.440 18	0.914 401 8
1 mile	63 360	5 280	1 760	1	160 934.72	1609.3472
1 centimeter	0.3937	0.032 808 33	0.010 936 111	0.000 006 213 699	1	0.01
1 meter	39.37	3.280 833	1.093 611 1	0.000 621 369 9	100	1

Units of Volume

Units	Cubic inches	Cubic feet	Cubic yards	Cubic centimeters	Cubic decimeters	Cubic meters
1 cubic inch	1	0.000 578 704	0.000 021 433 47	16.387 162	0.016 387 16	0.000 016.387 16
1 cubic foot	1 728	1	0.037 037 0	28 317.016	28.317 016	0.028.317 016
1 cubic yard	46 656	27	1	764 559.4	764.5594	0.764 559 4
1 cubic centimeter	0.016 023 38	0.000 035 314 45	0.000 001 307 94	1	0.001	0.000 000 001
1 cubic decimeter	61.023 38	0.035 314 45	0.001 307 943	1 000	1	0.001
1 cubic meter	61 023.38	35.314 45	1.307 942 8	1 000 000	1000	1

Units of Liquid Measure

Units	Fluid ounces	Liquid pints	Liquid quarts	Gallons	Milliliters	Liters	Cubic inches
1 fluid ounce	1	0.0625	0.031 25	0.007 812 5	29.5729	0.029 572 9	1.804 69
1 liquid pint	16	1	0.5	0.125	473.167	0.473 167	28.875
1 liquid quart	32	2	1	0.25	946.333	0.946 333	57.75
1 gallon	128	8	4	1	3785.332	3.785 332	231
1 milliliter	0.033 814 7	0.002 113 42	0.001 056 71	0.000 264 178	1	0.001	0.061 025 0
1 liter	33.8147	2.113 42	1.056 71	0.264 178	1000	1	61.0250
1 cubic inch	0.554 113	0.034 632 0	0.017 316 0	0.004 329 00	16.3867	0.016 386 7	1

Units of Mass

Units	Grains	Apothecaries' scruples	Pennyweights	Avoirdupois drams	Apothecaries' drams	Avoirdupois ounces
1 grain	1	0.05	0.014 666 67	0.036 571 43	0.016 666 7	0.002 285 71
1 apoth. scruple	20	1	0.833 333 3	0.731 428 6	0.333 333	0.045 714 3
1 pennyweight	24	1.2	1	0.877 714 3	**0.4**	0.054 857 1
1 avdp. dram	**27.343 75**	**1.367 187 5**	1.139 323	1	0.455 729 2	**0.0625**
1 apoth. dram	60	3	2.5	2.194 286	1	0.137 142 9
1 avdp. ounce	**437.5**	**21.875**	18.229 17	**16**	7.291 67	**1**
1 apoth. or troy ounce	**480**	**24**	**20**	17.554 28	**8**	1.097 142 9
1 apoth. or troy pound	**5760**	**288**	**240**	210.6514	**96**	13.165 714
1 avdp. pound	**7000**	**350**	291.6667	**256**	116.6667	**16**
1 milligram	0.015 432 356	0.000 771 618	0.000 643 014 8	0.000 567 383 3	0.000 257 205 9	0.000 035 273 96
1 gram	15.432.356	0.771 618	0.643 014 85	0.564 383 3	0.257 205 9	0.035 273 96
1 kilogram	15 432.356	771.6178	643.014 85	564.383 32	257.205 94	35.273 96

Units of Mass

Units	Apothecaries' or troy ounces	Apothecaries' or troy pounds	Avoirdupois pounds	Milligrams	Grams	Kilograms
1 grain	0.002 083 33	0.001 173 611 1	0.000 142 857 1	64.798 918	0.064 798 918	0.000 064 798 9
1 apoth. scruple	0.041 666 7	0.003 472 222	0.002 856 143	1295.9784	1.295 978 4	0.001 295 978
1 pennyweight	**0.05**	0.004 366 667	0.003 428 571	1555.1740	1.555 174 0	0.001 555 174
1 avdp. dram	0.056 966 146	0.004 747 178 8	**0.003 906 25**	1771.8454	1.771 845 4	0.001 771 845
1 apoth. dram	**0.125**	0.010 416 667	0.008 571 429	3887.9351	3.887 935 1	0.003 887 935
1 avdp. ounce	0.911 458 3	0.075 954 861	**0.0625**	28 349.527	28.349 527	0.028 349 53
1 apoth. or troy ounce	**1**	0.083 333 33	0.068 571 43	31 103.481	31.103 481	0.031 103 48
1 apoth. or troy pound	**12**	**1**	0.822 857 1	373 241.77	373.241 77	0.373 241 77
1 avdp. pound	14.583 333	1.215 277 8	**1**	**453 592.427 7**	**453.591 427 7**	**0.453 592 427 7**
1 milligram	0.000 032 150 74	0.000 002 679 23	0.00 002 204 62	**1**	0.001	**0.000 001**
1 gram	0.034 150 74	0.002 679 23	0.002 204 62	**1000**	**1**	**0.001**
1 kilogram	32.150 742	2.679 228 5	2.204 622 341	**1 000 000**	**1000**	**1**

*The figures in boldface type signify exact values.

Dimensional analysis involves five steps:

1. Write the units of the answer needed on the right side of the equation.
2. Set this equal to the quantity given, including the units, times some conversion factor(s).
3. Multiply the original quantity given by whatever factors necessary to convert its units to the units needed (see Table 6.6). Each factor will be a fraction in which the numerator (top) is equivalent to the denominator (bottom).
4. Check to see that all units have canceled with each other except the desired units of the answer.
5. Check the final answer for appropriate number of significant figures.

The dimensional analysis technique will be demonstrated with two examples using appropriate conversion factors.

EXAMPLE 6.9 An operator measured the depth of a liquid to be 233 cm. What is this depth expressed in inches?

$$233 \text{ cm} \times \frac{1.00 \text{ in}}{2.54 \text{ cm}} = 91.7 \text{ in}$$

Note that the cm units have canceled each other and the desired answer unit remains.

EXAMPLE 6.10 Bolts cost 3.27 cents each when bought in large lots. How much will a gross of bolts (144) cost? To solve this problem you would limit the number of significant figures to 3.

NOTE: See Chapter 19, "Physical Properties and Determinations," for temperature and other physical properties conversions.

CONVERSIONS BETWEEN ENGINEERING UNITS AND SI

Table 6.6 gives a listing of some conversion factors for converting U.S.-British units to SI units. This table is condensed in part from ASTM Standard for Metric Practice E 380-76. The letter E represents exponent followed by the exponent's sign and power of 10. For example, 2.567 234 E + 03 means 2567.234 and 2.569 544 E − 02 means 0.02569544.

TABLE 6.6 Some Conversion Factors for Engineering Units to SI Units

To convert from	To	Multiply by
abampere	ampere (A)	1.000 000 E + 01
abcoulomb	coulomb (C)	1.000 000 E + 01
abfarad	farad (F)	1.000 000 E + 09
abhenry	henry (H)	1.000 000 E − 09
abmho	siemens (S)	1.000 000 E + 09
abohm	ohm (Ω)	1.000 000 E − 09
abvolt	volt (V)	1.000 000 E − 08
acre foot (U.S. survey)	meter^3 (m^3)	1.233 489 E + 03
acre (U.S. survey)	meter^3 (m^3)	4.046 873 E + 03
ampere hour	coulomb (C)	3.600 000 E + 03

(Continued)

TABLE 6.6 Some Conversion Factors for Engineering Units to SI Units (*Continued*)

To convert from	To	Multiply by
are (French for area)	meter^2 (m^2)	1.000 000 E + 02
angstrom (Å)	meter (m)	1.000 000 E − 10
atmosphere (standard)	pascal (Pa)	1.013 250 E + 05
bar	pascal (Pa)	1.000 000 E + 05
barn	meter^2 (m^2)	1.000 000 E − 28
barrel (petroleum, 42 gal)	meter^3 (m^3)	1.589 873 E − 01
board foot	meter^3 (m^3)	2.358 737 E − 03
British thermal unit (Int)	joule	1.055 056 E + 03
British thermal unit (mean)	joule	1.055 056 E + 03
British thermal unit (therm)	joule	1.055 056 E + 03
Btu (Int) ft/h · ft^2 · °F	W/m·K	1.730 735 E + 00
Btu (Int) in/h · ft^2 · °F	W/m·K	1.442 279 E − 01
Btu (Int) in/s · ft^2 · °F	W/m·K	1.442 279 E + 02
Btu (International Table)/h	watt (W)	2.930 7112 E − 01
Btu (Internation Table)/ft^2 · h	W/m^2	3.154 591 E + 00
Btu (Internal Table)/h · ft^2 · °F	W/m^2·K	5.678 263 E + 00
bushel (U.S.)	meter^3 (m^3)	3.523 907 E − 02
caliber (inch)	meter (m)	2.540 000 E − 02
calorie (International Table)	joule (J)	4.186 800 E + 00
calorie (mean)	joule (J)	4.190 02 E + 00
calorie (thermochemical)	joule (J)	4.184 000 E + 00
calorie (Intern Table)/g · °C	J/kg·K	4.186 800 E + 03
calorie (thermochem)/g · °C	J/kg·K	4.184 000 E + 03
calorie (thermoc)/cm^2 · min	W/m^2	6.973 333 E + 02
calorie (thermoc)/cm^2 · s · °C	W/m-K	4.184 000 E + 02
carat (metric)	kilogram (kg)	2.000 000 E − 04
centimeter of mercury (0°C)	pascal (Pa)	1.333 22 E + 03
centimeter of water (4°C)	pascal (Pa)	9.806 38 E + 01
centipoise	pascal · s (Pa · s)	1.000 000 E − 03
centistokes	m^2/s	1.000 000 E − 06
circular mil	meter^2 (m^2)	5.067 075 E − 10
cup	meter^3 (m^3)	2.365 882 E − 03
day (mean solar)	second (s)	8.640 000 E + 04
degree (angle)	radian (rad)	1.745 329 E − 02
denier	kg/m	1.111 111 E − 07
dyne	newton (N)	1.000 000 E − 05
dyne · cm	N · m	1.000 000 E − 07
dyne · cm^2	pascal (Pa)	1.000 000 E − 01
electronvolt	joule (J)	1.602 19 E − 19
erg	joule (J)	1.000 000 E − 07
erg/cm^2 · s	W/m^2	1.000 000 E − 03
erg/s	watt (W)	1.000 000 E − 07
faraday (based on carbon-12)	coulomb (C)	9.648 70 E + 04
faraday (chemical)	coulomb (C)	9.649 57 E + 04
faraday (physical)	coulomb (C)	9.652 19 E + 04

(*Continued*)

TABLE 6.6 Some Conversion Factors for Engineering Units to SI Units (*Continued*)

To convert from	To	Multiply by
foot	meter (m)	3.048 000 E − 01
foot of water (39.2°F)	pascal (Pa)	2.988 98 E + 03
footcandle	lux (lx)	1.076 391 E + 01
footlambert	cd/m^2	3.426 259 E + 00
gal	m/s^2	1.000 000 E − 02
gallon (Canadian liquid)	meter^3 (m^3)	4.546 090 E − 03
gallon (U.K. liquid)	meter^3 (m^3)	4.546 092 E − 03
gallon (U.S. liquid)	meter^3 (m^3)	3.785 412 E − 03
gamma	tesla (T)	1.000 000 E − 09
gauss	tesla (T)	1.000 000 E − 04
gill (U.S.)	meter^3	1.182 941 E − 04
gram	kilogram (kg)	1.000 000 E − 03
gram-force/cm^2	pascal (Pa)	9.806 650 E + 01
horsepower (boiler)	watt (W)	9.809 50 E + 03
horsepower (electric)	watt (W)	7.460 000 E + 02
horsepower (metric)	watt (W)	7.354 99 E + 02
horsepower (water)	watt (W)	7.460 43 E + 02
horsepower (U.K.)	watt (W)	7.457 0 E + 02
hundredweight (long)	kilogram (kg)	5.080 235 E + 01
hundredweight (short)	kilogram (kg)	4.535 924 E + 01
inch of mercury (60°F)	pascal (Pa)	3.376 85 E + 03
inch of water (39.2°F)	pascal (Pa)	2.490 82 E + 02
inch of water (60°F)	pascal (Pa)	2.488 4 E + 02
inch^2	meter^2 (m^2)	6.541 600 E − 04
inch^3 (volume)	meter^3 (m^3)	1.638 706 E − 05
inch/second	m/s	2.54 0 000 E − 02
inch/second^2	m/s^2	2.54 0 000 E − 02
kayser	1 per meter	1.000 000 E + 02
kW·h	joule (J)	3.600 000 E + 06
knot (international)	meter/second	5.144 444 E − 01
langley	J/m^2	4.184 000 E + 04
liter	meter^3	1.000 000 E − 03
maxwell	weber (Wb)	1.000 000 E − 08
micron	meter (m)	1.000 000 E − 06
mil	meter (m)	2.54 0 000 E − 05
mile (international)	meter (m)	1.609 344 E + 03
millibar	pascal (Pa)	1.000 000 E + 02
millimeter of mercury (0°C)	pascal (Pa)	1.333 22 E + 02
oersted	ampere/meter	7.957 747 E + 01
ounce (avoirdupois)	kilogram (kg)	2.834 952 E − 02
ounce (troy or apothecary)	kilogram (kg)	3.110 348 E − 02
ounce (U.S. fluid)	meter^3	2.957 353 E − 05
peck (U.S.)	meter^3	8.809 768 E − 03
pica (printer's)	meter (m)	4.217 518 E − 03
point (printer's)	meter (m)	3.514 598 E − 04

(*Continued*)

TABLE 6.6 Some Conversion Factors for Engineering Units to SI Units (*Continued*)

To convert from	To	Multiply by
poise (absolute viscosity)	pascal second	1.000 000 E − 01
pound (lb avoirdupois)	kilogram	4.535 924 E − 01
pound (troy or apothecary)	kilogram	3.732 417 E − 01
quart (U.S. dry)	meter^3	1.101 221 E − 03
quart (U.S. liquid)	meter^3	9.463 529 E − 04
shake	second (s)	1.000 000 E − 08
slug	kilograms (kg)	1.459 390 E + 01
stokes (kinematic viscosity)	m^2/s	1.000 000 E − 04
tablespoon	meter^3	1.478 676 E − 05
teaspoon	meter^3	4.928 922 E − 06
ton (assay)	kilogram (kg)	2.916 667 E − 02
ton (long, 2240 lb)	kilogram (kg)	1.016 047 E + 03
ton (metric)	kilogram (kg)	1.000 000 E + 03
ton (refrigeration)	watt (W)	3.516 800 E + 03
ton (register)	meter^3 (m^3)	2.831 685 E − 00
ton (short, 2000 lb)	kilogram (kg)	9.071 847 E + 02
tonne	kilogram (kg)	1.000 000 E + 03
W·h	joule (J)	3.600 000 E + 03
W·s	joule (J)	1.000 000 E +00
yard	meter (m)	9.144 000 E − 01
yard^2	meter^2 (m^2)	8.361 274 E − 01
yard^3	meter^3 (m^3)	7.645 549 E − 01
year (calendar)	second (s)	3.153 600 E + 07

A gross has an infinite number of significant figures since it is a definition and the cost has 3 significant figures. Thus multiply 144 by $0.0327:

$$144 \text{ bolts} \times \$0.0327/\text{bolt} = \$4.7088$$

Although the calculated answer is $4.7088, the answer is limited to 3 significant figures ($4.71).

EXAMPLE 6.11 Calculate the number of seconds in 50.0 years.

$$50.0 \text{ years} \times \frac{365 \text{ days}}{1 \text{ year}} \times \frac{24 \text{ h}}{1 \text{ day}} \times \frac{60 \text{ min}}{1 \text{ h}} \times \frac{60 \text{ s}}{1 \text{ min}}$$

$$= 1.58 \times 10^9 \text{ seconds}$$

Note that the calculator answer (1.5769×10^9) was rounded to 3 significant figures because all factors were definitional (infinite) except for the limiting 3 significant figures in the 50.0 years.

CONVERSION OF MILLIMETERS TO DECIMAL INCHES (TABLE 6.7)

TABLE 6.7 Conversion Chart: Millimeters, Fractional Inches, and Decimal Inches

mm	In Frac.	In Dec.	mm	In Frac.	In Dec.	mm	In Frac.	In Dec.
.01		0.0004	8.3344	21/64	0.3281	21.4312	27/32	0.8437	57		2.244
.02		0.0008	8.7312	11/32	0.3437	21.8281	55/64	0.8594	58		2.283
.03		0.0012	9.000		0.3543	22.000		0.8661	59		2.323
.04		0.0016	9.1387	23/64	0.3594	22.2250	7/8	0.875	60		2.362
.05		0.0020	9.505	3/8	0.375	22.6219	57/64	0.8906	61		2.402
.06		0.0024	9.9719	25/64	0.3906	23.000		0.9055	62		2.441
.07		0.0028	10.000		0.3937	23.0187	29/32	0.9062	36		2.480
.08		0.0032	10.3187	13/32	0.4062	23.4156	59/64	0.9219	64		2.520
.09		0.0035	10.7156	27/64	0.4219	23.8125	15/16	0.9375	65		2.559
.10		0.004	11.000		0.4331	24.000		0.9449	66		2.598
.20		0.008	11.1125	7/16	0.4375	24.2094	61/64	0.9531	67		2.638
.30		0.012	11.5094	29/64	0.4531	24.6062	31/32	0.9687	68		2.677
.3969	1/64	0.0156	11.9062	15/32	0.4687	25.000		0.9843	69		2.717
.40		0.0158	12.000		0.4724	25.0031	63/64	0.9844	70		2.756
.50		0.0197	12.3031	31/64	0.4844	25.400		1.000	71		2.795
.60		0.0236	12.700	1/2	0.500	26		1.024	72		2.835
.70		0.0276	13.000		0.5118	27		1.063	73		2.874
.7937	1/32	0.012	13.0968	33/64	0.5156	28		1.102	74		2.913
.80		0.0315	13.4937	17/32	0.5312	29		1.142	75		2.953
.90		0.0354	13.8906	35/64	0.5469	30		1.181	76		2.992
1.000		0.0394	14.000		0.5512	31		1.220	77		3.031
1.1906	3/64	0.0469	14.2875	9/16	0.5625	32		1.260	78		3.071
1.5875	1/16	0.0625	14.6844	31/64	0.5781	33		1.299	79		3.110
1.9844	5/64	0.0781	15.000		0.5906	34		1.339	80		3.150
2.000		0.0787	15.0812	19/32	0.5937	35		1.378	81		3.189

2.3812	3/32	0.0937	15.4781	39/64	0.6094	36	1.417	82	3.228
2.7781	7/64	0.1094	15.875	8/8	0.625	37	1.457	83	3.268
3.000		0.1181	16.000		0.6299	38	1.498	84	3.307
3.175	1/8	0.125	16.2719	41/64	0.6406	39	1.535	85	3.346
3.5719	9/64	0.1406	16.6687	21/32	0.6562	40	1.575	86	3.386
3.9637	5/32	0.1502	17.000		0.6693	41	1.614	87	3.425
4.000		0.1579	17.0656	43/64	0.6719	42	1.654	88	3.465
4.3656	11/64	0.1719	17.4625	11/16	0.6875	43	1.693	89	3.504
4.7625	3/16	0.1875	17.8594	45/64	0.7031	44	1.732	90	3.543
5.000		0.1969	18.000		0.7087	45	1.772	91	3.583
5.1594	13/64	0.2031	18.2562	23/32	0.7187	46	1.811	92	3.622
5.5562	7/32	0.2187	18.6532	47/64	0.7344	47	1.850	93	3.661
5.9531	13/64	0.2344	19.000		0.748	48	1.890	94	3.701
6.000		0.2362	19.050	3/4	0.750	49	1.929	95	3.740
6.3500	1/4	0.250	19.4369	49/64	0.7656	50	1.969	96	3.780
6.7459	11/64	0.2658	19.8433	15/32	0.7812	51	2.008	97	3.819
7.000		0.2756	20.000		0.7874	52	2.047	98	3.858
7.1437	9/32	0.2812	20.2402	51/64	0.7969	53	2.087	99	3.898
7.5406	19/64	0.2969	20.6375	13/16	0.8125	54	2.126	100	3.937
7.9375	5/16	0.3125	21.000		0.8268	55	2.165		
8.000		0.315	21.0344	53/64	0.8281	56	2.205		

CONVERSION OF STANDARD DRILL BIT SIZES (TABLE 6.8)

TABLE 6.8 Drill Bit Sizes

Bit no.	Diam., in	Bit no.	Diam., in	Bit no.	Diam., in	Bit no.	Diam., in	Bit no.	Diam., in
1	0.2250	17	0.1730	33	0.1130	49	0.0730	65	0.0350
2	0.2210	18	0.1695	34	0.1110	50	0.0700	66	0.0330
3	0.2130	19	0.1660	35	0.1100	51	0.0670	67	0.0320
4	0.2090	20	0.1610	36	0.1065	52	0.0635	68	0.0310
5	0.2055	21	0.1590	37	0.1040	53	0.0595	69	0.0292
6	2.2040	22	0.1570	38	0.1015	54	0.0550	70	0.0280
7	0.2010	23	0.1540	39	0.0995	55	0.0520	71	0.0260
8	0.1990	24	0.1520	40	0.0980	56	0.0465	72	0.0250
9	0.9060	25	0.1495	41	0.0960	57	0.0430	73	0.0240
10	0.1935	26	0.1470	42	0.0935	58	0.0420	74	0.0225
11	0.9010	27	0.1440	43	0.0890	59	0.0410	75	0.0210
12	0.1890	28	0.1405	44	0.0860	60	0.0400	76	0.0200
13	0.1850	29	0.1360	45	0.0820	61	0.0390	77	0.0180
14	0.1820	30	0.1285	46	0.0810	62	0.0380	78	0.0160
15	0.1800	31	0.1200	47	0.0785	63	0.0370	79	0.0140
16	0.1770	32	0.1160	48	0.0760	64	0.0360	80	0.0135

REFERENCES

1. *Chemical Rubber Company Handbook of Chemistry and Physics*, 78th edition. 2608 pages, ISBN 0-8493-0595-0. CRC Press (1997–1998, edited annually).
2. *Perry's Chemical Engineers' Handbook*, Robert H. Perry and Don W. Green. ISBN 0-07-115448-5. McGraw-Hill, New York (1997).
3. "Laboratory Information Management Systems," R. D. McDowall. *Journal of Chemistry and Engineering* (22 March 1993): 32–36.
4. *Statistical Quality Control Methods,* Irving W. Burr. ISBN 0-8247-6344-0. Marcel Dekker (1976).
5. Current Good Manufacturing Practices (CGMP). Code of Federal Regulations, 21 CFR 160(b)(4). U.S. Department of Health and Human Services, Washington, DC (April 1991).
6. *Statistical Manual of the AOAC,* W. J. Youdan and E. H. Steiner. American Association of Official Analytical Chemists, Washington, DC (1975).
7. *Lange's Handbook of Chemistry,* 15th edition. Section 11. Edited by John A. Dean, ISBN 0-07-016384-7. McGraw-Hill, New York (1995).

CHAPTER 7
STANDARD OPERATING PROCEDURES

STANDARD OPERATING PROCEDURES (SOPs)

Standard operating procedures (SOPs) are comprehensive, job-specific procedures developed by companies for their own step-by-step procedures for daily operations and regulatory compliance. These procedures are designed for all personnel ranging from newly trained employees to long-term, experienced personnel. These SOPs should cover every aspect of the process operation. For example, most companies would have four or more general purpose reference documents in their SOPs. These documents would include: (a) Standard Operating Conditions and Limits (SOCL); (b) Pressure Release Devices (PRD); (c) Lock/Tag/Try (LTT); (d) Confined Space Entry (CSE); (e) job-related injuries, and many others. In addition, many companies will maintain log books on topics ranging from process equipment lubrication requirements to control valves and pump maintenance manuals. Companies are continuously upgrading their SOPs to comply with federal regulatory agencies, including the Food and Drug Administration (FDA), Drug Enforcement Agency (DEA) and OSHA, as well as international organizations such as the International Organization for Standardization. For example, SOP documents are mandated for employers to comply with FDA regulations like 21 CFR Parts 210 and 211. These regulations are referred to respectively as **Current Good Manufacturing Practices (CGMP)** for the manufacturing, processing, packing, or holding of drugs "Part 210 as general" and "Part 211 for finished pharmaceuticals." FDA regulation 21 CFR Part 211.100 states "Written production and process control procedures shall be followed in the execution of the various production and process control functions and shall be documented at the time of performance." All chemically related manufacturers rely heavily on SOPs since these companies tend to be regulation-driven.

STANDARD OPERATING CONDITIONS AND LIMITS (SOCLs)

Some process operations utilize a centralized control room(s), especially if heavily automated. The control room is normally operated continuously by one or more operators. **Control room operators** are responsible for monitoring the daily routine of running the process unit and their duties could include the total system's operation (alarms, overrides, off-specification, etc.). The operating procedures for most plants and processes are developed by the engineering group that designed and constructed the facility. There are operating procedures that are **equipment-specific** and developed by manufacturers, and there are **job-specific** operating procedures as established by SOPs for each job and establish personnel communication links. Job-specific information involves many different people and area-wide use; thus it is sometimes referred to as **standard operating conditions and limits (SOCLs)**. These SOCLs involve very specific procedures like lockout/tagout, confined space entry, pressure relief devices, control room jobs to even pump operations, and scheduled lubrication manuals. Federal regulations usually use the term "operating procedures" to refer to all procedures for an area.

GOOD LABORATORY PRACTICE (GLP)

Any company doing work covered by the **Toxic Substances Control Act (TSCA), Federal Insecticide, Fungicide and Rodenticide Act (FIFRA),** or **Good Laboratory Practice for Nonclinical Laboratory Studies** must comply with **Good Laboratory Practice (GLP)**. The Environmental Protection Agency is responsible for the TSCA and FIFRA, while the Food and Drug Administration reviews GLP. Noncompliance can result in complications ranging from criminal violations to delays in getting products to market. GLP is a compilation of procedures and practices designed to promote the quality and validity of all laboratory studies. Historically, the need for GLP started in the United States with the Food, Drug, and Cosmetic Act of 1938. This legislation has been amended several times since its inception by the FDA to encourage drug manufacturers to practice and follow stated GLP regulations. The most current Good Laboratory Practice regulations were codified in 1976 as Chapter 21 of the Code of Federal Regulations, Part 58.

CURRENT GOOD MANUFACTURING PRACTICE (CGMP)

Current Good Manufacturing Practice (CGMP) was developed before GLP and regulates the manufacturing and associated quahty control of drugs. For example, four sections that relate to SOPs found in Current Good Manufacturing Practice for Finished Pharmaceuticals (21 CFR 211 Subpart J "Records and Reports") are outlined below.

A. 21 CFR 211.180: General Requirements

1. Any production, control, or distribution record that is required to be maintained in compliance with this part and is specifically associated with a batch of a drug product shall be retained for at least 1 year after the expiration date of the batch or, in the case of certain OTC drug products lacking expiration dating because they meet the criteria for exemption under Sec. 211.137, 3 years after distribution of the batch.
2. Records shall be maintained for all components, drug product containers, closures, and labeling for at least 1 year after the expiration date or, in the case of certain OTC drug products lacking expiration dating because they meet the criteria for exemption under Sec. 211.137, 3 years after distribution of the last lot of drug product incorporating the component or using the container, closure, or labeling.
3. All records required under this part, or copies of such records, shall be readily available for authorized inspection during the retention period at the establishment where the activities described in such records occurred. These records or copies thereof shall be subject to photocopying or other means of reproduction as part of such inspection. Records that can be immediately retrieved from another location by computer or other electronic means shall be considered as meeting the requirements of this paragraph.
4. Records required under this part may be retained either as original records or as true copies such as photocopies, microfilm, microfiche, or other accurate reproductions of the originaI records. Where reduction techniques, such as microfilming, are used, suitable reader and photocopying equipment shall be readily available.
5. Written records required by this part shall be maintained so that data therein can be used for evaluating, at least annually, the quality standards of each drug product to determine the need for changes in drug product specifications or manufacturing or control procedures. Written procedures shall be established and followed for such evaluations and shall include provisions for:
    (a) A review of a representative number of batches, whether approved or rejected, and, where applicable, records associated with the batch.

(b) A review of complaints, recalls, returned or salvaged drug products, and investigations conducted under Sec. 211.192 for each drug product.

6. Procedures shall be established to ensure that the responsible officials of the firm, if they are not personally involved in or immediately aware of such actions, are notified in writing.

B. 21 CRF 211.182: Equipment Cleaning and Use Log

A written record of major equipment cleaning, maintenance (except routine maintenance such as lubrication and adjustments), and use shall be included in individual equipment logs that show the date, time, product, and lot number of each batch processed. If equipment is dedicated to manufacture of one product, then individual equipment logs are not required, provided that lots or batches of such product follow in numerical order and are manufactured in numerical sequence.

In cases where dedicated equipment is employed, the records of cleaning, maintenance, and use shall be part of the batch record. The persons performing and double-checking the cleaning and maintenance shall date and sign or initial the log indicating that the work was performed. Entries in the log shall be in chronological order.

C. 21 CFR 211.186: Master Production and Control Records

To ensure uniformity from batch to batch, master production and control records for each drug product, including each batch size thereof, shall be prepared, dated, and signed (full signature" handwritten) by one person and independently checked, dated, and signed by a second person. The preparation of master production and control records shall be described in a written procedure and such written procedure shall be followed. The Master production and control records shall include as a minimum those items listed in Table 7.1.

TABLE 7.1 Master Production and Control Records Information

(1) The name and strength of the product and a description of the dosage form
(2) The name and weight or measure of each active ingredient per dosage unit or per unit weight or measure of the drug product, and a statement of the total weight or measure of any dosage unit
(3) A complete list of components designated by names or codes sufficiently specific to indicate any special quality characteristic
(4) An accurate statement of the weight or measure of each component, using the same weight system (metric, avoirdupois, or apothecary) for each component. Reasonable variations may be permitted, however, in the amount of components necessary for the preparation in the dosage form, provided they are justified in the master production and control records
(5) A statement concerning any calculated excess of component
(6) A statement of theoretical weight or measure at appropriate phases of processing
(7) A statement of theoretical yield, including the maximum and minimum percentages of theoretical yield beyond which investigation according to 21 CFR 211.192 is required
(8) A description of the drug product containers, closures, and packaging materials, including a specimen or copy of each label and all other labeling signed and dated by the person or persons responsible for approval of such labeling
(9) Complete manufacturing and control instructions, sampling and testing procedures, specifications, special notations, and precautions to be followed

D. 21 CFR 211.188: Batch Production and Control Records

Batch production and control records shall be prepared for each batch of drug product produced and shall include complete information relating to the production and control of each batch. These records shall include: (a) an accurate reproduction of the appropriate master production or control

TABLE 7.2 Batch Production and Control Records Information

(1) Dates
(2) Identity of individual major equipment and lines used
(3) Specific identification of each batch of component or in-process material used
(4) Weights and measures of components used in the course of processing
(5) In-process and laboratory control results
(6) Inspection of the packaging and labeling area before and after use
(7) A statement of the actual yield and a statement of the percentage of theoretical yield at appropriate phases of processing
(8) Complete labeling control records, including specimens or copies of all labeling used
(9) Description of drug product containers and closures
(10) Any sampling performed
(11) Identification of the persons performing and directly supervising or checking each significant step in the operation
(12) Any investigation made according to 21 CFR 211.192
(13) Results of examinations made in accordance with 21 CFR 211.134

record, checked for accuracy, dated, and signed; and (b) documentation that each significant step in the manufacture, processing, packing, or holding of the batch was accomplished, including the 13 items listed in Table 7.2.

GENERAL GUIDELINES OF LOG BOOKS AND BATCH RECORDS

The above CGMP guidelines are provided as general rules. Every company or manufacturer will have its own log book/batch record procedure (chemical hygiene plan, SOPs, SOCLs, etc.) and format. The log book is the diary and record of the process operator. All data are recorded directly into the process log book and/or batch records depending on your company's SOPs. Data should not be recorded on stray sheets of paper, bench tops, or paper towels. It is imperative that the log book/batch sheet records be above question. Many federal compliance cases and patent disputes have been won or lost because of the credibility of the data contained in a process log book and batch records.

All data should be immediately entered in the log book/batch sheets. All raw data should be recorded neatly and directly in the log book/batch sheets. Mistakes should be "lined through" and remain legible. Mistakes should never be erased or totally obscured. (See Fig. 7.1 for the correct way to indicate a mistake in a log book.)

```
Incorrect   (original data must be always legible)
                12.3̶6̶   6
Correct     (correction should be legible, legitimized, and authorized)
                12.3̶4̶   6   transcription error
                            12/17/94 J.T.B.
```

FIGURE 7.1 Laboratory notebook corrections.

Information obtained from automatic recording devices, such as charts, should be noted and recorded in the log book with the appropriate reference indexed on the charts or recording paper for easy retrieval when needed. Charts and graphs can then be stored in separate, appropriate, secure locations.

Calculations of raw data which are carried out on other paper or on a calculator and the results must be recorded in the process control log book/batch records with the results. All calculations should be checked by either the person who performs the work or a competent coworker.

It should be understood that if raw data entries are transferred to a computer database, neither the electronically stored data nor its hard copies can be considered as legal substitutes for the original notebook data entries. If raw data are captured directly by a computer, for example from a chromatographic system connected directly to a computer, the process unit may elect to treat the electronic data or hard copy print out as raw data. If the magnetic media are treated as raw data, the company must retain the ability to display these data in a readable form for the entire period that the information is required by the appropriate agencies (this period could be as long as 30 years in some countries). In addition, there must be archives with appropriate conditions for orderly storage and retrieval of all raw data, documentation, protocols, and samples.

As already stated, data and calculated results (charging amounts, theoretical yield, percent yield, etc.) should be included on each batch and recorded in the process notebook. Do not forget to adhere to all rules dealing with significant figures (see Chapter 6, section on "Significant Figures") in your calculations.

Prudent manufacturing practice would suggest that all pertinent physical, chemical, and safety information also be included in the log book or made readily available. In most cases, this information should be gathered before the procedure is started.

INTERNATIONAL ORGANIZATION FOR STANDARDIZATION (ISO)

The **International Organization for Standardization (ISO),** based in Geneva, Switzerland, is a non-governmental organization established in 1947. The mission of this organization is to promote standardization and related activities for international commerce. More than 90 countries including all of Europe and the Americas are involved in this quality assurance standards program. The American National Standard Institute (ANSI) and the **American Society for Quality (ASQ)** have adopted these European guidelines for use in this country.

ISO 9000 is officially entitled "Quality Management and Quality Assurance Standards: Guidelines for Selection and Use." This standard sets forth the principal concepts and describes general guidelines, especially in purchaser-supplier contracts. ISO 9000 is a set of five individual, international standards for quality management and quality assurance (see Table 7.3), which cover all aspects of a company's operations from research and production to sales and services. ISO 9000 is the overall or basic standard.

TABLE 7.3 ISO 9000 Standards for Quality Management and Quality Assurance

ISO 9001 This standard provides a model for quality assurance in the design, production, and supply of products or services. It is most comprehensive of all the standards, and includes all of the requirements found in ISO 9002 and ISO 9003.
ISO 9002 A model for quality assurance in production and installation, however not research and development.
ISO 9003 A model for quality assurance when only final inspection and testing are required.
ISO 9004 A model dealing with guidelines for developing quality management.

FIGURE 7.2 ISO 9000 certification seal.

The **Registrar Accreditation Board (RAB)** is an affiliate of the American Society for Quality Control that accredits registrars for assessing, auditing, and certifying quality programs in the United States. These registrars will grant a Certificate of Compliance to any applicant organization whose fully evaluated system complies with the requirements of the selected ISO 9000 standard. Adoption of the ISO 9000 system requires that a manufacturer have documentation for managing an effective quality system. Documentation should be legible, readily identifiable, easily accessible, and include revision dates. All out-of-date documentation should be removed and disposed of according to set procedures. Table 7.4 shows some of the more commonly found deficiencies during ISO GLP audits.

TABLE 7.4 Common Deficiencies at ISO 9000 GLP Audit (courtesy Hewlett Packard Company)

Documents
1. Unapproved
2. Procedures do not correspond to practice
3. Documents not located according to distribution record
4. Approved suppliers list incomplete (ISO 9000 only)
5. Product release protocol faulty

Equipment
1. New/alternate equipment not qualified
2. Calibration incomplete
3. Performance measurement inadequate
4. Inadequate response to *Exceptions*

Material
1. Inadequate or *no* identification
2. Material from unapproved sources
3. Material poor in condition
4. Inadequate shelf life control
5. Inspection status not clear
6. Non-conforming material not controlled

People
1. Not trained
2. Unaware of requirements
3. Non-conforming with documented requirements
4. Training program/records inadequate
5. Organization/responsibilities non-conforming

Administration
1. Methodology incomplete or inconsistent
2. Non-conformance not dealt with (not identified, not documented, and/or not communicated)
3. Corrective action not identified or specified, action slow or not documented
4. Management reviews ineffective or infrequent
5. Deviation from documented procedures
6. Records incomplete or disorganized

STATISTICS

Since statistics are now used routinely in process quality control programs, a short introduction to statistics is presented in this chapter. Statistics are an absolute necessity in all scientific measurements for determining the accuracy and precision of data. Many new and specific mathematical terms are encountered in the study of statistics. For example, in the laboratory **accuracy** means the "correctness" of a given analysis, while **precision** indicates the "reproducibility" of an analytical procedure. The National Bureau of Standards described the difference between accuracy and precision: "accuracy" describes with closeness [of data] to the truth; "precision" describes the closeness of readings to *one another.*

An archery target (Fig. 7.3) provides a good analogy of these two often confused terms. If the arrows are all located directly in the bull's-eye, then both accuracy and precision are demonstrated. However, if all the arrows are clustered in the lower left quadrant and not near the bull's-eye, one would have good precision but poor accuracy.

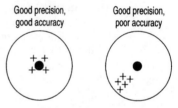

FIGURE 7.3 Difference between precision and accuracy.

A major source of error in process results can be the sampling procedure itself. The question to be asked is: does the process sample truly represent the material being tested? Many professional societies and governmental agencies (such as the Association of Official Analytical Chemists, American Society for Testing Materials and Environmental Protection Agency) have very specific instructions on obtaining, preparing, and storing representative samples. Consult Chapter 11, "Sampling," for information on proper sampling techniques and obtaining representative samples. In addition, Chapter 6, "Mathematics Review and Conversion Tables," discusses some of the mathematical concepts needed in statistical calculations. A potential source of error has been introduced with the advent of the electronic calculator which involves significant figures and statistical misuse.

STATISTICAL TERMINOLOGY

Some of the more frequently encountered terms in process statistics are defined as follows:

Absolute error is the difference between the true value and the measured value, with the algebraic sign indicating whether the measured value is above (+) or below (−) the true value.

Relative error is the absolute error (difference between the true and measured value) divided by the true value; usually expressed as a percentage.

Indeterminate errors are random errors that result from uncontrolled variables in an experiment and cannot normally be determined because they do not have a single source.

Determinate errors are those errors that can be ascribed to a particular cause and thus can usually be determined as being personal, instrumental, or method uncertainties.

Mean (m) describes the technique of "taking an average" by adding together the numerical values (x, y, z, etc.) of an analysis and dividing this sum by the number n of measurements used to obtain the mean. For example:

$$m = \frac{x + y + z}{n}$$

Median is the same data used for calculating the mean and can be displayed in an increasing or decreasing series and the "middle" value simply selected as the median. The advantage of median over the mean is that the median will always be one of the actual measurements. Of course, if the total number of measurements is an even number, there will not be a single middle value; thus the median will be the average of the two middle values.

Mode is the measurement value that appears most frequently in the series.

Deviation can be defined (as the name might imply) as how much each measured value differs from the mean. Mathematically, the deviation is calculated using the following equation, where d_x is the deviation, m represents the mean, and x stands for the measured value. The vertical bars ($||$) signify "absolute value"; thus the algebraic sign will always be positive:

$$d_R = |m - x|$$

Relative deviation (d_R) relates the deviation to the mean to indicate the magnitude of the variance. If the mean is a rather large number, then the deviation is not as critical as it would be in the case of a smaller mean. Mathematically, the relative deviation d_R can be calculated by dividing the deviation d by the mean m according to the following equation:

$$d_R = \frac{d}{m}$$

The d_R can be multiplied by 100 to express the percent relative deviation, or multiplied by 1000 to yield parts per thousand (ppt). These terms are used frequently in quantitative analysis to express precision of measurement.

$$\% \, d_R = \frac{dR}{m} \times 100$$

$$ppt \, d_R = \frac{dR}{m} \times 100$$

Average deviation (d_A) indicates the precision of all the measurements and is calculated by dividing the sum of all the individual deviations (dx, dy, dz, etc.) by the number n of deviations calculated.

$$d_A = \frac{dx + dy + dz}{n}$$

Standard deviation ($d_{s \text{ or } \sigma}$) is the most used of the deviation averaging techniques because it indicates *confidence limits*, or the *confidence interval* for analyzing all data. The standard deviation can be calculated in five steps:

1. Determine the mean m.
2. Subtract the mean from each measured data item.
3. Square each difference.
4. Find the average of the squared terms in Step 3.
5. Calculate the square root of the average found in Step 4 by dividing by one less than the actual number of measurements.

In other words, the standard deviation is calculated by taking the square root of the quotient of the sum of all the squared individual deviations divided by one less than the number of measurements ($n - 1$) used in the analysis. Statistically it has been determined that as the number of measurements n exceeds 30, the $n - 1$ term can be simplified to n.

$$d_{s \text{ or } \sigma} = \sqrt{\frac{dx + dy + dz}{n - 1}}$$

EXAMPLE 7.1 Calculate the standard deviation in these mass determinations: 36.78 mg, 36.80 mg, 36.87 mg, and 36.94 mg.

$$\text{Mean } m = \frac{36.78 + 36.80 + 36.87 + 36.94}{4} = 36.85$$

Measurement	Deviation $(m - x)$	Deviation squared $(m - x)^2$
36.78	0.07	0.0049
36.80	0.05	0.0025
36.87	0.02	0.0004
36.94	0.09	0.0081

$$\text{Standard deviation, } d_{s \text{ or } \sigma} = \sqrt{\frac{0.0159}{4-1}} = 0.07$$

Thus the answer should be reported as 36.85 ± 0.07 mg.

Relative standard deviation can be calculated by dividing the standard deviation by the mean. Like relative deviations, this ratio can be multiplied by 100 or 1000 to obtain the relative percent or parts per thousand deviation.

Confidence limit is the interval around an experimental mean within which the true result can be expected to lie with a stated probability. Given a large enough number of measurements, a symmetrically shaped distribution curve (**Gaussian**) can be drawn by plotting the number of times a specific numerical measurement is found versus the numerical value of the measurements. Most measurements should be clustered around the mean as shown in Fig. 7.4. In this particular distribution, the frequency of observations is plotted on the ordinate (y-axis) and the standardized variable Z, which is calculated by subtracting the mean value from the value of each result and dividing this difference by the standard deviation [$Z = (x - m)/d_s$] is plotted on the abscissa (x-axis). It can be shown that 68.3 percent of the area beneath this curve lies within one standard deviation d_s (also referred to as σ) either way from the mean. Approximately 95.5 percent of all values will be within ± 2 d_s and 99.3 percent within ± 3 d_s. These intervals are known as **confidence intervals**. Typically, laboratory data is "rejected" or "retained" in the final laboratory report based on these predetermined confidence limits.

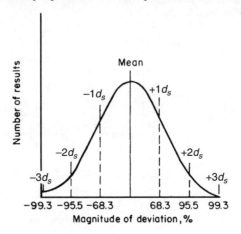

FIGURE 7.4 Gaussian distribution curve with confidence limits.

METHOD OF LEAST SQUARES

A technician or operator is frequently required to plot data on a graph. However, in many instances, the data does not fall on a straight line. One solution to this dilemma is simply to eyeball the best straight line using a ruler and as many data points as possible. A better approach involves using statistics to define the best straight line fit to the data. A straight-line relationship should be $y = mx + b$, where y represents the dependent variables, x represents the independent variable, m is the **slope** of the curve, and b represents the ordinate (y-axis) intercept. Mathematically, it has been determined that the best straight line through a series of data points is that line for which the sum of the squares of the deviations of the data points from the line is a minimum. This is known as the ***method of least squares***. The actual calculations involve the use of differential calculus, which is beyond the scope of this reference; however, most scientific calculators are capable of solving this type of calculation with minimum input from a technician. Normally, the calculator has a statistics function called **linear regression**. Since the actual procedure varies for different models, consult the manual for your scientific calculator for the specific details.

Statistics are not only useful for plotting graphs and rejecting unreliable laboratory or plant data, they are also routinely used in quality control programs. Control samples are routinely analyzed along with all other laboratory samples. These standards, or control samples, have usually been analyzed many times or have been purchased as a standard from a commercial source. The laboratory management establishes a confidence level (for example, three standard deviations) within which these control samples must fall, and a daily plot is maintained of the analytical results. With the central line normally representing the known concentration of the control, these quality control charts will indicate any sudden or even gradual trend from which the analytical results deviate. These control charts are also used in inter-laboratory or inter-plant comparisons and auditing. A typical quality control chart is shown in Fig. 7.5. Computer software programs are now commercially available that are specifically designed to generate analytical control charts for quality control programs. This software gives the technician the ability to generate instant statistics on any analytical data. For example, the number of samples or records, means, standard deviations, coefficients of variation, minima, and maxima are instantly displayed. These programs offer both accuracy and precision graphs using both fixed intervals and Shewhart X-bars. This computer software can plot reported value data, percent recovery data, and absolute differences. Control charting protocols specified by the American Society for Testing and Materials (ASTM), the Environmental Protection Agency (EPA), and Standard Methods are normally included in these software packages.

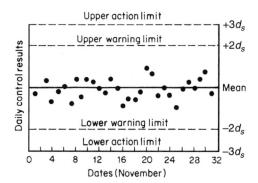

FIGURE 7.5 Quality control charting.

In quality control statistics, a **histogram** (frequency plot) is a chart that plots individual values versus the frequency of occurence. These charts, once created manually by quality control specialists, can be stored and displayed on a CRT instantaneously by almost any process employee. A **pareto chart** is another form of bar graph with the vertical axis usually the number of occurrences and the horizontal axis used to displace the least to most frequent. **Run charts** are yet another form of histogram with a process variable (for example, product thickness or strength) being plotted on the vertical axis and time plotted on the horizontal axis.

STATISTICAL PROCESS CONTROL

Statistical Process Control (SPC) packages are commercially available to assist engineering and process personnel in the daily operations to continuously monitor and refine production processes. The company's SOPs establish the desired "target," "mean," or "centerline" value for process personnel to evaluate process performance. SPC typically provides a graphical display of selected measurable quality variables with time and targets values. This is the type of process documentation that is required for ISO certification.

Many process plants use the **Western Electric Rules** to detect patterns of processes outside of normally acceptable statistical process control. The Western Electric rules have 11 general guidelines as shown in Table 7.5. Each analysis is tested against the Western Electric rules or company's SOPs to determine an out-of-normal conditions or results. The *centerline* refers to the *mean* or desired process result established by the company.

TABLE 7.5 Western Electric Rules for Statistical Process Control

Rule 1	One point 3 σ outside of centerline
Rule 2	Three consecutive points jumping 3 σ
Rule 3	Two of three points 2 σ above or below centerline
Rule 4	Four of five points 1 σ above or below centerline
Rule 5	Eight consecutive points above centerline
Rule 6	Eight consecutive points below centerline
Rule 7	Five consecutive points increasing in value
Rule 8	Five consecutive points decreasing in value
Rule 9	Fifteen consecutive points within 1 σ of centerline
Rule 10	Eight consecutive points outside 1 σ of centerline
Rule 11	After a 3 σ change, three cunsecutive points within ± 0.75 σ of the change point

A SPC chart is used to plot quality parameter points (pH, viscosity, concentrations, etc.) on one axis and time on the other axis. This type of chart (Fig. 7.6) is excellent for visualizing trends with time and provides the operator a means for monitoring processes. This continuous monitoring ensures that all systems are functioning properly and that the product will be within specifications. Some SPC guidelines use color coding to simplify the interpretation of these charts: for example, yellow code (1 σ or 1 standard deviations), orange (2 σ or 2 standard deviations), and red (3 σ or 3 standard deviations), and each level requires a different SOP-specified level of investigative or corrective action. Companies are becoming much more involved in quality control information, not only to improve production and quality (ISO) but to meet governmental regulations for being responsible for chemicals from "cradle to grave."

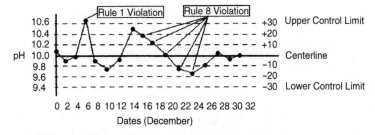

FIGURE 7.6 Graph of sample data (pH versus time) and the application of Western Electric rules to determine any out-of-normal conditions.

REFERENCES

1. "40 CFR Part 792: Toxic Substance Control Act (TSCA): Good Laboratory Practice Standards." Federal Register 54, Nr. 158, 34034–34052, U.S. Government Printing Office, Washington, DC (August 17, 1989).
2. "The OECD Principles of Good Laboratory Practice," Number 1. *Organization for Economic Cooperation and Development Series on Principles of Good Laboratory Practice and Compliance Monitoring.* Environmental Monograph No. 45, Paris (1992).
3. "General Requirements for the Competence of Calibration and Testing Laboratories," 3rd edition. ISO Guide 25, International Organization for Standardization, Case postale 56, CH-1211 Geneva 20, Switzerland (1990).
4. "USP Perspectives on Analytical Methods Validation." W. L. Paul, *Pharm. Technol.* 15(3):130–141 (1991).
5. "Good Laboratory Practice Part 2: Automating Method Validation and System Suitability Testing." Hewlett-Packard Product Notes, Pub No. 12-5091-3748E (1993).
6. "Current Good Manufacturing Practices (cGMP), Code of Federal Regulations, 21 CFR 160(b)(4)." U.S. Department of Health and Human Services, Washington, DC (April 1991).
7. "Current Good Manufacturing Practices (cGMP). 21 CFR 194(d)." U.S. Department of Health and Human Services, Washington, DC (April 1991).
8. "The EEC Guide to Good Manufacturing Practice for Medicinal Products." 111/2244/87-EN Rev. 3.0, European Economic Council (January 1993).
9. 1993 Annual Book of ASTM Standards, Vol. 14.01, ASTM Standard Practice E 685, pp. 731–735 and 1103–1187, American Society for Testing and Materials, Philadelphia, Pennsylvania (1993).
10. Youdan, W. J., and Steiner, E. H., *Statistical Manual of the AOAC,* American Association of Official Analytical Chemists, Washington, DC (1975).
11. *Industrial Hygiene Service Laboratory Manual,* Technical Report No. 78 Department of Health, Education and Welfare, PHS, CDC, Cincinnati and Washington, DC.
12. "Statistics in Quantitative Analysis" and "Sampling," H. A. Laitinen and W. E. Harris, in *Chemical Analysis,* 2nd edition, Chaps. 26 and 27. New York, McGraw-Hill Book Co. (1975).
13. "Experimental Statistics." M. G. Natrella, NBS Handbook 91, pp. 3.1–3.42 National Institute of Standards and Technology, Gaithersburg, MD (1963).

CHAPTER 8
LABORATORY GLASSWARE*

INTRODUCTION

For a number of years manufacturers have been fabricating glass laboratory equipment with ground-glass joints having standard dimensions and designed to fit each other perfectly. This feature eliminates the use of rubber, plastic, and cork stoppers as connections between different pieces of equipment. Thus, by judicious use and choice of the multitude of items available, equipment for varied procedures can be assembled more quickly and with less effort.

Ground-glass assemblies (1) give perfect seals for vacuum and moderate pressures, (2) completely eliminate stopper contamination, (3) extend temperature limits upward for use of the equipment, and (4) provide neater and more professional assemblies.

The interjoint items are identified by their names and joint-size numbers. Size is designated by two figures. The first indicates the approximate diameter of the larger tube in millimeters. The second designates the length of the ground surface. A 19/38 joint is about 19 mm in diameter and about 38 mm long. (See Table 8.1.) Reducing and expanding adapters permit the unlimited choice of joint designs and sizes. Interjoint glassware is designated as male or female connections, straight joint, ball joint, or any combination desired. (See Table 8.2.)

Use the ground-joint glassware exactly as you would use regular glassware with flexible connections such as stoppers and tubing, except that precautions must be observed. (These precautions will be enumerated later in this section.)

TABLE 8.1

Joint no.	OD of tube, mm	Joint no.	OD of tube, mm
5/20	5	34/45	32
7/25	6	40/50	37
10/30	8	45/50	42
12/30	10	50/50	47
14/35	12	55/50	52
19/38	17	60/50	57
24/40	22	71/60	68
29/42	27	103/60	100

TABLE 8.2

Ball or socket no.	Ball or socket no.
7/1	28/15
12/1	35/20
12/1.5	35/25
12/2	40/25
12/3	50/30
12/5	65/40
18/7	75/50
18/9	102/75
28/12	

---

*Glassware drawings in this chapter courtesy of Ace Scientific Supply Company, Linden, NJ.

Thousands of variations are available to the technician. Refer directly to the scientific glassware catalog to pick the exact item having the desired type and size of connector so that a perfect assembly can be made as desired.

Some of the advantages of ground-joint glassware are as follows:

1. Because no cork, rubber, or plastic stoppers or connections are used, contamination and discoloration of the chemicals are avoided.
2. Corks do not have to be selected, bored, or fitted—a time-saver.
3. Units can be assembled and reassembled again and again to carry out different operations.
4. Assembly of units is quick.
5. Broken parts are quickly replaced with duplicates which are standard and fit perfectly.
6. No impurities from broken cork, swollen rubber, or plastic enter the reaction.
7. Narrow tubes need not be used, as with cork and rubber stoppers, allowing full-width tubing.

GROUND-GLASS EQUIPMENT

A. Joints and Clamps

See Figs. 8.1 to 8.8.

FIGURE 8.1 Ground-glass joint with drip tube sealed to smaller end of the ground zone.

FIGURE 8.2 Stainless steel clamps for ground-glass joints are merely screwed together to obtain closure.

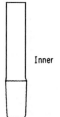

FIGURE 8.3 Full-length, male ground-glass joint.

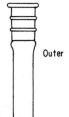

FIGURE 8.4 Full-length, female ground-glass joint.

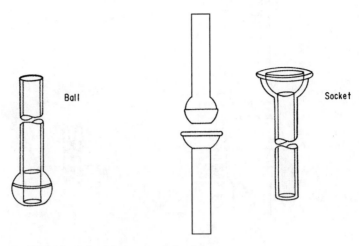

FIGURE 8.5 Ball-and-socket ground-glass joints.

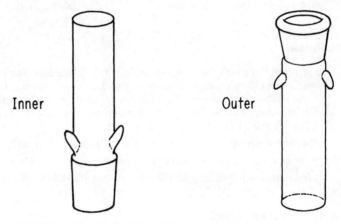

FIGURE 8.6 Ground-glass joints with hooks.

CARE OF GROUND-GLASS SURFACES

A. Lubrication

Although ground-glass joints usually seal well without the use of lubricants, it is generally advisable to lubricate ground-glass joints to prevent sticking and, therefore, to prevent breakage. Under some reaction conditions, it may be necessary not to use a lubricant, but under most conditions a lubricant should be used. Lubrication makes it easy to separate ground-joint ware and prevents leakage. Ground-glass joints must be kept clean and must be cleaned prior to lubrication. Dust, dirt, and particulate matter may score the surfaces and cause leakage.

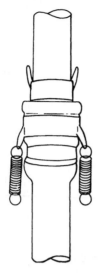

FIGURE 8.7 Tension hooks for holding ground-glass joints together securely (stainless steel tabs, hooks, and springs).

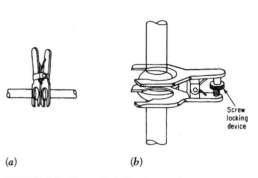

FIGURE 8.8 Clamps for ball-and-socket joints, (*a*) Spring-closed for smaller sizes, (*b*) Screw-locking device for larger sizes. The size number indicates the diameter, in millimeters, of the ball over which the clamp fits. (Size numbers: 7, 12, 12A, 18, 18A, 28, 35, 40, 50, 65, 75, 100.)

B. Choice of Lubricant

A lubricant must withstand high temperatures, high vacuum, and chemical reaction. Choice of a lubricant depends upon the conditions under which it is to operate. Lubricate *only* the upper part of the joint with a small amount of grease. *Always* clean ground-glass surfaces thoroughly. *Avoid contamination by contact with grease* by cleaning joint of flask containing the liquid *prior* to pouring.

The following lubricants may be used:

Silicone grease for high temperatures and high vacuum. It is easily soluble in chlorinated solvents.

Glycerin for long-term reflux or extraction. It is a water-soluble, organic-insoluble greasing agent.

Hydrocarbon grease for general laboratory use. It is soluble in most organic laboratory solvents.

C. Lubricating Ground-Glass Joints

When lubricating ground-glass joints, use the following guidelines:

1. Lubricate only the upper part of the inner joint.
2. Avoid greasing any part of the joint which may come in contact with vapor or liquid and cause contamination.
3. The choice of the lubricant depends upon the materials used in the glassware, and the effect that the conditions of the reaction and the materials have on the lubricant.
4. A properly lubricated joint appears clear, without striations.
5. Lubrication is required when the joints must be airtight and when the glassware contains strong alkaline solutions.

D. Components

See Figs. 8.9 to 8.19.

LABORATORY GLASSWARE 153

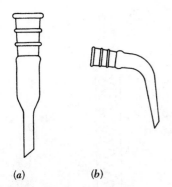

FIGURE 8.9 Drip adapters. (*a*) Straight. (*b*) Angled.

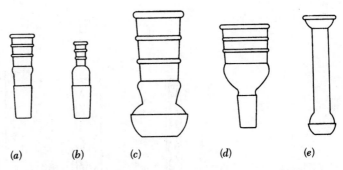

FIGURE 8.10 Typical adapters. (*a*) Straight. (*b*) Reducing. (*c*) Mixed—straight to ball. (*d*) Enlarging. (*e*) Ball-and-socket.

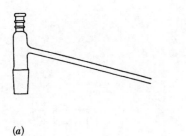

FIGURE 8.11 Adapters with simple distilling heads.

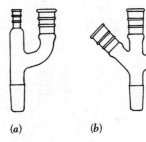

FIGURE 8.12 Parallel adapters: (*a*) three-way, (*b*) four-way.

FIGURE 8.13 (*a*) Gas-inlet-tube adapter. (*b*) Angled vacuum-connection adapter. (*c*) Straight vacuum-connection adapter.

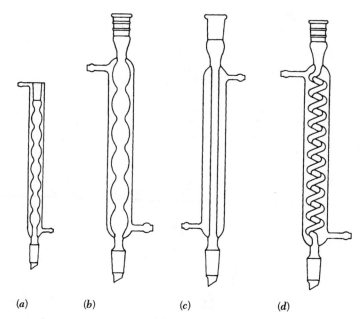

FIGURE 8.14 Various designs of condensers.

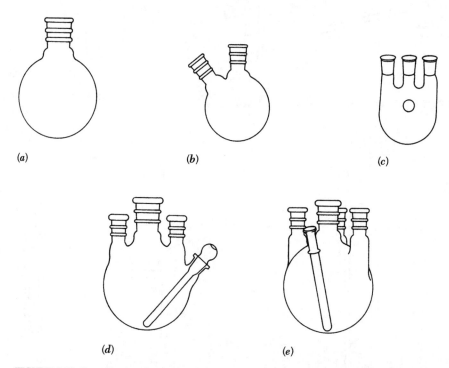

FIGURE 8.15 Round-bottomed reaction flasks fitted with different openings: (*a*) standard single-neck, (*b*) angled two-neck, (*c*) parallel three-neck, (*d*) three-neck and thermometer well, (*e*) four-neck and thermometer well.

LABORATORY GLASSWARE   155

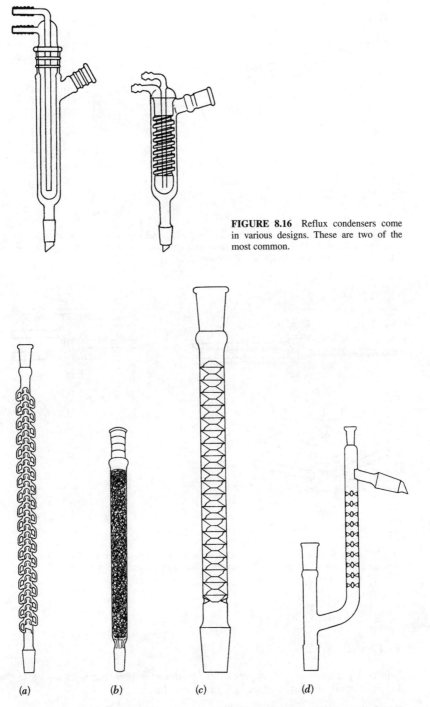

FIGURE 8.16 Reflux condensers come in various designs. These are two of the most common.

FIGURE 8.17 Fractional-distilling columns with atmospheric or vacuum istillation: (*a*) bubble cap, (*b*) stainless steel wire sponge, (*c*) Stedman screen packing, (*d*) Vigreux or Claisen.

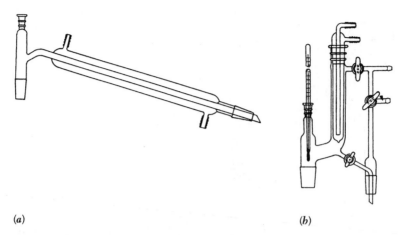

FIGURE 8.18 Distilling heads: (*a*) atmospheric Liebig type; (*b*) vacuum or atmospheric type with stopcock manifold allowing receiver distillate contents to be removed without breaking the vacuum. The reflux ratio is adjusted with the stopcock.

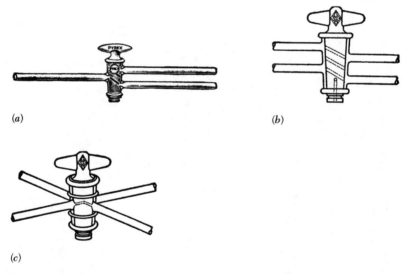

FIGURE 8.19 Stopcocks: (*a*) three-way, allows fluids to be channeled as desired or to cut off flow completely; (*b*) four-way oblique-bore with vent to bottom of plug; (*c*) four-way V-bore.

E. Assemblies

See Figs. 8.20 to 8.23.

Lubricating Stopcocks. A stopcock must be properly lubricated with a suitable stopcock grease. Too much lubricant may plug the bore or the tip of a buret.

CAUTIONS: Use recommended stopcock greases. Silicone greases should not be used. Ground surfaces *must be clean.*

LABORATORY GLASSWARE 157

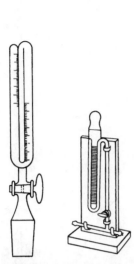

FIGURE 8.20 Manometer-type vacuum gauges used to indicate pressure in a closed system.

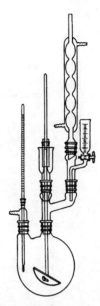

FIGURE 8.21 Reflux reaction apparatus with mixer, gas-inlet tube, thermometer, and addition funnel.

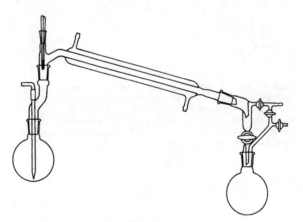

FIGURE 8.22 General distillation apparatus for use with vacuum or atmospheric pressure. Distillates are separable into fractions without interrupting vacuum distillation.

Procedure

1. Spread two circular bands of grease around the stopcock (Fig. 8.24).
2. Insert stopcock in buret and twist several times. The grease will spread out and the joint will be completely transparent.

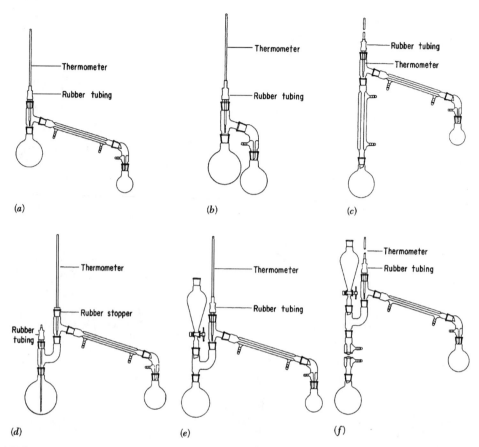

FIGURE 8.23 How-to suggestions for using ground-glass-joint assemblies: (*a*) Simple distillation, atmospheric pressure, or vacuum, (*b*) Distillation of viscous or high-boiling liquids or solids, (*c*) Fractional distillation, vacuum or atmospheric pressure (column should be packed with stainless steel sponge), (*d*) Steam distillation, or vacuum distillation with gas capillary bubbler, (*e*) Concentration of a solvent or extract, vacuum or atmospheric pressure. (*f*) Concentration of a solvent or extract, with fractional distillation, vacuum, or atmospheric pressure (column should be packed with stainless steel sponge).

STORAGE OF GLASSWARE

A. In Drawers

Glassware may be stored in sliding drawers. To prevent glass breakage, take heed of the following:

1. If many glass items of various shapes and forms are crowded in sliding drawers, breakage may occur.
2. If the drawer is too shallow for the items, breakage can occur when the drawer is closed.
3. If the drawer sticks, a sudden jerk can cause the glass items to contact each other violently.
4. Round flasks, bulbs, etc., can roll in the drawer. They should be cushioned and secured with padding.
5. Care should be exercised when removing or placing glass items in drawers.

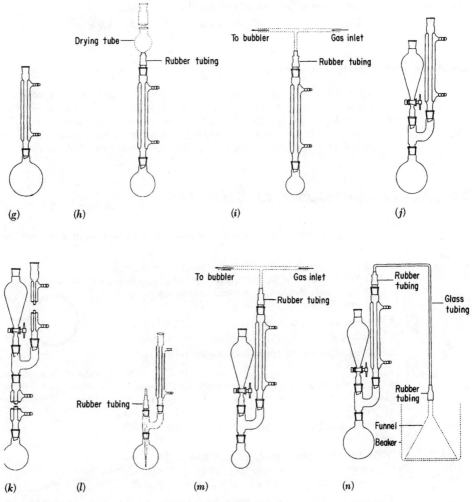

FIGURE 8.23 (*continued*) (*g*) Simple reflux, (*h*) Simple reflux, dry atmosphere, (*i*) Reflux, inert atmosphere. (*j*) Addition and reflux, high-boiling liquids, (*k*) Addition and reflux for low-boiling liquids. (*l*) Gas inlet and reflux, (*m*) Addition and reflux, inert atmosphere, (*n*) Addition and reflux with trapping of evolved gases. Safety suggestions: (1) Remember to recheck setups, especially when using a closed system. (2) Provide adequate ventilation, particularly when working with solvents or materials which could produce toxic vapors. (3) By and large, materials being used are flammable and should be handled with caution. (4) In handling glassware, follow the safety tips given. (5) Learn the location of safety equipment and be familiar with its use in case of emergency.

B. On Shelves and in Cabinets

When storing glassware on shelves and in cabinets:

1. There should be sufficient room.
2. No parts of the item should extend over the edge.
3. Items should be placed so that they cannot roll off or roll into other items.

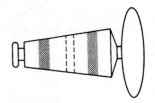

FIGURE 8.24 Greasing a stopcock.

4. Round-bottomed flasks should be seated on cork rings.

   5. Heavy, bulky, or cumbersome items should be stored at low levels.

C. Preparation for Long-Term Storage

When lab glassware is not to be used for a long period of time, take apart buret stopcocks, ground-glass joints, and flask stoppers to prevent sticking. Remove grease from joints. Loosen Teflon® stoppers and stopcocks slightly to prolong the life of the sealant material.

For easy storage and reuse, put a strip of thin paper between the ground-glass surfaces. Otherwise the ground joints may stick together and may be extremely difficult to separate.

ASSEMBLY OF GROUND-JOINT GLASSWARE

   1. Plan your assembly so that the working area will be uncluttered, with easy access to all components.

   2. Use as few clamps as possible to support the apparatus firmly.
   - **(a)** The precision of the ground-glass joints allows little room for misalignment.
   - **(b)** The joints themselves provide mechanical support and rigidity.

   3. Support all flasks with rings (Fig. 8.25) for stability, even though clamps may support the neck of the flask.

   4. *Always* assemble the apparatus from the bottom up.

FIGURE 8.25 Support ring; use with suitable clamp holders to support round-bottomed vessels, funnels, and other apparatus on support stands or frames. Available in variety of sizes and rod lengths.

NOTE: Assemble glassware so that any liquid flow always passes through the inner (male) joint. It should not flow into the joint (Fig. 8.26). This precaution keeps the joint surfaces free from liquid and prevents possible contamination from the lubricant.

   - **(a)** Fasten all clamps loosely at first, except the bottom clamp. *Use correct clamps of the proper size.* (See Figs. 8.27 and 8.28.)
   - **(b)** Gradually tighten the clamps as the apparatus goes into complete assembly, to accommodate the apparatus.
   - **(c)** *Be sure alignment is correct,* then finally tighten all clamps.

CAUTION: Always use stable ring stands, which *do not wobble,* or use a rigid frame network for support. (See Fig. 8.29.)

   - **(d)** Do not position your apparatus at an angle. Professional laboratory practices demand that vertically positioned items actually be exactly vertical.
   - **(e)** *Never force alignment* of ground-glass apparatus. This causes breakage, leakage, and improper function of the apparatus.

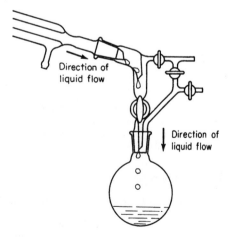

FIGURE 8.26 Direction of liquid flow in assembly with joints fitted with drip tips.

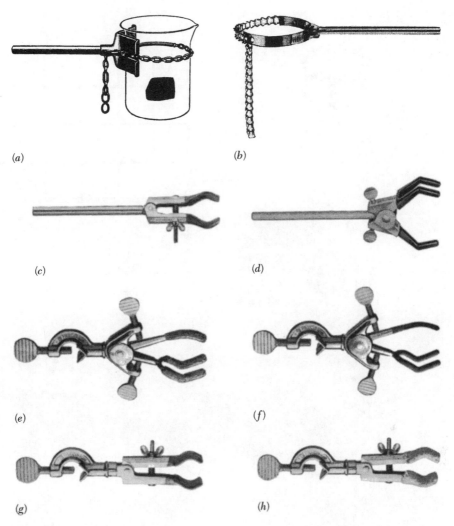

FIGURE 8.27 Clamps, (*a*) Beaker clamp, chain adjustable size, with spring tension, (*b*) Screw-type clamp with adjustable tension, (*c*) Utility clamp with long handle, (*d*) Trigrip, double-jaw, vinyl-covered clamp with long handle to hold equipment, (*e*) Double-jaw, three-prong, asbestos-covered clamp. (*f*) Double-jaw, three-prong, vinyl-covered clamp, (*g*) Fixed-position clamp with vinyl-covered jaws, (*h*) Fixed-position utility clamp with asbestos-covered jaws.

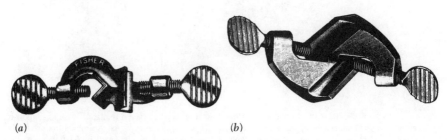

FIGURE 8.28 Clamp holders anchor rods securely to rods, support assemblies, and hold clamps to support assemblies. Available in parallel (*a*) or right-angle (*b*) direction.

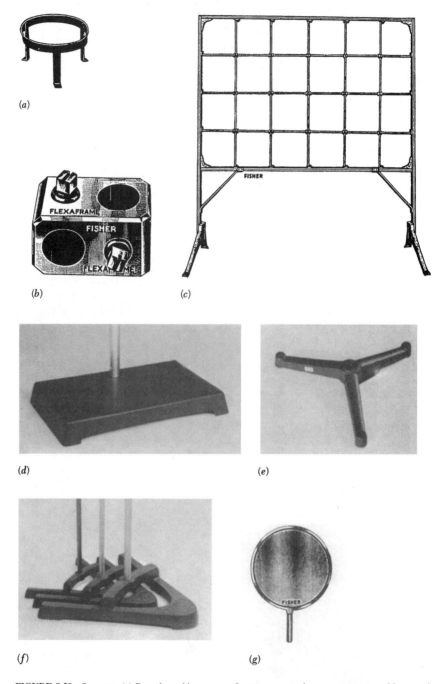

FIGURE 8.29 Supports. (*a*) Round, working-area-surface rest support base to support round-bottomed containers; available in a range of sizes. (*b*) Flexible frame can be assembled from rods of various lengths to provide the support requirements for the apparatus. Provides steady base without use of multiple support stands. (*c*) Right-angled rod clamps. (*d*) Rectangular single rod support base. Fairly sturdy on level areas. (*e, f*) Tripods offer sturdy support for apparatus; may have two vertical rods. (*g*) Support plate clamped to support base or support frame with suitable clamp holder is used to hold flat-bottomed containers.

A. Safety

1. *Always recheck your systems,* especially when you are working under reduced pressure or with extremely hazardous materials.
2. *Always check your ventilation system,* especially when you are working with toxic gases. *Use hoods* if necessary.
3. *Never use broken or cracked glassware.* You may lose your material or you may be injured by contact with the material.

B. Cutting Glass Tubing

When glass tubing is cut, sharp edges result. These can cause cuts and serious injury. Always use care when cutting glass tubing and when handling the cut pieces of tubing.

There are many sizes of tubing and glass rod (see Fig. 8.30 and Tables 8.3 and 8.4). Methods used for cutting them vary with their size.

Cutting Small-Diameter Tubing (25-mm OD or Less).

1. Scratch the tube or rod at the desired point with a three-cornered file or glass-cutter. Use only one or two strokes. Use considerable pressure. Do *not* saw (Fig. 8.31).
2. Wrap the tubing in a protective cloth to avoid cutting your hands. Place thumbs together opposite scratch.
3. Using little force, pull back on the tube and push thumbs outward quickly to break the glass. A straight, clean break should result.

C. Dulling Sharp Glass Edges

Sharp edges that result when glass tubing is cut can be dulled in the following ways (see Fig. 8.32).

1. Hold the glass article in the left hand and a piece of clean, new wire gauze (without asbestos) in the right hand.
2. Gently stroke the broken end with wire gauze while rotating the glass object.

End Finishing (Fire-Polishing). Fire-polishing the ends of tubing eliminates the sharp edges, prevents cuts, and enables the technician to insert the fire-polished tubing easily into corks, stoppers, and rubber or plastic tubing.

1. Smooth the sharp edges to prevent cuts.
2. Insert the end into the hot nonluminous portion of burner flame and rotate smoothly and evenly. (See Figs. 8.32 and 8.33.)

CAUTION: Wear gloves or keep hands well back from the end being heated. *It gets hot!*

D. Common Glassblowing Techniques

Making Points. Points are formed by pulling tubes to a smaller diameter (Fig. 8.35), and they can be used as handles for holding small pieces of tubing, and for closing them.

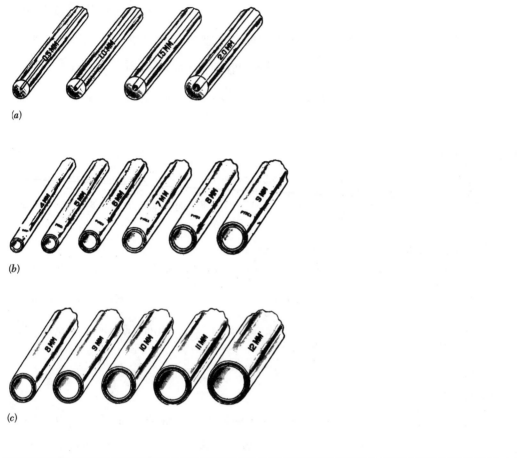

FIGURE 8.30 Glass tubing, (*a*) Capillary wall thickness and diameter of hole vary. Bore varies from ¼ to 3 mm diameter, OD from 5 to 10 mm, and combinations of bore to OD are available. (*b, c*) Tubing is soft, Pyrex, or Vycor® glass, is available from 2 to 178 mm OD. It normally comes in 4-ft lengths, (*d*) Glass-tubing sizer.

Procedure

1. Rotate the tube in the flame until it is pliable.
2. Remove it from the flame, while still rotating it, and slowly pull the ends about 8 in apart.
3. Melt the center of the smaller section until the ends are closed and sealed.

LABORATORY GLASSWARE

TABLE 8.3 Standard-Wall Glass Tubing

OD, mm	Wall, mm	OD, mm	Wall, mm
2	0.5	38	2.0
3	0.6	41	2.0
4	0.8	45	2.0
5	0.8	48	2.0
6	1.0	51	2.0
7	1.0	54	2.4
8	1.0	57	2.4
9	1.0	60	2.4
10	1.0	64	2.4
11	1.0	70	2.4
12	1.0	75	2.4
13	1.2	80	2.4
14	1.2	85	2.4
15	1.2	90	2.4
16	1.2	95	2.4
17	1.2	100	2.4
18	1.2	110	2.6
19	1.2	120	3.0
20	1.2	125	3.0
22	1.5	—	—
25	1.5	130	3.0
28	1.5	140	3.5
30	1.8	150	3.5
32	1.8	178	3.5
35	2.0		

TABLE 8.4 Standard Glass Rod

OD, mm
2
3
4
5
6
7
8
9
10
11
13
16
19
25
32
38

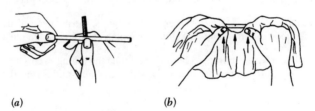

FIGURE 8.31 Proper method of (*a*) scratching, and (*b*) holding glass tubing for cutting.

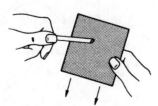

FIGURE 8.32 Dulling sharp edges with wire screen.

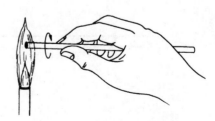

FIGURE 8.33 Fire-polishing (note rotation of the work).

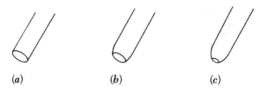

FIGURE 8.34 (*a*) Freshly cut tubing. (*b*) Tubing after polishing. (*c*) Tubing that has been heated too long.

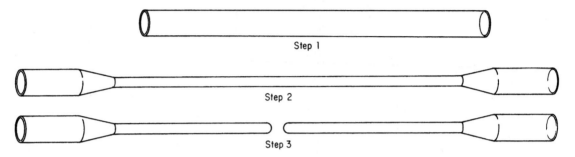

FIGURE 8.35 Making a "point."

Points are useful when it is necessary to work a larger piece of glass tubing which is too short to handle. The smaller tube or rod is fused to the end of the large tubing, and it can be centered to form an effective handle similar to a point.

Sealing the Ends of Capillary Tubes. The end of a capillary tube can be sealed by gently heating its tip in a small flame (see Fig. 8.36).

CAUTION: Use a small flame and heat only the tip until glass fuses.

Bending Glass Tubing. When bending glass, the diameter of the tubing and the distance between the tubes after the bend determine the length of tubing which must be heated. For small-bore tubing:

1. Use a wing-tip Bunsen burner and rotate the tubing evenly with both hands for uniform heating. Adjust the burner to give a nonluminous flame and a well-defined blue cone (see Fig. 8.37).

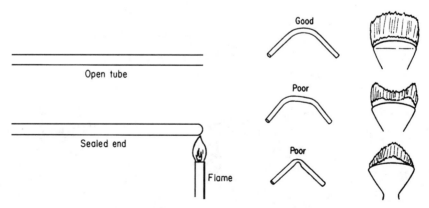

FIGURE 8.36 Sealing capillary tubes.

FIGURE 8.37 Effects of flame adjustment on bends.

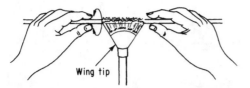

FIGURE 8.38   Work must be constantly rotated.

2. Hold the tube lengthwise in the flame and rotate it evenly with a back-and-forth motion until the glass becomes soft (Fig. 8.38).
3. After the glass is soft (it bends under its own weight), remove it from the flame.
4. Bend the glass to the desired shape, holding it in position until it hardens.

CAUTION: Wear gloves or keep hands well back from the end of the tube being heated. *It gets hot!*

HANDLING STOPPERS AND STOPCOCKS

A. Loosening Frozen Stoppers and Stopcocks

Laboratory glassware and reagent bottles have glass-to-glass connections which sometimes become frozen; To loosen stopcocks, stoppers, or any glass-to-glass connections that are frozen, you may use the following techniques.

CAUTION: Remember *glass is fragile.* It will break under severe thermal or mechanical shock. *Use care and handle gently. Do not apply too much force or subject to rough treatment.*

1. *Gentle tapping.* Gently tap the frozen stopcock or stopper with the *wooden* handle of a spatula. Tap so that the direction of the force will cause the stopper to come out. If you tap *too hard*, you will break the stopper. When tapping, hold the item in the left hand and gently tap with the wooden handle of a spatula held in the right hand.* Always work directly over a desk top covered with soft cushioning material to prevent falling stopper from breaking.
2. *Heating.* Try immersing the frozen connection in hot water. If that doesn't work, warm the stopcock housing or the stopper housing *gently* in the smoky flame of a gas-flame burner such as a Bunsen burner. Heat causes the housing to expand. Rotate constantly for even heating. When it is hot, gently tap with handle of spatula. Repeat several times, allowing the housing to expand and contract to break the frozen seal.
3. *Soaking.* Soak the frozen assembly in hot dilute glycerin and water or chlorinated solvent (CCl_4) and grease.
4. *Stopcock-removing clamp.* Affix this clamp (Fig. 8.39) to the frozen stopcock and housing and *gently* twist; the knurled nut applies pressure to separate the components.

CAUTION: Use care and use the stopcock-removing clamp with procedures 1 to 3 or combinations of these.

5. *Carbonated water method.* Frozen ground-glass joints are sometimes freed by immersing the frozen joint in a freshly opened container of carbonated water. The penetration of the liquid is evident by the appearance of the colored liquid between the joints. Allow to soak long enough

---

*Reverse if you are left-handed.

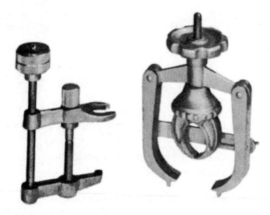

FIGURE 8.39 Clamps for removing frozen stopcocks and stoppers.

to allow fullest penetration. Then remove, rinse thoroughly with tap water, and dry—leaving the inner surface wet. Gently warm the outer joint over a smoky Bunsen flame while continually rotating the joint. After a full deposit of carbon has covered the outer joint, remove from the flame and attempt to twist the components apart (using asbestos gloves for protection against the hot glass). *Do not use force.* Repeat if necessary.

CAUTION: Never store alkaline reagents in ground-glass-stoppered flasks, bottles, or burettes. Ground-glass stoppers or stopcocks may stick.

REFERENCES

1. *The Laboratory Companion: A Practical Guide to Materials, Equipment, and Techniques.* Gary S. Coyne, ISBN 978-0-471-78086-1. Wiley Interscience, New Jersey (Nov. 2005).

CHAPTER 9
pH MEASUREMENT

ACIDS AND BASES

The simplest theory defines an acid as a substance that produces hydrogen ions (H⁺) when dissolved in water. Acids turn litmus paper red, whereas bases turn it blue. Acids react with active metals such as zinc and iron, dissolving the metals and producing hydrogen gas. Thus, according to **Arrhenius**, the properties of water solutions of acids are due to the hydrogen ion H⁺. The properties of acids in water solution are due to a hydrated hydrogen ion; that is, a hydrogen ion attached to at least one water molecule. The simplest species is the **hydronium** ion, H_3O^+:

$$HCl \rightarrow H^+(aq) + Cl^-(aq)$$

$$HNO_3 \rightarrow H^+(aq) + NO_3^-(aq)$$

The simplest acid/base theory defined a base as a substance that releases hydroxide ions (OH⁻) in water solution. Two common examples of bases are NaOH and $Mg(OH)_2$.

$$NaOH \rightarrow Na^+(aq) + OH^-(aq)$$

$$Mg(OH)_2 \rightarrow Mg^{2+}(aq) + 2OH^-(aq)$$

A. Strengths of Acids and Bases

Acids are classified as strong or weak depending on the number of H⁺ (or H_3O^+) ions they produce when dissolved in water. A **strong acid** ionizes almost 100 percent in aqueous solution; that is, nearly all the acid molecules break up into ions. Table 9.1 lists some common strong acids. Hydrochloric acid is a strong acid because HCl reacts almost completely with water to produce ions:

$$HCl + H_2O \rightarrow H_3O^+ + Cl^-$$

A **weak acid** ionizes only slightly in solution; most of the acid is present as molecules rather than ions. This can be indicated in an equation by using a double arrow:

$$HC_2H_3O_2 + H_2O \leftrightarrow H_3O^+ + C_2H_3O_2^-$$

In solutions of weak acids, generally less than 5 percent of the acid molecules exist in the form of ions. There are relatively few H⁺ (or H_3O^+) ions present in a solution containing a weak acid. Most acids are weak acids. Some common weak acids are listed in Table 9.2.

Just as there are strong and weak acids, there are strong and weak bases. Most common bases are ionic compounds of a metal from Group IA or IIA and the hydroxide ion. All of the hydroxides of the Group IA metals are strong bases. A strong base dissociates almost 100 percent into ions in water solution. A weak base, on the other hand, exists mainly in the form of molecules in solution. NaOH is a strong base and exists almost totally as sodium ions and hydroxide ions in water:

$$NaOH(s) \rightarrow Na^+(aq) + OH^-(aq)$$

TABLE 9.1	Some Common Strong Acids
Hydrobromic acid	HBr
Hydroiodic acid	HI
Hydrochloric acid	HCl
Nitric acid	HNO_3
Perchloric acid	$HClO_4$
Sulfuric acid	H_2SO_4

TABLE 9.2	Some Common Weak Acids
Acetic acid	$HC_2H_3O_2$
Carbonic acid	H_2CO_3
Phosphoric acid	H_3PO_4
Nitrous acid	HNO_2
Sulfurous acid	H_2SO_3
Hydrofluoric acid	HF

TABLE 9.3	Some Common Strong Bases
Strontium hydroxide	$Sr(OH)_2$
Barium hydroxide	$Ba(OH)_2$
Lithium hydroxide	LiOH
Sodium hydroxide	NaOH
Potassium hydroxide	KOH
Rubidium hydroxide	RbOH
Cesium hydroxide	CsOH

TABLE 9.4	Some Common Weak Bases
Ammonia	NH_3
Beryllium hydroxide	$Be(OH)_2$
Magnesium hydroxide	$Mg(OH)_2$
Calcium hydroxide	$Ca(OH)_2$
Zinc(II) hydroxide	$Zn(OH)_2$
Iron(III) hydroxide	$Fe(OH)_3$

Ammonia is a weak base, and, like weak acids, forms relatively few ions in water. Most of the ammonia is present as ammonia (NH_3) molecules in solution:

$$NH_3(aq) + H_2O(l) \rightarrow NH_4^+(aq) + OH^-(aq)$$

How strong is a strong acid? Is a sulfuric acid solution made by adding 10 mL of acid to 100 mL of water the same strength as one made by adding 10 mL of acid to a liter of water? The strength of acidic or basic solutions is often stated in terms of concentration of these solutions in moles of acid or base per liter of solution, or molarity (solutions and concentrations will be discussed in further detail later in this chapter). A 1 molar (1 M) solution of hydrochloric acid contains 1 mole of H^+ ions in each liter of solution. A 0.1 M HCl solution contains 0.1 mole of H^+ in each liter of solution. Experiments have shown that in pure water at 25°C, there are 1.0×10^{-7} moles of H^+ per liter; this is a hydrogen ion concentration of 1.0×10^{-7} M. In water, about 1 molecule in 500 million ionizes to hydrogen and hydroxide ions:

$$H_2O \rightarrow H^+ + OH^-$$

Pure water contains equal amounts of H^+ and OH^- and is neutral, that is, neither acidic nor basic. If a solution contains more H^+ than OH^-, it is acidic; if it contains more OH^- than H^+, it is basic.

B. The pH Scale

A simple mathematical notation for expressing the molar hydrogen ion concentration (H^+) that avoids exponential notation is the pH scale (see Fig. 9.1).

On this scale, a hydrogen ion concentration of 1×10^{-7} M becomes a pH of 7; a hydrogen ion concentration of 1×10^{-10} M becomes a pH of 10; and so on. By definition, pH is equal to the negative logarithm of the hydrogen ion concentration expressed in moles/liter:

$$pH = -\log [H^+]$$

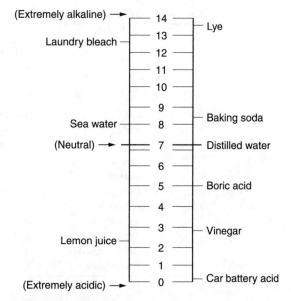

FIGURE 9.1 The pH scale.

EXAMPLE 9.1   To calculate the pH of solution that has a [H⁺] = 4 × 10⁻⁴ M, enter 4 × 10⁻⁵ on your hand calculator; press the "log" and then the "change-sign" key (or +/−) to get:

$$pH = -\log [4 \times 10^{-5}] = 4.4$$

For most acidic or basic solutions, the scale of pH values range from 0 to 14. If the pH = 7, the solution is neutral; if the pH is less than 7, the solution is acidic; if it is greater than 7, the solution is basic. Table 9.5 gives the relationship between acidic solutions, basic solutions, pH, and pOH. Mathematically, the sum of the pH and the pOH of an aqueous solution should be 14.

Note that pH decreases as a solution becomes more acidic. Table 9.6(a) gives approximate pH values for some common acid solutions and Table 9.6(b) gives approximate pH values for some common base solutions. The pH scale is set up mathematically so that each change of one pH unit corresponds to a tenfold change in the concentration of hydrogen ions in the solution. Thus, a liter of vinegar with a pH reading of 3 has ten times as many hydrogen ions as a liter of grape juice with a pH of 4. A liter of lemon juice with a pH of 2 has 100 times as many hydrogen ions as a liter of tomato juice that has a pH of 4.

pH THEORY

An **acid** is a substance that yields **hydrogen ions** when dissolved in water. **Bases** are substances that yield **hydroxyl ions** when dissolved in water. An acid that ionizes in dilute aqueous solution to produce many hydrogen ions (nearly complete ionization) is classified *as a strong* acid. An acid that ionizes slightly in water to produce few hydrogen ions is classified as a *weak* acid. A strong base ionizes in water nearly completely to produce many hydroxyl ions. A weak base ionizes slightly in water to produce few hydroxyl ions. The hydrogen ion is actually hydrated and can be represented

TABLE 9.5 Relationship Between Acidic Solutions, Basic Solutions, pH, and pOH

pH	H^+	H^+	OH^-	OH^-
0	10^0	1	0.00000000000001	10^{-14}
1	10^{-1}	0.1	0.0000000000001	10^{-13}
2	10^{-2}	0.01	0.000000000001	10^{-12}
3	10^{-3}	0.001	0.00000000001	10^{-11}
4	10^{-4}	0.0001	0.0000000001	10^{-10}
5	10^{-5}	0.00001	0.000000001	10^{-9}
6	10^{-6}	0.000001	0.00000001	10^{-8}
7	10^{-7}	0.0000001	0.0000001	10^{-7}
8	10^{-8}	0.00000001	0.000001	10^{-6}
9	10^{-9}	0.000000001	0.00001	10^{-5}
10	10^{-10}	0.0000000001	0.0001	10^{-4}
11	10^{-11}	0.00000000001	0.001	10^{-3}
12	10^{-12}	0.000000000001	0.01	10^{-2}
13	10^{-13}	0.0000000000001	0.1	10^{-1}
14	10^{-14}	0.00000000000001	1	10^0

TABLE 9.6(a) Approximate pH Values of Some Common Acid Solutions

Acids	Approximate pH
Hydrochloric acid, 1.0 M HCl	0.1
Hydrochloric acid, 0.1 M HCl	1.0
Sulfurous acid, 0.05 M H_2SO_3	1.5
Hydrochloric acid, 0.01 M HCl	2.0
Acetic acid, 1.0 M $HC_2H_3O_2$	2.4
Acetic acid, 0.1 M $HC_2H_3O_2$	2.9
Benzoic acid, 0.01 M C_6H_5COOH	3.1
Acetic acid, 0.01 M $HC_2H_3O_2$	3.4
Carbonic acid, saturated H_2CO_3	3.8
Boric acid, 0.033 M H_3BO_3	5.2

TABLE 9.6(b) Approximate pH Values of Some Common Base Solutions

Bases	Approximate pH
Sodium bicarbonate, 0.1 M $NaHCO_3$	8.4
Calcium carbonate, saturated $CaCO_3$	9.4
Iron(II) hydroxide, saturated $Fe(OH)_2$	9.5
Ammonium hydroxide, 0.01 M NH_4OH	10.6
Ammonium hydroxide, 0.1 M NH_4OH	11.1
Ammonium hydroxide, 1.0 M NH_4OH	11.6
Potassium hydroxide 0.01 M KOH	12.0
Calcium hydroxide, saturated $Ca(OH)_2$	12.4
Sodium hydroxide, 0.1 M NaOH	13.0
Sodium hydroxide, 1.0 M NaOH	14.0

as $H^+ \cdot H_2O$ or H_3O^+ commonly known as the hydronium ion. The ionization of HCl in water should be written as

$$HCl + xH_2O + yH_2O \rightarrow H^+ \cdot xH_2O + Cl^- \cdot yH_2O$$

But for simplicity it is represented by

$$HCl \xrightarrow{H_2O} H^+ + Cl^-$$

The relative strength of an acid or base is found by comparing the concentration of H^+ in solution with that of water. Pure water ionizes to a very small extent to produce a few hydrogen and hydroxyl ions in equilibrium with the water molecules.

$$H_2O \rightleftharpoons H^+ + OH^-$$

Each molecule of water ionizes to produce one H^+ and one OH^-. Pure water always contains equal amounts of each of these ions and is *neutral*.

Any solution that contains equal concentrations of H^+ and OH^- is **neutral**. A solution that contains an excess of OH^- over H^+ is **basic**.

At 25°C, the concentration of both the H^+ and OH^- ions in a liter of pure water amounts to only 1×10^{-7} moles each. The equilibrium expression for the dissociation of water is

$$H_2O \rightleftharpoons H^+ + OH^-$$

The dissociation constant k is

$$k = \frac{[H^+][OH^-]}{[H_2O]}$$

But, because the concentration of water (55.6 mol/L) is practically constant (55.6 − 0.0000001), it can be incorporated into the expression

$$k[H_2O] = [H^+][OH^-]$$

to give a new constant K_w:

$$K_w = [H^+][OH^-]$$

This new constant is called the ion-product constant for water, and

$$K_w = (1 \times 10^{-7})(1 \times 10^{-7})$$
$$= 1 \times 10^{-14}$$

At 25°C, in any aqueous solution, the product of the concentrations of the H^+ and OH^- ions will always be equal to this constant, 1×10^{-14}. *The concentration must be expressed in moles per liter.*

In a neutral solution the concentration $[H^+] = [OH^-] = 1 \times 10^{-7}$.

In an acid solution the hydrogen-ion concentration is greater than the hydroxyl-ion concentration, and the hydrogen-ion concentration is greater than 1×10^{-7}. The hydroxyl-ion concentration will be less than 1×10^{-7}. The product of the hydrogen-ion and the hydroxyl-ion concentration will always equal 1×10^{-14}.

In a basic solution the hydroxyl-ion concentration is greater than the hydrogen-ion concentration and will exceed 1×10^{-7}, while the hydrogen-ion concentration will be less than 1×10^{-7}. The product of the concentrations [H$^+$] and [OH$^-$] will always equal 1×10^{-14}. The hydrogen-ion concentration may vary from 1 mol/L in strongly acid solutions down to exceedingly small numbers in strongly basic solutions; for example, 4.2×10^{-12}. A simpler method of notation to avoid the use of awkward exponential numbers for expressing [H$^+$] is the use of the pH scale to express small hydrogen-ion concentrations. Keep in mind that pH decreases as a solution becomes more acidic.

The hydrogen-ion concentration expressed as a power of 10 is known as the pH. By definition, the pH is equal to the negative logarithm of the hydrogen-ion concentration, or

$$pH = -\log[H^+] = \log\frac{1}{[H^+]}$$

And, similarly, the pOH scale is often used to express the hydroxyl-ion concentration. pOH is defined as the negative logarithm of the hydroxyl concentration; that is,

$$pOH = -\log[OH^-] = \log\frac{1}{[OH^-]}$$

The sum of the pH and pOH is 14 because they originated from the hydrogen-ion and hydroxyl-ion concentrations (expressed in moles per liter). Thus if hydrogen-ion concentration [H$^+$] = 1×10^{-10} mol/L, then

$$pH = \log\frac{1}{1 \times 10^{-10}}$$
$$= \log 1 \times 10^{10} = \log 1 + \log 10^{10}$$
$$pH = 0 + 10 = 10$$

EXAMPLE 9.2 Calculate the pH of a solution that has [H$^+$] = 1×10^{-5} mol/L.

SOLUTION:
$$pH = \log\frac{1}{[H^+]}$$
$$= \log\frac{1}{1 \times 10^{-5}} = \log 1 \times 10^5$$
$$= \log 1 + \log 10^5$$
$$= 0 + 5 = 5$$

EXAMPLE 9.3 A solution has a hydrogen-ion concentration of 5×10^{-6}. What is the pH?

SOLUTION:
$$pH = \log\frac{1}{[H^+]}$$
$$= \log\frac{1}{5 \times 10^{-6}}$$
$$= \log\frac{1 \times 10^6}{5} = \log 1 + \log 10^6 - \log 5$$
$$= 0 + 6 - 0.70 = 5.3$$

ACID/BASE INDICATORS

The acidity or alkalinity of an aqueous solution in terms of its pH in the laboratory can be found by using an indicator which has a characteristic color at certain pH levels. For example, the common indicator phenolphthalein is colorless from 1 to 8.0 and red in the range from 9.8 to 14.0.

The indicator phenolphthaline is added to a solution of 0.1 M HCl, pH 1.0. The solution remains colorless. However, if phenolphthalein is added to a basic solution, pH = 10, the solution will turn red, indicating that the solution is basic and has a pH greater than 9.8.

NOTE: See Chapter 24, "Chemical Calculations and Concentration Expressions," for additional example problems.

Some acid-base indicators frequently show changes of color over a range of two pH units. Within a pH range that is included within its own interval of color change, such a substance provides a visual indication of the pH value of the solution (Table 9.7). The pH of common laboratory solutions is listed in Table 9.8.

Visual methods work fine in colorless solutions; the change in color of the indicator can be observed. However, in colored solutions indicators cannot be used, nor can we determine accurately in any solution the *exact* pH of the solution, regardless of its color, by purely visual means.

TABLE 9.7 Acid–Base Indicators

Name of indicator	pH range	Color change	Preparation
Methyl violet	0.2–3.0	Yellow to blue	0.05% dissolved in water
Cresol red	0.4–1.8	Red to yellow	0.1 g in 26 mL 0.01 m NaOH + 200 mL H_2O
Thymol blue	1.2–2.8	Red to yellow	Water + dil. NaOH
Orange IV	1.3–3.0	Red to yellow	H_2O
Benzopurpurin 4B	1.2–4.0	Violet to red	20% EtOH*
Methyl orange	3.1–4.4	Red to orange-yellow	H_2O
Bromphenol blue	3.0–4.6	Yellow to blue-violet	H_2O + dil. NaOH
Congo red	3.0–5.0	Blue to red	70% EtOH
Bromcresol green	3.8–5.4	Yellow to blue	H_2O + dil. NaOH
Methyl red	4.4–6.2	Red to yellow	H_2O +dil. NaOH
Chorphenol red	4.8–6.8	Yellow to red	H_2O + dil. NaOH
Bromcresol purple	5.2–6.8	Yellow to purple	H_2O + dil. NaOH
Litmus	4.5–8.3	Red to blue	H_2O
Alizarin	5.6–7.2	Yellow to red	0.1 g in MeOH[†]
Bromthymol blue	6.0–7.6	Yellow to blue	H_2O + dil. NaOH
Phenol red	6.6–8.2	Yellow to red	H_2O + dil. NaOH
Thymol blue	8.0–9.6	Yellow to blue	H_2O + dil. NaOH
o-Cresolphthalein	8.2–9.8	Colorless to red	0.04% in EtOH
Phenolphthalein	8.3–9.8	Colorless to red	70% EtOH
Thymolphthalein	9.4–10.5	Yellow to blue	70% EtOH
Alizarin yellow R	10.0–12.0	Yellow to red	95% EtOH
Indigo carmine	11.4–13.0	Blue to yellow	50% EtOH
Trinitrobenzene 135	12.0–14.0	Colorless to orange	70% EtOH

*Ethyl alcohol.
[†]Methyl alcohol.
Universal indicators are mixtures of several indicators and may be used for estimation of pH values.

TABLE 9.8 pH Values of Acids and Bases (Approximate)

Acids	pH	Bases	pH
Hydrochloric acid 1 N	0.1	Sodium bicarbonate 0.1 N	8.4
Sulfuric acid 1 N	0.3	Borax 0.1 N	9.2
Hydrochloric acid 0.1 N	1.1	Calcium carbonate, saturated	9.4
Sulfuric acid 0.1 N	1.2	Iron(II) hydroxide, saturated	9.5
Orthophosphoric acid 0.1 N	1.5	Sodium sesquicarbonate 0.1 M	10.1
Sulfurous acid 0.1 N	1.5	Magnesium hydroxide, saturated	10.5
Oxalic acid 0.1 N	1.6	Ammonium hydroxide 0.01 N	10.6
Hydrochloric acid 0.01 N	2.0	Potassium cyanide 0.1 N	11.0
Tartaric acid 0.1 N	2.2	Ammonium hydroxide 0.1 N	11.1
Formic acid 0.1 N	2.3	Ammonium hydroxide 1 N	11.6
Acetic acid 1 N	2.4	Sodium carbonate 0.1 N	11.6
Acetic acid 0.1 N	2.9	Potassium hydroxide 0.01 N	12.0
Benzoic acid 0.01 N	3.1	Trisodium phosphate 0.1 N	12.0
Alum 0.1 N	3.2	Calcium hydroxide, saturated	12.4
Acetic acid 0.01 N	3.4	Sodium metasilicate 0.1 N	12.6
Carbonic acid, saturated	3.8	Sodium hydroxide 0.1 N	13.0
Hydrogen sulfide 0.1 N	4.1	Potassium hydroxide 0.1 N	13.0
Arsenious acid, saturated	5.0	Sodium hydroxide 1 N	14.0
Hydrocyanic acid 0.1 N	5.1	Potassium hydroxide 1 N	14.0
Boric acid 0.1 N	5.2		

METHODS FOR DETERMINING pH

A. Colorimetric Determinations: Indicators

Principle
The hydronium-ion (hydrogen-ion) concentration of a solution can be measured by adding an indicator. An acid-base **indicator** is a complex organic compound that is a weak acid or a weak base. The molecular form of the indicator is different in color than that of the ionic form. The color changes are rapid and reversible and take place over a pH range of 2 units. The color of the solution indicates the pH of the solution.

Procedure
1. Choose the indicator that has a range covering the pH interval to be measured. Refer to the section on indicators.
2. Add several drops of indicator to the solution being tested.
3. Compare the color of the solution with the color range of the indicator (see Table 9.7 and Fig. 9.2).
4. Record the pH observed and determined by the color of the solution.

Preparation of Indicator Solutions.

Alizarin Yellow R: 0.01%   Dissolve 1.0 g of the dye, the sodium salt of 5-(*p*-nitrophenylazo) salicylate, in 200 mL 95% ethanol; dilute to 1 L with 95% ethanol.

Bromcresol Purple: 0.1%   Grind 1.0.g of the dye, dibromo-*o*-cresolsulfonephthalein, with 18.5 mL of 0.1 M NaOH in a mortar; dilute to 1 L with water.

FIGURE 9.2 The pH comparator determines the pH of clear, turbid, or naturally colored solutions by providing a sharp color difference with a wide selection of pH indicators.

Bromthymol Blue: 0.1%	Grind 1.0 g of the dye, dibromothymolsulfonephthalein, with 16.0 mL of 0.1 M NaOH in a mortar; dilute to 1 L with water.
Indigo Carmine: 0.1%	Dissolve 1.0 g of the dye, sodium indigo disulfonate, in 500 mL 95% ethanol; dilute to 1 L with distilled water. (Although this dye could be made up with water, an aqueous solution does not keep well.)
Methyl Orange: 0.1%	Dissolve 1.0 g of the dye in water; dilute to 1 L.
Methyl Red: 0.1%	Grind 1.0 g of the dye with 37 mL of 0.1 M NaOH in a mortar; dilute to 1 L with water.
Orange IV: 0.1%	Dissolve 1.0 g of the dye, the sodium salt of p-(p-anilinophenylazo) benzenesulfonic acid, in water; dilute to 1 L.
Phenolphthalein: 0.1%	Dissolve 1.0 g of the dye in 700 mL of 95% ethanol; dilute to 1 L with water.
Phenol Red: 0.1%	Grind 1.0 g of the dye, phenolsulfonephthalein, with 28.2 mL of 0.1 M NaOH in a mortar. Dilute to 1 L with water.

The Importance of Selecting the Proper Indicator. When color-changing indicators are used to indicate the equivalence point of acid–base titrations, the selection of the correct indicator for that particular system is most important. This is especially true when strong acid-weak base or weak acid-strong base titrations are carried out.

EXAMPLE 9.4 Weak Acid-Strong Base (see Fig. 9.3).

Methyl red (pH interval 4.4–6.2) should not be used for a weak acid-strong base titration, because the equivalence point is between pH 7 and pH 10.

Methyl orange (pH interval 3.1–4.4) should not be used for the weak acid-strong base titration.

Bromthymol blue (pH interval 6.0–7.6) would also give a false indication before the equivalence point was attained.

Phenolphthalein (pH interval 8.3–9.8) would be satisfactory. It would change color at the true equivalence point.

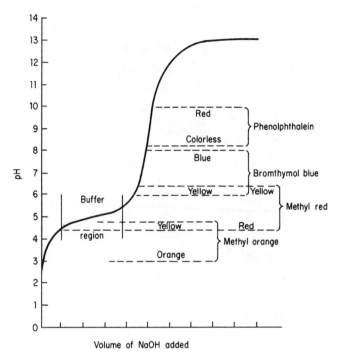

FIGURE 9.3 The pH range over which the indicator color changes determines the correct indicator for the procedure.

B. pH Test Paper

The pH range over which the color of the indicator changes determines the correct indicator for the procedure.

pH test papers are available for every value of pH; they cover both broad and narrow pH ranges (Fig. 9.4). The paper is impregnated with the indicator. A strip of test paper is wet with the liquid to be tested and then immediately compared with the standard color chart provided for each paper and range. The pH can be determined visually by comparison of the colors.

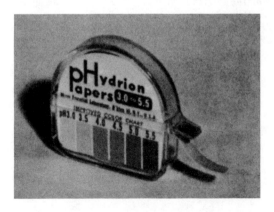

FIGURE 9.4 pHydrion® test paper determines pH directly by immediate comparison of the dipped paper with the color chart provided for each paper and range.

C. pH Meter

Fortunately, in the case of hydrogen-ion measurements, a convenient and direct determination of the pH can be made with an instrument known as the pH meter (Fig. 9.5). This instrument measures the concentration of the hydrogen ions in a solution by using a **calomel electrode** as a reference electrode and a **glass electrode** as "indicator" electrode (an electrode whose potential will vary with the concentration). The actual measurement of the hydrogen-ion concentration (or pH) is made by dipping the two electrodes into the same solution of unknown pH.

FIGURE 9.5 Fisher Accumet® Excel XL15 pH/Temperature/Millivolt Meter. (*Photo Courtesy of Thermo Fisher Scientific*)

Components. A pH meter (Fig. 9.6) is basically composed of a glass electrode, a calomel electrode, and a solid-state voltmeter. (Figure 9.7 shows a magnified drawing of the electrodes.)

1. The glass electrode has a fixed acid concentration inside. It is immersed in a solution of varying hydrogen-ion content. The electrode responds to the change in H_3O^+-ion concentration, yielding a corresponding voltage.
2. The calomel electrode is the reference electrode. Its voltage is independent of the H_3O^+-ion concentration.
3. The two electrodes constitute a galvanic cell whose electromotive force is measured by the solid-state voltmeter. The meter is calibrated to read in pH units, reflecting the H_3O^+-ion concentration.

Procedure

1. Rotate the switch to STANDBY. Allow to warm for 30 minutes.

CAUTION: When the instrument is not in use, keep on STANDBY.

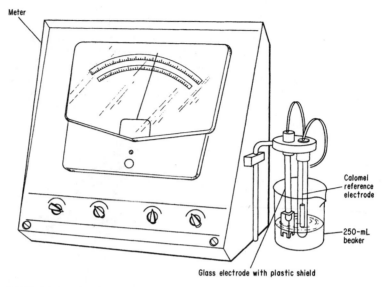

FIGURE 9.6 Parts of a pH meter.

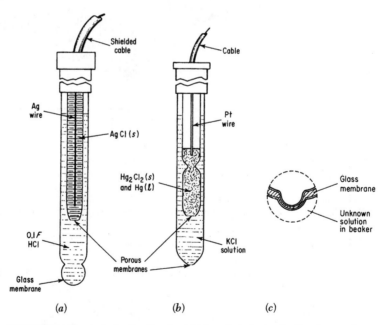

FIGURE 9.7 The electrode of a pH meter. (*a*) Glass electrode showing the glass membrane, which is fragile. Handle with care. Do not let it touch the bottom of the beaker. (*b*) Calomel electrode. (*c*) Enlarged detail of (*a*).

NOTE: Different manufacturers designate the neutral position of the selector switch with various markings, such as BAL for balance.

2. Raise the electrode from the storage solution in the beaker.

CAUTION: Always keep the electrodes in distilled water when not in use.

3. Rinse the electrodes thoroughly with distilled water. Blot with absorbent tissue.
4. Standardize against an appropriate buffer solution. (See Buffer Standardization of the pH Meter, below.)
5. Place the beaker of solution to be tested beneath the electrodes. This is the measurement of the pH of the solution.
6. Lower the electrodes carefully into the solution. Adjust temperature compensator to the temperature of the solution.
7. Rotate the selector knob to pH. Read the pH of the solution directly from the meter. Record the value.
8. When the determination is complete:
   (a) Switch to STANDBY. Raise the electrodes.
   (b) Rinse the electrodes with distilled water.
   (c) Store the electrodes in distilled water.

NOTE: Leave the instrument connected to the power line at all times, except when it is not to be used for extended periods. This will ensure stable, drift-free performance, and the slight temperature rise will eliminate humidity troubles. Component life also will be extended by elimination of repeated current surges. Turn selector switch to the balance position when the instrument is not in use.

Buffer Standardization of the pH Meter.*

Principle

All pH meters must be standardized daily by means of a buffer solution of known pH value. For maximum accuracy, use a buffer in the range of the sample to be tested. A **buffer solution** is one which tends to remain at constant pH. For example: use a pH 4 buffer for standardizing when work is to be done in the acid range; a pH 7 buffer when the work is near neutral: and a pH 9 buffer when the work is in the alkaline range.

The buffer temperature should be as close as possible to the sample temperature. Try to keep this temperature difference within 10°C.

Procedure

1. Determine the temperature of the buffer solution with a thermometer. Adjust the TEMPERATURE knob on the unit to that temperature.
2. Calibrate the meter by immersing the electrodes (Fig. 9.8) in a buffer solution of known pH.
   (a) Rotate the selector switch to pH.
   (b) Turn the adjustment knob to CALIBRATION to read the pH of the known buffer solution.

---

*This section reprinted by permission of Analytical Measurements, Chatham, N.J.

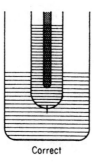
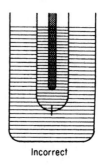

FIGURE 9.8 When using a pH meter, the electrode depth in solutions is important.

NOTE: Some manufacturers designate this position as ASYMMETRY CONTROL.

CAUTION: Do not allow the electrodes to touch the sides of the beaker.

3. Rotate the selector knob to STANDBY.
   (a) Raise the electrodes carefully.
   (b) Remove the buffer solution.
   (c) Rinse the electrodes thoroughly with distilled water.
   (d) Blot the electrodes with absorbent, lint-free tissue.

Electrode Maintenance. When not in use, the electrodes should be left soaking in water or buffer solution. Electrodes that have dried out should be soaked in water for several hours prior to use.

Keep the reference electrode reservoir filled, using saturated KCl solution. There should be some KCl crystals in the reservoir.

The fiber junction at the tip of the reference electrode may become plugged when used for long periods of time in samples containing suspended material. A plugged junction can usually be opened by boiling the tip of the electrode in dilute nitric acid.

Glass electrodes used in biological samples may start to drift due to the formation of a coating of protein. The coating may be removed by cleaning with a strong solution of detergent and water.

Hints for Precision Work. For accurate work, fresh buffer solution should be used. Buffer solution bottles should be capped when not in use to prevent evaporation and contamination. Never pour buffer solutions back into the bottle.

Between buffers and samples, rinse the electrodes with distilled water and blot with an absorbent tissue to prevent carryover of solution.

Immerse completely the pH sensitive bulb portion of the glass electrode into the solution to be measured. Be sure that the glass electrode does not touch the wall or bottom of the sample container.

Keep the electrodes and the samples measured at the same constant temperature. Temperature cycling of the electrodes will result in small drifts that may make reading of very small changes in pH difficult.

pH TITRATION*

A. Introduction

In an acid-base titration, the change in [H⁺] may be very large, that is, from 10^{-1} to 10^{-10}. This is a 100-million-fold change in concentration, and it would be rather inconvenient to plot such numbers. However, the change in pH is only from 1.0 to 11.0, and these numbers may conveniently be plotted. When the volume of titrant in milliliters is plotted against the pH, we obtain what is called a *titration curve*. (See Figs. 9.9 to 9.11.) See Chapter 24, "Chemical Calculations and Concentration

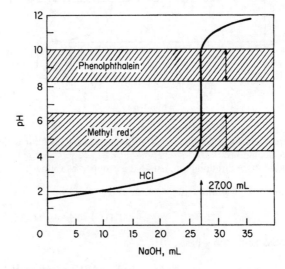

FIGURE 9.9 pH titration curve; strong acid versus strong base.

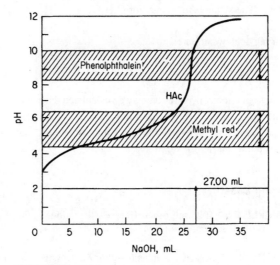

FIGURE 9.10 pH titration curve; weak acid versus strong base.

---

*This section reprinted by permission of Analytical Measurements, Chatham, N.J.

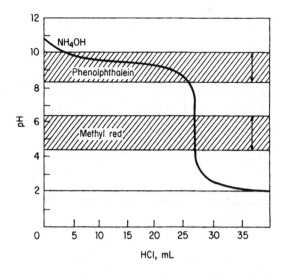

FIGURE 9.11 pH titration curve; strong acid versus weak base.

Expressions," and Chapter 25, "Volumetric Analysis," for additional information on pH titrations and calculations.

Acid–base titrations can usually be grouped as follows:

1. Strong acid–strong base:

$$H^+ + OH^- \rightarrow H_2O$$

2. Weak acid–strong base:

$$HAc + OH^- \rightarrow Ac^- + H_2O$$

NOTE: The symbol Ac^- stands for $C_2H_3O_2^-$.

3. Strong acid–weak base:

$$H^+ + NH_4OH \rightarrow NH_4^+ + H_2O$$

The hydrogen-ion concentration may be calculated during a titration from the amount of acid or base that has been added. We can therefore calculate the pH of any solution resulting from the reactions of type 1, 2, or 3. Let us now look at some sample calculations to see how the pH is computed at various points during the titrations.

B. Strong Acid–Strong Base

EXAMPLE 9.5 If 25 mL of a 0.10 M HCl solution is diluted to 100 mL with distilled water, what is the pH of the resulting solution?

SOLUTION: Since HCl is considered to be 100 percent ionized, the number of moles of H^+ in solution will be:

$$(0.025 \text{ L})(0.10 \text{ mol/L}) = 2.5 \times 10^{-3} \text{ mol}$$

which is now dissolved in a total volume of 100 m L (0.1 L).

$$[H^+] = \frac{2.5 \times 10^{-3} \text{mol}}{0.10 \text{ L}} = 25 \times 10^{-3} = 2.5 \times 10^{-2} \text{ mol/L}$$

The pH is calculated as follows:

$$pH = \log\frac{1}{[H^+]} = \log\frac{1}{2.5 \times 10^{-2}}$$
$$= \log\frac{1 \times 10^2}{2.5}$$
$$= \log 1 + \log 10^2 - \log 2.5$$
$$= 0 + 2 - 0.32 = 1.68$$

EXAMPLE 9.6 What will be the pH of the solution after 10 mL of 0.10 M NaOH has been added?

SOLUTION:

1. Original number of moles of HCl = (0.025 L)(0.10 mol/L) = 0.0025 mol.
2. Moles of NaOH added = (0.01 L)(0.10 mol/L) = 0.001 mol.
3. Moles of HCl remaining in solution = (0.0025 − 0.0010) = 0.0015 mol.
4. The 0.0015 mol of HCl is now dissolved in 110 mL of solution. The [H$^+$] = 0.0015 mol/0.11 L = 0.0136 or 1.36×10^{-3}.
5. The pH can now be calculated as

$$pH = \log\frac{1}{1.36 \times 10^{-2}} = \log\frac{1 \times 10^2}{1.36}$$
$$= \log 1 + \log 10^2 - \log 1.36$$
$$= 0 + 2 - 0.13 = 1.87$$

EXAMPLE 9.7 What will be the pH of the solution when 25 mL of NaOH has been added? This is the equivalence point, and we should expect that neutralization, $H^+ + OH^- \rightarrow H_2O$, has occurred, and the only H$^+$ in solution would be from the water, that is, 1×10^7 or a pH of 7.

SOLUTION:

1. Original number of moles of HCl = (0.025 L)(0.10 mol/L) = 0.0025 mol.
2. Moles of NaOH added = (0.025 L)(0.10 mol/L) = 0.0025 mol.
3. Moles of HCl remaining in solution = (0.0025 − 0.0025) = 0.00 mol.

This means that we have completely reacted the HCl, and only water and NaCl are in solution. Since neither Na$^+$ or Cl$^-$ reacts with the water, the source of H$^+$ in solution will be from the dissociation of water. Thus, $[H^+] = 1 \times 10^{-7}$ and the pH = 7.

EXAMPLE 9.8 What will be the pH of the solution after 30.0 mL of NaOH has been added?

SOLUTION:

1. Original number of moles of HCl = (0.025 L) (0.10 mol/L) = 0.0025 mol.
2. Moles of NaOH added = (0.030 L) (0.10 mol/L) = 0.003 mol.
3. Moles of NaOH in excess = (0.003 − 0.0025) = 0.0005 mol.

4. The 0.0005 mol of NaOH is now dissolved in 130 mL of solution. The [OH⁻] = 0.0005 mol/ 0.13 L = 0.0038 or 3.8×10^{-3}.

5. The [H⁺] can be found from the expression for the ion product of water:

$$K_w = 1 \times 10^{-14} = [H^+][OH^-]$$

$$1 \times 10^{-14} = [H^+] = [H^+](3.8 \times 10^{-3})$$

$$\frac{1 \times 10^{-14}}{3.8 \times 10^{-3}} = [H^+]$$

$$2.6 \times 10^{-12} = [H^+]$$

6. The pH can now be calculated:

$$pH = \log\frac{1}{2.6 \times 10^{-12}} = \log\frac{1 \times 10^{12}}{2.6}$$

$$= \log 1 + \log 10^{12} - \log 2.6$$

$$= 0 + 12 - 0.42 = 11.58$$

C. Weak Acid–Strong Base

EXAMPLE 9.9 If 25 ml of a 0.10 M HAc solution is diluted to 100 mL with distilled water, what is the pH of the resulting solution?

SOLUTION: Since HAc is ionized only to a slight extent, the [H⁺] and thus the pH can be calculated from the ionization constant.

$$HAc \rightarrow H^+ + Ac^-$$

$$K_i = \frac{[H^+][Ac^-]}{[HAc]}$$

Let $X = [H^+]$. Since the HAc ionizes to form equal amounts H⁺ and Ac⁻, the [Ac⁻] will also be equal to X. The concentration of the HAc is (0.025 L)(0.10 mol/L) = 0.0025 mol, which is now dissolved in 100 mL (0.1 L), or 0.0025 mol/0.1 L = 2.5×10^{-2} mol/L.

$$K_i = 1.8 \times 10^{-5} = \frac{(X)(X)}{(2.5 \times 10^{-2} - X)}$$

As X will be very small, we may disregard it when it is to be subtracted from the much larger 2.5×10^{-2}, with very little error. Thus,

$$K_i = 1.8 \times 10^{-5} = \frac{X^2}{2.5 \times 10^{-2}}$$

$$(2.5 \times 10^{-2})(1.8 \times 10^{-5}) = X^2$$

$$4.5 \times 10^{-7} = X^2$$

$$6.7 \times 10^{-4} = X = [H^+]$$

The pH is calculated as follows:

$$pH = \log \frac{1}{(6.7 \times 10^{-4})} = \log \frac{1 \times 10^4}{6.7}$$

$$= \log 1 + \log 10^4 - \log 6.7$$

$$= 0 + 4 - 0.83 = 3.17$$

EXAMPLE 9.10 What will be the pH of the solution after 10 mL of 0.10 M NaOH has been added?

SOLUTION:
1. Original number of moles of HAc = (0.025 L)(0.10 mol/L) = 0.0025 mol.
2. Moles of NaOH added = (0.01 L)(0.10 mol/L) = 0.001 mol.
3. Moles of HAc remaining in solution = (0.0025 − 0.001) = 0.0015 mol.
4. The 0.0015 mol of HAc is now dissolved in 110 mL of solution. The [HAc] = 0.0015 mol/0.11 L = 0.0136 = 1.36×10^{-2}.
5. Moles of acetate ion = acetate-ion concentration produced by the ionization of the remaining HAc (HAc → H^+ + Ac^-), which we designate X. In addition, there are the acetate ions produced during the titration.

$$HAc + OH^- \rightarrow Ac^- + H_2O$$

Each mole of NaOH forms a mole of Ac^-, and since we added 0.001 mol of NaOH, we form 0.001 mol of Ac^-. Therefore, the total Ac^- concentration is (X + 0.001 mol). We assume that X is small compared to the 0.001 mol of Ac^- produced in the reaction and approximate that the amount of Ac^- = 0.001 mol. Since the volume of the solution is now 110 mL (0.11 L), the [Ac^-] = 0.001 mol/0.11 L = 0.091 mol/L.

6. The [H^+] can be found from the ionization constant:

$$1.8 \times 10^{-5} = \frac{[H^+](9.1 \times 10^{-2})}{(1.36 \times 10^{-2})}$$

$$\frac{(1.8 \times 10^{-5})(1.36 \times 10^{-2})}{(9.1 \times 10^{-2})} = [H^+]$$

$$2.7 \times 10^{-6} = [H^+]$$

7. The pH can now be calculated as follows:

$$pH = \log \frac{1}{(2.7 \times 10^{-6})} = \frac{\log 1 \times 10^6}{2.7}$$

$$= \log 1 + \log 10^6 - \log 2.7$$

$$= 0 + 6 - .43 = 5.27$$

EXAMPLE 9.11 What will be the pH of the solution after 25 mL of 0.10 M NaOH has been added? This is the equivalence point.

SOLUTION:

1. Original number of moles of HAc = (0.025 L)(0.10 mol/L) = 0.0025 mol.
2. Moles of NaOH added = (0.025 L)(0.10 mol/L) = 0.0025 mol. Thus, the solution is neutralized and no HAc should remain.
3. Moles of Ac⁻ produced = moles of NaOH added = 0.0025 mol. Since the volume of solution is now 125 mL (0.125 L), the [Ac⁻] = 0.0025 mol/0.125 L = 0.02 mol/L.
4. The Ac⁻ formed will react with the water (hydrolyze):

$$Ac^- + H_2O \rightarrow HAc + OH^-$$

The [OH⁻] may be found from the ion product of water:

$$1 \times 10^{-14} = [H^+][OH^-]$$

If $[H^+] = X$, then $[OH^-] = \dfrac{1 \times 10^{-14}}{X}$ and the $[HAc] = \dfrac{1 \times 10^{-14}}{X}$ since the $[HAc] = [OH^-]$.

5. The [H⁺] can be found from the ionization constant:

$$1.8 \times 10^{-5} = \dfrac{X \, 0.02}{\dfrac{1 \times 10^{-14}}{X}}$$

$$1.8 \times 10^{-5} = \dfrac{X^2 \, 0.02}{1 \times 10^{-14}}$$

$$\dfrac{(1.8 \times 10^{-5})(1 \times 10^{-14})}{2 \times 10^{-2}} = X^2$$

$$9 \times 10^{-18} = X^2$$

$$3 \times 10^{-9} = X$$

6. The pH can be calculated as:

$$pH = \log \dfrac{1}{3 \times 10^{-9}} = \log \dfrac{1 \times 10^9}{3}$$
$$= \log 1 + \log 10^9 - \log 3$$
$$= 0 + 9 - 0.48 = 8.52$$

Note that the pH at the equivalence point is not 7 as in the case of a strong acid–strong base titration.

EXAMPLE 9.12 What will be the pH of the solution after 30 mL of NaOH has been added?

SOLUTION:

1. Original moles of HAc = (0.025 L)(0.10 mol/L) = 0.0025 mol.
2. Moles of NaOH added = (0.03 L)(0.10 mol/L) = 0.0030 mol.
3. Moles of NaOH in excess = (0.0030 − 0.0025) = 0.0005 mol.
4. The 5×10^{-4} mol of NaOH is now dissolved in 130 mL of solution. The [OH] = 5×10^{-4} mol/0.13 L = 3.8×10^{-3} mol/L.

5. Since the excess OH⁻ will drive the reaction

$$Ac^- + H_2O \rightleftharpoons HAc + OH^-$$

back to the left, the hydrolysis is repressed and the [OH⁻] can be assumed to be 3.8×10^{-3} mol/L with very little OH⁻ being formed by the hydrolysis reaction.

6. The [H⁺] may be found from the ion product of water:

$$1 \times 10^{-14} = [H^+](3.8 \times 10^{-3})$$

$$\frac{1 \times 10^{-14}}{3.8 \times 10^{-3}} = [H^+]$$

$$2.6 \times 10^{-12} = [H^+]$$

7. The pH can be calculated:

$$pH = \log \frac{1}{(2.6 \times 10^{-12})} = \log \frac{1 \times 10^{12}}{2.6}$$

$$= \log 1 + \log 10^{12} - \log 2.6$$

$$= 0 + 12 - 0.42 = 11.58$$

We have now calculated the pH during several steps of the titration of a strong or weak acid with a strong base. If we plot the pH versus mL of titrant (NaOH) for these, or similar, reactions we obtain a titration curve. The general form of such curves is given in Figs. 9.9 and 9.10.

D. Strong Acid–Weak Base

Even though we have not calculated the pH values for the titration of the third group of reactions (strong acid–weak base), the computations are similar and the general form of the titration curve is shown in Fig. 9.11.

BUFFER SOLUTIONS

Mixtures of weak acids and their salts or of weak bases and their salts are called **buffer solutions** because they resist changes in their hydrogen-ion concentration upon addition of small amounts of acid or base. Their pH tends to remain unchanged. Buffer solutions are used in the laboratory to standardize pH meters and whenever solutions having a definite pH are required. Tables 9.6 to 9.8 provide alternative procedures for preparing buffer solutions of any desired pH value.

Premixed, ready-to-use buffer solutions are also available and can be purchased in pint and quart containers. Also available, however, are vials or packets of concentrate (Fig. 9.12), which can be quantitatively transferred to specified volumetric flasks and diluted to the mark with distilled water. Thus, a stock of buffer-solution concentrates of any desired pH can be kept in a very small space.

A. Preparation of Buffer Solutions

Method 1.

Procedure
Prepare the stock solutions listed in Table 9.9 according to standard volumetric methods. Select the pH of the buffer solution desired. Mix the indicated volumes of the stock solutions required to prepare the desired buffer solution.

190 CHAPTER NINE

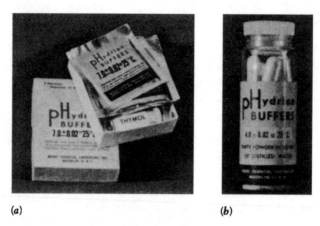

FIGURE 9.12 Buffer solution concentrates; tablets and envelopes (*a*), and capsules (*b*) for preparing standard pH solutions by using premeasured quantities to be dissolved in definite volume of distilled water.

TABLE 9.9 Preparation of Buffer Solutions, Method 1 (mixture diluted to 100 mL)

Desired pH, (25°C)	0.2 M KCl, mL	0.1 M potassium hydrogen phthalate, mL	0.1 M potassium dihydrogen phosphate, mL	0.2 M HCl, mL	0.1 M HCl, mL	0.1 M NaOH, mL	0.025 M borax, mL	0.05 M disodium hydrogen phosphate, mL	0.2 M NaOH, mL
1.0	25			67					
1.5	25			20.7					
2.0	25			6.5					
2.5		50			38.8				
3.0		50			22.3				
3.5		50			8.2				
4.0		50			0.1				
4.5		50				8.7			
5.0		50				22.6			
5.5		50				36.6			
6.0			50			5.6			
6.5			50			13.9			
7.0			50			29.1			
7.5			50			40.9			
8.0			50			46.1			
8.5							50		
9.0					15.2		50		
9.5					4.6		50		
10.0						8.8	50		
10.5						18.3	50		
11.0						22.7		50	
11.5						4.1		50	
12.0						11.1		50	
12.5	25					26.9			20.4
13.0	25								66.0

Method 2.

Procedure

Mix the two solutions below according to the instructions indicated in Table 9.10 to prepare 1 L of a buffer solution of the desired pH.

1. Dissolve 12.37 g of anhydrous boric acid, H_3BO_3, and 10.51 g of citric acid, $H_3C_6H_5O_7 \cdot H_2O$, in distilled water and dilute to 1 L in a volumetric flask. This makes a 0.20 M boric acid and a 0.05 M citric acid solution.
2. Dissolve 38.01 g of $Na_3PO_4 \cdot 12H_2O$ in distilled water and dilute to 1 L in a volumetric flask. This makes a 0.10 M tertiary sodium phosphate solution.

TABLE 9.10 Preparation of Buffer Solutions, Method 2

Desired pH	Solution 1, mL	Solution 2, mL
2.0	975	25
2.5	920	80
3.0	880	120
3.5	830	170
4.0	775	225
4.5	720	280
5.0	670	330
5.5	630	370
6.0	590	410
6.5	545	455
7.0	495	505
7.5	460	540
8.0	425	575
8.5	390	610
9.0	345	655
9.5	300	700
10.0	270	730
10.5	245	755
11.0	220	780
11.5	165	835
12.0	85	915

Method 3.

Procedure

This series of buffer solutions requires the preparation of five stock solutions. Large enough volumes of these solutions should be prepared to meet the need. The maximum volume of any one solution needed to prepare 1 L of a buffer solution of desired normality is 500 mL.

1. 0.1 N NaOH. Dissolve 4 g NaOH per liter of solution.
2. 0.1 M potassium hydrogen phthalate. Dissolve 20.42 g $KHC_8H_4O_4$ in enough distilled water to make 1 L of solution.
3. 0.1 M monopotassium phosphate. Dissolve 13.62 g KH_2PO_4 per liter of solution.
4. 0.1 M KCl. Dissolve 7.46 g per liter of solution.
5. 0.1 M boric acid. Dissolve 6.2 g H_3BO_3 in 0.1 M KCl (solution 4) to make 1 L of solution.

Mix the stock solutions according to the volumes indicated in Table 9.11 and then dilute the mixture to 1 L with distilled water.

TABLE 9.11 Preparation of Buffer Solutions, Method 3

Desired pH	Volume of each solution, mL				
	1	2	3	4	5
4	4	500		496	
5	226	500		274	
6	56		500	444	
7	291		500	209	
8	39			461	500
9	208			292	500
10	437			63	500

ION-SELECTIVE ELECTRODES

A conventional pH meter can be used to measure the voltaic potential of a specific ion in a solution by using a reference electrode (e.g., a calomel electrode) and a sensing electrode filled with a solution of the specific ion. This sensing electrode is usually referred to as an **ion-selective electrode** (ISE) for all ions in aqueous solutions except hydrogen. Use of a calomel reference electrode and a glass or "hydrogen" electrode for sensing the hydrogen ion concentration and determining the pH was discussed earlier in this chapter under pH Meter. The voltaic potential that develops between the reference electrode and the ISE is directly proportional to the concentration of the selective ion in solution. The actual measurement of the selective ion concentration (voltaic potential) is made by dipping both the reference and the selective-ion electrodes into the same solution.

An ISE analysis can be represented as a typical electrochemical cell with two half-cell components, the ion-selective electrode, and an external reference electrode. The ISE half-cell consists of an electrode with a porous conductive membrane specific for the analyte, an internal filling solution called the **electrolyte,** which contains a soluble salt of the analyte, and an internal reference electrode.

Figure 9.13 is a typical cell assembly for fluoride determinations using an ion-selective electrode equipped with an LaF_3 crystal membrane and a calomel or other reference electrode.

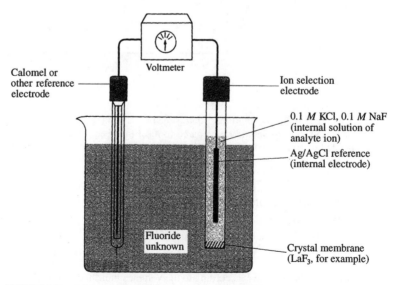

FIGURE 9.13 Schematic of cell assembly using a selective ion electrode.

The reference half-cell can be a number of different electrodes. A hydrogen electrode would be the ideal reference; however, it is a cumbersome and awkward electrode to use in the laboratory. A calomel electrode would be more convenient, and it can provide a standard reference potential of 0.2145 volts at 25°C under various operating conditions. The **calomel electrode** consists of elemental mercury and a KCl solution saturated with Hg_2Cl_2 (calomel). An analysis is performed by placing both the ISE and the reference electrodes into the solution to be analyzed. The cell electromotive force (emf) is the measurement of the potential difference across the ion-selective membrane and the reference electrode.

The magnitude of this difference is dependent on the analyte **activity.** Activity is a concentration parameter that describes the *effective* concentration of the analyte by taking into account the ionic strength of the total solution being analyzed. For solutions with low ionic strengths, activity, and concentrations are synonymous. The cell emf can be described briefly as the **Nernst equation:**

$$E = E^0 - \frac{0.059}{n} \log [\text{ion}]$$

where: E = the cell potential difference
E^0 = standard cell potential
n = signed charge on the selected ion
[ion] = molar concentration of ion

The Nernst equation can be simplified for ion-selective calibrations to

$$E_{obs} = K + K' \log C_{ion}$$

Both K and K' are constants for a specific ion and E_{obs} represents the observed voltage for the cell of a given ion concentration. If a plot is made of the observed voltages (E_{obs}) against the log of the specific ion concentrations (C_{ion}), a straight line will result.

A. Selective Ion Calibration and Calculation Example

1. Assume the following data were obtained using a silver ion electrode and a series of standard Ag^+ solutions. The silver ion concentrations will be expressed in molarity [M] and the E_{obs} expressed in volts.

[Ag^+]	log [Ag^+]	Volts
1.00×10^{-1}	-1	0.300
1.00×10^{-2}	-2	0.400
1.00×10^{-3}	-3	0.500
1.00×10^{-4}	-4	0.600

2. A linear plot of this molarity versus volts is shown in Fig. 9.14.
3. Assume that an unknown silver ion solution gave a meter reading of 0.46 volts.
4. The calibration graph at 0.46 volts corresponds to a log [Ag^+] reading of -2.5. (See calibration graph in Fig. 9.14.)
5. Taking the antilog of -2.5 gives a silver ion concentration of 0.00316 M Ag^+.

In most ion-selective calibration plots the $-\log C_{ion}$ is plotted on the x axis and the concentration actually decreases in going from left to right. If two or more ions of the same ionic charge are present

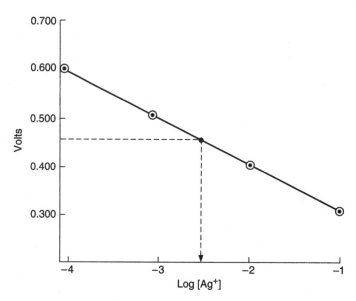

FIGURE 9.14 Typical calibration curve with cell voltage versus log of silver ion concentration.

in the solution, the potential developed at the membrane will depend on the activity of all ions present. For example, Li^+ ions are known to interfere slightly with Na^+ ion responses, and corrections should be determined for each ion and its matrix.

The most intensively studied material for selective ion electrodes has been glass membranes described earlier in this chapter with pH meters. Glass membranes were introduced in 1906 by Cremer for the measurement of the hydrogen ion. A glass consisting of 72.2% SiO_2, 6.4% CaO, and 21.4% Na_2O was found very effective for H^+ detection. The measurement of H^+ ion concentrations became the primary use of this glass-membraned ion-selective electrode. However, through systematic studies of glass composition it was discovered that other specific ions such as Na^+ could also be quantitatively determined. For example, a glass composition of 52.1% SiO_2, 19.1% Al_2O_3, and 28.8% Na_2O produces a **solid-state membrane** electrode that is 100,000 times more sensitive to Ag^+ than to H^+. Other types of **crystalline membrane** or solid-state membrane electrodes have been developed that can measure anions such as cyanide, thiocyanate, fluoride, plus larger cations such as the lanthanides and heavy metals.

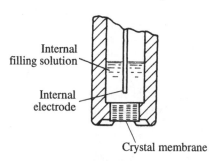

FIGURE 9.15 Construction of a crystal membrane electrode.

A third type of media known as **liquid-liquid membranes** offered a broader application on cation and anion concentration studies (NH_4^+, Cl^-, HCO_3^-, etc.). These liquid membrane electrodes tend to be specific for the alkali, alkaline earths, and many other cations. Liquid membranes with a polymer matrix used to stabilize the membrane have become quite popular. The liquid membrane consists of an organic, water-immiscible liquid phase incorporating the sites. Liquid membrane electrodes based on charged sites generally show selectivity for ions of opposite charge (see Fig. 9.15).

Another type of membrane called the **gas-permeable membrane** is now being used in the gas-sensitive electrodes. An ion sensor is in contact with an electrolyte that is separated from the sample by a gas-permeable membrane (e.g., CO_2, NO_x, etc.). Similarly, **enzyme substrate membranes** can be used in contact with ion sensors for the measurement of organic substances such as urea.

Although it is impossible to determine how many analytes can be measured with ion-selective electrodes, there are about 50 ions for which sensors are commercially available (see Table 9.12). ISEs are unique in several respects. They respond strictly to single-ion-activity changes; they may have a linear response over several orders of magnitude in activity (for example, ISE sensors for Ca^{2+} respond in the 10^{-7} to 1.0 mol/L range). ISEs can be made very small; such microelectrodes (tip diameter approximately 1 μm) can be used to measure ions in situ—for example, within the human body.

TABLE 9.12 Ion-Selective Electrodes

Electrode	Type	Concentration range, ppm	Temperature range	Interferences
Ammonium, NH_4^+	Gas sensing	0.01 to 17,000	0 to 50°C	Volatile amines
Bromide, Br^-	Solid state	0.40 to 79,900	0 to 80°C	S^{-2}, I^-, CN^-
Calcium, Ca^{2+}	Polymer membrane	0.02 to 40,100	0 to 40°C	Ag^+, Hg^{2+}, Cu^{2+}
Cyanide, CN^-	Solid state	0.2 to 260	0 to 80°C	S^{-2}, I^-, Br^-, Cl^-
Fluoride, F^-	Solid state	0.02 to saturated	0 to 80°C	OH^-
Sodium, Na^+	Glass	0.02 to saturated	0 to 80°C	Ag^+, H^+, Li^+, Ca^{2+}, K^+

REFERENCES

1. Byrne, T. P., "Ion-Selective Electrodes in Direct Potentiometric Clinical Analyzers." *Selective Electrode Rev.* 10:107–124 (1988).

2. Koryta, J., "Theory and Applications of Ion-Selective Electrodes. Part 8." *Anal Ckim Acta.* 233:1–30 (1990).

3. *The pH and Conductivity Handbook,* vol. 28, Omega Engineering Inc., P.O. Box 2284, Stamford, CT 06906.

4. Solsky, R. L., "Ion-Selective Electrodes." *Anal Chem.* 60:106A–113A (1988); *Anal Chem.* 62:21R–33R (1990).

CHAPTER 10
BASIC ELECTRICITY

INTRODUCTION

All employees should become familiar with OSHA's regulations dealing with electrical safety, 29 CFR 1910.331-.335. Electricity is the No. 1 cause of industrial fires and, according to recent National Safety Council statistics, is causing the accidental death of over 800 people annually. Thousands more have suffered disabling injuries due to electrical accidents. The No. 1 cause of work-related electrocutions has been contact with overhead power lines. The most effective way to avoid electrical hazards would be to de-energize the system or to maintain a safe distance. A general rule for unqualified persons and equipment working near overhead power lines would be *to maintain a minimum of 10 feet clearance from powerlines carrying 50 kilovolts or less.* (See Table 10.1, OSHA reference 29 CFR 1910.333 Table S-5)

The atomic theory of matter is the basis for understanding electricity. The atoms of all substances consist of electrons, protons, and neutrons. In solids, called **conductors**, the electrons can move easily from atom to atom. In substances, called **insulators,** the electrons are bound very tightly, each in its own atom, and do not move freely. Metallic elements conduct electricity, and most nonmetallic elements do not conduct electricity. Electrons flow from the more negative to the more positive points. The flowing of electrons through a conductor is referred to as **electricity** or an **electric current**. An electric current is most useful when it does work, and is not merely flowing through conductors. Work can be performed when the electrical energy is transformed into some other form of energy, such as mechanical, heat, or light energy.

ELECTRICAL SAFETY PRACTICES

Personal protective equipment (PPE) is certainly a safeguard against electrical hazards, but should not be used as the only means of protection. This specialized PPE consists of: (a) protective; leather gloves worn over rubber insulated gloves (see Table 10.2); (b) non-conductive hard hats; (c) eye protection equipped with side shields; (d) specialized filtered eye protection which should be worn around electrical welding arcs; and (e) non-conductive and specialized clothing (rubber sleeves) which should be worn in electrically hazardous areas.

Table 10.3 outlines some additional electrical safety practices that should be followed.

Table 10.4 lists some of the physiological effects on humans upon contact with a common 60 Hertz AC current:

Electricity will follow the path of least resistance. Never touch electrical wires or equipment without following the proper procedures. Remember, when handling live circuits, your body can act as a conductor of electricity and result in a fatal electric shock.

ELECTRICITY COMPARED TO FLUIDS

The principle of electricity can be compared to that of water (see Table 10.5). Water flows through pipes; electric current (I) flows through conductors. Water flows through pipes at varying rates, so many gallons per minute. Electrons also flow through conducting wires at a variable rate. A flow of billions

198 CHAPTER TEN

TABLE 10.1 Approach Distance for Qualified Employees—Alternating Current

Voltage range (phrase to phase)	Minimum approach distance
0–300 V	avoid contact
300–750 V	1.0 ft (30.5 cm)
750 V–2 kV	1.5 ft (46 cm)
2–15 kV	2.0 ft (61 cm)
15–37 kV	3.0 ft (91 cm)
37–87.5 kV	3.5 ft (107 cm)
87.5–121 kV	4.0 ft (122 cm)
121–140 kV	4.5 ft (137 cm)

TABLE 10.2 Glove Ratings for Electrical Work

Glove rating	Maximum use voltage
Class 0	1000 AC
Class 1	7600 AC
Class 2	17,000 AC
Class 3	26,500 AC

TABLE 10.3 Electrical Safety Guidelines and Practices

- Always maintain the recommended distance from power lines.
- Follow required lockout/tagout procedures when dealing with electricity.
- Use only non-conductive ladders (typically wood or fiberglass) and equipment, if possible, around electricity.
- Determine that all electrical equipment is properly grounded.
- Never bypass electrical safety features.
- Inspect all electrical tools, cords, and equipment before using.
- Determine that the electrical cord is rated for the load to be carried.
- Never exceed the capacity or deliberately overload a circuit.
- Avoid using electrical tools and equipment in wet conditions.
- Do not wear jewelry, watches, key chains, and other conductive materials while working around active electrical systems.
- Never probe blindly into activated circuits with tools or hands.
- Ground fault interrupts should be used whenever possible.
- Never drape or hang electrical extension cords over metal piping or ductwork.
- Report any electrical malfunctions immediately and discontinue use.

TABLE 10.4 Physiological Effects on Humans with Common Electrical Currents

- 0–1 milliamp range—usually harmless to humans. A slight tingling or involuntary movement to the electrical shock could cause a loss of balance and injury by falling.
- 5–15 milliamp range—can cause a loss of muscle control in humans. This loss of control may cause the person to not be able to release the source of current and increase the degree of injury.
- 75–300 milliamp range—enough current flow to cause death in humans.
- 2500 milliamps (2.5 amps) and greater—can cause immediate cardiac arrest and respiratory failure in humans. Internal organs can be permanently damaged and both internal and external burns can occur.

TABLE 10.5 Analogy of Electrical and Water Units of Measurement

Unit of electrical measurement and its symbol	Analog for water	Electrical representation	Water representation
Volt (electron pressure, emf, voltage) V Voltmeter	Pressure lb/in² (psi), pascal (Pa) (Pressure gauge)	Voltmeter The greater the difference in the quantity of electrons between the electrodes, the greater the electron pressure, the voltage.	The greater the height of the column of water (the greater the difference in the water level between the top level and the bottom) the greater the water pressure (lb/in²).
Ampere (flow of electrons, electric current A) Ammeter	Rate of flow of water: gal/min (gpm) L/min Water meter	Night light uses little current. Airport light uses great deal of current.	House faucet flow is 5 gal/min. (both at same pressure) Fire hydrant flow is 500 gal/min.
Ohm (resistance to flow of electrons) Ω R	Resistance of the pope to the flow of water.	110 V ac Dimmer (Variable resistance) Identical bulbs use the same voltage; increased resistance reduces the flow of current: reduces the light output.	Smooth, polished inside of pipe offers little resistance. Tube bends and corroded pipe offer resistance to water flow.

of electrons per second is equal to one **ampere**, the unit of electric current ($\cong 6 \times 10^{18}$ e/s = 1 A). Electric current (I) is measured in amperes (A).

Water in a pipe is under a definite pressure which can range from fractions of a pound per square inch to hundreds of thousands of pounds per square inch. Water can be in pipes and under pressure, yet stand still and not flow because the faucet is closed. Electricity can be in a conductor, under pressure, and not flow because the conductor (wire) does not make a complete circuit: the switch is off. The higher the electron pressure, the greater the tendency for electrons to flow. Electron pressure or **voltage** (E) is measured in volts (V).

In water systems, there is a certain amount of resistance to the flow of water through the pipes because of the turns, fittings, scale, and constrictions. A certain amount of power is required to overcome this resistance. Similarly, energy is required to overcome the energy bond by which electrons are attached to the atom in a conductor. This opposition to the flow of electricity is called **resistance** (R), and its unit of measurement is the **ohm** (Ω). When a current of 1 ampere is flowing through an electrical device as a result of a potential difference of exactly 1 volt, the device is defined as having a resistance of 1 ohm (Ω). Different substances offer a wide range of resistance to the flow of electrons, from extremely easy (metals like copper and silver), to reduced flow or complete blockage (nonmetals like: SiO_2 [glass] and rubber).

OHM'S LAW

Just as the energy obtainable from the flow of water is related to the fluid pressure, the amount of flow, and the resistance, the components of electricity are related by a law known as **Ohm's Law**, which states: Voltage (in volts) = current (in amperes) × resistance (in ohms)

$$E = I \times R$$

EXAMPLE 10.1 If an electric device operates at 110 V and has a resistance of 15.2 Ω the current required is 22 A.

$$110 \text{ volts} = I \times 15.2 \ \Omega$$

$$\frac{110 \text{ volts}}{15.2 \ \Omega} = I = 7.24 \text{ A}$$

To summarize our analogy between electricity flow and water flow, it is useful to visualize water flowing through a pipe from a water reservoir. The force making the water flow is caused by pressure (volts in electrical flow); the amount of water flowing would be given as quantity or gallons per unit time (amperes in electrical flow); the resistance to water flow is determined by the size of pipe, its length, and the roughness of its inner surface (ohms in electrical flow). Watts represent the power or driving force with the water flow taken times the pressure. The relation among power, current, and resistance is given by Ohm's law:

$$\text{Volts} = \text{amperes} \times \text{ohms}$$

$$\text{Volts} = \frac{\text{watts}}{\text{amperes}}$$

$$\text{Volts} = \sqrt{\text{watt} \times \text{ohms}}$$

$$\text{Ohms} = \frac{\text{volts}}{\text{amperes}}$$

$$\text{Ohms} = \frac{\text{volts}^2}{\text{watts}}$$

$$\text{Ohms} = \frac{\text{watts}}{\text{amperes}^2}$$

$$\text{Watts} = \frac{\text{volts}^2}{\text{ohms}}$$

$$\text{Watts} = \text{amperes}^2 \times \text{ohms}$$

$$\text{Watts} = \text{volts} \times \text{amperes}$$

EXAMPLE 10.2 What is the value of the resistance if a voltmeter reads 0.98 V when a current of 0.33 A is flowing through the circuit?

$$\text{Ohms} = \frac{\text{volts}}{\text{amperes}} = \frac{0.98 \text{ volts}}{0.33 \text{ amperes}} = 2.97 \, \Omega$$

CONDUCTORS

Conductors can be defined as any substance that will conduct electricity. **Conductivity** is defined as the reciprocal of resistance, C = 1/R and is expressed in reciprocal ohms (Ω^{-1} or **mho**). The most common forms of electrical conductors are wire(s) which may be a single strand or many strands braided together. Braided wire is usually used where the conductor must be flexible; single-strand wire is used where there will be no movement. The larger the diameter of the wire, the more easily the electrons will flow through it. The smaller its diameter, the more resistance it offers to the flow of electricity. Copper is the most commonly used conductor with aluminum being second. Wires (in the United States) are measured in thousandths of an inch (mils). A 1-mil wire has a diameter of 0.001 in; therefore a 100-mil wire has a diameter of 0.1 in. In normal usage, wires are designated by the **AWG (American Wire Gage)**, which is standard and specifies the diameter and area of wires (see Table 10.6). The larger the diameter and the larger the area of a wire, the more electricity it will conduct. The ease with which electrons flow in conductors, as noted before, depends upon the relative looseness of the atomic structure. The looser the bonds holding the electrons in the atom, the less the resistance to electron flow; and the more tightly the electrons are held, the higher the resistance.

TABLE 10.6 Wire Size and Resistance (standard annealed copper wire at 20°C)

AWG no.	Diameter, inch	Ohms/1000 ft	Ohms/meter
40	0.00315	1049.0	3.44
38	0.00397	659.6	2.16
36	0.00500	414.8	1.36
34	0.00631	260.9	0.855
32	0.00795	164.1	0.538
30	0.01003	103.2	0.338
28	0.01264	64.9	0.213
26	0.01594	40.8	0.134
24	0.02010	25.7	0.084
22	0.02535	16.1	0.053
20	0.03196	10.15	0.033
18	0.04030	6.385	0.021
16	0.05082	4.016	0.013
14	0.06408	2.525	0.0083
12	0.08081	1.588	0.0052
10	0.1019	0.999	0.0033
8	0.1285	0.628	0.0021
6	0.1620	0.395	0.0013
4	0.2043	0.249	0.00082
3	0.2294	0.201	0.00066
2	0.2576	0.159	0.00052
1	0.2893	0.126	0.00041
0	0.3248	0.100	0.00033

Wire size is important because of the current-carrying capacity and the voltage drop in the wire. All conductors, even wires, have a certain amount of resistance, and there will always be a power loss in the form of heat because of the current flowing through the conductor. That power loss is reflected in the voltage available at the end of the wire and is known as the voltage drop. This drop in voltage can be significant if the diameter of the wire is small (resistance is high) and the current flow is large (Fig. 10.1). Therefore, do not use long extension cords; your delivered voltage may be too low to allow the appliance to operate. The voltage drop can be measured with a voltmeter. When in doubt, use heavy extension cords to minimize the voltage drop.

CONDUCTIVITY OF VARIOUS METALS

Resistivity or specific resistance is reciprocal of conductivity. By comparing the resistivities of various metals, we can see which ones offer least resistance and therefore conduct electricity the best. Table 10.7 gives some of these values, and you can see that silver is the best conductor of all the metallic elements. Nickel is the worst of the group, but certain alloys of nickel, such as Nichrome, have even higher resistivity and are used as heating elements in hot plates, for example. Copper is used in electric wiring because it is the next best conductor to silver and costs considerable less.

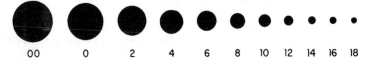

FIGURE 10.1 Actual diameters of different AWG sizes of wire, without insulation.

TABLE 10.7 Electric Resistivity of Metals

Metal	Resistivity, $\mu\Omega$-cm
Aluminum	2.6548
Copper	1.6730
Gold	2.35
Nickel	6.84
Silver	1.59
Tungsten	5.65

HOW TO READ A RESISTOR

Table 10.8 shows the color code normally used for identifying resistors. The colored band closest to the end identifies the first digit, the next colored band the second digit, the third band gives the multiplier in powers of 10, and a fourth band, if shown, identifies the tolerance.

ALTERNATING AND DIRECT CURRENT

There are two types of current: alternating and direct. The type normally found in industry and in homes all over the country is of the alternating type. In **alternating current, AC,** the electrons flow in one direction, then stop and reverse themselves, then stop and continues the cycle many

TABLE 10.8 Color Code for Resistor Values, in Ohms

Color	1st & 2nd bands	3rd band	4th band
Black	0	0	±20%
Brown	1	1	
Red2	2	2	±2%
Orange	3	3	
Yellow	4	4	
Green	5	5	
Blue	6	6	
Violet	7	7	
Gray	8	8	
White	9	9	
Gold		−1	±5%
Silver		−2	±10%
No Color			±20%

times a second. In the United States, the frequency of the cycling is 60 Hz (hertz); in European countries 50 Hz; one cycle per second equals one hertz.

Modern power stations generate between 11 and 14 kilo volts (kV) AC, which may be boosted by transformers as high as 275 kV for long-distance transmission. The power needed in various areas is brought to substations, where the voltage is reduced to around 2500 V, and then reduced further at secondary points to about 115–230 V AC where it is delivered to industry and residences. The figure "115 V AC" is rather flexible. Actually the voltage may range from 105 V to as high as 122 V, depending upon the conditions. Power companies at times arbitrarily reduce the voltage 5 to 10 percent to conserve energy (typically during very hot summer months). When this happens, disconnect as much equipment as possible to avoid damage to it.

Direct current, DC, is current that flows in only one direction. This is the type found in batteries and dry cells. The current is considered to flow from the negative terminal to the positive terminal, because the negative terminal of the battery has an excess of electrons and the positive terminal has a deficiency of electrons. Since AC equipment can be permanently damaged by connecting it to direct current and vice versa, it is wise to determine whether a piece of equipment is meant to operate AC or DC. This information is usually marked on the equipment: if the label states "AC only" or "DC only", the equipment must be hooked up only to AC or DC as required. A label stating AC/DC means that either type of current can be used.

FUSES

Fuses are safety devices used to limit the flow of current. Fuses are rated by the current they safely conduct: 5, 10, 20, 30 amps or more. When the current exceeds that stated amount, sufficient heat is generated in the fuse to cause the metal alloy in the fuse to melt; thus the circuit is broken, and electricity ceases to flow. The wiring in a particular circuit is of a definite size and rated to carry up to a certain amount of current safely. If too much current passes through this wiring, heat is generated and fire can result. The amount of current to be carried by a conductor is therefore limited by placing a fuse in the path of the electrons. The fuse will "blow" (burn out) when the current-carrying capacity of the fuse is exceeded. Thus a 15-A fuse will carry 15 A without blowing.

The ordinary fuse will carry its rated capacity indefinitely, but will blow when the current exceeds its rating. Electric monitors are rated by the number of amperes they draw when they are running continuously. A 15-A motor therefore draws 15 A. However, all electric motors draw excessive amounts of current when they are starting, as much as two to three times their normal rating; however, they drop down to their normal rating when they have reached full speed. It is during this starting period, the time of excess current flow, that fuses may blow, because their rating is based upon the motor's normal requirements. To prevent fuses from blowing every time a motor starts, time-lag fuses were developed. They will carry a large overload of current safely for a few seconds, but will blow just as quickly for continuous overloads as will an ordinary fuse. They have the appearance of ordinary fuses, but they are constructed differently and are used whenever electric motors are on the circuit.

CAUTION: Fuse panels are electrically "hot" (current flows through them). Be extremely careful when replacing fuses.

A. Type of Fuses

A visual inspection reveals which fuse has blown: the element has melted. Unscrew or unplug the fuse and replace it with a fuse having the same rating. Do not replace with a higher-rated fuse or insert a conductive material (coin, slug, etc.); larger currents can then be carried by the circuit. This may result in overheating and potentially cause a fire. Ferrule and cartridge type fuses are securely gripped by spring contacts, and may be "frozen" to the contacts because of corrosion. For these fuses, always use a fuse puller which is safe to use because your hands are away from "hot" contacts (live wires) and your hand motion is away from the box. See Fig. 10.2 for three types of common fuses.

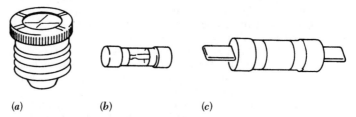

FIGURE 10.2 Three types of fuses, (*a*) Screw-type fuses, maximum 30 A. (*b*) Ferrule-type fuse, maximum 60 A. (*c*) Cartridge-type fuse, with knife-blade terminals; such fuses are usually made for currents greater than 60 A.

B. Procedure for Replacing Fuses

1. If practicable, throw the power switch to OFF to disconnect all power from the fuse box. Unfortunately, the power switch may control other important circuits, and total disconnect may not be practical.
2. Use a fuse puller to remove fuse.
3. Replace fuse with the fuse puller.

C. Fuse-Holding Blocks

Some fuse panels have fuse-holding blocks that are removable. The cabinet is fitted with copper bus bars, and the fuse blocks engage the bars to make contact. A red light goes on as a signal that a fuse block is burned out. To replace a fuse, remove the entire block which contains two fuses. Once the burned fuse is replaced, the entire block, made of insulating plastic, is pushed back into the cabinet. This piece of equipment is superior from a safety standpoint to the more traditional fuse box.

D. Determining the Cause for Blown Fuses

1. Replace the fuse. If the fuse does not blow again, there may have been a temporary overload, because the system is probably very near the capacity of the circuit.
2. If the fuse blows again, change some of the electrical units to another circuit.
3. If the fuse still blows, you probably have a short circuit, which you must isolate.
4. Disconnect all units from the circuit and replace the fuse. The fuse should not blow now because no units are connected and nothing is drawing current. If it does blow, call an electrician immediately.
5. Reconnect one of the units that was originally connected to the circuit; if the fuse does not blow, connect the second, then the third, etc. Eventually, reconnection of one of the units will blow the fuse.
6. Disconnect all the units except the last one that was connected. Replace the fuse. If the fuse blows again, that is the defective unit, and it should be serviced.

CIRCUIT BREAKERS

Circuit breakers, often used today instead-of fuses, perform the same function as a fuse, but operate in different way. They open the circuit in the event of an overload, but unlike fuses, they may be reset and reused. Circuit breakers look much like toggle switches (Fig. 10.3), but they usually have four positions: ON, OFF, TRIPPED, and RESET. When the circuit breaker is carrying current, the position is ON, and just like a fuse, it will carry its rated capacity indefinitely. However, even small overloads will cause the circuit breaker to trip, breaking the circuit. When this happens, push the handle to RESET, when you feel it engage, push the handle to ON. Circuit breakers are designed to carry small overloads for a short period of time, as for motor starting, but they trip on continuous overload.

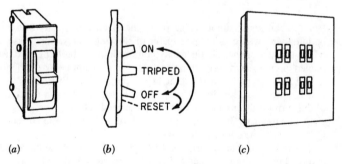

FIGURE 10.3 Circuit breaker, (*a*) Circuit breaker, (*b*) The operating positions of a circuit breaker, (*c*) Control panel for a circuit breaker.

SERIES AND PARALLEL CIRCUITS

Electric circuits are designed to be connected in series, parallel, or both. In **series circuits**, every electron flows through every electric device. An example is a string of old-fashioned Christmas-tree lights. These are connected in series; the electric circuit is broken when one bulb burns out, and electricity ceases to flow (see Fig. 10.4). In **parallel circuits**, each electric device uses its own current or

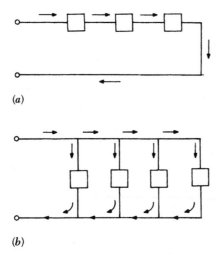

FIGURE 10.4 (*a*) Series and (*b*) parallel circuit diagrams.

number of electrons independently of the other. In the series circuit, when the electrons enter the conductor through one wire of the conductor, they must pass through every piece of equipment before returning to the other wire of the conductor. The current in the wire is therefore the same at all points. In the parallel circuit, the total current that enters the wire of the conductor at the plug is divided between the requirements of each electrical unit, each using its own amperage. Therefore, when several electric devices are connected in parallel, as when such devices are connected to the triple outlet of an extension cord, caution must be exercised that the sum of the amperes used by each unit does not exceed the safe carrying capacity of the extension cord. This concept must be understood in order to explain why fuses blow when additional units are added in parallel.

EXAMPLE 10.3 A circuit is fused for 20 A. An electric motor drawing 10 A and an electric heater drawing 7 A are connected to the same outlet. Thus the current through the circuit is 17 A. If another heater or unit that draws 5 A is also connected to the same outlet, then the total current will be 22 A, exceeding the safe current-carrying capacity of the wire conductors, and the fuse blows. The answer to this problem is to use another circuit, protected by another fuse. Never remove a fuse and replace it with a larger fuse or solid conductor (wire, coin, etc.). This removes the safety device, and if too much current is drawn through the wires, fire may result.

ELECTRICAL POWER

In a typical power supply (Fig. 10.5), the electric company brings in service of, say, 110 V to a building with wire conductors capable of carrying 100 A. This passes through the master switch (A), which can disconnect all electricity coming into the building. From there it is divided into smaller branch circuits, each fused to protect that circuit from being overloaded.

All the current, up to 100 A at 110 V, passes through A. Only 20 A of current can pass through fuse B which has been drawn from outlets B_1 and B_2. Fuse B (rated at 20 A) will blow when the sum of the currents through these two outlets exceeds 20 A.

A maximum of 30 A can pass through fuse C. When the current drawn from outlets C_1, C_2, and C_3 exceeds 30 A, only that fuse will blow.

Fuse D is capable of carrying 40 A and will pass all the current drawn from outlets D_1, D_2, D_3, and D_4, but it will blow when the current exceeds 40 A.

Only current drawn from outlet E_1 will pass through fuse E and the capacity of circuit E is 10 A. To summarize:

Total current available 100 A

Circuit B Capacity 20 A

Circuit C Capacity 30 A

Circuit D Capacity 40 A

Circuit E Capacity 10 A

Total of Circuits B, C, D, and E equals 100 A.

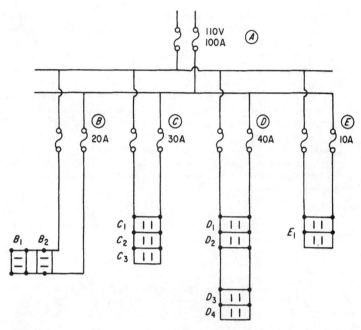

FIGURE 10.5 Current distribution in a typical installation.

The main fuse will blow when the total current drawn exceeds 100 A. If one of the subfuses, B, C, D, or E, is defective, and the current drawn through all circuits exceeds 100 A, then the main fuse, A, will blow.

SERVICING INOPERATIVE DEVICES

If a device does not work when connected to an electric outlet:

1. First check the outlet to see if electricity is available (if the outlet is "hot"), by plugging a desk lamp (which is known to work) into the socket.
2. Next check the electric plug to be sure that the wires are connected to the plug.
3. Check the cord connecting the plug to the unit to see if the wire is broken.
4. Examine the fuse in the electrical unit. It may have blown.
5. See that the switch is turned on.
6. If none of these steps results in operation, call an electrician.

ELECTRICAL TESTING INSTRUMENTS: VOLTMETER, AMMETER, CONTINUITY TESTER, AND MULTIMETER

Voltage is tested with a **voltmeter**. The two probes of the voltmeter are touched to the two terminals being checked. The voltage is read off the dial. The probes must touch the base wire, which is not covered with insulation The voltmeter must be connected to the two conductors when in use. Meters must be selected for the determination they are to make, the range to be covered, and the precision desired. *Exercise caution when you choose the meter. Do not overload.*

Current is tested with an **ammeter**, which must be inserted in the line so that the total amount of current can be read off the meter. The unit must be disconnected (one of the conductors either cut or be disconnected from itself at the junction), and the ammeter is inserted in series with the wire conductor. In work with energized circuits, one may use an external type of ammeter, which has a circular portion that can be opened, placed around a single wire, then closed. The current passing through the wire creates a magnetic field around the wire conductor, which actuates the needles of the ammeter and gives the reading in amperes of the current passing through the conductor.

A **continuity meter** is used for resistance measurement to determine whether or not a circuit is complete. Obviously, if the circuit is broken, then no current will flow during the continuity test.

The **multimeter** is a compact instrument that is used to measure the electrical characteristics of circuits (Table 10.9). Among the characteristics that can be measured, depending upon the instrument, are voltage, amperage, resistance of alternating and direct current, and output voltages. It can be extremely useful for determining the operating characteristics and for locating abnormalities in electrical systems.

TABLE 10.9 Steps for Operating a Multimeter

1. Do not overload.
2. Insert probes leads in proper outlets on the meter.
3. Set selector to proper range.
4. When measuring resistances, do not measure the components of units that have electricity flowing through them. This relatively high current will burn out the meter if it is used when the item is electrically "hot."

Note: Always follow the directions given in the instruction manual because of variations in design.

ELECTRIC MOTORS

Electric motors convert electric energy to mechanical energy and do work. They must be controlled with safe and positive-acting controls. Control means to govern or regulate, to start, stop, reverse, or alter the rate of revolution of the motor (normally rated in revolutions per minute or rpm's). Any piece of equipment that is used to accomplish this end is a control component or motor control.

A. Starting Motors

Small motors, fractional horsepower to 1/2 horsepower, can be started with a simple, knife-type, on-off switch. Many small motors use nothing more than a cord and plug with a toggle switch. For larger motors (1 to 10 hp), manual motor starters are used to give manual control and both overvoltage and undervoltage protection. In the event of a drastic voltage variation, the starter disconnects the motor from the line voltage, thus protecting the motor. Smaller motors may use 110 or 220 V single-phase AC, but larger motors invariably use 220-V three-phase AC. The contacts are made by manually pushing the START button, which is physically linked to the contacts (Fig. 10.6). Once the contacts are closed, the linkage is latched into position. Should an overvoltage or undervoltage condition develop, a safety component will unlatch the linkage, disconnecting the motor. When the STOP button is pushed, the linkage is released and the contacts disconnect.

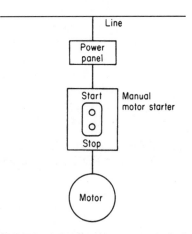

FIGURE 10.6 Manual motor starter. Push the START button to start and push the STOP button to stop.

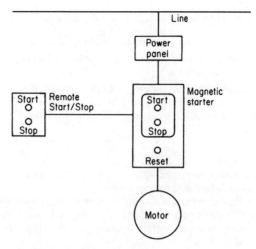

FIGURE 10.7 Magnetic motor starter. Push the START button to start and push the STOP button to stop.

B. Magnetic Starters

In contrast to the manual starter, which is normally located in the immediate vicinity of the motor, a magnetic starter (Fig. 10.7) can control a motor that is not in the immediate vicinity, as long as the magnetic starter is electrically connected to it. When the START button is pressed, it energizes a small coil in the starter which magnetically causes the contact points to close, starting the motor. Only a momentary push to start the motion is necessary. The contacts on closing also complete the electric circuit for the small coil, thus keeping the coil energized and the contacts closed. When the motor is to be stopped, the STOP button is pushed. This push breaks the electrical circuit of the energized small coil, causing all the contacts in the starter to disengage, including that of the small coil. Thus, the motor is stopped at that point. The cycle can be repeated at will. Magnetic starters also have overvoltage and undervoltage protection which automatically breaks the electrical circuit of the small coil, in effect causing the same result as pushing the STOP button. Magnetic starters may have several remote START-STOP switches as well as a set on the starter itself. When a magnetic starter becomes de-energized as a result of an over-voltage or undervoltage condition, the starter will not be operative again until the RESET button has been pushed.

C. Using Motors Correctly

Motors can be used for jobs that normally require twice the rated horsepower, but only for a short time. No motor can continually provide more than its rated horsepower. Most 60-Hz AC motors run at around 1725 to 1750 rpm. Special-purpose motors run at higher specified speeds. Motors develop heat when they run continuously, usually about 40°C (above room temperature). Therefore, motors running in a hot room may feel too hot to the touch, but if they are not running overloaded, they are running normally and are safe. Motors will be damaged if excessive current flows through them for a considerable time.

Here are a few general tips:

- Always use the proper motor for the job. Abusing motors by overloading them will damage them and burn them out.
- Chemicals and water can also cause motors to burn out. Keep them clean and away from possible spillage of water and chemicals.

Protect motors by following the simple instructions below:

- Three-phase motors: Use a correct fuse on each line.
- Single-phase motors: Use only one fuse on 115 V (the black ungrounded line) and two fuses on 230 V (one in each line).
- Thermal-overload devices carry a small overload (when starting the motor) for a short time without tripping, but trip quickly if a large overload is imposed. When they do trip, they can be reset by pressing the RESET button after a short time lapse.

D. Types of Motors

Capacitor motors are fitted with a cylindrical "capacitor" attached to the housing. They are used to start heavier loads, such as a vacuum pump.

Split-phase motors are usually fractional-horsepower motors, less than 1/3 hp. They have no brushes or commutators, but are not used for hard-to-start items such as compressors.

Repulsion-induction motors operate only on single-phase AC. They have the ability to start heavy-load items.

Universal motors operate on AC or DC, but they do not run at constant speed. The greater the load, the lower the speed. Rheostats can be used to control and vary their speeds, as in mixers.

Three-phase motors require three-phase AC. They cannot be run on single-phase AC. Their direction of rotation can be reversed by changing any two of the three leads to the motor.

Some larger motors, from about 1/2 hp and higher, have four leads; this means that the motor will run on 115 or 230-V AC, depending upon the connection. These are called dual-voltage motors. At the lower voltage, the coils are connected in parallel, which means that two leads will be joined at each terminal as in Fig. 10.8. At the higher voltage, the two coils of the motor are connected in series. Two of the coil terminals (one from each coil) are joined together and are not connected to the input voltage, as in Fig. 10.9.

FIGURE 10.8 Low voltage connection.

FIGURE 10.9 High voltage connection.

E. Belts, Pulleys, and Power Transmission

Motors may be coupled directly to the shaft of a piece of equipment with a rigid coupling for stirrers or with a flexible coupling to a machine. Rigid couplings should have the correct internal-diameter opening to fit the shaft of the motor and the machine. Force is not required to assemble the unit, provided the shaft is not scarred or burred. If the shafts are burred, smooth them with a fine file (on the motor while running) just enough to remove the burrs. Always tighten the locking screws firmly so that the shafts will not slip and become burred. Do not apply excessive pressure to tighten the locking screws.

A general rule for the machine to run at higher speeds is that the machine pulley is to be smaller than the motor pulley. For the machine to run at lower speeds, the machine pulley is to be larger than the motor pulley.

MAKING ELECTRICAL CONNECTIONS

You must strip the insulation from the end of electric wires before you can make a connection. Use care when removing the insulation.

1. Do not nick the copper wire. Bending of nicked wire leads to breaks.
2. Apply the knife or cutter at an angle when cutting away insulation to attach the wire to a terminal. This decreases the chance of nicking the wire and gives the correct insulation angle cut.

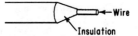

FIGURE 10.10 Correctly stripped wire for splicing.

3. Apply the knife at a greater angle when preparing the wire to be spliced to another wire (Fig. 10.10).

A. Electrical Terminal Connections

Faulty connections of electric wire to terminals of equipment are a major source of trouble and improper performance. Use care when making connections.

1. Strip away only sufficient insulation.
2. Bend the wire into a 3/4 circle with needle-nose pliers (Fig. 10.11).
3. Tighten the screw firmly, after you insert the loop in the correct direction (Fig. 10.12). Loop will open if it is inserted in the opposite direction and the screw is tightened.

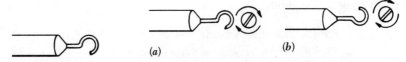

FIGURE 10.11 Loop for terminal connection.

FIGURE 10.12 (*a*) Loop closes when screw is tightened (correct). (*b*) Loop opens when screw is tightened (incorrect).

4. Solderless lug connections should be used on larger-diameter wires.

Braided wires: After stripping insulation, twist the braided wires with the fingers to make essentially one strand.

CAUTION: Avoid short circuits by making clean connections. Prevent wires from touching other terminals.

B. Splicing Wires Temporarily

Electric wires can be connected together easily with solderless connectors, provided there is no strain on the wires.

1. Strip the insulation from the end of the wires as shown correctly in Fig. 10.10.
2. Twist the two (or more) wires together as shown in Fig. 10.13. If one wire has a smaller diameter than the others, let it extend beyond (project some beyond) the others.
3. Slip the connector over the bare twisted wires and screw the insulating shell over the wires, tightening the connector securely by twisting. The insulating shell should extend over the insulation of the wires; then you will not have to tape the connection (see Fig. 10.14).

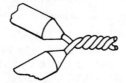 

FIGURE 10.13 Splicing wire together before securing with solderless connector.

FIGURE 10.14 The completed splice with insulating shell.

C. Continuous Wire

To connect one wire to another wire that is not to be cut, use the following procedure.
For solid-strand wires:

1. Strip the continuous wire for about 1/2 inch at the point desired as shown correctly in Fig. 10.10.
2. Wrap the end of the stripped wire to be connected to the continuous wire as in Fig. 10.15.
3. If it is a permanent connection, solder and then tape. If it is to be a temporary connection, merely tape with insulating tape.

FIGURE 10.15 Splicing wire (one uncut, one cut).

D. Splicing Wires Permanently

1. Strip the ends of the wires as shown correctly in Fig. 10.10.
2. Hook the two wires together as shown in Fig. 10.16.
3. Twist the wires in opposite directions as in Fig. 10.17.
4. Solder and then tape the twisted wires.

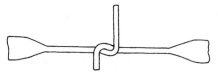

FIGURE 10.16 Hooking the wires.

FIGURE 10.17 Twisting the wires together.

E. Electrical Soldering

Procedure for soldering wires:

1. Use an electric soldering iron and flux-core solder.
2. Clean, by scraping, the wire or object to be soldered. Good soldering results will only occur if surfaces are clean.
3. Clean the soldering iron tip by allowing it to heat until the residue solder on the tip melts and then quickly wipe the tip firmly with a heavy piece of cloth (quickly so that the cloth does not burn). The tip will now be clean and shiny with a film of melted solder.
4. Apply the hot soldering iron to the surface to be soldered (Fig. 10.18).
5. Touch the solder to the area to be soldered. Do not touch the solder to the hot iron.
6. When the wire is hot enough to melt the solder, the solder will melt and flow evenly and smoothly, making good contact (Fig. 10.19). The joint should not be jiggled or moved until the solder has cooled and solidified. The surface of the solder will turn from shiny to dull when the solder has cooled.
7. Remove the iron when solder has flowed.

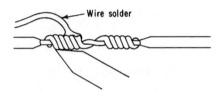

FIGURE 10.18 Correct method of soldering: Iron heats the wire not the solder.

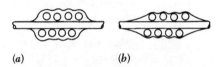

FIGURE 10.19 Soldering: (*a*) Incorrect hot solder is applied to cold wire, (*b*) Correct: solder is melted by hot wire providing a good flow of solder.

NOTE: You will get good results if:

1. The wire surfaces are clean.
2. The wires are heated to melt the solder. Never use the soldering iron to melt the solder directly.
3. The joint is never jiggled or moved until the solder has cooled and solidified. The joint will be weakened if this is done. The surface of the solder will turn from shiny to dull when the solder has cooled.

F. Taping

Tape electric connections to prevent short circuits and electric shocks.

1. Use plastic tape that is specially made for electrical work. Although it is thin, it possesses high insulating qualities.
2. Start the tape at one end of the insulation. Wrap spirally to the other insulation; then reverse and return to the start. Continue the process of wrapping until the taped section is as thick as original wire.

NOTE: Keep the tape stretched at all times.

GROUNDING ELECTRICAL EQUIPMENT

Normally the electricity available for use in the laboratory and plant has one side connected to the ground; the other side is "hot" at 110 V. If an electric device feels tingly when it is touched, the plug should be removed and the prongs reversed when it is replugged. This keeps the device at ground potential. In order to observe safety rules, the electric device should always be **grounded**; that is, an electric conductor, a wire, should be connected from the metal casing of the device to a water pipe or any metal object that is connected to the ground. This conductor will then carry any dangerous current to the ground instead of through any individual, thus affording positive protection against accidental electrocution by touching a shorted device.

Electricity is dangerous and can be fatal. It is not necessarily the high voltage that is dangerous. It is the number of amperes that can cause injury and death. Electricity is transported through electric wires that are covered by nonconducting materials (insulation). Most electric equipment is constructed of metal. When that metal comes in accidental contact with the wires carrying the electricity, a dangerous situation arises. When that happens, and you touch the normally neutral metal which is now "hot" (carrying current) because of the accidental contact, your body becomes the path for the electric current. It is the flow of the electric current through your body that is dangerous.

NOTES:

- The higher the voltage, the greater the number of milliamperes that can pass through your body under identical conditions. It is therefore more dangerous to touch a 230-V line than a 115-V line.

- The effectiveness of your contact with the hot wire and with the ground determines the number of milliamperes that pass through your body.
- When you stand on dry surfaces, the contact is poor and few milliamperes pass through you.

When you stand on wet surfaces, the contact is much better and many milliamperes may pass through you.

CAUTION: Avoid working with energized electric wires while standing on wet surfaces.

A. Summary of Safety Precautions

Wherever there are electric outlets, plugs, wiring, or connections, there is danger of electric shock.

1. Do not use worn wires; replace them.
2. Use connections that are encased in heavy rubber.
3. Ground all equipment: use three-prong plugs or pigtail adapters.
4. Do not handle any electric connections with damp hands or while standing on a wet or damp surface.
5. Do not continue to run motors after there has been a spill on the motor.

B. Importance of Grounding

Electric equipment is grounded for safety. Improperly grounded items can cause injury, fires, damage to other units, and death. When you refer to ground, you are referring to the earth, which is neutral and therefore safe. A "grounded piece of equipment" means that electrically the unit is at the same electric potential as the earth, which is what you want.

Grounding wires, called **grounds**, are electric-wire connecting to a point such as a water pipe buried in the ground or a ground rod which has been driven into the ground. Ground wires normally carry no electric current. They are connected to a metal component of the device, which itself is electrically connected to the motor, switch, case, cabinet, motor shell, base, or outlet box.

C. Properly Grounded Items

Should a motor become defective the insulation breaks, and the hot electric wire touches the casing. The casing becomes hot. Because the item is grounded properly, the current flows through the grounding wire to ground and not through you. As stated above, if the casing becomes hot, but because there is no ground, your body becomes the conductor when you touch the casing and electric shock results, causing injury or death.

D. Proper Grounding

1. Merely to attach an electric ground to a piece of apparatus is not sufficient. Be certain that every metal component in the apparatus is electrically grounded. Many times the mechanical connections and fittings on equipment may themselves be insulators and thus will insulate the other parts from the electrical ground.
2. Be certain that the electric ground connection is actually an electric ground and not itself insulated from ground. Consult an electrician to determine what is and what is not a "true ground". Then use that ground.

EXAMPLE 10.4 An electric motor may be grounded at the switch box but not be grounded itself. Should the motor become defective, with insulation breaking on the wires or coils, it will become hot because it is insulated from the ground. This is a dangerous situation. Safe installations are shown in Figs. 10.20 and 10.21.

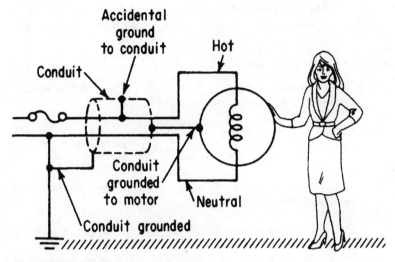

FIGURE 10.20 Safe installation, properly grounded.

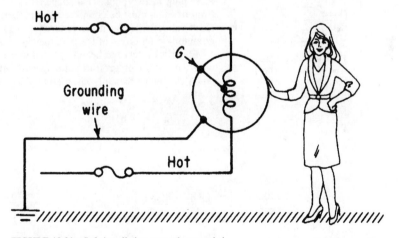

FIGURE 10.21 Safe installation, properly grounded.

E. How to Determine If Your Equipment Is Grounded

1. Complete your assembly.
2. Electrically disconnect your equipment from the power source.
3. Obtain a multimeter and set the selector to resistance, at the zero range.
4. Firmly press one probe onto your previously determined safe electric ground.

216 CHAPTER TEN

5. Then touch all pieces of metal of your assembly with the other probe. All readings should be zero, showing that all are electrically connected to each other.
6. Remove your multimeter, and connect your assembly to electric power.
7. Adjust the selector switch to the voltage used (for example, 115 V AC) on your multimeter.
8. Firmly press one probe of the multimeter to the electric ground, and then press the other probe to all metal surfaces. All readings should be zero volts.
9. Your equipment is electrically safe under normal conditions for use.

F. Ground Fault Circuit Interrupters

Most modern electric outlets are fitted with three openings, two for the electric wires and one for the grounding wire. These outlets look like Fig. 10.22.

FIGURE 10.22 This newer type of electrical outlet has a ground.

CAUTION: Just because the outlet looks as if it is a grounded outlet, do not assume it has been electrically grounded. Always test.

Older installations have only the two-opening receptacle (Fig. 10.23) and you must affix a grounding wire.

If your electrical device is fitted with a three-prong plug but your outlet receptacles are only the two-opening type, you first must use an adapter that converts from three-plug to two-plug. This adapter has a "grounding wire" hanging from it which must be connected to a proper ground according to local electrical code. Under no circumstances should you ever break off the grounding-plug prong of the plug merely to avoid locating an adapter.

FIGURE 10.23 The old type of electric outlet has no ground.

Electrical safety has been improved tremendously in the past decade with the advent of **ground fault circuit interrupters, GFCI**. These GFCI can be permanently installed in place of a receptacle and are also

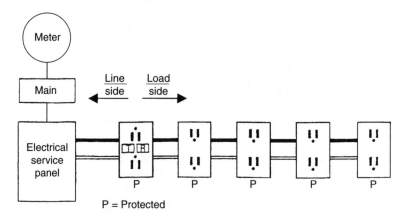

FIGURE 10.24 A typical GFCI installed as the first receptacle to protect all other receptacles in the circuit.

available as a combination circuit breaker/GFCI for installation in electrical service panels to protect all outlets on that circuit. GFCI are now required in most commercial and residential electrical circuits. These devices are designed to automatically interrupt potentially lethal currents caused by ground faults and improper grounding situations.

ELECTRIC CIRCUIT ELEMENTS

Standard symbols for the various elements encountered in electric circuits are given in Table 10.10.

TABLE 10.10 Standard Electric Circuit Symbols

Element	Symbol
Connecting wire (negligible resistance)	
Resistance (fixed)	
Variable resistance (two terminals) such as a dial box	
Rheostat or potentiometer (three terminals)	
Capacitance (fixed)	
Capacitance (variable)	
Inductance	
Inductance (iron core) or choke	
Transformer (or mutual inductance)	
Transformer (iron core)	
Autotransformer	
Battery (two or more cells in series)	
Voltaic cell	
Fuse	
Switch [single-pole single-throw (SPST)]	
Switch [double-pole double-throw (DPDT)]	
Tap key (SPST) momentary contact	
Galvanometer	
Ammeter	
Millimeter	
Voltmeter	
Wattmeter	

REFERENCES

1. Marsden, R. M., *Newnes Electronic Circuits Pocketbook*, Volume 1, Butterworth-Heinemann. ISBN 0-7506-0132-9, 1991.
2. Bolton, W., *Newnes Instrumentation and Measurement Pocketbook*, Butterworth-Heinemann. ISBN 0-7056-0039-X, 1991.
3. *The Flow and Level Handbook, The pH and Conductivity Handbook, The Temperature Handbook, The Data Acquisition Systems Handbook, The Pressure Strain and Force Handbook, The Electric Heaters Handbook,* Volume 28 Series. Omega Engineering, Inc. PO 4047, Stamford, CT 06907-0047.
4. McPartland, B., and McPartland, J., *National Electrical Code Handbook,* 23rd edition. Handbook based on National Fire Protection Association's 1997, National Electrical Code McGraw-Hill, Inc., 1997.

CHAPTER 11
SAMPLING

THE IMPORTANCE OF PROPER SAMPLING

If you were asked to sample a pizza, would an individual pepperoni slice, black olive, green pepper, or cheese slice truly represent the total pizza? A true representation of the pizza would require that this heterogeneous mixture be converted into a homogeneous mixture. The first step in obtaining an accurate analysis or assessment of a process is to take a representative sample. If the sample is not properly taken, then the analysis is worthless.

There are two general rules to follow in sampling:

1. Take a large initial sample.
2. Take a representative sample.

A large initial (**gross**) **sample** must be taken to provide a true composition of the entire raw material delivery, process stream, or final product. A single lump of coal taken from the top of each coal car may not represent the composition of the entire coal shipment. A **representative sample** is one that contains *all* the components in a raw material, process stream, or final product. Your company will have established detailed SOPs for sampling all process materials.

HETEROGENEOUS VERSUS HOMOGENEOUS MIXTURES

Raw materials and products must be sampled in order to conduct analyses for components or to determine their purity. The size of the lot to be sampled can range from a few grams to thousands of pounds, and yet the sample used in the analysis must represent as closely as possible the average composition of the total quantity being analyzed.

All matter is either homogeneous or heterogeneous. **Homogeneous** materials are uniform in composition, they have the same makeup throughout. In other words, if sampled, homogeneous materials will give the same analysis. Homogeneous materials may be pure substances or mixtures. Pure air, water, and chemicals are examples of homogeneous materials. **Heterogeneous** materials have a varied composition and will give different compositions depending on where they are sampled. All heterogeneous materials are mixtures; that is, they are materials whose components can keep their own identities and can usually be separated from each other by some physical means.

THE GROSS SAMPLE

The gross sample of the lot being analyzed is supposed to be a miniature replica in composition and in particle-size distribution. If it does not truly represent the entire lot, all further work to reduce it to a suitable laboratory size and all analytical procedures are a waste of time. The technique of sampling varies according to the substance being analyzed and its physical characteristics.

220 CHAPTER ELEVEN

BASIC SAMPLING RULES

1. The size of the sample must be adequate, depending upon what is being measured, the type of measurement being made, and the level of contaminants.
2. The sample must be representative and reproducible. In static systems multilevel sampling must be made.
3. Many companies use color coding on sample tags. For example, blue tags might be used for a routine laboratory analysis while green tags might be used for high priority laboratory samples.

SAMPLING GASES

Gases are primarily sampled in a process to determine the quality of a process stream and/or the quality of the final product. Gases are routinely sampled during process operations to determine any contaminants in process streams or emissions from the plant. These contaminants can be gaseous or particulate in nature. Where particulate contaminants have large particles, they can settle out immediately, but smaller particles tend to behave as a gas and settle many miles from the emissions source. The most common expression for atmospheric pollutants is micrograms per cubic meter ($\mu g/m^3$) of sampled air.

One device used to determine suspended atmospheric particulate pollutants is the high volume sampler. **High volume (Hi Vol) samplers** operate on the principle of filtration and are used to collect particulate ranging from 0.1 to 100 µm in diameter. A typical Hi Vol sampler is housed in a shelter and contains a vacuum-type air circulation blower. These samplers are equipped with glass fiber filters that are weighed before and after the sampling period. Most Hi Vol samplers are operated for 24 hours at an average sampling rate of about 1.4 m^3 of air per minute (approximately 2016 m^3/day). Since the accumulating particulate matter tends to slow the Hi Vol sampling rate, it is necessary to determine the average flow rate each time. Simply average the starting flow rate and the final flow rate. The mass of collected particulate can be determined by weight as shown below in the example problem. In addition, solvent extractions can be used to further identify organic and inorganic contaminants. Radioactive fallout can also be determined with this technique.

EXAMPLE 11.1 The following data was collected using a Hi Vol sampler, and calculates the particulate matter content in the sampled air.

Weight of filter (before) = 3.167 g
Weight of filter (after) = 3.567 g
Weight of total particulate = 0.400 g

Flow rate (start) = 1.81 m^3/min
Flow rate (finish) = 1.58 m^3/min
Flow rate (average) = 1.70 m^3/min

Time (start) = 4 PM 2/7/98
Time (finish) = 4 PM 2/8/98
Time (total—24 hours) = 1440 minutes

$$\text{Sample volume} = \frac{1.70 \text{ m}^3}{\text{minute}} \times 1440 \text{ minutes} = 2448 \text{ m}^3$$

$$\text{Particulate concentration} = \frac{0.400 \text{ g}}{2448 \text{ m}^3} = 163 \times 10^{-6} \text{ g/m}^3$$
$$= 163 \text{ µg/m}^3$$

A. Gas Sampling Techniques

Several sampling methods are available for collecting gas samples and gaseous pollutants. The principal methods that have been used are: (1) absorption in a liquid; (2) adsorption (held to the surface) on various solids; (3) condensing or freezing the pollutants; and (4) grab sampling.

Gas-liquid **absorption** sampling is the process by which a gaseous contaminant in air is removed by dissolving or reacting the contaminant with a liquid. A typical chemical absorption process would involve drawing a volume of air through a solution which reacts with the gaseous contaminant to form a nongaseous compound. It is not necessary to have 100 percent collection efficiency; however, the efficiency should be known and reproducible. In some circumstances, a sampling system having a relatively low collection efficiency (e.g., 40 to 50 percent) could be used provided that the desired sensitivity, reproducibility, and accuracy are obtainable.

A variety of devices have been used for sampling gaseous pollutants or process streams for analysis. One of the simplest and most common devices that has been used is an ordinary **gas washing bottle** which contains the absorbent, plus a gas dispersion tube for introducing the gaseous pollutant/components into an absorbing solution. A second type of gas absorber is the fritted-glass absorber. The fritted part of the dispersion tube is readily available in the form of a disc or cylinder of various pore sizes. The coarse and extra coarse frits provide good pollutant dispersion with a minimum pressure loss. Absorbers which use frits of approximately 50 μm or less pore size gradually become clogged with use. They may be cleaned by forcing an appropriate cleaning solution back and forth through the frit and then rinsing with distilled water in the same fashion. Various substances may be removed from the frits by cleaning with the appropriate solvent (e.g., hot hydrochloric acid for dirt and hot concentrated sulfuric acid containing sodium nitrite for organic matter, etc.).

The quantity of a particular gas that can be adsorbed by a given amount of adsorbent will depend on the following factors:

1. concentration of the gas in the immediate vicinity of the adsorbent,
2. the total surface area of the adsorbent,
3. the temperature of the system,
4. the presence of other molecules which may compete for a site on the adsorbent,
5. the characteristics of the substance being adsorbed (adsor-bate) such as weight, electrical polarity, chemical reactivity, and size and shape of the molecules,
6. the size and shape of the pores of the adsorbing medium, and
7 the characteristics of the adsorbent surface such as electrical polarity.

Ideal physical adsorption of a gas is favored by a high concentration of material to be adsorbed, a large adsorbing surface, freedom from competing molecules, low temperature, and a correct form of the adsorbate.

Most of the common adsorbents are granular in form and are supported in a column through which the gas to be sampled is drawn. Common adsorbents have the capacity to adsorb 10 to almost 100 percent of their weight. An ideal adsorbent should be of such size and form that it offers little or no resistance against flow. It should have a high adsorptive capacity, be inert and selective for the desired absorbate, resistant to breakage, deterioration and corrosion, be easily activated, and provide an easy release of adsorbate. Unfortunately, no single adsorbent possesses all of these characteristics so it becomes a matter of choosing the best adsorbent for the particular job.

B. Freeze-Out or Condensation

Air pollutants existing as gases can be trapped or removed by the freeze-out or condensation method. Trapping, in this discussion, implies the mechanism whereby a sample is collected. The term "removed" implies an air-cleaning mechanism used to remove unwanted gas contaminants from the gas stream. The method has a very high efficiency at relatively low flow rates. The method consists essentially of drawing air through collection chambers which have progressively lower temperatures. If the temperatures of the chambers are approximately equal to or less than the boiling point of the gaseous components of the air passing through it, these components will condense from the gaseous phase to the liquid phase. The condensate (liquid phase) is collected. The gaseous contaminants to be collected will determine the temperatures required in the collection chambers. The temperatures of the chambers can be controlled by using different immersion bath liquids. Contaminants with boiling points as low as $-195°C$ ($-383°F$) can be collected by this method. The type of freeze-out equipment required depends to a large extent on the application. The collection chambers themselves are placed in **Dewar flasks** which contain the cooling solutions as shown in Fig. 11.1.

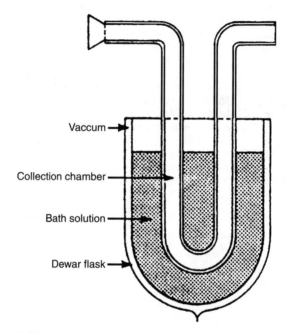

FIGURE 11.1 A Dewar used in the freeze-out or condensation method.

Among the collection chambers used, U-shaped and spiral-shaped tubes are the most common. Large radius bends should be designed into the tubes to facilitate smooth airflow and to prevent accumulation of ice at the bends. The collection efficiencies of most freeze-out systems are not very good. In order to efficiently condense a gas, it is necessary for the gas to come in contact with the cold surface of the collection chamber. Therefore, the efficiency of collection by freeze-out can be improved by: (1) filling the collection chamber with material which will increase the cold surface area; and (2) reducing the flow rate.

GRAB SAMPLING

A **grab sample** is a sample taken at a particular time within an interval of a few seconds to a minute. Sampling is generally used to determine the gaseous composition of the air or the effluent from a specific source at a particular instant. One can lengthen the sampling interval to obtain an average sample over the extended sampling time. The grab sampling techniques covered here apply to atmospheric sampling and source sampling. Grab samples are usually collected by using one of the following techniques: (1) use of an evacuated container; (2) purging (displacement of air); (3) displacement of a liquid; (4) inflation of a flexible bag; or (5) use of a syringe.

Grab sampling techniques are satisfactory when the gas of concern is to be collected in a liquid and the rate of absorption is slow. Under these conditions the absorbing solution is placed into the sampling apparatus, the sample is collected, and the system allowed to come to equilibrium before any analysis is made. Most grab sampling techniques utilize a minimum of equipment and require little special training. One common collection vessel for grab sampling is an evacuated round-bottom flask.

Safety requires that evacuated vessels made of fragile materials be placed in a protective container or wrapped with adhesive tape to reduce the hazards of implosion. An **implosion** is the sudden in-rush of gas into an evacuated space. It has the same consequences as an explosion.

Commercially available gas displacement collectors are available. These collectors are cylindrical glass tubes with stopcocks at each end (see Fig. 11.2). They can be evacuated to collect gas samples or use the liquid displacement technique described below. Metal containers of the same general design have been used, but they have been found to react with many samples. Their advantage is that they are virtually unbreakable. The stopcocks are opened and the gas is pulled from the source through the sampling train to flush out the "old air" that was present in the collector. After sampling, the collector should be held in place until the stopcocks have been closed (the stopcock next to the aspirating device is closed first) and the aspirating device has been removed.

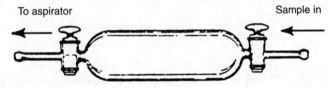

FIGURE 11.2 Gas displacement collector.

For grab sampling using the liquid displacement procedure typical containers are a glass tube with two stopcocks (Fig. 11.2), or any suitable container where the liquid can be displaced. The containers are filled with any liquid reagent which will not dissolve or react with the components of the gas sample. Some reagents which are commonly used are water, brine, and aqueous acidified sodium sulfate solution. To take the sample a sampling train is set up similar to the one shown in Fig. 11.3. The liquid is allowed to drain by opening the top stopcock, and then the bottom one. The **Orsat** analysis which has been used for many years for combustion gases utilizes this principle for sampling.

The inflation technique collects a sample by inflating a plastic bag. Plastics of various types have been used. Mylar® bags have been found satisfactory for aliphatic hydrocarbons, formaldehyde, ozone, SO_2, and NO_2. Saran® and various aluminized plastics have also been used. To take the sample, the sampling train is set up as shown in Fig. 11.4. A sample is collected by evacuating the air from the grab sample box which causes the flexible bag to inflate and draw the sample out of the duct. By reversing the flow of air from the pump which forces air into the box, the sample is expelled from the bag for analysis.

SAMPLING LIQUIDS

When liquids are pumped through pipes, a number of samples can be collected at various times and combined to provide the gross sample for analysis. Care should be taken that the samples represent a constant fraction of the total amount pumped and that all portions of the pumped liquid are sampled. Homogeneous liquid solutions can be sampled relatively easily, provided the material can be mixed

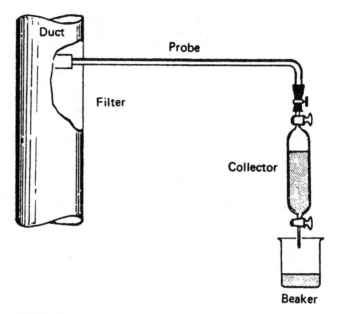

FIGURE 11.3 Grab sampling using a liquid displacement principle of collection.

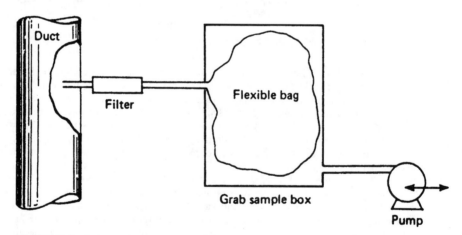

FIGURE 11.4 Grab sampling using a flexible bag.

thoroughly by means of agitators or mixing paddles. After adequate mixing, samples can be taken from the top and bottom and combined into one sample which is thoroughly mixed again; from this the final sample is taken for the analysis.

Liquid samplers called coliwasas are available in various forms. Most disposable **coliwasas samplers** are approximately 40 in long, made of chemical-resistant glass, have a liquid-tight inert plunger, and hold approximately 200 mL of liquid sample. Disposable glass coliwasas are prescored for safe disposal. Reusable glass coliwasas are manufactured of borosilicate glass and have ground-glass inner seals. This unscored-type of sampler can take samples to within 0.5 in of a drum bottom and can be cleaned with a special brush for reuse. Other disposable/reuseable coliwasas samplers are available in lengths up to 12 ft and fabricated of glass, stainless steel, Teflon, and other plastics.

Other liquid samples can be as simple a polyethylene **dipper** (100 mL to 1000 mL capacity) equipped with a custom length handle (3 ft to 12 ft). Bailers and drum samplers of various designs are readily available. Liquid thief samplers or "**bomb samplers**" made of stainless steel are available for corrosive liquid and can be used to sample liquids at any desired depth. A valve at the bottom of the sampler can be opened and closed by pulling a cord, thus permitting a sample to be taken. These bomb samplers with their cylindrical reservoir chambers fit easily into most openings.

SAMPLING HOMOGENEOUS SOLIDS

In the case of solid mixtures, sampling devices called thiefs and riffles are used. A solid **thief,** such as that illustrated in Fig. 11.5, generally consists of a tube with openings along its side and a removable rod which fits snugly into the tube. In use, the thief is plunged into the mixture with all the internal openings in the closed position. The inner tube is then rotated so that the holes match up, allowing material to enter the tube at several points. The inner tube is then rotated to shut the holes, and the sampler can be withdrawn. The sampled material can then be removed and the various segments examined individually or a composite (total) analysis performed. Soil samples are routinely taken using **augers** with core samples of 0.75 in diameter or greater, and lengths of normally 1 ft or more.

A **riffle** is a device which separates a given batch into several smaller amounts, thus permitting a single analysis to be representative of the entire batch. A riffle is illustrated in Fig. 11.6. Devices such as thiefs and riffles are useful in obtaining small samples without disturbing the homogeneity.

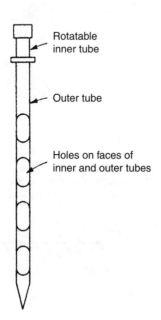

FIGURE 11.5 Thief probe sampler.

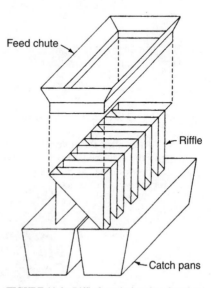

FIGURE 11.6 Riffle for reducing size of particulate samples.

SAMPLING NON-HOMOGENEOUS SOLIDS

The task of obtaining a representative sample from a lot of non-homogeneous solids requires that:

1. A gross sample be taken.
2. The gross sample be reduced to a representative laboratory-size sample by one of two methods: **coning and quartering**, or **rolling and quartering** as described below.

A. Coning and Quartering

When very large lots are to be sampled, a representative sample can be obtained by coning and quartering (Fig. 11.7). The first sample is formed into a cone, and the next sample is poured onto the apex of the cone. The result is then adequately mixed, and a new cone is formed. As each successive sample is added to the reformed cone, the total is mixed thoroughly and a new cone is formed prior to the addition of another sample.

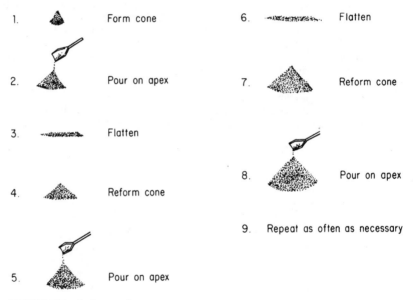

FIGURE 11.7 Coning samples.

After all the samples have been mixed by coning, the mass is flattened and a circular layer of material is formed. This circular layer is then quartered, and the alternate quarters are discarded. This is shown in Fig. 11.8. This process can be repeated as often as desired until a sample size suitable for analysis is obtained.

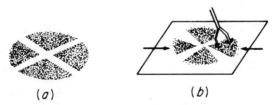

FIGURE 11.8 Quartering samples: select opposite quarters, discard other two quarters.

B. Rolling and Quartering

A representative cone of the sample is obtained by the coning procedure, and this sample is then placed on a flexible sheet of suitable size. The cone is flattened (Fig. 11.9), and then the entire mass of the sample is repeatedly rolled by pulling first one corner of the sheet over to the opposite corner, and then another corner over to its opposite corner (Fig. 11.10). The number of rollings required depends upon the size of the sample, the size of the particles, and the physical condition of the sample.

FIGURE 11.9 The cone is flattened after being formed.

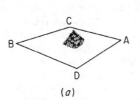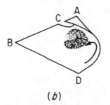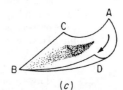

FIGURE 11.10 Rolling the sample.

To collect the sample, raise all four corners of the sheet simultaneously and collect the sample in the middle of the sheet (Fig. 11.11). The resulting rolled sample is flattened into a circular layer and quartered until a suitably sized sample is obtained.

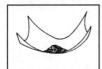

FIGURE 11.11 Collecting the rolled sample. (*a*) Roll A to B. (*b*) Roll C to D. (*c*) Caution: Do not roll A to D; mixing will not be as effective.

SAMPLING METALS

Figure 11.12 illustrates the problem with obtaining a representative sample from a solid metal sample. Notice how the composition of a solid can vary, in this case the carbon content in a piece of steel. Metals can be sampled by drilling the piece to be sampled at regular intervals from all sides, being certain that each drill hole extends beyond the halfway point. Additional samples can be obtained by sawing through the metal and collecting the "sawdust." All particles collected are then mixed thoroughly and quartered. Alternatively, the metal can be melted in a graphite crucible to provide the sample for analysis.

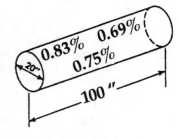

FIGURE 11.12 Variation of carbon content in a steel rod.

CRUSHING AND GRINDING

Crushing and grinding involves the reduction of the particle size of a material by physical means. Although the terms are used interchangeably, crushing is usually used in connection with coarse materials, while grinding is usually associated with the production of fine particles. The three main reasons for crushing and grinding are: (1) to increase chemical reactivity by increasing surface area; (2) to change physical characteristics; and (3) to improve mixing. See Chapter 15 "Gravimetric Analysis" for laboratory-scale grinding techniques, like mortar and pestle.

Where a reaction is confined to the surface of a solid, the rate of reaction will be a function of the exposed surface. To complete such a reaction in a minimum time, the surface area must be made as large as possible, and this can be accomplished by a grinding process. As an example, powdered coal is used where a high combustion rate is desired. Grinding may also alter the shape of a particle, for example, in the grinding of a pigment, where the grinding may flatten the particle to enable it to better cover or hide the substrate being painted. In the case where the pigment is ground in an oil base, the grinding also serves to provide a better mixture between the oil and the pigment.

The properties of a solid most directly related with particle-size reduction are hardness and grindability. **Hardness** has been described in Chapter 19 "Physical Properties and Determinations"— "Hardness" section, together with the Mohs system for measuring hardness. **Grindability** is a more general term and refers to the ease with which a particle of a material can be fractured so as to reduce its size. Although there is no well-accepted method for determining grindability, it can be measured by subjecting a sample to a grinding operation in a special grinder in which the energy input for a given size reduction can be measured: the less energy required, the greater the grindability. To describe a particular grinding operation, the reduction ratio is frequently used. The **reduction ratio** is the average particle size of the material before grinding, divided by the average particle size after grinding. Particle-size measurements are difficult to carry out in a reproducible manner because a uniform particle size and shape are rarely achieved and results must therefore be expressed in terms of a **particle-size distribution**. The results are also dependent on the particle property actually being measured and how the results are expressed, since most particles have a highly irregular geometry. The term particle size itself is therefore not precisely descriptive.

A. Classifying Particulate Sizes

Perhaps the most widely used description of particle-size distribution is by screen analysis. In a **screen analysis,** a sample is placed on a series of sieves, called a nest, and the nest is shaken with a circular motion to which is added an occasional bump (Fig. 11.13). The platform holding the sieves has a rotary motion, while the hammer on top provides a regular tapping action. The particles trickle down through the coarsest screen on top to the finest on the bottom. Each particle larger than the opening between the threads of the screen is retained by that particular sieve. In this way, a sequence of retentions is obtained and each retention can then be weighed. Thus, the weight percent retained on a particular sieve can be determined.

Standard sieves are made of woven wires that provide essentially square openings. There are three systems for designating sieve sizes: (1) **Standard**; (2) **Alternate**; and (3) **Tyler**. The physical sieve dimensions are given for the three systems in Table 11.1. Other types of sieves are available, such as those using silk bolting cloth (usually used for fine sieves, 150 to 400 mesh), electroplated wire mesh with square openings (usually for fine sieving, 150 down to 400 mesh), and punched sheet-metal sieves (usually used for coarse work, i.e., 0.25 in particles and larger). The sieve size is related to the number of holes per unit of length. A 10 mesh screen contains 10 holes per linear inch or 100 holes per square inch. A 200 mesh sieve would contain 40,000 (200×200) holes per inch. If a material is described as a 20/60 mesh range, this indicates that the material passes through a 20 mesh sieve, but not a 60 mesh sieve.

During the sieving of fine powders, the materials may have a tendency to ball, a phenomenon in which the particles seem to join together and form into fairly large spheres which do not pass

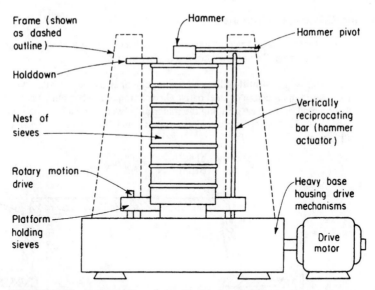

FIGURE 11.13 Automatic sieve shaker. The platform holding the sieves has a rotary motion while the hammer on top provides a regular tapping action.

through the openings in the sieve, even though the individual particle size is fine enough to pass easily through the sieve. In such a case, the sieving can be accomplished by dispersing the material in a liquid which does not dissolve or swell the particles. Balling is frequently associated with the presence of moisture and can sometimes be avoided by predrying the sample.

TABLE 11.1 Sieve Dimensions for Standard, Alternate, and Tyler Sieves

Standard	Alternate	Tyler
26.9 mm	1.06 inch	1.050 inch
19.0 mm	3/4 inch	0.742 inch
12.7 mm	1/2 inch	—
9.51 mm	3/8 inch	0.371 inch
8.00 mm	5/16 inch	2.5 mesh
4.00 mm	No. 5	5 mesh
2.00 mm	No. 10	9 mesh
1.00 mm	No. 18	16 mesh
841 micron	No. 20	20 mesh
595 micron	No. 30	28 mesh
420 micron	No. 40	35 mesh
297 micron	No. 50	48 mesh
250 micron	No. 60	60 mesh
177 micron	No. 80	80 mesh
149 micron	No. 100	100 mesh
125 micron	No. 120	115 mesh
105 micron	No. 140	150 mesh
74 micron	No. 200	200 mesh
44 micron	No. 325	325 mesh
37 micron	No. 400	400 mesh

Adapted from Robert H. Perry and Don W. Green (eds.), "Perry's Chemical Engineers' Handbook," 7th ed. (New York: McGraw-Hill Co., 1997).

B. Crushing and Grinding Equipment

The equipment used for crushing and grinding falls into four or more general types: (1) ball mill; (2) jaw crusher; (3) pan crusher; and (4) hammer mill.

Ball Mill.   A **ball mill** utilizes a rotating chamber which is loaded with the material to be ground, together with porcelain or steel balls. The rotation of the container causes the balls to roll around and impinge on each other, grinding the material between the balls. A ball mill is commonly used in the laboratory because of its convenience, but it can achieve large dimensions for industrial purposes as illustrated in Fig. 11.14.

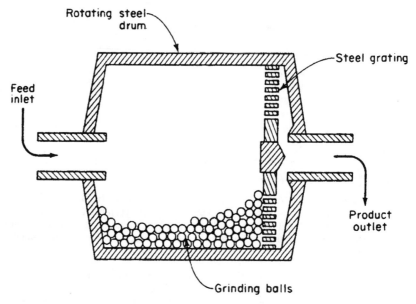

FIGURE 11.14   Industrial ball mill.

Jaw Crusher.   A **jaw crusher** is generally used to break up large particles of hard materials such as rocks. In a jaw crusher, a hardened-steel jaw moves back and forth against a stationary surface with the material to be crushed fed in at the top where the clearance is greatest, and moving to the bottom, where the clearance is minimal. While jaw crushers in large sizes are widely used in mineral processing, they are also available in small sizes suitable for laboratory and process operations.

Pan Crusher.   A **pan crusher** is used not only for crushing, but also for mixing liquids and solids. This type of apparatus, illustrated in Fig. 11.15, usually is available only in large sizes. It consists of two heavy wheels, called **mullers**, which roll around the inside of the pan, crushing the material with which it is filled against the side and bottom of the pan. Doctor blades are suitably positioned to scrape the sides and bottom of the pan as well as the sides of the muller wheels and to distribute the mix in such a way as to present new material continuously to the muller wheels.

Hammer Mill.   **Hammer mills**, of the type illustrated in Fig. 11.16, are available in a wide range of sizes. They consist of hammers which are free to swing as the rotor on which they are mounted rotates at relatively high speeds. The hammers swing out toward a grating in large mills, or against a wire-screen mesh in smaller mills. The impact action of the hammers breaks up the particles against the casing, and the openings in the casing automatically size the particles, since large particles stay in the mill until they can pass through the grating or screen. Hammer mills are frequently used in the chemical processes to produce very fine powders.

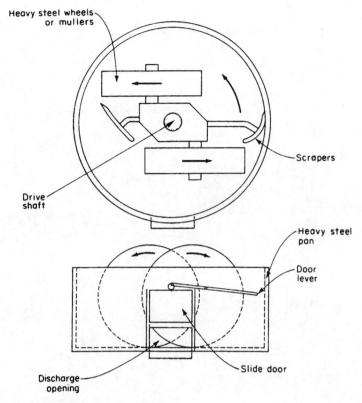

FIGURE 11.15 Pan crusher or "Muller."

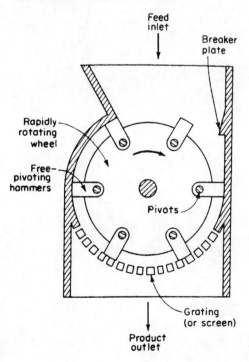

FIGURE 11.16 Hammer mill.

231

CHAPTER 12
LABORATORY FILTRATION

INTRODUCTION

Filtration is the process of removing material, often but not always a solid, from a substrate (liquid or gas) in which it is suspended. This process is a physical one; any chemical reaction is inadvertent and normally unwanted. Filtration is accomplished by passing the mixture to be processed through one of the many available sieves called **filter media.** These are of two kinds: surface filters and depth filters.

With the surface filter, filtration is essentially an exclusion process: particles larger than the filter's pore or mesh dimensions are retained on the surface of the filter; all other matter passes through. Examples are filter papers, membranes, mesh sieves, and the like. These are frequently used when the solid is to be collected and the filtrate is to be discarded. Depth filters, however, retain particles both on their surface and throughout their thickness; they are more likely to be used in industrial processes to clarify liquids for purification. A mat of Celite is an example. Filtration most commonly is used in one of four ways:

- **Solid-liquid filtration:** The separation of solid particulate matter from a carrier liquid
- **Solid-gas filtration:** The separation of solid particulate matter from a carrier gas
- **Liquid-liquid separation:** A special class of filtration resulting in the separation of two immiscible liquids, one of them water, by means of a hydrophobic medium
- **Gas-liquid filtration:** The separation of gaseous matter from a liquid in which it is usually, but not always, dissolved

In the laboratory, filtration is generally used to separate solid impurities from a liquid or a solution or to collect a solid substance from the liquid or solution from which it was precipitated or recrystallized. This process can be accomplished with the help of gravity alone (Fig. 12.1) or it can be speeded up by using vacuum techniques. (Refer to the section on vacuum filtration.) Vacuum filtration provides the force of atmospheric pressure on the solution in addition to that of gravity, and thus increases the rate of filtration.

The efficiency of filtration depends on the correct selection of the method to be used, the various pieces of apparatus available, the utilization of the filter medium most appropriate for the particular process, and the use of correct laboratory technique in performing the manipulations involved. During filtration (as previously mentioned) a liquid is usually separated from a solid by pouring the liquid through a sieve, usually filter paper. The liquid passes through the paper, but the solid is retained. Although the carrier liquid is usually relatively nonreactive, it is sometimes necessary to filter materials from highly alkaline or acidic carrier liquids or to perform filtration under other highly reactive conditions. A variety of filter media exists from which it is possible to select whichever one best fits the particular objectives and conditions of a given process. The most common filter media are:

- Paper
- Fiberglass "papers" or mats
- Gooch crucibles
- Sintered-glass (or fritted-glass) crucibles and funnels

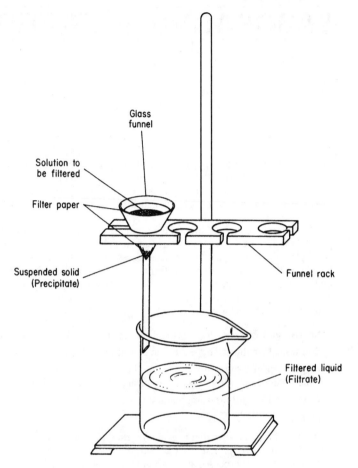

FIGURE 12.1 A gravity-filtration setup.

- Porous porcelain crucibles
- Monroe crucibles
- Millipore® membranes

All of those listed are available in various porosities, and their use will be discussed later in this section.

FILTRATION METHODS

There are two general methods of filtration: gravity and vacuum (or suction). During gravity filtration the filtrate passes through the filter medium under the combined forces of gravity and capillary attraction between the liquid and the funnel stem. In vacuum filtration a pressure differential is maintained across the filter medium by evacuating the space below the filter medium. Vacuum filtration adds the force of atmospheric pressure on the solution to that of gravity, with a resultant increase in the rate of filtration. The choice of method to be used depends upon the following factors:

1. The nature of the precipitate
2. The time to be spent on the filtration
3. The degree to which it is necessary to retain all the precipitate
4. The extent to which one can tolerate the contamination of the precipitate with the filtrate

We will discuss each type later in the chapter. Information applicable to both methods follows.

FILTER MEDIA

A. Paper

There are several varieties or grades of filter paper (Fig. 12.2) for special purposes; there are qualitative grades, low-ash or ashless quantitative grades, hardened grades, and even glass-fiber "papers." For a given filtration, you must select the proper filter paper with regard to porosity and residue. Some of this information is given in Table 12.1.

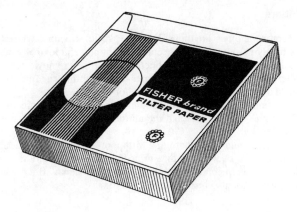

FIGURE 12.2 Filter paper comes in assorted sizes and porosities.

TABLE 12.1 Commonly Used Filter Papers*

W	S&S	RA	Porosity	Speed	Use for
Qualitative- or regular-grade papers					
4	604	202	Coarse	Very rapid	Gelatinous precipitates
1	595	271	Medium	Medium	Ordinary crystalline precipitates
3	602	201	Medium	Slow	Fine precipitates; used with Büchner funnels
Quantitative-grade papers (less than 0.1 mg ash)					
41	589 blue ribbon	...	Coarse	Very rapid	Gelatinous precipitates
40	589 white ribbon	...	Medium	Rapid	Ordinary crystalline precipitates
42	589 black ribbon	...	Fine	Slow	Finest crystalline precipitates

*Code W: Whatman; S&S: Schleicher and Shüll; RA: Reeve Angel.

Qualitative-grade papers will leave an appreciable amount of ash upon ignition (of the order of 0.7 to 1 mg from a 9-cm circle) and are therefore unsuitable for applications in quantitative analysis where precipitates are to be ignited on the paper and weighed. They are widely used for clarifying solutions, filtration of precipitates which will later be dissolved, and general nonquantitative separations of precipitates from solution.

Low-ash or ashless quantitative-grade papers can be ignited without leaving an ash. The residue left by an 11-cm circle of a low-ash paper may be as low as 0.06 mg; an ashless-grade paper typically leaves 0.05 mg or less from an 11-cm circle. In most analytical procedures, this small mass can be considered negligible.

Hardened-grade papers are designed for use in vacuum filtrations and are processed to have great wet strength and hard lintless surfaces. They are available in low-ash and ashless as well as regular grades.

Fiberglass papers are produced from very fine borosilicate glass and are used in Gooch, Büchner, or similar filtering apparatus to give a combination of very fine retention, very rapid filtration, and inertness to the action of most reagents to an extent not found in any cellulose paper.

All grades of filter paper are manufactured in a variety of sizes and in several degrees of porosity. Select the proper porosity for a given precipitate. If too coarse a paper is used, very small crystals may pass through, while use of too fine a paper will make filtration unduly slow. The main objective is to carry out the filtration as rapidly as possible, retaining the precipitate on the paper with a minimum loss.

B. Membrane Filters

Membrane filters are thin polymeric (plastic) structures with extraordinarily fine pores. These sheets of highly porous material are composed of pure, biologically inert cellulose esters or other polymeric materials. Such filters are distinctive in that they remove from a gas or liquid passing through them all particulate matter or microorganisms larger than the filter pores. With proper filter selection, they yield a filtrate that is ultra-clean and/or sterile.

Membrane filters are available in a wide variety of pore sizes in a number of different materials. The range of pore sizes and the uniformity of pore size in a typical filter are shown in Table 12.2 and Fig. 12.3.

When liquids pass thorugh a Millipore® membrane filter, all contaminants larger than the filter-pore size are retained on the surface of the filter, where they can be readily analyzed or counted. This is in sharp contrast with the action of a "depth" filter, which retains contaminants not only on its surface but also inside the filter matrix.

TABLE 12.2 Uniformity of Millipore® Filters

Filter pore size, μm	Maxium "rigid" particle to penetrate, μm	Filter pore size, μm	Maximum "rigid" particle to penetrate, μm
14	17	0.65	0.68
10	12	0.60	0.65
8	9.4	0.45	0.47
7	9.0	0.30	0.32
5	6.2	0.22	0.24
3	3.9	0.20	0.25
2	2.5	0.10	0.108
1.2	1.5	0.05	0.053
1.0	1.1	0.025	0.028
0.8	0.85		

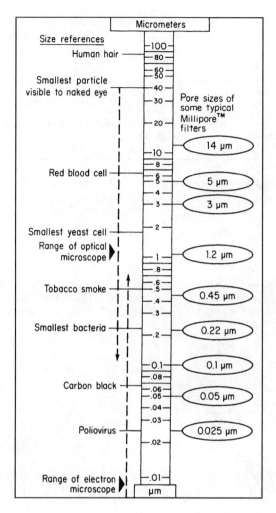

FIGURE 12.3 Scale comparing Millipore® filter-pore sizes with sizes of microbes and microparticles.

C. Fritted Glassware

The filtration of solids can be performed with funnels fitted with **fritted-glass** (also called **sintered-glass**) plates. Fritted glass is available in different porosities, and some of the problems encountered in using filter paper are minimized by using fritted-glass equipment. The grades of fritted glassware are listed in Table 12.3.

For optimum service, fritted ware must be maintained carefully. It is best to follow any manufacturer's instructions that may be enclosed with the equipment. The following paragraph and table comprise an example of such instructions.

Cleaning—A new fritted filter should be washed by suction with hot hydrochloric acid and then rinsed with water

TABLE 12.3 Grades of Fritted Ware

Designation	Nominal maximum pore size, μm
Extra-coarse	170–220
Coarse	40–60
Medium	10–15
Fine	4–5.5
Very fine	2–2.5
Ultrafine	0.9–1.4

before it is used. Clean all fritted filters immediately after use. Many precipitates can be removed from the filter surface simply by rinsing from the reverse side with water under pressure not exceeding 15 lb/in^2. Some precipitates tend to clog the fritted filter pores, and chemical means of cleaning are required. Suggested solutions are listed below.

Material	Cleaning solution
Fatty materials	Tetrachloromethane
Organic matter	Hot, concentrated cleaning solution
Mercury reside	Hot nitric acid
Silver chloride	Ammonia or sodium hyposulfite

See the section on vacuum filtration for a discussion of other types of crucibles and funnels.

FILTERING ACCESSORIES

A. Filter Supports

Some solutions that are to be filtered tend to weaken the filter paper, and at times the pressure on the cone of the filter will break the filter paper, ruining the results of the filtration. The thin woven textile disks shown in Fig. 12.4 are used to support the tip of the filter-paper cone.

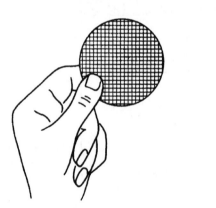

FIGURE 12.4 Filter-paper support.

They are approximately of the same thickness as the filter paper, and therefore ensure close contact of the reinforced paper with the funnel walls. They are folded along with the filter paper when it is formed into the normal conical shape, and they can easily be removed from the wet filter paper after the filtration has been completed, if desired. This is done when the filter paper and the collected precipitate are to be ashed.

B. Filter Aids

During filtration, certain gummy, gelatinous, flocculent, semicolloidal, or extremely fine particulate precipitates often quickly clog the pores of a filter paper, and then the filtering action stops. Normal analytical determinations require positive clarification at maximum filtration rates. Therefore, when a necessary filtration is impeded or halted by the presence of recalcitrant particles, filter aids are employed to speed up the process. Filter aids consist of diatomaceous earth and are sold under the trade names of Celite® or FilterAid; they are extremely pure and inert powderlike materials which form a porous film or cake on the filter medium. In use, they are slurried or mixed with the solvent to form a thin paste and are then filtered through a Büchner funnel (with the paper already in place on the funnel) to form a film or cake about 3 to 4 mm thick. The troublesome filtration slurry is then filtered through the cake, and the gummy, gelatinous, or finely divided particulate precipitate is caught in the cake, whereupon the filtration proceeds almost normally. An alternative procedure involves the direct addition of the filter aid directly to the problem slurry with thorough mixing. The filter aid then speeds up the filtration by forming a porous film or cake on the filter medium and yields filtrates of brilliant crystalline clarity. All the suspended matter has been retained on the funnel, and the filtration normally proceeds rapidly.

Limitations of Filter Aids.

1. They cannot be used when the object of the filtration is to collect a solid product, because the precipitate collected also contains the filter aid. Filter aids can be used only when the filtrate is the desired product.
2. Because they are relatively inert, they can be used in normally acidic and basic solutions; however, they cannot be used in strongly alkaline solutions or solutions containing hydrofluoric acid.
3. Filter aids cannot be used when the desired substance is likely to precipitate from the solution.

C. Wash Bottles

Two types of wash bottles are (1) the Florence flask (now largely a laboratory curiosity), which operates on breath pressure (the stream of water is directed by moving the tip with the fingers) and (2) the plastic wash bottle operated by squeezing, the "squeeze bottle" (Fig. 12.5).

The polyethylene plastic wash bottle has all but displaced the traditional glass Florence flask wash bottle. It is available in a variety of colors for content identification, eliminating the hazard of choosing the wrong bottle. The flow of the bottle can be increased by cutting off part of the tip.

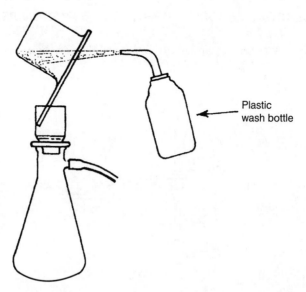

FIGURE 12.5   Washing a precipitate into a crucible with a squeeze bottle.

D. Filter Pump and Accessories

A water pump (Fig. 12.6) is used for suction filtrations, general vacuum manipulations, and pipet cleaning. Accessories are used to couple to water lines and prevent splashing to existing water. (See Water Aspirator in Chapter 5, "Pressure and Vacuum.")

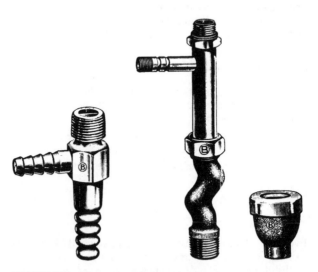

FIGURE 12.6 Filter pump and accessories.

MANIPULATIONS ASSOCIATED WITH THE FILTRATION PROCESS

Whether one uses gravity or vacuum filtration, three operations must be performed: **decantation, washing,** and **transfer** (Fig. 12.7).

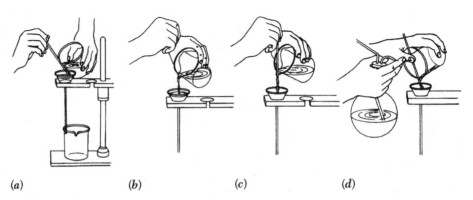

FIGURE 12.7 Gravity-filtering operation showing techniques for decantation and transfer of precipitates.

A. Decantation

When a solid readily settles to the bottom of a liquid and shows little or no tendency to remain suspended, it can be separated easily from the liquid by carefully pouring off the liquid so that no solid is carried along. This process is called *decantation*. To decant a liquid from a solid:

1. Hold the container (beaker, test tube, etc.) that has the mixture in it in one hand and have a glass stirring rod in the other (Fig. 12.7a).
2. Incline the beaker until the liquid has almost reached the lip (Fig. 12.7b).
3. Touch the center of the glass rod to the lip of the beaker and the end of the rod to the side of the container into which you wish to pour the liquid.
4. Continue the inclination of the beaker until the liquid touches the glass rod and flows along it into the second container. The glass rod enables you to pour the liquid from the beaker slowly enough that the solid is not carried along and also prevents the liquid from running back along the outside of the beaker from which it is being poured (Fig. 12.7c).

B. Washing

The objective of washing is to remove the excess liquid phase and any soluble impurities which may be present in the precipitate. Use a solvent which is miscible with the liquid phase but does not dissolve an appreciable amount of the precipitate.

Solids can be washed in the beaker after decantation of the supernatant liquid phase. Add a small amount of the wash liquid and thoroughly mix it with the precipitate. Allow the solid to settle. Decant the wash liquid through the filter. Allow the precipitate to settle, with the beaker tilted slightly so that the solid accumulates in the corner of the beaker under the spout (Fig. 12.8). Repeat this procedure several times.

Several washings with small volumes of liquid are more effective in removing soluble contaminants than a single washing using the total volume.

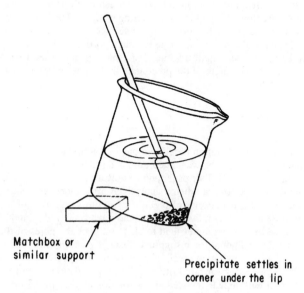

FIGURE 12.8 Supporting a beaker in a tilted position to allow the precipitate to settle prior to decantation.

FIGURE 12.9 Stirring rod with **rubber policeman** (a piece of rubber tubing with a flattened end) is used to remove traces of solids from containers and to speed up solutions of solids in liquids when used as a stirrer. Also used to prevent scratching the inside of a vessel.

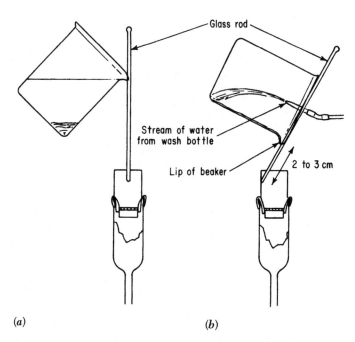

FIGURE 12.10 (*a*) Decantation for vacuum filtration. (*b*) Transferring the last portions of precipitate.

Transfer of the Precipitate. Remove the bulk of the precipitate from the beaker to the filter by using a stream of wash liquid from a wash bottle (Fig. 12.7*d*). Use the stirring rod (Fig. 12.9) to direct the flow of liquid into the filtering medium. The last traces of precipitate are removed from the walls of the beaker by scrubbing the surfaces with a rubber policeman attached to the stirring rod. All solids collected are added to the main portion in the filter paper (Fig. 12.10). If the precipitate is to be ignited, use small fragments of ashless paper to scrub the sides of the beaker; then add these fragments to the bulk of the precipitate in the filter with the collected solid.

GRAVITY FILTRATION

During gravity filtration the filtrate passes through the filter medium under the forces of gravity and capillary attraction between the liquid and the funnel stem (Fig. 12.1). The most common procedure involves the use of filter paper and a conical funnel (see Fig. 12.11). The procedure is slow, but it is highly favored for gravimetric analysis over the more rapid vacuum filtration because there is better retention of fine particles of precipitate and less rupturing or tearing of the paper. Moreover, gravity filtration is generally the fastest and most preferred method for filtering gelatinous precipitates because these precipitates tend to clog and pack the pores of the filter medium much more readily under the additional force supplied during a vacuum filtration.

Avoid accumulating precipitate on the filter paper during the early stages of the filtration process. This is necessary for rapid filtering, since the precipitate will be drawn into the pores of the paper, where it will impede the passage of solution and retard the rate of filtration. It is advisable to carry out precipitations in a beaker whenever possible because it has a pour spout to facilitate pouring of liquids without loss. The precipitate should be allowed to settle to the bottom of the beaker before

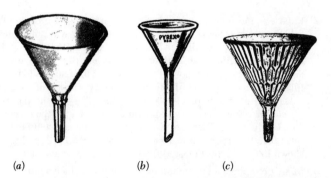

FIGURE 12.11 Class funnels, (*a, b*) varied size and stem length, (*c*) funnel with heavy-ribbed construction; raised ribs on inner surface facilitate rapid filtration.

filtration begins. The supernatant liquid phase, free of most of the suspended precipitate, is then poured onto the filter, leaving the precipitated solid essentially undisturbed. The bulk of the precipitate is not added until the last stages of the filtration as part of the *washing* process.

Procedure

Optimum filtering speed is achieved in gravity filtration by proper folding and positioning of the paper in the funnel (Figs. 12.12 and 12.13). If maximum speed is to be maintained, follow the suggestions given below.

1. Take maximum advantage of capillary attraction to assist in drawing the liquid phase through the paper. Use a long-stemmed funnel (Fig. 12.11*b*) and maintain a continuous column of water from the tip of the funnel stem to the undersurface of the paper. The tip of the funnel should touch the side of the vessel that receives the filtrate; this procedure aids the filtration and minimizes any loss of filtrate that might be caused by splashing. Accurate fit of the folded filter helps maintain an airtight seal between the funnel and the top edge of the wet filter paper.

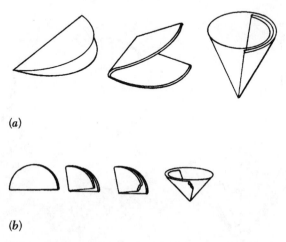

FIGURE 12.12 (*a*) Folding a filter paper, (*b*) Alternative method of folding filter paper. Steps in folding paper for use in filtering with a regular funnel. The second fold is not exactly at a right angle. Note the tear, which makes the paper stick better to the funnel.

FIGURE 12.13 Seating of a filter paper.

2. Expose as much of the paper as possible to provide free flow of liquid through the paper. If you fold the paper as shown in Fig. 12.12*b*, the paper will not coincide exactly with the walls of the funnel and the liquid will be able to flow between the paper and the glass. This type of fold will also help to maintain an airtight seal between the top edge of the filter paper and the funnel. Another way to expose most of the surface of the paper is to use a **fluted filter paper.** Fluting the filter paper increases the filtration in two ways. It permits free circulation of air in the receiving vessel and maintains pressure equalization. The solvent vapors present during the filtration of certain hot solutions can cause pressure to build up in the receiver and consequently decrease the speed of filtration. There are two ways of folding a flat circular piece of filter paper into a fluted filter. The first is shown in Fig. 12.14, the second in Fig. 12.15. The fluted filter in a funnel is shown in Fig. 12.16. Prepared fluted filters are also available commercially.

3. Many precipitates will spread over a wetted surface against the force of gravity; this behavior is known as **creeping,** and it can cause loss of precipitate. For this reason, be sure to pour the solution into the filter until the paper cone is no more than three-quarters filled. Never fill the cone completely. This precaution prevents loss of precipitate from both creep and overflow. It also provides an area near the top of the paper that is free of precipitate. By grasping this "clean" portion, you can remove the cone from the funnel and fold it for ignition (p. 263) without loss or contamination. See Fig. 12.17 for the best way to pour the **supernatant liquid** into the filter.

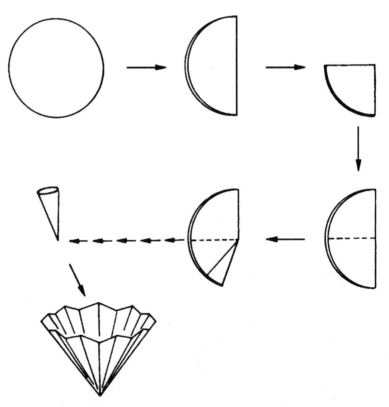

FIGURE 12.14 Folding a fluted filter. Fold the filter paper in half, then fold this half into eight equal sections, like an accordion. The fluted filter paper is then opened and placed in a funnel.

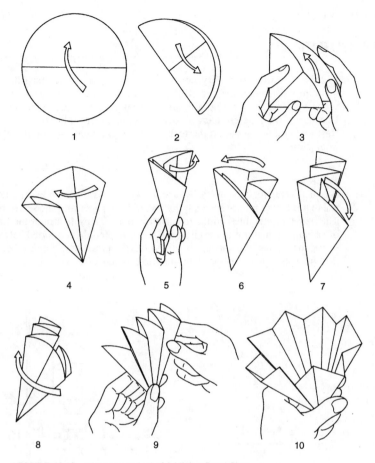

FIGURE 12.15 Alternate method of folding a fluted filter.

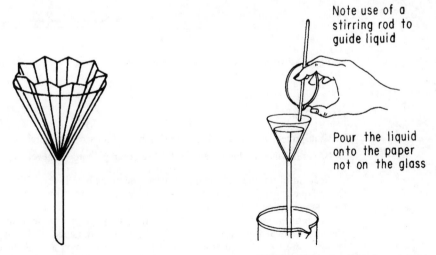

FIGURE 12.16 A fluted filter paper in a funnel.

FIGURE 12.17 Good filtration technique.

VACUUM FILTRATION

Vacuum filtration is a very convenient way to speed up the filtration process, but the filter medium must retain the very fine particles without clogging up. The vacuum is normally provided by a water aspirator, although a vacuum pump, protected by suitable traps, can be used. Because of the inherent dangers of flask collapse from the reduced pressure, thick-walled filter flasks should be used, and the technician should be always on the alert for the possibility of an implosion.

Procedure

A typical setup for carrying out a vacuum filtration is shown in Fig. 12.18. This illustration shows the use of a Büchner funnel in which the wetted filter paper must be seated before the suction is applied. The funnel or crucible is fitted to a suction flask. The side arm of the flask is connected to a source of vacuum such as a water aspirator. A water-trap bottle is inserted between the flask and the source of vacuum. When the vacuum is turned on, the pressure difference between the filter medium and the atmosphere helps to speed up the filtration process.

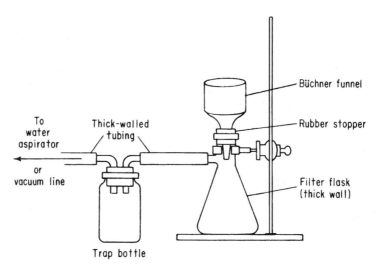

FIGURE 12.18 Complete vacuum-filtration assembly using a Büchner funnel.

CAUTIONS:

1. Wear protective glasses when the assembly is under reduced pressure.
2. Be careful that the liquid level in the "safety trap" bottle is never as high as the inlet tubes.

Vacuum filtration is advantageous when the precipitate is crystalline. It should not be employed for gelatinous precipitates because the added pressure forces the particles into the pores of the filter medium, clogging them so much that no liquid can pass through.

Vacuum filtration can be performed with filter paper as well as with the various crucibles. The common conical funnel paper is easily ruptured at the apex of the cone when under the added stress of the vacuum. To strengthen the cone at the apex, a small metal liner is often inserted (see Fig. 12.19).

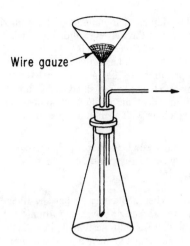

FIGURE 12.19 Funnel with a wire-gauze cone to be used as a support for the filter paper.

A. Equipment and Filter Media

Büchner Funnels. *Büchner funnels* (Fig. 12.20) are often used for vacuum filtration. They are not conical in shape but have a flat, perforated bottom. A filter-paper circle of a diameter sufficient to cover the perforations is placed on the flat bottom, moistened, and tightly sealed against the bottom by applying a slight vacuum.

When a Büchner funnel is used, the precipitate is allowed to settle, and the liquid phase is first decanted by pouring it down a stirring rod aimed at the center of the filter paper, applying only a light vacuum until sufficient solid has built up over the paper to protect it from breaking. The vacuum is then increased, and the remainder of the precipitate is added.

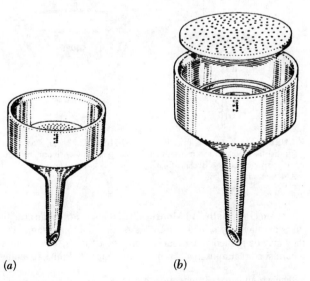

FIGURE 12.20 Büchner funnels, (*a*) Büchner suction funnel, (*b*) Büchner plain funnel with a removable plate, available in various sizes (14.5 to 308 mm).

The precipitate can be washed by adding small amounts of wash liquid over the surface of the precipitate, allowing the liquid to be drawn through the solid slowly with the vacuum. Precipitates cannot be dried or ignited and weighed in Büchner funnels.

Büchner funnels are not applicable for gravimetric analysis because they do permit rapid filtration of large amounts of crystalline precipitates. They are extremely useful in synthetic work, however. Precipitates can be air-dried by allowing them to stand in the funnel and drawing a current of air from the room through the precipitate with the vacuum pump or water aspirator. The last traces of water can be washed from the precipitate with a suitable water-miscible, volatile solvent.

NOTE: Solutions of very volatile liquids, ether, and hot solutions are not filtered very conveniently with suction. The suction may cause excessive evaporation of the solvent, which cools the solution enough to cause precipitation of the solute.

Wire-Gauze Conical Funnel Liner.   When a Büchner funnel and filter flask are not available, a serviceable apparatus can be constructed and used for vacuum filtration. A small piece of fine wire gauze is bent into the form of a cone that fits into the funnel (Fig. 12.19). The filter paper is then wetted and pressed against the side of the funnel to make a good seal with the glass. Vacuum should be applied gently so that a hole will not be torn in the paper. It may be necessary to use a double thickness of filter paper to prevent tearing. Wrap the flask with a towel before applying vacuum.

Crucibles.

Sintered-Glass Crucibles.   Glass crucibles with fritted-glass disks sealed permanently into the bottom end are available in a variety of porosities (Fig. 12.21). With care they can be used for quantitative analyses requiring ignition to a temperature as high as 500°C.

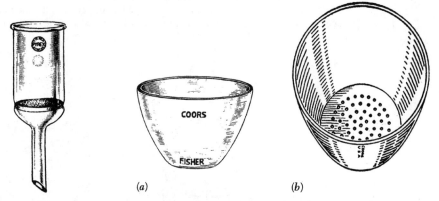

FIGURE 12.21 A funnel with a fritted-glass disk is used in vacuum filtration where the paper filter in a Büchner funnel would be attached.

FIGURE 12.22 (*a*) Porcelain crucible used for ignition of samples in analysis. (*b*) Gooch crucible with perforated bottom.

Porous Porcelain and Monroe Crucibles.   Porcelain crucibles (Fig. 12.22) with porous ceramic disks permanently sealed in the bottom are used in the same way as sintered-glass crucibles, except they may be ignited at extremely high temperatures. The Monroe crucible is made of platinum, with a burnished platinum mat serving as the filter medium. Advantages of these crucibles are:

- Their high degree of chemical inertness
- Their ability to withstand extremely high ignition temperatures

(a)   (b)

FIGURE 12.23   (a) Rubber adapter for suction filtration. (b) Adapter in use with a crucible in a suction flask.

These and other crucibles are seated in an adapter when filtering is performed (Fig. 12.23).

The Gooch Crucible. The Gooch crucible is a porcelain thimble with a perforated base. (See Fig. 12.22b.) The filter medium is either a mat of asbestos or a fiberglass paper disk. The mat is prepared by pouring a slurry of asbestos fiber suspended in water into the crucible and applying light suction.

Asbestos mats permit precipitates to be quantitatively ignited to extremely high temperatures without danger of reduction by carbon. With fiberglass mats, ignition temperatures above 500°C are not possible. Both filter media are resistant to attack by most chemicals.

The filter media used in these crucibles are quite fragile. Exercise extreme care when adding the liquid, so that the asbestos or glass paper will not be disturbed or broken, allowing the precipitate to pass through. Use a small, perforated porcelain disk over the asbestos mat or glass paper to deflect any stream of liquid poured into the crucible.

Platinum Crucibles. Platinum is useful in crucibles for specialized purposes. The chemically valuable properties of this soft, dense metal include its resistance to attack by most mineral acids, including hydrofluoric acid; its inertness with respect to many molten salts; its resistance to oxidation, even at elevated temperatures; and its very high melting point.

With respect to limitations, platinum is readily dissolved on contact with aqua regia and with mixtures of chlorides and oxidizing agents generally. At elevated temperatures, it is also dissolved by fused alkali oxides, peroxides, and to some extent hydroxides. When heated strongly, it readily alloys with such metals as gold, silver, copper, bismuth, lead, and zinc. Because of this predilection toward alloy formation, contact between heated platinum and other metals or their readily reduced oxides must be avoided. Slow solution of platinum accompanies contact with fused nitrates, cyanides, alkali, and alkaline-earth chlorides at temperatures above 1000°C; bisulfates attack the metal slightly at temperatures above 700°C. Surface changes result from contact with ammonia, chlorine, volatile chlorides, sulfur dioxide, and gases possessing a high percentage of carbon. At red heat, platinum is readily attacked by arsenic, antimony, and phosphorus, the metal being embrittled as a consequence. A similar effect occurs upon high-temperature contact with selenium, tellurium, and to a lesser extent, sulfur and carbon. Finally, when heated in air for prolonged periods at temperatures greater than 1500°C, a significant loss in weight due to volatilization of the metal must be expected.

The rules governing the use of platinumware are as follows:

- Use platinum equipment only in those applications that will not affect the metal. Where the nature of the system is in doubt, demonstrate the absence of potentially damaging components before committing platinumware to use.

- Avoid violent changes in temperature; deformation of a platinum container can result if its contents expand upon cooling.

- Supports made of clean, unglazed ceramic materials, fused silica, or platinum itself may be safely used in contact with incandescent platinum; tongs of Nichrome or stainless steel may be employed only after the platinum has cooled below the point of incandescence.

- Clean platinumware with an appropriate chemical agent immediately following use; recommended cleaning agents are hot chromic acid solution for removal of organic materials, boiling hydrochloric acid for removal of carbonates and basic oxides, and fused potassium bisulfate for the removal of silica, metals, and their oxides. A bright surface should be maintained by burnishing with sea sand.
- Avoid heating platinum under reducing conditions, particularly in the presence of carbon. Specifically, (a) do not allow the reducing portion of the burner flame to contact a platinum surface, and (b) char filter papers under the mildest possible heating conditions and with free access of air.

The Hirsch Funnel and Other Funnels. Hirsch funnels (Fig. 12.24), used for collecting small amounts of solids are usually made of porcelain. The inside bottom of the funnel is a flat plate with holes in it which supports the filter paper. Büchner and Hirsch funnels can also be obtained in glass with sintered-glass disks. Other funnels used like the **Hirsch** are the **Witt** and the **filter nail** (see Fig. 12.25).

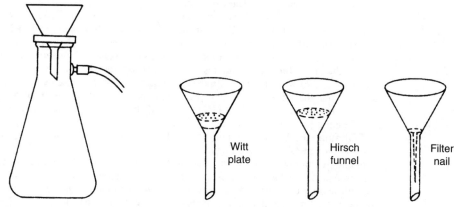

FIGURE 12.24 Suction filtration with a Hirsch funnel.

FIGURE 12.25 The Hirsch and other funnels for collecting small quantities of solids.

When using these funnels, a rubber ring forms the seal between the funnel and the filter flask, which is connected to the vacuum line or to the aspirator. (A rubber stopper or cork can also be used instead of the rubber ring to fit the funnel to the filter flask.)

B. Large-Scale Vacuum Filtration

Large amounts of material can be easily vacuum-filtered by using a tabletop Büchner funnel. The filtrate is collected in a vacuum flask placed alongside the filter on the tabletop (see Fig. 12.26). Alternatively, the vacuum flask can be located below the level of the funnel (Fig. 12.27).

FILTRATION FOR GRAVIMETRIC ANALYSIS

One of the most important applications of filtration is in gravimetric analysis, which pertains to the measurement of mass; the analysis therefore is completed by "massing" procedures. The substances which are massed are obtained (1) by forming an insoluble precipitate from the desired component;

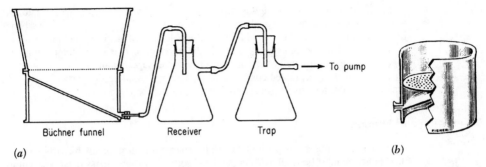

FIGURE 12.26 Vacuum filtration on a large scale. (*a*) Setup with tabletop Büchner funnel, (*b*) cutaway view of tabletop Büchner funnel (ID 56-308 mm).

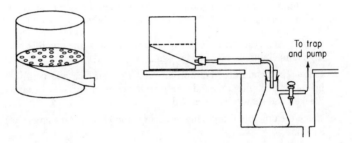

FIGURE 12.27 Alternative setup for vacuum filtration on a large scale.

from its weight, the percentage calculations can be made, given the mass of the original sample, or (2) by distilling off a volatile component; the nonvolatile residue is then massed. Both the volatile and nonvolatile portions can be massed, and calculations can be made from the data so obtained.

A precipitate suitable for analytical procedures should have the following characteristics:

- It should be relatively insoluble, to the extent that any loss due to its solubility would not significantly affect the result.
- It should be readily filterable. Particle size should be large enough to be retained by a filter.
- Its crystals should be reasonably pure, with easily removed solid contaminants.
- It should have a known chemical composition or be easily converted to a substance which does have a known composition.
- It should not be hygroscopic.
- It should be stable.

A. Aging and Digestion of Precipitates

Aging and digestion of precipitates frequently help to make them suitable for analytical procedures.

Freshly formed precipitates are aged by leaving them in contact with the supernatant liquid at room temperature for a period of time. There are frequently changes in the surface: a decrease in the total surface area or removal of strained and imperfect regions. Both effects are due to recrystallization because small particles tend to be more soluble than large ones, and ions located in imperfect and strained regions are less tightly held than is normal and therefore tend to return to the solution. On **aging** they are deposited again in more perfect form. These changes cause a beneficial decrease in adsorbed foreign ions and yield a more filterable as well as purer precipitate.

Heating during the aging process is called **digestion.** Increasing the temperature greatly enhances digestion. The precipitate is kept in contact with the supernatant at a temperature near boiling for a period of time. Flocculated colloids usually undergo rapid aging, particularly on digestion, and a major portion of the adsorbed contaminants may often be removed.

B. Filtration and Ignition

Eventually, the substances precipitated and prepared for filtration as the early steps in a gravimetric determination must be filtered, dried, and massed; that is, their mass must be accurately determined so that analytical calculations may be performed. Some precipitates are collected in tared crucibles and oven-dried to constant mass. The following procedures cover the handling of both types.

Preparation of Crucibles. All crucibles employed in converting a precipitate into a form suitable for weighing must maintain a substantially constant mass throughout the drying or ignition process; you must demonstrate that this condition applies before you start.

1. Inspect each crucible for defects, especially where the crucible has previously been subjected to high temperatures.
2. Place a porcelain crucible upright on a hard surface and gently tap with a pencil. You should hear a clear ringing tone indicating an intact crucible. A dull sound is characteristic of one that is cracked and should be discarded.
3. Clean the crucible thoroughly. Filtering crucibles are conveniently cleaned by backwashing with suction.
4. Bring the crucible to constant weight, using the same heating cycle as will be required for the precipitate. Agreement within 0.2 mg between consecutive measurements is considered constant mass.
5. Store it in a desiccator until it is needed.

Preparation of a Filter Paper. Fold the paper exactly in half; then make the second fold so that the corners fail to coincide for about 3 mm in each dimension. Tear off a small triangular section from the short corner to permit a better seating of the filter in the funnel. Open the paper so that a cone is formed and then seat it gently in the funnel with the aid of water from a wash bottle. There should be no leakage of air between paper and funnel, and the stem of the funnel should be filled with an unbroken column of liquid, a condition that markedly increases the rate of filtration.

Transfer of Paper and Precipitate to Crucible. When filtration and washing are completed, transfer the filter paper and its contents from the funnel to a tared crucible (Fig. 12.28). Use considerable care in performing this operation. The danger of tearing can be reduced considerably if partial drying occurs prior to transfer from the funnel.

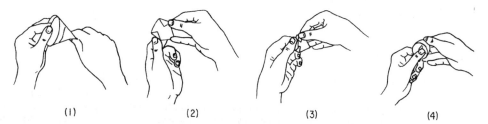

FIGURE 12.28 This is the way to transfer a filter paper and precipitate to a crucible.

First, flatten the cone along its upper edge; then fold the corners inward. Next, fold the top over. Finally, ease the paper and contents into the crucible so that the bulk of the precipitate is near the bottom.

Ashing of a Filter Paper. If a heat lamp is available, place the crucible on a clean, nonreactive surface; an asbestos pad covered with a layer of aluminum foil is satisfactory. Then position the lamp about 6 mm from the top of the crucible and turn it on. Charring of the paper will take place without further intervention; the process is considerably accelerated if the paper can be moistened with no more than one drop of strong ammonium nitrate solution. Removal of the remaining carbon is accomplished with a burner.

Considerably more attention must be paid to the process when a burner is employed to ash a filter paper. Since the burner can produce much higher temperatures, the danger exists of expelling moisture so rapidly in the initial stages of heating that mechanical loss of the precipitate occurs. A similar possibility arises if the paper is allowed to flame. Finally, as long as carbon is present, there is also the possibility of chemical reduction of the precipitate; this is a serious problem where reoxidation following ashing of the paper is not convenient.

In order to minimize these difficulties, the crucible is placed as illustrated in Fig. 12.29. The tilted position of the crucible allows for the ready access of air. A clean crucible cover should be located nearby, ready for use if necessary.

(a)   (b)

FIGURE 12.29 Ignition of a precipitate with access to air. (*a*) Start heating slowly from the side. (*b*) Do not let the flame enter the crucible.

NOTE: Always place the hot cover or crucible on a wire gauze—never directly on the desktop. The cold surface may crack the crucibles, and dirt, paint, etc., are easily fused into the porcelain, thus changing its mass.

Heating is then commenced with a small burner flame. This is gradually increased as moisture is evolved and the paper begins to char. The smoke that is given off serves as a guide to the intensity of heating that can be safely tolerated. Normally it will appear to come off in thin wisps. If the volume of smoke emitted increases rapidly, the burner should be removed temporarily; this condition indicates that the paper is about to flash. If, despite precautions, a flame does appear, it should be snuffed out immediately with the crucible cover. (The cover may become discolored owing to the condensation of carbonaceous products; these must ultimately be removed by ignition so that the absence of entrained particles of precipitate can be confirmed.) Finally, when no

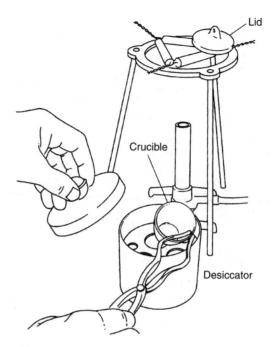

FIGURE 12.30 Cooling the crucible to constant mass.

further smoking can be detected, the residual carbon is removed by gradually lowering the crucible into the full flame of the burner. Strong heating, as necessary, can then be undertaken. Care must be exercised to avoid heating the crucible in the reducing portion of the flame.

Cooling the Crucible to Constant Mass. Place the warm crucible, contents, and lid in a **desiccator**,* containing an effective desiccant, to cool before determining its mass (Fig. 12.30). This procedure enables you to get balance readings to constant mass, especially in a humid atmosphere.

Summary.

1. Oxidize the paper completely to CO_2 and H_2O.
2. Record the mass of the previously prepared crucible.
3. Arrange the crucible according to Fig. 12.29a.
4. Increase the temperature *slowly* until all the black carbon residue is burned away.
5. Position the burner so that the reducing gases of the flame are *not* deflected into the crucible (Fig. 12.29b).
6. Reposition the crucible to expose fresh portions to the highest temperature of the burner.
7. Final ignition converts the precipitate to the anhydrous oxide: (a) Remove the crucible cover, (b) Ignite at red heat for 30 min with a Fisher, Meker, or other high-temperature burner; or you may use a muffle furnace.
8. Cool the crucible in a desiccator (Fig. 12.30).
9. Determine the mass of the cool crucible.
10. Repeat steps 7 to 9 until a constant mass is reached.

---

*For a complete discussion of desiccators, refer to Chapter 15, "Gravimetric Analysis."

REFERENCES

1. The Chemist's Companion, A Handbook of Practical Data, Techniques, and References. Arnold J. Gordon and Richard A. Ford, ISBN 0-471-31590-7, John Wiley & Sons, Inc. (1972).
2. Lange's Handbook of Chemistry, fifteenth edition, Section 11-Practical Laboratory Information, John A. Dean, editor, McGraw-Hill Inc., ISBN 0-07-016384-7 (1999).

CHAPTER 13
RECRYSTALLIZATION

INTRODUCTION

Recrystallization is a procedure whereby organic compounds which are solid at room temperature are purified by being dissolved in a hot solvent and reprecipitated by allowing the solvent to cool. The solvent may be a pure compound or a mixture, and the selection of the solvent depends upon a number of important factors. If the growth of the crystals is very fast and not selective, the precipitation process does not aid in purification. When the crystals grow very slowly and consist of pure compounds, the precipitation process is a purifying one; this second type of precipitation is usually defined as **crystallization.**

The crystallization process is very slow and requires relatively long periods of time to ensure that no impurities will be trapped in the crystal lattice as the crystal grows. Ordinary precipitation is a relatively fast process and occurs in minutes or hours. In this case, any impurities in the solution are actually trapped as the precipitate forms, resulting in an impure crystal.

In the laboratory a solid is purified by recrystallization by dissolving it in a hot solvent, filtering the solution, and then allowing the desired crystals to form in the filtrate, while the impurities remain in solution.

REQUIREMENTS OF THE SOLVENT

In general solvents should:

- Not react with the compound
- Form desirable, well-formed crystals
- Be easily removed from the purified crystals
- Have high solvency for the desired substance at high temperatures and low solvency for that substance at low temperatures
- Have high solvency for impurities

A. Solvency

The substance to be purified should be sparingly soluble in the solvent at room temperature, yet should be very soluble in the solvent at its boiling point. The solubility of a solute in a solvent is a function not only of the chemical structures of the solute and the solvent, but also of the temperature. In the majority of cases, the solubility of the solute in a solvent increases as the temperature increases, and in some cases the increase in solubility is very dramatic. This is the basis for the recrystallization method of purification. If the compound has been reported in the literature, its solubility in common solvents can be found in the reference. Normally, **polar organic compounds** (those which contain one or more —OH, —COOH, —CONH$_2$, —NH$_2$, or —SH functional groups) tend to dissolve in polar solvents such as water, the lower-molecular-weight alcohols, or combinations of them. **Nonpolar compounds** tend to dissolve in nonpolar organic solvents, such as benzene, the petroleum ethers, hexanes, chlorohydrocarbons, etc. (See Table 13.1.) The general rule regarding

TABLE 13.1 Solvent Polarity Chart*

Relative polarity	Compound formula	Group	Representative solvent compounds
Nonpolar	R—H	Alkanes	Petroleum ethers, ligroin, hexanes
↑ Increasing polarity ↓	Ar—H	Aromatics	Toluene, benzene
	R—O—R	Ethers	Diethyl ether
	R—X	Alkyl halides	Tetrachloromethane, chloroform
	R—COOR	Esters	Ethyl acetate
	R—CO—R	Aldehydes and ketones	Acetone, methyl ethyl ketone
	R—NH$_2$	Amines	Pyridine, triethylamine
	R—OH	Alcohols	Methanol, ethanol, isopropanol, butanol
	R—COHN$_2$	Amides	Dimethylformamide
	R—COOH	Carboxylic acids	Ethanoic acid
Polar	H—OH	Water	Water

*(See Chapter 23 "Organic Chemistry Review" for compound formulas information.)

solubility is that *like substances tend to dissolve in like substances,* but the molecule as a whole must be considered before making the decision. For example, a high-molecular-weight fatty acid, stearic acid, behaves more like a nonpolar substance than a polar one, because the —COOH group is not the major part of the molecule.

In general, the following points should be considered:

- A useful solvent is one that will dissolve a great deal of the solute at high temperatures and very little at low temperatures.
- If a solvent dissolves too much solute at low temperatures, it is unsuitable. You will be working with such a small volume of solvent that you will have a slush rather than a solution to filter. Furthermore, too much of the solute will not crystallize out at the low temperature and therefore much will be lost.
- If too much solvent is required to dissolve the solute even at its boiling point, it may be possible to recrystallize several grams, but extremely large volumes of solvent would be required to recrystallize several hundred grams.
- Quick tests of solubility are unreliable and are misleading because some solutes dissolve very slowly in boiling solvents. A quick observation may be misleading and cause you to reject the solvent as being unsatisfactory. Give the solute sufficient time to dissolve; otherwise you may use too much solvent because you will add additional quantities unnecessarily.
- The suitability of a solvent depends upon the establishment of equilibrium. Maximum solute will dissolve when equilibrium has been attained between the dissolved and solid solute.

B. Volatility

The **volatility** of a solvent determines the ease or difficulty of removing any residual solvent from the crystals which have formed. Volatile solvents may be removed easily by drying the crystals under vacuum or in an oven.

CAUTION: The temperature of the oven must be carefully watched and controlled so that the temperature is well below the melting point of the recrystallized compound or the flash point of the solvent.

Solvents with a high boiling point should be avoided, if possible. They are difficult to remove and the crystals usually must be heated mildly under high vacuum to remove such solvents.

Some common solvents and their boiling points are listed in Tables 13.2 and 13.3.

RECRYSTALLIZATION

TABLE 13.2 Common Water–Immiscible Solvents

Solvent	Boiling point, °C
Acetone	56.5
Methanol	64.7
Ethanol, 95%	78.1
Water	100
Dioxane	101
Acetic acid	118

TABLE 13.3 Common Water–Immiscible Solvents

Solvent	Boiling point, °C
Diethyl ether	34.6
Petroleum ether	40–60
Chloroform	61.2
Ligroin	65–75
Tetrachloromethane	76.7
Benzene	80.1
Ligroin	60–90

C. Solvent Pairs

Miscible solvents of different solvent power yield a mixture that gives a usable solvent system (Table 13.4).

Principle
The solute is soluble in one solvent but relatively insoluble in the second solvent.

NOTE: The solvents must be miscible in all proportions.

Procedure

1. Dissolve the solute in the minimum amount of the hot (or boiling) solvent in which it has maximum solubility.
2. Add the second solvent (in which the solute is relatively insoluble) dropwise to the boiling solution of the solute obtained in step 1, until the boiling solution just begins to become cloudy.
3. Add more of the solute-dissolving solvent (step 1) dropwise to the boiling solution until the solution clears up.
4. Allow the clear solution to slowly cool. Crystals of the solute should form.

NOTE: If crystals do not form and an oil separates from the solvent mixture, refer to the section on inducing crystallization.

TABLE 13.4 Solvent Pairs

Benzene–ligroin
Ether–acetone
Acetone–water
Ethanol–water
Methanol–water
Ether–petroleum ether
Acetic acid–water
Methanol–ether
Ethanol–ether
Methanol–methylene chloride
Dioxane–water

RECRYSTALLIZATION OF A SOLID

A. Selecting the Funnel

Gravity Filtration. Use either a short-stemmed or a stemless funnel. Long-stemmed funnels tend to cool the filtering solution, and crystallization then takes place in the stem, decreasing the flow rate and even clogging up the funnel.

Vacuum Filtration. Use a Büchner funnel, either porcelain or plastic. Jacketed Büchner funnels may be desirable to minimize any crystallization of the solute caused by evaporation under reduced pressure.

B. Heating the Funnel

Principle
If a hot recrystallization solution is poured through a cold funnel, the solvent cools and crystallization may sometimes take place in the funnel and its stem, clogging the funnel. Funnels (and thus solvents) can be heated or kept hot by the following procedures.

Method 1. Place a stemless or short-stemmed funnel in a beaker containing the pure solvent which is heated on a steam bath. Hot solvent can be poured through, and the reflux ring of the boiling solvent will heat the funnel (Fig. 13.1).

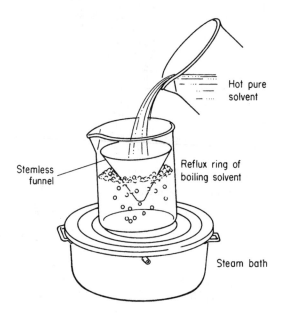

FIGURE 13.1 Heating and maintaining the temperature of a hot funnel.

Method 2. Place a funnel and fluted filter paper in the neck of an Erlenmeyer flask which is heated on a steam bath to reflux the pure solvent in it (Fig. 13.2). The reflux ring will heat the funnel. When the recrystallization procedure is to be started, the heated funnel and filter paper are transferred to a funnel support.

Method 3. Pass hot water or steam through a jacketed Büchner funnel during vacuum filtration (Fig. 13.3).

C. Receiver Selection

It is desirable to use an Erlenmeyer flask instead of a beaker to collect the filtered crystallizing solution because:

- The large opening of the beaker is conducive to catching dust and contaminating the product.
- The Erlenmeyer flask receiver can be easily stoppered, and the content can thus be stored without loss of the solvent by evaporation. If a beaker is used and all the solvent mother liquor evaporates, the process of recrystallization will be ruined. All the impurities dissolved in the mother liquor will crystallize out and coat the crystals that are to be purified.

D. General Procedure

Recrystallizing.

1. Select the most desirable solvent; refer to solubility tables.
2. Add the determined volume of solvent to the flask (no more than two-thirds the volume of the flask) and heat (Fig. 13.4). Add a few boiling stones if desired.

RECRYSTALLIZATION **261**

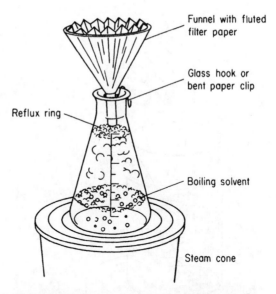

FIGURE 13.2 Alternate method of heating and maintaining the temperature of a hot funnel. The glass hook or bent paper clip ensures escape of vapor and prevents pressure from building up.

FIGURE 13.4 Heating the solvent for recrystallization.

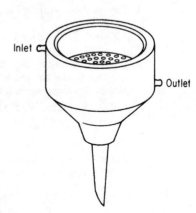

FIGURE 13.3 Jacketed Büchner funnel.

3. Add the minimum amount of hot solvent to the solute slowly to dissolve it. Boil, if necessary, to dissolve all the solute.

CAUTION: Do not add too much solvent. Stop adding solvent when only a small quantity of solute remains and further small additions of solvent do not dissolve that remaining solid. Usually, the insoluble material is an impurity. Always allow enough time after adding portions of solvent for the solute to dissolve, because some materials dissolve slowly.

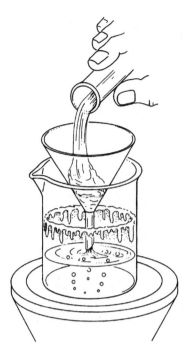

FIGURE 13.5 Preheating the funnel with hot solvent.

4. Preheat the filter funnel to prevent crystallization of the solid in the funnel (Fig. 13.5).

CAUTION: Observe *all fire-hazard cautions* because of volatile fumes.

5. Filter the boiling solution through the preheated funnel. Add the solution in small increments, and keep the filtrate hot and in a state of reflux to prevent premature crystallization. Steam baths are suitable for those solvents that have a boiling point lower than 100°C.
6. Collect the filtrate in a flask; allow it to stand and cool.
    (a) Cool or chill rapidly in a cooling bath for small crystals.
    (b) Cool slowly to get large crystals.
7. Filter die crystals from the mother liquor by gravity or suction. Further crystals can be obtained by evaporation and concentration of the mother liquor.

CAUTION: The later crystals may be impure as compared with those from the first crystallization.

8. Dry the crystals in a warm oven.

CAUTION: Be careful that the oven is not hot enough to melt the crystals.

Completing Crystallization. If the crystals are collected by filtration immediately after the solution has come to room temperature, or to a lower one by chilling, some of the crystals may not have been collected. Some materials require only a few minutes to crystallize out, while others may require days. The degree of completion of crystallization under the conditions of the experiment can be determined only by practice. It is always a good practice, when working with new substances, to bottle, label, and save the filtrate for a reasonable period of time, and observe if any more crystals come out of solution.

E. Preventing Crystallization of a Solute During Extraction

Should crystals of an organic substance begin to crystallize out of an organic solvent during extraction, add additional solvent in small portions until the crystals redissolve. This may happen because of the solubility of water in the organic solvent or because a water-soluble solvent such as alcohol, which was a part of the organic solvent mixture, was itself extracted by the water.

Washing Crystals. After all the crystals and solution have been transferred to the filter, some cold, fresh solvent (the same used in the recrystallization) should be poured over the crystals to wash them. If this is not done, any soluble impurities in the solvent that remained on the wet crystals will be deposited on those crystals when the solvent evaporates. Usually one washing (or possibly two) with cold solvent will free the crystals of any possible contamination from this source.

If vacuum filtration has been used and the crystals have been pulled down into a tight cake, the crystals can be washed on the filter, or, better yet, resuspended in a minimum amount of fresh solvent and refiltered. If, however, washing is done in the suction filter, first disconnect the vacuum, then carefully break up the cake gently with a rubber policeman and add fresh solvent to form a wash slurry. Take care not to tear the filter paper. Finally, reapply suction and pull the wash liquid through the cake.

Inducing Crystallization. When the solute fails to crystallize and remains as an oil in the mother liquor, one of the following techniques or some combination of them may be helpful in inducing crystals to form.

1. Scratch the oil against the side of the beaker with a glass stirring rod. Use a freshly cut piece of glass rod (not fire-polished) with a vertical (up-and-down) motion in and out of the solution. Seed crystals or nuclei may develop which will cause crystallization to take place.
2. "Seed" the oil with some of the original material, finely powdered, by dropping some into the cooled flask.
3. Cool the solution in a freezing mixture. Refrigerate for a long time.
4. Add crumbs of dry ice.
5. Let stand for a long period.
6. If a solvent pair has been used, oiling may be prevented and crystallization induced by adding a little more of the better solvent, or changing the solvent system.

DECOLORIZATION

Principle
Colored contaminants may sometimes be removed by adding finely powdered decolorizing charcoal, such as **Norit,** which adsorbs the contaminants. Soluble contaminants, not adsorbed, remain in solution in the mother-liquor filtrate.

A. General Procedure

1. Select the most desirable solvent; refer to solubility tables.
2. Place the substance to be purified in a suitably sized flask.
3. Add a determined volume of solvent (maximum two-thirds the volume of the flask) and a few boiling stones.
4. Add decolorizing carbon, 1 percent by weight of solute, if needed.
5. Boil until all crystals have dissolved.

6. Filter as quickly as possible through a fluted filter in the funnel (Fig. 13.6). Stemless funnels are best to use, because there is no stem in which crystallization can take place and clog the system. If necessary, warm the filter funnel to prevent crystallization of hot filtrate in the funnel.
7. Collect the filtrate in a flask; allow it to stand and cool. (a) Cool or chill rapidly in a cooling bath for small crystals, or (b) Cool slowly to get large crystals.
8. Filter the crystals from the mother liquor by gravity or suction filtration. Further crystals can be obtained by evaporation or concentration of the mother liquor.
9. Dry the crystals in a warm oven. Do not melt the crystals.

CAUTION: Observe all fire and toxic hazards because of volatile fumes.

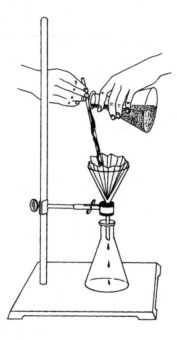

FIGURE 13.6 Experimental setup for recrystallization (decolonization filtration).

B. Additional Techniques and Hints

1. The decolorizing charcoal may be added after the solute has been dissolved in the minimum amount of solute. This is done to ensure complete solubilization of the solute.

CAUTION: Add the charcoal in small increments so that the solution does not froth and boil over.

2. If you are filtering with suction, use a heated or a jacketed Büchner funnel to prevent clogging caused by crystallization due to cooling. Use a layer of diatomaceous earth or Celite® to form a base on the paper in the Büchner funnel to trap and catch the finely divided particles of charcoal.

NOTE: Observe all safety precautions relating to filtration under reduced pressure. (See Chapter 12, "Laboratory Filtration.")

C. Batch Decolorization

Slightly colored solutions may be decolorized by adsorption on alumina (Fig. 13.7).

Procedure

1. Prepare a short column packed with activated alumina.
2. Pour the colored solution in the top of the packed column.
3. Collect the decolorized solution at the bottom. If this doesn't work, carbon decolorization will be necessary (or you are dealing with a colored substance).

FRACTIONAL CRYSTALLIZATION

Principle

You will often encounter mixtures of two compounds which have similar solubilities, and both will crystallize out of solution. They can be separated by fractional crystallization, which is a multistep crystallization repeated as many times as necessary.

Procedure

1. Isolate the mixture of compounds which are to be separated by fractional crystallization.
2. Select a solvent in which one compound is more soluble than the other compound.
3. Recrystallize the mixture using normal procedures. The result will also be a mixture, but there will be enrichment of the less soluble compound.
4. Repeat the crystallization as often as necessary until one pure product is obtained.

FIGURE 13.7 Decolorization by adsorption on alumina.

NOTE: "Seeding" a dilute solution with a pure crystal of the desired compound and then cooling the solution very slowly (without any mixture or agitation) may result in die crystallization of the desired compound, and leave the other compound in a supersaturated state.

CAUTION: Carry fractional crystallizations out in very dilute solution; otherwise, no purification takes place. Crystallization from a concentrated solution leads to total recovery of both compounds.

LABORATORY USE OF PURIFICATION BY FRACTIONAL CRYSTALLIZATION

Different substances have different solubilities in the same solvent at different temperatures. The changes in the solubilities with temperature are not the same. Advantage can be taken of such solubility relationships to effect the separation and purification of organic compounds and inorganic salts.

EXAMPLE 13.1 To separate KCl from NaCl, technicians take advantage of the marked change in the solubility of KCl with temperature as compared with NaCl (Fig. 13.8).

Procedure

1. Assume a mixture of 30 g NaCl and 50 g KCl in 100 g water at 100°C.
2. Cool the solution; precipitation of KCl starts to occur at about 70°C.

266 CHAPTER THIRTEEN

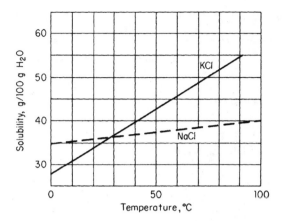

FIGURE 13.8 Typical solubility curves used in fractional crystallization.

3. At 0°C, about 20 g KCl will have crystallized out of solution; most of the NaCl will remain in solution because it has a greater solubility in water than KCl.
4. Filter the solution to obtain almost pure KCl.
5. Recrystallize the KCl in the minimum amount of water, cooling again to 0°C and filtering.

Procedure for Other Compounds

1. Locate in the literature the solubility-temperature data of the compounds, if they are available. Or determine the solubility of the compounds involved by experiment.
2. Graph the solubilities of the compounds in 100 g of solvent versus the temperature.
3. Use the KCl–NaCl separation example to develop a working procedure.

NONCRYSTALLIZATION: "OILING" OF COMPOUNDS

Sometimes crystallization does not take place as expected, and the solid may "oil out"; that is, the substance does not crystallize, but becomes a supercooled, amorphous liquid mass.

The substance may melt in the solvent instead of dissolving in it. **Oiling-out** occurs when the boiling point of the solvent is too high, and the melted solute is insoluble in the solvent.

Some substances are very difficult to crystallize, regardless of the boiling point of the solvent, and again they become oils instead of crystals, as the solvent cools. Some techniques that may be helpful were discussed in the section on inducing crystallization, earlier in this chapter.

Substances which have very low melting points are extremely difficult to crystallize, and crystal formation above the melting point is impossible. Obviously such substances must be purified by some other method.

REFERENCES

1. Gordon, Arnold J., and Ford, Richard A. *The Chemist's Companion, A Handbook of Practical Data, Techniques, and References,* John Wiley & Sons, Inc. ISBN 0-471-31590-7, 1972.
2. *Lange's Handbook of Chemistry,* 15th edition, Section 11—Practical Laboratory Information, John A. Dean, editor, McGraw-Hill, Inc. ISBN 0-07-016384-7, 1999.

CHAPTER 14
THE BALANCE

INTRODUCTION

Chemistry is a science of precision, a quantitative science. The most important single piece of apparatus available to the chemist is the balance. It is as important to the chemist as the microscope is to the biologist. Balances are mechanical devices used to determine the mass of objects. Because the mass to be determined ranges from kilograms to micrograms, the choice of the balance to be used for any determination is governed by the total mass of the object and the sensitivity desired. Therefore, the technician is always faced first with the decision of which balance to use. The precision required is the second decision.

All balances are expensive precision instruments, and you should use extreme care when handling and using them. Many kinds of balances are found in the chemical laboratory, ranging from rough measuring devices (the trip balances, the triple-beam balance) which are sensitive to 0.1 g to the analytical balances sensitive to fractions of a microgram.

Because balances are delicate instruments, the following comprehensive rules should be observed in caring for and using them. (These are general rules for all balances. Prudence will dictate which are not applicable in work with rougher measuring devices.)

1. Level the balance.
2. Inspect the balance to be certain that it is working properly. Use calibrated, undamaged masses.
3. Check the balance zero.
4. Be certain the beam is locked before removing or changing masses or objects to be massed.
5. Keep the balance scrupulously clean.
6. Work in front of the balance to avoid parallax errors.
7. Handle all masses and objects with forceps, never with fingers. Place the masses as close as possible to the center of the pans.
8. Avoid massing hot objects.
9. Release the locking mechanism slowly, avoiding jars.
10. Do not overload the balance.
11. Never place moist objects or chemicals directly on the balance pans.
12. Close the balance case (if part of the balance).
13. Triple count all masses to avoid error. Separate masses.
14. Record masses in notebook for addition. Never add mentally.

MEASUREMENT

To *measure,* by definition, is to determine the dimensions, capacity, or quantity of anything. A *measurement* is then the extent, capacity, or amount of something as determined by measuring. Any of our life experiences are studied by a system of measures and communicated to others by transferring a stimulus to an instrument which measures its intensity. For example, we transfer our feeling of temperature to others by comparing our body temperature with that of the surroundings and saying we are hot, cold, or comfortable. The stimulus is temperature and the instrument is our body's sense of feeling.

Laboratory instruments designed to measure are mere refinements of our body senses. For example, a balance permits us to determine the mass of an object more accurately than we could determine it by lifting.

In a quantitative determination we are accurately measuring some part of the whole, some constituent of the product. The amount of constituent can be measured by a volumetric, instrumental, or gravimetric technique, and the percent of a constituent must be a ratio of the amount of constituent to the amount of product. The amount of product is usually determined by mass, and the mass is measured by a balance. To distinguish the measurement of mass from the measurement of weight, technicians use the term *mass* as a verb and say that an object is *massed* rather than weighed.

DEFINITIONS OF TERMS

Mass An invariant measure of the quantity of matter in a object. The SI unit of mass is the kilogram, but in the laboratory gram quantities are more usual. Technicians properly use the term *mass* in discussing measurements made with a balance.

Weight The force of attraction exerted between an object and the earth. Weight equals mass times the gravitational attraction. Mass is proportional to weight, so we ordinarily interchange the terms, but the unit of weight is the newton.

Capacity The largest load on one pan for which the balance can be brought to equilibrium.

Precision (standard deviation) Degree of agreement of repeated measurements of the same quantity. It is a statistical value and is calculated:

$$S = \sqrt{\frac{\sum d^2}{f}}$$

where S = standard deviation
d = deviation between individual massing and average
$f = n - 1$

Precision and reproducibility are synonymous.

Readability The smallest fraction of a division at which the index scale can be read with ease.

Accuracy The agreement between the result of a measurement and the true value of the quantity measured.

The National Bureau of Standards described the difference between precision and accuracy: "Accuracy" has to do with closeness [of data] to the truth, "precision" only with closeness of readings to one another.

A. Factors Influencing Accuracy

1. Magnitude of the lever-arm error
2. Magnitude of error in scale indication due to variable load
3. Adjustment error of masses
4. Uniform value of divisions throughout the optical scale

5. Precision
6. Environmental factors

Factors 1 to 3 have no influence on the Mettler balance because of substitution.

Sensitivity   The change in load required to produce a perceptible change in indication. It is therefore a ratio and is not to be used to discuss the quality of a measurement.

ERRORS IN DETERMINING MASS

1. *Changes in moisture or CO_2 content.* Some materials take up H_2O or CO_2 from the air during the massing process. Such materials must be massed in a closed system.
2. *Volatility of sample.* Materials that are volatile at room temperature will lose mass while on the balance. Such materials must be massed in a closed system.
3. *Electrification.* An object carrying a charge of static electricity is attracted to various parts of the balance, and an error in mass may occur. An antistatic brush might help in such cases.
4. *Temperature.* If an object is warm relative to the balance, convection currents cause the pan to be buoyed up, and the apparent mass is less than the true mass. Determine mass at room temperature, if possible.
5. *Buoyancy.* This error is due to the weight of air displaced by the object on the pan and is generally quite small.

TYPES OF BALANCES

A. Equal-Arm Balances

Principle of Operation
The equal-arm analytical balance (Fig. 14.1) acts like a first-class lever. The addition of mass to one side of such a lever at rest (in equilibrium) will cause it to become unbalanced. The force at the point of load is a product of the mass involved and the horizontal distance from the fulcrum through

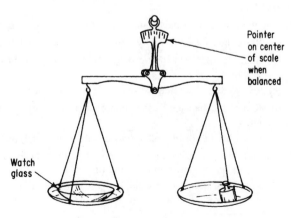

FIGURE 14.1   Equal-arm balance.

which it is acting. The lever again achieves its position of equilibrium when the force at the load site is exactly balanced on the opposite side of the fulcrum.

$$F_1 = F_2$$

where F_1 and F_2 are opposing forces.

Since the force is dependent on the distance from the fulcrum, it is essential that the pans of the balance be exactly equidistant from the fulcrum.

The massing operation on an equal-arm balance then consists of duplicating under load the equilibrium position of the unloaded balance.

Rider balances (Fig. 14.2),* Chainomatic® balances, and keyboard balances are examples of equal-arm balances.

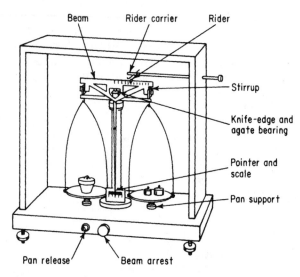

FIGURE 14.2 Controls and components of a laboratory analytical balance.

General Procedure

1. Find the rest point of the balance when there is *no load* on either pan.
   (a) Raise the balance beam by turning the operating knob into the free-swinging position.
   (b) Start the balance swinging 10 to 20 divisions by air current or by fanning gently with a piece of paper.
   (c) Record 3 to 5 consecutive swing points of the pointer.
   (d) Return the balance to the supported position by reversing the position of the operating knob.
2. Place the object to be massed (empty crucible) on the left pan of the balance. handle with forceps.
3. Transfer the appropriate masses with ivory-tipped forceps to the center of the right pan. Adjust the masses on the right pan to 10 mg light.
4. Move the rider to bring the swinging pointer to rest at the *original no-load rest point*.

---

*Although they are being replaced by more modern balances, these are still found in some laboratories.

5. Record the total mass needed to achieve step 4.
6. You have obtained the mass of the crucible.
7. Place the material to be massed in the crucible.
8. Repeat operations 2 through 5.
9. You have obtained the mass of the substance and the mass of the crucible combined.
10. Subtract the mass in step 5 from that in step 9 to get the mass of the material.

B. The Triple-Beam Balance

The capacity of a triple-beam balance is 2610 g with attachment masses. Its sensitivity is 0.1 g. See Fig. 14.3.

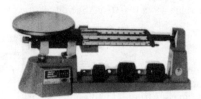

FIGURE 14.3 Two types of triple-beam balance.

Procedure

1. Observe all general massing procedures.
2. Slide all poises or riders to zero.
3. Zero the balance, if necessary, with balance-adjustment nuts.
4. Place the specimen on the pan of the balance.
5. Move the heaviest poise or rider to the first notch that causes the indicating pointer to drop; then move the poise back one notch, causing the pointer to rise.
6. Repeat procedure 4 with the next highest poise.
7. Repeat this procedure with the lightest poise, adjusting the poise position so that the indicator points to zero.
8. The mass of the specimen is equal to the sum of the values of all the poise positions, which are read directly from the position of the poises on the marked beams.

C. The Dial-O-Gram Balance

The capacity of a Dial-O-Gram® balance (Fig. 14.4) is 310 g. Its sensitivity is 0.01 g.

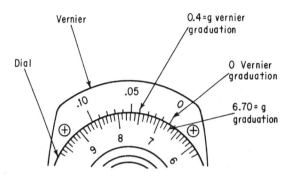

FIGURE 14.4 Dial-O-Gram control knob®.

Procedure

1. Observe all general massing procedures, sliding poises to zero.
2. Rotate the dial to 10.0 g.
3. Move the 200-g poise on the rear beam to the first notch which causes the pointer to drop; then move it back one notch.
4. Move the 100-g poise to the first notch which causes the pointer to drop.
5. Rotate the dial knob until the pointer is centered.
6. Add the values of the 200-g poise, the 100-g poise, and the dial reading. Each graduation of the dial reads 0.1 g, with a vernier breaking the value down to 0.01 g. (See discussion of verniers below.)

D. Two-Pan Equal-Arm Chainomatic Balances

Adjustment of the height of the chain in this type of balance causes changes in the mass applied to the right-hand pan. It eliminates the use of masses less than 0.1 g. It will apply masses from 0 to 100 mg (0.1 g) to the right-hand pan. See Fig. 14.5.

FIGURE 14.5 Ainsworth balance, Chainomatic® type.

Procedure

1. Use the procedure given under Two-Pan Equal-Arm Balances; except when adding masses of 100 mg or less, adjust the height of the chain indicator to get the same rest point as that of the original with *no load.*
2. The calibrated vernier gives the mass portion from 0 to 100 mg.

Using the Vernier on the Chainomatic Balance and Other Equipment.

Principle

A vernier is used to measure accurately a fraction of the finest division on the main scale of a measuring instrument.

Estimation Without Vernier. On the main scale, without a vernier, the sliding index indicates the portion on the scale corresponding to the measurement (Fig. 14.6).

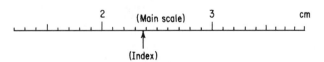

FIGURE 14.6 Example of scale (length).

1. The index points to a reading between 2.3 and 2.4 cm.
2. *Estimate* the index to be 8/10 of a division (8/10 of 0.1 cm, or 0.08 cm).
3. The *estimated* reading of the index pointer = 2.3 + 0.08 = 2.38 cm.

Exact Reading with Vernier.

EXAMPLE 14.1

1. The index in Fig. 14.7 points to a reading between 2.3 and 2.4 cm (the same as was obtained in the estimated procedure). The zero mark on the vernier equals the simple index pointer, indicating that the reading is between 2.3 and 2.4 cm.

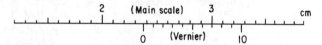

FIGURE 14.7 Example of vernier and scale (length).

2. The division of the vernier scale which coincides with a division of the main scale indicates the *exact* reading.
   (a) The vernier division which coincides is 7, which is exactly 0.07 cm.
   (b) The *accurate* reading is the sum of 2.3 cm and the exact 0.07 cm to give 2.37.

EXAMPLE 14.2

1. The vernier zero index (Fig. 14.8) is 2.7.
2. The vernier reading is 0.01 cm (division 1 on the vernier coinciding with division 2.8 on the main scale).
3. The exact reading is 2.7 + 0.01 = 2.71.

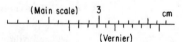

FIGURE 14.8 Additional example of vernier and scale (length).

E. Unequal-Arm Balance—Substitution

The principle of operation of the unequal-arm balance is substitution. The balance consists of an asymmetric beam. The maximum load is placed on both sides. On the shorter end are a pan and a full complement of masses. A counterpoise mass is used on the longer end to impart equilibrium to the system. When a load is placed on the pan, the analyst must remove an equivalent mass from the load side, within the range of the optical scale, to bring the balance into equilibrium. The total masses removed plus the optical-scale reading equal the mass on the pan. The Mettler single-pan balance operates on this principle. A diagram of its operating system appears in Fig. 14.9.

F. Single-Pan Analytical Balance

Basic Controls.

1. Pan-arrest control
   (a) Assures constant position of the beam between and during massing.
   (b) Protects the bearing surface from excessive wear and injury due to shocks.
2. Arrest control—three positions
   (a) Arrest position is used when removing or placing objects on the pan, when the balance is being moved, and when the balance is not in use.
   (b) Partial arrest position is used to obtain preliminary balance.
   (c) Release position is used when the final massings are being made.

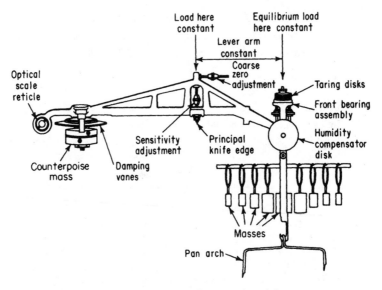

FIGURE 14.9 Diagram of the mass system of an unequal-arm balance.

3. Zero–adjust knob
   (a) Positions the optical scale to read zero when the pan is empty, because of minute changes in beam position.
4. Mass-setting knobs—two knobs which remove and replace masses from the beam
   (a) One knob removes masses in 1- to 9-g increments.
   (b) The second knob removes masses in 10-g increments, load limit 100 g.
5. Optical-scale adjustment
   (a) Turn knob positions the optical scale relative to a reference line so that the final mass can be obtained to 0.1 mg.

General Procedure
1. Check to see that the balance is level.
2. Zero the balance in arrest position, with pans clear and all mass readings at zero.
3. Mass the object.
   (a) Put the pan in arrest position.
   (b) Place the object on the pan.
   (c) Set to semiarrest position.
   (d) Adjust 1- and 10-g control knobs until the mass is within 1 g of the object's mass.
   (e) Return to arrest position, then to full release position with the arrest control.
   (f) Obtain the final mass by adjustment of the optical-system adjustment knob.

Types of Single-Pan Analytical Balances. Following is a list of the single-pan analytical balances most frequently encountered by technicians.

Mettler Model H-5 (See Fig. 14.10.)
Sartorius Series 2400 (See Fig. 14.11.)

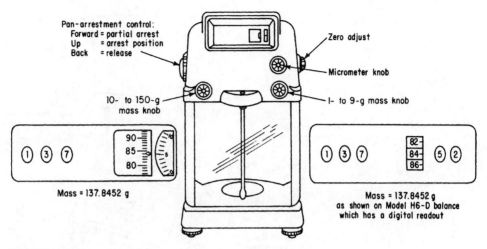

FIGURE 14.10 The Mettler HF Gram-atic single-pan analytical balance.

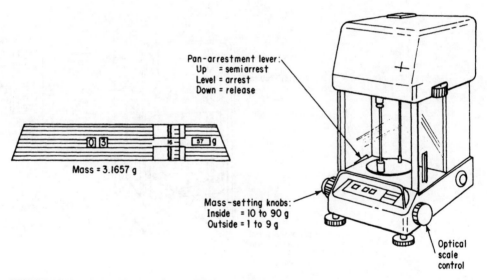

FIGURE 14.11 The Sartorius 2400 Series single-pan analytical balance.

Stanton Unimatic CL2 (See Fig. 14.12.)
Ainsworth Magni Grad Type 21 (See Fig. 14.13.)

G. Electronic Balances

The newest balances available are the electronic balances. These come in capacities from a few milligrams to kilograms. A single control bar turns the balance on to provide a digital readout of the mass (Fig. 14.14).

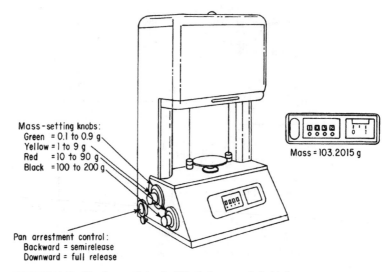

FIGURE 14.12 The Stanton Unimatic CL2 single-pan analytical balance.

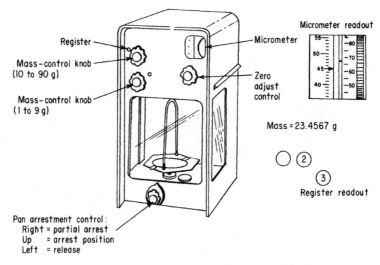

FIGURE 14.13 The Ainsworth Magni Grad Type 21 single-pan analytical balance.

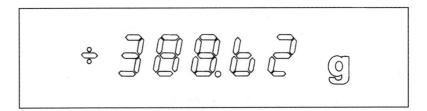

FIGURE 14.14 Digital readout from a new electronic balance.

Procedure

1. Place sample on pan, with balance turned on.
2. Press zero-set button.
3. Read digital readout for mass.

Electronic balances can be coupled directly to computers or recording devices if necessary.

H. Other Types of Balances

Various types of balances encountered in the laboratory are shown in Figs. 14.15 to 14.16.

FIGURE 14.15 Single-pan, top-loading, substitution-type, rapid-reading balance with varied sensitivities and optical micrometer scale.

FIGURE 14.16 Dial-reading torsion balance; allows rapid adjustment of the 10-g mass increment by rotation of the dial.

COMPUTERIZED BALANCES

Computerized analytical balances are the newest generation of balances. (See Fig. 14.17.) They offer highly precise weighing capability plus the convenience of modem electronics. The balances allow for 10-µg readability in a convenient-to-use design. A computerized balance automatically recalibrates itself whenever environmental conditions change and the instrument senses a drift. The balance can also be connected to a computer for automated weighing procedures. Software allows for calculations of statistics and reports results in the user's format. Some computerized balances are ISO/GLP-compliant with built-in printing and recording capabilities (for example, date, time, serial number, mass, sample number, and even a line for the operator's signature). All of this data can be communicated directly through a standard interface to any type of laboratory information management system.

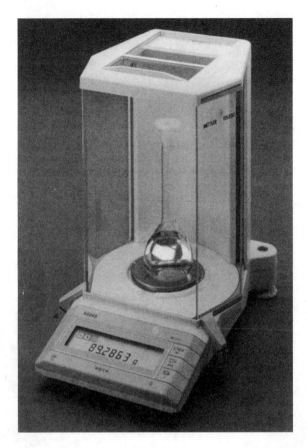

FIGURE 14.17 Computerized analytical balance offers 10-μg readability and automatic calibration. (*Courtesy Mettler-Toledo, Inc.*)

CHOOSING THE CORRECT BALANCE

A. Criteria for Choosing a Balance

1. Is the balance suitable for making the desired measurement?

EXAMPLE 14.3 Is a Mettler Model B-5 (four-place balance) suitable for massing 10 mg of material?

The accuracy in the optical scale is ±0.05 mg.

The error on a 10-mg sample is ±0.5 percent.

Conclusion: Analytical tests cannot tolerate 0.5 percent errors; therefore, a different balance must be chosen.

2. What balance is suitable to mass 10 mg of material?

Model B-5 accuracy in the optical scale is ±0.05 mg.

The error on a 10-mg sample is ±0.5 percent (not tolerable).

Model B-6 accuracy in the optical scale is ±0.02 mg.
The error on a 10-mg sample is ±0.2 percent (tolerable).

Conclusion: Model B-5 is not suitable. Model B-6 is suitable.

B. Model Calculation

Determination of chloride:

$$\% \, Cl = \frac{V \times N \times \text{milliequivalent mass} \times 100}{\text{mass of sample (g)}}$$

where V = volume of titrant (2 mL)
N = normality of titrant (1 N)

Milliequivalent mass of chloride = 0.035
Mass of sample, g = 0.010000 (no error assumed)
Assumption: The V and N of the titrant have no error.

$$\frac{2 \times 0.1 \times 0.035 \times 100}{0.010000} = 70.0\%$$

Four-place balance: $\dfrac{2 \times 0.1 \times 0.035 \times 100}{0.01015*} = 69.64\%$

possible error = ±0.37

Five-place balance: $\dfrac{2 \times 0.1 \times 0.035 \times 100}{0.01002*} = 69.76\%$

possible error = ±0.24

Microbalance: $\dfrac{2 \times 0.1 \times 0.035 \times 100}{0.01002*} = 69.98\%$

possible error = ±0.02

REFERENCES

1. R. M. Schoonover and F. E. Jones, "Air Buoyancy Correction in High Accuracy Weighing on Analytical Balance." Analytical Chemistry, 53:900 (1981).
2. Lange's Handbook of Chemistry, fifteenth edition, Section 11-Practical Laboratory Information, John A. Dean, editor, McGraw-Hill Inc., ISBN 0-07-016384-7 (1999).

*Inherent mass error in balance.

CHAPTER 15
GRAVIMETRIC ANALYSIS

INTRODUCTION

Gravimetric analysis refers to the isolation of a specific substance from a sample and ultimately weighing the substance in a pure or known form. This substance is usually isolated by precipitating it in some insoluble form, by depositing it as a pure metal in electroplating, or by converting it to a gas which is then quantitatively absorbed. Review Chapter 12, "Laboratory Filtration," for details of this isolation technique. It is necessary that (1) the sought substance be completely removed from the sample, (2) the moisture content and/or other volatile components be determined, and (3) the sample be representative of the material being analyzed.

CAUTION: Chips obtained solely from the surface may not represent the true composition of the object.

NOTE: Representative sampling techniques are discussed in Chapter 11, "Sampling."

HANDLING SAMPLES IN THE LABORATORY

The technician may analyze an already prepared sample or prepare a new sample which is to be tested, analyzed, or evaluated. Each sample should be completely identified, tagged, or labeled so that no question as to its origin or source can arise.

Some of the information which may be on the sample is:

1. The number of the sample
2. The notebook experiment-identification number
3. The date
4. The origin, for example, the technician's name, and cross-reference number
5. Weight or volume
6. Identifying code of the container
7. What is to be done with the sample, what determination is to be made, or what analysis is desired

Suitable containers are shown in Fig. 15.1.

A. Pretreatment of Samples

As it arrives at the laboratory, the sample often requires treatment before it is analyzed, particularly if it is in the form of a solid. One of the objectives of this pretreatment is to produce a material so homogeneous that any small portion removed for the analysis will be identical to any other portion. This usually involves reduction of the size of particles to a few tenths of a millimeter and thorough mechanical mixing. Another objective of the pretreatment is to convert the substance to a form in

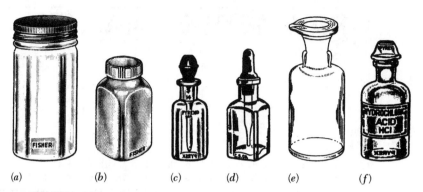

FIGURE 15.1 (*a*) and (*b*) Liquid- and solid-storage bottles for chemicals and samples; varied design, openings, and sizes. (*c–e*) Dropping bottles to dispense small volumes of liquids. (*f*) Liquid-storage bottle for laboratory acids, bases, reagents, and salts.

which it is readily attacked by the reagents employed in the analysis; with refractory materials particularly, this involves grinding to a very fine powder. Finally, the sample may have to be dried or its moisture content may have to be determined, because this is a variable factor that is dependent upon atmospheric conditions as well as the physical state of the sample.

B. Crushing and Grinding

In dealing with solid samples, a certain amount of crushing or grinding is sometimes required to reduce the particle size. Unfortunately, these operations tend to alter the composition of the sample, and for this reason the particle size should be reduced no more than is required for homogeneity and ready attack by reagents.

Ball or jar mills are jars or containers, usually made of porcelain, fitted with a cover and gasket which can be securely fastened to the jar. The jar is half filled with flint pebbles or porcelain or metal balls, and then enough of the material which is to be ground is added to cover the pebbles or balls and the voids between them. The cover is fastened securely to seal the mill hermetically, and the jar is revolved on a rotating assembly. The length of time for which the material is ground depends upon the fineness desired and the hardness of the material. The jar is then emptied into a coarse-mesh screen to separate the pebbles or balls from the ground material.

C. Changes in Sample

Several factors may cause appreciable alteration in the composition of the sample as a result of grinding. Among these is the heat that is inevitably generated. This can cause losses of volatile components in the sample. In addition, grinding increases the surface area of the solid and thus increases its susceptibility to reactions with the atmosphere. For example, it has been observed that the iron(II) content of a rock may be altered by as much as 40 percent during grinding—apparently a direct result of atmospheric oxidation of the iron to iron(III).

The effect of grinding on the gain or loss of water from solids is considered in a later section.

Another potential source of error in the crushing and grinding of mixtures arises from the difference in hardness of the components of a sample. The softer materials are converted to smaller particles more rapidly than the hard ones; any loss of the sample in the form of dust will thus cause an alteration in composition. Furthermore, loss of sample in the form of flying fragments must be avoided, since these will tend to be made up of the harder components.

D. Grinding Surfaces

A serious error can arise during grinding and crushing as a consequence of mechanical wear and abrasion of the grinding surfaces. For this reason only the hardest materials such as hardened steel, agate, or boron carbide are employed for the grinding surfaces. Even with these, contamination of the sample is sometimes encountered.

The **mortar** and **pestle** (Fig. 15.2), the most ancient of grinding tools, still find wide use in the analytical laboratory. These now come in a variety of sizes and shapes and are commonly constructed of glass, porcelain, agate, mullite, and other hard materials.

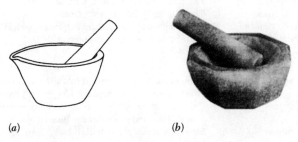

FIGURE 15.2 Mortars and pestles. (*a*) Porcelain mortar. (*b*) Agate mortar.

CAUTION:

1. *Always clean the mortar and pestle thoroughly both before and after grinding each sample.*
2. *Never grind two materials together unless specifically told to do so.*

Minor and some major explosions have occurred and fingers and eyes have been lost or burned because technicians have failed to observe these two simple rules.

E. Flame Tests for Identification of Elements

Technicians often make a preliminary examination of a sample to determine the presence of certain common elements. This examination may include a flame test, which is easily performed as follows.

1. Obtain a loop of platinum or Nichrome wire mounted in a glass rod. Clean it carefully by dipping it into a small amount of concentrated hydrochloric acid (in a test tube) and then heating it in the blue flame of a Bunsen-type burner until it is cherry red. If the wire loop is clean, the flame will not change color; if the flame shows any color at all, repeat the process until you are satisfied that the loop is free of any contaminant. Replace the acid frequently.
2. Pour a small amount of the powdered sample into a clean watch glass.
3. Heat the *clean* wire loop to cherry redness and dip it into the sample. Some of the powder will cling to the loop. Tap the rod lightly to dislodge any excess sample.
4. Place the wire loop plus the sample back in the flame and reheat it to cherry redness. As shown in Table 15.1, certain elements will lend characteristic colors to the flame.
5. Be sure to clean the wire loop thoroughly between samples.

TABLE 15.1 Flame Tests for Elements

Color of flame	Element indicated
Blue	
Azure	Lead, selenium, $CuCl_2$ (and other copper compounds when moistened with HCl); $CuBr_2$ appears azure blue, then is followed by green.
Light blue	Arsenic and some of its compounds; selenium.
Greenish blue	$CuBr_2$; arsenic; lead; antimony.
Green	
Emerald green	Copper compounds other than halides (when not moistened with HCl); thallium compounds.
Blue-green	Phosphates moistened with sulfuric acid; B_2O_3.
Pure green	Thallium and tellurium compounds.
Yellow-green	Barium; possibly molybdenum; borates (with H_2SO_4).
Faint green	Antimony and ammonium compounds.
Whitish green	Zinc.
Red	
Carmine	Lithium compounds (masked by barium or sodium), are invisible when viewed through green glass, appear violet through cobalt glass.
Scarlet	Calcium compounds (masked by barium), appear greenish when viewed through cobalt glass and green through green glass.
Crimson	Strontium compounds (masked by barium), appear violet through cobalt glass, yellowish through green glass.
Violet	Potassium compounds other than silicates, phosphates, and borates; rubidium and cesium are similar. Color is masked by lithium and/or sodium appears purple-red through cobalts glass and bluish-green glass.
Yellow	Sodium, even the most minute amounts, is invisible when viewed through cobalt glass.

MOISTURE IN SAMPLES

The presence of water in a sample represents a common problem that frequently faces the analyst. This compound may exist as a contaminant from the atmosphere or from the solution in which the substance was formed, or it may be bonded as a chemical compound, a hydrate. Regardless of its origin, however, water plays a part in determining the composition of the sample. Unfortunately, particularly in the case of solids, the water content is a variable quantity that depends upon such things as humidity, temperature, and the state of subdivision. Thus, the constitution of a sample may change significantly with environment and method of handling.

In order to cope with the variability in composition caused by the presence of moisture, the analyst may attempt to remove the water by drying prior to weighing samples for analysis. Alternatively, the water content may be determined at the time the samples are weighed out for analysis; in this way results can be corrected to a dry basis. In any event, most analyses are preceded by some sort of preliminary treatment designed to take into account the presence of water. There are many established tests used for this purpose.

FORMS OF WATER IN SOLIDS

It is convenient to distinguish among the several ways in which water can be held by a solid. Although it was developed primarily with respect to minerals, the classification of Hillebrand and his collaborators may be applied to other solids as well, and it forms the basis for the discussion that follows.

The **essential water** in a substance is that water which is an integral part of the molecular or crystal structure of one of the components of the solid. It is present in that component in stoichiometric quantities. Thus, the water of crystallization in stable solid hydrates (for example, $CaC_2O_4 \cdot 2H_2O$, $BaCl_2 \cdot 2H_2O$) qualifies as a type of essential water.

A second form is called **water of constitution**. Here the water is not present as such in the solid but rather is formed as a product when the solid undergoes decomposition, usually as a result of heating. This is typified by the processes

$$2KHSO_4 \rightarrow K_2S_2O_7 + H_2O$$
$$Ca(OH)_2 \rightarrow CaO + H_2O$$

Nonessential water is not necessary for the characterization of the chemical constitution of the sample and therefore does not occur in any sort of stoichiometric proportions. It is retained by the solid as a consequence of physical forces.

Adsorbed water is retained on the surface of solids in contact with a moist environment. The quantity is dependent upon humidity, temperature, and the specific surface area of the solid. Adsorption is a general phenomenon that is encountered in some degree with all finely divided solids. The amount of moisture adsorbed on the surface of a solid also increases with the amount of water in its environment. Quite generally, the amount of adsorbed water decreases as temperature increases, and in most cases it approaches zero if the solid is dried at temperatures above 100°C.

Equilibrium, in the case of adsorbed moisture, is achieved rather rapidly, ordinarily requiring only 5 or 10 minutes. This often becomes apparent to the chemist who weighs finely divided solids that have been rendered anhydrous by drying; a continuous increase in weight is observed unless the solid is contained in a tightly stoppered vessel.

A second type of nonessential water is called **sorbed water**. This is encountered with many colloidal substances such as starch, protein, charcoal, zeolite minerals, and silica gel. The amounts of sorbed water are often large compared with adsorbed moisture, amounting in some instances to as much as 20 percent or more of the solid. Interestingly enough, solids containing even this much water may appear to be perfectly dry powders. Sorbed water is held as a condensed phase in the interstices or capillaries of the colloidal solids. The quantity is greatly dependent upon temperature and humidity.

A third type of nonessential moisture is **occluded water.** Here, liquid water is entrapped in microscopic pockets spaced irregularly throughout the solid crystals. Such cavities often occur naturally in minerals and rocks.

Water may also be dispersed in a solid in the form of a solid solution. Here the water molecules are distributed homogeneously throughout the solid. Natural glasses may contain several percent of moisture in this form.

EFFECTS OF GRINDING ON MOISTURE CONTENT

Often the moisture content and thus the chemical composition of a solid is altered to a considerable extent during grinding and crushing. This will result in decreases in some instances and increases in others.

Decreases in water content are sometimes observed when one is grinding solids containing essential water in the form of hydrates; thus the water content of **gypsum**, $CaSO_4 \cdot 2H_2O$, is reduced from 20 to 5 percent by this treatment. Undoubtedly the change is a result of localized heating during the grinding and crushing of the particles.

Losses also occur when samples containing occluded water are reduced in particle size. Here, the grinding process ruptures some of the cavities and exposes the water so that it may evaporate.

More commonly perhaps, the grinding process is accompanied by an increase in moisture content, primarily because of the increase in surface area exposed to the atmosphere. A corresponding increase in adsorbed water results. The magnitude of the effect is sufficient to alter appreciably the composition of a solid. For example, the water content of a piece of porcelain in the form of coarse particles was zero, but after it had been ground for some time it was found to be 0.62 percent. Grinding a basaltic green-stone for 120 minutes changed its water content from 0.22 to 1.70 percent.

DRYING SAMPLES

Principle
Samples may be dried by heating them at 100 to 105°C or higher, if the melting point of the material is higher and the material will not decompose at that temperature. This procedure will remove the moisture bound to the surface of the particles.

Procedure
1. Label the beaker and the weighing bottle. (Remove cover from weighing bottle.)
2. Place the weighing bottle in the beaker, which is covered by a watch glass supported on glass hooks (Fig. 15.3).
3. Place in the oven for the required time at the temperature suggested.

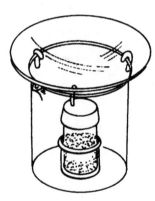

FIGURE 15.3 Arrangement for the drying of samples.

DRYING COLLECTED CRYSTALS

A. Gravity-Filtered Crystals

Gravity-filtered crystals collected on a filter paper may be dried by the following methods.

CAUTION: Be sure to label everything properly.

1. Remove the filter paper from the funnel. Open up the filter paper and flatten it on a watch glass of suitable size or a shallow evaporating dish. Cover the watch glass or dish with a large piece of clean, dry filter paper (secured to prevent wind currents from blowing it off) and allow the crystals to air-dry.

CAUTION: Hygroscopic substances cannot be air-dried in this way.

FIGURE 15.4 Pressing out excess moisture from the wet crystals is a method applicable to both gravity-filtered crystals and vacuum-filtered crystals as shown here.

FIGURE 15.5 The porous plate absorbs the excess water.

2. Press out excess moisture from the crystals by laying filter paper on top of the moist crystals and applying pressure with a suitable object. (See Fig. 15.4.)
3. Use a spatula to work the pasty mass on a porous plate (Fig. 15.5); then allow it to dry.
4. Use a portable infrared lamp to warm the sample and increase the rate of drying. Be sure the temperature does not exceed the melting point of the sample.
5. Use a desiccator filled with a desiccant (Fig. 15.6).

B. The Desiccator

A **desiccator** is a glass container filled with a substance which absorbs water (a **desiccant**); it is used to provide a dry atmosphere for objects and substances. Desiccators are employed to achieve and maintain an atmosphere of low humidity for the storage of samples, precipitates, crucibles, weighing bottles, and other equipment. Use a desiccator as follows:

FIGURE 15.6 Desiccator.

1. Remove the cover by sliding sideways as in Fig. 15.7.
2. Place the object to be dried on the porcelain platform plate.
3. Regrease the ground-glass rim with petroleum jelly or silicone grease if necessary.
4. Slide the lid back in position.

CAUTION: Hot crucibles should never be inserted immediately in the desiccator. Allow to cool in air for 1 minute prior to insertion. If this caution is not observed, the air will be heated in the desiccator when it is closed. On cooling, a partial vacuum will result. When the desiccator is opened, a sudden rush of air may spill the sample.

FIGURE 15.7 Removing the desiccator cover.

NOTE: **Vacuum desiccators** are equipped with side arms, so that they may be connected to a vacuum and the contents will be subject to a vacuum rather than to dried air. Vacuum-type desiccators should be used to dry crystals which are wet with organic solvents. Vacuum desiccators should not be used for substances which sublime readily.

A desiccator must be kept clean and its charge of desiccant must be frequently renewed to keep it effective. This is done as follows:

1. Remove the cover and the porcelain support plate.
2. Dump the waste desiccant in an appropriate waste receptacle.
3. Wash and dry the desiccator.
4. Refill with fresh desiccant (Fig. 15.8).
5. Regrease the ground-glass lid.
6. Replace the porcelain support.
7. Slide the lid into position on the desiccator.

C. Vacuum-Filtered Crystals

Vacuum-filtered crystals can be dried in one of two ways:

1. Remove the filter cake and use the procedures listed under drying gravity-collected crystals.
2. After all the crystals have been collected on the Büchner funnel, the funnel is covered loosely with an evaporating dish or a larger piece of filter paper (secured to prevent its blowing off). The vacuum system is then maintained to pull air through the moist crystals, which will dry after a short period of time.

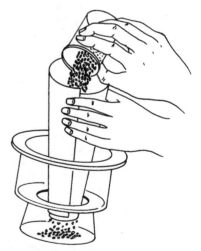

FIGURE 15.8 Filling the desiccator with fresh desiccant.

NOTE: Less time will be required to dry the crystals if, while continuing the vacuum system, the covering is removed periodically and the cake is mixed (careful!) and evened out with a spatula before the cover is resecured.

D. Drying Crystals in a Centrifuge Tube

When very small quantities of crystals are collected by centrifugation, they can be dried by subjecting them to vacuum in the centrifuge tube while gently warming the tube (Fig. 15.9). This procedure prevents any loss of the small quantity of crystals collected, as would occur if you tried to transfer the crystals out of the tube with a rubber policeman.

E. Abderhalden Drying Pistol

Some substances retain water and other solvents so persistently that drying them in ordinary desiccators at room temperature will not remove the solvents. In these cases, the **drying pistol** (Fig. 15.10) works very well. The substance to be dried is placed in a boat that is inserted in the pistol, which is then connected to a source of vacuum. Also in the contained volume is a pocket for an effective adsorbing agent, such as P_4O_{10} for water, solid KOH or NaOH pellets for acid gases, or thin layers of paraffin wax for the removal of organic solvents. The temperature at which the substance is to be dried is determined by the boiling point of the refluxing liquid.

FIGURE 15.9 Drying crystals in a centrifuge tube.

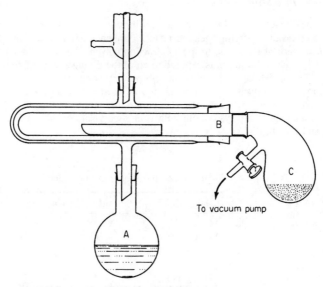

FIGURE 15.10 Abderhalden vacuum-drying apparatus ("pistol"). *A,* refluxing heating liquid; *B,* vacuum-drying chamber; *C,* desiccant.

DRYING ORGANIC SOLVENTS

Water can be removed from organic liquids and solutions by treating the liquids with a suitable drying agent to remove the water. Water is soluble to some extent in all organic liquids, and any organic solvent that has been used in a water-immiscible—organic-solvent extraction will contain water. Each organic solvent will dissolve its own characteristic percentage of water. For example, ethyl ether will contain about 1.5 percent water. To "dry" or dehydrate organic solvents, we use **drying agents**.

NOTE: The selection of drying agents must be carefully made. The drying agent selected should not react with the compound or cause the compound to undergo any reaction, but will *only* remove the water. The best drying agents are those which will react rapidly and irreversibly with water and will not react with or affect in any way the solvent or the solute dissolved in it.

Principle
Solid drying agents are added to wet organic solvents. They remove the water, and then the hydrated solid is separated from the organic solvent by decantation and filtration.

Procedure
1. Pour the organic liquid into a flask that can be stoppered. Add small portions of the drying agent, shaking the flask thoroughly after each addition. Add as much drying agent as is required.
2. Allow to stand overnight or for a predetermined time.
3. Filter the solid hydrate from the liquid with a funnel and filter paper.

NOTE: Several operations may be required. Repeat if necessary.

A. Efficiency of Drying Operations

The efficiency of a drying operation is improved if the organic solvent is repeatedly exposed to fresh portions of the drying agent, just as the efficiency of an extraction operation is improved by repeated exposure of the solute to fresh extraction solvent. The efficiency of the drying operation is lowest when the wet solution is exposed to all of the drying agent at one time.

Some dehydrating agents are very powerful and dangerous, especially if the water content of the organic solvent is high. These should be used only after the wet organic solvent has been grossly predried with a weaker agent. If you are in doubt as to the advisability of using a particular dehydrating agent, always consult with your supervisor or with specialists in the field. (See Tables 15.2 and 15.3.)

TABLE 15.2 Intensity of Drying Agents

High intensity	Moderate intensity	Low intensity
$Mg(ClO_4)_2$*	KOH	Na_2CO_3
Molecular sieves	NaOH	Na_2SO_4
Metallic sodium (Na)	K_2CO_3	$MgSO_4$
P_4O_{10}	$CaSO_4$	
H_2SO_4 (conc)	CaO, $CaCl_2$	

*$Mg(ClO_4)_2$ is the most efficient drying agent available, but is an explosion hazard with easily oxidized or acidic organic compounds.

TABLE 15.3 Characteristics of Drying Agents

Compound	Acidity	Comments
Calcium sulfate	Neutral	General use, commercially available as Drierite®, very fast.
Calcium chloride	Neutral	Reacts with N and O compounds, rapid; use for hydrocarbons and R-X.
Magnesium sulfate	Neutral	General use; rapid (avoid acid-sensitive compounds)
Sodium sulfate	Neutral	General use; mild; high-capacity; gross dryer for cold solutions.
Potassium carbonate	Basic	Use with esters, nitrites, ketones, alcohols (not for use with acidic compounds).
Sodium carbonate	Basic	Use with esters, nitriles, ketones, alcholes (not for use with acidic compounds).
Sodium hydroxide	Basic	Use only with insert compounds; very fast; powerful; good for amines.
Potassium hydroxide	Basic	Use only with inert compounds; very fast; powerful; good for amines.
Calcium oxide	Basic	Use for alcohols and amines; slow; efficient; not for use with acidic compounds.
Tetraphosphorus decoxide	Acidic	Use only with inert compounds (either, hydrocarbons, halides); fast; efficient.
Molecular sieves 3Å, 4Å	Neutral	General use; high-intensity; predry with common agent.
Sulfuric acid	Acidic	Very efficient; use for saturated hydrocarbons, aromatic hydrocarbons, and halides; reacts with olefins and basic compounds.

B. Classification of Drying Agents

1. Those which form compounds with water of hydration (the hydrates can be returned to the anhydrous form by suitable heating to remove the water): Na_2CO_3, Na_2SO_4, anhydrous $CaCl_2$, $ZnCl_2$, NaOH, $CaSO_4$, H_2SO_4 (95 percent) silica gel, CaO.

NOTE: Traces of acid remaining in wet organic-liquid reaction products are removed simultaneously with the water when *basic* drying agents are used: Na_2CO_3, NaOH, $Ca(OH)_2$.

2. Those which form new compounds by chemical reaction with water: metallic sodium, CaC_2, P_4O_{10}.

CAUTIONS: These dehydrating agents are extremely reactive and most efficient. *Handle with care.* They react with water to give NaOH, CaOH, and H_3PO_4, respectively.

1. *Do not use* them where either the drying agent itself or the product that it forms will react with the compound or cause the compound itself to undergo reaction or rearrangement.
2. *Use to dry* saturated hydrocarbons, aromatic hydrocarbons, ethers.
3. The compounds to be dried should not have functional groups, such as —OH^- and —$COOH^-$, which will react with the drying agent.
4. *Do not dry* alcohols with metallic sodium. *Do not dry* acids with NaOH or other basic drying agents. *Do not dry* amines (or basic compounds) with acidic drying agents. *Do not use* $CaCl_2$ to dry alcohols, phenols, amines, amino acids, amides, ketones, or certain aldehydes and esters.

C. Determining If Organic Solvent Is "Dry"

Drying agents will clump together, sticking to the bottom of the flask when a solution is "wet." They will even dissolve in very wet solutions if an insufficient amount of them has been added. Wet solvent solutions appear to be cloudy; dry solutions are clear. If the solution is "dry," the solid drying agent will move about and shift easily on the bottom of the flask.

FREEZE-DRYING: LYOPHILIZATION

Some substances cannot be dried at atmospheric conditions because they are extremely heat-sensitive materials, but they can be freeze-dried. Freeze-drying is a process whereby substances are subjected to high vacuum after they have been frozen, and under those conditions ice (water) will sublime. This leaves the non-sublimable material (everything but the water) behind in a dried state.

Commercial freeze-driers are available in some laboratories. They consist of a self-contained freeze-drying unit that will effectively remove volatile solvents. They may be simple ones, consisting merely of a vacuum pump, adequate vapor traps, and a receptacle for the material in solution. Others include refrigeration units to chill the solution plus more sophisticated instruments to designate temperature and pressure, plus heat and cold controls and vacuum-release valves.

Freeze-driers, as the name indicates, are usually used to remove all the volatile solvents or water, but they can be used to remove smaller amounts as required.

Freeze-drying procedures are excellent for drying or concentrating heat-sensitive substances. They differ from ordinary vacuum distillation in that the solution or substance to be dried must be frozen to a solid mass first. It is under these conditions that the water is selectively removed by sublimation, the ice going directly to the water-vapor state.

Procedure

1. Freeze the solution, spreading it out on the inner surface of the container (Fig. 15.11) to increase the surface area.
2. Apply high vacuum; the ice will sublime and leave the dried material behind.
    (a) Keep material frozen during sublimation. (Sublimation normally will maintain the frozen state.)

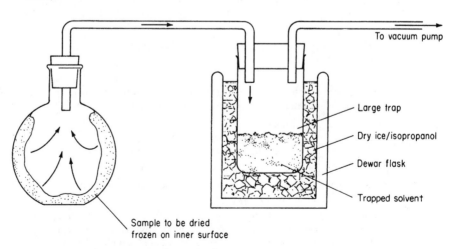

FIGURE 15.11 Setup for freeze-drying in the laboratory.

(b) Use dilute solutions in preference to concentrated solutions.

(c) Apply all safety procedures for working with high vacuum.

(d) Protect the vacuum pump from water with a dry-ice trap, and insert chemical gas-washing towers to protect the pump from corrosive gases.

A. Advantages of Freeze-Drying

1. Substances are locked in an ice matrix and cannot interact.
2. Oxidation is prevented because of the high vacuum.
3. The final product of freeze-drying is identical to the original product minus the water removed.

B. Freeze-Drying Corrosive Materials

The majority of freeze-drying operations involve solely the removal of water; however, many substances which are subjected to these procedures contain corrosive acids and bases. If these substances are to be freeze-dried, the unit must be protected against the corrosive vapors so that they do not attack the mechanical vacuum pump. *Always insert chemical gas towers to remove the corrosive vapors before they enter the pump.*

PREPARING THE SAMPLE FOR FINAL ANALYSIS

In order to complete many analyses, an aqueous solution of the sample is required; furthermore, the function to be determined must ordinarily be present in that solution in the form of a simple ion or molecule. Unfortunately many of the substances that are of interest can be converted to this form only by some treatment. For example, before the chlorine content of an organic compound can be determined, it is usually necessary to convert the element into a form that is amenable to analysis. Because this will require breaking of the carbon-chlorine bonds, the preliminary treatment of the sample is likely to be quite vigorous.

Various reagents and techniques exist for decomposing and dissolving analytical samples. Often the proper choice among these is critical to the success of an analysis, particularly where refractory substances are being dealt with.

A. Liquid Reagents Used for Dissolving or Decomposing Inorganic Samples

The most common reagents for attacking analytical samples are the mineral acids or their aqueous solutions. Solutions of sodium or potassium hydroxide also find occasional application.

Hydrochloric Acid. Concentrated hydrochloric acid is an excellent solvent for many metal oxides as well as those metals which lie above hydrogen in the electromotive series; it is often a better solvent for the oxides than the oxidizing acids. Concentrated hydrochloric acid is about 12 N, but upon heating hydrogen chloride is lost until a constant-boiling 6 N solution remains (boiling point about 110°C).

Nitric Acid. Concentrated nitric acid is an oxidizing solvent that finds wide use in attacking metals. It will dissolve most common metallic elements; aluminum and chromium, which become passive to the reagent, are exceptions. Many of the common alloys can also be decomposed by nitric acid. In this connection it should be mentioned that tin, antimony, and tungsten form insoluble acids when treated with concentrated nitric acid; this treatment is sometimes employed to separate these elements from others contained in alloys.

Sulfuric Acid. Hot concentrated sulfuric acid is often employed as a solvent. Part of its effectiveness arises from its high boiling point (about 340°C), at which temperature decomposition and solution of substances often proceed quite rapidly. Most organic compounds are dehydrated and oxidized under these conditions; the reagent thus serves to remove such components from a sample. Most metals and many alloys are attacked by the hot acid.

Perchloric Acid. Hot, concentrated perchloric acid is a potent oxidizing agent and solvent. It attacks a number of ferrous alloys and stainless steels that are intractable to the other mineral acids; it is frequently the solvent of choice. This acid also dehydrates and rapidly oxidizes organic materials.

CAUTION: Violent explosions result when organic substances or easily oxidized inorganic compounds come in contact with the hot, concentrated acid; as a consequence, a good deal of care must be employed in the use of this reagent. For example, it should be heated only in hoods in which the ducts are clean and free of organic materials and where the possibility of contamination of the solution is absolutely nil.

Perchloric acid is marketed as the 60 or 72 percent acid. Upon heating, a constant-boiling mixture (72.4 percent $HClO_4$) is obtained at a temperature of 203°C. Cold, concentrated perchloric acid and hot, dilute solutions are quite stable with respect to reducing agents; it is only the hot, concentrated acid that constitutes a potential hazard. The reagent is a very valuable solvent and is widely used in analysis. *Before it is employed, however, the proper precautions for its use must be clearly understood.*

Oxidizing Mixtures. More rapid solvent action can sometimes be obtained by the use of mixtures of acids or by the addition of oxidizing agents to the mineral acids. *Aqua regia,* a mixture consisting of three volumes of concentrated hydrochloric acid and one of nitric acid, is well known. Addition of bromine or hydrogen peroxide to mineral acids often increases their solvent action and hastens the oxidation of organic materials in the sample. Mixtures of nitric and perchloric acid are also useful for this purpose, as are mixtures of fuming nitric and concentrated sulfuric acids.

Hydrofluoric Acid. The primary use for this acid is the decomposition of silicate rocks and minerals where silica is not to be determined; the silicon, of course, escapes as the tetrafluoride. After decomposition is complete, the excess hydrofluoric acid is driven off by evaporation with sulfuric acid or perchloric acid. Complete removal is often essential to the success of an analysis because of the extraordinary stability of the fluoride complexes of several metal ions; the properties of some of these differ markedly from those of the parent cation. Thus, for example, precipitation of aluminum with ammonia is quite incomplete in the presence of small quantities of fluoride. Frequently, removal of the last traces of fluoride from a sample is so difficult and time-consuming as to negate the attractive features of this reagent as a solvent for silicates.

Hydrofluoric acid finds occasional use in conjunction with other acids to place certain steel samples into solution.

CAUTION: Hydrofluoric acid can cause serious damage and painful injury when brought in contact with the skin; it must be handled with respect.

B. Decomposition of Samples by Fluxes

Quite a number of common substances—such as silicates, some of the mineral oxides, and a few of the iron alloys—are attacked slowly, if at all, by the usual liquid reagents. Recourse to more potent fused-salt media, or fluxes, is then called for. Fluxes will decompose most substances by virtue of the high temperature required for their use (300 to 1000°C) and the high concentration of reagent brought in contact with the sample.

Where possible, the employment of a flux is avoided, for several dangers and disadvantages attend its use. In the first place, a relatively large quantity of the flux is required to decompose most substances—often 10 times the sample weight. The possibility of significant contamination of the sample by impurities in the reagent thus becomes very real.

Furthermore, the aqueous solution resulting from the fusion will have a high salt content, and this may lead to difficulties in the subsequent steps of the analysis. The high temperatures required for a fusion increase the danger of loss of pertinent constituents by volatilization. Finally, the container in which the fusion is performed is almost inevitably attacked to some extent by the flux; this again can result in contamination of the sample.

In those cases where the bulk of the substance to be analyzed is soluble in a liquid reagent and only a small fraction requires decomposition with a flux, it is common practice to employ the liquid reagent first. The undecomposed residue is then isolated by filtration and fused with a relatively small quantity of a suitable flux. After cooling, the melt is dissolved and combined with the rest of the sample.

Method of Carrying Out a Fusion. In order to achieve a successful and complete decomposition of a sample with a flux, the solid must ordinarily be ground to a very fine powder; this will produce a high specific surface area. The sample must then be thoroughly mixed with the flux; this operation is often carried out in the crucible in which the fusion is to be done by careful stirring with a glass rod.

In general, the crucible used in a fusion should never be more than half filled at the outset. The temperature is ordinarily raised slowly with a gas flame because the evolution of water and gases is a common occurrence at this point; unless care is taken there is the danger of loss by spattering. The crucible should be covered as an added precaution. The maximum temperature employed varies considerably depending upon the flux and the sample; it should be no greater than necessary, however, to minimize attack on the crucible and decomposition of the flux. The length of the fusion may range from a few minutes to 1 or 2 hours, depending upon the nature of the sample. It is frequently difficult to decide when the heating should be discontinued. In some cases, the production of a clear melt serves to indicate the completion of the decomposition. In others the condition is not obvious, and the analyst must base the heating time on previous experience with the type of material being analyzed. In any event, the aqueous solution from the fusion should be examined carefully for particles of unattacked sample.

When the fusion is judged complete, the mass is allowed to cool slowly; then just before solidification the crucible is rotated to distribute the solid around the walls of the crucible so that the thin layer can be readily detached.

Types of Fluxes. With few exceptions the common fluxes used in analysis (Table 15.4) are compounds of the alkali metals. Basic fluxes, employed for attack on acidic materials, include the carbonates, hydroxides, peroxides, and borates. The acidic fluxes are the pyrosulfates and the acid fluorides as well as boric oxide. If an oxidizing flux is required, sodium peroxide can be used. As an alternative, small quantities of the alkali nitrates or chlorates are mixed with sodium carbonate.

C. Decomposition of Organic Compounds

Analysis of the elemental composition of an organic substance generally requires drastic treatment of the material in order to convert the elements of interest into a form susceptible to the common analytical techniques. These treatments are usually oxidative in nature, involving conversion of the carbon and hydrogen of the organic material to carbon dioxide and water; in some instances, however, heating the sample with a potent reducing agent is sufficient to rupture the covalent bonds in the compound and free the element to be determined from the carbonaceous residue.

Oxidation procedures are sometimes divided into two categories. Wet ashing (or oxidation) makes use of liquid oxidizing agents such as sulfuric or perchloric acids. Dry ashing (or oxidation) usually implies ignition of the organic compound in air or a stream of oxygen. In addition, oxidations can be carried out in certain fused-salt media, sodium peroxide being the most common flux for this purpose.

TABLE 15.4 The Common Fluxes

Flux	Melting point, °C	Type of crucible used for fusion	Type of substance decomposed
Na_2CO_3	851	Pt	For silicates and silica-containing samples; alumina-containing sample; insoluble phosphates and sulfates.
Na_2CO_3 + an oxidizing agent such as KNO_3, $KCIO_3$, or Na_2O_2	Pt	Pt (not with Na_2O_2), Ni	For samples where an oxidizing agent is needed, that is, samples containing A, As, Sb, Cr, etc.
NaOH or KOH, KOH	318–380	Au, Ag, Ni	Powerful basic fluxes for silicates, silicon carbide, and certain minerals; man limitation, purity of reagents.
Na_2O_2	Decomposes	Fe, Ni	Powerful basic oxidizing flux for sulfides; acid-insoluble alloys of Fe, Ni, Cr, Mo, W, and Li; Pt alloys; Cr, Sn, Zn minerals.
$K_2S_2O_7$	300	Pt porcelain	Acid flux for insoluble oxides and oxide-containing samples.
B_2O_3	577	Pt	Acid flux for decomposition of silicates and oxides where alkali metals are to be determined.
$CaCO_3 + NH_4Cl$		Ni	Upon heating of the flux, a mixture of CaO and $CaCl_2$ is produced; used for decomposing silicates for the determination of the alkali metals.

In the sections that follow, we shall mention briefly some of the methods for decomposing organic substances prior to the analysis for the more common elements.

Combustion-Tube Methods. Several of the common and important elemental components of organic substances are converted to gaseous products when the material is oxidized. With suitable apparatus it is possible to trap these volatile compounds quantitatively and use them in analyzing for the element of interest. A common way to do this is to carry out the oxidation in a glass or quartz combustion tube through which is forced a stream of carrier gas. The stream serves to transport the volatile products to a part of the apparatus where they can be separated and retained for measurement; the stream may also serve as the oxidizing agent. The common elements susceptible to this type of treatment are carbon, hydrogen, oxygen, nitrogen, the halogens, and sulfur.

Figure 15.12 shows a typical combustion train for the determination of carbon and hydrogen in an organic substance. Figure 15.13 shows a movable combustion furnace. Oxygen is forced through

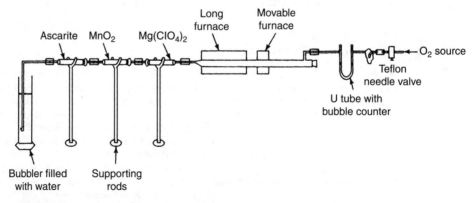

FIGURE 15.12 Combustion train with movable furnace.

FIGURE 15.13 Movable combustion furnace. (*Courtesy of Thermo Fisher Scientific.*)

the tube to oxidize the sample as well as to carry the products to the absorption part of the train. The sample is contained in a small platinum or porcelain boat that can be pushed into the proper position by means of a rod or wire. Ignition is initiated by slowly raising the temperature of that part of the tube which contains the sample. The sample undergoes partial combustion as well as thermal decomposition at this point, and the products are carried over a copper oxide packing that is maintained at a temperature of 700 to 900°C; this catalyzes the oxidation of the sample to carbon dioxide and water. Additional packing is often included in the tube to remove compounds that interfere with the determination of the carbon dioxide and water in the exit stream. Lead chromate and silver serve to remove halogen and sulfur compounds, while lead dioxide can be employed to retain the oxides of nitrogen.

The exit gases from the combustion tube are first passed through a massing tube packed with a desiccant that removes the water from the stream. The increase in mass of this tube gives a measure of the hydrogen content of the sample. The carbon dioxide in the gas stream is removed in the second massing tube packed with Ascarite® (sodium hydroxide held on asbestos). Because the absorption of carbon dioxide is accompanied by the formation of water, additional desiccant is contained in this tube.

Finally the gases are passed through a guard tube that protects the two massing tubes from contamination by the atmosphere.

Table 15.5 lists some of the applications of the combustion-tube method to other elements. A substance containing a halogen will yield the free element upon oxidation; this is frequently reduced to the corresponding halide prior to the analytical step. Sulfur finally yields sulfuric acid, which can be estimated by precipitation with barium ion or by alkalimetric titration.

Combustion with Oxygen in Sealed Containers.   A relatively straightforward method for the decomposition of many organic substances involves oxidation with gaseous oxygen is a sealed container. The reaction products are absorbed in a suitable solvent before the reaction vessel is opened. Analysis of the solution by ordinary methods follows.

A remarkably simple apparatus for carrying out such oxidations has been suggested by Schöninger (see Fig. 15.14). It consists of a heavy-walled flask of 300- to 1000-mL capacity fitted with a ground-glass stopper. Attached to the stopper is a platinum-gauze basket which holds from 2 to 200 mg of sample. If the substance to be analyzed is a solid, it is wrapped in a piece of low-ash filter paper cut in the shape shown in Fig. 15.14. Liquid samples can be massed in gelatin capsules which are then wrapped in a similar fashion. A tail is left on the paper and serves as an ignition point.

A small volume of an absorbing solution is placed in the flask, and the air in the container is then displaced by allowing tank oxygen to flow into it for a short period. The tail of the paper is ignited

TABLE 15.5 Combustion-Tube Methods for the Elemental Analysis of Organic Substances

Elements	Name of methods	Method of oxidation	Method of completion analysis
Halogens	Pregl	Sample combusted in a stream of O_2 gas over a red-hot Pt catalyst; halogens converted primarily to HX and X_2.	Gas stream passed through a carbonate solution containing SO_3^{2-} (to reduce halogens and oxyhalogens to halids); product, the halide ion X^2, determined by usual procedures.
	Grote	Sample combusted in a stream of air over a hot silica catalyst; produces are HX and X_2.	Same as above.
S	Pregl	Similar to halogen determination; combustion produces are SO_2 and SO_3.	Gas stream passed through aqueous H_2O_2, which converts sulfur oxides to H_2SO_4, which can then be determined.
	Grote	Similar to halogen determination; products are SO_2 and SO_3.	Similar to above.
N	Dumas	Sample oxidized by hot CuO to give CO_2, H_2O, and N_2.	Gas stream passed through concentrated KOH solution leaving only N_2, which is measured volumetrically.
C and H	Pregl	Similar to halogen analysis; products are CO_2 and H_2O.	H_2O adsorbed on a desiccant and CO_2 on Ascarite®; determined gravimetrically.
O	Unterzaucher	Sample pyrolized over C; O_2 converted to CO; H_2 used as carrier gas.	Gas stream passed over I_2O_5 ($5CO + I_2O_5 \rightarrow 5CO_2 + I_2$); liberated I_2 titrated.

and the stopper is quickly fitted into the flask; the container is then inverted as shown in Fig. 15.14; this will prevent the escape of the volatile oxidation products. Ordinarily the reaction proceeds rapidly, being catalyzed by the platinum gauze surrounding the sample. During the combustion, the flask is shielded to avoid damage in case of explosion.

After cooling, the flask is shaken thoroughly and disassembled; then the inner surfaces are rinsed down. The analysis is then performed on the resulting solution. This procedure has been applied to the determination of halogens, sulfur, phosphorus, and various metals in organic compounds.

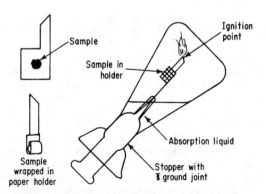

FIGURE 15.14 Schöninger apparatus for carrying out oxidation with gaseous oxygen in a sealed container.

Peroxide Fusion. Sodium peroxide is a strong oxidizing reagent which, in the fused state, reacts rapidly and often violently with organic matter, converting carbon to the carbonate, sulfur to sulfate, phosphorus to phosphate, chlorine to chloride, and iodine and bromine to iodate and bromate. Under suitable conditions the oxidation is complete, and analysis for the various elements may be performed upon an aqueous solution of the fused mass.

Once started, the reaction between organic matter and sodium peroxide is so vigorous that a peroxide fusion must be carried out in a sealed, heavy-walled, steel bomb. Sufficient heat is evolved in the oxidation to keep the salt in the liquid state until the reaction is completed; ordinarily the oxidation is initiated by passage of current through a wire immersed in the flux or by momentary heating of the bomb with a flame. Bombs for peroxide fusions are available commercially.

One of the main disadvantages of the peroxide-bomb method is the rather large ratio of flux to sample needed for a clean and complete oxidation. Ordinarily an approximate 200-fold excess is used. The excess peroxide is subsequently decomposed to sodium hydroxide by heating in water; after neutralization, the solution necessarily has a high salt content. This may limit the accuracy of the method for completion of the analysis.

The maximum size for a sample that is to be fused is perhaps 100 mg. The method is more suited to semimicro quantities of about 5 mg.

Wet-Ashing Procedures. Solutions of a variety of strong oxidizing agents will decompose organic samples. The main problem associated with the use of these reagents is the prevention of volatility losses of the elements of interest.

One wet-ashing procedure is the Kjeldahl method for the determination of nitrogen in organic compounds. Here concentrated sulfuric acid is the oxidizing agent. This reagent is also frequently employed for decomposition of organic materials where metallic constituents are to be determined. Commonly, nitric acid is added to the solution periodically to hasten the rate at which oxidation occurs. A number of elements are volatilized at least partially by this procedure, particularly if the sample contains chlorine; these include arsenic, boron, germanium, mercury, antimony, selenium, tin, and the halogens.

An even more effective reagent than sulfuric–nitric acid mixtures is perchloric acid mixed with nitric acid. A good deal of care must be exercised in using this reagent, however, because of the tendency of hot, anhydrous perchloric acid to react explosively with organic material. Explosions can be avoided by starting with a solution in which the perchloric acid is well diluted with nitric acid and not allowing the mixture to become concentrated in perchloric acid until the oxidation is nearly complete. Properly carried out, oxidations with this mixture are rapid and losses of metallic ions negligible.

CAUTION: *It cannot be too strongly emphasized that proper precautions must be taken in the use of perchloric acid to prevent violent explosions.*

Fuming nitric acid is another potent oxidizing reagent that is employed in the analysis of organic compounds. Its most important application is the analysis of the halogens and sulfur by the Carius method. The oxidation is carried out by heating the sample for several hours at 250 to 300°C in a heavy-walled sealed glass tube. Where halogens are to be determined, silver nitrate is added before the oxidation begins in order to retain them as the silver halides. Sulfur is converted to sulfate by the oxidation. A critical step in this procedure is that of forming a glass seal strong enough to withstand the rather high pressures that develop during the oxidation. Occasional explosions are almost inevitable, and a special tube furnace is ordinarily employed to minimize the effects of these.

Dry-Ashing Procedure. The simplest method for decomposing an organic sample is to heat it with a flame in an open dish or crucible until all the carbonaceous material has been oxidized by the air. A red heat is often required to complete the oxidation. Analysis of the nonvolatile components is then made after solution of the residual solid. Unfortunately a great deal of uncertainty always exists with respect to the recovery of supposedly nonvolatile elements when a sample is treated in this manner. Some losses probably arise from the mechanical entrainment of finely divided particulate matter in the hot convection currents around the crucible. In addition, volatile metallic compounds may be formed during the ignition. For example, copper, iron, and vanadium are appreciably volatilized when samples containing porphyrin compounds are heated.

In summary, the dry-ashing procedure is the simplest of all methods for decomposing organic compounds. It is often unreliable, however, and should not be employed unless tests have been performed that demonstrate its applicability to a given type of sample.

DETERMINING MASS OF SAMPLES

Store and dry samples in massing (weighing) bottles which have ground-glass contacting surfaces between the cover and the bottle (Fig. 15.15).

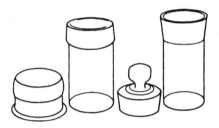

FIGURE 15.15 Typical massing bottles.

Procedure

1. Use a clean massing bottle fitted with a ground-glass cover.
2. Handle the bottle with suitable tongs or with a strip of lint-free paper (as illustrated in Fig. 15.16).
3. Do not touch the massing bottle with your fingers. Data will be significantly affected by the moisture and grease on your fingers.

FIGURE 15.16 Method of handling massing bottles.

A. Direct Mass Determination

1. Mass a clean receiving vessel or dish or a piece of glazed paper.
2. Transfer the desired quantity of substance into the receiving container with a clean spatula or by gently tapping the tilted massing bottle.
3. Mass the substance and the glazed paper or massing dish.
4. Calculate the mass of the sample by subtracting the mass of the paper or dish from the mass of the material and dish found in step 3. The difference in these two masses is the mass of the substance.

B. Mass Determination by Difference

1. Mass the special tared bottle which contains the sample.
2. Quantitatively remove the desired amount of the substance to the receiving container by gently pouring the material out of the massing bottle.
3. Remass the bottle.
4. Subtract the mass found in step 3 from the mass found in step 1. The difference in these two masses is the amount of material transferred.

Use *direct mass determination* when an exact quantity of substance is needed.

Use *mass determination by difference* (Fig. 15.17) when several samples of the same material are to be massed. This method is preferable when determining the mass of hygroscopic substances.

FIGURE 15.17 A convenient method to transfer a solid for massing by difference.

NOTE: Gently tapping the massing bottle or massing container enables you to better control the removal of the solid material without loss.

C. Gravimetric Calculations

The mass percent of a constituent is equal to the mass of the constituent divided by the sample mass and multiplied by 100.

For example, if a 1.000-g sample of limestone is found to contain 0.3752 g of calcium, it has

$$\frac{0.3752}{1.000} \times 100 \text{ or } 37.52\% \text{ calcium}$$

$$\frac{\text{Mass of substance}}{\text{Mass of sample}} \times 100 = \text{Percent by mass}$$

In most cases, however, the desired constituent is not massed directly but is precipitated and massed as some other compound. It is then necessary to convert the mass obtained to the mass in the desired form by using a **gravimetric factor**.

For example, a molecule of silver chloride is made up of one atom of silver and one atom of chlorine. The ratio of silver to chloride is as Ag/AgCl. Since the atomic mass of silver is 107.8 and that of chlorine is 35.5.

$$\frac{107.8 \text{ (atomic mass Ag)}}{107.8 \text{ (atomic mass Ag)} + 35.5 \text{ (atomic mass Cl)}} = 0.7526$$

This is called the *gravimetric factor*, and the percent of silver when weighed as AgCl is

$$\frac{\text{Mass of AgCl} \times 0.7526}{\text{Mass of sample}} \times 100$$

$$\text{or } \frac{\text{Mass of precipitate (AgCl)}}{\text{Mass of sample}} \times \text{gravimetric factor} \times 100 = \% \text{ Ag}$$

The factor for sodium when weighed as sodium sulfate is 2Na/Na$_2$SO$_4$ since there are 2 sodium atoms in sodium sulfate.

A general equation for gravimetric calculation is

$$\frac{\text{Mass of precipitate} \times \text{gravimetric factor} \times 100}{\text{Mass of sample}} = \text{percent of constituent}$$

Gravimetric factors are given in Table 15.6.

TABLE 15.6 Table of Gravimetric Factors

Sought	Massed	Factors
Na	Na$_2$SO$_4$	0.3237
K	K$_2$SO$_4$	0.4487
Ba	BaSO$_4$	0.5885
Ca	CaSO$_4$	0.2944
Cu	CuO	0.7988
Fe	Fe$_2$O$_3$	0.6994
Pt	Pt	1.000
Au	Au	1.000
Ag	Ag	1.000
H	H$_2$O	0.1119
C	CO$_2$	0.2729
S	BaSO$_4$	0.1374

MICRODETERMINATION OF CARBON AND HYDROGEN

Since organic compounds are characterized by the fact that they contain carbon and usually hydrogen, it can be seen that the ability to measure these elements accurately is of extreme importance. In spite of its importance there is no universal test for carbon and hydrogen. There are almost as many modifications as there are people running the test. Since the time of Pregl, however, the basic microcombustion technique has been the principal means for determining carbon and hydrogen. The technique is threefold:

1. Combustion of the organic material
2. Removal of interfering elements
3. Measurement of the carbon dioxide and water formed

Any modification or variation is involved with one of these three categories.

A. Combustion of Organic Material

The basic reactions in the combustion of organic material for the determination of carbon and hydrogen are

$$\text{Organic C} \xrightarrow[O_2]{\Delta} CO_2$$

$$\text{Organic H} \xrightarrow[O_2]{\Delta} H_2O$$

This combustion is usually carried out in a special tube. The tube packing, according to Pregl, contains silver, copper oxide, lead chromate, and lead dioxide. The combustion is carried out in oxygen; the copper oxide and lead chromate aid in the oxidation. Many investigators have described packings consisting of a variety of catalysts including cobaltic oxide, silver permanganate, silver vanadate, zirconium oxide, magnesium oxide, and silver tungstate.

Oxygen flow rates to carry out the oxidation have varied from 8 to >100 mL/min. At 50 mL/min a sample can be burned rapidly enough to effect complete combustion in 5 minutes and total sweep after 10 minutes, permitting the analyst enough time to prepare the next sample.

Aside from catalysts that are part of the combustion-tube packing, it is sometimes advantageous to cover the sample in the boat with catalysts to aid in the oxidation. This is especially true for organic compounds containing metals. Tungstic oxide appears to be a universal "clean" catalyst for most metal organics. A mixture of silver oxide and manganese dioxide has proved to be effective with highly chlorinated and sulfurized materials.

B. Removal of Interfering Substances

The interfering substances encountered in the determination of carbon and hydrogen are sulfur, the halogens, and nitrogen. Probably more has been published on removing these substances than on any other single area of carbon-hydrogen determination. Silver gauze or wire, maintained at 700°C, has been found effective for the removal of halogens (except fluoride) and sulfur products. The silver should be placed at the end of the combustion tube.

Removal of Sulfur. Sulfur in the carbon-hydrogen determination is converted to $SO_2 + SO_3$ and is absorbed by silver, forming Ag_2SO_4. Manganese dioxide will absorb all the sulfur dioxide from the products, thereby preventing interference with the carbon analysis. This is not quantitative. Again if this is the only mechanism employed for the removal of sulfur dioxide, no condensate is permissible or sulfuric acid will be formed.

Removal of Halogens (Except Fluorine). As previously mentioned, halogen can be removed by silver in the combustion tube. This is very effective. If halogen escapes from the tube, manganese dioxide placed in the train will absorb any chloride, and that chloride will not be absorbed on the magnesium perchlorate tube. However, it must be noted that no water may be permitted to condense prior to absorption.

Removal of Nitrogen Oxides. Nitrogen oxides formed during combustion of nitrogen compounds interfere with the carbon determination. Nitrogen oxides classically have been removed by absorption on lead dioxide maintained at 180°C. Since lead dioxide sets up an equilibrium with water and carbon dioxide (products of combustion), a weighed sample must be run prior to analysis of samples. When using this technique, one must not follow a high carbon-hydrogen analysis with a low one or vice versa. The analyst must also be absolutely sure of timing—one sample must follow the next in exactly the same time. Grades of lead dioxide are variable, and the analyst must exercise care in eliminating fines when packing the combustion tube.

Manganese dioxide is very effective when used for the absorption of nitrogen oxides. It is used externally between the magnesium perchlorate tube and the carbon dioxide absorption tube, normally at room temperature. It has a tendency to pick up carbon dioxide and release it; therefore, it must be dried prior to use. This material requires that no water condensation may take place prior to absorption: The formation of nitric acid will yield high hydrogen values. High-flow rates and external heating problems alleviate these problems. A simple hair dryer can be used to blow hot air on the connection between the combustion tube and absorption tube. This has proved to be very effective in preventing water condensation.

C. Measurement of Carbon Dioxide and Water

Water and carbon dioxide formed by the combustion of organic compounds have classically been collected on magnesium perchlorate and **Ascarite®** (NaOH on asbestos), respectively. This method is still extensively used today. One precaution when operating with a high flow rate is to have enough magnesium perchlorate at the exit end of the tube to remove all the water formed. Carbon dioxide is "absorbed" on Ascarite® as follows:

$$CO_2 + 2NaOH \rightarrow Na_2CO_3 + H_2O \tag{1}$$

D. Microprocedure for Determination of Carbon and Hydrogen in Organic Compounds

Principle
In this procedure, a known amount of organic matter is combusted in an atmosphere of oxygen. The water and carbon dioxide formed are absorbed by magnesium perchlorate and Ascarite®, respectively. The difference in the mass of the tubes before and after the combustion is measured, and the amounts of carbon and hydrogen are calculated (see Fig. 15.18).

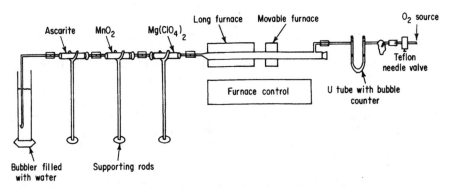

FIGURE 15.18   Schematic of equipment for analysis of carbon and hydrogen by combustion.

Scope
The method can be applied to all organic compounds containing carbon and hydrogen. Provision is made for the removal of interfering elements such as halogen, sulfur, and nitrogen. However, compounds containing large amounts of fluorine should not generally be analyzed by this method without prior treatment.

E. Total Oxidizable Carbon Analyzers

Total oxidizable carbon analyzers (TOCs) are completely automated, on-line analyzers for the detection of total oxidizable carbon in ultrapure water. These analyzers are extremely useful for continuous monitoring of organic impurity levels in type I reagent-grade water, USP purified water, water for injection, semiconductor water, and boiler feedwater in nuclear and fossil fuel power stations. TOCs require virtually no maintenance, use no reagents, are easy to install, and perform completely automated total oxidizable carbon analyses year after year. Their rugged construction provides exceptional reliability, allowing them to run for years without requiring calibration. Their calibration is traceable to NIST standards and can easily be verified. A typical oxidizable carbon analyzer is shown in Fig. 15.19.

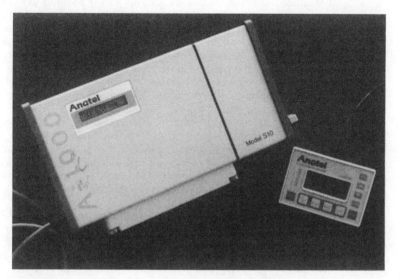

FIGURE 15.19 Total oxidizable carbon analyzer. (*Courtesy Anatel Corporation.*)

The analyzers detect organic impurities down to 0.05 ug/L in water from 0 to 100°C. They also measure the resistivity/conductivity and temperature of the water. The networkable sensors are available with a dedicated controller, self-configuring network firmware, a data logger, local or network printer, PC interface, opto-isolated digital inputs and outputs, and a bidirectional serial interface for a PLC or host computer.

KJELDAHL ANALYSIS

An important method for accurately determining elemental nitrogen in protein and other nitrogen-containing compounds is the *Kjeldahl analysis*. This total nitrogen determination is one of the oldest presently used analytical techniques. It was developed by German chemist Dr. Johan Kjeldahl in 1883. The method is based on the fact that digestion with sulfuric acid and various catalysts destroys nitrogen-containing organic materials, converting the nitrogen to ammonium acid sulfate. When the reaction mixture is made alkaline, ammonia is liberated and removed by steam distillation, collected, and titrated. A tremendous amount of work has been done on the selection of proper catalysts for the digestion. Various workers have used selenium, copper, mercury, and salts of each. Potassium sulfate added with the catalysts raises and controls the temperature of the reaction. Some workers have claimed that with the use of selenium, too high a temperature is achieved, thereby yielding low nitrogen results. However, it has been shown that with $HClO_4$ in the digestion mixture, selenium can be used effectively. It must be noted that any refractory nitrogen compounds, pyridines, etc., must be digested at ~370°C but at no higher than 400°C.

In the Kjeldahl determination of nitrogen, compounds containing N—O or N—N linkages must be pretreated or subjected to reducing conditions prior to analysis. Numerous agents have been utilized to effect this reduction. The N—O linkages are much easier to reduce than the N—N, and zinc or iron in acid is suitable for this purpose. There is no such general technique for the N—N linkages. Samples containing very high concentrations of halide can in some instances cause trouble because of the formation of oxyacids known to oxidize ammonia to N_2.

For nitrate-containing compounds, salicylic acid is added to form nitrosalicylic acid, which is reduced with thiosulfate. The ammonium hydrogen sulfate thus formed during digestion is reacted with NaOH to form free ammonia, which is distilled with steam. The ammonia is collected by passing it into boric acid solution and then is titrated with HCl. Some investigators have collected the ammonia in HCl and back-titrated with standard NaOH. The disadvantage of this technique is the need for two standard solutions and the critical loss of two components on the condenser. The pH at which the ammonium chloride complex is formed is 5.2. A mixture of methyl red and methylene blue has been shown to be very effective. Some use methyl purple, which is actually a mixture of methyl red and a blue dye.

The quantity of protein can be calculated from a knowledge of the percent of nitrogen contained in it. Although other, more rapid methods for determining proteins exist, the Kjeldahl method is the standard by which all other methods are based. The distillation of the ammonia gas after the Kjeldahl digestion is done in an apparatus such as that shown in Fig. 15.20. The long-necked Kjeldahl digestion flask is connected to a water-cooled condenser by means of a special trap that prevents any of the strongly alkaline mixture in the digestion flask from being carried over mechanically into the standardized titration solution.

The material is digested with sulfuric acid in the presence of a catalyst (mercury or selenium and potassium sulfate salt) to decompose it and convert the nitrogen to ammonium hydrogen sulfate as shown in the balanced equation below. The coefficients and subscripts (letters *a, b,* and *c*) represent moles in the balanced equations. A detailed procedure for determining the total nitrogen

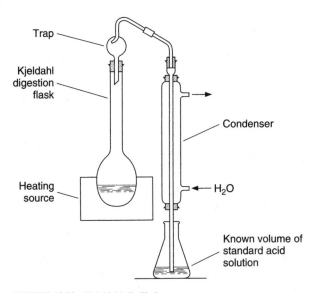

FIGURE 15.20 Kejeldahl distillation apparatus.

content by the Kjeldahl method can be found in the 11th edition of *Official Methods of Analysis of the AOAC* (1970).

$$C_aH_bN_c + \text{excess } H_2SO_4 \rightarrow aCO_2 + \tfrac{1}{2}bH_2O + cNH_4HSO_4 \tag{2}$$

The solution is cooled, concentrated alkali is added to make the solution alkaline, and the volatile ammonia is distilled into an excess of a standardized acid (usually HCl).

$$cNH_4HSO_4 + OH^- \rightarrow cNH_3 + cSO_4^{-2} \tag{3}$$

$$cNH_3 + (c+d)HCl \rightarrow cNH_4Cl + dHCl \tag{4}$$

Following distillation, the excess acid is back-titrated with standard base.

$$dHCl + dNaOH \rightarrow \tfrac{1}{2}H_2O + dNaCl \tag{5}$$

The percent nitrogen (%N) contained in a compound can be calculated from the weight of nitrogen analyzed by multiplying it by a gravimetric factor. The *gravimetric factor* for nitrogen is 14.00 and the gravimetric factor for various protein samples is established at 6.24 for gamma globulin and 6.25 for proteins in feeds.

$$\%N = \frac{(V\,HCl \times N\,HCl - V\,NaOH \times N\,NaOH) \times 14.00}{\text{sample weight}} \times 100 \tag{6}$$

where V HCl = volume of hydrochloric acid expressed in liters
N HCl = normality of hydrochloric acid
V NaOH = volume of sodium hydroxide expressed in liters
N NaOH = normality of sodium hydroxide
sample weight = mass of sample, as is, expressed in grams

A. Computerized Combustion Nitrogen Analyzers

Present-day nitrogen analyzers are based on the combustion principle, in which a sample is combusted at a high temperature in a pure oxygen atmosphere, interfering gases are scrubbed out, and the nitrogen is measured by thermal conductivity. Instruments utilizing the latest technology are microprocessor-driven and provide a host of advanced capabilities such as automatic calculations of nitrogen or protein as well as a high degree of automation (see Fig. 15.21).

Unattended operation becomes a reality and a convenience when the technician can weigh and load 50 samples or more in an autoloader and the instrument takes over analyzing and reporting the results automatically. A complete analysis can be performed in 3 to 10 minutes. There are no hazardous wastes to contend with, and very important, since the system is microprocessor-controlled, tens of thousands results can be stored. Those results can be displayed on high-resolution screens or printed on interfaced printers for reports. Table 15.7 shows a comparison of conventional Kjeldahl analysis and the more modern computerized combustion analyzers.

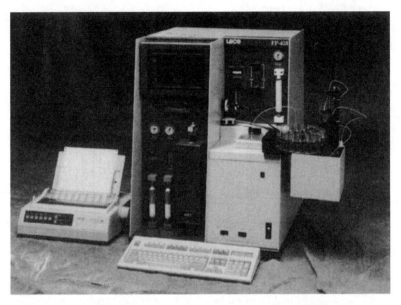

FIGURE 15.21 Computerized combustion nitrogen/protein analyzer. (*Courtesy Leco Corporation.*)

TABLE 15.7 A Comparison of Kjeldahl and Computerized Combustion Analyzers

Feature	Kjeldahl analysis	Computerized combustion analyzer
Analysis time	1–5 hours	3–9 minutes
Sample size	1–5 grams	0.01–500 mg
Precision	Approx. 0.5% RSD	<0.5% RSD
Hazardous chemicals	Mercury or selenium H_2SO_4, NaOH	Lead chromate
Automation (no. unattended)	None	Variable—23–125 analysis
Furance type	None	Vertical
Data system	None	For example, print results after each analysis for up to 200 samples. Method development. Linear regression. Store up to 32,000 results. Edit parameter after analysis.

B. Dumas Determination of Nitrogen

The Dumas method for nitrogen was introduced in 1831. It was not practical, however, until after Pregl adapted it to the micro scale. The Dumas method is based on combustion of the nitrogen-containing organic material in the presence of a catalyst at 780°C. The oxidation products are then passed over a reducing medium and the NO_x (various nitrogen oxides) are converted to N_2. The (various nitrogen oxides) following reactions show the process:

$$\text{Organic N compound} \xrightarrow[780° \text{ C}]{\text{CuO}} CO_2 + H_2O + NO_x + N_2 \tag{7}$$

$$NO_x \xrightarrow[780° \text{ C}]{\text{Cu*}} N_2 \tag{8}$$

---

*Metallic copper as reducing medium.

The nitrogen is then measured by bubbling the products into a solution of potassium hydroxide. The nitrogen displaces a corresponding volume of potassium hydroxide; the volume is measured, corrected for barometric pressure and temperature, and converted to mass. The percent nitrogen is then calculated. Samples containing a high concentration of methyl groups (CH_3), alkoxyl (OCH_3, etc.), and N-methyl (N-CH_3) compounds are known to release methane under normal conditions during the Dumas analysis. Cobaltic oxide has been shown to prevent the formation of methane.

Many investigators have attempted to automate the Dumas analysis. The Coleman analyzer programs the combustion and measures the nitrogen by means of a special nitrometer with a readout. Other techniques have applied thermal conductivity after a gas-chromatographic separation of the nitrogen. The automation of the Dumas analysis has been reasonably successful. The analyst must remember, however, that all samples are different and certain variables must be considered prior to analysis.

DETERMINATION OF HALOGENS

Many of the concepts covered later in discussing the determination of sulfur in organic compounds also apply to the determination of halogen. For simplicity, the term "halogen" will refer here only to chlorine, bromine, and iodine. Like most elemental analyses, the determination involves a three-step procedure:

1. Decomposition of the organic material and conversion of the halide to a measurable form
2. Removal of interfering ions
3. Measurement of the halogen-containing products

A. Decomposition of Organic Material

The basic means for decomposing organic compounds in an effort to measure total halogen is oxidation. Methods to carry out this procedure are:

1. Tube combustion (Carius method)
2. Bomb combustion, with peroxide (Parr method)
3. Flask combustion (Schöninger)
4. Pregl combustion methods (decomposition of the sample in an oxygen atmosphere at 680 to 700°C in the presence of a platinum catalyst)

The reaction for this oxidation is as follows:

$$\text{Organic } X \text{ compound} \xrightarrow[\Delta]{O_2} X_2 + CO_2 + H_2O \tag{9}$$

X stands for Br, Cl, or I.

$$X_2 \xrightarrow{\text{suitable absorbent}} 2X^- \quad \text{usually absorbed in a sodium carbonate solution.} \tag{10}$$

Hydrazine is sometimes used in the absorbent to aid in the reduction. It is essential when X is iodine.

B. Removal of Interfering Ions

The primary ions interfering with the halogen determination are the other halides present. For example, bromide will interfere with chloride and vice versa.

Cyanides and thiols also interfere; however, they are not present after combustion. Their presence must be noted, however, when determining water-soluble fractions; in this case the organic material is not destroyed.

Ion Exchange. An ion-exchange separation on Dowex 1 separates chlorine from bromine from iodine when the sample is eluted with various concentrations of sodium nitrate. Review Ion-Exchange Chromatography in Chapter 26, "Chromatography."

C. Other Methods of Determining Halogen

1. *Mohr method.* Formation and determination of red silver chromate.
2. *Volhard method.* Back-titrate excess silver nitrate with standard thiosulfate; iron(III) is the indicator. Solubility of silver chloride is a real problem.
3. *Colorimetry.* One of the most sensitive techniques for the determination of chloride ion involves the reaction which displaces thiocyanate ion from mercury(II) thiocyanate by chloride ion. The color formed when this reaction is carried out in the presence of iron(III) ion to form iron(III) thiocyanate is stable and proportional to the chloride content of the material being measured. The reaction is as follows:

$$2Cl^- + Hg(SCN)_2 + 2Fe^{3+} \rightarrow HgCl_2 + 2Fe(SCN)^{2+} \tag{11}$$

The technique can determine chloride down to 5 ppm.

DETERMINATION OF SULFUR

The determination of sulfur, like that of other elements in organic compounds, involves three basic steps:

1. Decomposition of the organic material and conversion of the sulfur to a measurable form
2. Removal of interfering ions
3. Measurement of the sulfur-containing products of combustion

Any modification or variation in a single method or between methods is involved with one of these three steps.

A. Decomposition of Organic Material and Conversion of Sulfur to a Measurable Form

Oxidation. The basic reaction for the determination of sulfur by an oxidative combustion is as follows:

$$\text{Organic S compound} \xrightarrow[O_2]{\Delta} CO_2 + H_2O + SO_3 + SO_2 \tag{12}$$

Increase of temperature yields a mixture that is 95 percent SO_2. The resulting oxides ($SO_2 + SO_3$) are converted to SO_4^{2-} as H_2SO_4 as follows:

$$SO_2 + SO_3 + H_2O_2 + H_2O \rightarrow 2H_2SO_4 \tag{13}$$

or

$$SO_2 + SO_3 + Br_2 + 3H_2O \rightarrow 2H_2SO_4 + HBr \qquad (14)$$

Tube Combustion (Dietert). In this high-temperature combustion the sulfur is converted by oxidation to 95 percent theoretical SO_2. The sample is burned with the aid of a catalyst (V_2O_5) and pure oxygen in an alundum tube maintained at 1300°C.

Parr Bomb. This method is based on the fact that Na_2O_2 (sodium peroxide) is a powerful oxidizing agent. If the organic matter is placed in a closed vessel and heated in the presence of peroxide, all the sulfur will be converted to the corresponding sodium salt.

The danger to the operator is the pressure that is built up within the bomb during combustion. Remember 1 g molecular weight of a substance occupies 22.4 L in the vapor state at standard temperature and pressure.

Sugar or benzoic acid is added to increase the burning capacity of the bomb and decrease time for combustion. Excess peroxide is used up in this manner also.

The melt from the cooled products of combustion is put into solution and inspected, and the analysis is performed after all H_2O_2 has been removed by boiling.

Schöninger Combustion. In the Schöninger combustion technique (see Fig. 15.14) the sulfur is converted by oxidation to SO_2 and SO_3 and subsequently oxidized to $—SO_4^{2-}$ with H_2O_2. In this technique, a sample is weighed directly onto a piece of filter paper or a capsule and placed in a platinum basket. A flask is completely filled with oxygen, and the ignited paper is placed in the flask. Complete combustion takes place within 20 seconds, and the products are absorbed. This method is useful for nonvolatile compounds only.

REFERENCES

1. Gordon, Arnold J., and Ford, Richard A., "Experimental Techniques," in *The Chemist's Companion, A Handbook of Practical Data, Techniques, and References*, John Wiley & Sons, Inc. ISBN 0-471-31590-7, 1972.
2. *Lange's Handbook of Chemistry*, 15th edition, Section 11—Practical Laboratory Information, John A. Dean, editor, McGraw-Hill. ISBN 0-07-016384-7, 1999.

CHAPTER 16
PREPARATION OF SOLUTIONS

INTRODUCTION

Part of a technician's job is to prepare the solutions needed for the various procedures performed in the laboratory. It is most important that directions be followed exactly and that all caution be observed. Consult Chapter 24 for more detailed chemical calculations and concentration expressions information.

CAUTION: Always recheck the label of the chemical that you are using. Use of the wrong chemical can cause an accident or ruin a determination.

GRADES OF PURITY OF CHEMICALS

Chemicals are manufactured in varying degrees of purity. Select the grade of chemical that meets the need of the work to be done. It is wasteful to use costly reagent grades when technical grades would be satisfactory. The various grades are listed and explained below.

A. Commercial or Technical Grade

This grade is used industrially, but is generally unsuitable for laboratory reagents because of the presence of many impurities.

B. Practical Grade

This grade does contain impurities, but it is usually pure enough for most organic preparations. It may contain some of the intermediates resulting from its preparation.

C. USP

USP grade chemicals are pure enough to pass certain tests prescribed in the U.S. Pharmacopoeia and are acceptable for drug use, although there may be some impurities which have not been tested for. This grade is generally acceptable for most laboratory purposes.

D. CP

CP stands for chemically pure. Chemicals of this grade are almost as pure as reagent-grade chemicals, but the intended use for the chemicals determines whether or not the purity is adequate for the purpose. The classification is an ambiguous one; read the label and use caution when substituting for reagent-grade chemicals.

E. Spectroscopic Grade

Solvents of special purity are required for spectrophotometry in the uv, ir, or near-ir ranges as well as in nmr spectrometry and fluorometry. Specifications of the highest order in terms of absorbance characteristics, water, and evaporation residues are given. The principal requirement of a solvent for these procedures is that the background absorption be as low as possible. Most of these chemicals are accompanied by a label which states the minimum transmission at given wavelengths. Residual absorption within certain wavelengths is mainly due to the structure of the molecule.

F. Chromatography Grade

These chemicals have a minimum purity level of 99+ mol % as determined by gas chromatography; each is accompanied by its own chromatogram indicating the column and parameters of the analysis. No individual impurity should exceed 0.2 percent.

G. Reagent Analyzed (Reagent Grade)

Reagent-analyzed, or reagent-grade, chemicals are those that have been certified to contain impurities in concentrations below the specifications of the Committee on Analytical Reagents of the American Chemical Society (ACS). Each bottle is identified by batch number. Use only reagent-analyzed chemicals in chemical analysis.

CAUTION: Be certain that the bottle has not been contaminated in previous use. Impurities may not have been tested for, and the manufacturer's analysis may possibly have been in error.

H. Primary Standard

Substances of this grade are sufficiently pure that they may serve as reference standards in analytical procedures. You may use them directly to prepare standard solutions by dissolving massed amounts in solvents and then diluting them to known volumes. Primary standards must satisfy extremely high requirements of purity; they usually contain less than 0.05 percent impurities.

CAUTION: Many compounds form hydrates with water. Some form more than one hydrate with water, combining with a different number of water molecules to form definite compounds. When using tables from reference handbooks, check *carefully* to confirm that the formula is the one that you seek and not another hydrate.

COMMON HAZARDOUS CHEMICALS

Consider all chemicals, reagents, and solutions as *toxic* substances. Many of the hazards of chemicals are not obvious or evident by smell, odor appearance, or immediately detectable by the organs of the body. See Fig. 16.1 for appropriate labels used on hazardous materials.

CAUTION: Never fill a receptacle with a material other than that called for by the label. Label all containers before filling them. Throw away contents of all unlabeled containers.

A. Incompatible Chemicals

Certain chemicals may react and create a hazardous condition. Separate storage areas should be provided for such **incompatible chemicals**. (See Chapter 3. "Chemical Handling and Hazard Communication," for additional information.)

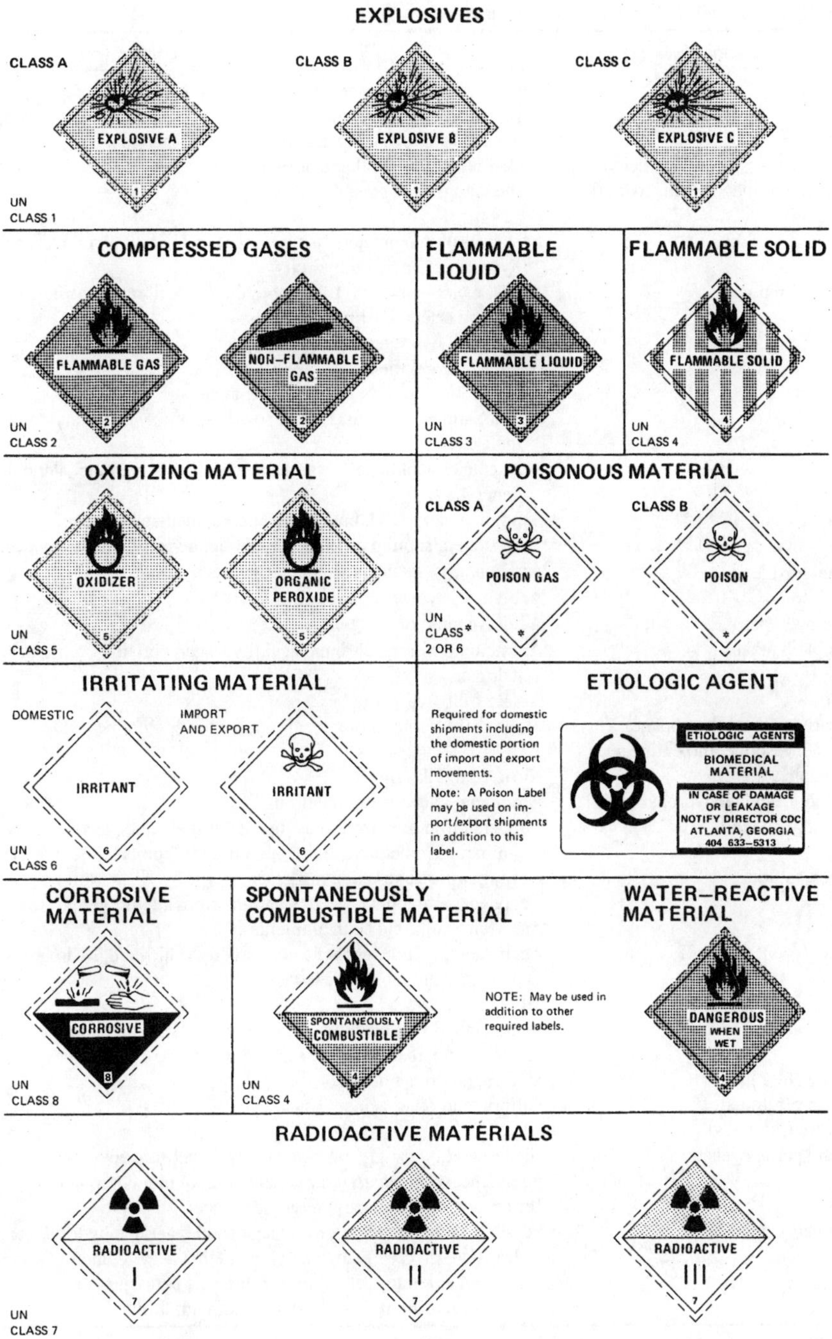

FIGURE 16.1 Warning labels for hazardous materials.

TABLE 16.1 Examples of Incompatible Chemicals

Chemical	Keep out of contact with
Acetic acid	Chromic acid, nitric acid, hydroxyl compounds, ethylene glycol, perchloric acid, peroxides, permanganates
Acetylene	Chlorine, bromine, copper, fluorine, silver, mercury
Alkaline metals, such as powdered aluminum or magnesium, sodium, potassium	Water, tetrachloromethane or other chlorinated hydrocarbons, carbon dioxide, the halogens
Ammonia, anhydrous	Mercury (in manometers, for instance), chlorine, calcium hypochlorite, iodine, bromine, hydrofluoric acid (anhydrous)
Ammonium nitrate	Acids, metals powders, flammable liquids, chlorates, nitrites, sulfur, finely divided organic or combustible materials
Aniline	Nitric acid, hydrogen peroxide
Bromine	Same as for chlorine
Carbon, activated	Calcium hypochlorite, all oxidizing agents
Chlorates	Ammonium salts, acids, metals powders, sulfur, finely divided organic or combustible materials
Chromic acid	Acetic acid, naphthalene, camphor, glycerin, turpentine, alcohol, flammable liquids in general
Chlorine	Ammonia, acetylene, butadiene, butane, methane, propane (or other petroleum gases), hydrogen, sodium carbide, turpentine, benzene, finely divided metals
Chlorine dioxide	Ammonia, methane, phosphine, hydrogen sulfide
Copper	Acetylene, hydrogen peroxide
Cumene hydroperoxide	Acids, organic or inorganic
Flammable liquids	Ammonium nitrate, chromic acid, hydrogen peroxide, nitric acid, sodium peroxide, the halogens
Fluorine	Isolate from everything
Hydrocarbons (butane, propane, benzene, gasoline, turpentine, etc.)	Fluorine, chlorine, bromine, chromic acid, sodium peroxide
Hydrocyanic acid	Nitric acid, alkali
Hydrofluoric acid, anhydrous	Ammonia, aqueous or anhydrous
Hydrogen peroxide	Copper, chromium, iron, most metals or their salts, alcohols, acetone, organic materials, aniline, nitromethane, flammable liquids, combustible materials
Hydrogen sulfide	Fuming nitric acid, oxidizing gases
Iodine	Acetylene, ammonia (aqueous or anhydrous), hydrogen
Mercury	Acetylene, fulminic acid, ammonia
Nitric acid (concentrated)	Acetic acid, aniline, chromic acid, hydrocyanic acid, hydrogen sulfide, flammable liquids, flammable gases
Oxalic acid	Silver, mercury
Perchloric acid	Acetic anhydride, bismuth and its alloys, alloys, alcohol, paper, wood
Potassium	Tetrachloromethane, carbon dioxide, water
Potassium chlorate	Sulfuric and other acids
Potassium perchlorate (see also Chlorates)	Sulfuric and other acids
Potassium permanganate	Glycerin, ethylene glycol, benzaldehyde, sulfuric acid
Silver	Acetylene, oxalic acid, tartaric acid, ammonium compounds
Sodium	Tetrachloromethane, carbon dioxide, water
Sodium peroxide	Ethyl or methyl alcohol, glacial acetic acid, acric anhydride, benzaldehyde, carbon disulfide, glycerin, ethylene glycol, ethyl acetate, methyl acetate, furfural
Sulfuric acid	Potassium chlorate, potassium perchlorate, potassium permanganate (or compounds with similar light metals, such as sodium, lithium)

SOURCE: Chemical Manufacturers Association, *Waste Disposal Manual.*

WATER FOR LABORATORY USE

Water is needed in the laboratory in various grades of purity depending upon the requirements of the procedures for which it is to be used.

A. Water-Purity Specifications

Solutions that conduct electric current are said to have **electric conductivity,** and the current is carried solely by the ions in solution. This electrical conductivity is also called the **specific conductance,** and it is measured in units of Siemens per centimeter (S/cm), which is the reciprocal of the resistance in ohms per centimeter (Ω/cm). This characteristic is used to specify the purity of the water. A micro siemens is one-millionth of a siemens. The resistance measurements require the use of an ac Wheatstone bridge, a conductivity cell, and a null indicator. Alternating current is normally used to avoid polarization of the electrodes.

Contrary to folk belief, rainwater is not absolutely pure water because it contains dissolved gases: O_2, N_2, CO_2, and oxides of nitrogen. As it falls on the ground and flows to city reservoirs, it dissolves minerals and other soluble substances. As a result, surface water can contain such positive ions as Na^+, K^+, Ca^{2+}, Mg^{2+}, Fe^{2+}, and Fe^{3+}, as well as such negative ions as CO_3^{2-}, HCO_3^-, SO_4^{2-}, NO_3^-, and Cl^-. Rainwater, after appropriate treatment, is often the source of city water. Depending upon the particular processes used in its purification and the efficiency of those processes, average city water normally contains dissolved inorganic substances, microorganisms, dissolved organic compounds of vegetable and plant origin, and particulate matter. This tap water contains sufficiently high levels of these impurities so that it cannot be used for testing procedures and analytical evaluations. It is also frequently "hard." Hard water contains appreciable quantities of Ca^{2+}, Mg^{2+}, Fe^{2+}, and Fe^{3+}; these minerals react with soap to form insoluble curds. Thus hard water does not lather readily. Soft water contains little mineral matter, lathers well, and is needed for laboratory use.

B. Softening Hard Water

Temporarily Hard Water.   Some waters are temporarily hard, because they contain the bicarbonates of calcium, magnesium, and/or iron. These bicarbonates can be removed simply by heating, thus converting the soluble bicarbonate to the insoluble carbonate, which can be filtered out. Therefore, temporarily hard water can be converted to soft water by boiling.

Permanently Hard Water.   When water contains sulfates or chlorides of calcium, magnesium, or iron, it is permanently hard water and is unsuitable for most laboratory work. It can be made soft in a number of ways. The first consists of adding substances that convert the undesirable soluble calcium, magnesium, and iron salts to precipitates which can be removed by suitable filtration. The most common water-softening substances are as follows:

Borax (sodium tetraborate), $Na_2B_4O_7$
Ammonium hydroxide, NH_4OH
Sodium carbonate, Na_2CO_3
Sodium hydroxide, NaOH
Potassium hydroxide, KOH
Trisodium phosphate, Na_3PO_4
Mixture of calcium hydroxide and sodium carbonate, $Ca(OH)_2$ and Na_2CO_3

Ion-Exchange Resins.   Soft water can also be obtained by passing hard water through a bed of special synthetic resins. These resins exchange their "soft" sodium ions for the waters' "hard" calcium, magnesium, and iron ions. As long as the resins will exchange their sodium ions for calcium ions, the water will be softened. However, when the resins will no longer effect this exchange because all the sodium ions are gone, the resin is said to be exhausted and it must be recharged. In the recharging process, sodium ions are replaced in the resin and the calcium ions are discarded. Water obtained in this way is called **deionized water**. For additional information on ion exchange, see Chapter 26, "Chromatography," section on ion-exchange chromatography.

C. Purifying Water by Reverse Osmosis

High-quality water of consistent purity can be obtained by the process of reverse osmosis whereby the dissolved solids are separated from feedwater by applying a pressure differential across a semipermeable membrane. This semipermeable membrane allows the water to flow through it, but prevents dissolved ions, molecules, and solids from passing through. The differential pressure forces the water through the membrane, leaving the dissolved particles behind, thus producing laboratory-grade water.

D. Distilled Water

Water purified by distillation is called **distilled water**. Electrically heated or steam-heated water stills (Fig. 16.2) distill raw water to give high-grade distilled water, venting volatile impurities. High-quality distilled water (Fig. 16.3) has a specific conductance of less than one microsiemens* per centimeter (μS/cm), which corresponds to about 0.5 ppm of a dissolved salts. Absolutely pure water has a conductivity of 0.55 μS/cm at 25°C. As mentioned before, this characteristic is used to specify the purity of the water.

FIGURE 16.2 Still for water. Flexible tube conducts distilled water to holding container.

---

*A more familiar name for this unit may be the *microohm*; the siemens is the SI unit of conductance.

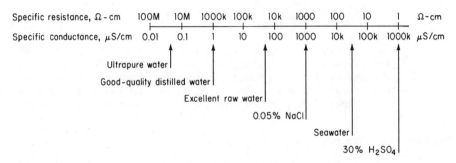

FIGURE 16.3 Specific conductances of various grades of water.

E. Demineralized Water

In the laboratory, the presence of even sodium ions may not be desirable or tolerated, and water that is free of all inorganic ions must be supplied. Demineralizers (Fig. 16.4) are used to remove mineral ions from water. Demineralizers contain several beds of resins and are packaged in cartridges; the water that is obtained this way is free of both mineral cations and anions.*

CAUTION: Demineralizers will not remove organic matter or nonelectrolytes from water. These must first be removed by distillation. You cannot substitute demineralized water for distilled water in every case. Check solution requirements.

F. Reagent-Grade Water

Reagent-grade water is water of the highest purity that is available from a practical standpoint; it is even purer than triple-distilled water, which can contain significant amounts of dissolved inorganic substances, organic substances, suspended particulates, and bacteria. These substances can interfere with or adversely affect analytical procedures employing atomic absorption photometry, chromatography, tissue culturing, etc.

Reagent-grade water is obtained from pretreated water, that is, water which has been distilled, deionized, or subjected to reverse osmosis; it is then passed through an activated carbon cartridge to remove the dissolved organic materials, through two deionizing cartridges to remove any dissolved inorganic substances, and through membrane filters to remove microorganisms and any particulate matter with a diameter larger than 0.22 micrometers (μm).

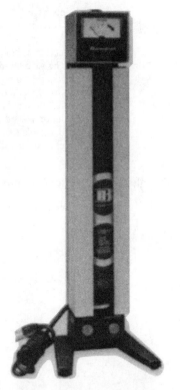

FIGURE 16.4 Demineralizer with dial to indicate the hardness of the water.

---

*The terms *deionized water* and *demineralized water* are sometimes used interchangeably. Make sure that the water you use meets the specifications named for conductivity and purity.

SOLUTIONS

Most chemical compounds found in the laboratory are not pure, but as previously explained, contain various percentages of impurities, depending on their grade. They are often used in solution form.

Solutions are mixtures that are characterized by homogeneity, absence of settling, and a molecular or ionic state of subdivision of the components. There are many kinds of solutions possible, because of the differences in the state of the **solute** (substance dissolved) and the **solvent** (the dissolving medium).

A. The Dissolution Process

When a solute is added to a solvent and enters the dissolved state, it will rapidly diffuse throughout the solvent, dispersing into separate atoms, ions, or molecules.

B. Solubility

Solvents have limited capacity to dissolve solutes, and that limit defines the **solubility** of the solute in the solvent (the maximum amount of solute that will dissolve in a fixed amount of solvent at a definite temperature). Pressure has an effect only upon gases; otherwise its effect is unimportant. When a two-component system contains the maximum quantity of solute in a solvent, the system is said to be **saturated**. An equilibrium is established between the pure solute and the dissolved solute. At the equilibrium point, the rate at which the pure solute enters the solution equals the rate at which the dissolved solute crystallizes out of the solution to return to the pure state. When a solution retains more than the equilibrium concentration of the dissolved solute (which happens under certain conditions), the solution is said to be **supersaturated**. Supersaturation is an unstable condition which can revert back to a stable one as a result of physical shock, decreased temperature, merely standing for a period of time, or some indeterminate factor.

C. Solubility Rules for Common Inorganic Compounds

Following are general solubility rules for inorganic compounds. See also Tables 16.2 and 16.3.

TABLE 16.2 Soluble Complex Ions

Cation	NH_3	CNS^-	Cl^-	CN^-	OH^-
Al^{3+}					$Al(OH)_4^-$
Ag^+	$Ag(NH_3)_2^+$			$Ag(CN)_2^-$	
Cd^{2+}	$Cd(NH_3)_4^{2+}$		$[CdCl_4]^{2-}$	$[Cd(CN)_4]^{2-}$	
Co^{2+}	$Co(NH_3)_6^{3+}$			$Co(CN)_4^{3+}$	
Cr^{2+}					$Cr(OH)_4^-$
Cu^{2+}	$Cu(NH_3)_4^{2+}$		$CuCl_4^{2-}$	$Cu(CN)_2^-$	
Fe^{2+}				$Fe(CN)_6^{4-}$	
Fe^{3+}		$Fe(CNS)^{2+}$	$FeCl_4^-$	$Fe(CN)_6^{3-}$	
Hg^{2+}	$Hg(NH_3)_4^{2+}$	$Hg(CNS)_4^{2-}$	$HgCl_4^{2-}$	$Hg(CN)_4^{2-}$	$Hg(OH)_4^{2-}$
Ni^{2+}	$Ni(NH_3)_6^{2+}$			$Ni(CN)_4^{2-}$	
Pb^{2+}			$PbCl_4^{2-}$		$Pb(OH)_3^-$
Zn^{2+}	$Zn(NH_3)_4^{2+}$			$Zn(CN)_4^{2-}$	$Zn(OH)_4^{2-}$

Arsenic (+3 or +5), antimony (+3 or +5), and tin (+2 or +4) react with Na_2S_x or $(NH_4)_2S$, to yield soluble AsS_4^{3-}, SnS_2^{2-}, and SbS_4^{4+}. Mercury will form HgS_2^{2-}. Concentrated NaOH or KOH yields AsO_2^-, AsO_4^{3+}, SnO_2^{2-}, SnO_3^{2-}, SbO_4^{4+}.

NO_3^-	All nitrates are soluble.
$C_2H_3O_2^-$	All acetates are soluble; $AgC_2H_3O_2$ is moderately soluble.
Cl^-	All chlorides are soluble except $AgCl$, $PbCl_2$, and $HgCl$. $PbCl_2$ is soluble in hot water, slightly soluble in cold water.
SO_4^{2-}	All sulfates are soluble except $BaSO_4$ and $PbSO_4$. Ag_2SO_4, Hg_2SO_4, and $CaSO_4$ are slightly soluble.
HSO_4^-	The bisulfates are more soluble than the sulfates.
CO_3^{2-}, PO_4^{3-}, CrO_4^{2-}, SiO_4^{2-}	All carbonates, phosphates, chromates, and silicates are insoluble, except those of sodium, potassium, and ammonium. An exception is $MgCrO_4$, which is soluble.
OH^-	All hydroxides (except sodium, potassium, and ammonium) are insoluble; $Ba(OH)_2$ is moderately soluble; $Ca(OH)_2$ and $Sr(OH)_2$ are slightly soluble.
S^{2-}	All sulfides (except sodium, potassium, ammonium, magnesium, calcium, and barium) are insoluble. Aluminum and chromium sulfides are hydrolyzed and precipitate as hydroxides.
Na^+, K^+, NH_4^+	All sodium, potassium, and ammonium salts are soluble. Exceptions: $Na_4Sb_2O_7$, $K_2NaCo(NO_2)_6$, K_2PtCl_6, $(NH_4)_2 \cdot NaCo(NO_2)_6$, and $(NH_4)_2PtCl_6$.
Ag^+	All silver salts are insoluble. Exceptions: $AgNO_3$ and $AgClO_4$; $AgC_2H_3O_2$ and Ag_2SO_4 are moderately soluble.

D. Increasing the Rate of Solution of a Solute in a Liquid

A solvent will only dissolve a limited quantity of solute at a definite temperature. (Refer to Chapter 13, "Recrystallization.") However, the rate at which the solute dissolves can be speeded up by the following methods:

1. Pulverizing, or grinding up the solid to increase the surface area of the solid on contact with the liquid.
2. Heating the solvent. This will increase the rate of solution because the molecules of both the solvent and the solute move faster.
3. Stirring.

Combinations of all three methods, when practical, enable one to dissolve solids more quickly.

COLLOIDAL DISPERSIONS

Colloidal dispersions are dispersions of particles that are hundreds and thousands of times larger than molecules or ions. Yet these particles are not large enough to settle rapidly out of solution, nor can they be seen under an ordinary microscope. These particles, which make up the dispersed phase in colloidal systems, are intermediate in size between those of a true solution and those of a coarse suspension. Thus, colloidal systems are somewhere between true solutions and coarse suspensions. However, they exhibit significant phenomena which are associated with adsorbed electric charges and enormous surface area. There are three types of colloidal systems: emulsions, gels, and sols.

Every colloidal system consists of two phases or portions:

1. The colloidal particles (the **dispersed phase**) are small aggregates or conglomerates of materials that are larger than the usual ion or molecule. These particles will remain suspended. They cannot be seen under the ordinary light microscope.
2. The dispersing medium (**dispersing phase**) is a continuous phase. It corresponds to the solvent of a solution.

TABLE 16.3 Solubilities of Inorganic Compounds in Water and Other Aqueous Solvents

Cations	Acetate	Borate	Bromide	Carbonate	Chlorate	Chloride	Chromate	Ferri-cyanide	Ferro-cyanide	Fluoride	Hydroxide	Iodide	Nitrate	Oxide	Phosphate	Sulfate	Sulfide	Sulfite
Aluminum	W	A	W		W	W	A		A	W	AB	W	W	A	A	W	7	A
Ammonium	W	W	W	W	W	W	W	W	W	W	W	W	W	W	W	W	W	W
Barium	W	A	W	A	W	W	A	W	W	2	1	W	W	1	A	i	W	A
Bismuth	W	A	3	A	W	5	A		A	A	A	A	2	A	A	A	7	
Cadmium	W		W	A	W	W	A			W	A	W	W	A	A	W	2	
Calcium	W	A	W	A	W	W	A	W	W	i	1	W	W	1	A	A	7	A
Chromium(III)	W	A	W	A	W	W	W			i	A	W	W	A	A	W	7	
Cobalt(II)	W	A	W	A	W	W	A	i	i	1	AB	W	W	A	A	W	2	A
Copper(II)	W	A	W	A	W	W	W	i	i	1	AB		W	A	A	W	2	A
Iron(II)	W	A	W	A	W	W		i	i	1	A	W	W	A	A	W	2	A
Iron(III)	W	A	W	A	W	W	W	W	i	1	A	W	W	A	A	W	A	A
Lead(II)	W	2	1	2	W	19	2	1	i	1	2B	1	W	A	2	i	2	2
Magnesium	W	A	W	A	W	W	W	W	W	A	A	W	W	A	A	W	7	W
Mercury(I)	1A		2	2	W	2	A	A		W		2	2	A	A	1		
Mercury(II)	W		A	A	W	W	A	i	i	i	A	i	W	A	A	W	i	A
Nickel	W	A	W	A	W	W	A	A	A	W	AB	W	W	A	A	W	2	A
Potassium	W	W	W	W	W	W	W	W	W	W	W	W	W	d	W	W	W	W
Silver	1	A	4	2	W	6	A	6	6	W		4	W	2	2	W	2	A
Sodium	W	W	W	W	W	W	W	W	W	W	W	W	W	d	W	W	W	W
Tin(II)	W	A			W	A		i	i	W	AB	W	W	A	A	1–8	A	A
Tin(IV)	W		W			W	W			W	AB	7d	W	A	A	W	A	
Zinc	W	A	W	A	W	W	A	A	i	A	AB	W	W	A	A	W	A	A

1 = only slightly soluble
2 = soluble in nitric acid, NHO_3
3 = soluble in hydrobromic acid, HBr
4 = soluble in potassium cyanide, KCN
5 = soluble in hydrochloric acid, HCl
6 = soluble in ammonium hydroxide, NH_4OH
7 = hydrolyzes
8 = soluble in sulfuric acid
9 = soluble in hot water

W = soluble in water
A = soluble in acid
B = soluble in water
d = decomposes in water
I = insoluble in water

A. Electrical Behavior

Since colloidal dispersions are either positively or negatively charged, they will migrate to an electrode of opposite charge. They can be coagulated by addition of highly charged ions of opposite sign: negatively charged particles are coagulated by positively charged ions, and positively charged particles by negatively charged ions.

B. Distinguishing between Colloids and True Solutions

Method 1. Colloidal dispersions, like true solutions, may be perfectly clear to the naked eye, but when they are examined at right angles to a beam of light (Fig. 16.5), they will appear turbid. This phenomenon is called the **Tyndall effect,** and it is caused by the scattering of light by the colloidal particles.

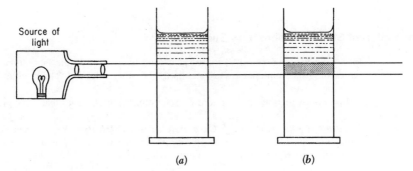

FIGURE 16.5 The Tyndall effect. (*a*) In a true solution a light beam is invisible. (*b*) In a colloidal solution a light beam is visible.

Method 2. The dispersed phase of a colloidal dispersion will pass through ordinary filter paper, but only the molecules and ions of a true solution will pass through semipermeable membranes. This selective passage of molecules and ions and retention of the particles of colloidal dispersions is called **dialysis**.

C. Distinguishing between Colloids and Ordinary Suspensions

1. Carefully examine the solution visually, or if possible, determine the transmission of light through the sample in a spectrophotometer.

   CAUTION: Do not shake the sample.

2. Shake the sample and compare results with those of step 1.
3. Ordinary suspensions become turbid and opaque on shaking. Particles tend to settle. Colloidal dispersions appear unchanged.

TABLE 16.4 Concentration of Desk Acids and Bases Used in the Laboratory

Reagent	Formula	Molecular mass	Molarity	Density, g/mL	% solute	Preparation*
Acetic acid, glacial	$HC_2H_3O_2$	60	17	1.05	99.5	
Acetic acid, dil.			6	1.04	34	Dilute 333 mL 17 M to 1 L.
Hydrochloric acid, conc.	HCl	36.4	12	1.18	36	
Hydrochloric acid, dil.			6	1.10	20	Dilute 500 mL 12 M to 1 L.
Nitric acid, conc.	HNO_3	63.0	16	1.42	72	
Nitric acid dil.			6	1.19	32	Dilute 375 mL 16 M to 1 L.
Sulfuric acid, conc.	H_2SO_4	98.1	18	1.84	96	
Sulfuric acid, dil.			3	1.18	25	Dilute 165 mL 18 M to 1 L.
Ammonium hydroxide, conc.	NH_4OH	35.05	15	0.90	58	
Ammonium hydroxide, dil.			6	0.96	23	Dilute 400 mL 15 M to 1 L.
Sodium hydroxide, dil.	NaOH	40.0	6	1.22	20	Dissolve 240 g in water, dilute to 1 L.

CAUTION: Always pour acid into water slowly with careful mixing.

D. Preparation of Standard Laboratory Solutions

Table 16.4 lists the concentration of acids and bases used in the laboratory and their preparation. Table 16.5 lists the preparation of standard laboratory solutions. These tables will save you calculation time.

Special instructions are required for the preparation of a carbonate-free NaOH solution.

Preparation and Storage of Carbonate-Free NaOH (0.1 N). Reagent-grade NaOH usually contains a considerable amount of Na_2CO_3, because it reacts with the CO_2 in the atmosphere. Carbonate-free NaOH (0.1 N) can be prepared as follows:

Procedure

1. Boil 1 L distilled water in a 2-L flask to free it of dissolved CO_2. Allow to cool (while covered with a watch glass), then transfer to a polyethylene bottle (or a rubber-stoppered hard-glass bottle).
2. Dissolve about 8 g of reagent-grade NaOH in 8 mL distilled water.*

CAUTION: This reaction is highly exothermic and evolves considerable heat. The solution is highly caustic and corrosive.

3. Filter the 50 percent NaOH solution through a Gooch crucible seated on a suction flask and catch the filtrate in a clean, hard-glass test tube. (See Fig. 16.6.)

NOTE: Highly caustic 50 percent NaOH solution will attack paper. The filter medium should be asbestos fiber or glass-fiber mat. To prepare the mat, stir acid-washed asbestos fibers in distilled water to make a suspension. With vacuum off, pour a small amount of the suspension through the crucible. Turn on suction and continue to pour the suspension into the crucible until a 1-mm-thick mat has been formed. When the crucible is held to the light, the perforations should be visible, but no light should pass through the openings.

---

*Only 4 g of NaOH are actually needed to prepare 1 L of 0.1 N NaOH, but we use 8 g because of the hygroscopic nature of NaOH and our uncertainty as to the water content of the NaOH.

TABLE 16.5 Preparation of Standard Laboratory Solutions

Name	Formula	Molecular mass	Concentration	Preparation
Acetic acid	CH_3COOH	60.05	1 M	Dilute 58 mL of glacial acetic acid to 1 L with water.
Aluminum chloride	$AlCl_3 \cdot 6H_2O$	241.43	0.05 M	Dissolve 12.1 g in distilled water and dilute to 1 L.
Aluminum nitrate	$Al(NO_3)_3 \cdot 9H_2O$	375.13	0.1 M	Dissolve 37.5 g $Al(NO_3)_3 \cdot 9H_2O$ in distilled water and dilute to 1 L.
			0.2 F^*	Dissolve 75 g of $Al(NO_3)_3 \cdot 9H_2O$ in distilled water and dilute to 1 L.
Aluminum sulfate	$Al_2(SO_4)_3 \cdot 18H_2O$	666.42	0.083 M	Dissolve 55 g of $Al_2(SO_4)_3 \cdot 18H_2O$ in distilled water and dilute to 1 L
			0.1 M	Dissolve 66.6 g $Al_2(SO_4)_3 \cdot 18H_2O$ in dissolved water and dilute to 1 L.
Ammonium acetate	$NH_4C_2H_3O_2$	77.08	1 M	Dissolve 77 g in distilled water and dilute to 1 L.
Ammonium carbonate	$(NH_4)_2CO_3$	96.10	0.5 M	Dissolve 48 g $(NH_4)_2CO_3$ in distilled water and dilute to 1 L.
Ammonium chloride	NH_4Cl	53.40	0.01 M	Dissolve 0.54 g NH_4Cl in distilled water and dilute to 1 L.
Ammonium hydroxide	NH_4OH	35.05	1 M	Dilute 67 ml conc. NH_4OH (15 M) to 1 L with distilled water.
Ammonium molybdate	$(NH_4)_2MoO_4$	196.01	0.5 M, 1 N	Dissolve 72 g MoO_3 in 200 mL of water, add 60 mL conc. NH_4OH, filter into 270 mL conc. HNO_3, add 400mL water, and dilute to 1 L. Dissolve 87 g $(NH4)_0Mo_7O_{24} \cdot 4H_2O$ and 240g NH_4NO_3 in about 800 mL of warm distilled water and 40 mL conc. (15 M) NH_4OH, and dilute to 1 L.
Ammonium nitrate	NH_4NO_3	80.04	0.1 M	Dissolve 8 g NH_4NO_3 in distilled water and dilute to 1 L.
Ammonium oxalate	$(NH_4)_2C_2O_4 \cdot H_2O$	142.11	0.25 M	Dissolve 35.5 g $(NH_4)_2C_2O_4 \cdot H_2O$ in distilled water and dilute to 1 L.
Ammonium persulfate	$(NH_4)_2S_2O_8$	228	0.1 M	Dissolve 22.8 $(NH_4)_2S_2O_8$ in distilled water and dilute to 1 L.
Ammonium sulfate	$(NH_4)_2SO_4$	132.14	0.1 M	Dissolve 13 g $(NH_4)_2SO_4$ in distilled water and dilute to 1 L.
Ammonium sulfide	$(NH_4)_2S$	68.14	1.0 M	Dilute 250 mL $(NH_4)_2S$ (22%) to 1 L with distilled water. Pass H_2S into 500 mL ice cold 1 M NH_4OH until saturated; then add 500 mL 1 M NH_4OH (use hood).
Ammonium sulfide, colorless	$(NH_4)_2S$	68.14	3 M	Saturate 200 mL NH_4OH with H_2S, add 200 mL NH_4OH, and dilute to 1 L with distilled water (use hood).
Ammonium thiocyanate	NH_4CNS	76.12	0.1 M	Dissolve 7.6 g NH_4CNS in distilled water and dilute to 1 L.
Antimony trichloride	$SbCl_3$	228.11	0.1 F^*	Dissolve 22.7 g $SbCl_3$ in 200 mL conc. HCl and dilute to 1 L with distilled water.
Aqua regia				Mix 1 part conc. HNO_3 with 3 parts conc. HCl.
Barium acetate	$Ba(C_2H_3O_2)_2 \cdot H_2O$	273.45	0.1 M	Dissolve 27.3 g $Ba(C_2H_3O_2)_2 \cdot H_2O$ in distilled water and dilute to 1 L (use hood).
Barium chloride	$BaCl_2 \cdot 2H_2O$	244.28	0.1 M	Dissolve 24.8 g $BaCl_2 \cdot 2H_2O$ in distilled water and dilute to 1 L.

(*Continued*)

TABLE 16.5 Preparation of Standard Laboratory Solutions (*Continued*)

Name	Formula	Molecular mass	Concentration	Preparation
Barium hydroxide	$Ba(OH)_2 \cdot 8H_2O$	315.48	0.02 N	Dissolve 3.1 g $Ba(OH)_2 \cdot 8H_2O$ in distilled water and dilute to 1 L.
			0.1 M, 0.2 N	Dissolve 32 g $Ba(OH)_2 \cdot 8H_2O$ in distilled water and dilute to 1 L.
Barium nitrate	$Ba(NO_3)_2$	261.35	0.1 M	Dissolve 26 g $Ba(NO_3)_2$ in distilled water and dilute to 1 L.
Bismuth chloride	$BiCl_3$	315.34	0.1 M	Dissolve 31.5 g $BiCl_3$ in 200 mL conc. HCl and dilute to 1 L with distilled water.
Bismuth nitrate	$Bi(NO_3)_3 \cdot 5H_2O$	485.07	0.1 M	Dissolve 49 g of $Bi(NO_3)_3 \cdot 5H_2O$ in 50 mL conc. HNO_3 and dilute to 1 L with distilled water.
Boric acid	H_3BO_3	61.83	0.1 M	Dissolve 6.2 g of H_3BO_3 in 1 L distilled water.
Bromine in carbon tetrachloride	Br_2	159.82	0.1 M	Dissolve 1 g liquid bromine in 100 g (63 mL) CCl_4 (use hood).
Bromine water	Br_2	159.82		Saturate 1 L distilled water with 10–15 g of liquid bromine (use hood).
Cadmium chloride	$CdCl_2$	183.31	0.1 M	Dissolve 18.3 g $CdCl_2$ in distilled water and dilute to 1 L.
Cadmium nitrate	$Cd(NO_3)_2 \cdot 4H_2O$	308.47	0.2 M	Dissolve 62 g $Cd(NO_3)_2 \cdot 4H_2O$ in distilled water and dilute to 1 L.
Cadmium sulfate	$CdSO_4 \cdot 4H_2O$	280.5	0.25 M	Dissolve 70 g $CdSO_4 \cdot 4H_2O$ in distilled water and dilute to 1 L.
Calcium acetate	$Ca(C_3H_3O_2)_2 \cdot H_2O$	176.2	0.5 M	Dissolve 88 g $Ca(C_2H_3O_2)_2 \cdot H_2O$ in distilled water and dilute to 1 L.
Calcium chloride	$CaCl_2$	111	0.1 M	Dissolve 11 g $CaCl_2$ in distilled water and dilute to 1 L.
	$CaCl_2 \cdot 6H_2O$	219.1	0.1 M	Dissolve 22 g $CaCl_2 \cdot 6H_2O$ in distilled water and dilute to 1 L.
Calcium hydroxide	$Ca(OH)_2$	74.1		Saturated: Shake 3–4 g reagent-grade CaO with distilled water and dilute to 1 L, then filter. Saturated solution (prepared every few days): Saturate 1 L distilled water with solid $Ca(OH)_2$; filter.
Calcium nitrate	$Ca(NO_3)_2 \cdot 4H_2O$	236.2	0.1 M	Dissolve 23.6 g $Ca(NO_3)_2 \cdot 4H_2O$ in distilled water and dilute to 1 L.
Chlorine water	Cl_2	70.91		Slightly acidify 225 mL 3% NaClO or 175 mL 5% NaClO with 6 mL acetic acid and dilute 1 L with distilled water. Chlorine water (prepared every few days): saturate 1 L distilled water with chlorine gas (use hood).
Chloroplatinic acid	$H_2PtCl_6 \cdot 6H_2O$	517.92	0.512 M, 0.102 N	Dissolve 26.53 g $H_2PtCl_6 \cdot 6H_2O$ in distilled water and dilute to 1 L.
Chromium(III) chloride	$CrCl_3$	158.36	0.167 M	Dissolve 26 g $CrCl_3$ in distilled water and dilute to 1 L.
Chromium(III) nitrate	$Cr(NO_3)_3 \cdot 9H_2O$	400.15	0.1 M	Dissolve 40 g $Cr(NO_3)_3 \cdot 9H_2O$ in distilled water and dilute to 1 L.
Chromium(III) sulfate	$Cr_2(SO_4)_3 \cdot 18H_2O$	716.45	0.083 M	Dissolve 60 g $Cr(SO_4)_3 \cdot 18H_2O$ in distilled water and dilute to 1 L.
Cobalt(II) chloride	$CoCl_2 \cdot 6H_2O$	237.93	0.1 M	Dissolve 24 g $CoCl_2 \cdot 6H_2O$ in distilled water and dilute to 1 L.
Cobalt(II) nitrate	$Co(NO_3)_2 \cdot 6H_2O$	2291.04	0.25 M	Dissolve 73 g $Co(NO_3)_2 \cdot 6H_2O$ in distilled water and dilute to 1 L.

(*Continued*)

TABLE 16.5 Preparation of Standard Laboratory Solutions (*Continued*)

Name	Formula	Molecular mass	Concentration	Preparation
Cobalt(II) sulfate	$CoSO_4 \cdot 7H_2O$	281.10	0.25 M	Dissolve 70 g $CoSO_4 \cdot 7H_2O$ in distilled water and dilute to 1 L.
Copper(II) chloride	$CuCl_2 \cdot 2H_2O$	170.48	0.1 M	Dissolve 17 g $CuCl_2 \cdot 2H_2O$ in distilled water and dilute to 1 L.
Copper(II) nitrate	$Cu(NO_3)_2 \cdot 3H_2O$	241.6	0.1 M	Dissolve 24 g $Cu(NO_3)_2 \cdot 3H_2O$ in distilled water and dilute to 1 L.
	$Cu(NO_3)_2 \cdot 6H_2O$	295.64	0.1 M	Dissolve 29.5 g $Cu(NO_3)_2 \cdot 6H_2O$ in distilled water and dilute to 1 L.
			0.25 M	Dissolve 74 g $Cu(NO_3)_2 \cdot 6H_2O$ in distilled water and dilute to 1 L.
Copper(II) sulfate	$CuSO_4 \cdot 5H_2O$	249.68	0.10 M	Dissolve 24.97 g $CuSO_4 \cdot 5H_2O$ in distilled water, add 3 drops conc. H_2SO_4, and dilute to 1 L.
Disodium phosphate	$Na_2HPO_4 \cdot 12H_2O$	358.14	0.1 M	Dissolve 35.8 g $Na_2HPO_4 \cdot 12H_2O$ in distilled water and dilute to 1 L.
Fehling's solution A				Dissolve 35 g $CuSO_4 \cdot 5H_2O$ in 500 mL distilled water.
Fehling's solution B				Dissolve 173 g $KNaC_4H_4O_6 \cdot 4H_2O$ (Rochelle salt) in 200 mL water, add 50 g solid NaOH in 200 mL water, and dilute to 500 mL with distilled water.
Hydrochloric acid	HCl	36.46	0.05 M	Dilute 4.1 mL of conc. HCl (37%) to 1 L with distilled water.
			0.1 M	Dilute 8.2 mL of conc. HCl (37%) to which distilled water.
			1.0 M	Dilute 82 mL of conc. HCl (37%) to 1 L with distilled water.
			3 M	Dilute 246 mL HCl (37%) to with distilled water.
			6.0 M	Dilute 492 mL of conc. HCl (37%) to 1 L with distilled water.
Hydrogen peroxide	H_2O_2	34.01	0.2 M	Dilute 23.3 mL 30% H_2O_2 to 1 L with distilled water. Hydrogen peroxide 3% is standard concentration commercially available.
Iodine in potassium iodide				Dissolve 5 g I_2 and 15 g KI in distilled water and dilute to 250 mL.
Iron(III) ammonium sulfate, saturated	$Fe_2(SO_4)_3(NH_4)_2SO_4 \cdot 24H_2O$	964.4		Dissolve about 1240 g $Fe_2(SO_4)_3(NH_4)_2SO_4 \cdot 24H_2O$ in distilled water and dilute to 1 L.
Iron(III) chloride	$FeCl_3 \cdot 6H_2O$	270.3	0.1 M	Dissolve 27.0 g $FeCl_3 \cdot 6H_2O$ in water containing 20 mL conc. HCl in distilled water and dilute to 1 L.
Iron(III) nitrate	$Fe(NO_3)_3 \cdot 6H_2O$	349.95	0.1 M	Dissolve 35 g $Fe(NO_3)_3 \cdot 6H_2O$ in distilled water and dilute to 1 L.
Iron(III) sulfate	$Fe_2(SO_4)_3 \cdot 9H_2O$	562.02	0.25 M	Dissolve 140.5 g $Fe_2(SO_4)_3 \cdot 9H_2O$ in 100 mL 3 M H_2SO_4 and dilute to 1 L with distilled water.
Iron(II) ammonium sulfate	$FeSO_4(NH_4)_2SO_4 \cdot 6H_2O$	392.14	0.5 M	Dissolve 196 g $FeSO_4(NH_4)SO_4 \cdot 6H_2O$ in water containing 10 mL conc. H_2SO_4 in distilled water and dilute to 1 L.
Iron(II) sulfate	$FeSO_4 \cdot 7H_2O$	278.02	0.1 M	Dissolve 28 g $FeSO_4 \cdot 7H_2O$ in water containing 5 mL conc. H_2SO_4, and dilute to 1 L with distilled water.
Lead(II) acetate	$Pb(C_2H_3O_2)_2 \cdot 3H_2O$	379.33	0.1 M	Dissolve 38 g $Pb(C_2H_3O_2)_2 \cdot 3H_2O$ in distilled water and dilute to 1 L.

(*Continued*)

TABLE 16.5 Preparation of Standard Laboratory Solutions (*Continued*)

Name	Formula	Molecular mass	Concentration	Preparation
Lead nitrate	$Pb(NO_3)_2$	331.2	0.05 M	Dissolve 16.6 g $Pb(NO_3)_2$ in distilled water and dilute to 1 L.
			0.1 M	Dissolved 33 g $Pb(NO_3)_2$ in distilled water and dilute to 1 L.
Lead nitrate (basic)	$Pb(OH)NO_3$	286.19	0.1 M	Dissolve 28.6 g of $Pb(OH)NO_3$ in distilled water and dilute to 1 L.
Lime water				See Calcium hydroxide.
Magnesium chloride	$MgCl_2 \cdot 6H_2O$	203.31	0.25 M	Dissolve 51 g $MgCl_2 \cdot 6H_2O$ in distilled water and dilute to 1 L.
Magnesium nitrate	$Mg(NO_3)_2 \cdot 6H_2O$	256.41	0.1 M	Dissolve 25.6 g of $Mg(NO_3)_2 \cdot 6H_2O$ in distilled water and dilute to 1 L.
Magnesium sulfate	$MgSO_4 \cdot 7H_2O$	246.48	0.25 M	Dissolve 62 g $MgSO_4 \cdot 7H_2O$ in distilled water and dilute to 1 L.
Manganese(II) chloride	$MnCl_2 \cdot 4H_2O$	197.91	0.25 M	Dissolve 50 g $MnCl_2 \cdot 4H_2O$ in distilled water and dilute to 1 L.
Manganese(II) nitrate	$Mn(NO_3)_2 \cdot 6H_2O$	287.04	0.25 M	Dissolve 72 g $Mn(NO_3)_2 \cdot 6H_2O$ in distilled water and dilute to 1 L.
Manganese(II) sulfate	$MnSO_4 \cdot 7H_2O$	277.11	0.25 M	Dissolve 69 g $MnSO_4 \cdot 7H_2O$ in distilled water and dilute 1 L.
Mercury(II) chloride	$HgCl_2$	271.5	0.25 M	Dissolve 68 g $HgCl_2$ in distilled water and dilute to 1 L.
Mercury(I) nitrate	$HgNO_3 \cdot H_2O$	280.61	0.1 M	Dissolve 28 g $HgNO_3 \cdot H_2O$ in 100 mL 6 F HNO_3 and dilute to 1 L with distilled water.
Mercury(II) nitrate	$Hg(NO_3)_2 \cdot H_2O$	332	0.02 M	Dissolve 6.6 g of $Hg(NO_3)_2 \cdot \frac{1}{2}H_2O$ in distilled water and dilute to 1 L.
Nickel(II) chloride	$NiCl_2 \cdot 6H_2O$	237.71	0.25 M	Dissolve 59 g $NiCl_2 \cdot 6H_2O$ in distilled water, dilute to 1 L.
Nickel(II) nitrate	$Ni(NO_3)_2 \cdot 6H_2O$	290.81	0.02 M	Dissolve 5.8 g $Ni(NO_3)_2 \cdot 6H_2O$ in distilled water, dilute to 1 L
			0.1 $F*$	Dissolve 29.1 g $Ni(NO_3)_2 \cdot 6H_2O$ in distilled water and dilute to 1 L.
Nickel(II) sulfate	$NiSO_4 \cdot 6H_2O$	262.86	0.25 M	Dissolve 66 g $NiSO_4 \cdot 6H_2O$ in distilled water, dilute to 1 L.
Nitric acid	HNO_3	63.01	1 M	Add 63 mL conc. HNO_3 (16 M) to distilled water and dilute to 1 L.
			3 M	Add 189 mL conc. HNO_3 (16 M) to distilled water to make 1 L.
			6.0 M	Add 378 mL conc. HNO_3 to distilled water to make 1 L.
Oxalic acid	$C_2H_2O_4 \cdot 2H_2O$	126.07	0.1 M	Dissolve 12.6 g $C_2H_2O_4 \cdot 2H_2O$ in distilled water and dilute to 1 L.
Potassium bromide	KBr	119.01	0.1 M	Dissolve 11.9 g KBr in distilled water and dilute to 1 L.
Potassium carbonate	K_2CO_3	138.21	1.5 M	Dissolve 207 g K_2CO_3 in distilled water and dilute to 1 L.
Potassium chloride	KCl	74.56	0.1 M	Dissolve 7.45 g KCl in distilled water and dilute to 1 L.
Potassium chromate	K_2CrO_4	194.20	0.1 $F*$	Dissolve 19.4 g K_2CrO_4 in distilled water and dilute to 1 L.
Potassium cyanide	KCN	65.12	0.5 M	Dissolve 33 g KCN in distilled water and dilute to 1 L (use hood).
Potassium dichromate	$K_2Cr_2O_7$	294.19	0.1 $F*$	Dissolve 29.4 g $K_2Cr_2O_7$ in distilled water and dilute to 1 L.

(*Continued*)

TABLE 16.5 Preparation of Standard Laboratory Solutions (*Continued*)

Name	Formula	Molecular mass	Concentration	Preparation
Potassium dihydrogen phosphate	KH_2PO_4	136.09	0.1 M	Dissolve 13.6 g of KH_2PO_4 in distilled water and dilute to 1 L.
Potassium ferricyanide	$K_3Fe(CN)_6$	329.26	0.167 M	Dissolve 55 g of $K_3Fe(CN)_6$ in distilled water and dilute to 1 L.
Potassium ferrocyanide	$K_4Fe(CN)_6 \cdot 3H_2O$	422.41	0.1 M	Dissolve 42.3 g of $K_4Fe(CN)_6 \cdot 3H_2O$ in distilled water and dilute to 1 L.
Potassium hydrogen phthalate	$KHO_4C_8H_4$	204	0.1 M	Dissolve 20.4 g $KHO_4C_8H_4$ in distilled water and dilute to 1 L.
Potassium hydrogen sulfate	$KHSO_4$	136.17	0.1 F*	Dissolve 13.6 g $KHSO_4$ in distilled water and dilute to 1 L.
Potassium hydrogen sulfite	$KHSO_3$	120.17	0.20 M	Dissolve 24.0 g $KHSO_3$ in distilled water and dilute to 1 L.
Potassium hydroxide	KOH	56.11	0.1 F*	Dissolve 5.6 g KOH in distilled water and dilute to 1 L.
			0.1 F*	Dissolve 56 g KOH in distilled water and dilute to 1 L.
			20%	Dissolve 250 g of KOH in distilled water and dilute to 800 mL.
			3 M, 3 N	Dissolve 168 g KOH in distilled water and dilute to 1 L.
Potassium iodate	KIO_3	214	0.10 M	(Acidified) Dissolve 21.4 g KIO_3 in distilled water, add 5 mL H_2SO_4, and dilute with distilled water to 1 L.
Potassium iodide	KI	166.01	0.1 M	Dissolve 16.6 g KI in distilled water and dilute to 1 L.
Potassium nitrate	KNO_3	101.11	01. M	Dissolve 10 g KNO_3 in distilled water and dilute to 1 L.
Potassium permanganate	$KMnO_4$	158.04	0.1 N	Dissolve 3.2 g $KMnO_4$ in distilled water, add 2 mL (18 M) H_2SO_4, and dilute to 1 L with distilled water.
			0.1 M	(Acidfied) Dissolve 15.8 g $KMnO_4$ in distilled water, add 10 mL (18 M) H_2SO_4, and dilute to 1 L.
			0.1 M	Dissolve 15.8 g $KMnO_4$ in distilled water and dilute to 1 L.
Potassium phosphate dibasic	K_2HPO_4	174.18	0.1 M	Dissolve 17.4 g K_2HPO_4 in distilled water and dilute to 1 L.
Potassium sulfate	K_2SO_4	174.27	0.1 M	Dissolve 17.4 g K_2SO_4 in distilled water and dilute to 1 L.
			0.25 M	Dissolve 44 g K_2SO_4 in distilled water and dilute to 1 L.
Potassium thiocyanate	KSCN	97.18	0.1 M	Dissolve 9.7 g KSCN in distilled water and dilute to 1 L.
Silver nitrate	$AgNO_3$	169.87	0.05 M	Dissolve 8.50 g $AgNO_3$ in distilled water and dilute to 1 L.
			0.10 M	Dissolve 17.0 g $AgNO_3$ in distilled water and dilute to 1 L.
			0.20 M	Dissolve 34.0 g $AgNO_3$ in distilled water and dilute to 1 L.
			0.50 M	Dissolve 85 g $AgNO_3$ in distilled water and dilute to 1 L.
			1.0 M	Dissolve 170 g $AgNO_3$ in distilled water and dilute to 1 L.

(*Continued*)

TABLE 16.5 Preparation of Standard Laboratory Solutions (*Continued*)

Name	Formula	Molecular mass	Concentration	Preparation
Sodium acetate	$NaC_2H_3O_2 \cdot 3H_2O$	136.08	0.50 M	Dissolve 68.0 g $NaC_2H_3O_2 \cdot 3H_2O$ in distilled water and dilute to 1 L.
			1 M	Dissolve 136 g $NaC_2H_3O_2 \cdot 3H_2O$ in distilled water and dilute to 1 L.
			1 M	Dissolve 82.0 g $NaC_2H_3O_2 \cdot 3H_2O$ in distilled water and dilute to 1 L.
			3 M, 3 N	Dissolve 408 g $NaC_2H_3O_2 \cdot 3H_2O$ in distilled water and dilute to 1 L.
Sodium arsenate	$Na_3AsO_4 \cdot 12H_2O$	424.07	0.1 M	Dissolve 42.4 g $Na_3AsO_4 \cdot 12H_2O$ in distilled water and dilute to 1 L.
Sodium bicarbonate	$NaHCO_3$	84.01	0.1 F*	Dissolve 8.4 g $NaHCO_3$ in distilled water and dilute to 1 L.
Sodium bromide	$NaBr$	102.9	0.1 F*	Dissolve 10.3 g $NaBr$ in distilled water and dilute to 1 L.
Sodium carbonate	Na_2CO_3	105.99	1 N	Dissolve 53 g Na_2CO_3 in distilled water and dilute to 1 L.
Sodium chloride	$NaCl$	58.44	0.1 M	Dissolve 5.9 g $NaCl$ in distilled water and dilute to 1 L.
Sodium cobaltinitrite	$Na_3Co(NO_2)_6$	404	0.08 M	Dissolve 25 g $NaNO_2$ in 75 mL water, add 2 mL glacial acetic acid, then 2.5 g $Co(NO_2)_2 \cdot 6H_2O$, and dilute with distilled water to 100 mL.
Sodium hydrogen phosphate	$Na_2HPO_4 \cdot 7H_2O$	268.1	0.05 M	Dissolve 13.4 g $Na_2HPO_4 \cdot 7H_2O$ in distilled water and dilute to 1 L.
Sodium hydrogen sulfate	$NaHSO_4$	120.0	1 M	Dissolve 120 g $NaHSO_4$ in distilled water and dilute to 1 L.
Sodium hydroxide	$NaOH$	40.0	0.1 M	Dissolve 4 g $NaOH$ in distilled water and dilute to 1 L.
			1 M	Dissolve 40.0 g $NaOH$ pellets in distilled water and dilute to 1 L. (*Caution*: Heat evolved; use goggles.)
			6.0 M	Dissolve 240 g $NaOH$ pellets in distilled water and dilute to 1 L. (*Caution*: Heat evolved; use goggles.)
Sodium iodide	NaI	149.9	0.1 F*	Dissolve 15 g NaI in distilled water and dilute to 1 L.
Sodium nitrate	$NaNO_3$	85.00	0.2 M	Dissolve 17 g $NaNO_3$ in distilled water and dilute to 1 L.
Sodium nitroprusside[1]	$Na_2Fe(CN)_5NO \cdot 2H_2O$	297.95	10%	Dissolve 100 g $Na_2Fe(CN)_5NO \cdot 2H_2O$ in 900 mL distilled water.
Sodium polysulfide	Na_2S_4			Dissolve 480 g $Na_2S \cdot 9H_2O$ in 500 mL water, add 40 g $NaOH$, and 16 mg powdered sulfur, mix well, and dilute to 1 L.
Sodium sulfate	Na_2SO_4	142.04	0.1 F*	Dissolve 14.2 g Na_2SO_4 in distilled water and dilute to 1 L.
Sodium sulfide	$Na_2S \cdot 9H_2O$	240.1	0.1 M	(Fresh) Dissolve 24 g $Na_2S \cdot 9H_2O$ in distilled water and dilute to 1 L.
			0.2 M	Dissolve 48 g $Na_2S \cdot 9H_2O$ in distilled water and dilute to 1 L.
Sodium sulfite	Na_2SO_3	126.04	0.1 M	Dissolve 12.6 g anhydrous Na_2SO_3 in distilled water and dilute to 1 L.

(*Continued*)

TABLE 16.5 Preparation of Standard Laboratory Solutions (*Continued*)

Name	Formula	Molecular mass	Concentration	Preparation
Sodium tetraborate (borax)	$Na_2B_4O_7 \cdot 10H_2O$	381.37	0.0025 M	Dissolve 9.5 g $Na_2B_4O_7 \cdot 10H_2O$ in distilled water and dilute to 1 L.
Sodium thiosulfate	$Na_2S_2O_3 \cdot 5H_2O$	248.18	0.1 M	Dissolve 24.8 g $Na_2S_2O_3 \cdot 5H_2O$ in distilled water and dilute to 1 L.
Strontium chloride	$SrCl_2 \cdot 6H_2O$	266.6	0.25 M	Dissolve 66.6 g $SrCl_2 \cdot 6H_2O$ in distilled water and dilute to 1 L.
Sulfuric acid	H_2SO_4	98.08	0.1 M	Add 5.6 mL conc. H_2SO_4 slowly to 500 mL distilled water and dilute to 1 L.

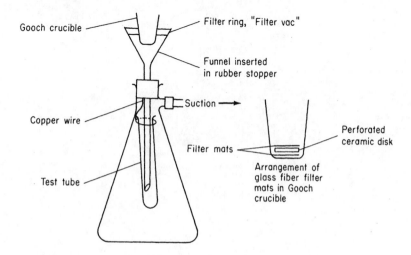

FIGURE 16.6 Assembly incorporating a Gooch crucible for filtering 50 percent NaOH.

4. Transfer the filtered 50 percent NaOH to a clean 10-mL graduated cylinder and add 6 mL of the solution to the prepared liter of cool boiled water. Shake well to ensure a uniform concentration of the resulting solution. The solution should be approximately 0.1 N, and, when stored against the CO_2 in the atmosphere, it should be protected with a drying tube of NaOH.

5. Standardize the solution by standard methods of analysis. (Refer to Chapter 25, "Volumetric Analysis.")

REFERENCE

1. Arnold J. Gordon and Richard A. Ford, *The Chemist's Companion, A Handbook of Practical Data, Techniques, and References.* John Wiley & Sons, Inc. ISBN 0-471-31590-7 (1972).

CHAPTER 17
PROCESS ANALYZERS

INTRODUCTION

Most industries now use online analyzers to measure and control chemical composition, concentration, and physical properties of process streams. These process analyzers can range from simple physical measurements (temperature, pressure, density, mass flow rate, viscosity, refractive index, conductivity, etc.) to complicated chemical instrumentation (pH, reduction/oxidation potentials, chromatography, spectroscopy, etc.).

PROCESS SENSORS

A. Temperature Sensors

When metals are heated, they expand; the amount of the expansion depends upon the temperature and the coefficients of expansion of the metals. When two metals having different coefficients of expansion are bonded together and heated, the strip of metal distorts and bends. This is the principle of thermostats that can act as electrical regulators for temperature control, known as **bimetallic thermometers**. In these thermometers, a rotary motion is developed by expansion of the bimetallic coil (see Fig. 17.1).

CAUTION: Do not attempt to use a dial-reading thermometer if the stem is bent or deformed. This causes binding of the helical coil on expansion, retarding the motion and giving inaccurate readings on the dial.

Filled thermometers (excluding mercury-glass thermometers) are composed of a bulb, a capillary tube, a pressure-activated and sensing mechanism, and a mechanical-lever amplifying pointer. The sensor may be a Bourdon type (see Chapter 5, "Pressure and Vacuum") or a bellows device, and the system operates on the principle of expansion of fluids (gas or liquid) by heat. As the fluid expands, it activates the sensor, which in turn moves the calibrated pointer or recording monitor or the electric relay part of the process analyzer, as shown in Fig. 17.2.

Thermistors are based on temperature-dependent electrical conductive properties of semiconductor materials. As the temperature increases in certain materials such as semiconductors, the electrical resistance decreases. They are extremely sensitive, and when heat is applied to them externally, they convert the change in heat to corresponding changes in voltage or current. They do not require reference or cold-junction compensation, which is required for thermocouples, and are available for a wide range of temperatures. They require a source of direct current incorporated in a simple bridge circuit with an indicating galvanometer. The standard thermistor bridge or telemetry circuit can provide sufficient voltage (without amplification) to actuate and operate signal-scanning or transmitting equipment (see Fig. 17.3).

Thermocouples are two dissimilar metals, usually in the form of wires, fused together at one end (the temperature-indicating end), that generates an electric current when that end is heated (Fig. 17.4). This is called the **Seebeck voltage**. The current is measured by a millivolt meter, called a **pyrometer**. There are many types of thermocouples that are suited to different temperature ranges and conditions. For example, a K-type thermocouple is suitable for a wide range of temperatures (−200 to 1350°C) but could not perform in cryogenic conditions, where an E-type would be better suited.

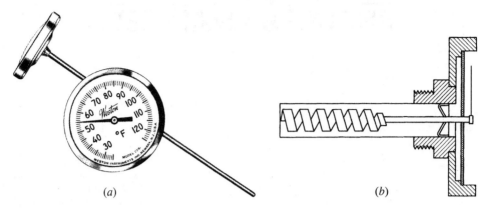

FIGURE 17.1 Bimetallic expansion thermometer: *(a)* dial-reading thermometer; *(b)* industrial temperature indicator with a helical bimetal element.

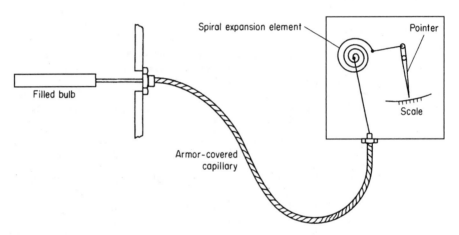

FIGURE 17.2 Liquid-filled remote indicating thermometer.

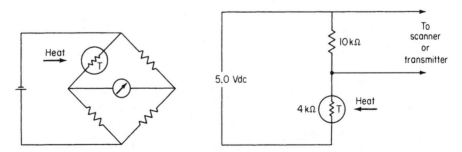

FIGURE 17.3 Typical thermistor temperature indicator circuit (simplified) and a typical thermistor telemetry circuit.

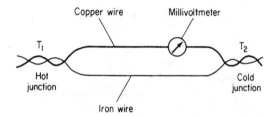

FIGURE 17.4 A simple thermocouple and pyrometer circuit.

Thermocouples are not as accurate as the thermistor. The thermocouple must be matched to a pyrometer because of the loop resistance of the thermocouple. The pyrometer is the indicator of the temperature signal of the thermocouple. Some pyrometers incorporate amplifier circuits to eliminate thermocouple resistance problems and to allow narrow-span indication. Various combinations of specially selected metals and alloys provide a broad range of thermocouples (see Table 17.1).

Thermocouple output voltages and allowable deviations are published in the *Annual Book of ASTM Standards* (NBS Monograph 125). Table 17.2 provides a listing of the more common ANSI thermocouple types and materials. Table 17.3 lists the ANSI thermocouple color codes used for identification purposes. Table 17.4 summarizes some of these data.

TABLE 17.1 Useful Range of Thermocouples

Base metals	Temperature, °F	EMF (mV)
Copper/Constantan	−300 to 750	−5.284 to 20.805
Iron/Constantan	−300 to 1600	−7.52 to 50.05
Chromel/Alumel	−300 to 2300	−5.51 to 51.05
Chromel/Constantan	32 to 1800	0 to 75.12
Platinum 10% Rh/Pt	32 to 2800	0 to 15.979
Platinum 13 % Rh/Pt	32 to 2900	0 to 18.636
Platinum 30% Rh/Pt 6% Rh	100 to 3270	0.007 to 13.499
Platinel 1813/Platinel 1503	32 to 2372	0 to 51.1
Iridium/Iridium 60% Rh 40%	2552 to 3326	7.30 to 9.55
Tungsten/Tungsten 26% Rh	60 to 5072	0.042 to 43.25
Tungsten 5% Rh/W 26% Rh	32 to 5000	0.042 to 38.45

TABLE 17.2 ANSI Thermocouple Types and Materials

Thermocouple type (ANSI)	Material
J	Iron/Constantan
K	Chromel/Alumel
T	Copper/Constantan
E	Chromel/Constantan
R	Platinum/Platinum 13% Rh
S	Platinum/Platinum 10% Rh
B	Platinum 6% Rh/Platinum 30% Rh
G*	Tungsten/Tungsten 26% Re
C*	Tungsten 5% Re/Tungsten 26% Re
D*	Tungsten 3% Re/Tungsten 25% Re

*Not ANSI symbols.

TABLE 17.3 ANSI Thermocouple Color Codes

Negative lead	Red
Type K, base metal	Yellow
Type J, base metal	Black
Type E, base metal	Purple
Type T, base metal	Blue

TABLE 17.4 Some Common Thermocouples and Their Characteristics

Noble metal thermocouples (for example: types B, R, and S) are all platinum or platinum/rhodium and are the most stable of all the common thermocouples. Type S is stable enough that it is specified as the standard for temperature calibrations between the antimony point (630.74°C) and the gold point (1064.43°C).

Type B thermocouples (platinum/6% Rh and platinum/30% Rh) generally should not be used below 50°C because they produce a double value millivolt/temperature curve between 0 and 42°C.

Types (C, D, and G) thermocouples all contain tungsten/rhenium and are especially suited for high temperatures, high vacuum, and reducing atmospheres. However, these thermocouples should not be used in an oxidizing atmosphere.

Type E thermocouples (chromel/constantan) can detect very small temperature changes and are ideally suited for temperatures below 0°C. Note that **Chromel** is nickel/10% chromium alloy.

Type J thermocouples (iron/constantan) are the least expensive of all the thermocouples, but impurities in the iron tend to make them less accurate. They are not recommended for use above 760°C because they can lose their calibration. The iron serves as the positive element and the **Constantan**, which is a copper-nickel alloy, acts as the negative element. Note that not all constantans (J versus T) are the same alloy.

Type N thermocouples (nicrosil/nisil) are nickel-based systems and offer better stability than most other base-metal (E, J, K, and T) thermocouple types. These newer alloys contain various mixtures of chromium, silicon, and magnesium). These newer thermocouples are replacing the older, less stable Chromel-Alumel combinations.

Type T thermocouples (copper/constantan) are well suited for temperatures below 0°T. The copper thermocouple lead offers the advantage of eliminating any necessary lead compensation, as found with most other thermocouples.

B. Resistance Temperature Detectors

Resistance Temperature Detectors (RTD) operate on the principle of change in electrical resistance in wire as a function of temperature. The element or temperature sensing unit can be either a wound wire or thin film. The element is normally pure platinum wound about a ceramic or glass core and hermetically sealed inside a capsule. The thin-film type consists of platinum being deposited as a film on a substrate and encapsulated. These RTD are very accurate and reproducible over a wide range of temperatures; a typical commercial unit has deviations of ±1.3°C at −200°C to ± 4.6°C at 850°C.

CONDUCTIVITY MEASUREMENTS

Conductivity is defined as the ability of a substance to conduct an electric current. It can be considered the reciprocal or inverse of *resistivity*. Process solutions can be classified into two classes: those that ionize completely (strong electrolytes) and those that ionize only partially (weak electrolytes). The magnitude of the current carried through the solution depends upon the number of ions present, which depends on the concentration of the electrolytes and its degree of ionization. Conductivity values are obtained from resistance measurements made with a conductivity bride (a Wheatstone bridge shown in Fig. 17. 5) by actually measuring the electrical resistance of a solution.

The base unit for resistance is the ohm (see "Ohm's Law," in Chapter 10, "Basic Electricity"). Conductivity is expressed in **mhos** (or ohm^{-1}) or **siemens**. The conductance of substance is normally carried out between two opposite, platinum electrodes of 1 cm^2, usually located exactly 1 cm apart.

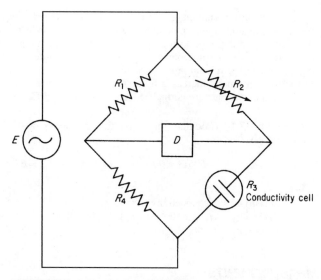

FIGURE 17.5 Schematic diagram of an alternating current (AC) Wheatstone bridge circuit for conductivity measurements.

For solutions of low conductivity, the electrodes can be placed closer together. **Conductivity** can be calculated using the following formula:

$$G = LA/l$$

where G = Conductivity in siemens
 L = Conductivity in siemens/cm
 A = Area normal to current flow in cm^2
 l = Length in cm between electrodes

Thus units of siemens/cm are produced. Prefixes of micro (μ) and milli (m) are commonly used with siemens. Potassium chloride standards are normally used to calibrate conductivity equipment following ASTM Standard method D1125-77. Table 17.5 gives some typical conductivity values for natural, laboratory, and process solutions.

TABLE 17.5 Conductivity of Various Aqueous Solutions

Aqueous solution	Conductivity @ 25°C
Ultrapure water	<0.1 µS/cm
Distilled water	0.5 µS/cm
Power plant boiler water	1.0 µS/cm
Mountain stream	1.0 µS/cm
Typical city water	50 µS/cm
Standard, 0.001 N KCl solution	146.93 µS/cm
Standard, 0.01 N KCl solution	1408.8 µS/cm
Standard, 0.1 N KCl solution	12,856 µS/cm
Standard, 1.0 N KCl solution	111,342 µS/cm
Potable water, maximum	1055 µS/cm
Typical ocean water	60 µS/cm
Sodium hydroxide, 10% NaOH	355 mS/cm
Sulfuric acid, 10% H_2SO_4	432 mS/cm

Conductivity measurements are used in the process industry as a quick, nondestructive online means of measuring ionic content of a sample stream. Conductivity's biggest disadvantage is that it is nonspecific and cannot distinguish between different ion types. Some ionic substances contribute very strong readings while other ionic substances generate very weak signals. Most organic compounds (sugars, petroleum products, soils, etc.) do not show much conductivity at all. Conductivity readings are normally increased by increased temperatures; thus temperature must be monitored. Most conductivity data are normalized and reported at 25°C.

EXAMPLE 17.1 Calculate the resistivity (in ohms) if a solution is found to be 60 microsiemens (μS).

$$60 \; \mu S = 60 \times 10^{-6} \; S = 6.0 \times 10^{-5} \; S = 6.0 \times 10^{-5} \; mho$$

Since:

Resistivity = 1/conductivity
Resistivity = $1/6.0 \times 10^{-5}$ mho = 1.7×10^4 ohms
Resistivity = 17,000 ohms

PRESSURE MEASUREMENTS

As was discussed in Chapter 5, "Pressure and Vacuum," there are three commonly used pressure scales: gauge pressure (psig), absolute pressure (psia), and vacuum scale (in Hg). Chapter 5 also discusses the three basic types of manometer used in chemical plants and laboratories: U-tube manometers, well manometers, and inclined manometers. Typically, a pressure gauge will contain a device (bellows, Bourdon tubes, diaphragms, etc.) that senses changes in pressure and converts it to mechanical energy. The most common pressure gauge is the **Bourdon gauge,** as shown in Chapter 5 in Fig. 5.9. The C-type Bourdon gauge gets it name from the C-shaped hollow tube used as the pressure sensor. In most automated systems, the Bourdon gauge is connected directly to a transmitter. The transmitter sends a signal to the controller connected to a pressure control valve and/or recorder. Diaphragm pressure gauges are normally composed of a fixed metal cup covered by a flexible metal plate. A pressure line enters the metal cup and the metal plate flexes with a mechanical linkage to a pointer to indicate the pressure.

PROCESS pH MEASUREMENTS

A specially designed voltmeter called a **pH meter** can be used to determine the acidity/basicity of any aqueous, conductive solutions. The device measures the potential differences between a pH-sensing electrode and a reference electrode immersed in the same solution. Chapter 9, "pH Measurement," covers pH meters, electrodes, buffers, calibrations, and so on, in great detail. Many process pH-sensing electrodes are fabricated with a special glass tip that is very sensitive to pH changes. Some of the newer pH meters use an antimony electrode instead of glass. However, antimony electrodes cannot be used in strong oxidizing or reducing media. The reference electrode can be one of two main types: (1) a **calomel reference electrode** contains a mixture of elemental mercury, mercury salts, and KCl; and (2) an **Ag/AgCl reference electrode** (see Fig. 17.6). Both reference electrode types produce a constant voltage and are independent of the pH of the solution. Most of the newer pH meters use electrodes that have been "combined" into one probe. Because pH determinations are sensitive to temperature changes, the pH meter requires some temperature correction for accuracy.

Figure 17.7 shows a picture of a new retractable pH sensor. This particular pH sensor can be housed in 316 SS and can be equipped with corrosion-resistant Hastelloy and titanium metal fittings. This model has a temperature compensation element located at the tip with the pH probe itself. A T-style handle that aids in the insertion and retraction into pressurized process environments is available.

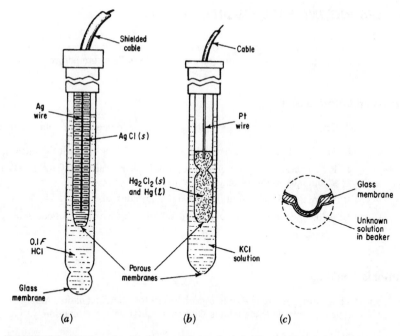

FIGURE 17.6 The electrodes of a pH meter. *(a)* glass electrode, *(b)* calomel electrode, and *(c)* enlarged detail of glass electrode.

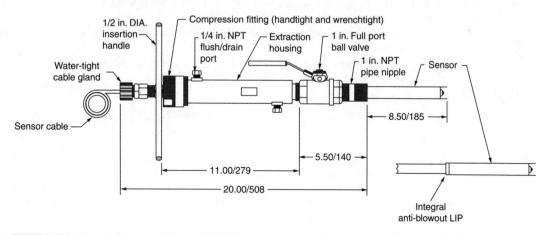

FIGURE 17.7 Retractable process pH sensor. *(TBX587 pH image courtesy of ABB Inc., www.abb.com/instrumentation.)*

All pH meters must be calibrated by placing the electrodes into a buffer solution of known pH. A **buffer solution** is any solution that resists changes in pH. The pH meter scale is adjusted (calibrated) to correspond to this known pH value. Standard buffer solutions at various pH values are available for pH meter calibration. For more accuracy, the pH meter must be calibrated with a standard buffer that has a pH value in the same range of the process stream being tested. Because pH readings are temperature-dependent, all pH calibrations should be corrected to the temperature of the process stream. Consult the Buffer Standardization of the pH Meter section in Chapter 9, "pH Measurement," for calibration details.

TROUBLESHOOTING PROCESS pH ELECTRODES

Restoring or rejuvenating pH electrode(s) can be done in several ways depending on the type of electrodes involved. Consult the manufacturer's literature before proceeding.

A. Potassium Chloride Soaking

A simple procedure calls for soaking the electrode in a warm, saturated potassium chloride solution:

1. Fill the reference electrode with its filling solution (consult the manufacturer's literature).
2. Prepare a soaking solution in a 100-mL beaker by adding 10 mL of a saturated potassium chloride solution to 90 mL of deionized water.
3. Warm the soaking solution until tepid; do not boil.
4. Allow the electrode to soak at a depth of 2 in for at least 3 hours.
5. Test the electrode.

B. Ammonia Soaking

This corrective procedure is especially useful for rejuvenating Ag/AgCl reference electrodes. This type of electrode tends to clog with silver chloride at the junction.

1. Empty the filling solution from the electrode.
2. Immerse the electrode in concentrated ammonia solution, but do not put ammonia inside the empty electrode. Ammonia will dissolve the silver chloride and permanently damage the silver/silver chloride reference.
3. Soak the electrode in the ammonia solution for 20 minutes; do not exceed this soaking period. Do not soak overnight.
4. Thoroughly rinse both the inside and outside of the empty electrode with deionized water. Empty the electrode and replace the filling solution.
5. Test the electrode.

C. Application of Vacuum

A water aspirator or vacuum pump can be used to dislodge obstructions from the junction of a reference electrode. Flexible tubing can be used to cover the end of the reference electrode and sufficient vacuum can be applied to draw some filling solution through the junction. Caution should be exercised in order not to apply too much vacuum and dislodge the reference electrode junction itself.

FLOW RATE MEASUREMENTS

There are many process flow measurements devices, with one of the simplest being the variable area flow meters, turbine flow meters, weirs, Coriolis mass flow meters, and so on. A **variable area flow meter,** as shown in Fig. 17.8, typically operates at constant pressure and can be used to measure the flow of gases, liquids, and even steam. The position of the float varies as the fluid flow varies. The gravity-operated type shown in Fig. 17.8 is sometimes called a **rotameter**. For better accuracy, the float is typically designed to spin or "rotate" under the influence of a fluid stream. An equilibrium is reached between the force of flowing fluids and the force of gravity. The numerical

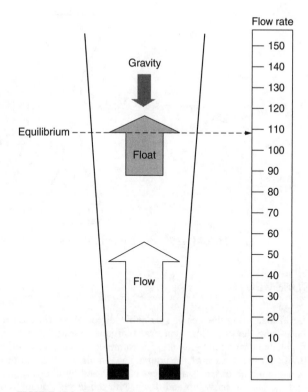

FIGURE 17.8 Variable area flow meter (rotameter).

scale must be calibrated using known gases, temperatures, and specific weights of floats. Other flow meter designs use impellers, rotating valves, differential pressure, ultrasonic measurements, and so on.

Flow transmitters use Bernoulli's principle and differential pressures in the process stream. As discussed in Chapters 5 and 18, Bernoulli's principle states that as the velocity of a fluid increases, the lateral pressure to the flow is decreased, resulting in the formation of a partial vacuum by the rapidly moving fluid. With the placement of a flow-restrictive device (orifice plates, flow nozzles, or venture tubes) in a process stream, the flow slows down and the pressure *increases* in the inlet side of the restriction. The fluid velocity through the restriction increases; thus the pressure on the outlet side of the restriction *decreases*. The difference between the inlet side pressure and outlet side pressure is measured by a **differential pressure (DP)** cell. Typically, the DP cell is connected to a transmitter that signals a flow controller.

Coriolis mass flow sensors have been commercially available for over 30 years. This mass flow rate–measuring device is based on the **Coriolis effect**. When fluid is moving through the sensor's tubes, forces are induced (see Fig. 17.9). These forces cause the flow tubes to twist in opposition to each other. This same twisting phenomena can be observed in a free garden hose that is allowed to flow rapidly. Mathematically, it happens that the mass flow rate is directly proportional to the twist angle and inversely proportional to the resonance frequency. The French engineer Gaspard Gustave de Coriolis (1792–1843) also discovered that the density of the fluid is inversely proportional to the square of the resonance frequency. The newest Coriolis meters now use resonating silicon microtubes and on-chip temperature sensors to measure mass flow rates, dose, chemical concentration, fluid density, and temperature. As shown in Fig. 17.9, a Coriolis force with flowing fluid on the lower tube bends the tube down. The Coriolis force on the upper tube with fluid flowing bends the tube up. As a result, the tubes twist and an optical device measures the magnitude of the tube twisting. The twisting is a function of the mass flow rate through the Coriolis meter.

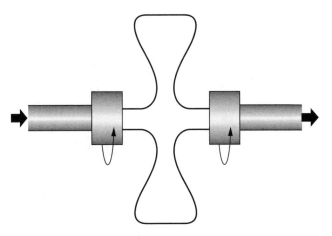

FIGURE 17.9 Coriolis mass flow sensor tube.

LEVEL MEASUREMENTS

Level measuring devices can be either direct or indirect. Basically, there are four types of level measuring devices: float type, radio frequency capacitance, ultrasonic/radar/microwave time of flight, and differential pressures measurements. Direct level measuring devices are in physical contract with the surface of the liquid. The simple direct level measuring instrument is a sight glass. A combination of floats and tapes can be used by allowing the float to rest on the liquid surface and the tape to move up and down with the level. Conductivity probes are also used to contact the liquid directly and indicate levels; such probes are especially useful with low- and high-level warning capabilities. Indirect level measuring devices rely on converting measurements of pressure changes, reflected light sources, radar, and/or sound waves, and so on. Differential pressure (DP) cells are routinely used as indirect level measuring devices. DP cells can be used to measure the hydrostatic pressure difference between two selected points on a pressurized vessel and converted to a level indicator.

MOISTURE MEASUREMENTS

Moisture measurements are normally found in two categories: absolute moisture methods and relative humidity methods. Absolute moisture methods can typically be directly calibrated, as in weight concentrations (for example, the Karl Fischer titration method), molar concentrations, or even in terms of dew-point temperatures. Loss in weight on heating (for example, moisture balances) is the most widely used example of an absolute method. The relative humidity methods are those calibrated in terms of percentage of saturation of moisture. The **dew-point method** is based on the fact that the relationship between water content and condensation points is well known. The dew-point method uses a visible, inert surface where the temperature can be well controlled and measured. By passing an unknown gas stream over the temperature-controlled surface, the onset of condensation can be observed and recorded. The original dew-point methods used visual observation, but now utilize electronic optical detection or thermocouple/thermistor temperature measurements. See Chapter 4 "Relative Humidity" section for additional dew-point calculation information. One of latest moisture-determining methods can now be done by using near-infrared spectroscopy (NIR). NIR simply bounces an infrared beam off of a sample and measures the change in energy intensity at a specific wavelength.

Various moisture analyzers utilize the high dielectric constant of water for determining its concentration. An alternating electric current is passed through a capacitor containing sample water between two fixed, capacitor plates.

REFERENCES

1. The Green Book, *Flow, Level, and Environmental Handbook and Encyclopedia*, 8th edition. Omega Engineering, Inc., P.O. Box 4047, Stamford, CT 06907–0047 (2009).
2. Enoksson, P., Stemme, G., and Stemme, E., "A Silicon Resonant Sensor Structure for Coriolis Mass-Flow Measurements." *J MEMS*. 6:119 (1997).
3. Miller, R. W., *Flow Measurement Engineering Handbook*. McGraw-Hill, Inc. (1996).
4. Spitzer, D. W., *Flow Measurements*. Instrument Society of America (1991).

CHAPTER 18
PLUMBING, VALVES, AND PUMPS

PLUMBING: TUBING VERSUS PIPING

Tubing has several advantages over conventional piping. First, tubing is generally easier to install, requiring only standard wrenches. In addition, tubing does not require threading, flaring, soldering, or welding. It is intrinsically safer to install because no torches, open flames, or sparks are associated with installation or maintenance. Compression fittings and tubing are inherently cleaner than piping because there are no fluxes, weld spatter, or need to boil out with acid/flushing cleaner. Tubing actually has a better strength-to-weight ratio than piping. It uses full wall thickness for containing pressure because no threading is necessary for the compression-type fittings. Because tubing weighs less than piping, it is less expensive to transport, easier to assemble, requires less support, and occupies less space. Tubing provides lower pressure drops because the bends are typically smoother and requires fewer sharp bends to design a system. The bending of tubing reduces the number of necessary connections and potential leak points (Fig. 18.1). In general, tubing systems connected by compression fittings tend to leak less than pipe connections.

GENERAL GUIDELINES FOR SELECTING TUBING*

1. The metal tubing must be constructed of a softer material than the fitting material. For example, brass fittings are not recommended for stainless steel tubing.
2. When the tubing and fittings are made of the same material, the tubing must be fully annealed.
3. Soft or pliable tubing (Tygon® or untreated polyvinyl chloride [PVC]) should always be used with an insert.
4. Tubing surface finish is vital for proper compression fitting seals. Do not use tubing with flat spots, visible weld seams, imperfections, or scratches.
5. Tubing that has been stretched or has lost its concentricity or ovality (roundness) should not be used.

Tubing can be easily cut, bent, and formed to meet any need. Secure connections that do not leak can be made, provided proper tools, fittings, and working procedures are used. Stainless steel is available as standard or degreased and passivated tubing. The degreased and passivated tubing has been thoroughly cleaned and degreased, treated with an acid solution for passivation, and then rinsed in distilled water and dried. Many new plastic materials are also available for high-pressure and chemical-resistant qualities.

---

*©2010 Swagelok Company.

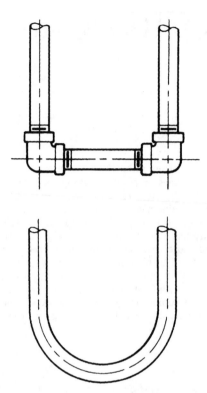

FIGURE 18.1 Pipe versus tubing bends. (©2010 Swagelok Company.)

METAL TUBING

Materials of construction are aluminum, brass, copper, carbon steel, and various alloys. Normally, copper tubing is used except when corrosion resistance and reactivity are of concern. Many different types of plastic tubes are available for use in a wide range of fluid applications. The most common types are nylon, polypropylene, Teflon® (TFE), and polypropylene.

It is suggested that Table 18.1 be used only to select possible materials for use. A more extensive investigation should be made of published material standards and solubility and/or compatibility considerations.

A. Cutting Tubing

Lay the metal tubing on a flat surface and slowly uncoil while holding the end flat against the surface (see Fig. 18.2). Do not uncoil more tubing than is necessary, because repeated uncoiling will distort, harden, and stiffen coiled tubing. Never attempt to straighten coiled copper tubing by stretching it. If tubing is stretched by no more than 1 percent (6 in. in 50 ft), an outside diameter reduction takes place and compression fitting sealing problems can occur.

The ends should be squared off at right angles to the tubing for secure connections. There are two methods to cut tubing:

- Use a tube cutter. Center the tubing between the two rollers and then rotate the cutter continuously around the tubing while very *slowly* applying pressure on the cutting wheel by twisting the knob screw (see Fig. 18.3).
- Use a hacksaw and guide blocks (see Fig. 18.4).

TABLE 18.1 Suggested Maximum Temperatures for Fittings, Tubing, and Seal Materials

Metals	°F	°C
Alloy 20	1200	649
Aluminum	400	204
Brass	400	204
Copper	400	204
Copper-Nickel (70/30)	700	371
Copper-Nickel (90/10)	600	316
Alloy B2	800	427
Alloy C-276	1000	537
Alloy 600	1200	649
Alloy 400	800	427
Nickel	600	316
Stainless Steel 304L	800	427
Stainless Steel 316-321-347	1200	649
Steel	375	190
Titanium	600	316
Plastics Delrin	195	91
Kel-F	200	93
Nylon	165	7
Perfluoroalkoxyl (PFA)	400	204
Polyethylene	140	60
Polypropylene	250	121
TFE	450	232
VESPEL	500	260
Elastomers Buna	255	124
Ethylene-Propylene	300	149
Kalrez	500	260
Neoprene	280	138
Silicone	450	232
Viton	400	204

Source: © 2010 Swagelok Company.

FIGURE 18.2 Uncoiling tubing.

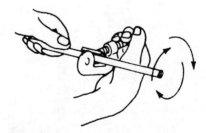

FIGURE 18.3 Cutting brass or copper tubing with a tube cutter.

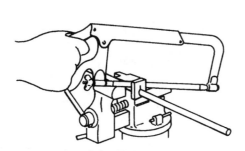

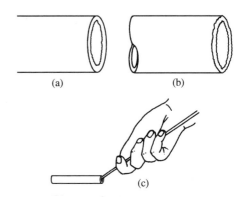

FIGURE 18.4 Hacksaw and guide blocks used to cut tubing.

FIGURE 18.5 (*a*) Burs formed inside tubing by action of the tube cutter. (*b*) Hacksaw burs form both inside and outside the tubing. (*c*) Removing inside burs with a file.

B. Removing Burs from Cut Tubing

When tubing is cut either with a tube cutter or with a hacksaw, burring of the tubing occurs. When a tube cutter is used, the burs are found inside (Fig. 18.5a); when a hacksaw is used, burs are found inside and outside (Fig. 18.5b). Burs can be easily removed by stroking the cut tubing gently with a flat file on the outside and with a rat-tailed file on the inside (Fig. 18.5c).

Another technique for deburring cut tubing utilizes a commercial deburring tool:

- To deburr outside diameter of tube, place the deburring tool over the end of the tube with blades on the inside. Rotate the tool back and forth, or in a clockwise direction.
- To deburr inside diameter, reverse the tool and insert blades inside the tube. Again, rotate the tool back and forth, or in a clockwise direction. Wipe the tube end clean with a cloth.

CAUTION: Do not place your fingers inside tool or near cutting edges.

BENDING TUBING

Tubing can be bent easily, and rounded bends without kinks can be obtained provided you do not attempt to bend the tubing too sharply. Bending to a radius of two or three times the tube diameter is possible; proper procedure must be used or difficulties such as flattening or wrinkling will occur, as shown in Fig. 18.6.

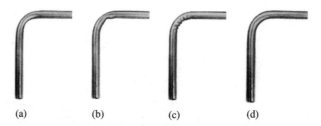

FIGURE 18.6 Types of tubing bends: (*a*) flattened bend (incorrect); (*b*) kinked bend (incorrect); (*c*) wrinkled bend (incorrect); (*d*) good 90° bend (correct). (©*2010 Swagelok Company.*)

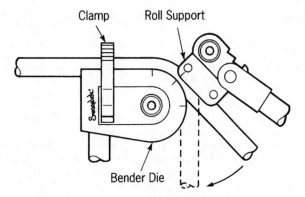

FIGURE 18.7 Tube benders. (©*2010 Swagelok Company.*)

Hand bending is accomplished by placing the thumbs toward each other, but far enough apart to allow length for making the bend. Hand bending to a small radius is very difficult, as flattening usually occurs. The knee can be used to form and bend tubing in small increments by shifting the knee to various points on the inside of the bend. Usually better bends can be produced by using a compression-type tube bend, as shown in Fig. 18.7.

A. Bends Near Tube Fittings

Bends located near fittings may be a source of leaks. Always leave a length of straight tube so that the deformed section at the bend does not enter the fitting. Figure 18.8 shows schematic and dimensions for a bend near a tube fitting.

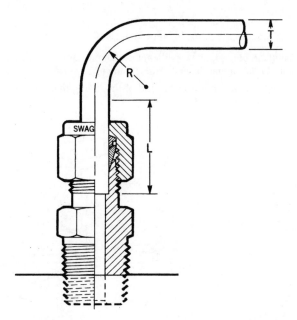

FIGURE 18.8 Tube bend near tube fitting. (©*2010 Swagelok Company.*)

TABLE 18.2 Recommended Dimensions (Metric) for Making Tubing Bends near Fittings

Tube O.D. "T" (mm)	3	6	8	10	12	14	16	18	20	22	25	32	38
Length* "L" (mm)	19	21	22	25	29	31	32	32	33	33	40	51	60
Length** "L" (mm)	16	17	18	20	24	25	25	25	26	27	33	47	55

*Recommended straight tube length.
**Absolute minimum straight tube length.

Table 18.2 shows the recommended bend radius and the recommended tube insert lengths (expressed in English units) for making tubing bends near fittings. "R" represents the radius of tubing bend as required or the minimum allowed for specified wall thickness and tube size as recommended by bender manufacturer. "L" represents the tube length required from the end of tube to beginning of bend. "T" represents tube outside the diameter.

CONNECTING TUBING MECHANICALLY

Many plant procedures require the use of soft copper tubing securely connected to gas cylinders, gas chromatography units, pressurized containers, cooling and heating containers, and so on. Copper tubing can easily be connected to itself or to any unit that has a standard pipe-thread fitting, and there are all conceivable types of connections and interconnections available. It requires only thought and ingenuity to make the connection.

The preparation of the copper tubing end is most important. The cut must be squared off, must be cleaned of inside and outside burs, and must not be deformed in any way. Once the edge has been prepared, the fitting connection is made. There are two basic types of connectors with which tubing can be mechanically connected together without the use of soldering equipment: (1) flared fittings require flaring of the tubing; (2) compression fitting do not. Both types provide secure connections that do not leak if the proper procedures are followed.

A. Flared Connections

Flaring requires the following procedure:

1. Cut the tubing to the proper lengths, making sure the ends are squared off and free of burs. Care must be taken that the end of the tubing to be connected is round and has not been deformed.

2. Slide the flange nut over the end of the tubing in such a way that the open end with the screw faces the connector (Fig. 18.9).

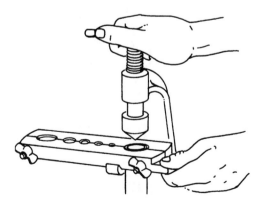

FIGURE 18.9 Flaring procedure.

3. Place the tubing in the flaring holder of the vise grip and tighten securely so that the tubing will not slip. The open end should be flush with the face of the holder having the countersunk recesses.
4. Slide the flaring tool over the holder with the flaring screw facing the open end of the tubing.
5. Screw down the flaring tool knob tightly, flaring the tubing.
6. Remove the flaring tool and tubing holder.

B. Connecting Flared Tubing

After the tubing has been flared, make the connection as shown in Fig. 18.10. The procedure is as follows:

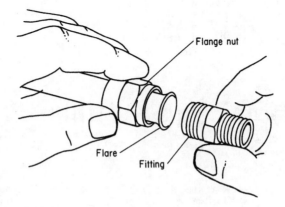

FIGURE 18.10 Assembling the flared tube and its fittings.

1. Select the desired connector.
2. Wipe clean the connector and flared tubing end.
3. Push the tubing and connector together and slide the flange nut to engage the connector fitting.
4. Hand-tighten the flange nut as much as possible; it should screw on easily.
5. Select appropriate end wrenches—one for the connector, the other for the flange nut.
6. Tighten the flange nut securely by using both wrenches in a counterclockwise motion to each other.

CAUTION: Do not exert excessive force.

COMPRESSION FITTINGS

Compression fitting (flareless-type connectors) does not require that the tubing be flared because they have a self-contained method (an expanding ferrule or ferrules) for providing leak-proof, all-metal connections. Figure 18.11 shows many of the commercially available tube fittings.

Compression fitting typically uses a nut, front ferrule, back ferrule, and body combination as shown in Fig. 18.12.

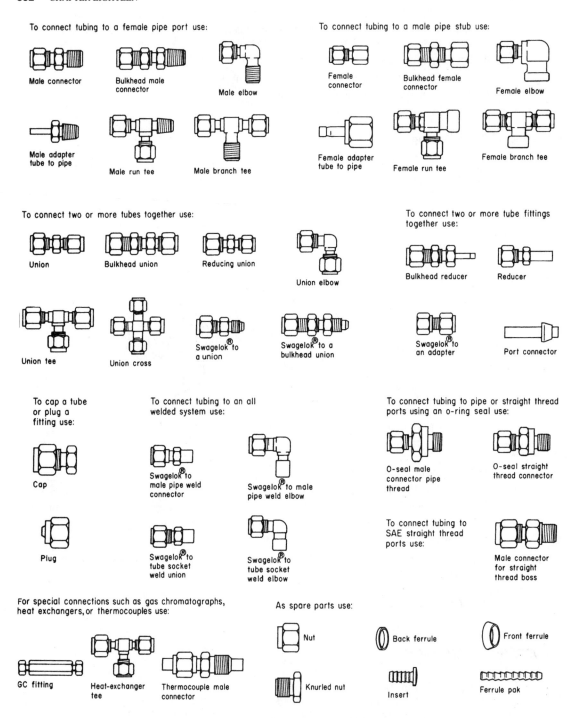

FIGURE 18.11 Tube fittings, available in all machinable materials. (©*2010 Swagelok Company.*)

PLUMBING, VALVES, AND PUMPS **353**

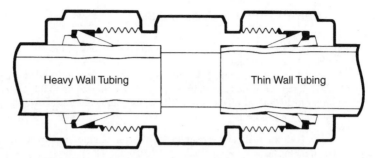

FIGURE 18.12 Compression fitting drawing. (©*2010 Swagelok Company.*)

The tubing must be prepared with the ends clean, free from burs, and circular, not deformed. To make the compression connections, follow these steps:

1. Swagelok compression fittings are shipped completely assembled, finger-tight, and ready for immediate use. Disassembly before use is not recommended because dirt and foreign materials can contaminate the fittings.
2. Tubing is inserted into the completely assembled fitting until it bottoms against the shoulder of the fitting. Before tightening the compression fitting, draw a line on the nut at the six o'clock position (see Fig. 18.13).
3. Holding the fitting body steady with a backup wrench, tighten the nut 1¼ turns from finger-tight. When properly tightened, the wrench should be at the nine o'clock position.

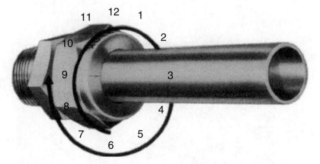

FIGURE 18.13 Compression fitting tightening sequence. (©*2010 Swagelok Company.*)

Tubing is available in various sizes up to 2 in. in diameter. All tubing of 1¼-in or greater diameter requires a **hydraulic swaging unit (HSU)**. Any steel or stainless steel Swagelok tube fitting must be made up using the HSU, never just wrenches. Torques are too high and tube variables make wrench pull-up unreliable.

CAUTION: Too much pressure will distort the fitting and cause breakage and leakage. See the manufacturer's specifications for details.

APPLYING PIPE THREAD TAPE OR SEALANTS

Pipe threads are still the most common end connections in the industry. Pipe threads *always* need a sealant. Tape should *never* be used on flared, coned, or tube compression fitting. Teflon tape that is wrapped around the male pipe threads is normally recommended for temperatures up to 450°F (232°C). Use ¼-in-wide tape on male tapered pipe threads up to 3/8-in diameter and ½-in-wide tape on larger male tapered pipe threads. The proper application is shown in Fig. 18.14 and is described in the following steps:

1. Clean both the male and female tapered threads, removing any dirt or previously used antiseize (antibinding) tape or compound.
2. Beginning at the first thread, wrap tape in the direction of the male tapered thread spiral and join with a slight overlap. Two wraps are suggested for stainless steel tapered pipe threads.
3. Make sure that the tape does not overhang the first thread, as the tape could shred and get into the fluid system.
4. Cut off any excess tape. Draw the free end of the tape around the threads tautly so that it conforms to the threads. Press in firmly at the overlay point. The connection is now ready.

Several different pipe sealants or "dopes" are available. The primary purposes of a sealant are to lubricate and seal the assembly. Sealants can also prevent galling and seizing of pipe threads. Anaerobic pipe thread sealants are commercially available with pressure ratings up to 10,000 psig. Many contain Teflon in a semiliquid paste in a squeezable plastic tube. These sealants can be used from approximately −65°F (−53°C) to 350°F (180°C). These anaerobic pastes seal the threads when the air is excluded. All gas systems require a 24-hour curing period for most of these liquid sealants. It should be noted that this type of sealant is not recommended for certain applications: do not use it on any plastic pipe except Teflon, and do not use it with halogens, oxygen, ozone, hydrazine, or nitrogen dioxide. Do not use such sealant with high concentrations of acids or bases. It also is not recommended for some food, cosmetic, drug, and/or water applications. Always consult the manufacturer for a compatibility data sheet before using these liquid sealants.

 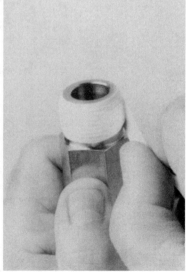

FIGURE 18.14 Applying Teflon tape. (*©2010 Swagelok Company.*)

PIPES AND FITTINGS

Galvanized pipe assemblies can be constructed by cutting the pipe to size, threading the ends, and then using any of the pipe fittings (Fig. 18.15) to make the required assembly. The fittings are available in standard pipe sizes starting with 1/16-in ID. MPT pipe thread is *tapered*. It must be used with fittings that have the same thread. It cannot be used with parallel threaded machine fittings.

Pipes and fittings are connected by welded, threaded, flanged, flared, compressed, or sweated joints, as shown in Fig. 18.16. In general, welded and flanged joints are used for iron pipes

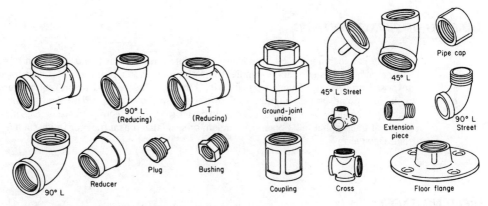

FIGURE 18.15 Typical galvanized iron pipe fittings.

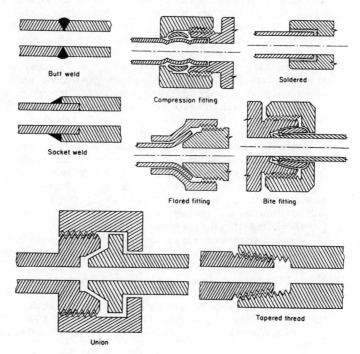

FIGURE 18.16 Commonly used pipe joints.

of large diameter, usually 2 in or larger, whereas threaded joints are used for small diameters. Flanged, flared, compressed, or sweated joints are used for small-diameter copper and brass tubing.

Steel pipes once were classified into three categories according to their wall thickness as standard, extrastrong, and double extrastrong. Pipes are now classified into 10 categories or schedule numbers according to wall thickness by the **American Standards Association**. The approximate diameter of a pipe is called the **nominal** diameter. For example, a Schedule 40 nominal 2-in diameter steel pipe has an actual external diameter of 2.375 in and an internal diameter of 2.067 in.

TRACING OF PIPING SYSTEMS

Tracing is a technique used in industry to maintain tubes, pipes, and vessels in a process system at temperatures usually above ambient temperatures. Water pipes must be heat-traced to prevent exposed process equipment from freezing. Heat tracing is used to prevent process gases from condensing and process liquids from solidifying in the system. Many fluid streams require heat tracing to reduce viscosity and make them easier to pump. Electrical resistance heating can be used for heat tracing, but most plants use steam, or specialized heat transfer fluids such as Therminol® and Dowtherm®.

WELDING SAFETY

The Voluntary Industry Standards for Chemical Process Industries Technical Workers does not list welding as a specific competency for process technicians. However, process technicians are certainly involved in process maintenance. A core competence was identified for these individuals to work directly with maintenance personnel in tasks ranging from scheduled maintenance to plant turnaround activities. There are approximately 600,000 welders and flame cutters currently employed in the United States, so it is very likely that a process technician would assist a maintenance technician in welding tasks. The American Welding Society (AWS) lists more than 100 different welding processes. One of the more common welding processes is arc welding, which uses an electric arc struck between two carbon electrodes or two tungsten electrodes or between a single electrode and the metal itself. The three main types of arc welding are shielded metal-arc welding, gas **tungsten-arc welding** (TIG), and gas **metal-arc welding** (MIG).

Welding can create many safety and personal safety hazards for the maintenance and process operators. These arcs and flames produce potentially toxic gases, fumes, particles, and ultraviolet light. The most common toxic gases and fumes are ozone, oxides of nitrogen, carbon monoxide, carbon dioxide, and metal vapors. As with any workplace assignment, consult the manufacturer's instructions and the appropriate material safety data sheets (MSDS) to determine the potential hazards and determine the permissible exposure limits (PEL) for the materials involved. A checklist of general welding safety rules is provided in Table 18.3.

Eye and face protection is critical in welding and cutting operations. Welding helmets, face shields, goggles, and safety glasses are available to protect the operator's eyes from heat, intense light, sparks, and spattered metal. However, *all* employees exposed to this type of radiation should have eye protection available. Filtered lenses are used to reduce harmful radiation that can burn an exposed individual's eyes and lead to blindness. As a general rule, operators should go to the lightest shade that gives a good view of the welding procedure without going below the recommended minimum shade. OSHA regulation 29 CFR 1910.252 (b)(2)(ii)(H) provides a guide for selecting the proper shade number with various welding operations and is summarized in Table 2.3.

TABLE 18.3 Checklist (√) General Welding Safety Rules

- ☐ 1. Read section on "Respirators" in Chapter 2.
- ☐ 2. Review section on "Fire Safety" in Chapter 2. Fire extinguishers must be present during all welding, cutting, and brazing operations.
- ☐ 3. Review section on "Aerial Lifts" in Chapter 2.
- ☐ 4. Review section on "Handling Compressed Gas Cylinders" in Chapter 4.
- ☐ 5. Review section on "Working in Confined Spaces" in Chapter 2.
- ☐ 6. Review section on "Personal Protective Equipment (PPE)" in Chapter 2. In addition, always keep outer clothing as free of oil and grease as possible, especially when using oxygen during welding. Wear flame-resistant clothing that fully covers the arms and legs. Woolen clothing is not as ignitable as cotton clothing. Special flame-resistant gloves, leggings, aprons, shoes, and boots are available for welders. Clothing should always be buttoned at the neck to prevent hot metal from getting inside.
- ☐ 7. Some welding processes can generate high levels of noise. Air carbon arcs and plasma arc cuttings can damage hearing. Always wear proper hearing protection. Always wear flame-resistant ear plugs and a hat during a welding or cutting procedure to protect the ears and head from burns.
- ☐ 8. Always wear the proper eye protection as shown in Table 2.1 and post signs to designate the welding area.
- ☐ 9. Before welding or cutting, perform an atmospheric test for flammable substances or substances that might produce toxic vapors if exposed to excessive heat.
- ☐ 10. Empty all pockets of ignitable or flammable materials before welding. Butane cigarette lighters are especially dangerous around welding and cutting operations.
- ☐ 11. Always blank or disconnect any connections to a drum or vessel before welding and monitor these vessels or connections for possible flammable vapors.
- ☐ 12. Housekeeping is critical while welding. Do not block escape routes and keep areas free of tripping hazards.
- ☐ 13. Employees in areas adjacent to welding zones must be provided with eye and face protection and flameproof screens if necessary.
- ☐ 14. Check all electrical and compressed gas welding equipment before use for defective or worn parts.
- ☐ 15. Remove all arc electrodes from their holders and disconnect the welder from the power source when not in use.
- ☐ 16. Always shut off the fuel and oxygen gases at both the torch valves and compressed gas cylinders.
- ☐ 17. Fluxes can be very corrosive to human tissue: avoid any contact and wash effective area with soap and water immediately.
- ☐ 18. Hot welded areas should be designated as such to prevent coworkers from accidentally coming into contact with these areas.

A. Hot Work Permits

Hot work is defined as any job activity that produces or uses flames, heat, or sparks that could act as a source of ignition. Welding certainly qualifies as a hot work activity. A hot work permit (Table 18.4) ensures that all potential fire hazards have been controlled and/or identified and that the welding will occur in a fire safe area. Remember, welding, cutting, and brazing operations should not be allowed in areas not authorized by hot work permits. In addition, hot work permits should not be allowed for areas with potentially explosive atmospheres, such as improperly cleaned tanks or equipment, and areas near large quantities of exposed, readily ignitable materials.

TABLE 18.4 Conditions That Hot Works Permits Must Identify

- (√) The location of hot work.
- (√) The materials being welded.
- (√) Dates that permit the authorized work.
- (√) Authorizing signature.

358 CHAPTER EIGHTEEN

B. Fire Prevention from Welding Activities

It is a good safety practice to employ **fire watchers** (maintenance technicians or process operators) during all welding and cutting operations. These trained individuals are responsible for watching for potential fires. Table 18.5 presents a checklist for fire watchers.

TABLE 18.5 Checklist (√) for Fire Watchers

☐ 1. Fire watchers must be trained in the use and location of all fire extinguishing equipment.
☐ 2. Fire watchers must be trained in sounding fire alarms and evacuation procedures.
☐ 3. A fire watch should be maintained for at least 30 minutes after completion of welding or cutting operations to detect and extinguish possible smoldering fires.

A fire watch is **mandatory** whenever any of the following conditions exists:

☐ 4. Appreciable combustible materials, in building construction or contents, closer than 35 ft to the point of operation.
☐ 5. Appreciable combustibles are more than 35 ft away but are easily ignited by sparks.
☐ 6. Wall or floor openings within a 35 ft radius expose a combustible material in adjacent areas including concealed spaces in walls or floors.
☐ 7. Combustible materials are adjacent to opposite sides of metal partitions, walls, ceilings, or roofs and are likely to be ignited by conduction or radiation.

PLUMBING SOLDERING

Most soldering tasks in a process operation would be assigned to specialized maintenance technicians (electricians, plumbers, and electronics technicians) except in perhaps some smaller or non-union plants. When permanent tubing installations are desired, the tubing can be connected to standard fittings (Fig. 18.17) by soldering.

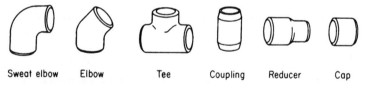

Sweat elbow   Elbow   Tee   Coupling   Reducer   Cap

FIGURE 18.17 Typical copper-pipe fittings to be used with solder.

The normal soldering procedure is as follows:

1. All surfaces should be cleaned to remove oxides and then lightly scraped with sandpaper to roughen the surface.
2. Use a torch to heat the fitting or the tube enough so that the solder will melt and flow when solder wire is touched to the joint. Do *not* heat the solder.
3. Allow the melted solder to flow evenly throughout the heated surfaces to be connected while maintaining secure contact with sufficient pressure.
4. Remove the flame and wipe away excess solder with a towel while maintaining enough force to keep components together.
5. Allow soldered connections to cool. Then test for leaks.

TABLE 18.6 Common Solder Alloys

Identification	% Composition	mp, °F
Woods metal	12.5 Sn, 25% Pb, 50% Bi, 12.5% Cd	165
Eutectic, T-L	63% Sn, 37% Pb	361
ASTM 60A	60% Sn, 40% Pb	370
ASTM 50A	50% Sn, 50% Pb	417
Eutectic, T-S	96.5% Sn, 3.5% Ag	430
ASTM 40A	40% Sn, 60% Pb	460

Soldering is a form of welding that joins materials below their melting points with a filler metal (see Table 18.6) having a melting point below 450°C (840°F).

The filler metal (solder) is drawn between closely fitted surfaces by capillary action. Soldering can be performed by dipping, using electric soldering irons, or using a torch. Soft solders are low-melting alloys (usually below 200°C), usually tin (Sn) and lead (Pb). The Environmental Protection Agency and other governmental agencies have restricted the use of these solders, which contain potentially toxic heavy metals. Hard solder usually requires temperatures of between 600 and 800°C.

FLUID FLOW

To understand fluid flow, you must first understand the definition of fluids. Typically, fluids are considered to be liquids, such as water, oil, blood, and so forth, but a **fluid** is more accurately defined as any material that can take the shape of its container. This means that gases such as air, propane, refrigerants, and so forth are classified as fluids the same as liquids. Another definition of a fluid is a substance that changes shape when a sheer stress is applied, no matter how small. Solids retain a fixed shape until a threshold force is reached to change the solid shape. Fluids have no such threshold and will change shape if any sheer force is applied. Fluids are useful for many applications and make life possible. Water is used as a coolant in cars and many industrial processes that require cooling. However, the fluid in this case works as part of an entire system where heat is added to the fluid in one part of the journey, then taken away in another part. Fluids by themselves are not terribly interesting, but fluids often act as a critical piece in an engineering system. The study of fluids used in such a system is referred to as **fluid mechanics**, where the physical process of fluid flowing through a system is analyzed and understood. At first glance, this would appear to be a straightforward study; however, many situations arise that complicate the fluid flow process and require a deeper understanding of the mechanics of fluid flow. Fluids in a system are often called **working fluids**, implying that they actually do some "work." You don't often think of a fluid being pumped through a system as doing work; however, consider the steam engine used in the early days of the industrial revolution. The working fluid is water that undergoes a phase transformation to become a gas via boiling. This increases the pressure that drives a piston.

A common way of characterizing a flowing fluid is by determining whether the flow is laminar or turbulent. **Laminar flow** can be described as a well-ordered flow pattern, sort of like cars on a highway traveling in their own lanes. **Turbulent flow**, by contrast, appears to have no order (see Fig. 18.18).

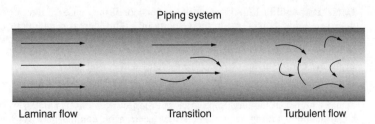

FIGURE 18.18 Graphic display of laminar flow to turbulent flow.

The transition from laminar to turbulent flow is difficult to predict and can often occur very suddenly. Many studies have been performed in an attempt to quantify this transition. Osborne Reynolds (1842–1912) performed most of the ground-breaking work on the transition between laminar and turbulent flows and developed a dimensionless parameter used to classify the flow condition. This parameter is called the **Reynolds number (Re)** and is defined by the equation:

$$\mathbf{Re} = \frac{DV\rho}{\mu}$$

where: D = diameter of pipe, m or ft
 V = average velocity of flowing fluid, m/s or ft/s
 ρ = fluid density, kg/m³ or lb/ft³
 μ = viscosity of fluid, kg/m·s or lb/ft·s

For fluid flow in pipes, laminar flow exists for Reynolds numbers less than **Re** = 2300. Turbulent flow is possible for flows greater than 2300, but not necessary. Turbulent flow is often induced by the roughness on the inside of pipes or flow conduit. For very smooth pipes, laminar flow can exist for Reynolds numbers as high as 40,000. The surface roughness is primarily what slows a fluid down as it is being pumped through a system.

EXAMPLE 18.1 What is the Reynolds number for water flowing through a 0.30-in ID pipe at 6 ft/s at room temperature, given that the density of water is 62.4 lb/ft³ and the viscosity of water is 1 centipoise (cP)?

NOTE: 1 cP = 0.672 × 10⁻³ lb/ft·s, thus μ = 0.672 × 10⁻³ lb/ft·s.

$$D = 0.30 \text{ in} \times \frac{1 \text{ ft}}{12 \text{ in}} = 0.025 \text{ ft}$$

$$\mathbf{Re} = \frac{DV\rho}{\mu} = \frac{0.025 \text{ ft} \times 6 \text{ ft/s} \times 62.4 \text{ lb/ft}^3}{0.672 \times 10^{-3} \text{ lb/ft·s}}$$

$$\mathbf{Re} = 14000 \ (\textit{turbulent})$$

Characterizing the type of flow (whether it is turbulent or laminar) and understanding the phase of the fluid (whether its gas, liquid, or mixed), all help in understanding the system. From sophisticated engineered systems to simple piping networks, an understanding of the fluid inside the system is important.

BERNOULLI'S PRINCIPLE

Daniel Bernoulli in 1738 introduced the concept of the conservation of energy for fluid flows. He discovered that an increase in the velocity of a fluid increases its kinetic energy while decreasing its static energy. A simple example of this energy balance is demonstrated by discharging a can of compressed air and noticing that the can walls get colder. **Bernoulli's principle** is derived by using the conservation of energy principles and relates internal pressure to speed of a moving fluid. Suppose you have a large pipe attached to a smaller pipe with a liquid flowing through the system, a common occurrence in fluid mechanics systems (see Fig. 18.19). The pressure is actually higher in the large pipe (V_1), but the flow velocity is smaller (V_2) when compared to the small-diameter piping section. This may seem a bit counterintuitive, but in the small pipe, the

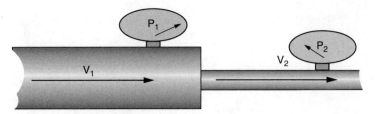

FIGURE 18.19 Bernoulli's principle.

fluid is moving faster, giving it larger kinetic energy and less potential energy (pressure). The large pipe, in contrast, has more potential energy (pressure) and less kinetic energy, as evidence by the slower flow speed.

PROCESS VALVES

Valves typically perform one of four basic functions: (isolation, regulation, direction, and protection. Gas valves, smaller-volume liquid valves, and plumbing issues are covered extensively in Chapter 4, "Handling Compressed Gases," and in this chapter. There are several different categories of values used in industrial applications. Process valves are primarily used for regulating and/or shutting off the flow of fluids. The most common process valve is the gate valve. A **gate valve** works exactly like a gate to the fluid flow (see Fig. 18.20). When the valve is fully opened, the fluid flows straight through the opening. This straight-through design causes very little friction or pressure drop within the valve. As the valve is closed, the gate moves perpendicularly across the flowing fluid until it meets the bottom or seat of the valve. The fluid is now completely blocked by the gate. This closing gate does not provide an accurate regulation of fluid flow but is primarily used in the fully open or fully closed position. Gate valves are not recommended to be used to throttle flows.

By design, a **globe valve** causes the flowing fluid to change its direction abruptly while flowing through the valve seating mechanism (see Fig. 18.17). This design creates much greater friction than the gate value and an appreciable pressure drop; however, the rate of flow can be better regulated. There are four common plug-type discs used in globe valves: plug, needle, ball, and composition. Figure 18.21 shows a plug disc being used to throttle or restrict the flow.

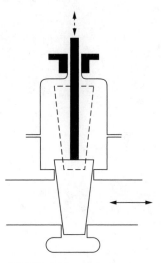

FIGURE 18.20 Typical gate valve. (*Courtesy* Handbook for Chemical Technicians, *Strauss and Kaufman,* ©*The McGraw-Hill Companies.*)

Unlike the gate or globe valve, the ball valve does not lift the flow controlling or throttling mechanism out of the process stream. A sphere (ball) with an opening bored through it is rotated, as shown in Fig. 18.22. The ball valve offers very little restriction to flow and can be fully opened or closed with a simple 90° turn. In the open position, the bored opening lines up perfectly with the inner diameter of the flow process pipe. Because the seats are typically made of plastic-coated materials, ball valves are not recommended for high-temperature or -pressure operations. Some ball valves are designed with multiple ports, which allow an operator to switch process flows quickly and still maintain the net positive suction head (NPSH, described in the next section), on the pump. Figure 18.23 shows a ball valve equipped with additional safety capabilities.

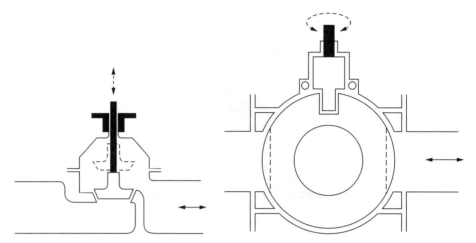

FIGURE 18.21 Typical globe valve. (*Courtesy* Handbook for Chemical Technicians, *Strauss and Kaufman,* ©*The McGraw-Hill Companies.*)

FIGURE 18.22 Typical ball valve. (*Courtesy* Handbook for Chemical Technicians, *Strauss and Kaufman,* ©*The McGraw-Hill Companies.*)

FIGURE 18.23 Process ball valve equipped with flanges and latch and lock mechanism. (©*2010 Swagelok Company.*)

Valves come in the same schedules as the fittings and the pipes with which they are to be used, and the nominal size of the fitting is the same as the nominal size of the pipe it fits. Current safety regulations require that valves be "locked out" during system maintenance (see the lockout/tagout section in Chapter 2, "Chemical Plant and Laboratory Safety," for more details).

Metering valves are designed to regulate (see Chapter 4 "Handling Compressed Gases") or adjust the rate of flow. They usually have a long, finely tapered stem tip and a fine-pitch stem thread

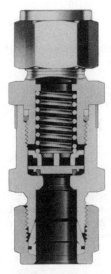

FIGURE 18.24 Check valve. (© 2010 Swagelok Company.)

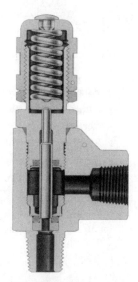

FIGURE 18.25 Safety relief valve. (©2010 Swagelok Company.)

and open in 5 to 10 turns. Valves of this type are not recommended for isolation service and are generally designed so they cannot be shut off.

Direction (check) valves are designed to have three or more connections, allowing flow to be directed toward two or more systems. **Check valves** (Fig. 18.24) are a type of direction valve because they are held closed by a spring. These valves open automatically when the inlet pressure exceeds the outlet pressure by enough to overcome the spring force. When the flow stops, the check valve closes to prevent flow in the reverse direction. Other more conventional check valve designs include the swing check, lift check, ball check, and stop check mechanisms, all of which are designed to close when the process flow is reversed. See Fig. 18.20 for examples of some typical check valve designs.

Relief valves protect a fluid system from excessive pressure. Pressure increase due to uncontrolled reactions or unexpected surges of pressure can be relieved by means of a **safety relief valve** (Fig. 18.25) installed in the gas line or cylinder. The valve is held closed by a spring as the system operates at its normal (designated) pressure at 150°F. When the pressure increases to a set point of the valve, it opens automatically and remains open until the system pressure decreases to below the set point. A **frangible disc** also serves the purpose of a safety relief valve, but cannot be reset. A metal disc is installed to burst at a predetermined pressure and to release the entire contents of the cylinder. A third type of relief assembly is the fusible plug. **Fusible plugs** do not operate on pressure but on temperature changes. These plugs are normally designed to melt at 212°F for acetylene tanks and generally around 165°F for other gases. Not all gas cylinders are equipped with safety valves. Certain compressed gases such as fluorine and the Class A poisons are examples.

FLASH ARRESTORS

A safety device required on all flammable gas and oxygen lines is a **flash arrestor**. This inline safety device prevents flashbacks in fuel gas and oxygen systems. Flash arrestors have built-in check valves to stop gas flow-back into the cylinder. Most of these flash arrestors are designed to absorb excess heat and minimize ignition sources. In addition, many are designed to produce shock waves under flashback conditions and automatically activate a safety shutoff valve.

PUMPS

According to the American National Standards Institute (ANSI) pump standards, there are six major types of pumps:

- **Kinetic pumps:** centrifugal, regenerative turbine, and special effects
- **Rotary pumps:** vane, piston, flexible member, lobe gear, circumferential piston, and screw
- **Vertical pumps**
- **Sealless centrifugal pumps**
- **Reciprocating power pumps**
- **Direct action (steam) pumps**

Pumps are primarily used in the process industry to transfer fluids or to increase pressure. Some specialized pumps have been equipped with an **Explosion-Proof Motor (XPRF)**, a motor that is totally enclosed to prevent sparks or arcs within its housing from igniting surrounding vapor or gas. This designation is also given to motors designed to withstand minor explosions within the motor/pump system itself.

A. Kinetic Pumps

Centrifugal pumps operate on the centrifugal principle of simply throwing fluids in an outward direction by using a rotating impeller. The fluid usually enters the pump near the center of the rotating impeller; by adhering to the moving impeller blades, the fluid is thrown toward the outer casing of the pump. The big advantage to this type of pump is its ability to handle fluids with entrained solids. Some centrifugal pumps have impellers made of rigid materials, whereas the self-priming types have flexible, rubber-like (elastomeric) impellers (see Fig. 18.26). Centrifugal pumps have certain advantages over other pump types: they are low maintenance, inexpensive, and compact and have easy-to-change moving parts (impellers rather than pistons). However, centrifugal pumps tend not to pump fluids against high pressures.

Centrifugal pumps require that fluids be pushed (not pulled or sucked) into the impellers. They must be primed to establish a **net positive suction head (NPSH)**. If this NPSH is not established, then the pump will cavitate. The NPSH is an engineering function of the pump and can be calculated using the following general equation (see Fig. 18.27).

$$NPSHa = Ha + Hs$$

where: Hs = Suction head (+) or suction lift (−)
Ha = Pressure on the liquid surface in the supply tank

NOTE: 1 atmosphere = 33.7 ft of water; see Chapter 5, "Pressure and Vacuum."

Cavitation occurs when the suction pressure drops below the NPSH or air pockets surround the impeller. Theoretically, this condition occurs when the NPSH available (NPSHa) does not exceed the NPSH required (NPSHr). Cavitation can be easily recognized by the pump's excessive noise and should be corrected immediately, as it can cause severe damage to a running pump. Another advantage to centrifugal pumps is their ability to operate with as much as 100 percent **slip** (no fluid flow) for short periods of time. Most positive-displacement–type pumps cannot accommodate as much slip.

The term **dead head** describes the ability of a pump to continue to run without damage when all discharge lines are closed. Normally, only centrifugal pumps are recommended for this duty. The term **prime** is used to describe a charge of liquid required to begin pumping action when the liquid source level is lower than the pump. The priming charge can be manually added to the pump's intake

PLUMBING, VALVES, AND PUMPS **365**

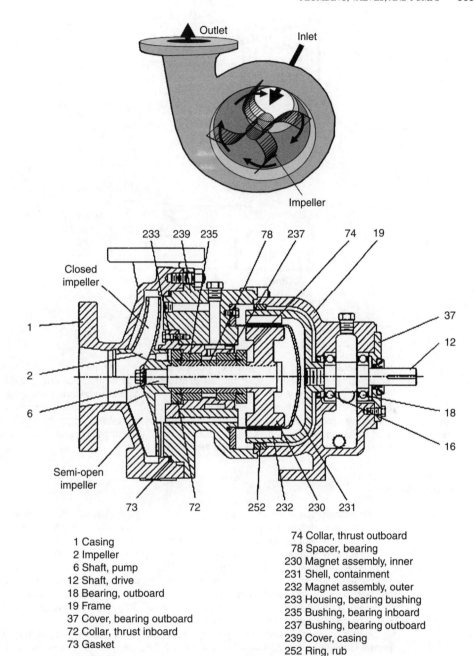

1 Casing	74 Collar, thrust outboard
2 Impeller	78 Spacer, bearing
6 Shaft, pump	230 Magnet assembly, inner
12 Shaft, drive	231 Shell, containment
18 Bearing, outboard	232 Magnet assembly, outer
19 Frame	233 Housing, bearing bushing
37 Cover, bearing outboard	235 Bushing, bearing inboard
72 Collar, thrust inboard	237 Bushing, bearing outboard
73 Gasket	239 Cover, casing
	252 Ring, rub

FIGURE 18.26 Magnetic-driver centrifugal pump. (*Courtesy of the Hydraulic Institute, Parsippany, N.J.*)

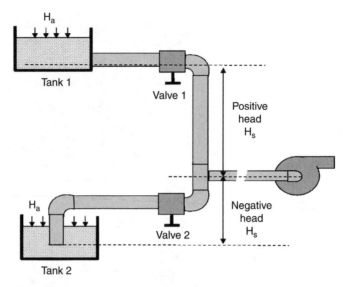

FIGURE 18.27 Net positive suction head (NPSH).

line or held by a foot valve. In many pumps, the priming liquid provides a seal and lubricant for the pumping mechanism. There are **self-priming pumps** designed to draw liquids up from below the pump inlet (suction lift). Pumps that require priming are described as requiring flooded suction. **Flooded suction** can best be defined as the gravitational flow of liquids into a pump inlet from an elevated source needed for priming. It is not necessary on self-priming pumps. A **foot valve** is a valve used at the point of fluid intake to retain fluids in the system and prevent loss of prime. Such a valve basically contains a check valve with a built-in strainer.

B. Rotary Pumps

A **rotary pump** combines the rotary motion with a positive displacement of the fluid. Rotary pumps use either gears or vanes to produce the positive displacement. Such pumps have a great advantage over centrifugal pumps because they require no priming.

Some rotary pumps work as **gear pumps**, as shown in Fig 18.28b. This type of pump traps fluid between teeth or two or more rotating gears. Gear pumps are good for low pulsation applications and the new models are often magnetically driven. As the gears rotate in opposite directions, fluid is picked up by each tooth and is delivered to the exit side of the pump. Thus rotary pumps are designed to deliver fluids at high pressures and constant rates. Rotary pumps are usually better at handling heavy or viscous liquids than are centrifugal or reciprocating pumps.

Sliding vane and flexible vane pumps are included in the rotary pump family. The **sliding vane pumps** have spring-loaded vanes that rotate inside the pump casing to move the fluids. As the impeller rotates, it moves the fluid from the inlet port to the outlet port. The spring loading provides a tight seal against the casing even as the vanes wear down from use. These sliding vanes require periodic replacement as shown in Fig. 18.28a.

Another type of rotary pump uses flexible vanes. These **flexible vanes pumps** do not use springs to provide the sealing tension, but use vanes made of pliable materials. These flexible vanes displace fluids just as the sliding vane pumps do. Wear is a problem with flexible-type vanes, and these pumps also require periodic maintenance.

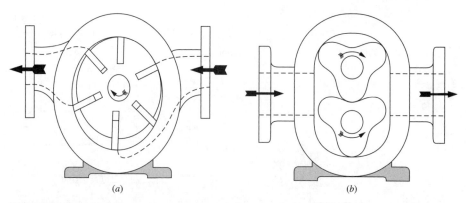

FIGURE 18.28 Rotary pumps: (*a*) sliding vane pump; (*b*) gear pumps. (*Courtesy of the Hydraulic Institute, Parsippany, N.J.*)

The screw-type or **progressive cavity pump** also qualifies as a rotary pump. These pumps use a screw shaft on a rotary motor to force liquids down a cavity and out a discharge line (see Fig. 18.29). Unlike centrifugal pumps, displacement pumps cannot tolerate slip and severe damage can occur if the fluid flow is stopped while the pump is operating. The screw-type volume-displacement pump does not have to be primed and can even pump solid/liquid (**slurries**) mixtures. When started, the pump will prime itself, provided that it is in reasonably good condition and that the gear drive is not too far above the surface of the solution. The problem of using a clean pump, free of contaminants, always faces the operator. Contaminants can be removed from the pump and transfer lines by purging the system with water, a hydrocarbon solvent, or purging it with compressed air or nitrogen gas. Be certain to use a purging method that is compatible with the pump, transfer lines, residual chemicals, and contaminants to be removed. This pump uses a screw-type rotor turning inside an elastomeric stator. Fluids are trapped between the tightly fitting rotor and stator and are forced through the pump by the screw-type rotation. Progressive cavity pumps are excellent for high-viscosity or abrasive fluids and/or slurries (particulate-containing fluids). Progressive cavity pumps are sometimes referred to as screw-type rotary or Moyno® pumps.

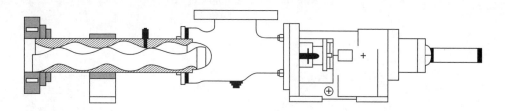

FIGURE 18.29 Progressive cavity (single-screw) pump. (*Courtesy of the Hydraulic Institute, Parsippany, N.J.*)

C. Reciprocating Pumps

A **reciprocating pump** is classified as a positive-displacement pump and usually requires a series of one-way valves to allow fluids to flow only in one direction. All reciprocating pumps use a back-and-forth motion to draw liquids into a cylinder and then a reverse motion to displace the liquid from a cylinder (see Fig. 18.30).

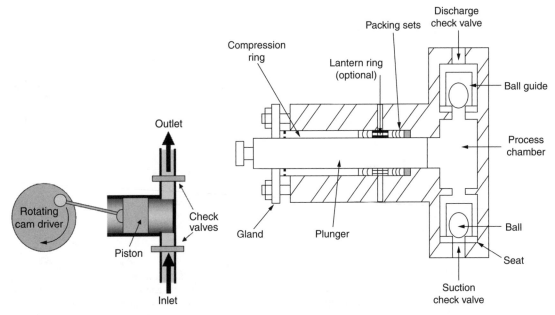

FIGURE 18.30 Reciprocating pump. (*Courtesy of the Hydraulic Institute, Parsippany, N.J.*)

Check valves are designed to allow fluids to flow only in one direction. They are primarily used in discharge lines and inlet lines to prevent reverse flow in reciprocating pumps (see Fig. 18.31). Reciprocating pumps use a piston, plunger, or diaphragm to displace fluids. The piston, plunger, or diaphragm is used to draw fluids into a cylinder and force them out.

Some reciprocating pumps are steam-powered, but they can be electrically powered or air-driven (**pneumatic**, pronounced "new-MAT-ic"). Generally, steam-powered reciprocating pumps are used to pump volatile and flammable liquids because of their basic fireproof and explosion-proof properties. They are also well suited for high-pressure systems.

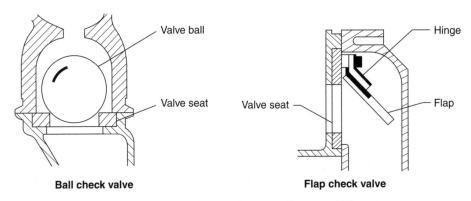

FIGURE 18.31 Check valves. (*Courtesy of the Hydraulic Institute, Parsippany, N.J.*)

D. Metering Pumps

Metering pumps are available in a wide range of designs and capacities, providing adjustable flow ranges from fractions of a milliliter per minute to hundreds of liters per minute. They are self-priming, capable of lifting a solution to heights equivalent to about 20 ft of water. Their output stroke is adjustable and they are made of a variety of materials (plastics, stainless steel, etc.). The selection is determined by the chemical properties and physical requirements of the process.

When solutions and slurries must be pumped or transferred while at the same time be protected against any possible contamination, the answer may be a peristaltic pump. A **peristaltic pump** (pronounced "pair-is-TALL-tick") uses flexible tubing and a series of rollers to create a worm-like fluid movement (Fig. 18.32). Motor-driven rollers are used to push the fluid through the tubing. These pumps tend to be easy to clean and provide few sources of contamination, and thus are excellent for food and pharmaceutical applications. This series of metal rollers actuated by an electric motor gently forces any liquid through the tubing. Solutions never come in contact with any portion of the pump; therefore, no contamination of the solution results and the pump does not have to be cleaned. These pumps can be used to move corrosive liquids, to pump several solutions at the same time, and to feed and mix solutions.

FIGURE 18.32 Watson-Marlow Sci-Q 400 U/D series variable speed peristaltic pump. (*Courtesy of Thermo Fisher Scientific, Inc.*)

E. Variable-Frequency Drive (VFD) Pumps

A variable-frequency drive (VFD) is an electronic controller that adjusts the speed of an electric motor by modulating the electric power being delivered. Motors used in this application must be of a special service type because the frequency generated by the VFD is not a perfect sine wave. The huge advantage to this new pump technology is that continuous control of the pump motor speed can be matched to the demand for the output. The operator can fine-tune the desired process flow and reduce energy costs and equipment maintenance and basically eliminate the need for inline flow control devices. These variable frequency drives enable pumps to accommodate fluctuating demand, which usually can be done at lower speeds and draw less energy.

F. General Pump Maintenance

Bearing Lubrication. Most process pumps use a rotating shaft for operation. These shafts must be supported with a bearing. **Bearings** come in five different types: single-row deep-groove, angular-contact, spherical roller, single-row tapered, and journal thrust. Bearings with rolling contact (ball or tapered rollers) are called antifriction whereas those designed to have sliding contact are called journal bearings. Journal bearings have a lubricant groove on the inside surface that serves as a small reservoir. All bearings must be properly lubricated in accordance with the manufacturer's service manual or company's standard operating procedure (SOP) of lubricating schedules. If bearings are not lubricated properly, they can overheat, rust, or corrode and cause the shaft to eventually seize (bind) and stop. This is even true for a pump that is not in use. An idle pump can develop rusted or corroded bearings in a very short period.

The type of lubricant should be found in the manufacturer's service manual. Oils are commonly used for light-to-moderate duty pumps in high-rpm applications. If no SOP or service manual is available, a filtered mineral oil of SAE 10 to 20 weight can be used. Greases are generally used to lubricate bearings under heavy loads and low-to-moderate rpm. If no SOP or service manual is available, a soda-soap or lithium-compounded grease with mineral oil base might be acceptable. Many synthetic lubricants are available, but those used must be free of abrasive contaminants. Pumps that are used to handle water should use lubricants that repel water. If dirt and contamination are found on the bearings, it is good practice to flush the bearings, housing, and sump with flushing solvent before adding new lubricant. Proper lubrication is so important to the service of a pump that many are equipped with oil reservoirs, grease cup lubricators, drip feeders, and even pressurized systems.

CAUTION: Too much oil or grease can damage a bearing. Overfilling can rupture lubricant seals, allowing contaminants to enter the bearing.

General Pump Maintenance Guidelines. All pumps are driven directly by a motor or some mechanical source. Most pumps (except those that are magnetic-coupled and peristaltic) have a shaft that transfers this work from the drive source to the pump itself. Sealing devices are required to prevent leakage of fluids along this moving shaft. The seal must be tight enough to prevent major leakage, but not cause damage to the expensive shaft. In fact, some pumps are designed to tolerate slight fluid leaks, which serve as lubricants. Some pumps are designed with external supply lines to the seal area for cooling and lubricating. Many pumps are equipped with **packing glands**, as shown in Fig. 18.33.

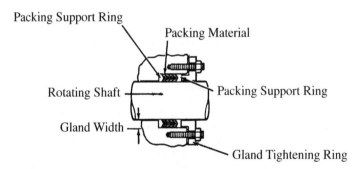

FIGURE 18.33 Pump: packing glands, gaskets, and seals. (©*2010 Swagelok Company.*)

In this design, a gland-adjustment nut or collar is placed on the outer end of the packing ring. Various materials, called **compression packings**, made of yarns, ribbons, asbestos, and so forth, are braided together in different shapes to improve lubrication retention around bearings and other moving parts. As the packing ring wears down and leakage increases, the nut or collar is adjusted to reduce the leakage. This compression packing material must be replaced periodically.

Square braid packings can generally carry a large percentage of lubricant and are used for high-speed/low-pressure rotary pump service. Lattice braid packings are relatively dense and suitable for reciprocating and centrifugal pumps. **Braid-over-braid packings** (also called *round braid*) are recommended for high-pressure, slow-speed applications with expansion joints, valve stems, and so on. Some older packing materials used asbestos packing materials on high-temperature pumps because they resist hardening and heat. Always consult the manufacturer's service manual or the company's SOP before servicing any equipment. Manufacturers and vendors now offer free troubleshooting advice over the Internet for most process equipment. There are three types of materials used to prevent the pump from leaking:

- **Packing seal.** Multiple flexible rings are mounted around the pump shaft and compressed together by tightening gland nuts. Some leakage can occur with this type of seal as it serves to lubricate the rotating shaft.
- **Lip seal.** This seal consists of a flexible rubber ring with the inner edge held closely against the rotating shaft by a spring.
- **Mechanical seal.** This seal consists of a rotating part and a stationary part with highly polished touching surfaces. Mechanical seals can be damaged by dirt and grit in the pumping liquids.

There are **sealless pumps**, usually equipped with a magnetic drive. This type of pump does not contain the conventional drive shaft and thus requires no seals. They use magnetic force to transmit power from the motor to the pump (as shown in Fig. 18.26). Table 18.7 provides a summary listing of some general pump maintenance guidelines.

TABLE 18.7 General Pump Maintenance Guidelines

The following maintenance and service duties are required:
- Routinely inspect all plumbing, insulation, tracer lines, pumps, and vessels, for leaks or other potential problems.
- Lightly touch motors and pumps for temperature and vibration checks.
- Routinely inspect instrument air, cooling systems, and utility lines for proper operation.
- Perform scheduled lubrication: adding oil, filling oil reserves, and greasing moving parts, etc.
- Flush, clean, and prepare pumps and turbines for maintenance. Carefully remove and properly dispose of any residual chemicals or contaminants.
- Disassemble and properly adjust pump packings, gaskets, and seals.

REFERENCES

1. American National Standards Institute/Hydraulic Institute (ANSI/HI) Pump Standards. The current release includes 26 currently available standards and a master index. http://www.pumps.org.
2. *Tube Fitter's Manual*, F. J. Callahan, Swagelok Company, 31400 Aurora Road, Solon, Ohio, 44139 (1993). This is a comprehensive treatment on tubing material compatibility and proper installation, operation, and maintenance of tubing, fittings, and valves.
3. The Green Book, *Flow, Level, and Environmental Handbook and Encyclopedia*, 8th edition, Omega Engineering, Inc., P.O. Box 4047, Stamford, CT 06907–0047 (2009).
4. Strauss, Howard J., and Kaufman, Milton, *Handbook for Chemical Technicians*, 1st edition, McGraw-Hill, Inc., ISBN 0-07-062164-0 (1976).

CHAPTER 19
PHYSICAL PROPERTIES AND DETERMINATIONS

PHYSICAL TESTING

A. Introduction

Physical testing is a very important part of running a chemical process operation. The purity of incoming raw materials and the quality of production at various stages can be determined by measuring the physical properties of materials. Some of the more common physical properties are listed in Table 19.1.

B. Physical and Chemical Changes

Substances, either elements or compounds, can undergo two types of change. They can undergo **physical change**, where the identity of the substance is not lost but the shape or form has been changed by the application of heat or pressure. The original shape, form, or state of matter always can be attained after infinite physical changes; the process is reversible. Examples are the melting of ice, freezing of water, melting of iron, vaporization of water, and so forth. **A chemical change** occurs when new substances are formed which are entirely different in composition and properties than the original substances. Chemical bonds have been made or broken; there is a change in energy and also a change in composition. Examples are the burning of fuel, the rusting of iron, the cooking of foods, and so forth. When a piece of iron rusts, undergoing a chemical change, it forms flaky brown rust (iron oxide), which does not conduct electricity, is not attracted by a magnet, is not shiny, and has no tensile strength. Each substance is characterized by its own physical properties. Comparison of the properties of a newly formed substance with the properties of the original substance determines whether or not a new substance is formed by chemical change, or whether there has been a physical change. Possible evidence that a chemical reaction has taken place would be if one or more of the phenomena shown in Table 19.2 has occurred.

TEMPERATURE MEASUREMENTS

A. Introduction

The measurement and reporting of temperatures are very important since temperature has a direct effect on many physical properties (for example, volume and density) or even the state of materials. Since the density of a material changes with temperature, both the density and temperature should be stated in any scientific measurements. The English temperature scale is known as the **Fahrenheit** scale and is found in daily use in the United States, both domestically and industrially. The temperature scale used in engineering applications is called the **Rankine** (°R) temperature scale. A degree on the Rankine temperature scale is equivalent to a degree on the Fahrenheit scale. However, most foreign countries use the Celsius scale, and almost all laboratory and scientific data in this country use the Celsius scale. The **Celsius** (°C) scale is also called the centigrade scale. Another temperature

TABLE 19.1 Some of the More Common Physical Properties

Color	Solubility	Tensile strength
Density	Specific gravity	Melting point
Boiling point	Elasticity	Thermal conductivity
Viscosity	Surface tension	Optical rotation
Malleability	Refractive index	Electrical conductivity
Specific heat	Freezing point	Hardness
Ductility	Bulk modulus	Crystal structure

TABLE 19.2 Evidence of a Chemical Reaction

- The evolution or absorption of heat
- The formation of a precipitate
- The evolution of a gas
- The change in color of a solution
- The emission of an odor
- The development or loss of fluorescence
- The emission of light (chemiluminescence)

scale in common use in the scientific community is called the **Kelvin** (K) or **Absolute** temperature scale. A degree on the Kelvin temperature scale is equivalent to a degree on the Celsius scale. This temperature scale is used almost exclusively when dealing with gases and is discussed in detail in Chapter 5, "Pressure and Vacuum," and Chapter 4, "Handling Compressed Gases." The relationship of the four temperature scales is shown in Fig. 19.1

All four temperature scales use the freezing and boiling points of water as their reference points. Pure water freezes at 32°F and boils at 212°F on the Fahrenheit scale. The Celsius scale has pure water freezing at 0°C and boiling at 100°C. The most common thermometer is the glass thermometer (Table 19.3). However, these glass thermometers can range from the low cost, general purpose type to the very accurate and certified American Society of Testing Materials (ASTM) type. These ASTM thermometers are available with a calibration certification from the National Institute of Standards with a range of −112 to 572°F (−80 to 300°C). A glass thermometer should never be subjected to rapid temperature changes, for example removing it from a boiling water bath to an ice bath may cause the glass to break or the liquid filling to separate. There are two types of glass thermometers, the **total immersion** type is designed to indicate the correct temperature when the bulb and entire liquid column are immersed in the test solution. The **partial immersion** type has a line around the stem to indicate the depth of immersion required for a correct reading.

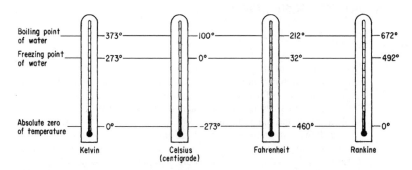

FIGURE 19.1 Relationship of the four temperature scales.

TABLE 19.3 Representative Thermometers

ASTM No.	Test	Temperature range	Scale division degree	Stem immersion	Length, ±5 mm
1C	General use	−20 to +150°C	1	76 mm	322
1F	General use	0 to +302°F	2	76 mm	322
2C	General use	−5 to +300°C	1	76 mm	390
2F	General use	+20 to +580°F	2	76 mm	390
3C	General use	−5 to +400°C	1	76 mm	413
3F	General use	+20 to +760°F	2	76 mm	413
5C	Cloud and pour	−38 to +50°C	1	108 mm	231
5F	Cloud and pour	−36 to +120°F	2	108 mm	231
6C	Low cloud and pour	−80 to +20°C	1	76 mm	232
6F	Low cloud and pour	−112 to +70°F	2	76 mm	232
7C	Low distillation	−2 to +300°C	1	Total	386
7F	Low distillation	+30 to +580°F	2	Total	386
8C	High distillation	−2 to +400°C	1	Total	386
8F	High distillation	+30 to +760°F	2	Total	386
9C	Pensky-Martens, low range tag closed tester	−5 to +110°C	0.5	57 mm	287
9F	Pensky-Martens, low range tag closed tester	+20 to 230°F	1	57 mm	287
10C	Pensky-Martens high range	+90 to +370°C	2	57 mm	287
10F	Pensky-Martens high range	+200 to +700°F	5	57 mm	287
11C	Open flash	−6 to +400°C	2	25 mm	308
11F	Open flash	+20 to +760°F	5	25 mm	308
12C	Gravity	−20 to +102°C	0.2	Total	420

B. Temperature Scale Conversions

An inspection of the two temperature scales will show that there are 180 degrees between the freezing and boiling point of water on the Fahrenheit scale and only 100 degree difference on the Celsius scale. Notice that by convention, the Kelvin scale does NOT require the degree (°) sign. See Table 19.4 for Fahrenheit/Celsius temperature conversion chart. The four temperature scales can be calculated or converted by selecting the appropriate formula below:

$$°F = 1.8°C + 32$$
$$°C = \frac{(°F - 32)}{1.8}$$
$$K = °C + 273$$
$$°R = °F + 460$$

EXAMPLE 19.1 Since normal body temperature is 98.6°F, what is body temperature expressed in Rankine degrees, Celsius degrees, and Kelvin degrees?

$$°R = °F + 460$$
$$°R = 98.6°F + 460 = 558.6°R$$
$$°C = (°F - 32)/1.8$$
$$°C = (98.6°F - 32)/1.8$$
$$°C = (66.6)/1.8 = 37°C$$
$$K = °C + 273$$
$$K = 37°C + 273 = 310 \text{ K}$$

TABLE 19.4 Temperature-Conversion Table

−140 to 20			21 to 55			56 to 90			91 to 340			350 to 690			700 to 1040		
°C		°F	°C		°F	°C		°F	°C		°F	°C		°F	°C		°F
−95.5	**−140**	−220	−6.1	**21**	69.8	13.3	**56**	132.8	32.8	**91**	195.8	177	**350**	662	371	**700**	1292
−90.0	**−130**	−202	−5.6	**22**	71.6	13.9	**57**	134.6	33.3	**92**	197.6	182	**360**	680	377	**710**	1310
−84.4	**−120**	−184	−5.0	**23**	73.4	14.4	**58**	136.4	33.9	**93**	199.4	188	**370**	698	382	**720**	1328
−78.9	**−110**	−166	−4.4	**24**	75.2	15.0	**59**	138.2	34.4	**94**	201.2	193	**380**	716	388	**730**	1346
−73.3	**−100**	−148	−3.9	**25**	77.0	15.6	**60**	140.0	35.0	**95**	203.0	199	**390**	734	393	**740**	1364
−67.6	**−90**	−130	−3.3	**26**	78.8	16.1	**61**	141.8	35.6	**96**	204.8	204	**400**	752	399	**750**	1382
−62.2	**−80**	−112	−2.8	**27**	80.6	16.7	**62**	143.6	36.1	**97**	206.6	210	**410**	770	404	**760**	1400
−56.6	**−70**	−94	−2.2	**28**	52.4	17.2	**63**	145.4	36.7	**98**	208.4	216	**420**	788	410	**770**	1418
−51.1	**−60**	−76	−1.7	**29**	84.2	17.8	**64**	147.2	37.2	**99**	210.2	221	**430**	806	416	**780**	1436
−45.5	**−50**	−58	−1.1	**33**	86.0	18.3	**65**	149.0	37.8	**100**	212.0	227	**440**	824	421	**790**	1454
−40.0	**−40**	−40	−0.6	**31**	87.8	18.9	**66**	150.8	43	**110**	230	232	**450**	842	427	**800**	1472
−34.4	**−30**	−22	0	**32**	89.6	19.4	**67**	152.6	49	**120**	248	236	**460**	860	432	**810**	1492
−28.9	**−20**	−4	0.6	**33**	91.4	20.0	**68**	154.4	54	**130**	266	243	**470**	878	438	**820**	1506
−23.3	**−10**	14	1.1	**34**	93.2	20.6	**69**	156.2	60	**140**	284	249	**480**	896	443	**830**	1526
−17.8	**−0**	32	1.7	**35**	95.0	21.1	**70**	158.0	66	**150**	302	254	**490**	914	449	**840**	1544
−17.2	**1**	38.8	2.2	**36**	96.8	21.7	**71**	159.8	71	**160**	320	260	**500**	932	454	**850**	1562
−16.7	**2**	35.6	3.8	**37**	98.6	22.2	**72**	161.6	77	**170**	338	266	**510**	950	460	**860**	1580
−16.1	**3**	37.4	3.3	**38**	100.4	22.8	**73**	163.4	82	**180**	356	271	**520**	968	466	**870**	1598
−15.6	**4**	39.2	4.9	**39**	102.2	23.3	**74**	165.2	88	**190**	374	277	**530**	986	471	**880**	1616
−15.0	**5**	41.0	4.4	**40**	104.0	23.9	**75**	167.0	93	**200**	392	282	**540**	1004	477	**890**	1634
−14.4	**6**	42.8	5.0	**41**	105.8	24.4	**76**	168.8	99	**210**	210	288	**550**	1022	482	**900**	1652
−13.9	**7**	44.6	5.6	**42**	107.6	25.0	**77**	170.0	100	**212**	414	293	**560**	1040	488	**910**	1670
−13.3	**8**	46.4	6.1	**43**	109.4	25.6	**78**	172.4	104	**220**	428	299	**570**	1058	493	**920**	1688
−12.8	**9**	48.2	6.7	**44**	111.2	26.1	**79**	174.2	110	**230**	446	304	**580**	1076	499	**930**	1706
−12.2	**10**	50.0	7.2	**45**	113.0	26.7	**80**	176.0	116	**140**	464	310	**590**	1094	504	**940**	1724
−11.7	**11**	51.8	7.8	**46**	114.8	27.2	**81**	177.8	121	**250**	482	316	**600**	1112	510	**950**	1742
−11.1	**12**	53.6	8.3	**47**	116.6	27.8	**82**	179.6	127	**260**	500	321	**610**	1130	516	**960**	1760
−10.6	**13**	55.4	8.9	**48**	118.4	28.3	**83**	181.4	132	**270**	518	327	**620**	1148	521	**970**	1778
−10.0	**14**	57.2	9.4	**49**	120.2	28.9	**84**	183.2	138	**280**	536	332	**630**	1166	527	**980**	1796
−9.4	**15**	59.0	10.0	**50**	122.0	29.4	**85**	185.0	143	**290**	554	338	**640**	1184	532	**990**	1814
−8.9	**16**	60.8	10.6	**51**	123.8	30.0	**86**	186.8	149	**300**	572	343	**650**	1202	538	**1000**	1832
−8.3	**17**	62.6	11.1	**52**	125.6	30.6	**87**	188.6	154	**310**	590	349	**660**	1220	543	**1010**	1850
−7.8	**18**	64.4	11.7	**53**	127.4	31.1	**88**	190.4	160	**320**	608	354	**670**	1238	549	**1020**	1868
−7.2	**19**	66.2	12.2	**54**	129.2	31.7	**89**	192.2	166	**330**	626	360	**680**	1256	554	**1030**	1886
−6.7	**20**	68.0	12.8	**55**	131.0	32.2	**90**	194.0	171	**340**	644	366	**690**	1274	560	**1040**	1904

*The conversion table may be used for converting degrees Fahrenheit to degrees Celsius or vice versa. Boldface numbers in the center refer to the known temperature in either Celsius or Fahrenheit. Equivalent temperature is found in the appropriate left or right column.

Source: SGA Scientific, Inc., Bloomfield, N.J.

C. Thermometer Corrections

Occasionally, the liquid (mercury, alcohol, etc.) inside a thermometer separates because of mishandling or transit. "Shaking down" a large thermometer does not work as with small capillary medical thermometer and usually results in breakage. The liquid can normally be reunited by: (1) cooling the thermometer bulb in a salt/ice solution (0°F) or an acetone/dry ice bath, or (2) heating the thermometer slowly until the liquid enters the expansion chamber (located at opposite end of the bulb).

MELTING-POINT DETERMINATIONS

A. Introduction

The melting point of a crystalline solid is the temperature at which the solid substance begins to change into a liquid. Pure organic compounds have sharp melting points. As a rule, contaminants lower the melting point and extend it over a long range. The temperature of the melting point and the sharpness of the melting point are criteria of purity. The melting-point range is the temperature range between which the crystals begin to collapse, melt, and the material becomes completely liquid.

The majority of organic compounds melt at convenient temperatures which range from about 50 to 300°C, and their melting points are useful aids in identifying the compounds as well as indicating their purity. Impurities **always** lower the melting point of a pure substance. Many compounds have the same melting point, yet mixtures of different compounds having the same melting point will melt at lower temperatures. This depression is a characteristic feature of mixed melting points and is extremely useful when one is trying to identify a compound; the melting point may be as much as 50°C lower than that of the pure compound.

1. As a rule, samples which melt at the same temperature and whose melting point is not depressed by admixture are usually considered to be the same compound.
2. Narrow-range melting points are indicative of the relative purity of a compound. Acceptably pure compounds have a 1°C range; normal commercially available compounds have a 2 to 3°C range. Extremely pure compounds have a 0.1 to 0.3°C range.
3. A wide melting-point range indicates that the compound is impure and contaminated.
4. Some substances will decompose, discolor, soften, and shrink as they are being heated. The operator must be able to recognize and to distinguish such behavior from that at the true melting point of a compound.
5. Some substances tend to shrink and soften prior to reaching their melting point. Others may release solvents of crystallization, but the melting point is reached only when the solid substance begins to change to a liquid, that is, when the first drop of liquid becomes visible.
6. When substances decompose upon melting, discoloration and/or charring usually takes place. The **decomposition point** is usually taken as the best estimate temperature and the value listed is followed by the letter "d," for example: 298°C d.

There are several procedures for determining melting points.

B. Capillary-Tube Method

Procedure

1. Obtain commercially available capillary tubes sealed at one end.
2. Fill the capillary tube with the powdered (carefully ground with a mortar and pestle, if necessary) compound to a height of 3 to 4 mm:
   (a) Scrape the powder into a pile.
   (b) Push the powder into the open end of the capillary tube.

(c) Shake the powder to the bottom of the tube by tamping lightly against the desktop or by gently rasping the tube with a file.

(d) Pack the powder tightly.

3. Attach the capillary to the thermometer with a rubber band, and immerse in an oil bath (Fig. 19.2). *NOTE:* Heated oil expands. Hot oil can swell and loosen the rubber band, causing the capillary tubing to fall off the thermometer. Be sure the rubber band is placed well above the oil.
4. Heat the oil bath quickly to about 5°C below the melting point, stirring continuously.
5. Now heat slowly; raise the temperature about 1°C/min, mixing continuously.
6. Record the temperature when fusion is observed, and record the melting-point range.
7. Discard the capillary after the determination has been made.

NOTE: Step 4 enables you to save a great deal of time. Therefore prepare two samples of the compound. Determine the approximate melting point first. Then allow the bath to cool about 15°C below that point and insert the second tube. Reheat slowly to obtain the melting point. Stir vigorously and constantly so that the temperature reading will not lag behind the actual temperature of the heating fluid.

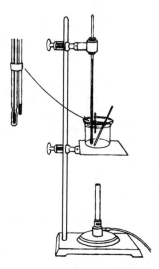

FIGURE 19.2 Laboratory setup for determining melting points.

C. Thiele Tube Method

An alternate method makes use of the Thiele melting-point apparatus (Fig. 19.3). The Thiele tube is a glass tube so shaped that when heat is applied by a micro-burner or Bunsen burner to the side arm, that heat is distributed to all parts of the vessel by convection currents in the heating fluid. No stirring is required.

Procedure

1. Obtain commercially available capillary tubes sealed at one end.
2. Fill the capillary tube with the powdered (carefully ground with a mortar and pestle, if necessary) compound to a height of 3 to 4 mm:

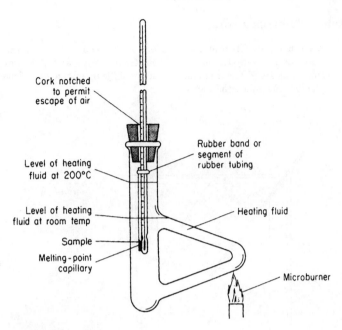

FIGURE 19.3 Thiele tube melting-point apparatus.

- **(a)** Scrape the powder into a pile.
- **(b)** Push the powder into the open end of the capillary tube.
- **(c)** Shake the powder to the bottom of the tube by tamping lightly against the desktop or by gently rasping the tube with a file.
- **(d)** Pack the powder tightly.

3. Attach the capillary to the thermometer with a rubber band, and immerse in an oil bath (Fig. 19.2). *NOTE:* Heated oil expands. Hot oil can swell and loosen the rubber band, causing the capillary tubing to fall off the thermometer. Be sure the rubber band is placed well above the oil, A Thiele tube filled with oil is used instead of an open beaker as shown in Fig. 19.3.

CAUTION: *Never heat a closed system. Always vent the Thiele tube.*

4. Do not heat the bath too fast; the thermometer reading will lag behind the actual temperature of the heating fluid.
5. Hot oil can swell and loosen the rubber band, causing the capillary tubing to fall off the thermometer. Be sure the rubber band is placed well above the oil.

Never determine the melting point by observing the temperature at which the melted substance solidifies when the bath cools. The substance may have decomposed, forming a new substance with a different melting point, or the substance may have changed into another crystalline form having a different melting point. Multiple melting points may be run simultaneously if the melting points of the different substances differ by 10°C. Identify tubes to avoid mistakes.

NOTE: For those substances which tend to decompose, inserting the capillary into the heating bath when the temperature is only a few degrees below the melting point may work.

D. Electric Melting-Point Apparatus

This is an electrically heated metal block which is controlled by a variable transformer or a rheostat. The block is equipped with a thermometer inserted into a close-fitted hole bored into the block's center (Fig. 19.4). The temperature reading of the thermometer indicates the temperature of the metal block on which the solid melts.

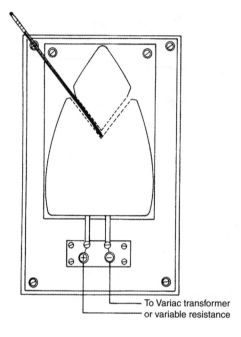

FIGURE 19.4 Electric melting-point apparatus.

Procedure

1. Obtain commercially available capillary tubes sealed at one end.
2. Fill the capillary tube with the powdered (carefully ground with a mortar and pestle, if necessary) compound to a height of 3 to 4 mm:
   (a) Scrape the powder into a pile.
   (b) Push the powder into the open end of the capillary tube.
   (c) Shake the powder to the bottom of the tube by tamping lightly against the desktop or by gently rasping the tube with a file.
   (d) Pack the powder tightly.
   (e) Place the filled capillary tube into the opening provided in the melting-point apparatus.
3. Follow the heating procedure detailed under Capillary-Tube Methods for determining melting point, raising the heat quickly to about 5°C below the melting point of the substance, then increasing the heat slowly.
4. When the determination is complete, turn off the electricity.

BOILING-POINT DETERMINATIONS

A. Introduction

The boiling point of a liquid is indicated when bubbles of its vapor arise in all parts of the volume. This is the temperature at which the pressure of the saturated vapor of the liquid is equal to the pressure of the atmosphere under which the liquid boils. Normally, boiling points are determined at standard atmospheric pressure: 760 Torr or 1 atm. The boiling point of a liquid is influenced by the atmospheric pressure. If the prevailing atmospheric pressure is less than one atmosphere, then the liquid will boil at a lower temperature. Appendix B, "Physical Properties of Water," shows the relationship between temperature and water's vapor pressure.

B. Test Tube Method

Procedure

1. Obtain a ring stand, clamps, thermometer, test tube, and burner. (See Fig. 19.5.)
2. Clamp a test tube containing 2 to 3 mL of the compound on a stand.
3. Suspend a thermometer with the bulb of the thermometer 1 in above the surface of the liquid.
4. Apply heat gently until the condensation ring of the boiling liquid is 1 in above the bulb of the thermometer.
5. Record the temperature when the reading is constant.

C. Capillary-Tube Method

Boiling points of liquids can be determined in microquantities by the Capillary Tube Method.

Procedure

1. Attach a small test tube (5 mm diameter) to a thermometer with a rubber band as shown in Fig. 19.6.
2. Use a pipette to introduce a few milliliters of the liquid whose boiling point is to be determined into the attached test tube.
3. Drop in a short piece of capillary tubing (sealed at one end) so that the open end is down.
4. Begin heating a beaker half-filled with mineral oil or dibutyl phthalate. A Thiele tube as shown in Fig. 19.3 can be used instead of the beaker.
5. Heat until a continuous, rapid, and steady flow of vapor bubbles emerges from the open end of the capillary tube, then stop heating.
6. The flow of bubbles will stop and the liquid will start to enter the capillary tube. Record this temperature. It is the boiling point.

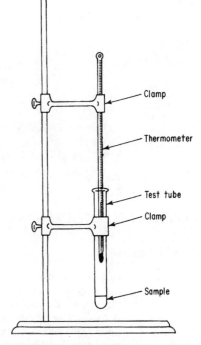

FIGURE 19.5 Laboratory setup for determining boiling points.

Notes:

1. Heating cannot be stopped before step 6 is reached. If heating is stopped below the boiling point, the liquid will enter the capillary tube immediately because of atmospheric pressure. If this

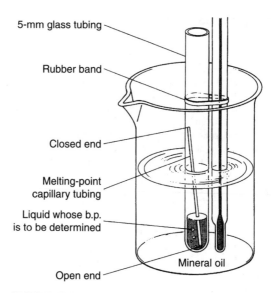

FIGURE 19.6 Capillary-tube or micro boiling-point determination.

happens, discard the capillary tube, add more liquid, and insert a new capillary tube before restarting procedure.

2. Heated oil expands. Hot oil will swell and loosen the rubber band, causing the tubing to fall off the thermometer. Be sure that the rubber band is placed well above the oil.

FLASH POINT DETERMINATIONS

A. Introduction

Generally speaking, fire and explosion are the most prevalent hazards to chemical operators. Raising the temperature of almost any liquid increases its vapor pressure (increases the concentration of vapor above the liquid), but does not result in combustion until its ignition temperature is reached. A combustible material will spontaneously start to burn when its temperature is raised to its **ignition point** in the presence of air. No source of ignition is necessary to start a fire above the ignition point of a liquid. With liquids, it is often difficult to reach the ignition point temperature by simple heating, but the vapors above the liquid can be readily ignited by a spark or flame, after which the heat generated will continue to maintain the liquid at a temperature sufficient to supply the flame with more vapor until the liquid is consumed. As the temperature of the liquid is raised in the presence of air, a concentration of vapor in air is achieved which will result in ignition, but for which the concentration is too low to result in sufficient heat being generated to sustain the combustion. The vapor simply ignites in a momentary flash. The liquid temperature at which this flash is first observed as the liquid is being slowly heated is known as the **flash point**. As the temperature of the liquid is raised above the flash point, ignition will cause longer and more pronounced flashes until the combustion becomes self-sustaining. Thus the flash point is a very important measure of the ease with which a liquid can ignite and explode or burn. The lower the flash point, the more flame-hazardous the material. Figure 19.7 shows the numerical scales used by OSHA and NFPA with the relationship between flash point and boiling point of flammable liquid.

PHYSICAL PROPERTIES AND DETERMINATIONS   **383**

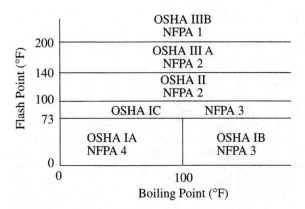

FIGURE 19.7 OSHA and NFPA classification of flammable liquids.

B. Open-Cup and Closed-Cup Method

In determining the flash point, a sample of the liquid contained in a metal cup is slowly heated under carefully stipulated conditions with the temperature of the liquid noted at frequent intervals. Periodically, a small flame is brought in contact with the vapors at the top of the cup, a procedure that is repeated until a flash is observed. the temperature at which the first ignition occurs is called the **flash point**.

Two types of cups are used, open and closed, as illustrated in Fig. 19.8a and 19.8b, respectively. While both types of cups can be used on almost any combustible liquid, the open cup is used for less volatile liquids (i.e., having a high flash point, such automotive lubricating oil), while the closed cup is used for more volatile liquids (i.e., having a low flash point, such as ethyl alcohol).

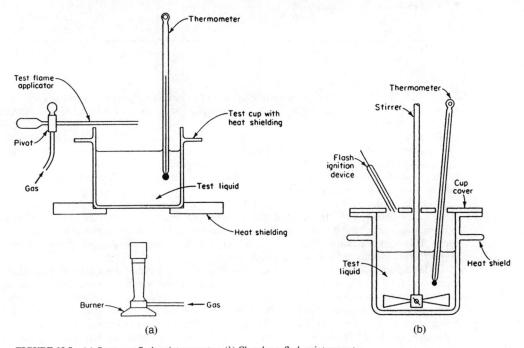

FIGURE 19.8 (*a*) Open-cup flash point apparatus. (*b*) Closed-cup flash point apparatus.

From the literature, the point of ether (diethyl ether) is given as −20°F (−28.9°C), indicating that this compound has the potential for igniting at temperatures well below room temperature. Since it is a volatile liquid, the closed-cup flash point is used. Under the conditions described, the liquid will flash only momentarily. However, if the concentration of vapor above the liquid is too high before an igniting flame is brought in contact with it, the mixture of vapor and air can burn violently in the form of an explosion. The same would be true of a combustible gas, such as hydrogen or methane, when the concentration of the combustible gas in air is sufficiently high. The explosive concentrations of various vapors and gases when mixed with air are given in Table 19.5. Any concentration of vapor or liquid below the **lower explosive limit (LEL)** or above the **upper explosive limit (UEL)** is free of explosion hazard. However, the converse is true in that any concentration that lies between these limits represents a mixture which will explode if ignited. These concentrations are therefore referred to as the explosive limits for the particular mixture.

TABLE 19.5 Flash Points and Explosion Limits of Some Common Chemicals

Name	Flash Point °F		Explosive Limits	
	Closed Cup	Open Cup	Lower % in Air	Upper % in Air
Acetone	0	15	2.12	13.0
Benzene	12	—	1.4	8.0
Diethyl ether	−20	—	1.7	48.0
Ethanol	55	—	3.3	19.0
n-Hexane	−7	—	1.3	6.9
Methanol	54	60	6.0	36.5
Petroleum ether	−50	—	1.4	5.9
Toluene	40	45	1.3	7.0
o-Xylene	63	75	1.0	—

EXAMPLE 19.2 The partial pressure of ethyl alcohol vapor in a mixture of vapor in air at atmospheric pressure is determined to be 145 torr. If the atmospheric pressure is found to be 750 torr, is the mixture explosive?

The concentration of ethyl alcohol vapor at an atmospheric pleasure of 770 torr, is $145/770 \times 100 = 18.8$ percent. Since this is within the explosive limits for ethyl alcohol as given in the literature (see Table 19.5, 3.28 to 19 percent), the mixture is explosive.

At a total pressure of 750 torr, the concentration is $145/750 \times 100 = 19.3$ percent, which is outside the explosive limits, and the mixture is *not explosive*.

Up to this point, we have considered combustion of a flammable mixture initiated by the application of a flame or spark. However, if such a flammable mixture is simply heated, a temperature is reached at which ignition automatically takes place (i.e., without the application of an ignition agent). This is known as **auto-ignition**, and the auto-ignition temperature is the temperature at which such a flammable mixture will spontaneously ignite.

Under certain conditions, a flammable material can ignite without the apparent application of heat in a process known as **spontaneous combustion.** In spontaneous combustion, the ignition temperature is reached as a result of heat generated by a slow chemical reaction where the heat cannot be dissipated fast enough and the temperature of the system rises. Corn and wheat dust can undergo spontaneous combustion as a result of the heat generated by the fermentation of various sugars and starches under conditions such that the heat accumulates until the auto-ignition temperature is reached.

DENSITY

A. Definition

The density of any substance can be found by dividing the mass of that substance by the volume that it occupies. For example:

$$\text{Density} = \frac{\text{mass of the substance}}{\text{volume of the substance}}$$

Density is expressed in the following units.

Grams per cubic centimeter	g/cm^3
Grams per milliliter	g/mL
Grams per cubic foot	lb/ft^3

The density of water at 4°C is 1.000 g/cm^3 = 1.000 g/mL; therefore, the terms milliliters and cubic centimeters are usually interchangeable. (However, in the U.S. customary system the density of water at 4°C = 62.4 lb/ft^3.)

EXAMPLE 19.3 Calculate the density of an unknown liquid if 47.8 g occupy 22.6 mL.

$$D = \frac{\text{mass}}{\text{volume}} = \frac{47.8 \text{ g}}{22.6 \text{ mL}} = 2.12 \text{ g/mL}$$

B. Determining Solid Densities

Regularly Shaped Solids. A symmetrical solid with a volume that can be determined using measured dimensions.

Procedure

1. Determine the mass of the object.
2. Measure the object and obtain relevant dimensions.
3. Calculate the volume, using the appropriate mathematical formula

Cube	($V = L^3$)
Rectangle	($V = L \times W \times H$)
Sphere	($V = 4/3 \pi r^3$)
Cylinder	($V = \pi r^2 \times h$)

4. Divide the mass by the volume.

EXAMPLE 19.4 Calculate the density of a rectangle weighing 33.67 g with a length of 2.34 cm, width of 3.55 cm, and a height of 10.77 cm.

$$\text{Mass} = 33.67 \text{ g}$$
$$\text{Volume} = L \times W \times H$$
$$= 2.34 \text{ cm} \times 3.5 \text{ cm} \times 10.77 \text{ cm}$$
$$= 89.5 \text{ cm}^3$$
$$\text{Density} = \frac{\text{mass}}{\text{volume}} = \frac{33.67 \text{ g}}{89.5 \text{ cm}^3} = 0.376 \text{ g/cm}^3$$

Irregularly Shaped Solids. A solid with an unsymmetrical shape where the volume cannot be determined from measured dimensions.

Procedure

1. Determine the mass of the object.
2. Determine the volume by water displacement (Fig. 19.9).
   (a) Use a graduated cylinder containing a measured and recorded amount of water (original volume).
   (b) Submerge the weighed solid completely in the graduated cylinder containing the water and record the larger volume reading (final volume).
   (c) Subtract the original volume from the final volume and obtain the volume of the object.
   (d) Determine the density by dividing the mass of the object by the volume.

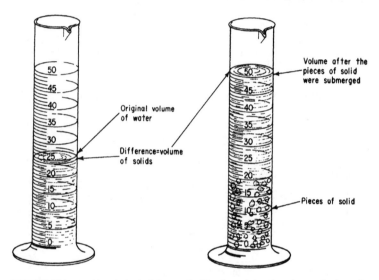

FIGURE 19.9 Archimedes' principle—method for determining the volume of an irregular solid.

EXAMPLE 19.5 A graduated cylinder is partially filled with water (density = 1.00 g/mL). An irregular shaped piece of metal weighing 60.98 g is then immersed in the water, causing the water level to change from 50.6 to 55.8 mL. Calculate the density of the unknown metal.

$$\text{Mass of sample} = 60.98 \text{ g}$$
$$\text{Volume of sample} = 55.8 \text{ mL} - 50.6 \text{ mL} = 5.20 \text{ mL}$$
$$\text{Density} = \frac{\text{mass}}{\text{volume}} = \frac{60.98 \text{ g}}{5.20 \text{ mL}} = 11.7 \text{ g/mL}$$

C. Determining Liquid Densities

Density-Bottle Method Procedure.

1. Use a calibrated-volume liquid container fitted with thermometer, or determine the volume of the container by using distilled water and reference tables listing the density of water at different temperatures. (See Appendix B, "Physical Properties of Water.")

2. Determine the mass of the empty container.
3. Determine the mass of the filled container (filled to the top of the capillary) and record the temperature. The density bottle is placed in a thermostatically controlled bath. There is some loss of liquid from the density bottle during the time that it is removed from the bath, wiped dry, and massed (especially with very volatile liquids because of evaporation of the liquid).
4. Obtain the mass of the liquid by subtraction.
5 Divide the mass by the volume of the calibrated container.

CAUTION: Take care to exclude air bubbles, and carry out operations as quickly as possible.

Float Method. A group of weighted, hermetically sealed glass floats can be used to determine the density of a liquid. When such floats are dropped into a liquid and thoroughly wetted, they will float in a more dense liquid, sink in a less dense liquid, or remain suspended in a liquid which has equal density (Fig. 19.10). By experimentally determining which float remains suspended, the density of the liquid can be determined by reading the density of that float.

Procedure
1. Fill a beaker, graduated cylinder, or any glass container with the liquid to be tested.
2. Add floats and note which float remains suspended.
3. Read the density of the liquid from that float.

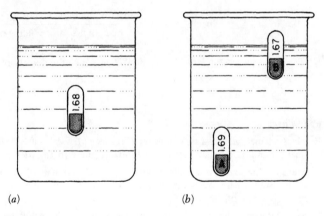

FIGURE 19.10 Float method for determining the density of liquids.

D. Determining Gas Densities

All gases have a much lower density than liquids or solids, and the determination of the density of a gas is more difficult because of the extremely small mass involved.

E. Dumas Method

The Dumas method involves the direct determination of mass of a known volume of gas in a calibrated sphere of known volume. The sphere of known volume is fitted with a capillary tube and stopcock, whereby the gas is introduced into the flask.

Procedure

1. The volume of the flask is verified by filling it with a liquid of known density and determining the mass of the filled sphere.
2. The flask is emptied, cleaned, dried, and evacuated.
3. The gas to be tested is introduced into the flask, and the mass of the flask and gas together is measured.

$$\text{Density of gas} = \frac{\text{mass of gas}}{\text{volume of gas}}$$

NOTE: The density calculated is the density of the gas at the temperature at which the determination is made. A thermometer should be hung in the immediate vicinity and read several times during the procedure.

SPECIFIC GRAVITY

A. Introduction

When metric units are used, specific gravity has the same numerical value as density; specific gravity is the mass of a substance divided by the mass of an equal volume of water. Specific gravity is expressed by a number; since it is a ratio, it has no units.

$$\text{Specific gravity} = \frac{\text{mass of the same volume of substance}}{\text{mass of the same volume of water}}$$

NOTE: The water standard is taken at 4°C, at which the density of distilled water is 1.000 g/mL.

B. Pyknometer Method

Procedure

1. Measure the mass of a **pyknometer** (Fig. 19.11), a calibrated-volume groundglass vessel fitted with a closure and a thermometer.
2. Measure the mass when the pyknometer is filled with distilled water. Subtract the mass obtained in step 1 from this value. This gives the mass of the water.
3. Repeat the procedure when the pyknometer is filled with the unknown liquid. Subtract the mass obtained in step 1 from this quantity. This gives the mass of the equal volume of the unknown liquid.
4. Divide the mass obtained in step 3 by the mass obtained in step 2 to get the specific gravity. (All mass measurements should be made at the same temperature.)

FIGURE 19.11 Pyknometer.

C. Hydrometer Method

A hydrometer is a glass container, weighted at the bottom, having a slender stem which is calibrated to a standard (Fig. 19.12). The depth to which the container will sink in a liquid is a measure of the specific gravity of the liquid. Specific gravity is read directly from the calibrated scale on the stem of the hydrometer. Figure 19.13 shows a hydrometer placed into a cylinder containing a liquid sample.

FIGURE 19.12 Glass hydrometer for specific gravity determination. It has a weighted tip and graduated, direct-reading tube.

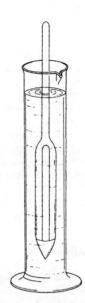

FIGURE 19.13 Hydrometer in use.

D. Specific Gravity Conversions

Hydrometers are calibrated in specific gravity units (the mass of a liquid divided by the mass of an equal volume of water taken at certain temperatures, such as 60 or 70°F), or in arbitrary units as degrees **Baumé** (Bé), degrees A.P.I. (American Petroleum Institute), or degrees **Brix** (also called **Fisher**). Decrees Brix are arbitrarily graduated so that 1° Brix = 1 percent sugar in solution. Hydrometer readings given in specific gravities can be converted to these units by the following formulas. See Table 19.6 for direct conversion of degrees Baume' to specific gravity.

Liquids lighter than water:

$$°Bé = \frac{140}{\text{sp gr } 60°F/60°F} - 130$$

$$°A.P.I = \frac{141.5}{\text{sp gr } 70°F/60°F} - 131.5$$

$$°Brix = \frac{400}{\text{sp gr } 60°F/60°F} - 400$$

Liquids heavier than water:

$$°Bé = 145 - \frac{145}{\text{sp gr } 60°F/60°F}$$

VISCOSITY

A. Definition

Viscosity is the internal friction or resistance to flow that exists within a fluid, either liquid or gas. Fluid flow in a pipe line has a greater velocity at the center than next to the metal surfaces, partly because of the friction between the fluid and the boundary surfaces. This causes the adjacent outer layers to move more slowly. The slower-moving layers in turn retard the motion of center layers. In the study of oils and organic liquids, viscosity is very significant. In industry "heavier" oils and liquids have higher viscosities, not greater densities.

The metric (SI system) measure of the ability to flow is expressed as **poise** (P), defined as dyne-seconds per square centimeter. Viscosities are usually tabulated in **centipoises** (cp). **Fluidity** is the reciprocal of viscosity, and in the SI system the unit for fluidity is called the **rhe**.

$$1 \text{ rhe} = 1/\text{poise} \text{ or } 1 \text{ poise}^{-1}$$

B. Viscosity Standards

References in handbooks and the literature provide viscosity data for many *pure* solvents and chemical substances. However, viscosity measurements normally require that viscosity-determining equipment be standardized against materials of certified viscosity. These viscosity standards are available in a very wide range of viscosities, and they are certified as permanent viscosity standards. Most of them are fluid silicones and are accurate to within 1 percent of the stated viscosity value. Viscosities range from 5 to over 100,000 cp. There are various industrial methods for determining viscosity.

TABLE 19.6 Conversion Degrees Baumé American to Specific Gravity at 60°F (15.55°C)

°Bé	0°	1°	2°	3°	4°	5°	6°	7°	8°	9°
For liquids *lighter than water*, degrees Baumé = $\frac{140}{\text{sp gr}} - 130$										
10	1.000	0.993	0.986	0.979	0.972	0.966	0.959	0.952	0.946	0.940
20	0.933	0.927	0.921	0.915	0.909	0.903	0.897	0.892	0.886	0.880
30	0.875	0.870	0.864	0.859	0.854	0.848	0.843	0.838	0.833	0.828
40	0.824	0.819	0.814	0.809	0.804	0.800	0.795	0.791	0.786	0.782
50	0.778	0.773	0.769	0.765	0.761	0.757	0.753	0.749	0.745	0.741
60	0.737	0.733	0.729	0.725	0.722	0.718	0.714	0.711	0.707	0.704
70	0.700	0.696	0.693	0.690	0.686	0.683	0.680	0.676	0.673	0.670
80	0.667	0.664	0.660	0.657	0.654	0.651	0.648	0.645	0.642	0.639
90	0.636	0.633	0.631	0.628	0.625	0.622	0.619	0.617	0.614	0.611
For liquids *havier than water*, degrees Baumé = $145 - \frac{145}{\text{sp gr}}$										
0	1.000	1.007	1.014	1.021	1.028	1.036	1.043	1.051	1.058	1.066
10	1.074	1.082	1.090	1.098	1.107	1.115	1.124	1.133	1.142	1.151
20	1.160	1.169	1.179	1.188	1.198	1.208	1.218	1.229	1.239	1.250
30	1.261	1.272	1.283	1.295	1.306	1.318	1.330	1.343	1.355	1.368
40	1.381	1.394	1.108	1.422	1.436	1.450	1.465	1.480	1.495	1.510
50	1.526	1.543	1.559	1.576	1.593	1.611	1.629	1.648	1.667	1.686
60	1.706	1.726	1.747	1.768	1.790	1.812	1.835	1.859	1.883	1.908

C. Small-Bore Tube Method

The flow of the liquid through a small-bore tube, such as is found in a thermometer, can be measured with a graduated cylinder and a stopwatch (Fig. 19.14). A constant hydrostatic pressure (head) is maintained by constant feed and overflow. The volume of liquid that passes through the capillary tube is collected in a graduated cylinder, and the time required is measured with the stopwatch.

$$\text{Coefficient of absolute viscosity} = \frac{\text{volume collected}}{\text{time}}$$

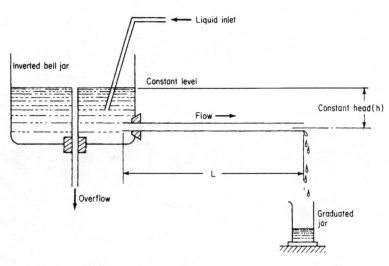

FIGURE 19.14 Small-bore tube viscometer.

D. Saybolt Viscometer Method

The Saybolt viscometer (Fig. 19.15) has a container for liquids with a capacity of 60 mL, fitted with a short capillary tube of special length and diameter. The liquid flows through the tube, under a falling head, and the time required for the liquid to pass through is measured in seconds. If temperature is a critical factor, the viscometer is kept at constant temperature in a temperature-controlled bath.

E. The Falling-Piston Viscometer Method

The liquid to be tested is placed in the test cylinder, and the falling piston is raised to a fixed, measured height (Fig. 19.16). The time that is required for the piston to fall is a measure of the viscosity. The higher the viscosity, the more time it takes for the piston to drop to the bottom.

F. Rotating Concentric Cylinder Viscometer Method

Two concentric cylinders, which are separated by a small annular space, are immersed in the liquid to be tested (Fig. 19.17). One cylinder rotates with respect to the other, and liquid in the space rotates in layers. A viscous force tends to retard the rotation of the cylinder when the viscometer is in motion. The torque on the inner cylinder, which is caused by the viscous force retarding its rotation, is measured by a torsion wire from which the cylinder is suspended. These are sometimes referred to by a manufacturer's name, Brookfield® viscometer.

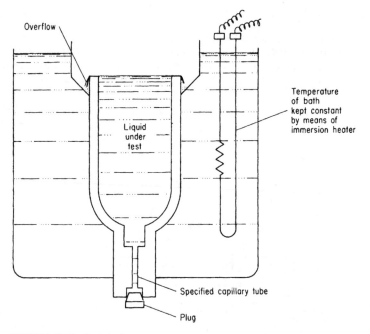

FIGURE 19.15 Saybolt viscometer.

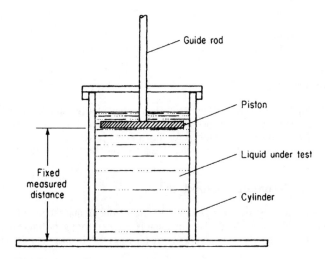

FIGURE 19.16 Falling piston viscometer.

G. Falling-Ball Viscometer Method

A ball will fall slowly through a viscous liquid. At first the ball accelerates, but then it will fall with a constant velocity. The technician measures the time required for the ball to fall a known distance, after the condition of uniform viscosity has been achieved. The cylinder must have a large enough

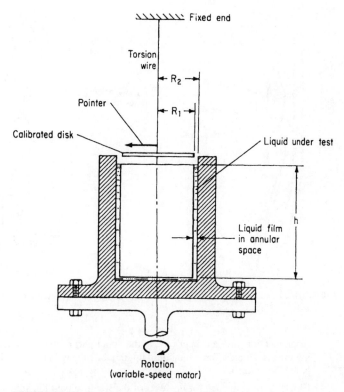

FIGURE 19.17 Rotating concentric cylinder viscometer.

diameter so that (1) no eddy currents are set up and (2) the cylinder surface will not affect the fall of the ball. Commercial falling-ball viscometers are fitted with a ball-release device, and the time required for the ball to descend is measured with a stopwatch.

H. Ostwald Viscometer Method

The capillary-type viscometer (Fig. 19.18) is frequently used in the process industry. These capillary viscometers have a sample reservoir, an upper bulb with a mark and a lower bulb with a mark. The capillary is the actually working part of the viscometer in determining the flow rate of a liquid. Large capillaries are necessary for highly viscose liquids and small capillaries are needed for low-viscosity liquids. Since viscosity is very temperature-dependent, the liquid sample must be allowed to reach temperature equilibrium in a constant temperature bath before analysis. A stopwatch can be used to time the drainage of the liquid from one mark to the second mark. This time can be compared to a previously analyzed viscosity standard, and the viscosity determined using the following procedure.

Procedure

1. Wash the viscometer and make sure it is absolutely clean.
   (a) Wash thoroughly with soap and water.
   (b) Rinse at least five times with distilled water.
   (c) Finally, introduce sufficient distilled water into the large round bulb, or reservoir (Fig. 19.18).
2. Allow the distilled water to come to thermal equilibrium in a constant temperature bath.

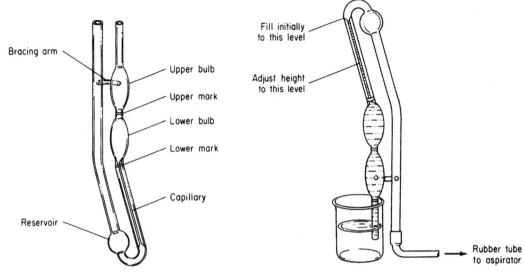

FIGURE 19.18 Ostwald-Cannon-Fenske capillary viscometer.

FIGURE 19.19 Procedure for filling viscometer.

3. Apply suction with a rubber tube to the upper part of the viscometer. This is best done by inverting the viscometer. Draw the liquid up into the tube with the two bulbs to a level above the second bulb (Fig. 19.19).
4. Clock the time needed for the level of the water to pass the signal markings. Make several determinations, using a stopwatch.
5. Drain the viscometer and dry it completely.
6. With a pipette, add an appropriate volume of the solution to be tested to the reservoir, that is, the same volume as in step 1(c).
7. Clock the time needed for the level of the liquid under test to pass the signal markings.
8. Calculate the relative viscosity of the test liquid by comparing the average time required for its flow against that of water or some other viscosity reference standard 25°C.

I. Determining Kinematic Viscosities

Kinematic viscosity is equal to the absolute viscosity divided by the mass density, and most laboratories measure kinematic viscosity. The SI unit of kinematic viscosity is the **stokes** (St) and is equal to 1 mL/s. A viscometer used for this purpose is calibrated in kinematic centistokes (2 to 10,000) and Saybolt universal seconds (33 to 46,000). Kinematic viscosity is defined as the ratio of viscosity to density. The stoke can be calculated by dividing the viscosity expressed in poises by the density of the substance at the temperature. Units of centipoises and centistokes appear quite often in the literature and represent 0.01 poise and/or 0.01 stokes, respectively.

$$\text{Stoke} = \frac{\text{gram}}{\text{second} \times \text{centimeter} \times \text{density}} = \frac{\text{poise}}{\text{density}}$$

To obtain the Saybolt universal viscosity (SSU) equivalent to a kinematic viscosity at temperature "t," multiply the equivalent Saybolt universal viscosity at 100°F by $1 + (t - 100)\, 0.000064$. These

TABLE 19.7 Specific Gravities and Viscosities of Some Common Substances

Substance	Temperature C°	Specific gravity $H_2O = 1$ @ 60°F	Temperature °C	Viscosity centistokes	Viscosity Saybolt universal seconds
Acetic acid, 10%	15	1.014	15	1.35	31.7
Acetic acid. 50%	15	1.061	15	2.27	33.0
Acetic acid, glacial	15	1.055	15	1.34	31.7
Ethanol	40	0.772	37.8	1.2	31.5
Methanol	20	0.79	15	0.74	—
Automotive SAE-5W	15.6	0.88–0.94	−17.8	1295 max	6M–max
SAE-20	15.6	0.88–0.94	98.9	5.7–9.6	45–58
SAE-30	15.6	0.88–0.94	98.9	9.6–12.9	58–70
SAE-50	15.6	0.88–0.94	98.9	16.8–22.7	85–110
SAE-90	15.6	0.88–0.94	98.9	14–25	74–120
Corn oil	15.6	0.924	100	8.6	54
Cocoanut oil	15.6	0.925	37.8	29.8–31.6	140–148
n-Decane	20	0.73	37.8	1.001	31
Ethylene glycol	15.6	1.125	21.1	17.8	88.4
Gasoline (a)	15.6	0.74	15.6	0.88	—
Gasoline (b)	15.6	0.72	15.6	0.64	—
Glycerine	20	1.260	20.3	648	2950
Olive oil	15.6	0.91–0.92	37.8	43.2	200
Petroleum ether	15.6	0.64	15.6	31 (est)	1.1
Sucrose solution 86.4 brix	15.6	1.459	37.8	1750 cp	—
Sucrose solution 82.3 brix	15.6	1.431	37.8	380 cp	—
Sucrose solution 78.4 brix	15.6	1.405	37.8	160 cp	—
Sodium hydroxide, 30%	15.6	1.33	18.3	10.0	58.1
Sulfuric acid, 100%	20	1.839	20	14.6	76
Toluene	20	0.866	20	0.68	—

centistokes to Saybolt seconds equivalents (at various temperatures) can be found in the viscosity section of most critical tables. Table 19.7 lists the specific gravity and viscosity of some common substances.

SURFACE TENSION

A. Introduction

The cohesion of the molecules of a liquid is manifested in the phenomenon called **surface tension**. The molecules in the interior of the liquid are subjected to balanced forces between them. The molecules at the surface of the liquid are subjected to unbalanced forces, because they are attracted to the molecules below them. As a result, the surface of the liquid appears to resemble that of an elastic membrane, causing liquid surfaces to contract. This inward pull on those surface molecules results in surface tension, the tendency of a liquid to form drops and the resistance to expansion of the surface area. As a result, in the absence of any external influence, a small liquid sample will assume the spherical shape of a drop.

The surface tension of a liquid is important in determining whether the liquid will "wet" a solid. When the contact angle is more than 90°, the liquid tends to "ball up," i.e., form droplets on the solid, much like mercury on glass, and does not wet, or spread over, the surface. A material with a low surface tension tends to wet solids, and this is important in detergents (where the dirt particles must be wetted before the liquid can wash them away), and in emulsifiers (where the particles of one liquid must be wetted by the other liquid before the former can be dispersed in the latter). In surface-tension analyses, it must be remembered that it is only the nature of the surface itself that is important. Thus, water will not spread on a piece of glass that has even an extremely thin layer of oil on it, whereas it will spread readily on a clean glass surface.

Many materials, such as soaps and alcohols, have the property of lowering the surface tension of a liquid. Materials have been developed which show this property to a pronounced degree. Because they have such an effect on the surface properties of a liquid, as a group they are known as **surface-active materials.**

Liquid molecules also have attraction for other substances: this property is called **adhesion**. When there is an adhesive force between liquids and the surfaces of containers (tubes, beakers, etc.), the liquid is said to "wet" the surface. The property is called **capillary** and is related to surface tension. Table 19.8 lists the surface tension of several common liquids measured at 20°C.

TABLE 19.8 Surface Tension of Some Common Liquids at 20°C

Liquid	Surface Tension (dynes/cm)
Acetone	23.7
Aniline	42.9
Benzene	28.9
n-Butyl alcohol	29.2
Carbon disulfide	32.3
Carbon tetrachloride	25.4
Ethyl alcohol	21.2
Ethyl ether	17.0
Glycerin	63.4
Isobutyl alcohol	23.0
Isopropyl alcohol	21.7
Mercury	470
Methyl alcohol	22.6
n-Octyl alcohol	27.5
n-Propyl acetate	24.3
n-Propyl alcohol	23.8
Pyridine	38.0
Tetrachloroethylene	31.7
Toluene	28.5
Water	72.9
m-Xylene	28.9
o-Xylene	30.1
p-Xylene	28.4

B. Fluid Surface-Tension Measurement

Fluid surface tensiometers are instruments which measure both static (equilibrium) and dynamic (non-equilibrium) fluid surface tension. It is the only commercially available instrument that completely replicates ASTM Designation D 3825-90, "Standard Test Method for Dynamic Surface Tension by the Fast-Bubble Technique." Internal programs allow the user to calculate exactly the surfactant diffusion times.

REFRACTIVE INDEX

A. Introduction

The **refractive index** of a liquid is the ratio of the velocity of light in a vacuum to the velocity of light in the liquid. The refractive index of a liquid is a constant for that liquid, and its determination furnishes us with both a method of identifying a substance and a method for determining the purity of substances. Since the angle of refraction varies with the wavelength of the light, the measurement of refractive index requires that light of a known wavelength be used, usually that of the yellow sodium D line with a wavelength of 5890 Å. A typical refractive index would be:

$$n_D^{20} = 1.4567$$

Where the superscript indicates the Celsius temperature and the subscript indicates that the sodium D line was used as a reference. Refractive index values can be reported to four decimal places quite accurately. Small amounts of impurities have a significant effect on the experimental value, and, in order to match the established reported refractive indices, substances must be very carefully purified.

B. Specific Refraction

The refractive index decreases with temperature because the density decreases, resulting in fewer molecules per unit volume. However, a quantity called the **specific refraction** is independent of the temperature and may be calculated by the equation:

$$r = \frac{(n^2 - 1)}{(d)(n^2 + 2)}$$

where
  r = specific refraction
  d = density
  n = refractive index

C. Refractometers

The refractive index is easily determined with a refractometer. The refractometer compares the angles at which light from an effective point source passes through the test liquid and into a prism whose refractive index is known. The refractive index of the liquid is read directly from a scale within the refractometer.

General precaution with any refractometer:

- Only a few drops of sample are necessary.
- If a mixture to be analyzed contains volatile components, close the prisms quickly to avoid selective evaporation.
- Never touch the prisms with any hard objects. Always clean the prism surfaces with a soft tissue moistened with isopropyl alcohol or suitable lens cleaner.
- After cleaning and drying, place the prisms in the locked position to prevent dirt and dust from collecting on them.

General Refractometer Procedure (always consult the manufacturer's instructions or company's SOP for proper operational procedure):

1. Adjust the constant temperature bath to the desired temperature. Connect to the refractometer with a small circulating pump. Figures 19.20 and 19.21 show schematics of an Abbé refractometer.
2. Twist the knurled locking screw counterclockwise to unlock the prism assembly. Lower the bottom part of the hinged prism until it is parallel with the benchtop.
3. Clean the upper and lower prisms with soft, non-abrasive, absorbent, lint-free cotton wetted with methyl alcohol, allow them to air dry.
4. Place a drop of the standard test solution of known refractive index at that temperature on the prism. Check the thermometer in the well of the prism assembly.
5. Close the prism assembly. Lock by twisting the knurled knob.
6. Set the magnifier index on the scale to correspond with the known refractive index of the standard.
7. Look through and turn the compensator knob until the colored, indistinct boundary seen between the light and dark fields becomes a sharp line.
8. Adjust the knurled knob at the bottom of the magnifier arm until the sharp line exactly intersects the midpoint of the cross hairs in the image (Fig. 19.22). *NOTE:* If necessary, recheck and reposition the index on the magnifier arm to read as in step 6.
9. The refractometer is standardized.

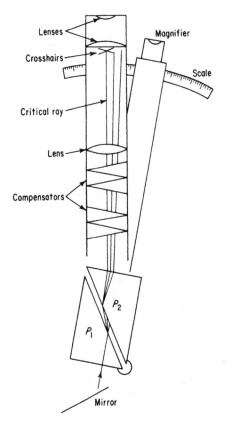

FIGURE 19.20 Schematic for the Abbé refractometer.

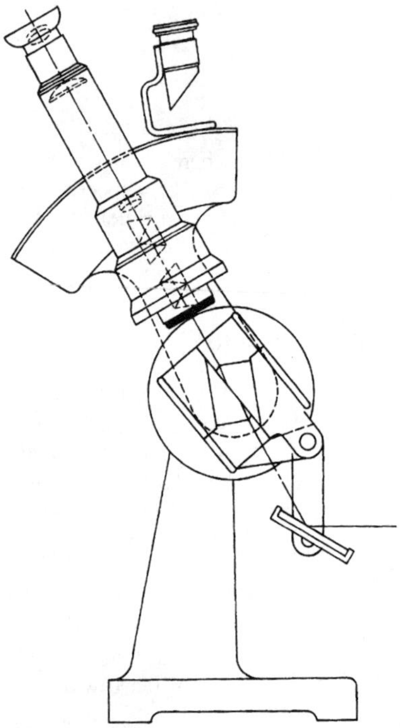

FIGURE 19.21 Drawing of the Abbé refractometer.

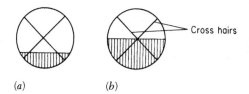

FIGURE 19.22 Adjustment of the refractometer (a) Incorrect. (b) Correct.

10. Repeat steps 2 through 5.
    (a) Open the prism assembly (two prism faces).
    (b) Clean the prism halves.
    (c) Place a drop of the sample on the lower prism half.
    (d) Close the prism assembly
11. Look through the eyepiece and move the magnifier arm until the sharp line exactly intersects the midpoint of the crosshairs on the image.
12. Use index of refraction table as shown in Table 19.9 to evaluate sample's possible identity or purity.

13. Read the refractive index from the magnifier-index pointer.
14. Clean the assembly's prisms and lock them together.

TABLE 19.9 Index of Refraction for Some Common Substances

	Refractive index, n_D^{20}
Gases:	
Air	1.00029
Ammonia	1.00038
Carbon dioxide	1.00045
Carbon monoxide	1.00034
Chlorine	1.00077
Methane	1.00044
Water (vapor)	1.00025
Liquids:	
Acetic acid	1.3716
Acetone	1.3588
Benzene	1.5011
Carbon disulfide	1.6319
Carbon tetrachloride	1.4601
Chlorobenzene	1.5241
Ethyl alcohol	1.3611
Glycerin	1.4746
Methanol	1.3288
Toluene	1.4961
Water	1.33299
Plastics:	
Cellulose acetate	1.48
Cellulose nitrate	1.50
Melamine	1.59
Methyl methacrylate	1.49
Nylon	1.53
Polyethylene	1.52
Polypropylene	1.52
Polystyrene	1.59
Urea	1.55
Solids:	
Amber	1.55
Borax, fused	1.46
Fluorite	1.43
Glass, crown	1.53
Rock salt	1.53
Quartz	1.56

POLARIMETRY

A. Introduction

Ordinary white light vibrates in all possible planes which are perpendicular to the direction of propagation, and it consists of many different wavelengths. Sodium light is monochromatic light, having only one frequency, but it still vibrates in all possible planes (Fig. 19.23a).

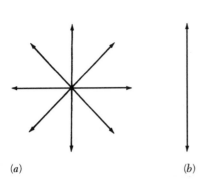
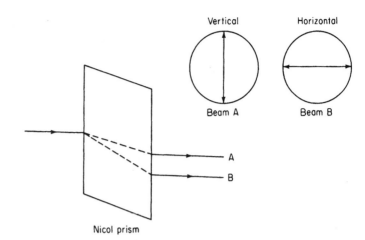

FIGURE 19.23 (*a*) Light polarization, unpolarized light vibrating in all possible directions. (*b*) Light wave vibrating in only one plane (polarized light).

FIGURE 19.24 Generation of two beams of polarized light through a Nicol prism.

When light which is vibrating in all possible planes is passed through a Nicol prism, two polarized beams of light (Fig. 19.23b) are generated (Fig. 19.24). One of these beams passes through the prism, while the other beam is reflected so that it does not interfere with the plane-polarized transmitted beam. If the beam of plane-polarized light is also passed through another Nicol prism, it can pass through only if the second Nicol prism has its axis oriented so that it is parallel to the plane-polarized light. If its axis is perpendicular to that of the plane polarized light, the light will be absorbed and will not pass through.

Some substances have the ability to bend or deflect a plane of polarized light, and that phenomenon is called **optical rotation**. Only asymmetric molecules can do this, and the magnitude and direction of the deflection of the plane of polarized light which passes through an asymmetric substance in solution can be measured in a polarimeter. A carbon having four different groups bonded to it is said to be an asymmetric carbon atom. Unpolarized light from the light source passes through a polarizer. Only polarized light is transmitted. This polarized light passes through the sample cell. If the sample does not deflect the plane of light, its angle is unchanged. If an optically active substance is in the sample tube, the light is deflected. The analyzer prism is rotated to permit maximum passage of light and is then said to be lined up. The degree (angle) of rotation α is measured (Fig. 19.25).

Factors that affect the angle of rotation α are:

1. Concentration (the greater the concentration, the greater the angle of rotation)
2. Solvent
3. Temperature
4. Wavelength of the polarized light
5. Nature of the substance
6. Length of sample tube

B. Direction of Rotation of Polarized Light

The direction of rotation is indicated by a (+) for **dextrorotation** (to the right) and a (−) for **levorotation** (to the left). To avoid confusion and to correlate structure and nomenclature, the symbols *D* and *L* are used without regard to the direction of optical rotation. Substances are labeled *D* or *L* with regard to the configuration of the asymmetric carbon atom.

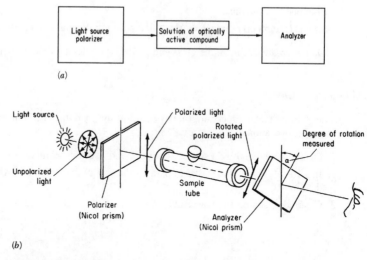

FIGURE 19.25 Components of a polarimeter.

C. Specific Rotation

The angle of rotation can be combined in a formula with the sample concentration and length of the sample tube to provide a value called **specific rotation** $[\alpha]_D^{25}$ which includes notation for the temperature and wavelength of the polarized light. Table 19.10 gives specific rotation angles for some common sugars in aqueous solution.

$$[\alpha]_D^{25} = \frac{\text{angle of rotation } \alpha}{\text{concentration} \times \text{length of sample tube}}$$

where concentration is expressed in grams of solute per milliliter of solvent, length is expressed in decimeters, $D = D$ line of sodium, and $25 = 25°C$. *NOTE:* the temperature is usually between 20 and 25°C unless otherwise stated.

TABLE 19.10 Specific Rotation of Sugars in Aqueous Solution

Sugar	$[\alpha]_D^{25}$
Galactose	+83.9°
Glucose ($\alpha + \beta$)	+52.7°
α–Glucose	+112°
β–Glucose	+18.7°
Mannose	+14.1°
Lactose	+52.4°
Maltose	+136.0°
Sucrose	+66.5°

D. Causes of Inaccurate Measurements

1. Solid particles suspended in the solution. Always filter solutions properly and avoid dust.
2. Entrapped air bubbles.
3. Strains (improperly annealed or strained glass) in the glass end plates.

Do not create strains in the glass end plates by tightening them with excessive pressure.

E. How to Use a Polarimeter

General Polarimeter Procedure (always consult the manufacturer's instructions or company's SOP for proper operational procedure):

1. Obtain a polarized light source, analyzer, and polarimeter tube.
2. In a 10-mL volumetric flask, prepare a solution* of the compound whose optical activity is to be determined.
3. Clean, dry, and partially assemble a polarimeter tube (Fig. 19.26).

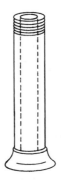

FIGURE 19.26 Polarimeter sample tube.

FIGURE 19.27 Filling a polarimeter tube.

4. With a dropper fill the tube completely to overflow (Fig. 19.27).
5. Slide an end glass on the tube so that no air is entrapped. Close the end of the polarizer tube. Do not screw on too tightly.
6. Position the polarimeter tube in the polarimeter.
7. Turn on the light source and allow it to warm up.
8. Rotate the analyzer tube until the two halves of the image that are viewed through the eyepiece match exactly.
9. Read the dial and record the magnitude of rotation.
10. Obtain a *blank solvent reading* by repeating steps 3 through 9 with pure solvent blank in a clean polarimeter tube.
11. The difference between readings 9 and 10 is α (Fig. 19.28).

---

*Temperature change causes a change in solution concentration due to the contraction or expansion of the solution. Compensate for this effect by making up the solution at room temperature (the temperature of the polarimeter) or use a polarimeter with a thermostat.

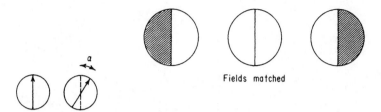

FIGURE 19.28 Patterns seen in adjustment of a polarimeter analyzer to determine the angle of rotation α.

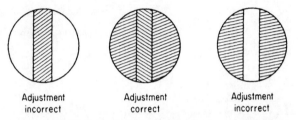

FIGURE 19.29 Split-field image changes in the polarimeter as the analyzer is rotated.

Newer types of polarimeters use a double-field optical system. The split image enables the operator to match the light intensity more accurately. As the plane of polarized light is rotated, the split field changes as the analyzer is rotated (Fig. 19.29).

F. Specific Rotation Calculations

If the angle of rotation is plotted on the y axis of a graph against the concentration in grams per milliliter on the x axis, the specific rotation can be calculated from the slope of the line obtained (Fig. 19.30).

$$[\alpha]_D^{25} = \text{slope} = \frac{\text{change in angle of rotation}}{\text{change in concentration}}$$

EXAMPLE 19.6 The specific rotations of α–glucose and β-glucose in water are given in Table 2.9 as +112° and +18.7°, respectively. Given the glucose (α + β mixture)-specific rotation value of 52.7°, calculate the equilibrium concentrations of α- and β-glucose.

(% α-glucose)(sp rotation) + (% β-glucose)(sp rotation) = mixture rotation

$$(x)(+112°) + (1 - x)(+18.7°) = 52.7°$$
$$x = 0.36$$

Therefore: α-glucose = 36% and β-glucose = 64%.

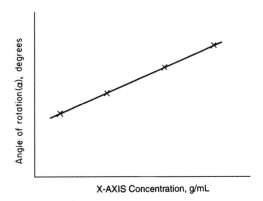

FIGURE 19.30 Graphing to obtain specific rotation.

RHEOLOGY

A. Young's Modulus and Elasticity

Rheology is the study of the deformation or flow of matter, especially metals, alloys, and plastics. Production rates of finished extruded and molded goods can often be increased by altering the formulation of the plastic mixture. The property of **elasticity** in a body is its ability to return to its original condition after the force which caused a distortion is removed. We do not normally think of glass or steel as being highly elastic bodies, but nevertheless, there are more elastic materials such as rubber, which return to their original shape once a distorting force has been removed. When a force, generally referred to as a stress, is applied to an elastic body, the distortion, or strain, which it causes is directly proportional to the stress. Conversely, if an elastic body is strained, an internal stress will be developed. The fact that the stress and strain are directly proportional is known as **Hooke's law.** Expressing this mathematically: k = stress/strain. The value of k is a constant for any given material and is known as the modulus of elasticity.

If too great a stress is applied to a body, it can be permanently distorted even after removal of the stress, indicating that there is a limit beyond which the proportionality expressed by Hooke's law does not hold. This stress is known as the **elastic limit** of the material. If a stress is applied to a body beyond the elastic limit, eventually the body will rapture. The stress at which this occurs is known as the breaking stress. The type of elasticity which is usually of greatest interest is the change in length of a body under a tensile or compressive load. When the stress is expressed as force per unit area and the strain as a change in length per unit length, we see a special form of the **elastic modulus.**

Young's modulus (Y) is expressed mathematically by the following equation. The uppercase Greek letter delta (Δ) indicates a change, so that the symbol ΔL means the change in length.

$$Y = \frac{\text{stress}}{\text{strain}} = \frac{F/A}{\Delta L/L}$$

In English units, force is expressed in pounds, area in square inches, and length in inches; therefore, the unit for Young's modulus is lb/in^2.

$$Y = \frac{\text{lb/in}^2}{\text{in/in}} = \frac{\text{lb}}{\text{in}^2}$$

EXAMPLE 19.7 A copper core 0.0160 in. in diameter and 36 in long is placed under a tensile force of 1.65 lb. This causes an elongation of 0.0170 in. What is Young's modulus for copper?

The cross-sectional area of wire

$$= \frac{d^2}{4} = \frac{(3.14)(0.0160)^2}{4} = 0.000201 \text{ in}^2$$

$$Y = \frac{\text{stress}}{\text{strain}} = \frac{F/A}{\Delta L/L} = \frac{1.63 \text{ lb}/0.000201 \text{ in}^2}{0.0170 \text{ in}/36 \text{ in}}$$

$$= 17.4 \times 10^6 \text{ lb/in}^2$$

Table 19.11 gives the elastic modulus for copper wire as 15.80×10^6 lb/in^2.

TABLE 19.11 Elastic (Young's) Modulus for Some Materials

Material	Young's Modulus	
	lb/in$^2 \times 10^6$	dyn/cm$^2 \times 10^{11}$
Aluminum, rolled	9.85	6.91
Brass, cold-rolled	13.09	9.02
Copper, rolled	18.06	12.46
Copper, wire	15.80	11.10
Iron, cast	13.0	9.1
Iron, wrought	27.5	19.4
Tungsten, drawn	51.49	35.50

B. Bulk Modulus and Compressibility

Solids can be placed under tensile or compressive stresses, but there would be little meaning in everyday experience in applying a tensile force to a liquid. Nevertheless, liquids display a high degree of elasticity when subjected to a compressive stress. When pressure is applied to a given volume of liquid, the volume tends to become smaller, and like Hooke's law for solids, the ratio of the change in pressure to the change in volume per unit volume is a constant and is known as the bulk modulus. The **bulk modulus** (B) is defined by the following equation:

$$B = -\frac{\Delta p}{\Delta V/V}$$

The minus sign appears because an increase in pressure causes a decrease in the volume of the liquid. Table 19.12 lists the bulk modulus for a variety of common liquids. As in the case of Young's

TABLE 19.12 Bulk Modulus for Various Liquids at 20°C

Liquid	Bulk modulus lb/in^2 ($\times 10^5$)	Bulk modulus dyne/cm$^2 \times (10^{10})$
Acetone	1.15	0.79
Aniline	3.21	2.21
Benzene	1.54	1.06
Butyl alcohol	1.74	1.20
Carbon tetrachloride	1.41	0.97
Ethyl alcohol	1.31	0.90
Ethylene glycol	3.92	2.70
Mercury	3.53	2.43
Methyl alcohol	1.19	0.82
Water	3.17	2.18

modulus, the units for bulk modulus are lb/in². Compressibility is the reciprocal of bulk modulus and has unit of in²/lb.

EXAMPLE 19.8 A vessel holding 1000 gal of water is subjected to a pressure of 5000 lb/in². At this pressure, what is the volume of the water? Table 19.12 gives the bulk modulus of water as 3.17×10^5 lb/in².

$$B = -\frac{\Delta p}{\Delta V / V}$$

$$\Delta V = -\frac{\Delta p \; V}{B} = -\frac{5.000 \text{ lb/in}^2 (1000 \text{ gal})}{3.17 \times 10^5 \text{ lb/in}^2} = -15.8 \text{ gal}$$

The final volume is therefore

$$V - \Delta V = 1000 - 16 \text{ gal} = 984 \text{ gal}$$

C. Hardness

Hardness is another property of matter which frequently has to be described in chemical work. The **hardness** of a substance is fundamentally a measure of its ability to abrade or indent another substance. An arbitrary scale of hardness, based on the ability of 10 selected minerals to scratch each other, is known as the **Mohs scale**. The Mohs scale shown in Table 19.13 is based on the following minerals, with talc the softest and diamond the hardest. The hardness of various materials can also be obtained based on the ability of one to scratch the other. These relative hardnesses for a number of important chemical materials and minerals have been determined. For metals and plastic materials, special systems for measuring hardness have been developed.

TABLE 19.13 Hardness Scale (Mohs Scale)

Material	Mohs scale value
Talc	1.0
Gypsum	1.8
Calcite	3.0
Fluorite	4.0
Apatite	5.0
Feldspar	6.0
Quartz	7.0
Topaz	8.0
Corundum	9.0
Diamond	10

EXAMPLE 19.9 Based on the assigned Mohs scale of hardness values shown in Table 19.13, asphalt has a value of 1.5, aluminum metal a 2.5, and iron a 4.5 value.

REFERENCES

1. *Chemical Rubber Company Handbook of Chemistry and Physics*, 78th edition, CRC Press, Inc. ISBN 0-8493-0595-0 (1997–1998).
2. Perry, Robert H., and Green, Don W., *Perry's Chemical Engineering Handbook*, McGraw-Hill Book Co. ISBN 0-07-115448-5 (1997).
3. *Lange's Handbook of Chemistry*, 15th edition, Section 11, John A. Dean, editor, McGraw-Hill Inc. ISBN 0-07-016384-7 (1999).
4. Gordon, Arnold J., and Ford, Richard A., *The Chemist's Companion, A Handbook of Practical Data, Techniques, and References*, John Wiley & Sons, Inc. ISBN 0-471-31590-7 (1972).

CHAPTER 20
EXTRACTION

INTRODUCTION

Solutes have different solubilities in different solvents, and the process of selectively removing a solute from a mixture with a solvent is called **extraction**.

The solute to be extracted may be in a solid or in a liquid medium, and the solvent used for the extraction process may be water, a water-miscible solvent, or a water-immiscible solvent. The selection of the solvent to be used depends upon the solute and upon the requirements of the experimental procedure.

SOXHLET EXTRACTION

A **Soxhlet extractor** (Fig. 20.1) can be used to extract solutes from solids, using any desired volatile solvent, which can be water-miscible or water-immiscible. The solvent is vaporized. When it condenses, it drops on the solid substance contained in a thimble and extracts soluble compounds. When the liquid level fills the body of the extractor, it automatically siphons into the flask. This process continues repeatedly as the solvent in the flask is vaporized and condensed.

Procedure
Set up apparatus as illustrated in Fig. 20.1.

1. Put the solid substance in the porous thimble, and place the partially filled thimble in the Soxhlet inner tube.
2. Fill the flask one-half full of extracting solvent.
3. Assemble the unit.
4. Turn on the cooling water.
5. Heat.
6. When the extraction is complete, turn off the heat and cooling water.
7. Dismantle the apparatus, and pour the extraction solvent containing the solute into a beaker. Isolate the extracted component by evaporation.

EXTRACTION OF A SOLUTE USING IMMISCIBLE SOLVENTS

Principle
A solute may be soluble in many solvents that are immiscible. When a solution of that solute in one of two immiscible solvents is shaken vigorously with the other immiscible solvent, the solute will be distributed between the two solvents in such a manner that the ratio of the concentrations (in moles per liter) of the solute is constant. This ratio is called the **distribution coefficient**, and it is independent of the volumes of the two solvents and the total concentration of the solute.

This type of extraction transfers a solute from one solvent to another. It can be used to separate reaction products from reactants and to separate desired substances from others in solution. The separatory funnel is used for this purpose. **Immiscible solvents,** which are incapable of mixing with

408 CHAPTER TWENTY

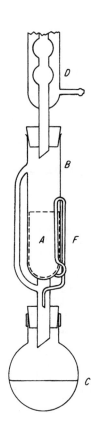

FIGURE 20.1 Soxhlet extractor. D = reflux condenser, B = body of extractor, A = extraction thimble, F = liquid return siphon, C = extracting solvent.

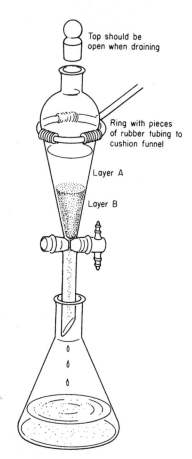

FIGURE 20.2 The extraction procedure: separating immiscible liquids by means of a separatory funnel.

each other to attain homogeneity and will separate from each other into separate phases, *must be used*. **Miscible solvents,** which are capable of being mixed in any ratio without separation into two phases, cannot be used.

NOTE: Multiple extractions with smaller portions of the extraction solvent are more effective than one extraction with a large volume.

The choice of the extraction solvent determines whether the solute remains in the separatory funnel or is in the solvent which is drawn off. *The solvent that has the greater density will be the bottom layer.* Thus the less dense extraction solvent remains in the separatory funnel, and the more dense extraction solvent is drawn off.

A. Using Extraction Solvent of Higher Density

Use a separatory funnel with a ring stand or other support. (See Fig. 20.2.)

  1. Use a clean separatory funnel, lubricating the barrel and plug of the funnel with a suitable lubricant. (Refer to procedures for the lubrication of a stopcock, Chapter 8.)

2. Pour the solution to be extracted into the funnel, which should be large enough to hold at least twice the total volume of the solution and the extraction solvent.
3. Pour in the extraction solvent; close with the stopper.
4. Shake the funnel *gently*.
5. Invert the funnel and open the stopcock slowly to relieve the pressure built up.
6. Close the stopcock while the funnel is inverted, and shake again.
7. Repeat steps 5 and 6.
8. Place the funnel in a ring-stand support and allow the two layers of liquid to separate. Remove stopper closure.
9. Open the stopcock slowly and drain off the bottom layer.
10. Repeat operation, starting at step 3, with fresh extraction solvent as many times as desired.
11. Combine the lower layers that have been drawn off.

Cautions and Techniques

1. Separator funnels are very fragile and expensive (especially if they are graduated). If one is to be supported in an iron ring stand, pad the ring with rubber tubing (cut longitudinally). This prevents breakage if the funnel bumps against the ring or the ring support when it is removed or inserted.
2. Always be certain that the stopcock is in a closed position before the funnel is returned to the normal vertical position and that it is securely seated, not floating loosely.
3. Always hold the stopper securely seated in the funnel. When the funnel is shaken vigorously to mix the two solvents, pressure inside the funnel increases as a result of the additive effect of the two partial vapor pressures of the two immiscible solvents. If $NaHCO_3$ is used and there are acids present, the pressure may also be increased by the resultant evolution of CO_2 gas. The high pressure is reduced to atmospheric pressure by holding the inverted funnel in both hands, holding the stopper securely in place with one hand and opening the stopcock with the other (Fig. 20.3). The venting procedure is repeated as many times as is necessary, until no further pressure buildup can be detected. This state is signaled by the disappearance of the audible "whoosh" of the escaping vapors.

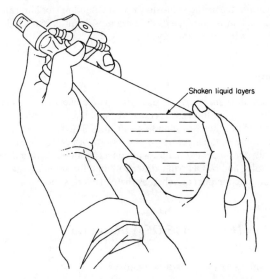

FIGURE 20.3 How to hold, shake, and vent a separator funnel.

4. Traces of insoluble material often collect at the interfaces between the two insoluble liquids. It is extremely difficult to separate the layers without taking some of this material along, but it can be easily removed by filtration when the extraction is completed, or at an even later stage, while washing or drying the extract.

EXTRACTION PROCEDURES IN THE LABORATORY

Extraction procedures are used to separate, purify, and analyze substances; three principal methods follow.

A. Using Water

Water is a polar solvent, and polar substances are soluble in it; examples of such substances are inorganic salts, salts of organic acids, strong acids and bases, low-molecular-weight compounds, carboxylic acids, alcohols, polyhydroxy compounds, and amines. Water will extract these compounds from any immiscible organic solvents that contain them.

B. Using Dilute Aqueous Acid Solution

Dilute hydrochloric acid (between 5 and 10 percent HCl) will extract basic substances such as organic amines, cyclic nitrogen-containing ring compounds, alkaloids, etc. The basic compound is converted to the corresponding hydrochloride, which is soluble in the aqueous solution, and is therefore extracted from the immiscible organic solvent. After the acid extraction is completed, the organic solvent is extracted with water to remove any acid that might be left in the organic solvent.

C. Using Dilute Aqueous Basic Solution

Dilute NaOH or 5 percent $NaHCO_3$ will extract acidic solutes from an immiscible organic solvent by converting the acidic solute to the corresponding sodium salt, which is soluble in water. After the basic extraction is completed, the organic solvent is extracted with water to remove any base that might be left in the organic solvent.

D. Selective Extraction

Extraction procedures may be used to separate phenols from carboxylic acids because, even though phenols are acidic, the carboxylic acids are about 10^5 times more acidic than phenols. Phenols are not converted to the corresponding salt by $NaHCO_3$, whereas carboxylic acids are converted to the corresponding salt. Therefore an extraction with $NaHCO_3$ solution will extract the carboxylic acid as the salt. An NaOH extraction will convert both phenols and carboxylic acids to the corresponding sodium salts, and NaOH will extract both into the aqueous phase. The NaOH is a strong enough base to deprotonate the phenol, whereas the $NaHCO_3$ is too weak.

CONSIDERATIONS IN THE CHOICE OF A SOLVENT

1. Like substances tend to dissolve like substances.
2. Organic solvents tend to dissolve organic solutes (see Table 20.1).

TABLE 20.1 Common Organic Solvents

Lighter than water	Heavier than water
Diethyl ether	Chloroform
Benzene	Ethylene dichloride
Petroleum ether	Methylene chloride
Ligroin	Tetrachloromethane
Hexane	

TABLE 20.2 Characteristics of Various Solvents Used for Extraction of Aqueous Solutions

Compound	Characteristics
Diethyl ether	Is generally a good solvent; absorbs 1.5 percent water; has a strong tendency to form peroxides.
Methylene chloride	May form emulsions; is easily dried.
Petroleum ethers (pentanes, hexanes, etc.)	Are easily dried; are poor solvents for polar compounds.
Benzene	Tends to form amulsions.
Ethyl acetate	Is good for polar compounds; absorbs large amounts of water.
2-butabol	Is good for highly polar compounds; dries easily.
Tetrachloromethane	Is good for nonpolar compounds; is easily dried.
Chloroform	Is easily dried; tends to form emulsions.
Diisopropyl ether	Tends to form peroxides.

3. Water tends to dissolve inorganic compounds and salts of organic acids and bases.
4. Organic acids, soluble in organic solvents, can be extracted into water solutions by using bases (NaOH, Na_2CO_3, or $NaHCO_3$).

See Table 20.2 for information regarding various solvents.

A. Diethyl Ether

Diethyl ether is a commonly used organic solvent because

1. It is easily removed from solutes because of its high volatility.
2. It is cheap.
3. It is an excellent high-power solvent.

However, it has disadvantages:

1. It is a fire hazard.
2. It is toxic.
3. It is highly soluble in water.
4. It is poorly recoverable because of its high volatility.
5. It is an explosion hazard because peroxides form.

PEROXIDES IN ETHER

The safety of using diethyl ether can be increased by detecting and removing any peroxides that form.

A. Detection of Peroxides in Ethers

Ethers tend to form peroxides on standing, and these peroxides can cause severe and destructive explosions. Always test ethers for peroxides before distilling them, either in concentrating solutions or purifying the ethers. Any one of three tests can be used.

Test 1. Colorless ferrothiocyanate changes to red ferrithiocyanate.

1. Prepare reagent.
    (a) Dissolve 9 g $FeSO_4 \cdot 7H_2O$ in 50 mL 18 percent HCl.
    (b) Add a little granular zinc.
    (c) Add 5 g sodium thiocyanate. When the red color fades, add an additional 12 g of sodium thiocyanate.
    (d) Decant clear supernatant from excess zinc into fresh storage bottle.
2. Add ether to be tested dropwise to reagent in a test tube. The solution will turn red if peroxides are present. This test is sensitive to 0.001 percent.

Test 2

Method 1. Colorless iodide changes to yellow (brown) iodine.

1. Prepare 10 percent aqueous potassium iodide (KI) solution.
2. Add 1 mL of KI solution to a sample of the ether to be tested. Let the mixture stand 1 min.
3. Appearance of yellow color indicates peroxides.

Method 2.

1. Prepare 10 percent KI solution in glacial acetic acid (100 mg KI/mL).
2. Add 1 mL of liquid to be tested to 1 mL of the solution.
3. A yellow color indicates a low peroxide concentration; a brown color indicates a high peroxide concentration.

Test 3

1. Prepare a test solution containing the following proportion of substances (a 0.1 percent solution):

    1 mg $Na_2Cr_2O_7 \cdot 2H_2O$
    1 mL H_2O
    1 drop dilute H_2SO_4

2. Add a few drops of the solution containing peroxides to the test solution. The development of a blue color indicating presence of perchromate ion in the organic layer is a positive test for peroxides.

B. Removal of Peroxides from Ethers

There are several convenient methods that can be used:

1. Pass the ether through a column containing activated alumina.

    CAUTION: Do not allow alumina to dry out. Elute or wash the alumina with 5 percent aqueous $FeSO_4$.

2. Store the ether over activated alumina.

CAUTION: See the preceding text.

3. Shake ether with a concentrated $FeSO_4$ solution (100 g $FeSO_4$ + 42 mL conc. HCl + 85 mL H_2O).

NOTE: Some ethers produce aldehydes when so treated. Remove them by washing with 1 percent $KMnO_4$ followed by 5 percent aqueous NaOH extraction to remove any acids formed; follow again with water wash.

4. Wash ethers with sodium metabisulfite solution: sodium pyrosulfite ($Na_2S_2O_5$). This substance reacts stoichiometrically with ethers.
5. Wash ethers with cold triethylenetetramine (make the mixture 25 percent by mass of ether).
6. In cases of water-soluble ethers, reflux with 0.5 percent (by mass) CuCl and follow by distillation of the ether.

RECOVERY OF THE DESIRED SOLUTE FROM THE EXTRACTION SOLVENT

After extraction, the extraction solvent and the solute that it now contains are processed to recover the *solute* wanted and, if practical, to reclaim the extraction solvent for economic reasons. (Refer to Chapter. 21, "Distillation and Evaporation.")

A. Emulsions

Frequently, when aqueous solutions are extracted with organic solvents, or organic solutions are extracted with aqueous solutions, emulsions form instead of two separate and distinct phases. **Emulsions** are colloidal suspensions of the organic solvent in the aqueous solvent or suspensions of the aqueous solvent in the organic solvent as minute droplets. This situation occurs when the solute may act as a detergent or soap or when viscous and gummy solutes are present. It also may happen if the separatory funnel containing the two solvents is shaken especially vigorously. Once emulsions form, it may be a very long time before the components separate.

B. Breaking Emulsions

Emulsion formation can be minimized by gently swirling the separatory funnel instead of shaking it vigorously, or the funnel may be gently inverted many times to achieve extraction.

Emulsions caused by too small a difference in the densities of the water and organic layer can be broken by the addition of a high-density organic solvent such as tetrachloromethane (carbon tetrachloride). Pentane can be added to reduce the density of the organic layer, if so desired, especially when the aqueous layer has a high density because of dissolved salts. Saturated NaCl or Na_2SO_4 salt solutions will increase the density of aqueous layers.

"**Salting out**" also helps to separate the layers. In simple extractions, the distribution coefficient of the extracting solvent may be increased by the addition of a soluble inorganic salt (NaCl or Na_2SO_4) to the water layer. The salt dissolves in the water layer and decreases the solubility of organic liquids in it. The mixture of immiscible liquids in the separatory funnel may form a **homogeneous solution** on shaking and may not separate to form two separate and distinct layers. This condition may be caused by the presence of a mutual solvent, such as alcohol or dioxane. In these cases, add NaCl or Na_2SO_4 crystals or small increments of a saturated solution of these salts, and shake again. Normally, two layers will begin to separate.

One or more of the following techniques may also be of value in breaking emulsions:

1. Add a few drops of silicone defoamer.
2. Add a few drops of dilute acid (if permissible).
3. Draw a stream of air over the surface with a tube connected to a water pump.
4. Place the emulsion in a suitable centrifuge tube and centrifuge until the emulsion is broken.
5. Filter by gravity or with a Büehner funnel (using an aspirator or pump, for higher vacuum).
6. Add a few drops of a detergent solution.
7. Allow the emulsion to stand for a time.
8. Place the emulsion in a freezer.

EXTRACTION BY DISPOSABLE PHASE SEPARATORS

When many extractions must be made, speed is important. A new method utilizes the 1PS® separator manufactured by Whatman, Inc. 1PS is a high-speed, disposable medium providing complete separation of immiscible aqueous solutions from organic solvents. The 1PS form of separatory funnel is so effective and yet so low-cost that it can be discarded freely after use. An important advantage of 1PS (over conventional, fragile, and expensive separatory funnels) is that 1PS simultaneously, with phase separation, filters out most if not all solids in the organic phase.

The 1PS can be used with solvents that are either lighter or heavier than water. If heavier than water, the solvent passes directly through the apex of the 1PS cone; if lighter than water, the solvent passes through the walls of the cone.

It is not necessary when using 1PS to wait until the two phases have settled out into separate layers; 1PS will separate drops of one phase suspended in the other just as well as if settling out had occurred. It will not, however, separate the components of a stable emulsion.

1PS can be used *flat* under suction to separate solvents heavier than water, provided that the pressure differential does not exceed 70 mmHg.

1PS is unaffected by mineral acids to 4 N. It will tolerate alkalies to 0.4 M.

Successful separations have been achieved at 90°C. However, surface tension is inversely proportional to temperature; temperature limits for a given separation, therefore, must be experimentally determined.

Phase separation *will not occur* if the interfacial surface tension is lowered by addition of polar compounds to the system. Surface tension is also lowered in the presence of surface-active agents, i.e., surfactants, above certain proportions.

Procedure
Disposable Phase Separator 1PS separates aqueous and organic solvent phases in a simple conical filter funnel. (See Fig. 20.4.)

1. Fold the paper in the normal way and place it in a conical filter funnel.
2. Pour the mixed phases directly into the funnel. It is not normally necessary to allow the phases to separate cleanly before pouring.
3. Allow the solvent phase to filter completely through the paper.
4. If required, wash the retained aqueous phase with a small volume of clean solvent. This is normally necessary when separating lighter-than-water solvents in order to clean the meniscus of the aqueous phase.

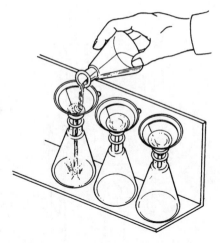

FIGURE 20.4 Extraction with the Disposable Phase Separator.

Do not allow the aqueous phase to remain in the funnel for a long time after phase separation, or it will begin to seep through. This effect is caused by evaporation of the solvent from the pores of the paper, creating a local vacuum which in turn draws the water through.

CONTINUOUS LIQUID-LIQUID EXTRACTIONS

When a solute is to be transferred from one solvent into another, a procedure which requires many extractions with large volumes of solvent because of its relative insolubility, continuous liquid-liquid extractions may be used. Specially designed laboratory glassware is required, because the extracting solvent is continually reused as the condensate from a total reflux. Choose the solvent in which the solubility of the desired solute to be extracted is most favorable, and one in which the impurities are least soluble. The solvents must be immiscible.

A. Higher-Density-Solvent Extraction

In this method the extracting solvent has a higher density than the immiscible solution being extracted. The condensate from the total reflux of the extracting heavier solvent is diverted through the solution to be extracted, passes through that solution, extracting the solute, and siphons back into the boiling flask (Fig. 20.5). Continuous heating vaporizes the higher-density solvent, and the process is continued as long as is necessary.

B. Lower-Density-Solvent Extraction

The extracting solvent has a lower density than the immiscible solution being extracted. The condensate from the total reflux of the extracting lower-density solvent is caught in a tube (Fig. 20.6). As the tube fills, the increased pressure forces some of the lower-density solvent out through the bottom. It rises through the higher-density solvent, extracting the solute, and flows back to the boiling flask. Continuous heating vaporizes the low-density solvent, and the process is continued as long as is necessary.

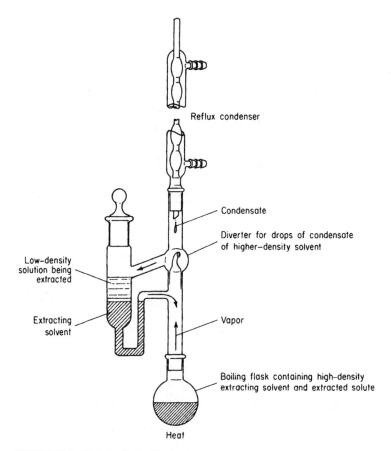

FIGURE 20.5 High-density liquid extractor.

ULTRAFILTRATION

Dialysis of small sample volumes for buffer exchange and desalting is time-consuming; solutes diffuse at different rates, and often enormous quantities of dialysate are needed. Ultrafiltration with the Immersible Molecular Separator® is a better way.

The Immersible Molecular Separator consists of a Pelicon membrane mounted on a cylindrical plastic core. The membrane can handle substances with molecular weights up to 10,000. To use the equipment, simply dip it in the sample solution, attach the tubing to a suitable vacuum source, and process. Solutes are removed automatically through the membrane into the vacuum trap, while macromolecules (with molecular weights over 10,000) are retained in the original vessel. Adsorption of protein to the membrane is negligible.

Figure 20.7 depicts a typical setup for constant-volume buffer exchange. The Immersible Molecular Separator is placed in a stoppered test tube or vessel containing the sample. The immersible tubing is attached to a vacuum source with a trap, while the line from the replacement solution reservoir is connected to the sample container. During filtration, the vacuum created by the departing filtrate draws in an equal amount of new solution to mix with the sample.

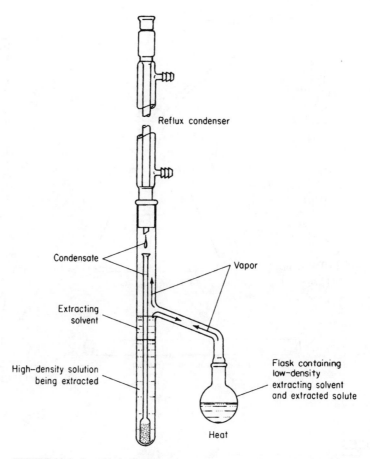

FIGURE 20.6 Low-density liquid extractor.

KUDERNA-DANISH SAMPLE CONCENTRATOR

When trace quantities of materials are extracted, it is necessary to evaporate most of the solvent to concentrate the desired compound for further analysis. This can be difficult to do without losing the desired compound. A current technique of removing large volumes of solvent without appreciable loss of the desired compound is a **Kuderna-Danish concentrator**. A typical concentrator is shown in Fig. 20.8. The bottom receiving vessel is generally held firmly in place by small springs to prevent component loss. The extract is placed in the flask, and the air-cooled Snyder column is attached as shown. The apparatus is placed in a special heating block or a steam bath so that the collector is in the steam and hot water. If the solvent boils at a low temperature, be careful not to increase the amount of heating at any time or an explosion may result. If fresh solvent is added to wash the Snyder column, be certain to add a fresh boiling chip. With a fatty compound and/or large volumes of solvent, use two **Snyder columns** (placed one on top of the other in series) to prevent the sample from foaming out the top.

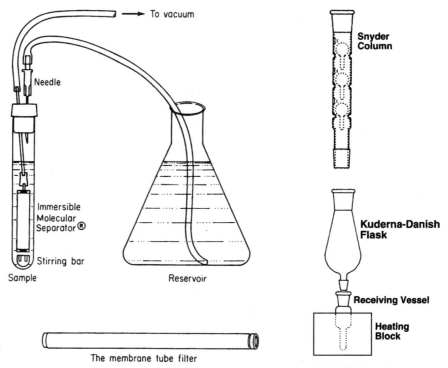

FIGURE 20.7 Ultrafiltration: dialysis in a test tube.

FIGURE 20.8 Kuderna-Danish sample concentrator.

REFERENCES

1. Robins, Lanny A., and Cusack, Roger W., Section 15, "Liquid-Liquid Extraction Operations and Equipment," *Perry's Chemical Engineers' Handbook*, 7th edition. McGraw-Hill, Inc. ISBN 0-07-048841-5 (1997).

2. Dean, John A., Section 11, "Practical Laboratory Information," *Lange's Handbook of Chemistry*, 15th edition. McGraw-Hill, Inc. ISBN 0-07-016384-7 (1999).

3. Gordon, Arnold J., and Ford, Richard A., "Experimental Techniques Section," *The Chemist's Companion, A Handbook of Practical Data, Techniques, and References*, John Wiley & Sons, Inc. ISBN 0-471-31590-7 (1972).

CHAPTER 21
DISTILLATION AND EVAPORATION

INTRODUCTION

Distillation is a process in which the liquid is vaporized, recondensed, and collected in a receiver. The liquid that has not vaporized is called the **residue**. The resultant liquid, the condensed vapor, is called the **condensate** or **distillate**.

Distillation is used to purify liquids and to separate one liquid from another. It is based on the difference in the physical property of liquids called **volatility**. Volatility is a general term used to describe the relative ease with which the molecules may escape from the surface of a pure liquid or a pure solid. The vapor pressure of a substance at a given temperature expresses this property. (See Fig. 21.1.)

A. Vapor Pressure

A volatile substance is one which exerts a relatively high vapor pressure at room temperature. A nonvolatile substance is one that exerts a low vapor pressure. The more volatile a substance, the higher its vapor pressure and the lower its boiling point. The less volatile a substance, the lower its vapor pressure and the higher its boiling point.

All liquids and solids have a tendency to vaporize at all temperatures, and this tendency varies with temperature and the external pressure that is applied. When a solvent is enclosed, vaporization will take place until the partial pressure of the vapor above the liquid has reached the vapor pressure at that temperature. Further evaporation of the liquid can be accomplished by removing some of the vapor above it, which in turn reduces the vapor pressure over the liquid.

SIMPLE DISTILLATION

An experimental setup for simple distillation is shown in Figs. 21.2 and 21.3. The glass equipment may be standard and require corks or may have ground-glass fitted joints. To be sure your setup is correct, follow the checklist below:

1. The distilling flask should accommodate twice the volume of the liquid to be distilled.
2. The thermometer bulb should be slightly below the side-arm opening of the flask. The boiling point of the corresponding distillate is normally accepted as the temperature of the vapor. If the thermometer is not positioned correctly, the temperature reading will not be accurate. If the entire bulb of the thermometer is placed too high, above the side arm leading to the condenser, the entire bulb will not be heated by the vapor of the distillate and the temperature reading will be too low. If the bulb is placed too low, too near the surface of the boiling liquid, there may be a condition of superheating, and the thermometer will show too high a temperature.
3. All glass-to-glass or glass-to-cork connections should be firm and tight.
4. The flask, condenser, and receiver should be clamped independently in their proper relative positions on a steady base.
5. The upper outlet for the cooling water exiting from the condenser should point upward to keep the condenser full of water.

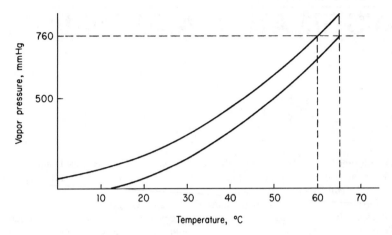

FIGURE 21.1 Dependence of vapor pressure on temperature.

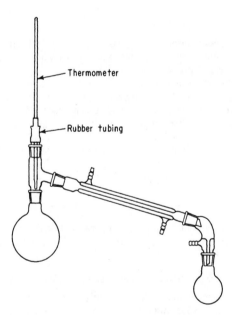

FIGURE 21.2 Apparatus for simple distillation at atmospheric pressure.

Procedure

1. Pour the liquid into the distilling flask with a funnel that extends below the side arm.
2. Add a few boiling stones to prevent bumping.
3. Insert the thermometer.
4. Open the water valve for condenser cooling.

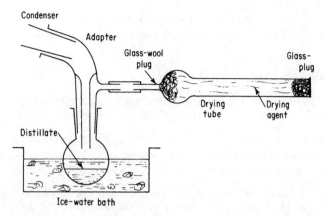

FIGURE 21.3 Protecting a distillate from atmospheric moisture.

5. Heat the distilling flask until boiling begins; adjust the heat input so that the rate of distillate is a steady two to three drops per second.
6. Collect the distillate in the receiver.
7. Continue distillation until only a small residue remains. Do not distill to dryness.

B. Distillation of Pure Liquids

Distillation can be used to test the purity of liquids or to remove the solvent from a solution.
The experimental setup for pure liquids is the same as that shown in Figs. 21.2 and 21.3.

1. The composition of the condensate is necessarily the same as the original liquid and is the same as the residue.
2. The composition does not change (see Fig. 21.4).

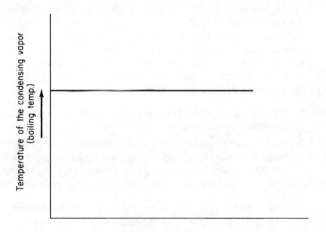

FIGURE 21.4 Distillation curve of a pure liquid.

3. The boiling temperature remains constant throughout the distillation.
4. Distillation establishes only the purity and boiling point of the pure liquid.

C. Distillation of a Solution

This process effects the separation of the nonvolatile dissolved solids because they remain in the residue and the volatile liquid is distilled, condensed, and collected.

1. The temperature of the distillate is constant throughout because it is pure.
2. The temperature of the boiling solution increases gradually throughout the distillation because the boiling solution becomes saturated with the nonvolatile solids.

When a nonvolatile substance is in the liquid being distilled, the temperature of the distilling liquid (the **head temperature**) will be the same as that of the pure liquid, since the vapor being condensed is uncontaminated by the impurity. The temperature of the pot liquid will be higher, because of the decreased vapor pressure of the solution containing the nonvolatile solute. The temperature of the pot liquid will continue to increase as the volatile component distills away, further lowering the vapor pressure of the solution and increasing the concentration of the solute.

CAUTION: When evaporating a solution to recover the solute or when using electric heat or burners to distill off large volumes of solvent to recover the solute, do not evaporate completely to dryness. The residue may be superheated and begin to decompose.

D. Distillation of a Mixture of Two Liquids

Principle
Simple distillation of a mixture of two liquids will not effect a complete separation. If both are volatile, both will vaporize when the solution boils and both will appear in the condensate. The more volatile of the two liquids will vaporize and escape more rapidly and will form a larger proportion of the distillate. The less volatile constituent will concentrate in the liquid that remains in the distilling flask, and the temperature of the boiling liquid will rise. The more volatile of the two liquids will appear first in the distillate. When the difference in the volatilities of the two liquids is large enough, the first distillate may be almost pure. The last of the distillate collected will be richer in the less volatile component. If there is sufficient difference in the volatilities of the two liquids, the last of the distillate may be almost pure. The distillate collected between the first portion and the last portion will contain varying amounts of the two liquids. (See Fig. 21.5.)

You may separate two liquids with different volatilities by *changing the receiver* several times during the distillation, thus collecting several portions of distillate.

1. The first portion collected while the boiling temperature is near that of the more volatile liquid may contain that liquid with little impurity.
2. The last portion collected when the distillation temperature is nearly equal to the boiling point of the less volatile liquid may contain that liquid and little of the other. Intermediate portions will contain *both liquids in varying proportions.* You may redistill each of the intermediate portions collected in the receiver to separate them further into their pure components.

The lower curve of the diagram in Fig. 21.5 gives the boiling points of all mixtures of these compounds. The upper curve gives the composition of the vapor in **equilibrium** with the boiling liquid phase. The vapor phase is much richer in the more volatile component, benzene, than the liquid phase with which it is in equilibrium. Therefore, the first few drops of vapor that condense will be richer in the more volatile component than in the less volatile one, toluene.

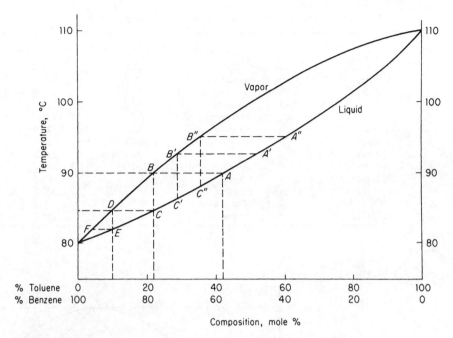

FIGURE 21.5 Boiling-point-composition diagram for the system benzene–toluene.

E. Rate of Distillation (No Fractionation)

The rate of distillation is controlled by the rate of the input of heat. In normal distillations a rate of about 3 to 10 mL/min, which corresponds roughly to one to three drops per second, is average. When liquids are to be separated by fractional distillation (see Fractional Distillation, this chapter), the rate may be a great deal smaller, depending upon the difficulty of the separation.

F. Concentration of Large Volumes of Solutions

When it becomes necessary to distill off large volumes of solvent to recover very small quantities of the solute, it is advisable to use a large distilling flask at first. (Never fill any distilling flask over one-half full.) When the volume has decreased, transfer the material to a smaller flask and continue the distillation. This minimizes losses caused by the large surface area of large flasks. If the solute is a high-boiling substance, the walls of the flask act as a condenser, making it difficult to drive the material over.

AZEOTROPIC DISTILLATION

Azeotropic mixtures distill at constant temperature without change in composition. Obviously, one cannot separate azeotropic mixtures by normal distillation methods.

Azeotropic solutions are nonideal solutions. Some display a greater vapor pressure than expected; these are said to exhibit *positive deviation*. Within a certain composition range such mixtures boil at temperatures higher than the boiling temperature of either component; these are **maximum-boiling azeotropes** (see Table 21.1 and Fig. 21.6).

TABLE 21.1 Maximum-Boiling-Point Azeotropic Mixtures

Component A		Component B		bp of azeotropic mixture, °C	% of B (by mass) in mixture
Substance	bp, °C	Substance	bp, °C		
Water	100.0	Formic acid	100.8	107.1	77.5
Water	100.0	Hydrofluoric acid	19.4	120.0	37.0
Water	100.0	Hydrochloric acid	−84.0	108.6	20.22
Water	100.0	Hydrobromic acid	−73.0	126.0	47.6
Water	100.0	Hydriodic acid	−35.0	127.0	57.0
Water	100.0	Nitric acid	86.0	120.5	68.0
Water	100.0	Sulfuric acid	10.5 (mp)	338.0	98.3
Water	100.0	Perchloric acid	110.0	203.0	71.6
Acetone	56.4	Chloroform	61.2	64.7	80.0
Acetic acid	118.5	Pyridine	115.5	130.7	65.0
Chloroform	61.2	Methyl acetate	57.0	64.8	23.0
Phenol	181.5	Aniline	184.4	186.2	58.0

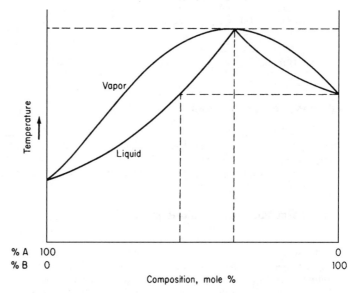

FIGURE 21.6 Maximum-boiling-point azeotrope.

Mixtures that have boiling temperatures much lower than the boiling temperature of either component exhibit *negative deviation*; when such mixtures have a particular composition range, they act as though a third component were present. In Fig. 21.7, the minimum boiling point at Z is a constant boiling point because the vapor is in equilibrium with the liquid and has the same composition as the liquid does. Pure ethanol (bp 78.4°C) cannot be obtained by fractional distillation of aqueous solutions that contain less than 95.57 percent of ethanol because this is the azeotropic composition; the boiling point of this azeotropic mixture is 0.3° lower than that of pure ethanol.

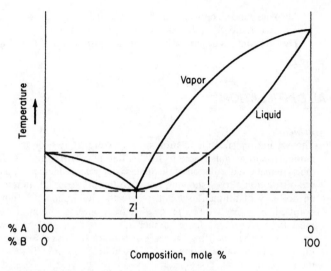

FIGURE 21.7 Minimum-boiling-point azeotrope.

Other examples are:

1. A mixture of 32.4 percent ethyl alcohol and 67.6 percent benzene (bp 80.1°C) boils at 68.2°C.
2. A ternary azeotrope (bp 64.9°C) is composed of 74.1 percent benzene, 18.5 percent ethyl alcohol, and 7.4 percent water. (See also Table 21.2.)

TABLE 21.2 Minimum-Boiling-Point Azeotropic Mixtures

Component A		Component B		bp of azeotropic mixture, °C	% of A (by mass) in mixture
Substance	bp, °C	Substance	bp, °C		
Water	100.0	Ethyl alchol	78.3	78.15	4.4
Water	100.0	Isopropyl alcohol	82.4	80.4	12.1
Water	100.0	n-Propyl alcohol	97.2	87.7	28.3
Water	100.0	$tert$-Butyl alcohol	82.6	79.9	11.8
Water	100.0	Pyridine	115.5	92.6	43.0
Methyl alcohol	64.7	Methyl iodide	44.5	39.0	7.2
Ethyl alcohol	78.3	Ethyl iodide	72.3	63.0	13.0
Methyl alcohol	64.7	Methyl acetate	57.0	54.0	19.0
Ethyl alcohol	78.3	Ethyl acetate	77.2	71.8	31.0
Water	100.0	Butyric acid	163.5	94.4	18.4
Water	100.0	Propionic acid	140.7	100.0	17.7
Benzene	80.2	Cyclohexane	80.8	77.5	55.0
Ethyl alcohol	78.3	Benzene	80.2	68.2	32.4
Ethyl alcohol	78.3	Toluene	110.6	76.7	68.0
Methyl alcohol	64.7	Chloroform	61.2	53.5	12.5
Ethyl alcohol	78.3	Chloroform	61.2	59.4	7.0
Ethyl alcohol	78.3	Methyl ethyl ketone	79.6	74.8	40.0
Methyl alcohol	64.7	Methylal	42.2	41.8	18.2
Acetic acid	118.5	Toluene	110.6	105.4	28.0

Absolute ethyl alcohol can be obtained by distilling azeotropic 95.5% ethyl alcohol with benzene. The water is removed in the volatile azeotrope formed.

The procedure is described earlier under Distillation of Pure Liquids.

FRACTIONAL DISTILLATION

Principle

The separation and purification of a mixture of two or more liquids, present in appreciable amounts, into various fractions by distillation is **fractional distillation**. It consists essentially in the systematic redistillation of distillates (fractions of increasing purity). Figures 21.8 through 21.10 are distillation curves showing how two liquids separate. Fractionations can be carried out with an ordinary distilling flask, but, where the components do not have widely separated boiling points, it is a very tedious process. A fractionating column (Fig. 21.11) is essentially an apparatus for performing a large number of successive distillations without the necessity of actually collecting and redistilling the various fractions. The glass column is filled with pieces of glass, glass beads, metal screening, or glass helices (Fig. 21.12). Some columns are more efficient than others.

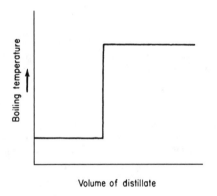

FIGURE 21.8 Curve for the ideal separation of two components.

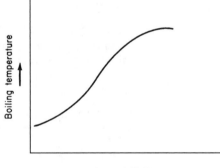

FIGURE 21.9 Curve for the actual separation, in a simple distillation apparatus, of two components that have boiling points close together.

The separation of mixtures by this means is a refinement of ordinary separation by distillation. Thus a series of distillations involving partial vaporization and condensation concentrates the more volatile component in the first fraction of distillate and leaves the less volatile component in the last fraction or in the residual liquid. The vapor leaves the surface of the liquid and passes up through the packing of the column. There it condenses on the cooler surfaces and redistills many times before entering the condenser. Each minute distillation causes a greater concentration of the more volatile liquid in the rising vapor and an enrichment of the residue which drips down through the column in the less volatile components.

By means of long and efficient distillation columns (see Fig. 21.11), two liquids may be completely separated.

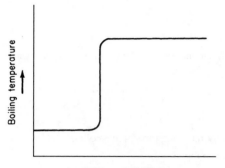

FIGURE 21.10 Curve for the actual separation of two liquids, which have boiling points close together, when the separation is carried out in a fractional distillation apparatus with an efficient column.

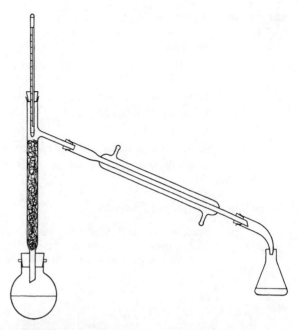

FIGURE 21.11 Fractional distillation apparatus, for use under vacuum or at atmospheric pressure. Column should be packed.

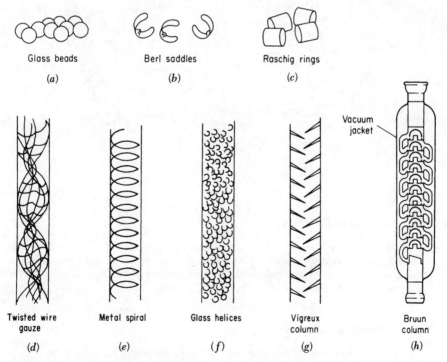

FIGURE 21.12 Packings for fractionating columns. Some (*a,b,c*) are shown loose; others (*d,e,f*) are shown in place. In the Vigreux (*g*) and Bruun (*h*) columns, the packings are built in.

427

Procedure

1. Select the type of fractionating column to be used, one that offers a large surface contact inside (see Fig. 21.12).
2. Set up the equipment.
3. Open the inlet cooling-water valve.
4. Apply heat.
5. Keep a large volume of liquid condensate continually returning through the column.
6. *Distill slowly* to effect efficient separation (See *Caution,* p. 437.)

A. Heating Fractionating Columns

Sometimes a fractionating column must be heated in order to achieve the most efficient fractionation of distillates. This may be accomplished by wrapping the column with heating tape or a resistance wire, such as Nichrome wire, controlled by variable transformers (see Fig. 21.13).

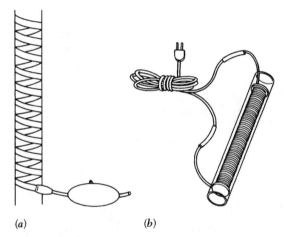

FIGURE 21.13 Methods of heating fractionating columns. (*a*) Using heating tape. (*b*) Nichrome wire heater.

B. Efficiency of Fractionating Columns

We measure the efficiency of a fractionating column in terms of the number of theoretical plates that it contains. A column with one theoretical plate is one in which the initial distillate has a vapor composition that is at equilibrium with the original solution. It is impossible to operate fractionating columns at equilibrium.

C. Bubble-Cap Fractionating Columns

Bubble-cap fractionating columns (Fig. 21.14) have definite numbers of trays or plates and are fitted with either bubble caps or sieve perforations—or modifications of these two—to enable the achievement of intimate vapor-liquid dispersion. However, because fractionating columns cannot

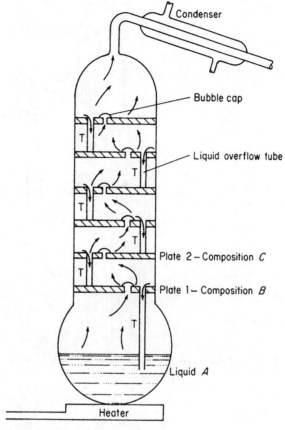

FIGURE 21.14 Bubble-cap fractionating column.

be operated at equilibrium (practically), the number of theoretical plates is always lower than the number of actual plates, depending upon the rate of distillation, reflux ratio, and other factors.

CAUTION: If you heat the pot too vigorously and remove the condensed vapor too quickly, the whole column will heat up uniformly and there will be no fractionation. The fractionating column will become flooded by the returning condensate.

D. Total-Reflux–Partial-Takeoff Distilling Heads

Exercise good judgment in the control of the amount of heat applied, and, for truly effective fractionation, use a total-reflux–partial-takeoff distilling head as shown in Fig. 21.15. With the stopcock S completely closed, all condensed vapors are returned to the distilling column, a total reflux condition. With the stopcock partially opened, the number of drops of condensate falling from the condenser which returns to the fractionating column can be adjusted. The ratio of the number of drops of distillate allowed to pass through stopcock S into the receiver to the number of drops of reflux is called the **reflux ratio**. With an efficient column, reflux ratios as high as 100 to 1 can be used to effectively separate compounds that have very close boiling points.

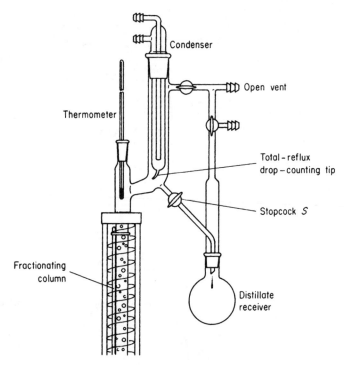

FIGURE 21.15 Total-reflux–partial-takeoff still head.

VACUUM DISTILLATION

Principle

Many substances cannot be distilled satisfactorily at atmospheric pressure because they are sensitive to heat and decompose before the boiling point is reached. Vacuum distillation, distillation under reduced pressure, makes it possible to distill at much lower temperatures. The boiling point of the material is affected by the pressure in the system. The lower the pressure, the lower the boiling point; the higher the pressure, the higher the boiling point.

A. Nomographs

Nomographs are special graphs that enable the technician to determine more accurately the boiling points at different pressures; they also provide a method of converting boiling points to the pressure desired.

Procedure for using the nomograph (see Fig. 21.16):

1. Select the desired boiling point at the reduced pressure.
2. Use a transparent plastic ruler and connect that boiling point (column A) with the given corresponding pressure (column G). It will intersect column B at a definite point. Record that point.
3. Using the point obtained from column B in step 2, select the new pressure desired on column C. Align the plastic ruler with these two points and read the corresponding temperature for the boiling point at the new pressure where the ruler intersects column A.

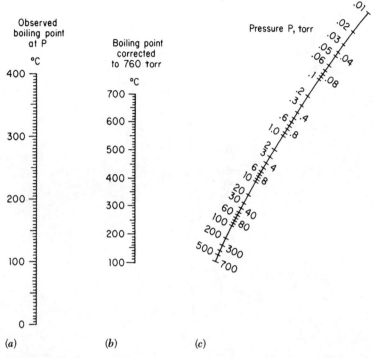

FIGURE 21.16 Nomograph for pressure and temperature at various boiling points.

EXAMPLE 21.1 A liquid boils at 200°C at 10 torr pressure. What would be (1) the normal boiling point at 760 torr and (2) the boiling point at 1 torr pressure?

SOLUTION Given: bp = 200°C (column A) at 10 torr (column C).

1. Intersection point in column B is 350°C, the boiling point at 760 torr pressure (corrected).
2. The line connecting the points 350°C on column B, and 1 torr pressure on column C intersects column A at 150°C. Thus 150°C is the boiling point at 1 torr pressure.

CAUTION: Glass equipment may collapse under reduced pressure. Use *safety glasses* and a *safety shield* as well as special equipment shown in Fig. 21.17.

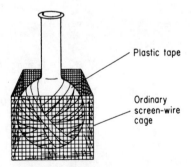

FIGURE 21.17 Safety precautions for a large flask in a vacuum system.

B. General Requirements

The experimental setup is as shown in Fig. 21.18.

1. *A source of vacuum.* Efficient water pumps, aspirators, will theoretically reduce the pressure in the system to the vapor pressure of the water passing through the pump. In practice, the pressure is usually about 10 mm higher. (*a*) Oil mechanical vacuum pumps. (*b*) Use rubber pressure

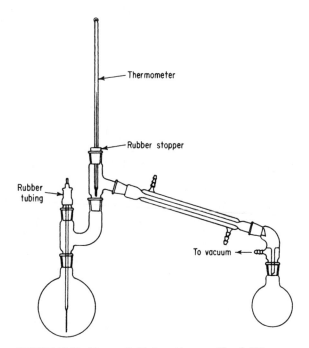

FIGURE 21.18 Vacuum distillation with gas-capillary bubbler.

tubing. (*c*) The entire distillation system should be airtight, free from leaks. (*d*) Lubricate all joints and connections.

2. *Safety trap* to protect manometer and vacuum source from overflow-liquid contamination (Fig. 21.19). The trap must be correctly connected. Vapors condense on the sides of the trap and fall to the bottom if they are not solidified.

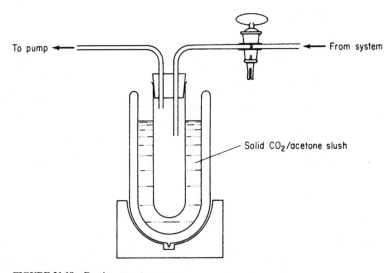

FIGURE 21.19 Dry-ice vapor trap.

Dry ice will freeze your skin; *do not handle with bare fingers or hands.*

(a) Crush dry ice in a cloth towel with a hammer.

(b) Use a scoop to fill the Dewar flask after the trap has been inserted.

(c) Add solvent, acetone, or isopropanol in small increments until the trap is filled and the liquid level is near the top.

3. *Pressure gauge* (manometer). Exercise great care when allowing air into the evacuated system. It must be done slowly to avoid breakage when the mercury column rises to the top of the closed tube. (Refer to the section on manometers in Chapter 5, "Pressure and Vacuum.")

4. *Manostat* (pressure regulator). To maintain constant pressure in the system, it automatically opens and closes needle valves, permitting air to enter or keeping the system airtight because of vacuum variations.

5. *Capillary air inlet* (Fig. 21.20).

6. *Special vacuum distillation* flasks to minimize contamination of the distillate caused by frothing of the boiling solution.

7. *Heating baths*, *electric mantles*, and *fusible alloy* or *sand baths*.

8. *Special distilling heads* (Fig. 21.21) to permit removal of distillate fractions without interrupting the distillation.

FIGURE 21.20 Capillary gas- or air-inlet tube.

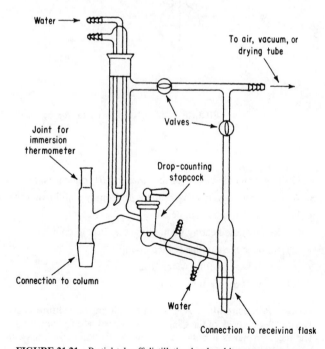

FIGURE 21.21 Partial-takeoff distillation head and its component parts.

C. Assemblies for Simple Vacuum Distillation and Fractionation

Simple Vacuum Distillation.

Procedure

A *Claisen flask* for use in this process is shown in Fig. 21.22.

1. Fill the Claisen flask (A) one-third full.
2. Apply vacuum; adjust the capillary air inlet (C) with the pinch clamp (D).
3. Heat bath to about 20°C higher than the temperature at which the material will distill.

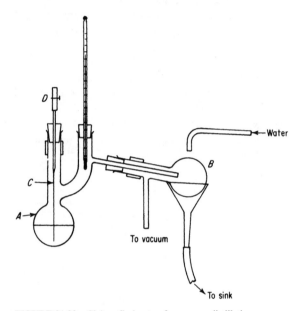

FIGURE 21.22 Claisen flask setup for vacuum distillation.

4. Cooling water flowing over the receiver (B) condenses the vapors to give a distillate.
5. A safety trap prevents any condensate from contaminating the suction pump or manometer.
6. When distillation is completed:
   (a) Remove the heating bath; allow the flask to cool.
   (b) Remove the capillary pinch clamp.
   (c) Cut off the cooling water.
   (d) Turn off the vacuum pump.

Vacuum Fractionation.   A **Claisen fractionating column** is shown in the apparatus in Fig. 21.23. The distillation neck of the Claisen flask serves as a fractionating column because of the indentations. The receiver makes it possible to remove distillate fractions without interrupting the distillation or breaking the vacuum.

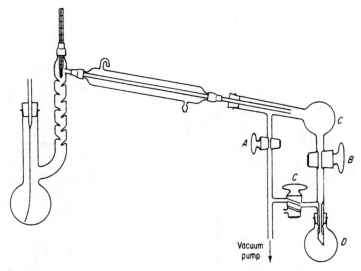

FIGURE 21.23 Claisen apparatus for fractional distillation in a vacuum.

Procedure
Follow the steps and observe the cautions previously listed. To remove distillate fractions during distillation:

1. Close stopcock B.
2. Close stopcock C.
3. Reverse rotation stopcock C so that air flows into flask D.
4. Gently remove flask D, empty into bottle, and replace.
5. Rotate stopcock C 180° so that D is now under vacuum.
6. Open stopcock B to allow collected distillate to drain into D.

Complete the fractional distillation and disconnect the equipment as described above.

D. Modified Distillate Receiver

Collecting individual distillate fractions without removing the receiver (as in the procedure above) can be accomplished with the modified receiver. The receiver is adjusted to collect the particular fraction desired and then rotated to meet the need. This method does not introduce any possible problems that might occur in the procedure if the vacuum distilling pressure is affected when the receiver is removed to isolate the distillate fraction. (See Fig. 21.24.)

A more complex apparatus is shown in Fig. 21.25.

STEAM DISTILLATION

Principle
Steam distillation is a means of separating and purifying organic compounds by volatilization. The organic compound must be insoluble or slightly soluble in water. When steam is passed into a mixture of the compound and water, the compound will distill with the steam. In the distillate, this distilled compound separates from the condensed water because it is insoluble in water.

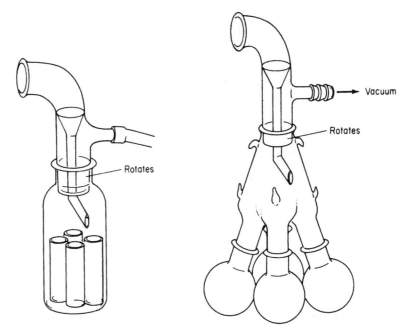

FIGURE 21.24 Modified receivers for collection of distillate fractions.

Most compounds, regardless of their normal boiling point, will distill by steam distillation at a temperature below that of pure boiling water. For example, naphthalene is a solid with a boiling point of 218°C. It will distill with steam and boiling water at a temperature less than 100°C.

Some high-boiling compounds decompose at their boiling point. Such substances can be successfully distilled at low temperature by steam distillation.

Steam distillation can be used to rid substances of contaminants because some water-insoluble substances are steam-volatile and others are not.

When it is desirable to separate nonvolatile solids from high-boiling organic solvents, steam distillation will remove all solvents (water-insoluble).

Procedure

1. Place the compound or mixture in the distilling flask with a little water. Pass cooling water through the condenser (see Fig. 21.26). A Claisen flask may be substituted for the round-bottomed flask. The Claisen still head helps to prevent any contamination of the distillate caused by spattering of the steam-distilled mixture. If there is no readily available source of piped steam, the steam can be generated in an external steam generator (Fig. 21.27) and then passed into the mixture to be steam-distilled.

CAUTION: Always equip the steam generator with a safety tube to prevent explosions.

NOTE: If only a small amount of steam is needed, for instance, to rid a substance of a small amount of steam-volatile impurity, water can be combined with the material directly in the distilling flask; the flask is then directly heated with a Bunsen flame or any other suitable heat source. The long steam-inlet tube is replaced with a stopper.

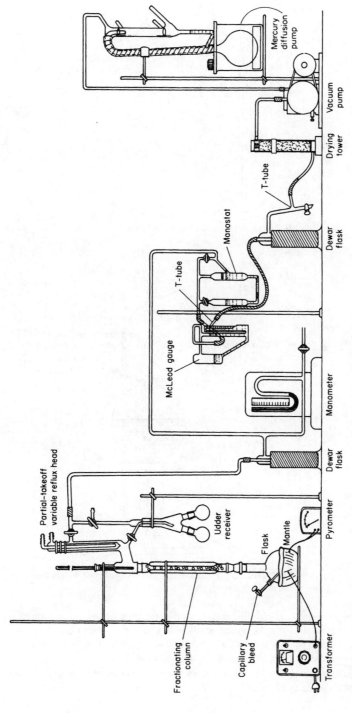

FIGURE 21.25 Apparatus setup using a diffusion pump for high-vacuum fractional distillation.

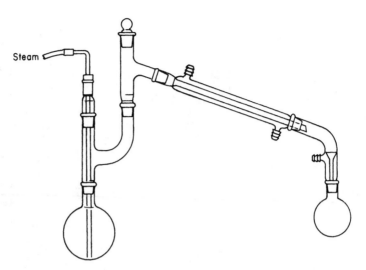

FIGURE 21.26 Apparatus for steam distillation.

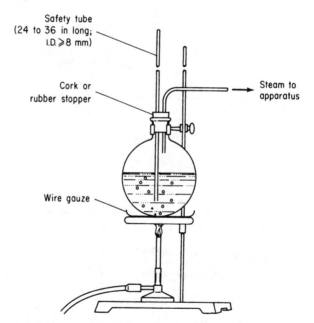

FIGURE 21.27 External steam generator equipped with boiling chips and a safety tube.

2. Pass steam into the distilling flask with the steam outlet below the surface of the liquid. The distilling flask *itself* may be heated gently with a burner. If steam is available from a laboratory steam line, insert a water trap (Fig. 21.28) in the entering steam line to trap condensed water. Otherwise, the condensed water may fill up the distilling flask.
3. Continue passing steam into the flask until no appreciable amount of water-insoluble material appears in the condensate.

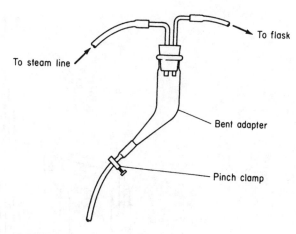

FIGURE 21.28 Water trap.

CAUTION: Steam will cause severe burns. Handle with care!

DANGER: If the substance crystallizes in the condenser, it will close the tube. Steam pressure could build up when the tube closes and cause an explosion. *Use care! Drain the condenser of cooling water. The crystals will melt and pass into the distillate.* When the tube is clear, slowly pass the cooling water through the condenser.

4. Always disconnect the steam-inlet tube from the flask.

A. Steam Distillation with Superheated Steam

The use of superheated steam can increase the proportion of the low-vapor-pressure component in the distillate, and, at the same time, reduce the amount of steam condensate in the distilling flask. The distilling flask is surrounded by a heating bath which is heated to the same temperature as the superheated steam; this minimizes any cooling of the steam which could occur before it enters the flask. (See Fig. 21.29.)

A commercially available metal superheater, heated by a Meker burner, is shown in Fig. 21.30.

REFLUXING

The reflux procedure allows you to heat a reaction mixture for an extended period of time without loss of solvent. The condenser, which is fixed in a vertical position directly above the heated flask, condenses all vapors to liquid. Because none of the vapors escape, the volume of liquid remains constant.

Reflux procedures are carried out in neutral, acid, or basic solution, depending upon the reaction.

Typical operations include hydrolysis-saponification of acid amides, esters, fats, nitriles, substituted amides, and sulfonamides. Hydrolysis-saponification is used to split organic molecules (which were made by combination of two or more compounds) into the original compounds.

Experimental setups are shown in Figs. 21.31 to 21.34.

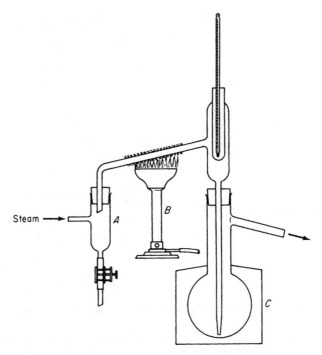

FIGURE 21.29 Distillation apparatus for use with superheated steam.

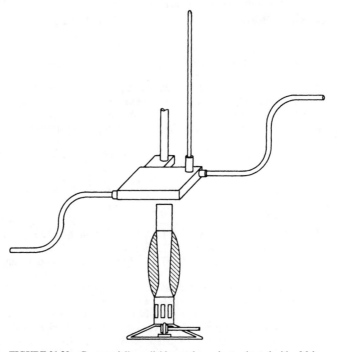

FIGURE 21.30 Commercially available metal superheater, heated with a Meker burner.

DISTILLATION AND EVAPORATION 441

FIGURE 21.31 Simple reflux condenser.

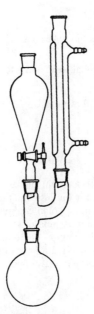

FIGURE 21.32 Addition tube and reflux condenser for use with high-boiling liquids.

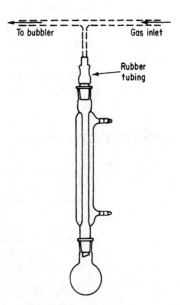

FIGURE 21.33 Apparatus for refluxing in an inert atmosphere.

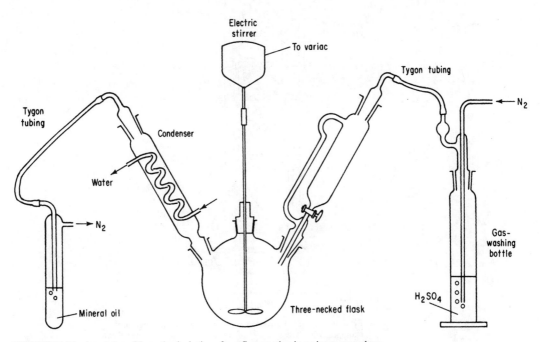

FIGURE 21.34 Apparatus with mechanical stirrer for reflux reaction in an inert atmosphere.

Procedure

1. The water inlet to the condenser is the lower one. The water outlet to the condenser is the upper one.
2. Fill the heating flask no more than half full; add a few boiling stones.
3. Turn on the cooling water.
4. Heat to reflux for the desired period of time.

PROCESS DISTILLATION

Large process distillation units usually contain **horizontal plates** or **trays** spaced from 1 to 3 ft apart. These plates serve as a means for the less volatile component vapors to condense and flow back down to a lower plate while the more volatile components pass through openings in the plates as vapor to higher levels in the column. A pipe or flue that conveys vapors or condensed liquids in a distillation tower is called a **downcomer**. There are over a dozen different designs for these trays (bubble cap, sieve, flexitrays, float-valve trays, uniflex trays, cascade trays, Turbogrid®, ripple trays, etc.), but they all serve the same purpose. Figure 21.35 is primarily used on a laboratory or

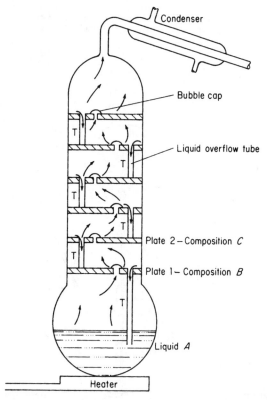

FIGURE 21.35 Fractionating tower shown with six trays equipped with bubble caps for the gases to rise and overflow tubes (T) for the liquids to return.

small-scale distillation. A process distillation would be designed to have a flow of continuous feed stock enter the tower at about midpoint and to provide outlets for the distilled products at various tray levels. The condenser removes the less volatile gaseous components and typically returns this condensed liquid back as reflux to the fractionating tower.

Packing materials are also used in commercial distillation units to replace the tray and bubble caps system. The distillation packings can consist of anything that provides surface area for the hot liquid and vapor to equilibrate. Packings consisting of **Raschig rings, Lessing rings, Berl saddles, partition rings**, copper mesh, iron mesh, and many other inert solids (see Fig. 21.36) are found in use. Most of these packing materials are placed in the distillation column in a random order. Some of the newer packings are placed in the distillation tower in a stacked or more ordered arrangement. These stacked packings generally cause lower pressure drops within the tower.

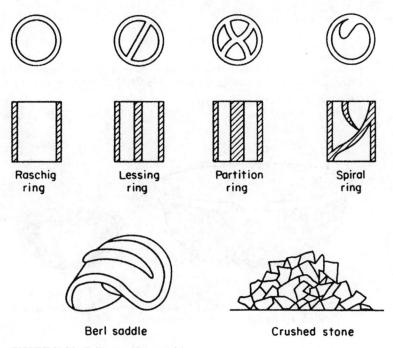

FIGURE 21.36 Column packing materials.

EVAPORATION OF LIQUIDS

A. Small Volumes

Evaporation of solvents is necessary at times to concentrate solutions and to obtain crystallization of solutes.

Method 1.

1. Pour the small volume of solution into the watch glass placed over a beaker of water (Fig. 21.37).
2. Boil the water. The heat transfer through the steam formed evaporates the solvent of the solution.

FIGURE 21.37 Evaporation over a water bath.

FIGURE 21.38 Alternate procedure using evaporating dish instead of watch glass.

Method 2.
Use an evaporating dish instead of a watch glass (Fig. 21.38). Evaporating dishes come in various sizes and are made of various materials (Fig. 21.39). Use the appropriate one.

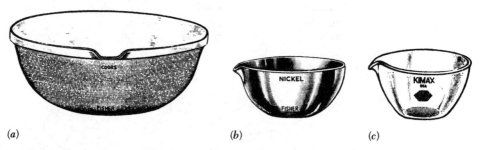

(*a*)   (*b*)   (*c*)

FIGURE 21.39 Evaporating dishes: (*a*) porcelain with heavy rim, (*b*) nickel, (*c*) crystallizing dish to hold or contain solutions from which solids are expected to crystallize.

B. Direct Heating of Evaporating Dishes

You can speed up the evaporation of water by directly heating the evaporating dish with a Bunsen burner, diffusing the heat with a wire gauze.

When the volume is very low, transfer the evaporating dish to the top of a beaker of boiling water. The steam acts as the heating agent. (See Fig. 21.39.)

C. Transferring Residues from Watch Glasses or Evaporating Dishes

When the water has been evaporated to the desired volume or the desired concentration, the material can be transferred to an appropriate container, such as a smaller evaporating dish, by rinsing with distilled water. The watch glass or evaporating dish is rinsed with a wash bottle, using a back-and-forth motion (Fig. 21.40).

FIGURE 21.40 Watch glass is rinsed with a wash bottle using a back-and-forth motion.

D. Large Volumes

Concentrate solutions by boiling off the desired volume of solvent.

Method 1.
1. Pour the solution that is to be concentrated by boiling off solvent into a suitably sized beaker that is covered by a watch glass resting on glass hooks (Fig. 21.41).
2. Heat the solution to evaporate the solvent.

Method 2.
When you need to concentrate a solution by evaporating the solvent, you can accelerate the process by directing a stream of air (or nitrogen gas, for easily oxidized substance) gently toward the surface of the liquid. This stream of gas will remove the vapors that are in equilibrium with the solution.

FIGURE 21.41 Arrangement for the evaporation of liquids.

CAUTION: If toxic or flammable solvent vapors should be involved, conduct the evaporation in a hood.

Compressed air sometimes contains oil and water droplets. Filter the air by passing it through a cotton-filled tube and a tube filled with a drying agent, such as anhydrous calcium chloride.

E. Evaporation Under Reduced Pressure

Solvents can be evaporated more quickly by evaporating them at reduced pressure and gently heating (Fig. 21.42). Refer to Fig. 21.43 for trap bottle.

Procedure

1. Place the solution to be concentrated in a round-bottomed flask or suction flask.

CAUTION: Do not use Erlenmeyer flasks having volumes larger than 125 mL. There is a danger of implosion and collapse.

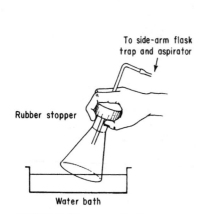

FIGURE 21.42 Evaporation of solvent under reduced pressure at elevated temperature.

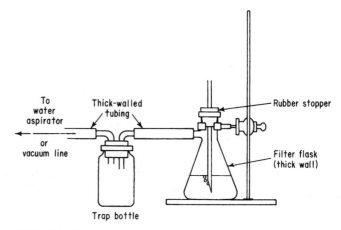

FIGURE 21.43 Evaporation of solvent under reduced pressure at room temperature.

2. Connect with rubber tubing to a safety trap which in turn is connected to a water aspirator.
3. Apply vacuum by turning on water.
4. *Swirl* the flask to expose large areas of the liquid and speed evaporation.

NOTE: *Swirling* technique helps suppress bumping.

5. The flask cools as the solvent evaporates.
6. Heat the flask by immersing in a warm-water bath.

F. Evaporation Under Vacuum

Water Aspirator.

1. Place liquid in a flask (fitted with a capillary air-inlet tube, Fig. 21.43) that is connected to a water aspirator with tubing.
2. Turn on the water aspirator to apply vacuum; adjust the capillary in the flask.
3. Gently apply heat with a warm-water bath.
4. When evaporation is completed, disconnect the tubing from the water aspirator *before* turning off the water.

Mechanical Vacuum Pump.

1. Place liquid in a flask (fitted with a capillary air-inlet tube) connected to a dry-ice trap by tubing.
2. Connect the outlet of the trap to a vacuum pump.
3. Turn on vacuum pump; adjust capillary on flask to a fine-air-bubble stream.
4. Gently apply heat with a warm-water bath.
5. Disconnect the tubing, connecting the flask to the vapor trap *before* turning off the vacuum pump.

ROTARY EVAPORATOR

Rotary evaporators (Fig. 21.44) provide a very rapid means to evaporate solvents and concentrate solutions. The flask rotates while the system is under vacuum (see section on the water aspirator in Chapter 5, "Pressure and Vacuum"), providing a very large surface area for evaporation. The walls of the flask are constantly rewetted as the flask rotates, minimizing superheating and bumping. Heat is supplied to the flask by steam bath, oil bath, heating mantle, or other heat source to meet the need.

Rotary evaporators can be used for evaporation and vacuum drying of powders and solids, and for low-temperature distillation of heat-sensitive substances. Substances can be degassed and distilled under inert atmospheres. The rotating flask ensures good mixing and good heat transfer from the heating bath.

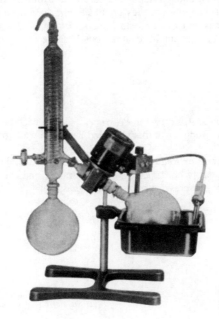

FIGURE 21.44 Rotary evaporator.

SUBLIMATION

The vapor pressure of solids increases as the temperature increases, and, because of this, some solids can go from the solid to the vapor state without passing through the liquid state. This phenomenon is called **sublimation**, and the vapor can be resolidified by reducing its temperature. The process can be used for purifying solids if the impurities in the solids have a much lower vapor pressure than that of the desired compound.

The advantages of sublimation as a purification method are:

1. No solvent is used.
2. It is a faster method of purification than crystallization.
3. More volatile impurities can be removed from the desired substance by subliming them off.
4. Substances that contain occluded molecules of solvation can sublime to form the nonsolvated product, losing the water or other solvent during the sublimation process.
5. Nonvolatile or less volatile solids can be separated from the more volatile ones.

The disadvantage of sublimation as a purification method is that the process is not as selective as crystallization because the vapor pressures of the sublimable solids may be very close together.

A. Atmospheric and Vacuum Sublimation

A sublimation point is a point at which the vapor pressure of a solid equals the applied pressure: it is a constant property like a melting point or a boiling point. Many liquids evaporate at temperatures below their boiling points, and some solids sublime (evaporate) below their melting points. If a substance sublimes below its melting point at atmospheric pressure, its melting point must be determined under pressure, in a sealed capillary tube.

Many solids do not develop enough vapor pressure at 760 mmHg (atmospheric pressure) to sublime, but they do develop enough vapor pressure to sublime at reduced pressure. For this reason, most sublimation equipment is constructed with fittings making it adaptable for vacuum connections. Furthermore, the use of vacuum is advantageous because the lower temperatures required reduce thermal decomposition.

B. Methods of Sublimation

Simple Laboratory Procedure at Atmospheric Pressure. Gently heat the sublimable compound in a container that has a loosely fitting cover that is chilled with cold water or ice (Fig. 21.45).

Methods Useful at Atmospheric or Reduced Pressure. The sublimation equipment illustrated in Figs. 21.46 to 21.49 can be easily constructed in the laboratory and can be used for sublimation procedures at normal atmospheric pressure or at reduced pressure.

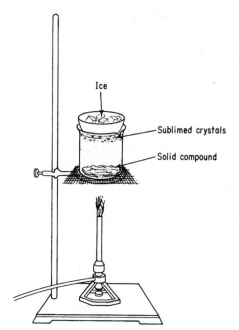

FIGURE 21.45 Simple laboratory setup for sublimation.

DISTILLATION AND EVAPORATION **449**

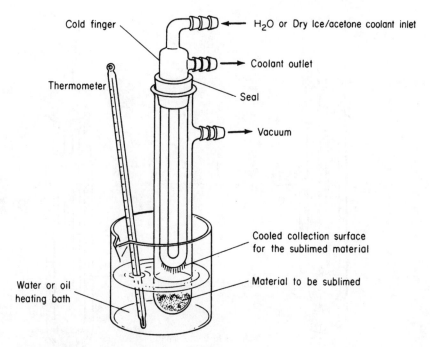

FIGURE 21.46 Cold-finger sublimation apparatus.

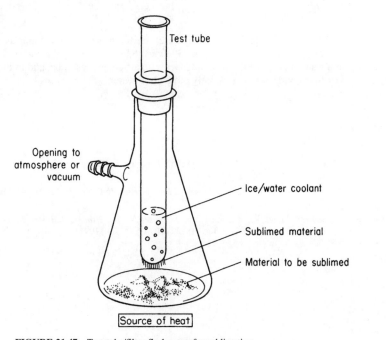

FIGURE 21.47 Test-tube/filter-flask setup for sublimation.

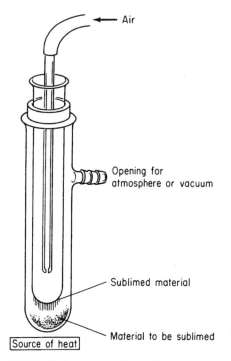

FIGURE 21.48 Air-cooled test-tube/filter-flask setup.

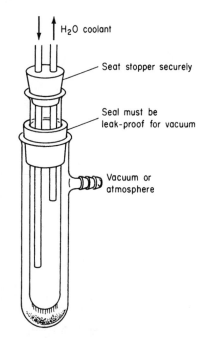

FIGURE 21.49 Test-tube/filter-flask setup with continuous water coolant.

The apparatus in Fig. 21.46 is based on the use of a cold finger as a condenser. Figure 21.47 demonstrates the use of a coolant-filled test tube as a condenser; the coolant here is a mixture of ice and water. In Fig. 21.48 the condenser is an air-cooled test tube, and in Fig. 21.49 the coolant for the test-tube condenser is constantly circulating water.

REFERENCES

1. Perry, Robert H., and Green, Don W., *Perry's Chemical Engineering Handbook*, McGraw-Hill Book Co. ISBN 0-07-115448-5 (1997).
2. Peters, Max S., *Elementary Chemical Engineering*, 2nd edition. McGraw-Hill Book Co. ISBN 0-07-049586-6 (1984).
3. Sawers, James R., and Eastman, Margaret M. R., *Process Industry Procedures and Training Manual*. 1st edition. McGraw-Hill, Inc. ISBN 0-07-054277-5 (1996).

CHAPTER 22
INORGANIC CHEMISTRY REVIEW

MATTER

All matter is made up of atoms and atoms are the smallest building blocks of nature. Atoms differ from one another because they contain different numbers of subatomic particles: protons, electrons, and neutrons (Table 22.1). The number of protons in the nucleus is called the **atomic number** (Z). The **mass number** represents the total number of protons and neutrons in the nucleus. Atoms must be neutral, therefore they contain as many electrons (− charges) as they have protons (+ charges). The protons and neutrons are located in the center (nucleus) of the atom and the electrons spin outside the nucleus. Pure substances can be either elements or compounds, depending on the atoms they contain. **Elements** are pure substances that contain only one type of atom. An atom is the smallest particles that can represent an element. As shown in Table 22.1, hydrogen, helium, lithium, beryllium, boron, carbon, calcium, and lead are examples of different elements (atoms).

Currently, there are 109 identified and named elements (Table 22.2). These 109 elements can be put together in millions of combinations to form molecules. A **molecule** can be defined as any group of atoms bonded together with no charge, for example water (H_2O), table salt (NaCl), and sugar ($C_6H_{12}O_6$). A molecule is the smallest particle that can represent a compound. A **compound** is any pure substance made up of two or more atoms that are chemically joined. Thus, H_2O, NaCl, and $C_6H_{12}O_6$ are all examples of compounds.

THE PERIODIC TABLE

The number of protons that an atom contains identifies that particular atom and no other element can have this number of protons (**atomic number**). Hydrogen has one proton and an atomic number of one; oxygen has eight protons and its atomic number is eight. An atom with 82 protons is lead (Pb). The atom with the largest number of protons has an atomic number of 109. Every atom also has a **mass number**, which is the number of protons plus the number of neutrons in the atom. An atom that has six protons and eight neutrons would have a mass number of 14. The atom with atomic number six is carbon, so this particular atom is called carbon-14. There is another carbon atom that has six protons in the nucleus but only six neutrons.

Each of the 109 elements has an atomic mass made up of the total mass of the subatomic particles in its atom. Table 22.2 is a listing of the elements in alphabetical order with their atomic numbers, average atomic masses, and their symbols. Elements listed with their atomic masses in brackets [ ] are radioactive and this number represents the average of the most stable atoms.

An arrangement of all the elements by their similar chemical and physical properties is known as the **periodic table**. Table 22.3 shows the elements placed in horizontal rows (periods) and vertical columns (families) based on their properties and similar electron arrangements.

A different symbol is used to identify each of the 109 elements. These symbols are derived from the names of the elements using the first letter of the element in most cases. If two elements begin with the same letter, then a second letter is used. In this table, the elements are arranged in order of increasing atomic numbers, beginning with hydrogen (Z = 1) and ending with meitnerium (Z = 109). Instead of being in one continuous line, the elements are placed in columns because, as the atomic number increases, certain properties are found to repeat. The elements in the first column of the table (except for hydrogen) are all soft, silvery metals. The elements in the last column are all gases at

TABLE 22.1 Subatomic Particles (Protons, Neutrons, and Electrons) in Various Atoms

Atomic number	Symbol	Mass number	Number of protons	Number of neutrons	Number of electrons
1	H	1	1	0	1
2	He	4	2	2	2
3	Li	7	3	4	3
4	Be	9	4	5	4
5	B	11	5	6	5
6	C	12	6	6	6
20	Ca	40	20	20	20
82	Pb	207	82	125	82

TABLE 22.2 Atomic Masses of Elements

Element	Symbol	Atomic number	Atomic mass
Actinium	Ac	89	[227]
Aluminum	Al	13	26.98
Americium	Am	95	[243]
Antimony	Sb	51	121.75
Argon	Ar	18	39
Arsenic	As	33	74.92
Astatine	At	85	[210]
Barium	Ba	56	[137.34]
Berkelium	Bk	97	[249]
Beryllium	Be	4	9.01
Bismuth	Bi	83	208.9
Bohrium	Bh	107	[262]
Boron	B	5	10.81
Bromine	Br	35	[79.91]
Cadmium	Cd	48	112.40
Calcium	Ca	20	40.08
Californium	Cf	98	249
Carbon	C	6	12.01
Cerium	Ce	58	140.12
Cesium	Cs	55	132.91
Chlorine	Cl	17	35.453
Chromium	Cr	24	52.00
Cobalt	Co	27	58.93
Copper	Cu	29	63.54
Curium	Cm	96	[247]
Dubnium	Db	105	[262]
Dysprosium	Dy	66	162.50
Erbium	Er	68	167.26
Einsteinium	Es	99	[254
Europium	Eu	63	151.96
Fermium	Fm	100	[253]
Fluorine	F	9	19.00
Francium	Fr	87	[223]
Gadolinium	Gd	64	157.25
Gallium	Ga	31	69.72
Germanium	Ge	32	72.59
Gold	Au	79	196.97
Hafnium	Hf	72	178.49
Hassium	Hs	108	[265]
Helium	He	2	4.00
Holmium	Ho	67	164.93
Hydrogen	H	1	1.01
Indium	In	49	114.82
Iodine	I	53	126.90

(Continued)

TABLE 22.2 Atomic Masses of Elements (*Continued*)

Element	Symbol	Atomic number	Atomic mass
Iridium	Ir	77	192.2
Iron	Fe	26	55.85
Krypton	Kr	36	83.80
Lanthanum	La	57	138.91
Lawrencium	Lr	103	257
Lithium	Li	3	6.94
Lutetium	Lu	71	174.97
Magnesium	Mg	12	24.31
Manganese	Mn	25	54.94
Meitnerium	Mt	109	[266]
Mendelevium	Md	101	[258]
Mercury	Hg	80	200.59
Molybdenum	Mo	42	95.94
Neodymium	Nd	60	144.24
Neon	Ne	10	20.18
Neptunium	Np	93	[237]
Nickel	Ni	28	58.71
Niobium	Nb	41	92.91
Nitrogen	N	7	14.01
Nobelium	No	102	[256]
Osmium	Os	76	190.2
Oxygen	O	8	16.00
Palladium	Pd	46	106.4
Phosphorus	P	15	30.97
Platinum	Pt	78	195.09
Plutonium	Pu	94	[242]
Polonium	Po	84	[210]
Potassium	K	19	39.10
Praseodymium	Pr	59	140.91
Promethium	Pm	61	[147]
Protactinium	Pa	91	[231]
Radium	Ra	88	[226]
Radon	Rn	86	[222]
Rhenium	Re	75	186.2
Rhodium	Rh	45	102.91
Rubidium	Rb	37	85.47
Ruthenium	Ru	44	101.07
Rutherfordium	Rf	104	[261]
Samarium	Sm	62	150.35
Scandium	Sc	21	44.96
Seaborgium	Sg	106	(263)
Selenium	Se	34	78.96
Silicon	Si	14	28.09
Silver	Ag	47	107.87
Sodium	Na	11	22.99
Sulfur	S	16	32.06
Strontium	Sr	38	87.62
Tantalum	Ta	73	180.95
Technetium	Tc	43	[99]
Tellurium	Te	52	127.60
Terbium	Tb	65	158.92
Thallium	Tl	81	204.37
Thorium	Th	90	232.04
Thulium	Tm	69	168.93
Tin	Sn	50	118.69
Titanium	Ti	22	47.90
Tungsten	W	74	183.85
Uranium	U	92	238.03
Vanadium	V	23	50.94
Xenon	Xe	54	131.30
Ytterbium	Yb	70	173.04
Yttrium	Y	39	88.91
Zinc	Zn	30	65.37
Zirconium	Zr	40	91.22

TABLE 22.3 The Periodic Table

Filled Shells	I	II													III	IV	V	VI	VII	0
																				2 He Helium 4.00260 2
	1 H Hydrogen 1.0079 -1 +1 1																			
	3 Li Lithium 6.941 2-1 +1	4 Be Beryllium 9.01218 2-2 +2													5 B Boron 10.81 2-3 +3	6 C Carbon 12.011 2-4 +2 +4 -4	7 N Nitrogen 14.0067 2-5 +1+2 +3+4 +5 -3	8 O Oxygen 15.9994 2-6 -2	9 F Fluorine 18.99840 2-7 -1	10 Ne Neon 20.179 2-8
	11 Na Sodium 22.98977 2-8-1 +1	12 Mg Magnesium 24.305 2-8-2 +2					Transition Elements								13 Al Aluminum 26.98154 2-8-3 +3	14 Si Silicon 28.086 2-8-4 +2 +4	15 P Phosphorus 30.97376 2-8-5 +3 +5 -3	16 S Sulfur 32.06 2-8-6 +4 +6 -2	17 Cl Chlorine 35.453 2-8-7 +1+5 +7 -1	18 Ar Argon 39.948 2-8-8
2	19 K Potassium 39.098 -8-8-1 +1	20 Ca Calcium 40.08 -8-8-2 +2	21 Sc Scandium 44.9559 -8-9-2 +3	22 Ti Titanium 47.90 -8-10-2 +3 +4	23 V Vanadium 50.9414 -8-11-2 +2+3 +4+5	24 Cr Chromium 51.996 -8-13-1 +2+3 +6	25 Mn Manganese 54.9380 -8-13-2 +2+3 +4+7	26 Fe Iron 55.847 -8-14-2 +2 +3	27 Co Cobalt 58.9332 -8-15-2 +2 +3	28 Ni Nickel 58.70 -8-16-2 +2	29 Cu Copper 63.546 -8-18-1 +1 +2	30 Zn Zinc 65.38 -8-18-2 +2	31 Ga Gallium 69.72 -8-18-3 +3	32 Ge Germanium 72.59 -8-18-4 +2 +4	33 As Arsenic 74.9216 -8-18-5 +3+5 -3	34 Se Selenium 78.96 -8-18-6 +4+6 -2	35 Br Bromine 79.904 -8-18-7 +1+5 -1	36 Kr Krypton 83.80 -8-18-8
2-8	37 Rb Rubidium 85.4678 -18-8-1 +1	38 Sr Strontium 87.62 -18-8-2 +2	39 Y Yttrium 88.9059 -18-9-2 +3	40 Zr Zirconium 91.22 -18-10-2 +4	41 Nb Niobium 92.9064 -18-12-1 +3 +5	42 Mo Molybdenum 95.94 -18-13-1 +6	43 Tc Technetium (97) -18-13-2 +4+6 +7	44 Ru Ruthenium 101.07 -18-15-1 +3	45 Rh Rhodium 102.9055 -18-16-1 +3	46 Pd Palladium 106.4 -18-18-0 +2 +4	47 Ag Silver 107.868 -18-18-1 +1	48 Cd Cadmium 112.40 -18-18-2 +2	49 In Indium 114.82 -18-18-3 +3	50 Sn Tin 118.69 -18-18-4 +2 +4	51 Sb Antimony 121.75 -18-18-5 +3 +5	52 Te Tellurium 127.60 -18-18-6 +4+6 -2	53 I Iodine 126.9045 -18-18-7 +1+5 +7 -1	54 Xe Xenon 131.30 -18-18-8
2-8-18	55 Cs Cesium 132.9054 -18-8-1 +1	56 Ba Barium 137.34 -18-8-2 +2	57-71 See Lanthanides	72 Hf Hafnium 178.49 -32-10-2 +4	73 Ta Tantalum 180.9479 -32-11-2 +5	74 W Tungsten 183.85 -32-12-2 +6	75 Re Rhenium 186.207 -32-13-2 +4+6 +7	76 Os Osmium 190.2 -32-14-2 +3 +4	77 Ir Iridium 192.22 -32-15-2 +3 +4	78 Pt Platinum 195.09 -32-17-1 +2 +4	79 Au Gold 196.9665 -32-18-1 +1 +3	80 Hg Mercury 200.59 -32-18-2 +1 +2	81 Tl Thallium 204.37 -32-18-3 +1 +3	82 Pb Lead 207.2 -32-18-4 +2 +4	83 Bi Bismuth 208.9804 -32-18-5 +3 +5	84 Po Polonium (209) -32-18-6	85 At Astatine (210) -32-18-7 -1	86 Rn Radon (222) -32-18-8
2-8-18-32	87 Fr Francium (223) -18-8-1 +1	88 Ra Radium 226.0254 -18-8-2 +2	89-103 See Actinides	104 Rf Rutherfordium (261) -32-10-2	105 Db Dubnium (262) -32-11-2	106 Sg Seaborgium (263) -32-12-2	107 Bh Bohrium (262) -32-13-2	108 Hs Hassium (265) -32-14-2	109 Mt Meitnerium (266) -32-15-2	110 (269)	111 (272)	112 (277)	(113)	(114)	(115)	(116)	(117)	118 (118)

Lanthanides	57 La Lanthanum 138.9055 -18-9-2 +3	58 Ce Cerium 140.12 -19-9-2 +3 +4	59 Pr Praseodymium 140.9077 -21-8-2 +3	60 Nd Neodymium 144.24 -22-8-2 +3	61 Pm Promethium (145) -23-8-2 +3	62 Sm Samarium 150.4 -24-8-2 +3	63 Eu Europium 151.96 -25-8-2 +2 +3	64 Gd Gadolinium 157.25 -25-9-2 +3	65 Tb Terbium 158.9254 -26-9-2 +3	66 Dy Dysprosium 162.50 -28-8-2 +3	67 Ho Holmium 164.9304 -29-8-2 +3	68 Er Erbium 167.26 -30-8-2 +3	69 Tm Thulium 168.9342 -31-8-2 +3	70 Yb Ytterbium 173.04 -32-8-2 +2 +3	71 Lu Lutetium 174.97 -32-9-2 +3
Actinides	89 Ac Actinium (2,7) -18-9-2 +3	90 Th Thorium 232.0381 -18-10-2 +4	91 Pa Protactinium 231.0359 -20-9-2 +4 +5	92 U Uranium 238.029 -21-9-2 +3+4 +5+6	93 Np Neptunium 237.0482 -22-9-2 +3+4 +5+6	94 Pu Plutonium (244) -24-8-2 +3+4 +5+6	95 Am Americium (243) -25-8-2 +3+4 +5+6	96 Cm Curium (247) -25-9-2 +3	97 Bk Berkelium (247) -27-8-2 +3 +4	98 Cf Californium (251) -28-8-2 +3	99 Es Einsteinium (252) -29-8-2 +3	100 Fm Fermium (257) -30-8-2 +3	101 Md Mendelevium (258) -31-8-2 +3	102 No Nobelium (259) -32-8-2 +2 +3	103 Lr Lawrencium (262) -32-9-2 +3

KEY: +1 +3 Au Gold 196.9665 -32-18-1 — Atomic Number, Symbol, Name, Oxidation States, Atomic Weight, Electron Configuration

Note: Atomic weights are those of the most commonly available long-lived isotopes on the 1973 IUPAC Atomic Weights of the Elements. A value given in parentheses denotes the mass number of the longest-lived isotope.

room temperature and do not readily enter into chemical reactions. Because of their similar properties, the elements in a column are referred to as a **group**, or **family**. The horizontal rows of the table are called **periods**.

The elements in Group 1 (IA) of the periodic table are known as the **alkali metals**; those elements located in Group 2 (IIA) are called **alkaline earth metals**, and those in Group 17 (VIIA) are known as the **halogens** (salt-formers). The unreactive gases in Group 18 (VIIIA) are the **noble gases**. The elements in the center of the table, Group 3 (IIIB) through 12 (IIB) are called the **transition elements**. These transition elements are all metals and you may recognize many from everyday use (copper, silver, tin, iron, and so forth).

Note the heavy stair step line toward the right side of the table. This line generally separates the metals on the left from the nonmetals on the right. **Metals** tend to lose electrons in a chemical reaction and **nonmetals** tend to gain electrons in a chemical reaction. All of the elements, except aluminum, that lie along this line are called semi-metals or **metalloids**. These elements have some properties of both nonmetals and metals. They are brittle like nonmetals, yet semi-lustrous solids like metals which are shiny. Because metalloids do not conduct heat and electricity as well as metals, but better than nonmetals, they are called semiconductors. Silicon and germanium are widely used in semiconductor devices. Metallic character is a periodic property; that is, it repeats in a regular way. Look at periods 2 and 3, for example. From left to right in the second period, the character of the elements changes from metallic to semi-metallic to nonmetallic; the same trend repeats in the third period, as well as in the fourth, fifth, and sixth. Other properties, such as density and size of atoms, also vary in a periodic fashion with increase in atomic number.

The periodic table is a useful tool in chemistry, providing a quick reference for many properties of the chemical element. The table is particularly useful in predicting how elements will bond with each other to form compounds.

CHEMICAL BONDING

Except for the noble gases, substances do not normally exist as single atoms. Most substances are groups of atoms bonded together. **Chemical bonds** are the attractive forces that hold the atoms together. The way in which atoms interact with each other is related to the number and arrangement of electrons in these atoms. Electrons exist in atoms at different distances from the nucleus and, therefore, at different energy levels. The electrons involved in bonding are those in the outer energy level of the atoms, and are called the **valence electrons**. The number of valence electrons that an atom has available for bonding in "A" group elements on the periodic table can be determined from their position in the periodic table. Each atom of the Group "A" elements has the number of valence electrons that corresponds to its Roman numeral group number. For example, elements of Group IA have one valence electron, Group IIA have two valence electrons, Group VIIA have seven, and finally Group VIIIA have eight valence electrons. The elements in "B" groups of the periodic table, otherwise known as the transition metals, are capable of having more than one set number of valence electrons. Because of this, the periodic table will not be used to determine the number of valence electrons for Group "B" atoms as it will for Group "A" atoms. Group "B" atoms will be addressed later in this chapter.

The goal of every atom is to reach a lower level of energy in order to become more stable. The noble gases are the most stable atoms on the periodic table because they have eight electrons in their outer energy shell. Helium is an exception with two electrons. It has been discovered that atoms tend to lose, gain, or share enough electrons to achieve an octet of eight electrons in their outer energy shell. The goal of all atoms is to have the same number of valence electrons as their nearest noble gas.

A. Ionic Bonds

Sodium is a highly reactive metal. It reacts so rapidly with air that it must be stored under oil or kerosene. It also reacts violently with water, producing so much heat in the process that it melts. Chlorine is greenish-yellow gas that is extremely irritating to the eyes and the respiratory tract. If a piece of

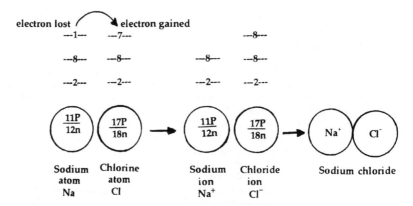

FIGURE 22.1 Metal element reacting with a nonmetal element.

sodium metal is dropped into a container of chlorine gas, a violent reaction occurs, leaving a white solid. The white solid is sodium chloride, ordinary table salt. We can explain this reaction in the following way (Fig. 22.1). Sodium, in Group IA, has one valence electron; by losing this one electron, it will have the same number of electrons as neon, the noble gas just preceding it on the periodic table. Chlorine is in Group VIIA and has seven valence electrons; by gaining one electron, chlorine will have the same number of electrons as its nearest noble gas, argon. In this reaction, both sodium and chlorine form ions that are more stable than their respective atoms. All atoms tend to become much more stable when they take on the same electron arrangement as a noble gas element.

An **ion** is an atom (or group of atoms) that has an electrical charge. As a neutral atom, sodium has 11 protons and 11 electrons; when it loses its one valence electron, it becomes positively charged and is represented as Na^+. A neutral chlorine atom has 17 protons and 17 electrons; when it gains one valence electron, it becomes negatively charged and is represented as Cl^-. Because the sodium and chloride ions have opposite charges, they attract each other. In each particle of table salt, there are millions of sodium and chloride ions. The ions arrange themselves in an orderly fashion that is repeated in three dimensions, making up a crystal of sodium chloride. The forces holding the ions together in the crystal are **ionic bonds**. Because the ratio to sodium to chloride ions is one to one, we write the formula for sodium chloride as NaCl. **Formulas** are used to represent compounds, just as symbols are used to represent elements.

Magnesium metal reacts with bromine, a reactive reddish-brown liquid, to produce a stable crystalline substance known as magnesium bromide. Magnesium is a Group IIA element and has two valence electrons; if it loses these two valence electrons, it will have the same electron arrangement as neon, the noble gas nearest to it. Magnesium forms the ion, Mg^{2+}. Bromine, in Group VIIA, has seven valence electrons. By gaining one electron it will have the electron structure of the noble gas krypton. Bromine forms the bromide ion, Br^-. Because magnesium has a positive two charge, while bromide is only negative one, two bromides ions are needed to combine with one magnesium ion. The formula for the resulting compound, magnesium bromide, is $MgBr_2$. The positive and negative charges must be equal in a compound.

Generally, the elements on the far left of the periodic table react with those on the far right (excluding the noble gases) to form stable crystalline solids. These crystalline solids are held together by ionic bonds and the compounds are known as ionic compounds. Note that elements on the far left of the periodic table are metals and they tend to give up electrons forming positive charged ions. Elements on the far right are nonmetals which have a tendency to take on electrons forming a negative charged ion. All ionic compounds contain a positive charged ion(s) and a negative charged ion(s). Ionic compounds made up of a metal and a nonmetal tend to be brittle, fairly hard, with high melting points. Ionic bonds are strong forces of attraction and account for the hardness and high melting temperatures of these substances. Pure sodium chloride, for example, melts at 808°C.

B. Covalent Bonds

Ionic bonds are formed between two opposite charged ions such as a metal and a nonmetal. Bonds can also form between similar atoms or even identical atoms. In these cases, the atoms are usually nonmetals and share their electrons. The shared electrons constitute a **covalent bond** and the resulting particle is a molecule. In electron sharing, two nuclei attract the same electrons and this attractive force holds the two atoms together.

The simplest covalent molecule is that of hydrogen, H_2. The hydrogen atom, with just one electron, needs one more electron to have the noble-gas electron structure of helium. If only other hydrogen atoms are around, two hydrogen atoms will share a pair of electrons. Only seven elements are known to be diatomic: hydrogen, nitrogen, oxygen, fluorine, chlorine, bromine, and iodine. Covalent bonds also form between different atoms. Hydrogen and chlorine react to form hydrogen chloride. Both hydrogen and chlorine atoms each need one electron to have the same number of electrons as their nearest noble gases. Thus compounds like HCl, HBr, and H_2O can also form using covalent bonds.

MOLECULAR AND STRUCTURAL FORMULAS

The **molecular formula** represents the compound description. It gives the exact number of each kind of atom in the compound. The molecular formula for ethylene glycol (antifreeze) is $C_2H_6O_2$, indicating that one molecule contains two carbon atoms, six hydrogen atoms, and two oxygen atoms.

Another type of descriptive formula is the **structural formula**. It gives the positions of every atom in the molecule. For ethylene glycol, the structural formula is:

$$\begin{array}{c} H \quad H \\ | \quad\;\; | \\ H-C-C-H \\ |\quad\;\; | \\ OH\;\;OH \end{array}$$

A. Molecular Weights

The simplest molecule is hydrogen with a formula of H_2. Since hydrogen molecules contain two hydrogen atoms and they each have an atomic weight of 1.0079 amu, hydrogen has a molecular weight of 2.0158 amu. Using a periodic table, **molecular weights** can be calculated for any compound by adding *all* the atomic weights found in the formula. The terms molecular weight and formula weight can be used interchangeably. Remember to add the mass for *each* atom present. The amu abbreviation stands for atomic mass units, but we will simply substitute grams for this theoretical mass unit. Chapter 24, "Chemical Calculations and Concentration Expressions," gives more details on molecular weight and other chemical calculations.

EXAMPLE 22.1 Using the periodic table, calculate the molecular weights of the following compounds:

(a) Chlorine (Cl_2) $Cl \times 2 = 35.45 \text{ g} \times 2 = \mathbf{70.906 \text{ g}}$

(b) Magnesium sulfide (MgS) $Mg = 24.305 \text{ g} \times 1 = 24.305 \text{ g}$
$S = 32.060 \text{ g} \times 1 = \underline{32.060 \text{ g}}$
$\mathbf{56.365 \text{ g}}$

(c) Sucrose ($C_6H_{12}O_6$)

$$C \times 6 = 12.011 \text{ g} \times 6 = 72.066 \text{ g}$$
$$H \times 12 = 1.0078 \text{ g} \times 12 = 12.094 \text{ g}$$
$$O \times 6 = 15.999 \text{ g} \times 6 = \underline{95.994 \text{ g}}$$
$$\mathbf{180.154 \text{ g}}$$

(d) Sodium sulfate pentahydrate ($Na_2SO_4 \cdot 5 H_2O$)

$$Na \times 2 = 22.990 \text{ g} \times 2 = 45.980 \text{ g}$$
$$S \times 1 = 32.064 \text{ g} \times 1 = 32.064 \text{ g}$$
$$O \times 4 = 15.999 \text{ g} \times 4 = 63.996 \text{ g}$$
$$H_2O \times 5 = 18.000 \text{ g} \times 5 = \underline{90.000 \text{ g}}$$
$$\mathbf{232.04 \text{ g}}$$

WRITING CHEMICAL FORMULAS

A **chemical formula** is a combination of symbols that represents a compound. The formula shows the elements present in the compound. Writing the formula for an ionic compound is simply a matter of balancing the charge on the positive ion with that of the negative ion. To write the formula for the compound of aluminum and oxygen, we need to know the charges that aluminum and oxygen take on as ions. Aluminum, being a Group IIIA element and a metal, will give up three electrons and take on a 3+ charge, Al^{3+}. Oxygen is a Group VIA element and a nonmetal which means it will take on two electrons resulting in a 2− charged ion, O^{2-}. The sum of the charges making up the compound must be zero, so we find the lowest common multiple of 3 and 2, which is 6. It is necessary to have two Al^{3+} and three O^{2-} ions to form a neutral compound. The numbers of atoms are shown as subscripts, giving the formula Al_2O_3. Notice that the symbol for the positive ion is always written first.

To successfully write correct chemical formulas, it is important to know what charge an atom will take on. Remember metals will always be positive and nonmetals negative. The metals give up their valence electrons so that Group IA atoms will take on a 1+ charge, Group IIA a 2+, Group IIIA a 3+, and so on. Nonmetals want eight electrons in their outer shell; therefore the charge that a nonmetal atom takes on will be whatever number they need to have eight electrons in their outer shell. For example, oxygen is in Group VIA having six electrons and needing only two more to achieve an octet of eight. Therefore an oxygen atom will take on a 2− charge. Atoms of Group VA elements will take on a 3− charge and Group VIIA will take on a 1− charge. These predictions are made directly from the periodic table.

Ca^{2+} and Fe^{3+} are examples of **monatomic** (one atom) positive ions. These ions contain one atom and are all called **cations** because of their positive charge. Ions that contain only one atom but with a negative charge are called **anions**. The charges of some ions cannot be determined using the group number in the periodic table. This is particularly true of the transition elements as was stated earlier. Table 22.4 provides a list of some common cations.

Some of the more common anions are listed in Table 22.5. Compounds that contain ions with more than one atom are called polyatomic ions and are discussed later in this chapter. Negative monatomic ions are named by replacing the ending of the element name with "**ide**," as shown by the examples in Table 22.5.

NAMING INORGANIC COMPOUNDS

Because there are approximately 100,000 compounds possible by combining all the elements (except carbon), there must be a systematic way of naming them. All compounds containing elements other than carbon are called **inorganic**. There are many compounds, however, that are still identified by

TABLE 22.4 Some Common Cations

Name	Formula	Valence
Aluminum	Al^{3+}	+3
Ammonium	NH_4^+	+1
Antimony	Sb^{3+}	+3
Arsenic(V) (arsenic)	As^{5+}	+5
Arsenic(III) (arsenious)	As^{3+}	+3
Barium	Ba^{2+}	+2
Bismuth	Bi^{3+}	+3
Cadmium	Cd^{2+}	+2
Calcium	Ca^{2+}	+2
Chromium(III) (chromic)	Cr^{3+}	+3
Chromium(II) (chromous)	Cr^{2+}	+2
Cobalt(III) (cobaltic)	Co^{3+}	+3
Cobalt(II) (cobaltous)	Co^{2+}	+2
Copper(II) (cupric)	Cu^{2+}	+2
Copper(I) (cuprous)	Cu^+	+1
Hydrogen	H^+	+1
Iron(III) (ferric)	Fe^{3+}	+3
Iron(II) (ferrous)	Fe^{2+}	+2
Lead(IV) (plumbic)	Pb^{4+}	+4
Lead(II) (plumbous)	Pb^{2+}	+2
Lithium	Li^+	+1
Magnesium	Mg^{2+}	+2
Manganese(III) (manganic)	Mn^{3+}	+3
Manganese(III) (manganic)	Mn^{2+}	+2
Mercury(II) (mercuric)	Hg^{2+}	+2
Mercury(I) (mercurous)	Hg^+	+1
Nickel(III) (nickelic)	Ni^{3+}	+3
Nickel(II) (nickelous)	Ni^{2+}	+2
Potassium	K^+	+1
Silver	Ag^+	+1
Sodium	Na^+	+1
Strontium	Sr^{2+}	+2
Tin(IV) (stannic)	Sn^{4+}	+4
Tin(II) (stannous)	Sn^{2+}	+2
Zinc	Zn^{2+}	+2

TABLE 22.5 Some Common Monatomic Anions

Name	Formula	Charge
Chloride	Cl^{1-}	−1
Oxide	O^{2-}	−2
Nitride	N^{3-}	−3
Fluoride	F^{1-}	−1
Sulfide	S^{2-}	−2
Carbide	C^{4-}	−4
Hydride	H^{1-}	−1
Phosphide	P^{3-}	−3
Selenide	Se^{2-}	−2
Telluride	Te^{2-}	−2
Iodide	I^{1-}	−1
Bromide	Br^{1-}	−1

their common names. Examples are water (H_2O), ammonia (NH_3), vinegar ($HC_2H_3O_2$), and baking soda ($NaHCO_3$). The systematic name for baking soda is sodium hydrogen carbonate. In this section, simple rules for naming inorganic compounds are introduced. **Organic chemistry** (carbon-containing) compounds are discussed in Chapter 23, "Organic Chemistry Review."

NAMING BINARY COMPOUNDS

The names of compounds that are composed of ions are derived from the names of the ions. Positive monatomic ions have the name of the element:

Ag^+, silver ion Zn^{2+}, zinc ion

If an element forms more than one positive ion, the charge on the ion is indicated as a Roman numeral in parentheses after the name of the element (see Table 22.6):

Fe^{2+}, iron(II) ion Sn^{2+}, tin(II) ion
Fe^{3+}, iron(III) ion Sn^{4+}, tin(IV) ion (read as tin four ion)

TABLE 22.6 Root Names of Multiple Charged Metals

English name	Latin name	Lower charge	Higher charge
Copper	Cuprum	Cuprous, Cu+1	Cupric, Cu+2
Cobalt	—	Cobaltous, Co+2	Cobaltic, Co+3
Iron	Ferrum	Ferrous, Fe+2	Ferric, Fe+3
Lead	Plumbum	Plumbous, Pb+2	Plumbic, Pb+4
Tin	Stannum	Stannous, Sn+2	Stannic, Sn+4
Arsenic	—	Arsenious, As+3	Arsenic, As+5
Manganese	—	Manganous, Mn+2	Manganic, Mn+3
Mercury	—	Mercurous, Hg+1	Mercuric, Hg+2
Chromium	—	Chromous, Cr+2	Chromic, Cr+3
Nickel	—	Nickelous, Ni+2	Nickelic, Ni+3

A. The "ous/ic" Nomenclature System

In this nearly obsolete system, sometimes called the Latin system, the ending "-ous" denotes the lower oxidation state (valence or combining power) of the metal, "-ic" the higher oxidation state (valence or combining power of the element). It should also be noted that the English word for the metal must be changed to its Latin form:

PbO, Plumb*ous* oxide   [The lead (Latin name: plumbum) cation has a charge of +2, thus an "ous" ending]

PbO_2, Plumb*ic* oxide   [The lead (Latin name: plumbum) cation has a charge of +4, thus an "ic" ending]

$FeCl_3$, Fer*ric* chloride   [The iron (Latin name: ferrum) cation has a charge of +3, thus "ic" ending]

$FeCl_2$, Fer*rous* chloride   [The iron (Latin name: ferrum) cation has a charge of +2, thus "ous" ending]

Compounds that contain only two elements are called **binary compounds**. The binary compounds above are all ionic; they contain a metal and nonmetal. Binary compounds composed of two elements are named by first naming the more positive element and then adding the name of the more negative element, Modified to end in "**ide**."

To name compounds composed of positive and negative monatomic ions, simply drop the word "ion" and combine the names:

$CaBr_2$, calcium bromide

Cu_2S, copper(I) sulfide or cuprous sulfide

CuS, copper(II) sulfide or cupric sulfide

Fe_2O_3, iron(III) oxide or ferric oxide

Compounds composed of two nonmetals can sometimes form a number of different compounds, depending on the number of atoms present. Because of this, a prefix system is used to indicate the number of atoms involved in the bond. *Rule: Compounds containing two different nonmetals bonded together ALWAYS require prefixes in their names* (Table 22.7). The following are several examples:

CO, carbon monoxide* N_2O, dinitrogen oxide

CO_2, carbon dioxide   NO, nitrogen oxide

SF_6, sulfur hexafluoride N_2O_5, dinitrogen pentoxide

TABLE 22.7 Prefixes for Naming Compounds

Prefix	Number of atoms	Prefix	Number of atoms
Mono-	1	Penta-	5
di-	2	hexa-	6
tri-	3	hepta-	7
tetra-	4	octa-	8
nona-	9	deca-	10

NAMING COMPOUNDS CONTAINING POLYATOMIC IONS

A. Naming Covalent Compounds

Covalent compounds involve *two nonmetallic elements*. To name these compounds, use the appropriate Greek prefixes found in Table 22.7 "di-", "tri-", "tetra-," etc., to indicate the number of atoms of each element involved and use the ending "-ide."

CO_2, carbon *di*oxide

HCl, hydrogen chloride   (*rule does not apply: not two nonmetal elements*)

H_2S, hydrogen sulfide   (*rule does not apply: not two nonmetal elements*)

PCl_3, phosphorus *tri*chloride

Cl_2O, *di*chlorine oxide

N_2O_4, *di*nitrogen *tetra*oxide

A polyatomic ion is a charged particle that contains more than one atom, for example the nitrate ion, NO_3^-. Table 22.8 is a listing of some common polyatomic ions and their charges. Compounds

---

*The use of the monoprefix is being discouraged. The more accepted name would be to eliminate the mono prefix. Thus, CO should be called carbon oxide. NO is simply nitrogen oxide, not mononitrogen monoxide.

TABLE 22.8 Common Polyatomic Ions

Name	Formula	Valence
Acetate	$C_2H_3O_2^-$	−1
Ammonium	NH_4^+	+1
Bicarbonate	HCO_3^-	−1
Bisulfate	HSO_4^-	−1
Bisulfide	HS^-	−1
Bisulfite	HSO_3^-	−1
Bromate	BrO_3^-	−1
Bromide	Br^-	−1
Bromite	BrO_2^-	−1
Carbonate	CO_3^{2-}	−2
Chlorate	ClO_3^-	−1
Chloride	Cl^-	−1
Chlorite	ClO_2^-	−1
Chromate	CrO_4^{2-}	−2
Cyanide	CN^-	−1
Cyanate	OCN^-	−1
Dichromate	$Cr_2O_7^{2-}$	−2
Dihydrogen phosphate	$H_2PO_4^{2-}$	−1
Ferricyanide	$Fe(CN)_6^{3-}$	−3
Ferrocyanide	$Fe(CN)_6^{4-}$	−4
Fluoride	F^-	−1
Hydrogen carbonate	HCO_3^{1-}	−1
Hydrogen phosphate	HPO_4^{2-}	−1
Hydrogen sulfate	HSO_4^{1-}	−1
Hydroxide	OH^-	−1
Hypobromite	BrO^-	−1
Hypochlorite	ClO^-	−1
Hypoiodite	IO^-	−1
Iodide	I^-	−1
Nitrate	NO_3^-	−1
Nitrite	NO_2^-	−1
Oxalate	$C_2O_4^{2-}$	−2
Perchlorate	ClO_4^-	−1
Perbromate	BrO_4^-	−1
Periodate	IO_4^-	−1
Permanganate	MnO_4^-	−1
Phosphate	PO_4^{3-}	−3
Phosphite	PO_3^{3-}	−3
Sulfate	SO_4^{2-}	−2
Sulfide	S^{2-}	−2
Sulfite	SO_3^{2-}	−2
Thiocyanate	SCN^-	−1
Thiosulfate	$S_2O_3^{2-}$	−2

containing polyatomic ions are named by simply combining the name of the positive ion with that of the negative ion. An example is $CaCO_3$, which is named calcium carbonate. Note that the name of the polyatomic ion is not changed. The procedure for writing a formula for a compound containing a polyatomic ion is the same as that for monatomic ions. For example, what is the formula for ammonium carbonate? The ammonium ion has the formula NH_4^+ and the carbonate ion is CO_3^{2-}. To make the sum of the charges equal zero, use two ammonium ions. To indicate that two ammonium ions are used, put them in parentheses. The formula for ammonium carbonate is $(NH_4)_2CO_3$.

EXAMPLE 22.2 Using Tables 22.4 and 22.8, name the following polyatomic compounds:

$(NH_4)_3PO_4$, ammonium phosphate
$Ni(OH)_2$, nickel(II) hydroxide
$CuSO_4$, copper(II) sulfate
$Cr_2(C_2O_4)_3$, chromium(III) oxalate
$Fe(CN)_3$, iron(III) cyanide

NAMING ACIDS

All of the common acids contain hydrogen. The names of these acids are related to the names of the ionic compounds that have already been discussed. Binary acids contain hydrogen and a second element. These acids are named by using the prefix **hydro-** and changing the ionic compound ending from "**-ide**" to "**-ic**." Thus, HCl is hydrochloric acid, HBr is hydrobromic acid and H_2S is hydrosulfuric acid.

Another group of acids, the oxygen-containing acids, contain hydrogen and a polyatomic ion. Their names are derived from the polyatomic ion with which the hydrogen is combined. In naming the acids, the "**-ite**" ending of the polyatomic ion is changed to "**-ous**" and the "**-ate**" ending becomes "**-ic**."

In naming acids, the word **acid** is always used as a part o the name.

HYDRATES

The moisture present in a material can be free or bound. Th **free-moisture** content of a material is that water which can be removed without causing a chemical or physical change in the material. **Bound-moisture** is that water which is associated with the chemical structure of the material. Hydrated compounds like magnesium sulfate pentahydrate ($MgSO_4 \cdot 5H_2O$) would fall into this classification. A dot is normally used between the compound and the number of water molecules present in a hydrate. When calculating the molecular weight of hydrates, the mass of the attached water molecule(s) must be included. Free moisture can be removed by simple evaporative heating, while bound moisture may require more complex procedures.

TABLE 22.9 Common Polyatomic Ions and Common Acids

Common polyatomic ions	Acids
NO_2^-, nitrite	HNO_2, nitrous acid
NO_3^-, nitrate	HNO_3, nitric acid
SO_3^{2-}, sulfite	H_2SO_3, sulfurous acid
SO_4^{2-}, sulfate	H_2SO_4, sulfuric acid
ClO^-, hypochlorite	$HClO$, hypochlorous acid
ClO_2^-, chlorite	$HClO_2$, chlorous acid
ClO_3^-, chlorate	$HClO_3$, chloric acid
ClO_4^-, perchlorate	$HClO_4$, perchloric acid
PO_3^{3-}, phosphite	H_3PO_3, phosphorous acid
PO_4^{3-}, phosphate	H_3PO_4, phosphoric acid

COMMON OR TRIVIAL NAMES

Some **common** or **trivial names** and formulas for some inorganic compounds are given in Table 22.10.

TABLE 22.10 Common or Trivial Names

Name	Formula
Alumina	Al_2O_3
Ammonia	NH_3
Aqua fortis	HNO_3
Aqua regia	$HCl + HNO_3$
Baking soda	$NaHCO_3$
Battery acid, oleum	H_2SO_4
Bicarbonate	HCO_3^{1-}
Bisulfate	HSO_4^{1-}
Bluestone	$CuSO_4 \cdot 5\,H_2O$
Borax	$Na_2B_4O_7 \cdot 10\,H_2O$
Brimstone	S
Calomel	Hg_2Cl_2
Caustic potash	KOH
Caustic soda	NaOH
Chile saltpeter	$NaNO_3$
Cream of tartar	$KHC_4H_4O_6$
Diamond	C
Dipping acid	H_2SO_4
Dolomite	$CaCO_3 \cdot MgCO_3$
Epsom salts	$MgSO_4 \cdot H_2O$
Fuming sulfuric acid	$H_2SO_4 + \text{excess } SO_3$
Galena	PbS
Gypsum	$CaSO_4 \cdot 2\,H_2O$
Hypo	$Na_2S_2O_3$ or $Na_2S_2O_3 \cdot 5\,H_2O$
Iron rust	Fe_2O_3
Laughing gas	N_2O
Lime	CaO or $Ca(OH)_2$
Limestone	$CaCO_3$
Lye	NaOH
Marble	$CaCO_3$
Milk of magnesia	$Mg(OH)_2$
Monohydrogen phosphate	HPO_4^{2-}
Muriatic acid	HCl (aq)
Oleum	$H_2SO_4 + \text{excess } SO_3$
Oil of vitriol	H_2SO_4 (concentrated)
Phosphine	PH_3
Plaster of paris	$CaSO_4 \cdot \tfrac{1}{2}H_2O$
Potash	K_2CO_3
Pyrites (fool's gold)	FeS_2
Quartz	SiO_2
Quicksilver	Hg
Sal ammoniac	NH_4Cl
Salt	NaCl
Saltpeter	KNO_3
Slaked lime	$Ca(OH)_2$
Stop bath, vinegar	$HC_2H_3O_2$
Washing soda	$Na_2CO_3 \cdot 10\,H_2O$

REFERENCES

1. *Lange's Handbook of Chemistry*, 15th edition, John A. Dean, editor, "Section 3—Inorganic Compounds," McGraw-Hill, Inc. ISBN 0-07-016384-7 (1999).
2. *Handbook of Chemistry and Physics*, 83rd edition or current edition, David R. Lide, editor-in-chief, CRC Press LLC. ISBN 0-8493-0483-0 (2003).
3. *The Merck Index*, 13th edition or current edition, Maryadele J. O'Neil, senior editor, Merck & Co., Inc. ISBN 0911910-13-1 (2001).

CHAPTER 23
ORGANIC CHEMISTRY REVIEW

NAMING ORGANIC COMPOUNDS

This chapter is designed to review the nomenclature rules for naming organic compounds by the **International Union of Pure and Applied Chemistry** (IUPAC) system. The IUPAC names will be routinely given for each compound, and common names, if used, will be stated in parenthesis. An **organic** compound can be defined as any substance that contains carbon. Carbon has four electrons in its valence shell and needs to share four additional electrons to achieve stability according to the octet rule. Therefore, many organic compounds will contain carbon atoms with four single covalent (sharing) bonds. However, carbon atoms are capable of forming double and triple bonds. Molecules having only carbon and hydrogen atoms are known as **hydrocarbons**. These hydrocarbon molecules having single, double, and triple bonds are known as the alkane, alkene, and alkyne families, respectively. Most hydrocarbons are found in: (a) linear structural compounds called aliphatic with varying numbers of single, double, and triple bonds; (b) cyclic hydrocarbons that form ring structures with primarily single bonds; and (c) structures classified as aromatics because of the specific C6H6 formula with alternating single and double bonds in a circular structure.

ALKANES

The simplest hydrocarbon molecules, known as alkanes, have carbon and hydrogen atoms bonded to each other by only single bonds. A hydrocarbon containing all single bonds between adjacent carbon atoms is referred to as saturated. **Saturation** refers to the fact that all bonding sites between adjacent carbon atoms are filled (saturated) with hydrogen atoms. The general formula for an **alkane** is C_nH_{2n+2} where n represents the number of carbon atoms and 2n+2 represents the number of hydrogen atoms in the hydrocarbon chain. These simple "straight-chained" hydrocarbons are referred to in the petroleum industry as **paraffins**. Crude oil used in the petroleum industry consists mostly of hydrocarbons mixed with sulfur, nitrogen- and oxygen-containing compounds.

Three basic IUPAC rules for naming alkanes are as follows:

A. IUPAC Rule 1

Name the alkane by selecting the name of the longest continuous chain of carbons found in the structure using Table 23.1.

B. IUPAC Rule 2

Each carbon-hydrogen group (called carbon branches) not counted in the continuous chain is named as an alkyl group. An alkyl (al'kil) group is simply a hydrocarbon with one hydrogen atom removed, as shown in Table 23.2. The name is derived by dropping the "ane" ending on the parent hydrocarbon and adding "yl." The "~" symbol used in Table 23.2 is used to show the point of attachment for this branch (alkyl group).

TABLE 23.1 The IUPAC Names of the First 20 Simple Alkanes, Number of Carbons, and Physical Properties

Name	Carbons	Formula	State	mp (°C)	bp (°C)	Sp. Gravity
Methane	1	CH_4	gas	−182.6	−161.7	0.4240
Ethane	2	C_2H_6	gas	−172.0	−88.6	0.5462
Propane	3	C_3H_8	gas	−187.1	−42.2	0.5824
Butane	4	C_4H_{10}	gas	−135.0	−0.5	0.5788
Pentane	5	C_5H_{12}	liquid	−129.7	36.1	0.6264
Hexane	6	C_6H_{14}	liquid	−94.0	68.7	0.6594
Heptane	7	C_7H_{16}	liquid	−90.5	98.4	0.6837
Octane	8	C_8H_{18}	liquid	−56.8	125.6	0.7028
Nonane	9	C_9H_{20}	liquid	−53.7	150.7	0.7179
Decane	10	$C_{10}H_{22}$	liquid	−29.7	174.0	0.7298
Undecane	11	$C_{11}H_{24}$	liquid	−25.6	195.4	0.7404
Dodecane	12	$C_{12}H_{26}$	liquid	−9.6	216.3	0.7493
Tridecane	13	$C_{13}H_{28}$	liquid	−6.0	230	0.7568
Tetradecane	14	$C_{14}H_{30}$	liquid	5.5	251	0.7636
Pentadecane	15	$C_{15}H_{32}$	liquid	10.0	268	0.7688
Hexadecane	16	$C_{16}H_{34}$	liquid	18.1	280	0.7749
Heptadecane	17	$C_{17}H_{36}$	liquid	22.0	303	0.7767
Octadecane	18	$C_{18}H_{38}$	solid	28.0	308	0.7767
Nonadecane	19	$C_{19}H_{40}$	solid	32.0	330	0.7776
Eicosane	20	$C_{20}H_{42}$	solid	36.8	343	0.7777

TABLE 23.2 Naming Alkyl Groups

Parent	Alkyl name	Formula
methane	methyl	~CH_3
ethane	ethyl	~CH_2CH_3
propane	propyl	~$CH_2CH_2CH_3$
butane	butyl	~$CH_2CH_2CH_2CH_3$
pentane	pentyl	~$CH_2CH_2CH_2CH_2CH_3$

The alkyl group name should precede the name of the longest continuous chain. If similar alkyl groups are attached (branched) to the same parent chain, then the prefixes shown in Table 23.3 should be used to indicate the number of groups.

TABLE 23.3 IUPAC Prefixes for Multiple Alkyl Branches

2 similar groups = di
3 similar groups = tri
4 similar groups = tetra
5 similar groups = penta
6 similar groups = hexa
7 similar groups = hepta
8 similar groups = octa
9 similar groups = nona
10 similar groups = deca

C. IUPAC Rule 3

The carbon atoms on the longest continuous chain are numbered in an order which will locate the attached alkyl groups using the smallest set of numbers possible. These alkyl branches are normally listed in alphabetical order. If two of the same alkyl groups are bonded to the same carbon, the number is listed twice, for example:

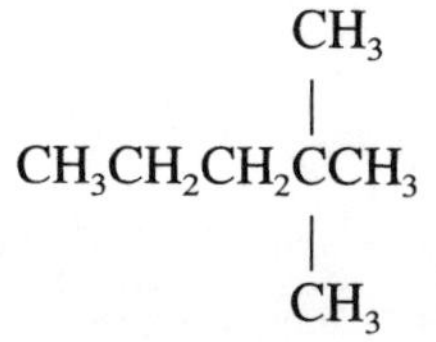

$$CH_3CH_2CH_2C(CH_3)_2CH_3$$ (with two CH₃ branches on the same carbon)

CH₃
|
CH₃CH₂CH₂CCH₃
|
CH₃

2,2-dimethylpentane

ALKENES

Hydrocarbon molecules containing a double carbon bond(s) are known as alkenes. The general formula for an **alkene** is C_nH_{2n}. Hydrocarbons which contain double and/or triple bonds are sometimes referred to as unsaturated. The IUPAC rules for naming alkenes are similar to those for naming alkanes:

1. Determine the base name by selecting the longest continuous chain that contains the double bond. Change the ending of the alkane name which corresponds to the number of carbon atoms in the chain from "ane" to "ene." For instance, if the longest chain has five carbon atoms (pentane), the base name for the alkene would be pentene.

2. Number the chain so as to include both carbons of the double bond and begin numbering at the end of the chain nearest the double bond. Designate the location of the double bond by indicating the number of the first carbon atom of the double bond before the parent chain name:

$$CH_3—CH_2—HC=CH—CH_3 \qquad H_2C=CH—CH_2—CH_3$$

2-pentene   1-butene

(not 3-pentene)   (not 3-butene)

3. The locations of alkyl groups are indicated by the number of the carbon atom to which they are attached:

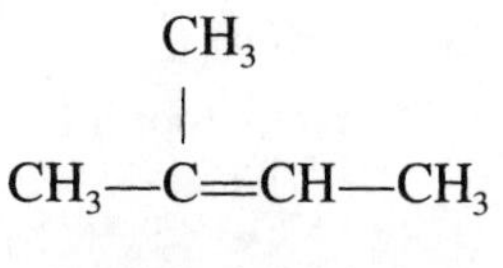

CH₃
|
CH₃—C=CH—CH₃

2-methyl-2-butene

H₃C   CH₃
 | |
CH₃—C=C—CH₂—CH₃

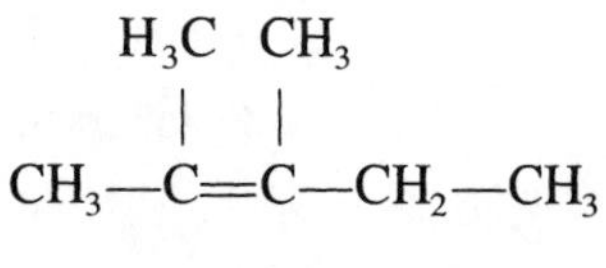

2,3-dimethyl-2-pentene

ALKYNES

Hydrocarbon molecules that contain triple bond(s) between the carbon atoms are called **alkynes**. The general formula for alkynes is C_nH_{2n-2}. The IUPAC rules for naming alkynes are similar to those for alkenes, except that the parent alkane (longest continuous chain containing the triple bond) name ending is changed from "ane" to "yne." The numbering system is exactly the same as alkenes, the triple bond is designated the lowest possible number, and the constituents or alkyl groups are numbered according to the triple bond. For example:

$$HC \equiv CH \qquad CH_3-CH_2-C \equiv C-CH_2-CH_3$$

ethyne   3-hexyne

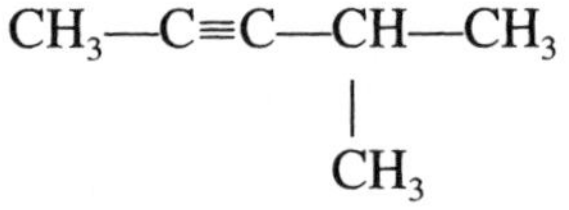

$$CH_3-C \equiv C-CH-CH_3$$
$$|$$
$$CH_3$$

4-methyl-2-pentyne

CYCLIC HYDROCARBONS

The carbon atoms normally associated in a straight chain can form a ring by connecting the end carbons. Cycloalkanes have two fewer hydrogen atoms than the same straight chained alkane; therefore the general formula for a **cycloalkane** is C_nH_{2n}. The petroleum industry describes this group of cyclic hydrocarbons as naphthene or **naphthenic compounds**. Cyclic molecules are named using the IUPAC system by attaching the prefix "cyclo" to the beginning of the parent chain name of the organic molecule, for example:

$CH_3CH_2CH_3$

propane

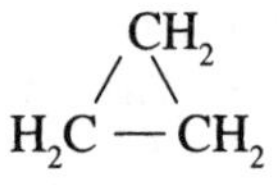

cyclopropane

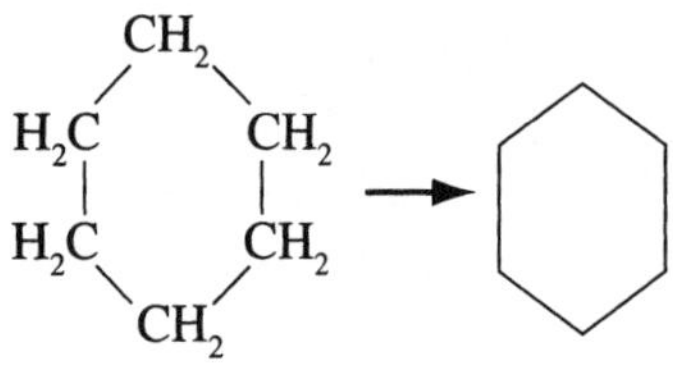

cyclohexane

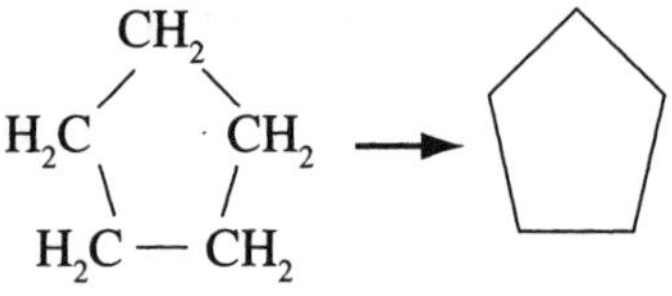

cyclopentane

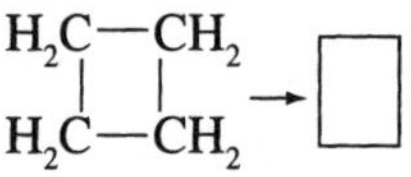

cyclobutane

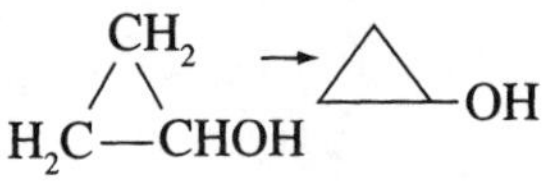

cyclopropanol

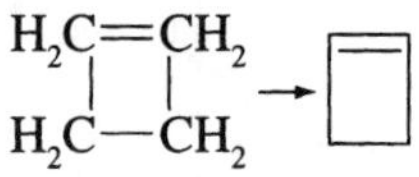

cyclobutane

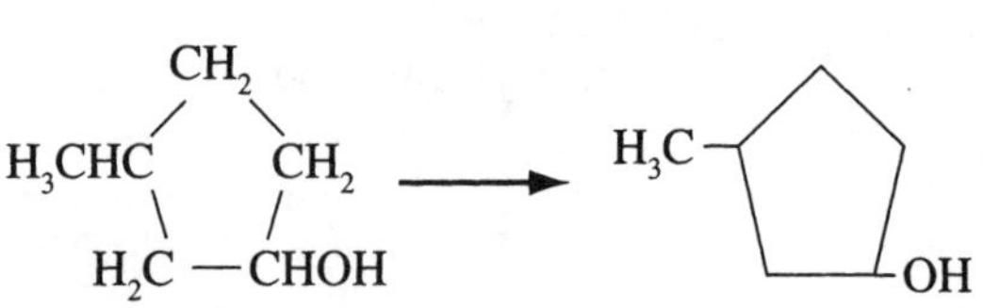

3-methylcyclopentanol

Note that numbers must be assigned when more than one substituent exists; otherwise carbon number "one" will always bear the single functional group.

Cyclic compounds can be represented by geometrical figures as demonstrated.

ISOMERS

The naming of organic compounds is complicated because of isomers. An isomer is a compound with the same formula but different structures. Table 23.4 indicates how the number of possible isomeric structures increases rapidly as the number of carbons and multiple bonds increases. Butane (formula: C_4H_{10}) is the first saturated hydrocarbon to have two isomeric structures as shown below:

$$CH_3CH_2CH_2CH_3 \qquad \begin{array}{c} CH_3CHCH_3 \\ | \\ CH_3 \end{array}$$

butane (normal butane)   2-methylpropane (isobutane)

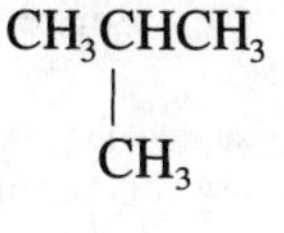

Unlike the single-bonded alkanes, there is no rotation about the double bond of the alkenes. This loss of rotation about the double bond creates stereoisomers with some alkenes. A **stereoisomer** is defined as a compound with identical groups on the same side (**cis**) of the double bond and (**trans**) if the identical groups are on opposite sides of the double bond.

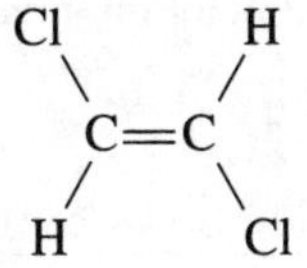

$$\begin{array}{c} Cl \qquad\quad H \\ \diagdown \quad \diagup \\ C = C \\ \diagup \quad \diagdown \\ H \qquad\quad Cl \end{array} \qquad\qquad \begin{array}{c} Cl \qquad\quad Cl \\ \diagdown \quad \diagup \\ C = C \\ \diagup \quad \diagdown \\ H \qquad\quad H \end{array}$$

trans-dichloroethene 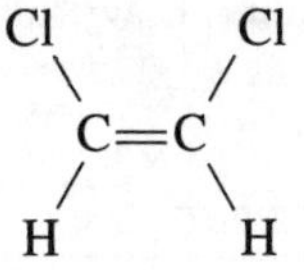   cis-dichloroethene

TABLE 23.4 Possible Number of Isomers with Increasing Number of Carbons in a Compound

Formula	Number isomers
CH_4	1
C_2H_6	1
C_3H_8	1
C_4H_{10}	2
C_5H_{12}	3
C_8H_{18}	18
$C_{10}H_{22}$	75

AROMATIC COMPOUNDS

An **aromatic** compound is one which contains a benzene molecule (C_6H_6). Aromatic compounds are sometimes referred to as **arenes**. Benzene is unique because it is a cyclic molecule with alternating double and single bonds. This alternating single- and double-bond configuration gives the benzene ring a great deal of stability and many compounds are found in nature with this structure. Because of its abundance, benzene is used as the IUPAC parent name for most molecules containing a six carbon-membered alternating double- and single-bonded ring. A shorthand method of writing the benzene structure is a circle inserted into the hexagon structure to represent the alternating bonds as shown below. This circle inside of the ring indicates that the electrons in the double bonds are actually spread over all six carbon atoms in the ring. When benzene is a substituent on another molecule, it is called phenyl (fen'uhl). Again, the lowest numbers possible are used to locate two or more substituents bonded to the ring.

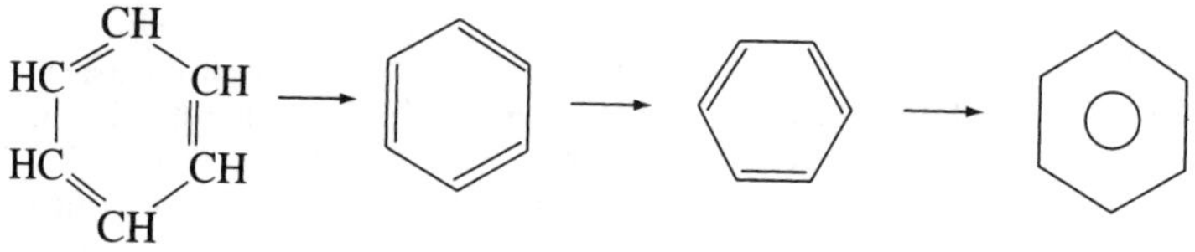

Benzene

Oil refineries divide crude oil mixtures into various fractions based on their boiling-point differences (see Chapter 21, "Distillation and Evaporation").

1. **Straight-run gasoline:** the most volatile liquid hydrocarbons and generally blended with ordinary motor fuels.
2. **Naphtha:** naphthenic oils and waxes plus light oils, used mainly for solvents.
3. **Kerosene:** a light oil used for solvents and heating oils.
4. **Gas oil:** a mixture of light and heavy oils which will ultimately be converted in a cracking process to gasoline and fuel oil.
5. **Residuum:** heavy lubricating oils, motor oil, grease, petroleum jelly, waxes, asphalt, and tar.

TABLE 23.5 The Common and IUPAC Names for Aromatic Compounds

IUPAC name	Structure	Common name(s)	mp (°C)	bp (°C)	Sp. Gr.
benzene	C_6H_6	benzol, phene	5.5	80.1	0.8787
methyl benzene	$CH_3C_6H_5$	toluene, toluol	−95	110.6	0.8669
ethyl benzene	$C_2H_5C_6H_5$		−94.97	136.2	0.8670
1,4-dimethyl benzene	$(CH_3)_2C_6H_4$	p-xylene	13.26	138.4	0.8611
1,2-dimethyl benzene	$(CH_3)_2C_6H_4$	o-xylene	−25.18	144.4	0.8802
1,3-dimethyl benzene	$(CH_3)_2C_6H_4$	m-xylene	−47.87	139.1	0.8642
hydrobenzene	C_6H_5OH	phenol	41.0	182.6	1.082
nitrobenzene	$C_6H_5NO_2$		5.7	210.9	1.199

ORGANIC FAMILIES AND THEIR FUNCTIONAL GROUPS

A. Alkyl Halides

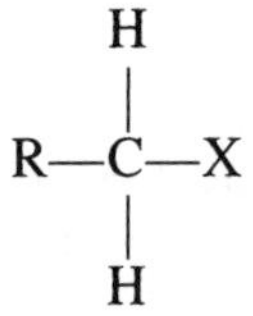

$$R-\underset{\underset{H}{|}}{\overset{\overset{H}{|}}{C}}-X$$ where: X = any halogen (F, Cl, Br, or I)
R = any alkyl or aryl group

Using the IUPAC nomenclature system, alkanes with a halogen atom as a substituent are named as a haloalkane. The general formula is RCH_2X. The parent alkane name is preceded by the halogen which is present and the number of the carbon to which the halogen is bonded.

$$Cl-CH_3 \qquad \underset{Br}{CH_3CH_2-CHCH_3} \qquad \underset{Cl}{CH_3-C=CH_2}$$

chloromethane   2-bromobutane   2-chloro-1-propene

TABLE 23.6 The IUPAC Names of Common Organic Halides

IUPAC name	Structure	Common name(s)	mp (°C)	bp (°C)	Sp. Gr.
chloromethane	ClCH$_3$	methyl chloride	−97	−23.7	0.920
dichloromethane	Cl$_2$CH$_2$	methylene chloride	−96	40.8	1.336
trichloromethane	HCCl$_3$	chloroform	−63.5	61.2	1.489
tetrachloromethane	CCl$_4$	carbon tetrachloride	−23	76.8	1.575
tribromomethane	HCBr	bromoform	7.8	149.5	2.865
chlorobenzene	C$_6$H$_5$Cl	phenyl chloride	−45.6	132	1.1058
dichloroethene	ClCH$_2$CH$_2$Cl	ethyiene dichloride	−35.5	83.8	1.238
1-chloro-4-toluene	CH$_3$C$_6$H$_5$Cl	p-to!yl chloride	7.5	162	1.0697
1,1,2 trichloroethane	Cl$_2$CCHCl	trichloroethylene	−86	87	1.477

ALCOHOLS

Any compound with the general formula R—OH, where R represents any hydrocarbon group and ~OH a hydroxyl group, is considered to be an **alcohol**. This alcoholic functional group is referred to as a carbinol (C—OH) group. If the hydroxyl group is attached to an aryl group (benzene ring), it is called a **phenol**. Consider alcohols, phenols, and ethers as derivatives of water. Absolute alcohol is another name for pure, water-free ethanol. Denatured alcohol is ethanol that has been made unfit for human consumption by adding small portions of methanol, benzene, and/or aviation gasoline. The contaminants do not change the chemical properties of the ethanol and no alcohol tax is required. Table 23.7 lists some common simple (one OH group) alcohols and their IUPAC names.

IUPAC naming

1. Select the longest continuous carbon chain that contains the OH group as the parent chain.
2. Drop the "e" ending of the alkane parent name and replace it with suffix "ol."
3. The "longest" chain must include the carbon bearing the OH group.

Alcohols can be classified as primary, secondary, or tertiary. Primary alcohol means that only one carbon atom is directly attached to the carbinol carbon.

$$\underset{H}{\overset{H}{R-\underset{|}{\overset{|}{C}}-OH}}$$

where R is any H or carbon-containing group. The R groups can be similar or different as indicated by R′ and R″.

Secondary alcohol means that two carbon atoms are directly attached to the carbinol carbon.

$$\begin{array}{c} R' \\ | \\ R-C-OH \\ | \\ H \end{array}$$

Tertiary alcohol means that three carbon atoms are directly attached to the carbinol carbon.

$$\begin{array}{c} R' \\ | \\ R-C-OH \\ | \\ R'' \end{array}$$

Alcohols with two OH groups are called **glycols** or **diols** and those with three OH groups are called **triols**. Table 23.8 gives several examples.

TABLE 23.7 The Common and IUPAC Names for Simple Alcohols

IUPAC name	Structure	Common name(s)	mp (°C)	bp (°C)	Sp. Gr.
methanol	CH_3OH	methyl alcohol or wood alcohol	−97.7	64.7	0.792
ethanol	CH_3CH_2OH	ethyl alcohol or grain alcohol	−114	78.3	0.789
1-propanol	$CH_3CH_2CH_2OH$	n-propyl alcohol or propyl alcohol	−126	97.2	0.804
2-propanol	$CH_3CHOHCH_3$	isopropyl alcohol or rubbing alcohol	−88.5	82.3	0.786
1-butanol	$CH_3CH_2CH_2CH_2OH$	n-butyl alcohol	190	117	0.810
2-butanol	$CH_3CH_2OHCH_2CH_3$	sec-butyl alcohol		99.5	0.808
2,2-dimethylpropanol	$(CH_3)_3COH$	t-butyl alcohol	25	82.5	0.789
1-pentanol	$CH_3CH_2CH_2CH_2OH$	amyl alcohol	−78.5	138.0	0.817
1-decanol	$CH_3(CH_2)_8CH_2OH$	n-decyl alcohol	6	232.9	0.829

TABLE 23.8 The Common and IUPAC Names for Diols and Triols

IUPAC name	Structure	Common name(s)	mp (°C)	bp (°C)	Sp. Gr.
1,2-ethanediol	CH_2OHCH_2OH	ethylene glycol, antifreeze	−13	197	1.116
diethylene glycol	$HOC_2H_4OC_2H_4OH$		−8	245	1.118
1,2-propanediol	$CH_2CHOHCH_2OH$	propylene glycol	—	188	1.038
triethylene glycol	$HOC_2H_4OC_2HOC_2H_4OH$		−7	287	1.125
dipropylene glycol	$(CH_3CHOHCH_2)O$		—	232	1.025
1,3 butylene glycol	$C_4H_8(OH)_2$		—	208	1.006
1,2,3-propanetriol	$HOCH_2CHOHCH_2OH$	glycerol, glycerin	20	290	1.261

AMINES

Amines are compounds with organic groups attached to a central nitrogen atom and can be classified in one of four groups. Most amines have a strong decaying fish odor. Consider amines as derivatives of ammonia (NH_3). The general formula is R—NH_2. Table 23.9 gives some common amines and their IUPAC names.

$$\begin{array}{cccc} H-N-H & R-N-H & R-N-H & R-N-R \\ | & | & | & | \\ H & H & R & R \\ \text{ammonia} & \text{primary} & \text{secondary} & \text{tertiary} \\ & \text{amine} & \text{amine} & \text{amine} \end{array}$$

Notice that the classification of amines is based on the number of organic groups attached to the nitrogen atom, not those attached to a carbon as was the case with primary/secondary/tertiary alcohols. A **quarternary** amine is defined as an amine which has four alkyl or aryl groups attached to the central nitrogen atom, for example, diethyldimethylammonium chloride.

Amines are very often called by common names as alkyl derivatives of ammonia with the word "ammonia" being changed to "amine." A capital "N" preceding the name of an amine indicates the branches are located on the nitrogen atom, not the carbon chain.

$$CH_3-NH_2 \qquad CH_3CH_2-\overset{\overset{H}{|}}{N}-CH_2CH_3$$
(methylamine)   (N,N-diethylamine)

The IUPAC name for the amine functional group is called araino (—NH_2, uh meen'o). It is considered the substituent and its position on the chain is indicated by the lowest numbers.

$$CH_3CH_2\overset{\overset{CH_3}{|}}{C}HCH_2\underset{\underset{NH_2}{|}}{C}HCH_3 \qquad \text{5-methyl-3-aminohexane}$$

$$H_2N-CH_2CH_2CH_2CH_2CH_2CH_2-NH_2 \qquad \text{1,6-diaminohexane}$$

ETHERS

Ethers are defined as any compound with an oxygen atom linked by single bonds to two carbon-containing (R) groups. The general formula is R—O—R′.

Symmetrical ether—both R and R′ groups are identical.

Mixed ether—the two R groups are different (thus unsymmetrical).

CH₃—O—CH₃   methoxymethane or (dimethyl ether)

CH₃—O—CH₂CH₃   methoxyethane or (ethyl methyl ether)

Common names are assigned to ethers by naming the two alkyl groups bonded to the oxygen and adding the word "ether." The IUPAC system names ethers as alkoxyalkanes, alkoxyalkenes, and alkoxyalkynes by selecting the longest continuous hydrocarbon chain as the parent and adding the alkoxy (RO—) prefix. Table 23.10 gives some ethers by their common and IUPAC names.

TABLE 23.9 The Common and IUPAC Names for Amines

IUPAC name	Structure	Common name(s)	mp (°C)	bp (°C)	Sp. Gr.
aminomethane	CH₃NH₂	methylamine	−92.5	−6.5	0.699
aminoethane	CH₃CH₂NH₂	ethylamine	−80.6	16.6	0.689
1-aminopropane	CH₃CH₂CH₂NH₂	propylamine	−83.0	48.7	0.719
1-aminobutane	CH₃CH₂CH₂CH₂NH₂	n-butylamine	−50.5	76	0.740
1-aminopentane	CH₃CH₂CH₂CH₂CH₂NH₂	n-amylamine	−55.0	104	0.766
2-amino-1-ethanol	₂HNCH₂CH₂OH	ethanolamine		171	1.022
1-aminobenzene	C₆H₅NH₂	aniline	−6	184	—
aminotoluene	CH₃C₆H₄NH₂	p-toluidene	44	200	—

TABLE 23.10 The Common and IUPAC Names for Some Ethers

IUPAC name	Structure	Common name	mp (°C)	bp (°C)	Sp. Gr.
methoxymethane	CH₃OCH₃	dimethyl ether, methyl ether	−140	−24.9	0.661
methoxyethane	CH₃OCH₂CH₃	methyl ethyl ether		7.9	0.697
ethyoxyethane	CH₃CH₂OCH₂CH₃	ether, diethyl ether	−116	34.6	0.714
propoxypropane	CH₃CH₂CH₂OCH₂CH₂CH₃	di-n-propyl ether	−122	90.5	0.736
methoxybenzene	C₆H₅OCH₃	anisole	−37.3	154	0.994
phenoxybenzene	C₆H₅OH₅C₆	diphenyl ether	26.9	259	1.072

CAUTION: Ethers tend to be chemically unstable. Aliphatic ethers react slowly with air to produce peroxides. All ether containers should be dated when received, purchased only in iron containers to reduce production of peroxides, and the excess discarded properly after opening the container. Ethers tend to be volatile, have low solubility in water, and have very low flash points. Do not discard excess ethers down drain; consult a supervisor and/or MSDS for proper disposal.

ALDEHYDES

Aldehydes are defined as any compound that contains a carbonyl group (C=O, kar′buh nil) with one or more hydrogen atoms.

General formula:
$$R-\overset{H}{\underset{}{C}}=O$$

The IUPAC nomenclature rules state that the "e" ending on the parent hydrocarbon's name be deleted and the "al" suffix be added. The word aldehyde starts with the prefix letters "al." Since it will always be the first carbon in the chain, no locator numbers are needed to designate the carbon to which the carbonyl is bonded. Table 23.11 gives both the common and IUPAC names for some aldehydes.

TABLE 23.11 The Common and IUPAC Names for Some Aldehydes

IUPAC name	Structure	Common name	mp (°C)	bp (°C)	Sp. Gr.
methanal	H_2CO	formaldehyde	−92	−21	0.815
ethanal	H_3CHCO	acetaldehyde	−123	20.8	0.781
trichloroethanal	Cl_3CHCO	chloral	−57.5	97.8	1.512
propanal	H_3CH_2HCO	propionaldehyde	−81	48.8	0.807
butanal	$H_3CCH_2CH_2HCO$	n-butylaldehyde	−97	74.7	0.817
pentanal	$H_3CCH_2CH_2CH_2HCO$	n-valeraldehyde	−92	103.7	0.819
benzenecarbonal	C_6H_5CHO	benzaldehyde	−56	179	1.046

CAUTION: The simplest aldehyde is formaldehyde and is a gas at room temperature. It is very soluble in water and is sold as a 37 percent aqueous solution called **formalin**. Formaldehyde has been used as a very effective preservative and bactericide, but is a carcinogen and should not be used without proper precautions.

KETONES

Ketones are structurally very similar to aldehydes. Ketones also have a carbonyl functional group, but the difference is that it is located on a carbon other than the first carbon of the chain. The general formula is R—CO—R′. The R′ is to designate that this R group can have a different structure (i.e., number of carbons and hydrogens) than R group itself.

The carbon to which the carbonyl is bonded must be assigned the lowest possible number. The parent alkanes ending is changed by dropping the "e" and adding the suffix "one." The word ketone ends in the letters "one." Table 23.12 lists some common ketones by their IUPAC and common names.

TABLE 23.12 The Common and IUPAC Names for Some Ketones

IUPAC name	Structure	Common name	mp (°C)	bp (°C)	Sp. Gr.
2-propanone	CH_3—CO—CH_3	dimethyl ketone, acetone	−95	56.1	0.7915
2-butanone	CH_3—CO—CH_2CH_3	methyl ethyl ketone, MEK	−86	79.6	0.805
2-petitanone	CH_3—CO—$CH_2CH_2CH:_3$	methyl propyl ketone	−77.8	102.1	0.812
3-pentanone	CH_3CH_2—CO—CH_2CH_3	diethyl ketone	−42.0	101.7	0.814
6-undecanone	$CH_3CH_2CH_2CIH_2CH_2$—CO—$CH_2CH_2CH_2CH_2CH_3$		14.6	228	0.826
chloro-2-propanone	CH_3COCH_2Cl	chloroacetone	−44.5	119	1.162
cyclohexanone	$CH_2(CH_2CH_2)_2CO$		—	156.7	0.949
benzophenone	$C_6H_5COC_6H_5$		48	305.4	1.083

CARBOXYLIC ACIDS

Carboxylic acid are classified as compounds that contain both the carbonyl group and the hydroxyl groups (*carb*onyl + hyd*roxyl*). These organic acids tend to be weak acids and have unpleasant odors.

General formula: R—COOH

TABLE 23.13 The Common and IUPAC Names for Some Carboxylic Acids

IUPAC name	Structure	Common name	mp (°C)	bp (°C)	Sp. Gr.
methanoic acid	HCOOH	formic acid	8.4	100.5	1.220
ethanoic acid	CH_3COOH	acetic acid, vinegar	16.6	118	1.049
propanoic acid	CH_3CH_2COOH	propionic acid	−22	141	0.992
bntanoic acid	$CH_3CH_2CH_2COOH$	butyric acid	−4.7	162.5	0.959
pentanoic acid	$CH_3CH_2CH_2CH_2COOH$	valeric acid	−34.5	187.0	0.939
hexanoic acid	$CH_3CH_2CH_2CH_2CH_2COOH$	caproic acid	−1.5	205	0.929
chloroethanoic acid	$CH_2ClCOOH$	chloroacetic acid	63	189.5	1.37
2-hydroxypropanoic acid	$CH_3CHOHCOOH$	lactic acid	18	—	1.249
benzoic acid	C_6H_5COOH	benzenecarboxylic acid	121.7	249	1.266

The IUPAC rules state that the "e" be deleted from the parent hydrocarbon and the suffix "oic" and word "acid" be added. No location numbers are needed to designate the carbon containing the carboxylic functional group since it will always be the first carbon. Table 23.13 gives the IUPAC name of several carboxylic acids, the structure, the common name, and physical properties.

AMIDES

Amides are derived from a carboxylic acid in which the OH group has been replaced by an amine (NH_2) group. The general formula is R—$CONH_2$. Location numbers are used to locate branches on carbon atoms of the parent hydrocarbon and a capital "N" prefix is used to indicate branches on the nitrogen atom of the amine group. The IUPAC system requires selecting the longest carbon chain as the parent carboxylic acid, then dropping the "oic" ending and adding suffix "amide."

$CH_3CH_2CH_2CH_2CONH_2$ (pentanamide)

$CH_3CONHCH_2CH_3$ (N-ethylacetamide)

ESTERS

Esters are compounds produced by the reacting of an alcohol and a carboxylic acid. The ester's common name is derived by naming the alkyl group of the alcohol followed by the organic anion of the acid molecule. The general formula is R—CO—OR. Table 23.14 shows some common esters.

$$CH_3CO—OH + CH_3OH \longrightarrow CH_3CO—O—CH_3 + H_2O$$

(acetic acid) + (methyl alcohol) $\longrightarrow$ (methyl acetate) + (water)

The IUPAC nomenclature rules require that the acid's "ic" ending be changed to "ate" and the name of the alkyl or aryl (R) group of the alcohol must precede this name.

EXAMPLE 23.1

pentyl ethanoate (pentanol + ethanoic acid)

ethyl propanoate (ethanol + propanoic acid)

methyl butanoate (methanol + butanonic acid)

TABLE 23.14 The Common and IUPAC Names for Some Esters

IUPAC name	Structure	Common name	mp (°C)	bp (°C)	Sp. Gr.
methyl methanote	CH_3OOCH	methyl formate	−99	32	0.974
ethyl methanoate	C_2H_5OOCH	ethyl formate	−80.5	54	0.906
ethyl ethanoate	$C_2H_5OOCCH_3$	ethyl acetate	−83.6	77.1	0.901
butyl ethanoate	$C_4H_9OOCCH_3$	n-butyl acetate	−77.9	126.5	0.882
t-butyl ethanoate	$(CH_3)_3COOCCH_3$	t-butyl acetate	—	97	0.896
pentyl ethanoate	$C_5H_{11}COOCCH_3$	n-amyl acetate	−70.8	147.6	0.879

AMINO ACIDS

An amino acid is a compound that contains at least one amino group (~NH_2) and at least one carboxylic acid group (~COOH). There are 20 different amino acids found in the human body (Table 23.15).

TABLE 23.15 The Common Amino Acids and Their Abbreviations

Name	Abbreviation	Formula
Glycine	Gly	$CH_2(NH_2)COOH$
Alanine	Ala	$CH_3CH(NH_2)COOH$
Valine	Val	$(CH_3)_2CHCH(NH_2)COOH$
Leucine	Leu	$(CH_3)_2CHCH_2CH(NH_2)COOH$
Isoleucine	Ile	$CH_3CH_2CH(CH_3)CHCH(NH_2)COOH$
Proline	Pro	$(CH_2)_3CH(NH)COOH$
Phenylalanine	Phe	$(C_6H_5)CH_2CH(NH_2)COOH$
Tryptophane	Trp	$(C_8H_5N)CH_2CH(NH_2)COOH$
Methionine	Met	$CH_3SCH_2CH_2CH(NH_2)COOH$
Aspartic Acid	Asp	$HOOCCH_2CH(NH_2)COOH$
Glutamic Acid	Glu	$HOOCCH_2CH_2CH(NH_2)COOH$
Lysine	Lys	$H_2N(CH_2)_4CH(NH_2)COOH$
Histidine	His	$(H_3N_2C_3)CH_2CH(NH_2)COOH$
Serine	Ser	$HOCH_2CH(NH_2)COOH$
Threonine	Thr	$CH_3CH(OH)CH(NH_2)COOH$
Tyrosine	Tyr	$HO(C_6H_5)CH_2CH(NH_2)COOH$
Cysteine	Cys	$HSCH_2CH(NH_2)COOH$
Asparagine	Asn	$H_2NCOCH_2CH(NH_2)COOH$
Glutamine	Gln	$H_2NCOCH_2CH_2CH(NH_2)COOH$

POLYMERS

Polymers are very large organic molecules with molecular weights ranging from 100 to 500,000. **Polymers** are made by joining (bonding) many small individual molecules known as **monomers** together until the desired polymer is formed. Molecules composed of a two monomeric chain are called **dimers**; a three monomeric chain is called a **trimer**, a four monomeric chain is called a **tetramer**, and a several monomeric chain is called a **polymer**. Some polymers are naturally occurring such as starch, proteins, and natural rubber. Other polymers are synthetic or man-made polymers: teflon, nylon, polyester, rubber, polyvinyl chloride, and so forth.

monomer   polymer

$$\begin{array}{c} F \quad\quad F \\ \diagdown \quad \diagup \\ C = C \\ \diagup \quad \diagdown \\ F \quad\quad F \end{array} \longrightarrow (CF_4)_x$$

tetrafluoroethylene   polytetrafluoroethylene
(Teflon®)

There are basically two types of polymers. The first is a polymer made of one type of monomer and is called a homo-polymer. The second is a polymer made of two or more different monomers and these are called copolymers and the process is known as **copolymerization**. Starch would be considered a homopolymer because it is composed of strictly glucose monomers. A protein is a copolymer because it can contain any combination of 20 or so different amino acid monomers.

Polymers can be produced either by an addition reaction or a condensation reaction. An **addition** reaction involves monomers that contain at least one carbon-carbon double bond in their structure. To produce an addition-type polymer, the process begins with some type of an initiator which breaks the carbon-carbon double bond of the monomer molecules present. As the double bond is broken, one carbon atom becomes electron rich, resulting in a negative charge. The second carbon atom of the bond becomes electron poor, resulting in a positive charge. The polymer begins to form as the negative end of one monomer bonds with the positive end of another monomer. The process continues until the depletion of monomer present in the reaction vessel.

A second type of copolymerization reaction is known as **condensation**. This reaction requires monomers with different functional groups so that when they react, chain growth takes place. Another characteristic of a condensation reaction is the production of another small molecule in addition to the copolymer. A typical small molecule produced is water or hydrogen chloride. The polymer chain size, molecular weight, physical property, and so on are determined by the final number of monomeric units in the polymer, no matter which polymerization process takes place. The strength and durability of a polymer is also dependent on the beginning monomers as well as the final size and molecular weight.

Thermoplastic materials are softened by the application of heat and can be reworked many times. **Thermosetting** materials are resistant to softening with the application of heat and tend to decompose with extensive heating.

EXAMPLE 23.2 Polyvinyl chloride is an important thermoplastic material in the process industry. To produce polyvinyl chloride, acetylene and hydrogen chloride gas are reacted (as shown in step 1). The vinyl chloride molecules are then caused to undergo polymerized by being treated with a special peroxide catalyst. This process produces a long, linear-chain molecule (polymer) as shown in step 2. The subscript "n" shows that an indefinite number of vinyl chloride molecules (monomers) can be connected in the polyvinyl chloride polymer. Vinyl chloride can be polymerized in the presence of other monomers, like vinyl acetate, to produce copolymers with different properties.

(step 1)

$$HC \equiv CH + HCl \longrightarrow Cl-CH=CH_2$$

(step 2)

$$(Cl-CH=CH_2)_n \longrightarrow$$

$$\begin{array}{c} -CH-CH_2-CH-CH_2-CH-CH_2- \\ \quad | \quad\quad\quad\quad\quad | \quad\quad\quad\quad\quad | \\ \quad Cl \quad\quad\quad\quad\; Cl \quad\quad\quad\quad\; Cl \end{array}$$

SUMMARY OF ORGANIC FAMILIES AND THEIR FUNCTIONAL GROUPS

Although carbon-hydrogen bonds are predominant in organic molecules, carbon can also covalently bond to other atoms such as oxygen, nitrogen, chlorine, etc. These different constituents are known as functional groups. All organic compounds are classified into families (Table 23.16) by their functional groups.

TABLE 23.16 Organic Families and Their Functional Groups

Formula	Functional group	Family
C_nH_{2n+2}	single bond	alkane
C_nH_2n	double bond	alkene
C_nH_{2n-2}	triple bond	alkyne
C_6H_6	benzene	aromatic
R—X	halogen	alkyl halide
R—OH	hydroxyl	alcohol
R—C—NH_2	amino	amine
R—O–R'	oxygen	ether
R—C=O | H	carbonyl (terminal)	aldehyde
R—C=O | R	carbonyl (non-terminal)	ketone
R—C=O | OH	carbonyl + hydroxyl	carboxylic acid
R—C=O | NH_2	carboxyl + amin	amide
R—C=O | OR'	carboxyl + carbinol	ester
H OH | | R—C—C=O | NH_2	carboxylic acid + amine	amino acid

ORGANIC REACTIONS

A. Cracking

The petroleum industry converts a large portion of all high-molecular weight hydrocarbons into gasoline by thermal or catalytic cracking. The high-molecular weight hydrocarbons are not volatile enough for gasoline, but the shorter-chained molecules are ideal. **Thermal cracking** is the exposure of high-molecular weight hydrocarbons to high temperatures and pressures, which causes cleavage of the long carbon chains. In **catalytic cracking**, high-molecular weight hydrocarbon chains can be broken by exposure to solid catalysts (for example, silicates) even at low temperatures and pressure.

$$C_3CH_2CH_2CH_2CH_2CH_2CH_2CH_3 \longrightarrow 2\ C_3CH_2CH_2CH_3$$

B. Alkylation

After cracking processes had been developed, the petroleum industry found an abundance of alkenes available. Alkylation processes were developed to cause alkenes to react with branched hydrocarbons to primarily produce high octane gasolines. High octane gasoline mixtures contain an abundance of highly branched hydrocarbons, like trimethyl pentane, commonly called **octane**. Both cold and hot sulfuric acids have been found to be excellent catalysts for the polymerization or alkylation of olefins. A typical industrial process would involve coupling isobutane with butene to produce iso-octane (2,2,5-trimethylpentane) as shown below:

$$CH_3CHCH_3CH_3 + CH_2=CHCH_2CH_3$$

$$\longrightarrow CH_3-\underset{\underset{CH_3}{|}}{\overset{\overset{CH_3}{|}}{C}}-CH_2-\underset{\underset{CH_3}{|}}{CH}-CH_3$$

C. Fermentation

Many important chemical processes involve fermentation. Examples include ethyl alcohol production in the brewery industry and penicillin in the pharmaceutical industry. Yeast is a fermenting agent and produces two important enzymes known as invertase and zymase. Note that any organic ending in "ase" is an enzyme. An **enzyme** is a catalyst which causes a reaction to take place without being consumed or modified itself. If the sugar concentration is too high, fermentation will not take place. It is normal to dilute the sugar solution with water to approximately 10 percent by weight. In addition, small amounts of sulfuric acid and ammonium sulfate are added to adjust the pH and add nutrients for the yeast. The invertases cause the sucrose to **hydrolyze** (react with water) and form two simpler sugars, glucose and fructose, as shown:

$$C_{12}H_2O_{11} + H_2O \longrightarrow C_6H_{12}O_6 + C_6H_{12}O_6$$

sucrose water glucose fructose

The zymase changes the simple sugars into ethyl alcohol (ethanol) and carbon dioxide as shown below:

$$C_6H_{12}O_6 \longrightarrow 2\ C_2H_5OH + 2\ CO_2$$

glucose or fructose ethyl alcohol carbon dioxide

REFERENCES

1. *Lange's Handbook of Chemistry*, 15th edition, John A. Dean editor, Section 1—Organic Compounds. McGraw-Hill, Inc. ISBN 0-07-016384-7 (1999).

2. *Handbook of Chemistry and Physics*, 83rd edition or current edition, David R. Lide, editor-in-chief. CRC Press LLC. ISBN 0-8493-0483-0 (2003).

3. *The Merck Index*, 13th edition or current edition, Maryadele J. O'Neil, senior editor. Merck & Co., Inc. ISBN 0911910-13-1 (2001).

CHAPTER 24
CHEMICAL CALCULATIONS AND CONCENTRATION EXPRESSIONS

MATERIAL BALANCE

In any chemical process, an overall accountability of pounds of feed stocks to final pounds of products must be made. This comparison study is referred to as a **material balance**. Material balance data are normally presented in moles/hour or barrels/day and should match within ±5 percent.

A chemical equation is a written statement that uses symbols and formulas to describe the changes that occur in a chemical reaction. It is a shorthand description of a chemical change and represents a material balance. For example, methane reacts with oxygen (combustion) to form carbon dioxide and water. The unbalanced chemical equation for this reaction is

$$CH_4(g) + O_2(g) \rightarrow CO_2(g) + H_2O(l) + \text{energy}$$

The substances that react with one another, the **reactants**, are written on the left side of the arrow, and the substances *produced*, the **products**, are on the right. The arrow shows the direction of the reaction and is read as "yields," "gives," or "produces." The plus (+) simply means "and." In addition to the essential plus sign and the arrow notation used in chemical equations, other symbols that specify physical state are often used. The (s) following the symbol for carbon indicates that carbon is a solid in this reaction; the (g) means gas. In addition, the symbol (l) is used to represent a liquid and (aq) to represent an aqueous solution, a substance dissolved in water.

The word that describes material (mass) balance between the reactants and the products is called **stoichiometry**. Chemical engineering deals with both mass balance and energy balance in any process operation. Notice that energy would be considered a product in the methane combustion reaction shown above. **Unit operations** deal primarily with the physical changes (mass and energy) that take place, as opposed to chemical changes, in a process.

MOLECULAR RELATIONSHIPS FROM EQUATIONS

A valid chemical equation must satisfy two conditions. It must be consistent with (a) experimental facts and (b) the law of conservation of mass. To satisfy the first condition, only the reactants and products actually involved in the reaction are shown in the equation, and the correct symbols and formulas for these substances must be used. In the reaction

$$C(s) + O_2(g) \rightarrow CO_2(g)$$

oxygen is represented as O_2, showing that it occurs as a diatomic gas. Recall from Chapter 22, "Inorganic Chemistry Review," that six other elements exist as **diatomic gases**: H_2, N_2, F_2, Cl_2, Br_2, and I_2. To satisfy the second condition, there must always be the same number of each kind of product atoms as there are reactant atoms because the law of conservation of mass states that no mass is lost in a chemical reaction.

In the above reaction, there is one atom of carbon on each side of the equation and there are two atoms of oxygen on each side. The equation satisfies the law of conservation of mass. The equation is therefore said to be balanced. The **law of conservation of mass** is obeyed (also referred to as material balance). If an equation is not balanced, it is brought into balance by placing **coefficients** in front of the appropriate symbols or formulas. The methane combustion equation shown earlier is not balanced.

$$CH_4(g) + O_2(g) \rightarrow CO_2(g) + H_2O(l) + \text{energy}$$

There are four atoms of hydrogen on the reactant's side of the arrow, but only two hydrogen atoms show on the product's side of the equation. In addition, the oxygen atoms are not balanced with two atoms (one molecule) on the left and a total of three oxygen atoms being shown on the product's side of the equation. By placing a "2" in front of the O_2 and H_2O this can be corrected.

$$CH_4(g) + 2\,O_2(g) \rightarrow CO_2(g) + 2\,H_2O(l) + \text{energy}$$

Now there are four hydrogen atoms and four oxygen atoms on each side of the equation; it is balanced. Note that we did not try to balance the equation by changing the formulas of any reactant or product. For example, water is H_2O, not H_2O_2. That change would have balanced the equation, but it would not represent the experimental facts, that methane gas reacts with oxygen gas to produce only carbon dioxide and water. The formula H_2O_2 represents hydrogen peroxide not water.

Equations are used to represent what is reacting and what is being formed in a chemical reaction. The relative numbers of reactant particles and resulting particles are indicated by the coefficients of the formulas which represent these particles. The particles may be atoms, molecules, or ions.

EXAMPLE 24.1 Ammonia (NH_3) burns in oxygen (O_2) to yield nitrogen (N_2) and water (H_2O). Write the balanced equation for this reaction.

$$NH_3 + O_2 \rightarrow N_2 + H_2O \ (\textit{unbalanced})$$

Balanced chemical equation:

$$4\,NH_3 + 3\,O_2 \rightarrow 2\,N_2 + 6\,H_2O$$

Four molecules of ammonia react with three molecules of oxygen to give two molecules of nitrogen and six molecules of water. When ammonia reacts with oxygen as in this equation, three molecules of oxygen are always consumed for every four molecules of NH_3 and always yield two molecules of nitrogen and six molecules of water. All atoms are accounted for—every nitrogen, hydrogen, and oxygen atom is counted in the reactants and the products.

When the reaction equation is properly balanced, the number of atoms of any element in a molecule of a reactant or product can be found as follows: multiply the coefficient of the molecule by the subscript of the elements in the molecule. When there is no subscript for an element, the value for that subscript is always 1. In a balanced equation, the number of atoms on the reactant side will always equal the number of atoms on the product side.

EXAMPLE 24.2 Calculate the number of atoms contained on the reactants and products side of this balanced equation:

$$4\,NH_3 + 3\,O_2 \rightarrow 2\,N_2 + 6\,H_2O$$

Reactants:

4×1 nitrogen gives 4 nitrogens.

4×3 hydrogens give 12 hydrogens.

3×2 oxygens give 6 oxygens.

Products:

2 × 2 nitrogens give 4 nitrogens.

6 × 2 hydrogens give 12 hydrogens.

6 × 1 oxygen gives 6 oxygens.

All reactant atoms are found in the products.

MASS RELATIONSHIPS FROM EQUATIONS

Balanced chemical equations tell what the reactants are and what the products are and indicate the relative masses of the reactants and the products. All atoms and all molecules have definite masses, and the actual masses of the atoms and the molecules are proportional to their atomic and molecular masses. **Molecular masses** are determined by adding the atomic mass of each atom in a formula (see Chapter 22, "Inorganic Chemistry Review").

EXAMPLE 24.3 Phosphoric acid reacts with potassium hydroxide to yield potassium phosphate and water.

$$H_3PO_4 + 3\ KOH \rightarrow K_3PO_4 + 3\ H_2O$$

(1 molecule) (3 molecules) → (1 molecule) (3 molecules)

(98 g) (3 × 56 g) → (212 g) (3 × 18 g)

(98 g) (168 g) → (212 g) (54 g)

(266 g) → (266 g)

Every molecule of phosphoric acid that reacts with three molecules of potassium hydroxide, one molecule of potassium phosphate and three molecules of water are formed.

As detailed below, every 98 g of H_3PO_4 reacts with 168 g KOH, and 212 g of K_3PO_4 and 54 g H_2O are formed (refer to Table 22.2 for atomic masses).

Calculating the molecular mass of H_3PO_4:

3 hydrogens = 3 × 1.008 = 3.024
1 phosphorus = 1 × 30.98 = 30.98
4 oxygens = 4 × 16.00 = 64.00
Molecular mass = 98 g

Molecular mass of KOH:

1 potassium = 1 × 39.1 = 39.1
1 oxygen = 1 × 16.00 = 16.00
1 hydrogen = 1 × 1.008 = 1.008
Molcular mass = 56 g

Molecular mass of K_3PO_4:

3 potassium atoms = 3 × 39.1 = 117
1 phosphorus atom = 1 × 30.98 = 30.98
4 oxygen atoms = 4 × 16.00 = 64.00
Molecular mass = 212 g

Molecular mass of H_2O:

2 hydrogen atoms = 2 × 1.008 = 2.016
1 oxygen atom = 1 × 16.00 = 16.00
Molecular mass = 18 g

Total mass of the reactants = total mass of the products

98 g H_3PO_4 + 168 g NaOH = 266 g
212 g K_3PO_4 + 54 g H_2O = 266 g

This is the conservation of mass. In normal chemical reactions:

Mass of the reactants = mass of the products

Mass is neither created nor destroyed.

MOLE RELATIONSHIPS FROM CHEMICAL EQUATIONS

Chemical equations give us a count of the atoms and molecules involved in a chemical change. In the reaction,

$$C + O_2 \rightarrow CO_2$$

one atom of carbon reacts with one molecule of carbon dioxide (one atom of carbon and two atoms of oxygen). The equation also specifies mass relationships. We know from Table 22.2 that one atom of carbon has a mass of 12.01 atomic mass units (amu) and that one atom of oxygen has a mass of 16.00 amu. The reaction states that 12.01 amu of carbon reacts with 32.00 (2 × 16.00 amu) of oxygen to produce a molecule of carbon dioxide with a mass of 44.01 amu [12.01 amu + 2(16.00 amu)]. Because atomic masses are relative masses, any other units could be used and the mass relationship would still be valid. Thus, if 12.01 lb of carbon is reacted with 32.00 lb of oxygen, 44.01 lb of carbon dioxide is produced; likewise, 12.01 g of carbon reacting with 32.00 g of oxygen produces 44.01 g of carbon dioxide. These mass relationships can be used to calculate amounts of reactants and products in chemical reactions.

As an example, let's calculate the amount of oxygen that would be needed to completely combine with 36.0 g of carbon. From the equation we know that 12.0 g of carbon requires 32.0 g of oxygen; therefore, 36.0 g/12.0 g, or three times as much oxygen (96.0 g). Using the factor-label method, the problem would be set up as follows:

$$36.0 \text{ g C} \times \frac{32.0 \text{ g } O_2}{12.0 \text{ g C}} = 96.0 \text{ g } O_2$$

The amount of carbon dioxide which would be formed could be calculated the same way:

$$36.0 \text{ g C} \times \frac{44.0 \text{ g } CO_2}{12.0 \text{ g C}} = 132 \text{ g } CO_2$$

For chemical reactions in which the coefficients are not in a one-to-one ratio, it is convenient to use a quantity called a mole for calculations of this type. The mole to a chemist is much like a dozen is to a farmer. It is an important unit of measurement. A mole is defined, in terms of mass, as the atomic mass of an element or molecular mass of a compound. A mole in terms of particles, atoms or molecules, is defined as 6.02×10^{23} atoms or molecules. This very large number (6.02×10^{23}) is called **Avogadro's number** and is important in chemical calculations. Chemists use Avogadro's number of atoms and molecules for the same reason that farmers sell their produce by the dozen. One egg is not practical or convenient while one atom or molecule would be insignificantly small to deal with in a chemical process. By taking Avogadro's number of atoms or molecules, we literally have enough matter to measure. In other words, 6.02×10^{23} atoms of any element are equivalent to that element's atomic mass or 6.02×10^{23} molecules of any compound are equivalent to a compound's molecular mass (sum of the atomic masses for every atom of the formula). For example, in every 12.0 g of carbon there are 6.02×10^{23} atoms of carbon. In every 16.0 g of oxygen there are 6.02×10^{23} atoms of oxygen (however, there are 32.0 g of oxygen molecules since oxygen is diatomic, O_2) and 44.0 g of carbon dioxide contains 6.02×10^{23} molecules of carbon dioxide.

The numerical coefficients of a balanced chemical equation represents the number of moles for each substance in the reaction. Let's consider the following example and how a mole is used in calculations involving chemical equations.

$$CH_4 + 2\,O_2 \rightarrow CO_2 + 2\,H_2O$$

VOLUME RELATIONSHIPS FROM EQUATIONS

When volatile or gaseous compounds are formed, volume relationships are determined by the coefficients of the reactants and products. Review Chapter 5 "Pressure and Vacuum," for more details.

1. Every mole of gas that reacts or is formed occupies 22.4 L at Standard Temperature and Pressure conditions (STP) of 0°C and 1.0 atmosphere.
2. For every fraction of a mole of gas that reacts or is formed as in rule 1, that same fraction of 22.4 L of gas is involved.

$$\text{Hydrogen} + \text{Oxygen} \rightarrow \text{Water}$$

$$2\,H_2(g) + O_2(g) \rightarrow 2\,H_2O(g)$$

Notice that two volumes of hydrogen react with one volume of oxygen to give two volumes of water vapor.

$$\text{Zinc} + \text{Hydrochloric acid} \rightarrow \text{Zinc chloride and Hydrogen}$$

$$Zn(s) + 2\,HCl(aq) \rightarrow ZnCl_2(aq) + H_2(g)$$

Note that 1 gram-atomic mass of zinc will yield 22.4 L of hydrogen (1 g mole); 2 g moles of HCl were needed. The use of ratio and proportion will provide the answer of reactant or product mass or volume relationships.

EXAMPLE 24.4 How many grams of water will be produced when 1000 g of methane is burned (reacted with oxygen)?

First, convert 1000 g of methane to moles of methane.

The molecular mass of methane, CH_4:

$$(1 \times 12.0\text{ g}) + (4 \times 1.01\text{ g}) = 16.0\text{ g}$$

$$1000\text{ g CH}_4 \times \frac{1\text{ mol CH}_4}{16.0\text{ g CH}_4} = 62.5\text{ mol CH}_4$$

Second, using the balanced chemical equation, relate moles of methane to moles of water:

$$62.5\text{ mol CH}_4 \times \frac{2\text{ mol H}_2O}{1\text{ mol CH}_4} = 125\text{ mol H}_2O$$

Finally, to obtain the mass of water, convert moles of water to grams of water:

$$125 \text{ mol H}_2\text{O} \times \frac{18.0 \text{ g H}_2\text{O}}{1 \text{ mol H}_2\text{O}} = 2250 \text{ g H}_2\text{O}$$

PERCENT YIELD

The above calculations assume that the reaction was complete; that is, all of the reactants were used up and the maximum amount of product was produced. In other words, we calculated a theoretical yield. In practice, the amount of product obtained in pure form is almost always less than the theoretical amount. Some product is always lost in the process of separating and purifying it. We express product loss in terms of **percent yield**.

$$\text{Percent yield} = \frac{\text{actual yield}}{\text{theoretical yield}} \times 100$$

EXAMPLE 24.5 In the previous problem, 2250 g of water should have been produced (theoretical yield). If a process operator found only 1800 g of water (actual yield) in the reaction vessel, what would the percent yield of the reaction be?

$$\text{Percent yield} = \frac{1800 \text{ g}}{2250 \text{ g}} \times 100 = 80\%$$

SOLUTION TERMINOLOGY

The word **solution** means a homogeneous mixture of two or more substances. A solution is made up of two constituents. The major one is known as the **solvent**, which is the substance doing the dissolving, and the minor one is the **solute**, which is the substance (one or more) being dissolved.

Solute + solvent = solution

The most common solutions are those in which the solvent is a liquid and the solute is either a soluble solid, soluble gas, or miscible liquid.
Examples:

solid/liquid	sugar dissolved in water
gas/liquid	carbon dioxide dissolved in water
liquid/liquid	ethyl alcohol dissolved in water

A few solid/solid solutions (brass, which is zinc in copper) and gas/gas solutions (air, which is oxygen and other gases in nitrogen) are known to exist.

The term, **miscible**, which was used earlier, means that the liquid solute dissolves totally in all proportions with the liquid solvent; for example, grain alcohol in water. In contrast, oil and water are totally **immiscible**; that is, they are completely insoluble in each other, forming two layers, regardless of the proportions used.

The dissolving of a solid solute in a liquid solvent is dependent on several factors: temperature affects both the speed and the extent of the dissolution process. Elevated temperatures generally cause solubilities and the rate of dissolution to increase, allowing more solute to dissolve in a given solvent at a faster rate. Particle size affects only the rate at which the solute dissolves in the solvent. Because the solution process occurs at the surface of the solid solute, increasing its surface area will increase the rate of dissolution.

CONCENTRATION EXPRESSIONS

Since solutions are mixtures of solutes and solvents, a variety of possible compositions exists. When describing solutions, one must also discuss concentrations. The concentration of a solution is the amount of solute in a given amount of solvent or solution. Several terms exist for expressing concentrations: (1) percent; (2) parts per million and parts per billion; (3) molarity, and (4) normality.

A. Percent

Solutions, whose concentrations are in percentages, refer to the percent solute dissolved in a total solution. **Percent** can be defined as part divided by whole times 100. Part can be represented by solute and whole is normally represented by solution, thus:

$$\text{Percent} = \frac{\text{part}}{\text{whole}} \times 100$$

Solutions, whose concentrations are in percentages, refer to the percent solute dissolved in a total solution.

A **percent by weight/weight** (% wt/wt) solution is one in which the solute and solvent are measured in mass.

$$\% \text{ (wt/wt)} = \frac{\text{grams solute}}{\text{grams solute} + \text{grams solvent}} \times 100$$

A **percent by weight/volume** (% wt/v) solution is one in which the solute is measured by mass and the solvent is measured by volume. As a rule, when this concentration unit is used, it is understood that the volume is 100 mL. The equation used to calculate % wt/v follows:

$$\% \text{ (wt/v)} = \frac{\text{grams solute}}{100 \text{ mL of solvent}} \times 100$$

A **percent by volume/volume** (% v/v) solution is one in which the solute and solvent are both measured by volume.

$$\% \text{ (v/v)} = \frac{\text{volume of solute}}{\text{volume of solution}} \times 100$$

EXAMPLE 24.6 A solution is prepared using 100 mL of acetone in 1.00 L of pure water. The density of acetone is 0.790 g/mL and of water is 1.00 g/mL. Calculate the % (wt/wt), % (wt/v), and % (v/v).

% (wt/wt)

Mass of acetone: 100 mL × 0.790 g/mL = 79.0 g

Mass of water: 1.00 L = 1000 mL × 1.00 g/mL = 1000 g

$$\% \text{ (wt/wt)} = \frac{79.0 \text{ g acetone}}{79.0 \text{ g acetone } + 1000 \text{ g water}} \times 100$$

$$= 7.32\%$$

It should be pointed out that % wt/wt solutions can be prepared using liquid solutes as well as solids provided the density of the liquid is known. Since density equals the mass per volume, the mass of any given volume of a substance can be calculated by knowing the substance's density (mass = density × volume).

% (wt/v)

Mass of acetone: 79.0 g

Volume water: 1000 mL

$$\% \text{ (wt/v)} = \frac{79.0 \text{ g acetone}}{1000 \text{ mL water}} \times 100 = 7.90\%$$

Notice that the "g" unit in the numerator does not cancel with the "mL" unit in the denominator. This is why the volume used for % wt/vol solutions will always be 100 mL.

% (v/v)

Volume of acetone: 100 mL

Volume water: 1000 mL

$$\% \text{ (v/v)} = \frac{100 \text{ mL acetone}}{10.0 \text{ mL acetone } + 1000 \text{ mL water}} \times 100$$

$$= 9.90\%$$

B. Parts per Million

The term **parts per million (ppm)** refers to a solution which has one part solute for every million parts solution, and a part per billion is one part solute for every billion parts solution. Concentration units ppm and ppb can be calculated in the same manner as percentage except for the factor. For percentage the factor is one hundred, for parts per million the factor is one million, and for parts per billion the factor is one billion:

$$\text{ppm} = \frac{\text{grams solute}}{\text{grams solute } + \text{ grams solvent}} \times 1{,}000{,}000$$

$$\text{ppb} = \frac{\text{grams solute}}{\text{grams solute } + \text{ grams solvent}} \times 1{,}000{,}000{,}000$$

CHEMICAL CALCULATIONS AND CONCENTRATION EXPRESSIONS   **491**

Another method for calculating ppm and ppb concentration units is the ratio method using the definitions below. The following example demonstrates the "part/whole" method for calculating the concentration of ethylene glycol in terms of ppm and ppb.

$$1.0 \text{ ppm} = \frac{1.0 \text{ μg}}{1.0 \text{ mL}} \quad \text{or} \quad \frac{1.0 \text{ mg}}{1.0 \text{ L}}$$

$$1.0 \text{ ppb} = \frac{1.0 \text{ ng}}{1.0 \text{ mL}} \quad \text{or} \quad \frac{1.0 \text{ μg}}{1.0 \text{ L}}$$

EXAMPLE 24.7 How many ppm (wt/wt) of ethylene glycol (density = 1.11 g/mL) would be present in a solution in which 10.0 mL of ethylene glycol were diluted to 1.00 L with water (density = 1.00 g/mL)?

1. Calculate the grams of ethylene glycol in 10.0 mL:

$$10.0 \text{ mL ethylene glycol} \times \frac{1.11 \text{ g}}{1.00 \text{ mL}} = 11.1 \text{ g ethylene glycol}$$

2. Calculate the grams of water in 1.00 L:

$$1.00 \text{ L water} \times \frac{1000 \text{ mL}}{1.00 \text{ L}} \times \frac{1.00 \text{ g}}{1.00 \text{ mL}} = 1000 \text{ g water}$$

3. Calculate the ppm:

$$\text{ppm} = \frac{\text{parts}}{\text{whole}} \times 10^6 = \frac{11.1 \text{ g}}{1011.1 \text{ g}} \times 10^6 = 11{,}000 \text{ ppm}$$

C. Molarity

Molarity (M) is defined as the moles of solute per liter of solution. Molarity is also referred to as **molar**. One mole of any substance equals its formula or molecular weight. In most cases, the formula weights can be read directly from the manufacturer's label or MSDS of the chemical being used.

$$M = \frac{\text{moles of solute}}{\text{liter of solution}}$$

EXAMPLE 24.8 If 265 g of sodium fluoride (NaF) are dissolved in enough deionized water to produce 355 mL of solution, what is the molarity of the solution?

Sodium fluoride label reads:

Formula Weight (F.W.) = 41.988 g/mol.

1. Convert the grams of NaF to moles of NaF:

$$265 \text{ g NaF} \times \frac{1 \text{ mol NaF}}{41.998 \text{ g NaF}} = 6.31 \text{ mol NaF}$$

2. Convert the volume from mL to L:

$$355 \text{ mL} \times \frac{1.0 \text{ L}}{1000 \text{ mL}} = 0.355 \text{ L}$$

3. Calculate the molarity:

$$\frac{6.31 \text{ mol NaF}}{0.355 \text{ L}} = 17.8 \text{ M NaF solution}$$

EXAMPLE 24.9 If 200 mL of acetone were dissolved in enough water to make 8.00 L of solution, what would be the molarity? The density of acetone is 0.790 g/mL and the molecular weight is 58.08 g/mole.

$$200 \text{ mL} \times \frac{0.790 \text{ g}}{1.00 \text{ mL}} \times \frac{1.00 \text{ mol}}{58.08 \text{ g}} = 2.72 \text{ moles of acetone}$$

$$\text{Molarity} = \frac{2.72 \text{ moles}}{8.00 \text{ L solution}} = 0.340 \text{ M}$$

D. Normality

Normality (N) is defined as the number of gram-equivalent weights of solute per liter of solution.

$$N = \frac{\text{gram-equivalent weights of solute}}{\text{liter of solution}}$$

A **gram-equivalent weight** is defined as the number of grams necessary to provide exactly one mole of hydrogen ions (H^+) in an acid, one mole of hydroxide ions (OH^-) in a base, or one mole of electrons in an oxidation/reduction reaction (see Redox section below). An equivalent is simply one mole of H^+ ions, one mole of OH^- ions, or one mole of electrons.

(Acids)   1 mol HCl = 1 eq H^+

1 mol H_2SO_4 = 2 eq H^+

1 mol H_3PO_4 = 3 eq H^+

(Bases)   1 mol NaOH = 1 eq H^-

1 mol $Ba(OH)_2$ = 2 eq H^-

1 mol $Al(OH)_3$ = 3 eq H^-

(Redox)

1 mol CO → CO_2 = 2 eq (where: C^{2+} changes to C^{4+})

1 mol Fe → Fe^{3+} = 3 eq (where: Fe^0 changes to Fe^{3+})

1 mol MnO_4^- → Mn^{2+} = 5 eq (where: Mn^{+7} changes to Mn^{2+})

EXAMPLE 24.10 Calculate the normality of a solution containing 166.8 g H_3PO_4 in enough water to produce 5.0 L of solution.

$$166.8 \text{ g } H_3PO_4 \times \frac{1 \text{ mol}}{98.0 \text{ g}} \times \frac{3 \text{ eq}}{\text{mol}} = 5.12 \text{ eq } H_3PO_4$$

$$N = \frac{5.12 \text{ eq}}{5.0 \text{ L}} = 1.02 \text{ N } H_3PO_4$$

NOTE: *This would be a 0.34 M H_3PO_4 solution.*

REDOX REACTIONS

The term **redox** is used to indicate oxidation/reduction type reactions. **Oxidation** is the loss of electrons in a reaction and reduction is the gaming of electrons in a reaction. An oxidation cannot occur in a reaction without the simultaneous occurrence of a reduction reaction. An **oxidizing agent** is any substance that gains electrons in a chemical reaction. A **reducing agent** is any substance that gives up electrons in a reaction. In the following redox reaction, the Fe atom changed from a metal with charge of 0 to a Fe^{3+} ion which has a +3 charge. Iron was oxidized (lost electrons), therefore it is the reducing agent. The oxygen went from a charge of 0 to a charge of −2, therefore it is the oxidizing agent since it gained electrons.

$$4 \text{ Fe} + 3 \text{ O}_2 \rightarrow 2 \text{ Fe}_2O_3$$

DILUTIONS

Chemicals are frequently purchased in concentrated form to eliminate the shipment and cost of the solvent, (refer to Table 24.1). If a different concentration of the solution is desired, the concentrated solution can be diluted with additional solvent. A dilution formula that can be used with any method of expressing concentrations (molarity, percent, etc.) is as follows:

$$C_o \times V_o = C_f \times V_f$$

where
  C_o = original concentration of stock solution
  V_o = volume of stock solution that will be diluted
  C_f = concentration of final (diluted) solution
  V_f = volume of final (diluted) solution

The dilution formula can be arranged to solve for any one of the four variables.

EXAMPLE 24.11 An operator has a concentrated solution of 12.0 M HCl and needs 75.0 gal of 2.50 M HCl.

$$C_o \times V_o = C_f \times V_f$$

$$(12.0 \text{ M HCl})(V_o) = (2.50 \text{ M HCl})(75.0 \text{ gal})$$

$$V_o = \frac{(2.50 \text{ M HCl})(75.0 \text{ gal})}{(12.0 \text{ M HCl})}$$

$$V_o = 15.6 \text{ gal}$$

Therefore 15.6 gal of the stock 12.0 M HCl solution would be diluted with enough water to achieve a total volume of 75.0 gal and thus produce the desired 2.50 M HCl final concentration.

Water needed = 75.0 gal − 15.6 gal = 59.4 gal

NEUTRALIZATION AND TITRATION

If an acid is added to a base, a neutralization reaction can occur. **Neutralization** is the reaction of an acid with a base to produce a salt and water. (Review Chapter 9, "pH Measurement.")

Acid + Base → Salt + Water

For example, if equivalent amounts of nitric acid and sodium hydroxide react, the salt, sodium nitrate, and water are formed.

$$HNO_3 + NaOH \rightarrow NaNO_3 + H_2O$$

If hydrochloric acid is reacted with an equivalent amount of sodium hydroxide, sodium chloride and water are formed. These reactions are called neutralization reactions, because the properties of the acid and the base are destroyed, or neutralized. The proton (H^+) of the acid reacts with the hydroxide (OH^-) of the base, forming water. If two hydroxides were available from the base, it would take two protons to completely neutralize them as in the following example:

$$2\ HCl + Ca(OH)_2 \rightarrow CaCl_2 + 2\ H_2O$$

TABLE 24.1 Concentrations and Densities of Some Commercial Acids and Bases

Compound	Density g/mL	Concentration % by weight	mol. wt.	Molarity
Ammonium hydroxide NH_4OH	0.90	27.0	17.03	14.3
Acetic acid $HC_2H_3O_2$	1.05	99.5	60.05	17.4
Hydrofluoric acid HF	1.15	48.0	20.01	
Hydrochloric acid HCl	1.18	37.0	36.46	11.8
Nitric acid HNO_3	1.41	70.0	63.01	15.6
Perchloric acid $HClO_4$	1.66	72.0	100.47	
Phosphoric acid H_3PO_4	1.69	85.0	98.00	14.7
Sulfuric acid H_2SO_4	1.83	95.5	98.08	17.8

Acid/base neutralization reactions are used in titrations to determine the concentrations of unknown solutions. **Titration** is an analytical method in which a solution of known concentration (see Table 24.1) is reacted with a solution of unknown concentration in order to find the concentration of the unknown solution. A very useful equation* in acid/base neutralization reactions is:

$$C_A \times V_A = C_B \times V_B$$

where
C_A = concentration of the acid
V_A = volume of the acid
C_B = concentration of the base
V_B = volume of the base

---

*It should be noted that this equation is only valid for acids and bases that produce equal numbers of H^+ ions and OH^- ions.

EXAMPLE 24.12 Calculate the concentration of a hydrochloric acid solution if 27.77 mL are required to neutralize 45.44 mL of a 0.2233 M NaOH solution.

$$HCl + NaOH \rightarrow NaCl + H_2O$$

$$C_A \times V_A = C_B \times V_B$$

$$C_A = \frac{C_B \times V_B}{V_A}$$

$$C_A = \frac{0.2233 \text{ M} \times 45.44 \text{ mL}}{27.77 \text{ mL}}$$

$$C_A = 0.3654 \text{ M}$$

REFERENCES

1. Brown, Theodore, Lemay, L. H. Eugene, Jr., Bursten, Bruce E., and Burdge, Julia R., *Chemistry: The Central Science*, 9th edition. Prentice Hall. ISBN 0-13-038168-3 (2003).

2. "Section 3—Inorganic Compounds," *Lange's Handbook of Chemistry*, 15th edition, John A. Dean, editor. McGraw-Hill, Inc. ISBN 0-07-016384-7 (1999).

3. *Handbook of Chemistry and Physics*, 83rd edition or current edition, David R. Lide, editor-in-chief. CRC Press LLC. ISBN 0-8493-0483-0 (2003).

CHAPTER 25
VOLUMETRIC ANALYSIS

INTRODUCTION

A volumetric method is one in which the analysis is completed by measuring the volume of a solution of established concentration needed to react completely with the substance being determined. A review of Chapter 9, "pH Measurement," and Chapter 24, "Chemical Calculations and Concentration Expressions," is recommended.

A. Definitions of Terms

Units of Volume

Liter (L): A volume equal to 0.001 m^3.

Milliliter (mL): One-thousandth of a liter.

Cubic centimeter (cm^3): Can be used interchangeably with milliliter without effect.

Titration. A process by which a substance to be measured is combined with a reagent and quantitatively measured. Ordinarily, this is accomplished by the controlled addition of a reagent of known concentration to a solution of the substance until reaction between the two is judged to be complete; the volume of reagent is then measured.

Back Titration. A process by which an excess of the reagent is added to the sample solution and this excess is then determined with a second reagent of known concentration.

Standard Solution. A reagent of known composition used in a titration. The accuracy with which the concentration of a standard solution is known sets a limit on the accuracy of the test.

Standard solutions are prepared by one of the following methods:

- Carefully measuring a quantity of a pure compound and calculating the concentration from the mass and volume measurements
- Carefully dissolving a massed quantity of the pure reagent itself in the solvent and diluting it to an exact volume
- Using a prestandardized commercially available standard solution

Equivalence Point. The point at which the standard solution is chemically equivalent to the substance being titrated. The equivalence point is a theoretical concept. We estimate its position by observing physical changes associated with it in the solution.

End Point. The point where physical changes arising from alterations in concentration of one of the reactants at the equivalence point become apparent.

Titration Error. The inadequacies in the physical changes at the equivalence point and our ability to observe them.

Meniscus. The curvature that is exhibited at the surface of a liquid which is confined in a narrow tube such as a burette or a pipette.

B. Typical Physical Changes During Volumetric Analysis

- Appearance or change of color due to the reagent, the substance being determined, or an indicator substance
- Turbidity formation resulting from the formation or disappearance of an insoluble phase
- Conductivity changes in a solution
- Potential changes across a pair of electrodes
- Refractive index changes
- Temperature changes

Reading the Meniscus. Volumetric flasks, burettes, pipettes, and graduated cylinders are calibrated to measure volumes of liquids. When a liquid is confined in a narrow tube such as a burette or a pipette, the surface is bound to exhibit a marked curvature, called a **meniscus**. It is common practice to use the bottom of the meniscus in calibrating and using volumetric ware. Special care must be used in reading this meniscus. Render the bottom of the meniscus, which is transparent, more distinct by positioning a black-striped white card behind the glass. (See Fig. 25.1.)

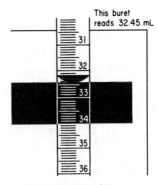

FIGURE 25.1 Useful technique for reading a meniscus.

Procedure

Location of the eye in reading any graduated tube is important:

1. With the eye above the meniscus, too small a volume is observed.
2. With the eye at the same level as the meniscus, the correct volume is observed.
3. With the eye below the meniscus, too large a volume is observed.

The eye must be level with the meniscus of the liquid to eliminate **parallax errors** (Fig. 25.2). Read the top of the black part of the card with respect to the graduations on a burette.

C. Washing and Cleaning Laboratory Glassware

General Rules

1. Always clean your apparatus immediately after use, if possible. It is much easier to clean the glassware *before* the residues in them become dry and hard. If dirty glassware cannot be washed immediately, put it in water to soak.
2. Handle glassware carefully while cleaning it, especially bulky flasks and long, slender columns.

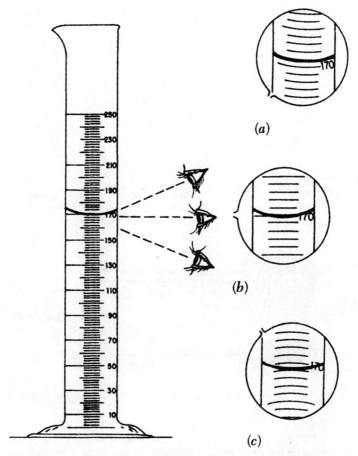

FIGURE 25.2 Avoiding parallax error in reading a meniscus. (*a*) Eye level too high; volume too high; (*b*) Eye level correct; volume correct; (*c*) Eye level too low; volume too low.

CAUTION: Round-bottomed flasks are especially fragile. When they are full, never set them down on a hard, flat surface while holding them by the neck. Even gentle pressure can cause a star crack, which is difficult to see in a wet flask. Cracked glassware is a hazard. Check each round-bottomed flask carefully before it is used. Support round-bottomed flasks *only* on ring stands or on cork rings.

3. Rinse off all soap or detergent residue after washing glassware to prevent any possible contamination later.

Most pieces of laboratory glassware can be cleaned by washing and brushing with a detergent or a special laboratory cleaning product called Alconox. After they have been thoroughly cleaned, they are rinsed with tap water and finally with a spray of distilled water. If the surface is clean, the water will wet the surface uniformly. On soiled glass the water stands in droplets (Fig. 25.3).

Use brushes carefully and be certain that the brush has no exposed sharp metal points that can scratch the glass and cause it to break. Such scratches are a frequent and unexpected cause of breakage when the glassware is heated. Brushes come in all shapes and sizes (Fig. 25.4). Using the correct brush makes the cleaning job much easier.

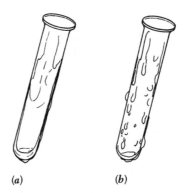

FIGURE 25.3 Water spreads out smoothly and evenly on clean glass (*a*), but stands in droplets on soiled glass (*b*).

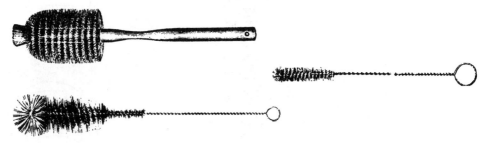

FIGURE 25.4 Brushes in a varied selection of sizes and designs for cleaning laboratory equipment.

CAUTION: Never exert excessive force on the brush. You may force the metal through the bristles to scratch the glassware, causing breakage and injury to yourself.

Cleaning Volumetric Glassware

1. Always rinse volumetric glassware equipment three times with distilled water after you have emptied and drained it. This prevents solutions from drying on the glassware, causing difficulty in cleaning.
2. Dry volumetric glassware at room temperature, *never in a hot oven.* Expansion and contraction may change the calibration.
3. The glass surfaces should be wetted *evenly.* Spotting is caused by grease and dirt. *Removing grease:* Rinse and scrub with hot detergent solution followed by adequate distilled-water rinses. *Removing dirt:* Fill or rinse with dichromate cleaning solution. Allow to stand for several hours, if necessary. Follow with multiple distilled-water rinses.

Pipettes. Pipettes may be cleaned with a warm solution of detergent or with cleaning solution. Draw in sufficient liquid to fill the bulb to about one-third of its capacity. While holding it nearly horizontal, carefully rotate the pipette so that all interior surfaces are covered. Drain inverted, and rinse thoroughly with distilled water. Inspect for water breaks and repeat the cleaning cycle as often as ne*cessary.*

Burettes. Thoroughly clean the tube with detergent and a long brush. If water breaks persist after rinsing, clamp the burette in an inverted position with the end dipped in a beaker of cleaning solution. Connect a hose from the burette tip to a vacuum line. Gently pull the cleaning solution into the burette, stopping well short of the stopcock. Allow to stand for 10 to 15 minutes, and then drain. Rinse thoroughly with distilled water and again inspect for water breaks. Repeat the treatment if necessary.

After some use, the grease lubricant on stopcocks tends to harden; small particles of grease can break off, flow to the tips of burettes, and clog them. The grease may be cleaned out by inserting a fine, flexible wire into the burette tip and breaking up the plug. Flush out the particles with water.

CAUTION: The wire must be flexible and not too thick. Otherwise, you may break or split the tip. Alternative method of cleaning burette tips:

1. Carefully heat the plugged tip of the burette with a match (Fig. 25.5), warming the grease and melting it; use several matches if necessary. Water pressure or air (blowing through the cleaned

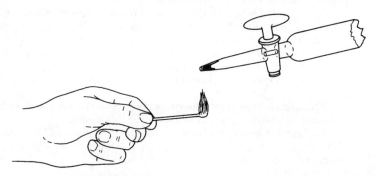

FIGURE 25.5 Cleaning a burette tip.

top of the burette) will push the melted grease out.

CAUTION: Heat cautiously; the tip may break.

2. Wipe surface soot from tip.

CAUTION: Handle burettes and pipettes very carefully. They are longer than you think. Their tips are especially fragile and are easily broken on contact with sinks or water faucets. Protective rubber mats help prevent breakage and chipping.

Automatic Pipette and Tube Cleaners. Apparatus designed to clean and dry many pipettes or tubes at one time is available. It consists of a stainless steel or high-density polyethylene cylinder which utilizes intermittent siphon action, setting up a turbulent "fill-and-empty" action cycle. Either hot or cold water can be used, and some units are fitted with an electric heater to provide heated air through the cylinder after the washing cycles.

Cleaning Glassware Soiled With Stubborn Films and Residues. When you cannot completely clean glassware by scrubbing it with a detergent solution, more drastic cleaning methods must be used.

Dichromate–Sulfuric Acid Cleaning Solution

CAUTION: Prepare and handle this cleaning solution with extreme care. Avoid contact with clothing or skin.

Dissolve 92 g $Na_2Cr_2O_7 \cdot 2H_2O$ (sodium dichromate) in 458 mL H_2O, and cautiously add with stirring 800 mL concentrated H_2SO_4.* The contents of the flask will get very hot and become a semisolid red mass. When this happens, add just enough sulfuric acid to bring the mass into solution.

---

*Potassium dichromate may also be used, but it is less soluble than the sodium compound.

Allow the solution to cool before attempting to transfer it to a soft-glass bottle. After the glassware has been cleaned with a detergent and rinsed carefully, pour a small quantity of the chromate solution into the glassware, allowing it to flow down all parts of the glass surface. Pour the solution back into its stock bottle. Then rinse the glassware, first with tap water and then with distilled water, until the glass surface looks clean.*

The solution may be reused until it acquires the green color of the chromium (III) ion. Once this happens, it should be discarded.

Dilute Nitric Acid Cleaning Solution. Films that adhere to the inside of flasks and bottles may often be removed by wetting the surface with dilute nitric acid, followed by multiple rinses with distilled water.

Aqua Regia Cleaning Solution. *Aqua regia* is made up of three parts of concentrated HCl and one part of concentrated HNO_3. This is a very powerful, but extremely dangerous and corrosive, cleaning solution. *Use in a hood with extreme care.*

Alcoholic Potassium Hydroxide *or* Sodium Hydroxide Cleaning Solution. Add about 1 L ethanol (95 percent) to 120 mL H_2O containing 120 g NaOH or 105 g KOH.

This is a very good cleaning solution. Avoid prolonged contact with ground-glass joints on interjoint glassware because the solution will etch glassware and damage will result. This solution is excellent for removing carbonaceous materials.

Trisodium Phosphate Cleaning Solution. Add 57 g Na_3PO_4 and 28.5 g sodium oleate to 470 mL H_2O.

This solution is good for removing carbon residues. Soak glassware for a short time in the solution, and then brush vigorously to remove the incrustations.

Nochromix Cleaning Solution. This is a commercial oxidizer solution, but it contains no metallic ions. The powder is dissolved in concentrated sulfuric acid, yielding a clear solution. The solution turns orange as the oxidizer is used up. *Use with care.*

Metal Decontamination of Analytical Glassware. Where metal decontamination is desired, in chelometric titration, treat acid-washed glassware by soaking it in a solution containing 2 percent NaOH and 1 percent disodium ethylenediamine tetraacetate for about 2 hours; follow this bath with a number of rinses with distilled water.

Ultrasonic Cleaning. Ultrasonic cleaning units generate high-frequency sound waves (sound waves of higher frequencies than those detectable by the human ear) that penetrate deep recesses, turn corners, and pass through barriers. Laboratory glassware and optical equipment as well as narrow-bore pipettes, manometers, and similar items are easily cleaned with such units. The items are immersed in the cleaning solution, and the power is turned on.

The cleaners are usually fitted with stainless steel tanks and come in various sizes. They can clean a load of badly soiled pipettes and assorted glassware faster and better than any other method of cleaning. The principle on which ultrasonic cleaners work is **cavitation**, the formation and collapse of submicron bubbles (bubbles smaller than 1 μm in diameter). These bubbles form and collapse about 25,000 times each second with a violent microscopic intensity which produces a scrubbing action. This action effectively frees every surface of the glassware of contaminants, because the glassware is immersed in the solution and the sound energy penetrates wherever the solution reaches.

Fluidized-Bath Method. A fluidized bath may be used for cleaning polymeric and carbonaceous residues from laboratory glassware and metallic apparatus.

1. Establish a temperature in the fluidized medium of approximately 500°C (932°F).
2. Place the workpieces and hardware to be cleaned in a basket, and lower the basket into the fluidized medium.
3. Adjust heater control. Heat is transmitted to the immersed parts, and thermal breakdown of the plastic polymer commences almost immediately. Cleaning is effected rapidly. Be sure the fumes

---

*As many as 10 washings with demineralized or distilled water are necessary to remove traces of the chromium ions.

that are emitted from the surface of the bath are removed from the work area by means of a fume hood and extractor fan.

4. Remove the basket and parts from the bath, and allow them to cool to room temperature. Metals can be quenched in water.

5. Any noncombustible residues may be easily removed by blowing with an air jet or by brushing lightly.

D. Drying Laboratory Glassware

After glassware has been thoroughly cleaned and rinsed, it must be dried.

Drainboards and Drain Racks. Drainboards and drain racks are used for draining and drying various sizes and shapes of glassware. The supports have pins and pegs anchored in an inclined position to ensure drainage. Some drainboards are fitted with hollow pegs and a hot-air blower to speed up the drying process. Place glassware securely on rack. Do not allow pieces to touch each other and cause accidental breakage.

Dryer Ovens. Dryer ovens are designed for efficient high-speed drying of glassware. They come in various sizes and power ratings; some of them are fitted with timers.

Quick Drying. If it is necessary to dry the insides of flasks or similar vessels, gently warm them over a Bunsen flame and then gently pass a stream of compressed air through a glass tube that leads to the bottom of the flask until the item is dry.

Rinsing Wet Glassware with Acetone. Water-wet glassware can be dried more quickly by rinsing the item with several small portions of acetone, discarding the acetone rinse after each time. Then place the item in a safety oven or gently pull air through the glassware by connecting a pipette with rubber tubing to an aspirator and inserting the pipette in the glassware.

CAUTION: Never heat thick-walled glassware in a flame. It will break (Fig. 25.6).

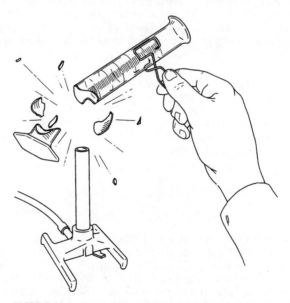

FIGURE 25.6 Heavy, thick-walled glassware will break if heated in a flame.

TOOLS OF VOLUMETRIC ANALYSIS

Pipettes, burettes, and volumetric flasks are standard volumetric equipment. Volumetric apparatus calibrated **to contain** a specified volume is designated **TC**, and apparatus calibrated **to deliver** a specified amount, **TD**.

Only clean glass surfaces will support a uniform film of liquid; the presence of dirt or oil will tend to cause breaks in this film. The appearance of water breaks is a sure indication of an unclean surface. Volumetric glassware is carefully cleansed by the manufacturer before being supplied with markings, and in order for these to have meaning, the equipment must be kept equally clean when in use.

As a general rule, the heating of calibrated glass equipment *should be avoided.* Too rapid cooling can permanently distort the glass and cause a change in volume.

A. Volumetric Flasks

Volumetric flasks are calibrated to contain a specified volume when filled to the line etched on the neck.

Directions for the Use of a Volumetric Flask. Before use, volumetric flasks should be washed with detergent and, if necessary, cleaning solution. Then they should be carefully and repeatedly rinsed in distilled water; only rarely need they be dried. Should drying be required, however, it is best accomplished by clamping the flasks in an inverted position and employing a mild vacuum to circulate air through them.

Introducing Standard Directly into a Volumetric Flask. Direct preparation of a standard solution requires that a known mass of solute be introduced into a volumetric flask. In order to minimize the possibility of loss during transfer, insert a funnel into the neck of the flask. The funnel is subsequently washed free of solid.

Dilution to the Mark. After introducing the solute, fill the flask about half full and swirl the contents to achieve solution. Add more solvent, and again mix well. Bring the liquid level almost to the mark, and allow time for drainage. Then use a medicine dropper to make such final additions of solvent as are necessary. Firmly stopper the flask and invert repeatedly to ensure uniform mixing.

NOTES:

1. If, as sometimes happens, the liquid level accidentally exceeds the calibration mark, the solution can be saved by correcting for the excess volume. Use a gummed label to mark the actual position of the meniscus. After the flask has been emptied, carefully refill to the mark with water. Then, using a burette, measure the volume needed to duplicate the actual volume of the solution. This volume, of course, should be added to the nominal value for the flask when the concentration of the solution is calculated.
2. For exacting work, the flask should be maintained at the temperature indicated on the flask.

B. Pipettes

Pipettes (pipets, pipettors, chemical droppers) are designed for the transfer of known volumes of liquid from one container to another. Pipettes that deliver a fixed volume are called **volumetric** or **transfer pipettes** (Fig. 25.7). Other pipettes, known as **measuring pipettes**, are calibrated in convenient units so that any volume up to maximum capacity can be delivered (Fig. 25.8). Table 25.1 gives the capacities and tolerances for some volumetric pipettes and Table 25.2 shows the calibration units for some measuring pipettes.

Certain accessories are useful in working with pipettes Fig. 25.9.

VOLUMETRIC ANALYSIS **505**

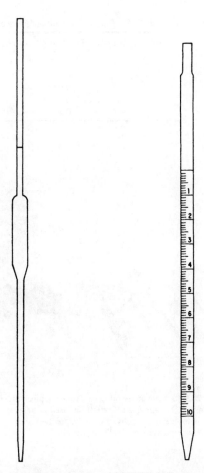

FIGURE 25.7 Volumetric, or transfer pipette.   **FIGURE 25.8** Measuring pipette.

TABLE 25.1 Volumetric Transfer Pipettes Tolerances, ± mL*

Capacity, mL	Class A	Class B
0.5	0.006	0.012
1	0.006	0.012
2	0.006	0.012
3	0.01	0.02
4	0.01	0.02
5	0.01	0.02
10	0.02	0.04
15	0.03	0.06
20	0.03	0.06
25	0.03	0.06
50	0.05	0.10
100	0.08	0.16

*Accuracy tolerances for volumetric transfer pipettes are given by ASTM standards E969 and Federal Specification NNN-P-395.

TABLE 25.2 Calibration Units of Measuring Pipettes

Capacity, mL	Divisions, mL	Capacity, mL	Divisions, mL
0.1	0.01	5	0.1
0.2	0.01	10	0.1
1	0.01	25	0.1
1	0.1	50	0.2
2	0.1		

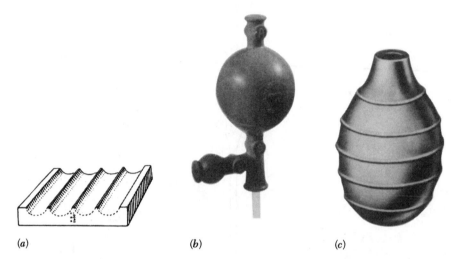

(a) (b) (c)

FIGURE 25.9 Pipette accessories, (*a*) Porcelain pipette rest to prevent contamination and to prevent pipettes from rolling and breaking. (*b*) Pipette filler used to transfer sterile, corrosive, and toxic liquids safely. Easily controlled by squeezing, it delivers quickly, precisely, and safely. (*c*) Pipette filler. Squeeze first, then immerse in liquid; release pressure gradually as needed.

Directions for the Use of a Pipette. The following instructions pertain specifically to the manipulation of transfer pipettes, but with minor modifications they may be used for other types as well. Liquids are usually drawn into pipettes through the application of a slight vacuum. Use mechanical means, such as a rubber suction bulb or a rubber tube connected to a vacuum pump. See Fig. 25.9*b* and *c*.

Procedure

1. Clean pipette thoroughly and rinse with distilled water.
2. Drain completely, leaving no rinse-water drops inside. If the pipette is wet with water, rinse three times with the solution to be used in the analysis.
3. Keep the tip of the pipette below the surface of the liquid.
4. Draw the liquid up in the pipette using the pipette bulb or suction (aspirator) (Fig. 25.10).

CAUTION: *Always* use an aspirator, pipette bulb, and/or safety flask when working with liquids (Figs. 25.9*b* and *c* and 25.10). *Never pipette any solution by mouth.*

5. Disconnect the suction when the liquid is above the calibration mark. *Quickly remove the suction unit and immediately place the index finger of the hand holding the pipette over the exposed end of the pipette to the closed end.*
6. Release pressure on the index finger to allow the meniscus to approach the calibration mark.

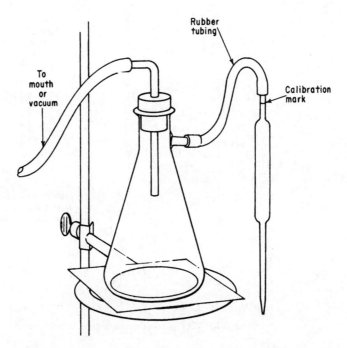

FIGURE 25.10 Using a pipette with a safety flask and water aspirator.

7. At the mark, apply pressure to stop the liquid flow, and drain the drop on the tip by touching it to the wall of the liquid-holding container.
8. Transfer the pipette to the container to be used and release pressure on the index finger. Allow the solution to drain completely; allow a time lapse of 10 seconds, or the period specified on the pipette. Remove the last drop by touching the wall of the container (Figs. 25.11c and d and 25.12).
9. The calibrated amount of liquid has been transferred. *Do not blow out the pipette.* In the case of color-coded pipettes, a frosted ring indicates complete blowout.

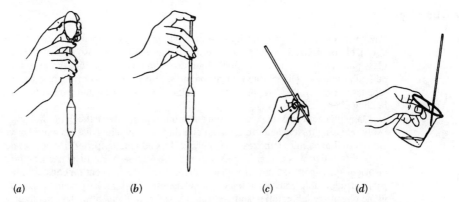

FIGURE 25.11 Technique for using a volumetric pipette. (*a*) Draw liquid past the graduation mark. (*b*) Use forefinger to maintain liquid level above the graduation mark. (*c*) Tilt pipette slightly and wipe away any drops on the outside surface. (*d*) Allow pipette to drain freely.

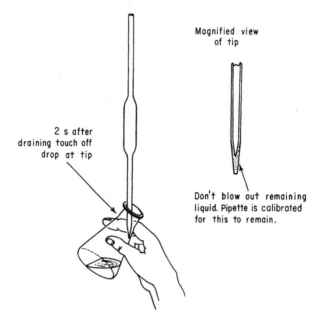

FIGURE 25.12 Correct technique for draining a pipette.

NOTES:
1. The liquid can best be held at a constant level in the pipette if the forefinger is slightly moist (use distilled water). Too much moisture, however, makes control difficult.
2. It is good practice to avoid handling the pipette by the bulb.
3. Pipettes should be thoroughly rinsed with distilled water after use.

C. Other Pipettes

Pipettes come in a range of designs, from single-piece glass pipettes, to disposable, to more complex adjustable or electronically operated forms. **Pasteur pipettes**, also known as droppers, are plastic or glass pipettes used to transfer small amounts of liquids, but are not graduated or calibrated for any particular volume. Disposable transfer pipettes, also known as **Beral pipettes**, are similar to Pasteur pipettes. However, they are made from a single piece of plastic and their bulb can serve as the liquid-holding chamber.

Many pipettes types work by creating a partial vacuum above the liquid-holding chamber and selectively releasing this vacuum to draw up and dispense. Pipettes that dispense between 1 and 1000 μL are referred to as micropipettes (see Table 25.4), and macropipettes dispense a greater volume range of liquids. Micropipettes and macropipettes are generally air-displacement pipettes and positive-displacement pipettes. Generally, piston-driven, air-displacement pipettes are micropipettes that dispense an adjustable volume of liquid from a disposable tip. Air-displacement pipettes are less common in the chemistry laboratories and are used for small volumes of more volatile or viscous substances, such as DNA.

TABLE 25.3 Tolerances of Micropipets (Eppendorf)

Capacity, µL	Accuracy, %	Precision, %
10	1.2	0.4
40	0.6	0.2
50	0.5	0.2
60	0.5	0.2
70	0.5	0.2
100	0.5	0.2
250	0.5	0.15
500	0.5	0.15
600	0.5	0.15
900	0.5	0.15
1000	0.5	0.15

TABLE 25.4 Burette Accuracy Tolerances, ± mL*

Capacity, mL	Subdivisions, mL	Class A and Precision Grade	Class B and Standard Grade
10	0.05	0.02	0.04
25	0.10	0.03	0.06
50	0.10	0.05	0.10
100	0.20	0.10	0.20

*Class A burets conform to specification ASTM E694 for standard taper stopcocks and to ASTM E287 for Teflon or polytetrafluoroethylene stopcock plugs.

D. Burettes

Burettes, like measuring pipettes, deliver any volume up to their maximum capacity. Burettes of the conventional type (Fig. 25.13a) must be manually filled. Others with side arms are filled by gravity. For more accurate work, Schellbach burettes (Fig. 25.14) are employed. These have a white background with a blue stripe and can be read at the point of least magnification. When unstable reagents are employed, a burette with a reservoir bottle and pump may be employed. Table 25.5 shows typical burette capacities and accuracy tolerances.

E. Directions for the Use of a Burette

Before being placed in service, a burette must be scrupulously clean. In addition, it must be established that the stopcock is liquid-tight.

NOTE: Grease films that appear unaffected by cleaning solution may yield to treatment with such organic solvents as acetone or benzene. Thorough washing with detergent should follow such treatment.

F. Filling

Make certain that the stopcock is closed. Add 5 to 10 mL of solution and carefully rotate the burette to wet the walls completely; allow the liquid to drain through the tip. Repeat this procedure two more times. Then fill the burette above the zero mark. Free the tip of air bubbles by rapidly rotating the stopcock and allowing small quantities of solution to pass. Finally, lower the level of the solution to, or somewhat below, the zero mark; after allowing about a minute for drainage, take an initial volume reading.

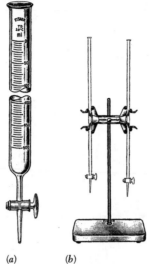
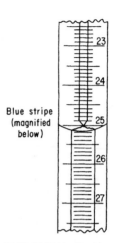

FIGURE 25.13 Burettes, (*a*) Single-dispensing burette with graduated etched scale, standard taper, and stopcock. Such burettes are available with volumes from 10 to 1000 cm^3. The newer burettes are equipped with plastic (e.g., Teflon®) stopcocks, (*b*) Titrating assembly and stand with white base for easy observation of color changes. (*Fisher Scientific Company.*)

FIGURE 25.14 The blue line makes it easy to read a Schellbach burette.

G. Holding the Stopcock

Always push the plug into the barrel while rotating the plug during a titration. A right-handed person points the handle of the stopcock to the right, operates the plug with the left hand, and grasps the stopcock from the left side as shown in Fig. 25.15.

Procedure

1. Test the burette for cleanliness by clamping it in an upright position and allow it to drain. *No water drops should adhere to the inner wall. If they do, reclean the burette.*
2. Grease the stopcock with clean grease, *after cleaning it.*
   (a) Remove the stopcock.
   (b) Clean the barrel and stopcock with a swab soaked in an appropriate solvent: petroleum ether, diethyl ether, acetone, etc.
   (c) Regrease as shown in Fig. 25.16. Improperly applied grease will spread and obstruct the holes, making *recleaning* necessary. Greasing is only necessary with ground glass stopcocks, newer burettes equipped with plastic stopcocks do not typically require grease.

PERFORMING A TITRATION

1. Use a setup such as that shown in Fig. 25.13*b*. Rinse the cleaned burette three times with the solution to be used, draining completely each time.

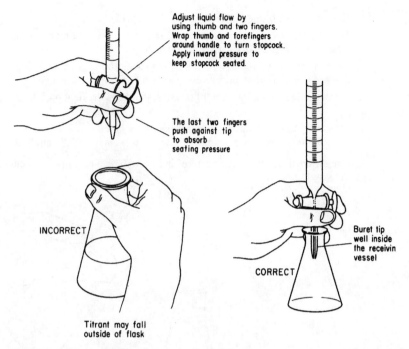

FIGURE 25.15 Preferred method for manipulating a stopcock and using a burette during titration.

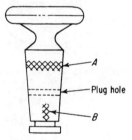

FIGURE 25.16 How to grease a stopcock.

2. Fill the burette above the zero graduation.
3. Drain slowly until the tip is free of air bubbles and completely filled with liquid and the meniscus of the liquid is at the zero graduation.
4. Add the titrant to the titration flask slowly, swirling the flask with the right hand until the end point is obtained. To avoid error, with the tip well within the titration vessel, introduce solution from the burette in increments of a milliliter or so. Swirl (or stir) the sample constantly to ensure efficient mixing. Reduce the volume of the additions as the titration progresses; in the immediate vicinity of the end point, the reagent should be added a drop at a time. When it is judged that only a few more drops are needed, rinse down the walls of the titration. Allow a minute or so to elapse between the last addition of reagent and the reading of the burette. Keep your eye level with the meniscus for all readings.

5. Rinse the walls of the flask frequently (Fig. 25.17).
6. For precision work, volumes of less than one drop can be rinsed off the tip of the burette with wash water (Fig. 25.18). (See Splitting a Drop of Titrant.)
7. Near the end point, the trail of color from each drop is quite long (Fig. 25.19). The end point is reached when the color change does not disappear after 30 seconds.
8. Allow the burette to drain for 30 seconds; then read the final position of the meniscus.

CAUTION: Never allow reagents to remain in burettes overnight. The stopcock may "freeze" because of prolonged contact, especially with bases such as KOH and NaOH.

9. The difference between the "before" and "after" readings on the burette is the volume of liquid delivered.

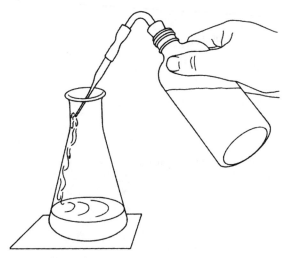

FIGURE 25.17 Rinse the walls of the flask frequently during a titration.

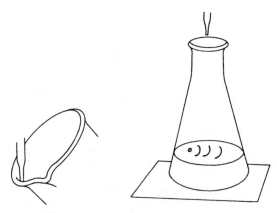

FIGURE 25.18 Removal of an adhering drop.

FIGURE 25.19 As the end point is approached, the color trail from each drop gets larger and persists longer.

Splitting a Drop of Titrant. There are times in precision titration when one drop is more than that is needed to reach the equivalence point, and only part of a drop is needed. In these cases you may use method 1 or 2; method 1 is preferable.

1. Carefully open the stopcock so that only part of a drop appears, and then close the stopcock. Carefully direct a small stream of water from the wash bottle on the tip of the burette to wash the part drop into the solution. Mix the solution thoroughly before adding another part drop.
2. Quickly but carefully spin the closed stopcock 180°. A small stream of liquid will shoot out.

CAUTION: Spin quickly, but carefully. Hold the stopcock securely so that it does not become loose in the socket. Be sure that the stopcock is closed after the spin.

A. Titration Curves

The end point can be readily determined by using an indicator and observing color changes. Sometimes it is necessary to determine the end point or equivalence point by using a potentiometric measurement on a titration curve. These titration curves are generated by plotting pH (or cell potential) against volume of titration used. The inflection point on the curve corresponds to the equivalence point and represents the maximum rate of change of pH or cell potential per unit of volume of titrant. A typical pH titration curve appears in Fig. 25.20.

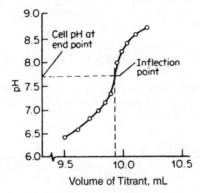

FIGURE 25.20 A normal titration curve.

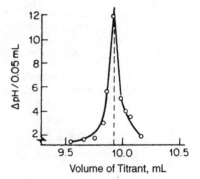

FIGURE 25.21 A first-derivative titration curve.

Another common practice of displaying titration curves involves plotting the change in pH (or cell potential) per increment (usually 0.05 mL) of titrant. This type of plot is referred to as a first *derivative titration curve*. A typical first-derivative curve is shown in Fig. 25.21. Many automatic titration equipped with microprocessors can display titration curves in normal, first-derivative, and second-derivative modes.

ACID-BASE TITRATIONS AND CALCULATIONS

A common reaction in chemistry involving an acid and a base is called a *neutralization* and can be represented by this simple equation:

$$\text{acid} + \text{base} \rightarrow \text{salt} + \text{water}$$

The concentration of an unknown solution of acid or base can be determined by reacting a measured quantity of the unknown solution with a measured volume of an appropriate acid or base of known concentration. This process is called an *acid-base titration*. In a typical titration, a measured volume of the unknown solution is placed in a flask, and a solution of known concentration is added (measured) from a burette until an equivalent amount of both the acid and base are present. The point at which the acid and base have been added in equivalent amounts is called the *equivalence point*. Usually the equivalence point of a titration is determined by using an acid-base indicator. These *acid-base indicators* have characteristic colors in acidic and basic solutions (see Chap. 9, "pH Measurement," for details) and are selected to undergo these color changes at the specific equivalence point or pH.

In a typical acid-base titration, the concentration of an unknown sodium hydroxide solution, NaOH, can be determined by titration with a known concentration of hydrochloric acid, HCl, using a phenolphthalein indicator.

Phenolphthalein is colorless in an acidic solution and is pink in a basic solution. The chemical reaction is shown by the following equation:

$$NaOH + HCl \rightarrow NaCl + H_2O$$

A *simple acid-base equation* for calculating concentrations is as follows:

$$C_{acid} \times V_{acid} = C_{base} \times V_{base}$$

where

C_{acid} = concentration of the acid
V_{acid} = volume of the acid
C_{base} = concentration of the base
V_{base} = volume of the base

Acids and bases can also be standardized (concentration accurately determined) by using a primary standard. A primary standard is usually a solid acid or base compound of high purity, stability, and known molecular weight as shown in Table 25.5.

TABLE 25.5 Acid-Base Primary Standards

Compound	Formula	Molecular wt.
Primary acids		
Potassium hydrogen phthalate	$KHC_8H_4O_4$	204.23
Sulfamic acid	HNH_2SO_3	97.10
Oxalic acid dihydrate	$HOOC-COOH(H_2O)$	126.07
Benzoic acid	C_6H_5COOH	122.13
Primary bases		
Sodium carbonate	Na_2CO_3	106.0
Tris(hydroxylmethyl)aminomethane	$(CH_2OH)_3CNH_2$	121.4

A. Procedure for Standardization of a Base or Acid

1. Review the section in this chapter entitled Performing a Titration.
2. Using a volumetric pipette, transfer exactly 25.00 mL of standardized acid solution into a clean 25-mL Erlenmeyer flask. If a primary acid standard were used instead of the standardized acid solution, a very accurately weighed quantity would be added to the flask with enough water to dissolve the solid. Add three drops of phenolphthalein indicator solution to the Erlenmeyer flask.

3. Titrate the standardized acid solution slowly by releasing a basic solution from a filled (starting volume reading known) burette through the stopcock. Swirl the Erlenmeyer flask gently as the base solution is added. A white piece of paper under the flask will allow the indicator color change to be more visible. As the base is added to the acid, a faint pink color will develop at the point of contact. As the end point (equivalence point) is approached, this pink coloration will persist for longer and longer periods. When this occurs, add the base drop by drop until the end point is reached. This is indicated by the first drop of base which causes the entire solution to become faintly pink and to remain pink for at least 30 seconds. Record the final (end point) burette reading to the nearest 0.01 mL.

4. Calculate the concentration of the basic solution using the simple acid-base equation given above.

B. Back Titrations

In a **back titration** an excess (more than required) of the titrant is added to the sample. The excess is then "back titrated." Why must we sometimes use this technique? The reaction between the titrant and the sample may take place too slowly. The reaction will occur faster if an excess of titrant is used. For example, some materials, such as marble or limestone, $CaCO_3$, are dissolved in the same solution, HCl, that would normally be used to titrate them. It would take too long to titrate solid $CaCO_3$ directly, so an excess of the titrant, HCl, is used to react with the sample quickly and the excess *unreacted* HCl is then determined. An example of a back titration calculation would be as follows.

EXAMPLE 25.1 Assume that exactly 50.00 mL of 0.2000 M HCl were added from a pipette to an 0.8000-g sample of limestone and the flask was gently swirled until the sample had completely reacted. The *excess* HCl was then titrated with 0.1000 M KOH, 36.00 mL being required. Calculate the percent $CaCO_3$ in the limestone.

$$? \% \ CaCO_3 = \frac{g \ CaCO_3}{g \ sample} \times 100$$

SOLUTION Reacted HCl = (0.0500 L × 0.2000 M) − (0.03600 L × 0.100 M KOH)
Reacted HCl = 0.00640 mole HCl

$$\frac{(0.00640 \text{ mole HCl}) \times \frac{1 \text{ mole } CaCO_3}{2 \text{ moles HCl}} \times \frac{100.90 \text{ g } CaCO_3}{\text{mole } CaCO_3}}{0.8000 \text{ g sample}} \times 100$$

$$\% \ CaCO_3 = \frac{(0.0064 \text{ mole HCl})(0.5 \text{ mole } CaCO_3/\text{HCl})(100.90 \text{ g } CaCO_3)}{0.8000 \text{ g}} \times 100$$

$\% \ CaCO_3 = 40.36\%$

OXIDATION-REDUCTION TITRATIONS AND CALCULATIONS

The acid-base reactions and calculations in the previous section assumed that all reactions are one-to-one. In reality, molar quantities of the reactants in a chemical equation cannot always be reduced to a one-to-one ratio. For example, the following common acid-base reaction is not a simple one-to-one reaction.

$$H_2SO_4 + 2NaOH \rightarrow Na_2SO_4 + 2H_2O$$

Two moles of sodium hydroxide are required to neutralize one mole of sulfuric acid. It should be obvious that the simple acid-base equation which assumes a one-to-one reaction ratio will not yield the correct answer if these two substances are titrated together. Table 25.6 lists some common titrimetric oxidizing and reducing reagents.

TABLE 25.6 Common Titrimetric Oxidizing and Reducing Agents

Agent	Formula	Product	Subscript, s
Oxidizing			
Permanganate (in acid)	MnO_4^-	Mn^{2+}	5
Dichromate (in acid)	$Cr_2O_7^{2-}$	Cr^{3+}	6
Chlorite (in acid)	ClO_3^-	Cl^-	6
Peroxide (in acid)	H_2O_2	H_2O	2
Nitrate (in dilute acid)	NO_3^-	NO	3
Ferric (in acid)	Fe^{3+}	Fe^{2+}	1
Reducing			
Iodide (acid or neutral)	I^-	I_2	1
Oxalate (in acid)	$C_2O_4^{2-}$	CO_2	2
Thiosulfate (in acid)	$S_2O_3^{2-}$	$S_4O_6^{2-}$	1
Sulfur dioxide (in acid)	SO_2	SO_4^{2-}	2
Ferrous (in acid)	Fe^{2+}	Fe^{3-}	1
Zinc (in acid)	Zn	Zn^{2+}	2

A mathematical formula that will always yield a correct answer even with different equation-balancing ratios is the **stoichiometric equation** given below. **Stoichiometry** calculations involve converting the mass of one substance (reactant or product) in a chemical reaction into equivalent terms for all other substances (reactant or product) in the balanced reaction.

$$N_A \times V_A = N_B \times V_B$$

where

N_A = normality of substance A
V_A = volume of substance A
N_B = normality of substance B
V_B = volume of substance B

Normality and molarity have a simple mathematical relationship. The subscript (s) term is detailed in Table 25.7.

$$N = M \times s$$

where

N = normality
M = molarity
s = subscript on the H^+ ion, OH^- ion, or the change in the charge on the ion

TABLE 25.7 Stoichiometric Relationship of Compounds as Acids, Bases, or Redox Reagents

Compound	Formula	Subscript, s
H^+ ions		
Hydrochloric acid	HCl	1
Nitric acid	HNO_3	1
Potassium hydrogen phthalate	KHC_8O_4	1
Sulfuric acid	H_2SO_4	2
Phosphoric acid	H_3PO_4	3
OH^- ions		
Sodium hydroxide	NaOH	1
Potassium hydroxide	KOH	1
Calcium hydroxide	$Ca(OH)_2$	2
Aluminum hydroxide	$Al(OH)_3$	3
Redox agent change		
Ferric to ferrous	Fe^{3+}/Fe^{2+}	1
Stannic to stannous	$Sn^{3+}Sn^{2+}$	2
Oxalate to carbon dioxide	$C_2O_4^{2-}/CO_2$	2
Permanganate to manganous	MnO_4^-/Mn^{2+}	5

EXAMPLE 25.2 A 4 M H_2SO_4 solution would have what normality?

$$N = M \times s$$
$$N = 4 \times 2$$
$$N = 8$$

EXAMPLE 25.3 A 0.4 M $KMnO_4$ solution would have what normality?

$$N = M \times s$$
$$N = 0.4 \times 5$$
$$N = 2.0$$

CHELATES AND COMPLEXO-METRIC TITRATIONS

Chelates are a special class of coordination compounds that result from the reaction of a metal ion and a ligand that contains two or more donor groups. The result is an ion that has different properties than the parent metal ion. Chelates containing two groups that coordinate are called **bidentate**; those that have a lone pair, **monodentate**. Various chelating agents (chelating ligands) are used to remove troublesome ions from water solutions, as in water softening. In photographic processes, the thiosulfate ion removes silver ions from the film by the formation of the soluble $[Ag(S_2O_3)_2]^{3-}$ ion.

A chelate is a cyclic coordination complex in which the central metal ion is bonded to two or more electron-pair donors from the same molecule or ion, forming five- or six-membered rings in which the metal ion is part of the ring structure. These chelate ring structures are very stable rings, even more stable than complexes involving the same metal ion and same molecules or ion ligands which do not form rings. Aminopolycarboxylic acids and polyamines are excellent chelating agents;

TABLE 25.8 Commonly Used Chelating Agents

Abbreviation	Name	Formula of the anion
CDyTA	1,2-diamino-cyclohexane-tetraacetic acid	$\begin{array}{c}CH_2\\H_2C\diagup\diagdown CH-N(CH_2COO^-)_2\\H_2C\diagdown\diagup CH-N(CH_2COO^-)_2\\CH_2\end{array}$
DTPA	Diethylene-triamine-pentaacetic acid	$^-OOC-CH_2-N\left(CH_2-CH_2-N\diagup^{CH_2COO^-}_{CH_2COO^-}\right)_2$
EDTA	Ethylenediamine-tetraacetic acid	$(^-OOC-CH_2)_2N-CH_2-CH_2-N(CH_2-COO^-)_2$
EGTA	Ethylene glycol bisaknoethyl ether tetraacetic acid	$(^-OOCCH_2)_2N-CH_2-CH_2-O-CH_2-CH_2-O$ $(^-OOCCH_2)_2N-H_2C-CH_2$
EEDTA	Ethyl ether diaminetetraacetic acid	$(^-OOCCH_2)_2=N-CH_2-CH_2-O$ $(^-OOCCH_2)_2=N-H_2C-CH_2$
HEDTA	N′-hydroxyethyl-ethylenediamine-triacetic acid	$-OOCCH_2$ $N-CH_2-CH_2-N(CH_2COO^-)_2$ $HO-CH_2CH_2$
MEDTA	1-methylethylene-diaminetetra-acetic acid	CH_3 $(^-OOCCH_2)_2N-CH-CH_2-N(CH_2COO^-)_2$
NTA	Nitriloacetic acid	$N(CH_2COO^-)_2$
Penten	Pentaethylene-hexamine	$H_2N-CH_2-CH_2-NH-CH_2-CH_2-N-H$ CH_2 $H_2N-CH_2-CH_2-NH-CH_2-CH_2-NH-CH_2$
Tetren	Tetraethylene-pentamine	$NH(CH_2-CH_2-NH-CH_2-CH_2-NH_2)_2$
Trien	Triethylene-tetraamine	$H_2NCH_2-CH_2-NH-CH_2-CH_2-NH-CH_2-CH_2$

of these, EDTA, ethylenediaminetetraacetic acid, is among the best. These chelating agents form complexes with other substances that are only slightly dissociated. Some of the commonest are listed in Table 25.8.

The ethylenediaminetetraacetic acid ion reacts with practically every metal in the periodic table to form stable one-to-one five-membered chelate ring complexes. EDTA can be used to titrate metals in neutral or alkaline solutions, and titration curves can be drawn. When the negative logarithm of the metal-ion concentration is plotted versus the volume of EDTA titrant added, the curves show a sharp break at the equivalence point (Fig. 25.22).

The equivalence point for complexometric titrations can be shown by colored indicators, the *metallochromic indicators*. These indicators, although they form complexes with metal ions, are also acid-base indicators. One form of the indicator has a specific color, and the complex has another. The

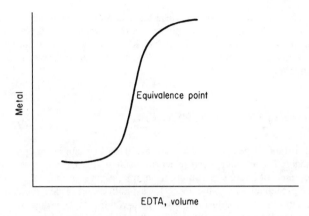

FIGURE 25.22 Titraion curve of an EDTA complexometric titration.

point of color change on the addition of the EDTA is the equivalence point. Common complexometric indicators are Eriochrome Black T, Eriochrome Blue Black B, and Calmagite.

Complexometric titrations can also be monitored by potentiometric and coulometric procedures.

KARL FISCHER METHOD FOR WATER DETERMINATION

The determination of water is one of the most widely practiced analyses in industry. The Karl Fischer method relies on the very specific reaction between water and a reaction mixture devised by German chemist Karl Fischer in 1935. The original mixture contained pyridine, sulfur dioxide, and iodine in methanol. It reacts quantitatively with water. The chemical reaction is believed to occur as follows:

$$C_5H_5N + C_5H_5N\text{—}SO_2 + C_5H_5N\text{—}I_2 + H_2O \rightarrow \text{¼ } C_5H_5N\text{—}HI + C_5H_5N\text{—}SO_3$$

There is a secondary reaction with the methanol:

$$C_5H_5N\text{—}SO_3 + CH_3OH \rightarrow C_5H_5NH\text{—}O\text{—}SO_2\text{—}OCH_3$$

Two main techniques, volumetric and coulometric, have been developed since the discovery by Karl Fischer. Pyridine (C_5H_5N) has a very objectionable odor and newer quantitative aqueous methods have been developed without pyridine. Both the volumetric and coulometric methods use the same reaction principle outlined above. Volumetric titrations are preferentially used for the determination of large amounts of water (1 to 100 mg). Coulometry is a micromethod and particularly suitable for the 10 μg to 10 mg water range. Elemental iodine in the Karl Fischer reaction can be generated coulometrically with 100 percent efficiency. Coulometric procedures can provide an absolute instrumental technique, and the analysis requires very little calibration or standardization. Coulometric instruments calculate the water content automatically from the generated titration current.

An entirely automated titrimeter will determine moisture in the range from 1 μg/mL to 100 percent water content. The instrument combines a burette, a sealed titration vessel, a magnetic stirrer, and a pump system for changing solvent. Liquid samples are injected through a septum; solid samples are inserted through a delivery opening in the titration head. The titration is kinetically controlled: the speed of titrant deliver is adjusted to the expected content of water. Typically 50 mg of water are titrated in less than 1 minute with a precision of better than 0.3 percent. An optional **pyrolysis** (high-temperature heating in the absence of air or oxygen) device is available for moisture determinations in solid samples. A complete line of coulometric and volumetric Karl Fischer titration control standards is commercially available. Karl Fischer water standards are indispensable to standardize and demonstrate the reliability of these titrators according to ISO 9000, GMP, GLP, and U.S. FDA guidelines.

ELECTRONIC PIPETTES AND DISPENSERS

Electronic dispensing systems are available that can be programmed to operate in many different modes. The technician can select from standard pipetting, repetitive dispensing of single or multiple volumes, titration, serial dilution, sample mixing, or manual mode for customized sample dispensing procedures. Even the speed for aspiration and dispensing can be independently set. Automatic pipettes come in adjustable- and fixed-volume models that cover the range from 0.5 to 5000 µL and are ISO 9000 certified. The adjustable-volume pipettes can be recalibrated to ensure optimum performance, particularly when pipetting viscous or nonaqueous liquids. The newest models of pipettes are completely autoclavable at 121°C and are constructed of durable chemical- and UV-resistant materials to extend the life of the pipette. This type of automatic pipette comes with disposable plastic tips to minimize the cleaning and contamination problem associated with conventional glass pipettes. These disposable tips are certified to be pyrogen-free, DNA-free, ATP-free, and sterile by the manufacturer.

REFERENCES

1. Christian, Gary D., *Analytical Chemistry*, 6th edition. John Wiley and Sons, Inc. ISBN 978-04-7121472-4 (2003).
2. "Section 11.6—Volumetric Analysis," *Lange's Handbook of Chemistry*, 15th edition, John A. Dean, editor. McGraw-Hill, Inc. ISBN 0-07-0163384-7 (1999).
3. Fischer, K., "New Methods for the Volumetric Determination of Water Contents in Liquids and Solids," *Angew Chemie* 48:394 (1935).
4. Mitchell, J., and Smith, D. M., *Aquammetry,* John Wiley, New York, 1977 (vol. 1), 1983 (vol. 2).
5. Scholz, E., "Karl Fischer Reagents without Pyridine," *Fres Z Anal Chem* 309:123 (1981). German.

CHAPTER 26
CHROMATOGRAPHY

INTRODUCTION

The term **chromatography** was coined in 1906 by a Russian botanist, Mikhail Tswett, after the words *chromatus* and *graphein,* meaning "color" and "to write." Tswett's coinage was appropriate for his technique for separating colored plant pigments. Tswett discovered that by washing the compounds through a column packed with an adsorbent medium (calcium carbonate) the least-adsorbed pigments were washed through the column quickly, while the strongly adsorbed pigments were immobilized by their attraction to the column packing. (In all chromatographic processes, one medium is fixed in the system and is called the **stationary phase**, while a second medium flows through the fixed medium and is called the **mobile phase**.) Adsorption in this column process is directly related to the affinity of the solute for either the stationary adsorbent or the flowing solvent, and the process is generally referred to as **column chromatography**. Later applications of this adsorption-desorption process showed that colorless substances could also be separated by this technique, but the term "chromatography" remained in use. Today, all types of chromatographic processes are used primarily for the nondestructive separation of complexed mixtures.

ADSORPTION CHROMATOGRAPHY

Adsorption chromatography typically uses silica gel or alumina as the stationary (solid) phase and organic solvents for the mobile (liquid) phase. **Adsorption chromatography** occurs when the sample components transfer from the mobile phase to the stationary phase where they are selectively adsorbed on the surface. This chromatographic process has the advantages of being simple and applicable over a wide range of temperatures, but is best for resolving heat-labile substances at low temperatures. The main disadvantages are that many substances undergo chemical changes with these very active adsorbents, separations tend to be concentration-dependent, and the adsorbents have low capacities for adsorbing solutes.

PARTITION CHROMATOGRAPHY

A second and more versatile type of chromatography called **partitioning chromatography** is based on the differential distribution of the sample components between the two phases. Partition chromatography differs from adsorption chromatography in that two liquid phases are used in which the sample components vary in their degrees of solubility. The term *partition* was first encountered in Chapter 20, "Extraction," when describing the distribution of components between two immiscible liquids. The solubility and polarity of the individual components dictate the ultimate distribution of these components in the mobile and stationary liquids. The stationary phase consists of a thin layer of liquid coated on a porous inert solid and the mobile phase can be a pure liquid or a combination of liquids. One of the first applications of this technique used water adsorbed into silica gel as the stationary phase and a mixture of butanol and chloroform as the mobile phase to separate complexed mixtures of amino acids. In addition to silica gel, other common adsorbents are cellulose, starch, diatomaceous earth, and even powdered rubber. The versatility of this system is much greater than that of adsorption chromatography, and it tends to be less concentration-dependent.

THIN-LAYER CHROMATOGRAPHY

A. Introduction

Thin-layer chromatography (TLC) is a simple, rapid, and inexpensive method for analyzing a wide variety of materials ranging from inorganic ions to high-molecular-weight biological compounds. Thin-layer chromatography is widely used in the chemical, biological, pharmaceutical, and medical industries, not only for quality control and research analyses, but for the preparative separation and isolation of compounds of interest. TLC actually began with the Dutch biologist Beyerinck, in 1889, when he allowed a drop of a mixture of hydrochloric and sulfuric acids to diffuse through a thin layer of gelatin. In 1942, Békésy used a layer of adsorbent between glass plates held apart by cork gaskets. This was filled with an adsorbent slurry and used in the same manner as column chromatography. Thin-layer chromatography is actually a subdivision of liquid chromatography in which the mobile phase is a liquid and the stationary phase is situated as a thin layer on the surface of a flat plate. TLC is grouped with paper chromatography under the term **planar liquid chromatography** because of the flat geometry of the paper or layer stationary phases. Although paper chromatography has made valuable contributions in separation techniques, paper is not universally satisfactory for all separations. Because of its inertness to chemicals, glass is the universal support for thin-layer chromatography. Glass plates of different sizes, shapes, and thickness are used to suit the users, particular needs. Pyrex brand glass plates are available for cases where high-temperature heating of the plate is necessary for making compounds visible.

In thin-layer chromatography, an adsorbent (a powder of at least about 50 m^2/g specific surface area) is applied to a supporting plate in a thin layer. Generally, a binding agent is used to adhere the adsorbent to the support, although some work is done without a binder using very finely divided adsorbent which clings to the support and forms a rather soft layer. The mixture of adsorbent and binder is applied as a thin slurry and the excess moisture is removed under varying conditions depending on the adsorbent, the binder, and the desired degree of activity.

Silica gel is the most commonly used adsorbent for thin-layer chromatography, accounting for an estimated 70 to 80 percent of all TLC plates used. Most commercial silica gel produced and sized for TLC use has consistently reproducible pore size, particle size, particle distribution, surface area, and impurity levels. With all of these variables carefully controlled, the most important factor influencing TLC separations is moisture. The most nearly ideal condition for general TLC silica gel separations is considered to be at the level of 11 to 12 percent water. Good practice then suggests that the TLC plates be dried at 70 to 80°C for 30 minutes and then allowed to cool in an open, clean environment. This is particularly advisable if the TLC plates have been on hand for some time, two to three months, or exposed to questionable conditions. If the environment suffers from high humidity or an extraordinary concentration of chemical fumes, it may be advisable to store the plates after drying in a closed desiccators. In no case is it recommended that temperatures above 110°C be used for drying. Silica gel comes with a wide variety of binders and additives such as:

Silica gel G	10 percent gypsum (calcium sulfate) as the binder
Silica gel H	Silicon dioxide/aluminum oxide binder
Silica gel F	Fluorescent indicator added, a subscript indicates the wave-length
Silica gel P	Preparative TLC
Silica gel R	Specially purified adsorbent
Silica gel R.P.	A silanized gel for reverse-phase work

B. Preparation of the Plate for Thin-Layer Chromatography

1. Before preparing the slurry, treat the mixing flask and glass rod with a hydrophobic substance, such as dimethyldichlorosilane, thus rendering them hydrophobic. Dissolve 2 mL of the dimethyldichlorosilane in 100 mL of toluene and thoroughly wash the mixing vessel and glass rod with it.

Use this solution as a rinse to waterproof both the mixing vessel and rod (to be used for preparing the slurry) and all other glassware to be used. Finally, rinse all glassware with methanol and distilled water prior to use.

2. Prepare a slurry of the adsorbent. In the case of aluminum oxide or silica gel, use Table 26.1 or 26.2 to determine the proportions of the slurry.

Other adsorbents used for TLC are cellulose and polyamides; some of them contain binders that make them stick to the glass.

TABLE 26.1 Slurry Preparation with Aluminum Oxide

Film thickness, µm	Mass of adsorbent, g	Water, mL
150	3.0	4.0
250	5.0	7.0
375	7.5	10.0
500	10.0	13.0

TABLE 26.2 Slurry Preparation with Silica Gel

Film thickness, µm	Mass of adsorbent, g	Water, mL
150	3.0	6.0
250	5.0	7.0
375	7.5	15.0
500	10.0	18.0

3. Put the required mass of adsorbent into the mixing flask and then add the specified volume of water. Shake thoroughly for about 5 s.

4. Place a clean, dry, glass plate (Fig. 26.2) on a paper towel and pour the slurry evenly across the carrier plate near the bottom edge. Use the glass rod to spread the slurry with a smooth steady motion to the top edge of the plate (but not over it); slide the rod, don't roll it. A commercial spreader (Fig. 26.3) may also be used.

CAUTION: This operation may be repeated to obtain a smooth coating, but cannot be repeated once the slurry hardens.

5. Without allowing further flow of the coating, dry the coated plate in an oven at 89 to 90°C for about 1 h. This permits the plate to be handled and at the same time activates the coating. Store dried plates in a suitable desiccator.

C. Thin-Layer Chromatography Procedure

Standard TLC procedure involves three steps.

1. The substance to be separated into fractions is spotted on the edge of the plate with a micropipette in such a manner as to yield a minimum area. Better separation and development are obtained with small sample spots.

2. The prepared solvent mixture is placed in the bottom of a developing tank (Fig. 26.4) and the plate (or plates) is positioned in the tank with the upper part of the carrier plate leaning against the side of the tank. The tank is securely covered with a glass plate and the developing begins as the solvent rises up the plate by capillary attraction.

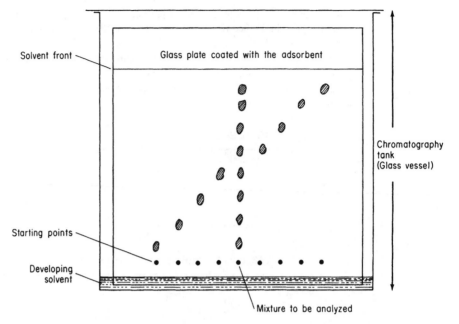

FIGURE 26.1 Separation by thin-layer chromatography.

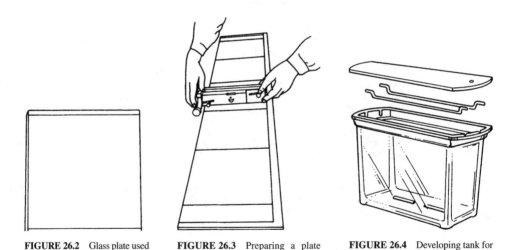

FIGURE 26.2 Glass plate used in thin-layer chromatography.

FIGURE 26.3 Preparing a plate using a commercial spreader.

FIGURE 26.4 Developing tank for thin-layer chromatography.

An alternate method uses a sandwich technique. A blank plate and the spotted plate are placed channel-to-channel and clamped together. They are placed in the trough (Fig. 26.5), which contains the solvent in a vertical position, with the level of the solvent lower than the level of the spotting.

The TLC plate for this variation is prepared in the normal manner, but a strip about ½ in wide is scraped off the sides and upper edge. An inverted-U cardboard frame is placed over the cleared strip, and the cover plate placed over the U frame. Clamps securely hold the sandwich together. Only about 15 mL of solvent is needed to fill the trough for developing the chromatogram.

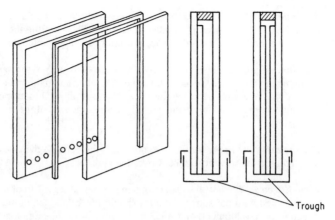

FIGURE 26.5 Assembly showing normal and saturated sandwich chamber.

3. The spots must be located. Once the solvent front has reached the desired level, the plate is removed and allowed to dry. Now the location of the spot must be determined. This location can be used as a criterion for the identification of the substance, particularly if controls have been set up; the intensity of the spot is a quantitative measure of the concentration of the substance. Chromophoric substances can be located visually; colorless substances require other means.

Some substances fluoresce under UV light, and irradiation of the plate will indicate the position of the spot. Other plates are coated with fluorescent materials; the spot will obscure this fluorescence when the plate is irradiated. Spray reagents, selected to react with the spot (Table 26.3), reveal its location. The spray must be applied uniformly to the dried plate. Finally, exposure of the plate to chemical vapors can also reveal the location of the spot.

TABLE 26.3 Spraying Reagents

Type	Application
Strong acid [H_2SO_4; H_3PO_4 (heated to 120°C)]	Natural substances
Strong acids [H_2SO_4; H_3PO_4 with 0.5–1.0% aldehyde (such as anisaldehyde)]	Natural substances
Antimony(V) chloride (20%) and carbon tetrachloride* (80%) heated to 20°C	Resins, terpenes, and oils
Iodosulfuric acid solution [1.0 N iodine, 16% H_2SO_4 (1:1)]	Organic nitrogenous compounds
Iodine 0.5% solution in chloroform or alcohol	Organic nitrogenous compounds
Antimony(III) chloride (25%) in chloroform; may yield fluorescence when heated	Carotenoids, steroid glycosides

*Tetrachloromethane

D. Tips on Technique for Thin-Layer Chromatography

Precoated, Commercially Available Plates. Precoated aluminum sheets are commercially available, and they offer certain advantages over glass plates. Aluminum (or rigid plastic) sheets can be cut easily with scissors, and they are stiff enough to stand without supports. The coatings are abrasion-resistant and uniform.

Activation of Plates. All prepared or precoated plates should be stored and protected from environmental contamination. They must be "activated" by heating them in an oven before use and then cooled and stored in a suitable desiccator prior to being spotted and developed.

Spotting. Capillary tubes or micropipettes are used to spot the sample solutions. The capillary is positioned in the desired location over the TLC plate and momentarily touched to the plate with due care to avoid disturbing the coating. The solvent is allowed to evaporate. The procedure is repeated until the whole sample has been spotted, keeping the size of the spot as small as possible for better separation.

Placing the Plate in the Tank. Do not touch the plate on the sides. Hold it by its edges and place it squarely in the solvent surface. Cover the tank securely.

Edge Effects. The best results are obtained when at least 1 in of the outer edge of the carrier plate is not coated with the slurry. Samples should not be spotted closer than 1 in from the edge in the direction of development, and no closer than 1 in to the edge that is parallel to the development. Edge effects such as distortion of the solvent boundary or distortion of the spot shape can result from disregarding these tips.

GAS CHROMATOGRAPHY

Gas chromatography (GC) is one of the fastest and most useful separation techniques available in the laboratory (Fig. 26.6). Gas chromatographic analysis is basically limited to organic compounds that are volatile and not thermally labile (decomposable). There are two modes of gas chromatography: **gas-solid (adsorption) chromatography** and **gas-liquid (partition) chromatography**. Gas-liquid chromatography (GLC) is used more extensively than gas-solid chromatography (GSC). Both types of gas chromatography require that the sample be converted into or exist in the vapor state and be transported by an inert carrier gas through a column packed with either a liquid phase coated on a solid support (GLC) or simply a solid adsorbent with no liquid-phase coating (GSC).

A sample is injected into a heated block where it is immediately vaporized and swept as a concentrated vapor into a column. Separation occurs as the various compound vapors are selectively adsorbed by the stationary phase and then desorbed by fresh carrier gas. This sorption-desorption process occurs repeatedly as the compounds move through the column toward a detector. The compounds will be eluted from the column with those having a high affinity for the column packing being slower than those with little affinity.

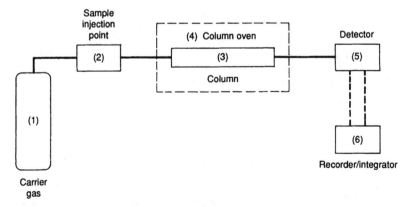

FIGURE 26.6 Block diagram of a gas chromatograph.

INSTRUMENT DESIGN AND COMPONENTS

A typical gas chromatograph consists of the following: (1) a carrier gas supply, (2) sample injection port, (3) column, (4) column oven, (5) detector, and (6) a recorder/integrator system. A block diagram showing these main GC components is shown in Fig. 26.6.

A. Carrier Gas

The **carrier gas** is used to transport the sample molecules from the injection port to the detector and provide the means for partitioning the sample molecules from the stationary phase. The most common carrier gases are helium and nitrogen. These gases are supplied in high-pressure tanks which require a two-stage pressure regulator for reducing the inlet gas pressure and controlling the gas velocity through the column. This gas must be of high purity with minimal moisture or other contaminants present to reduce erroneous detector signals. Various commercial gas purifiers are available for removing carrier-gas contaminants, especially oxygen and water. Refillable in-line traps are also available that contain molecular sieve 5A, charcoal, etc., for removing oxygen, moisture, and hydrocarbon contaminants. These purifiers should be checked and reconditioned periodically, especially if drifting baseline is being experienced. If hydrogen gas is being used for an FID detector or as the carrier gas, a hydrogen generator is strongly recommended (over a cylinder of compressed hydrogen) for high purity and laboratory safety.

B. Sample Injection

A sample inlet system must be provided that allows liquid samples in the range of 1 to 10 μL to be injected with a microsyringe (see section on Microliter-Syringe-Handling Techniques) through a **septum** into a block that is heated to a temperature in excess of the compounds' boiling points. The liquid sample is immediately vaporized as a "plug" and swept through the column by the carrier gas. Septum (plural: *septa*) is a self-sealing rubber or elastomer material that forms a leakproof entrance into the injection port. Septa are manufactured in discs ranging in size from 5 to 16 mm to fit specific injection inlets and are available in very expensive high-puncture-tolerance/low-bleed/preconditioned types or in less expensive simple rubber stock. Some Teflon-faced septa are available to reduce septum-bleeding contamination into the column at higher temperatures. A leaky septum is one of the most common sources of trouble in gas chromatography; therefore, it should be changed periodically.

CAUTION: Remember to always turn off a thermal conductivity detector before disrupting the carrier-gas flow by removing the septum or changing gas cylinders.

Gas and liquid samples can also be introduced using a **gastight valve and calibrated volume loop** system. These valves have multiport arrangements and can have sample loops for injecting liquid or gas samples. A typical six-port valve with sample purge, sample volume loop, and column connection is shown in Fig. 26.7.

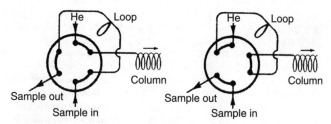

FIGURE 26.7   Typical six-port GC sample valve. (*Courtesy Varian Associates, Inc.*)

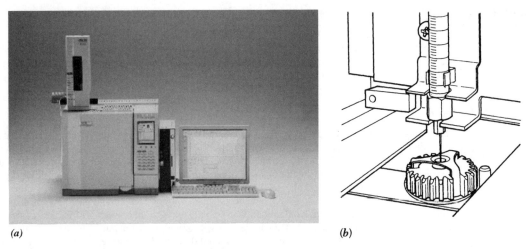

(a)   (b)

FIGURE 26.8 (a) Gas chromatograph equipped with an automatic sampler and workstation (*model GC-2010 Plus, courtesy Shimadzu Scientific Instruments.*); (b) Automatic sampler with microliter syringe.

Samples can be introduced using a programmable injector for the continuous operation of chromatographic systems as shown in Fig. 26.8a and b. These automatic injectors use the same type of microsyringes but are capable of a higher degree of reproducibility than a manual technique. This unattended operation releases the operator for other duties and allows 24-hour operation of a GC system equipped with an automated integration/recorder system.

Special injector splitters are used with capillary columns (see section Columns below), which usually require samples of less than 1 µL. These injector splitters mix the vaporized sample and split (ratio between 1/10 and 1/1000) the original injection volume by venting the excess. Other options include on-column injection and special glass inserts to line the interior of the metal injection ports to prevent the metal-catalyzed decomposition of certain types of compounds.

C. Columns

Two basic types of columns are currently being used: packed and capillary. Packed columns will usually have 1000 to 3000 plates per meter (see section Column Efficiency in this chapter), while capillary columns can exceed 4000 plates per meter. **Packed columns** are normally made of copper, stainless steel, or glass, with common bores of 1.6, 3.2, 6.4, or 9.5 mm and lengths of 1 to 3 m. Glass columns and glass injection-port liners are necessary when dealing with labile compounds that might react or decompose on contact with metal surfaces, especially at elevated temperatures. These columns have been "packed" with a coated, sieved (ranging from 60 mesh to 120 mesh) inert solid support (see section Solid Supports). The solid support is coated (usually 1 to 10 percent by weight) with a highly viscous liquid phase (see section Liquid Phases).

Capillary columns do not contain solid support coatings and simply have the liquid phase (less than 1 µm thick) coated directly onto the interior walls of the column. This wall-coated open tubular (WCOT) technique provides an open, unrestricted carrier-gas path through the typical 0.25-mm-diameter column. Since capillary columns present very little restriction to gas flow, they can be made extremely long (50 to 150 m) for greater compound resolution. A newer form of capillary column technology called **support-coated open tubular** (SCOT) has recently been developed. A layer of solid support is adsorbed onto the interior of the capillary tubing walls and the liquid phase applied. The primary advantage of this surface-area-increasing technique is to increase the sample capacity of the column, and, in some cases, sample splitting is not necessary as was the case with conventional capillary columns.

D. Column Oven

Operating the column at a constant temperature **(isothermal)** during an analysis is critical for reproducible results. However, it may be necessary to change the column temperature in a reproducible way **(temperature programming)** during an analysis to separate components having greatly differing boiling points and/or polarities in a reasonable period of time. Good resolution is obtained for low-boiling components at moderate oven temperatures; however, the analysis will require an excessive amount of time to elute any high-boiling compounds, and the chromatographic peaks will be too broad for proper quantitative interpretation. Temperature programming can reduce this difficulty by allowing the low-boiling compounds to elute at initially low temperatures, and the column oven is then increased at a reproducible rate to elute the high-boiling compounds in a reasonable time. A typical temperature program would consist of the following steps as shown in Fig. 26.9:

Point A	Initial column temperature
Point B	Final column temperature
Zone 1	Post-injection period
Zone 2	Temperature-programming rate (°C/min)
Zone 3	Upper temperature period
Zone 4	Automatic cool-down
Zone 5	Isothermal recovery to point A

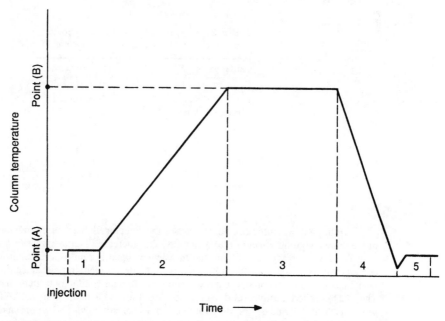

FIGURE 26.9 GC oven temperature programming profile. Point A = initial column temperature; point B = final column temperature.

E. Detectors

Each chromatograph has a detection device at the exit of each column to monitor the gas composition. There are many types of detectors: thermal-conductivity, flame-ionization, electron-capture, photo-ionization, Hall electrolytic-conductivity, flame-photometric, thermionic-specific, and coulometric detectors are the most common. Some chromatographic systems have dual-channel detectors to analyze a single sample by two dissimilar detectors simultaneously. This technique will provide the operator with additional information on the identification of compounds. This dual-detector capability is also necessary for stabilizing the detector baseline when temperature programming is being used. As the column temperature is deliberately increased in the programming cycle, some of the liquid phase *bleeds* from the column and increases the detector background signal. A matched column with the same bleed rate is installed to produce an opposite, but identical, signal to the detector; the two opposing signals cancel each other, and the baseline remains relatively straight.

Thermal-Conductivity Detectors (TCDs). Some thermal-conductivity (hot-wire) detectors use a thin filament of metal, while other TC types use thermistors. In thermal conductivity detectors, a tungsten filament is heated by a continuous current flow and cooled by the carrier gas as it exits from the column.

CAUTION: Thermal-conductivity detectors can be damaged by turning on the filament current without proper carrier-gas flow.

Hydrogen and helium are the best TC carrier gases because of their very high thermal conductivities; however, hydrogen is flammable. Hydrogen has the highest thermal conductivity (53 cal/°C mol) and produces the greatest sensitivity of all carrier gases. Helium is also an excellent TC carrier gas (42 cal/°C mol) and is much safer than hydrogen. Most other gases have thermal conductivities of less than 10 cal/°C mol, as shown in Table 26.4.

TABLE 26.4 Thermal Conductivities of Some Common Carrier Gases and Solutes (in calories/°C mol)

Carrier gases	Solutes
Hydrogen 53	Acetone 4.0
Helium 42	Benzene 4.2
Nitrogen 8	Butane 5.7
	Cyclohexane 4.2
	Chloroform 2.5
	Ethane 7.4
	Ethanol 5.3
	Nonane 4.5

The elution of a sample component causes the detector filament or thermistor to heat up because of the diluted cooling effect of the carrier gas. This increase in temperature (resistance) is measured by a Wheatstone bridge circuit and the imbalance signal is sent to a recorder/integrator. Thermal-conductivity detectors are the most universal of all chromatographic detectors because they can detect any gaseous compound that has a different thermal conductivity than that of the carrier gas. The TCD is called a **universal detector** because it responds to all analytes. TCDs are not the most sensitive of the GC detectors; they routinely have sensitivities of ~10 ppm and linear ranges of 4 orders of signal magnitude. A major advantage is that this detector is nondestructive and can be used for preparatory work. A typical thermal-conductivity detector is shown in Fig. 26.10.

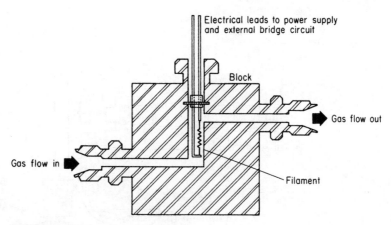

FIGURE 26.10 Thermal-conductivity detector.

Flame-Ionization Detector. The most widely used detector in gas chromatography is the flame-ionization detector (FID). The FID is extremely sensitive to all organic compounds, but is not classified as a universal detector like the TCD. Hydrogen and air are mixed in a burner to produce a very hot (approximately 2100°C) flame that can ionize carbon-containing compounds. A collector electrode with a dc potential is placed above the flame to measure its conductivity. As the column effluent passes through the burner jet, certain compounds are ionized and create a current flow that is proportional to the concentration of the carbon atoms in the flame. These generated currents are of only ~10^{-12} amps and require an electrometer for amplification of the signal. The FID is very sensitive and can detect concentration in the range of 100 ppb in some cases. It also has an excellent linear range and can handle signal magnitude changes of 6 orders or more. In addition, FID has excellent signal stability compared to other types of detectors and is not as sensitive to carrier-gas changes or other instrument parameter changes. The organic molecules are converted into positively charged ions and are then attracted to the negatively charged collector ring above the flame jet. The ion current is roughly proportional to the number of carbon atoms present; therefore each compound requires a response factor. The analyzed compounds are destroyed in passing through the flame. Therefore FIDs are not routinely used in preparative scale work. A cross-sectional view of a typical FID is shown in Fig. 26.11.

FID detectors are much more sensitive (perhaps ×1000) than TC detectors. To set the FID flows, begin with all the manifold block flows off and the carrier gas off. Measure each flow separately with a bubble meter attached to the FID vent tube. First set, and then turn off, each of the detector gas flows at the flow manifold block. Set the carrier-gas flow last, since it can be turned off without changing the flow settings. After setting the detector flows, shut off all manifold block flows until ready to light the flame. If flows are set correctly, the FID flame will be easy to ignite. Proper ignition should result with a slight audible "pop." To check the flame, hold a mirror or shiny metallic surface near the exhaust to observe moisture or condensation. Do not place paper or other combustible materials into the flame to check for ignition as this practice can grossly contaminate the detector. As stated earlier, flame ionization detectors are primarily limited to organic compounds and do not respond very well to air, water, or most inorganic compounds (Table 26.5). In addition, there are several very specialized GC detectors: electron capture detectors, nitrogen-phosphorous detectors, flame photometric detectors, etc.

F. Recorder/Integrator

The signal from the detector can be sent to a servo (potentiometric) strip-chart recorder where the magnitude of the voltage is measured and recorded. The recorder (*x* axis) draws a straight line when only carrier gas is passing through the detector, and any eluting compounds cause the recorder to

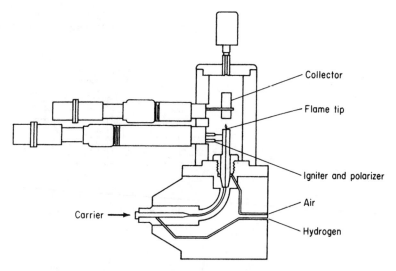

FIGURE 26.11 Typical flame-ionization detector.

TABLE 26.5 Gaseous Substances Giving Little or No Response in the Flame-Ionization Detector

He	CS_2	NH_3
Ar	COS	CO
Kr	H_2S	CO_2
Ne	SO_2	H_2O
Xe	NO	$SiCl_4$
O_2	N_2O	$SiHCl_3$
N_2	NO_2	SiF_4

respond in proportion to the quantity of that compound. These strip-chart recorders generally have selectable voltage ranges of 1 mV to 5 V. Gas chromatographs normally use 10-mV signals or less; however, extended-voltage-range recorders permit other laboratory uses (LC, atomic absorption spectroscopy, etc.). The y axis of the recorder measures the retention time for each eluting compound from its time of injection. Most GC recorder systems allow for different chart speeds (usually ranging from 1 to 30 cm/min) to improve manual integration measurements and/or to conserve paper. Some newer gas chromatographs have computerized integration capabilities that automatically record the retention time to the thousandth of a minute, if desired, and the proportional detector response of each compound. Figure 26.12 shows an electronic integrator displaying retention time (RT), area responses for each peak, and the chromatogram. This type of integrator can be used in both gas and liquid chromatography.

GENERAL GC OPERATIONAL PROCEDURES

The following GC instrumentation conditions should be recorded with each analysis:

- Sample identification name or number
- Date and time

FIGURE 26.12 Gas chromatograph dedicated workstation and computer terminal. (*Model GC-2010 Plus, courtesy Shimadzu Scientific Instruments.*)

- Sample size
- Sample dilution factor(s), if any
- Carrier-gas flow rate
- Attenuator setting and range setting, if applicable
- Detector sensitivity setting or bridge current
- Column packing, length, and concentration
- Column oven temperature
- Injector temperature
- Detector temperature
- Chart speed
- Recorder chart range in millivolts
- Appropriate integrator settings

Instrument Start-Up Procedure

1. Check septum.
2. Turn on gases first.
3. Adjust flow rates.
4. Turn on power to the instrument.
5. Adjust temperatures.
6. Turn on the detector.
7. Turn on recorder/integrator.

8. Inject standards.
9. Inject samples.
10. Record all GG conditions and data.

Instrument Shutdown Procedure

1. Turn off detector.
2. Turn off recorder/integrator.
3. Turn off or reduce temperatures.
4. Turn off all gases.
5. Turn off all power. (Some GC manufacturers recommend that their instruments be maintained at elevated temperatures with a slight carrier-gas flow while not in use.)

A. Microliter-Syringe-Handling Techniques

In gas chromatography, the quantitative results can usually be no more accurate than the sample reproducibility. Since very small liquid sample volumes are normally injected (usually 1 µL or less), the accuracy, reproducibility, and technique of the operator are critical. General guidelines for using a microliter syringe (Fig. 26.13) are as follows:

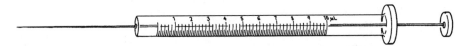

FIGURE 26.13 A microliter syringe.

1. The GC microliter syringe and plunger should be dry and clean. Most glass syringes can be cleaned by first washing with cold chromic acid solution followed by rinsing with distilled water. The syringe should be blown dry with compressed air and the plunger wiped with lint-free paper.
2. To ensure the most quantitative transfer of sample, the syringe's interior surface and plunger should first be wetted with the sample. Check the syringe for air bubbles. Air can be removed by *slowly* drawing liquid into the syringe and *rapidly* expelling the liquid. This filling and emptying procedure should be repeated until no air bubbles are visible.
3. Draw at least twice as much sample into the syringe as desired. Invert the syringe and slowly depress the plunger until the desired volume is obtained. Before attempting to inject the sample, wipe the needle clean with a lint-free tissue, using a quick motion and being careful not to siphon sample from the needle or transfer body heat by handling the syringe excessively.
4. One hand should be used to quickly guide the syringe needle and to prevent its bending while passing through the GC septum. The other hand should be used to control the syringe barrel and the thumb to prevent the plunger from being expelled by the GC's gas pressure.
5. The needle should be inserted as far into the GC injection port as possible, the plunger depressed as smoothly and quickly as possible, and the syringe withdrawn immediately while the plunger is still depressed.
6. The recorder should be started and marked as the sample is injected so that retention times can be measured starting with the injection. Obviously, if an electronic integrator is available, it should be turned on at the start of the injection.

CAUTION: Injection ports are usually very hot.

B. Gas Chromatographic Fittings and Valves

See Compression Fittings in Chapter 18 "Plumbing, Valves, and Pumps" and references listed at the end of this chapter.

SOLID SUPPORTS

Solid supports provide the surface area for a liquid phase to be exposed to a mobile gas phase. The ideal solid support should be inert, not pulverize readily, and have a high surface area (greater than 1 m^2/g). Solid supports are normally graded into **mesh sizes.** For example, an 80 mesh means a screen with 80 holes per linear inch. The solid particles are sorted based on their ability to pass through the screen mesh openings. Mesh ranges (60/80) are used to indicate the largest to smallest solid-support particles. Naturally occurring silicates and **diatomaceous earth**, which is composed of the skeletons of thousands of single-celled plants, are used extensively as GC solid-support materials. The type of support derived from diatomaceous earth is known by the trade name Chromosorb® and comes with various chemical and physical properties. These solid supports can be **silanized** by treatment with dimethyldichlorosilane or other silanes to reduce the surface activity of the silicates in the diatomaceous earth. Acid-washing treatments are also beneficial in reducing the tailing effects caused by solid-support adsorption.

Chromosorb P® is a pink-colored material, with a rather high surface area (4 to 6 m^2/g), has the highest liquid-phase holding capacity or adsorptivity, but is the least inert of the Chromosorbs.

Chromosorb W® is a white-colored material prepared by mixing with sodium carbonate flux at about 900°C. This material tends to be more rugged than Chromosorb P, but has a very low surface area (approximately 1 m^2/g).

Chromosorb G® combines the high surface area of P and ruggedness and inertness of W.

Chromosorb T® is not made from diatomaceous earth but is actually a fluorocarbon polymer (Teflon®). This solid support is much more inert than the Chromosorbs and is used, when coated, to separate very polar compounds such as water without the typical adverse tailing effects.

Porapak® is a trade name for stryrene/vinyl benzene cross-linked polymers. This polymer is formed into porous beads which serve as both the liquid phase and the solid support. The Porapaks are excellent for high-temperature (usually 250°C maximum) separations of polar mixtures. These polymeric beads can also be coated with conventional liquid phases, usually of less than 5 percent by weight.

Table 26.6 shows the typical and maximum liquid loadings for specific GC support materials.

LIQUID PHASES

Selecting the proper liquid phase for a particular GC separation can be the most difficult task in operating such a system. Generally, the liquid phase must have the following characteristics:

1. Nonvolatile or have boiling point of at least 100°C greater than the maximum column-operating temperature
2. Thermally stable, not decompose with heat
3. Good solubility for the sample components

Historically, hundreds of substances have been used as GC liquid phases, but the most important ones number fewer than 50. These liquid phases can be classified *as polar, nonpolar,* or of *intermediate polarity.* The old laboratory expression, "likes dissolve likes," can be readily applied in selecting GC liquid phases. *Polar* substances are usually better separated on *polar* columns and *nonpolar* substances on *nonpolar* columns. Table 26.6 gives a classification of common compounds or functional groups listed in decreasing order of polarity.

TABLE 26.6 Typical and Maximum Loadings for Solid Supports

Support material	Typical loading, w/w%	Maximum loading, w/w%	
		Typical liquid phase	Sticky or gum phases*
Anakrom	3–10	25	20
Chromosorb G	2–6	12	7
Chromosorb P	10–30	35	20
Chromosorb T	1–2	6	3
Chromosorb W	3–10	25	20
Chromosorb 101, 102 through 108	3–10	25	10
Gas Chrom Q	3–10	25	20
Porapak N, Q, Q-S, R, S	1–5	8	4
Porapak T	1–2	6	4
Teflon®	1–2	6	3
Tenax	1–4	5	4

*OV-1, SE-30, SE-52, OV-275, DEGS. Apiezon L are typical sticky or gum phases.
SOURCE: Varian Associates, Inc.

Some typical GC liquid phases are listed in Table 26.7, classified as to their degree of polarity. This table should be used in conjunction with and as a guide to Table 26.6, classification of solute polarities.

COLUMN EFFICIENCY

Fractional distillation and chromatographic processes can be compared in their respective abilities to separate mixtures. A basic review of distillation theory and techniques from any general chemistry or organic chemistry laboratory manual would be beneficial in understanding this section on GC column efficiency. *Resolution* is defined as the degree of separation between adjacent peaks on a chromatogram. For symmetrical peaks (gaussian-shaped), the resolution R can be calculated with the equation that follows. For symmetrical peaks, a resolution R calculation of 1.0 corresponds to approximately a 2 percent peak overlap. Figure 26.14 shows a chromatogram with two unsymmetrical peaks and a technique for estimating the resolution.

$$R = 2(t_2 - t_1)/w_1 + w_2$$

where

t_2 and t_1 are retention times for peaks 2 and 1, respectively
w_1 and w_2 represent the estimated peak base widths for peaks 1 and 2

A simple mathematical relationship can be drawn between the number of theoretical plates (number of transfer equilibria) in a fractional distillation and the resolution of components in a chromatographic column. In a chromatographic column, the number of theoretical plates n can be calculated by the equation given below. A better name for this value is *column efficiency* since it depends on column flow rate, temperature, column length, and even the compound itself. The n term is unitless; therefore t_R and W must be in the same units of length (usually mm). The greater the number of theoretical plates n, the more efficient and better the resolution in the column. Figure 26.15 shows a typical chromatogram and the appropriate measurements.

$$n = 16(t_R/W)^2$$

where

t_R = retention time
W = width of peak

TABLE 26.7 Solute Classification

Class	Solute
Class I (most polar)	Water
	Glycol, glycerol, etc.
	Amino alcohols
	Hydroxy acids
	Polyphenols
	Dibasic acids
Class II (polar)	Alcohols
	Fatty acids
	Phenols
	Primary and secondary amines
	Oximes
	Nitro compounds with α-H atoms
	Nitriles with α-H atoms
	NH_3, HF, N_2H_4, HCN
Class III (intermediate)	Ethers
	Ketones
	Aldehydes
	Esters
	Tertiary amines
	Nitro compounds with no α-H atoms
	Nitriles with no α-H atoms
Class IV (low polarity)	$CHCl_3$
	CH_2Cl_2
	CH_3CHCl_2
	CH_2ClCH_2Cl
	$CH_2ClCHCl_2$, etc.
	Aromatic hydrocarbons
	Olefinic hydrocarbons
Class V (nonpolar)	Saturated hydrocarbons
	CS_2
	Mercaptans
	Sulfides
	Halocarbons not in Class IV, such as carbon tetrachloride

SOURCE: Gow-Mac Instrument Co. Reproduced by permission.

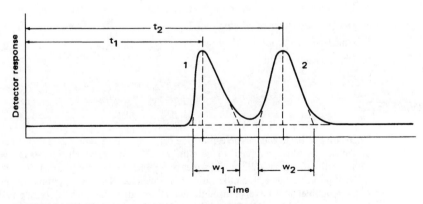

FIGURE 26.14 Chromatogram with two unsymmetrical peaks.

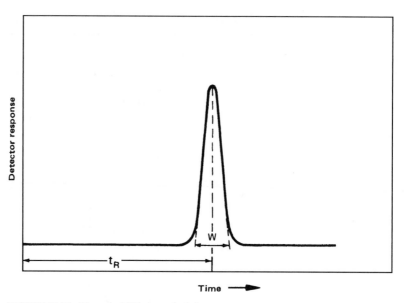

FIGURE 26.15 Theoretical GC-plate calculations.

Another term used to describe column efficiency is **Height Equivalent to a Theoretical Plate (HETP)**. This value can be calculated with the equation given below, with L representing length of column expressed in centimeters. Obviously, the smaller the HETP (length of column required to produce the equivalent of a distillation plate), the better the resolution of mixture components.

$$\text{HETP} = \frac{L}{n}$$

In general, resolution can be improved by one or more of the following:

1. Increase the retention time t_R.
2. Increase the column length.
3. Use smaller-diameter stationary phase.
4. Decrease diameter of column.
5. Optimize the carrier-gas flow rate.
6. Reduce sample size.

FLOWMETERS

There are two basic ways to determine the carrier-gas flow rate in a gas chromatograph. The simplest and most versatile technique is to use a **bubble flowmeter**, which consists of a reservoir of soap solution, a calibrated glass tube, and a squeeze bulb as shown in Fig. 26.16. A rubber tube connects the GC exit port to a side arm on the calibrated flowmeter. The GC column effluent causes the soap film to form bubbles that traverse through the calibrated column. The flow rate is determined by timing how long the bubble takes to travel the calibrated column. For example, assuming that a stopwatch is

TABLE 26.8 Liquid-Phase Classification

Class	Liquid phase
Class A (I)	FFAP
	20M-TPA
	Carbowaxes
	UCONs
	Versamid 900
	Hallcomid
	Quadrol
	THEED (tetrahydroxyethylenediamine)
	Mannitol
	Diglycerol
	Castorwax
Class B (II)	Tetracyanoethyl pentaerythritol
	Zony E-7
	Ethofat
	β,β-Oxydipropionitrile
	XE-60 (nitrile gum)
	XF-1150
	Amine 220
	Epon 1001
	Cyanoethyl sucrose
Class C (III)	All polyesters
	Dibutyl tetrachlorophthalate
	SAIB (sucrose acetate isobutyrate)
	Tricresyl phosphate
	STAP
	Benzyl cyanide
	Lexan
	Propylene carbonate
	QF-1 (silicone, fluoro-1)
	Polyphenylether
	Dimethylsulfolane
	OV-17 (50% phenyl silicone)
Class D (IV&V)	SE-30
	SF-96
	DC-200
	Dow 11
	Squalane
	Hexadecane
	Apiezons
	OV-1 (methyl silicone)

SOURCE: Gow-Mac Instrument Co.

started as a bubble passes the 0-mL calibration mark, stopped at the 10-mL mark, and found to take exactly 15 seconds (0.25 min), the flow rate would be 10 mL/0.25 minute or 40 mL/minute. Bubble flowmeters are normally calibrated in 0–2, 0–10, and 0–60 mL-volume ranges that are useful for capillary and standard GC applications. This type of flowmeter is also available in digital models, which provide more convenience and accuracy.

Another device used to determine carrier-gas flow rate is called a **rotameter**. Again a calibrated glass tube is used, but fitted with a ball that rises in the tube as the column effluent flow rises. This type of flowmeter is normally built into the instrument directly in the carrier-gas stream between

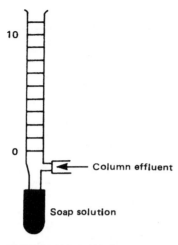

FIGURE 26.16 Bubble flowmeter.

the gas source and the injector port. This type of flowmeter usually contains a valve for adjusting the carrier-gas flow rate through the column. Rotameters must be calibrated for the different carrier gases; otherwise the numerical scale is only in arbitrary units.

Flow measurements can now be determined instantly and accurately with state-of-the-art microprocessor-based flowmeter and/or electronic mass flowmeters. Some of these automatic devices use a heated sensing element and the thermodynamic heat-conduction principles to determine the true mass flow rate in chromatographic or industrial processes, or any other gas-dependent system. Most of these electronic flowmeters are calibrated for N_2, He, H_2, Air, and 5 percent Ar/CH_4 mixtures. The larger units have a working range of 1.0 to 500 mL/min and an accuracy of ±2 percent of reading or 0.05 mL. They can display volume in mL/min, linear velocity in cm/sec, or split ratio. Some units are microflowmeters (0.10 to 20 mL/min range) and are excellent for optimizing low-capillary-column work. Compliance with ISO 9000, GLP, and other stringent quality control protocols makes it necessary to utilize these electronic flowmeters.

QUALITATIVE ANALYSIS

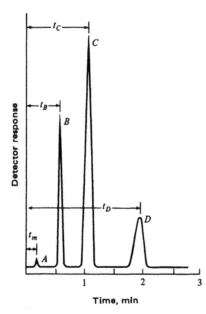

FIGURE 26.17 Retention times on a chromatogram.

The time required for each compound to pass through the column is called the **retention time** and is characteristic of that compound and can be used to identify each substance in a mixture (Fig. 26.17). Peak A in Fig. 26.17 represents air and indicates that even air has a column retention time or at least residence time. The retention time for component B is represented by t_B, the retention time for component C is represented by t_c, and so forth for this mixture. The actual retention time is measured from the time of injection to the peak of the component. Caution should be exercised in assuming that a retention time definitely identifies a compound; several compounds can have identical retention times. The identification of compounds using gas chromatographic peaks is performed by comparing the retention time of each eluted compound with the retention time of a pure standard under identical conditions. In other words, a known sample retention time is determined immediately before or after the unknown analysis on the same column using identical column temperature, flow rate, sample size, etc.

A third technique of peak identification is called **spiking**, in which a known compound is deliberately added to the unknown mixture and the size of the original peak is observed (Fig. 26.18). If the unknown peak increases in size after the spiking, it would indicate that the two compounds have the same retention time.

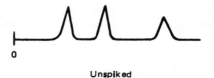

Unspiked

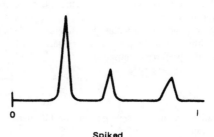

Spiked

FIGURE 26.18 Chromatograms of unspiked and spiked samples.

CAUTION: *It should be stressed that since several compounds can have identical retention times, one should be cautious about assuming the identity of a compound using GC retention data only.*

INTEGRATION TECHNIQUES

The concentration of each eluding compound is directly proportional to the area under the recorded peak. At one time, chromatographers had to cut out each peak on the chromatogram, weigh the paper and assume the mass to be proportional to the compound's concentration. However, there are several alternative methods currently being used to determine this area response: peak height, height times one-half base width, height times half-width, planimeter, electronic integrators, and computers. A discussion and summary of the precision of these various integration techniques is given in Table 26.9.

TABLE 26.9 Precision of Various Integration Techniques

Type	Precision
Electronic integrator/computer	0.5% or better
Height times half-width	3%
Triangulation	4%
Planimeter	4%
Peak height	5%
Cut and weigh	5%

A. Cut and Weigh

This integration technique is obviously one of the more tedious and time-consuming integration techniques. Determining the respective masses of the actual chromatographic peaks can be determined quite accurately from the recorder paper. The integrated area in the chromatographic peak should be proportional to the detector response to that component. The real disadvantages are (1) the accuracy with which the technician can cut out the peak area(s) is questionable and (2) the chromatographic record is destroyed in the process.

B. Peak Height

This integration procedure is the quickest and least accurate of these quantifying techniques. The assumption is made that if a compound's concentration is proportional to its area response, then its peak height should be directly proportional to the compound's concentration. Peak-height measurements are more prone to errors from GC temperature and flow variations and injection techniques than are the area-measuring techniques. Capillary chromatography lends itself to peak-height integration techniques since the peaks are very narrow and closely spaced; however, area determinations are generally the preferred method.

C. Triangulation

This integration technique assumes that the GC peak is a triangle and the area can be calculated using the formula:

$$\text{Area} = \tfrac{1}{2} \text{ base} \times \text{height}$$

The disadvantage with this technique is that it requires that all chromatographic peaks return to the baseline for accurate measurement. Many chromatograms will not provide this type of ideal peak separation. This triangulation technique is shown in Fig. 26.19. The *half-width method* is a slight modification of this technique in that the peak height is measured and ½ base width is assumed to be the width at half this measured peak height. This technique is demonstrated in Fig. 26.20. The advantage with this modification comes from eliminating the necessity to measure the rounded corners of the triangle at the baseline.

D. Planimeter

This is a mechanical device used to trace out the perimeter of geometric shapes or peaks on the chromatogram and, by conversion factor, converting this distance to area. Figure 26.21 explains

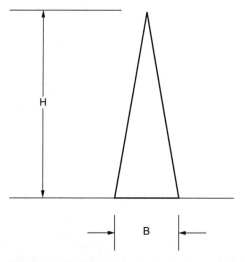

FIGURE 26.19 Triangulation method using one-half base (B) times height (H) method.

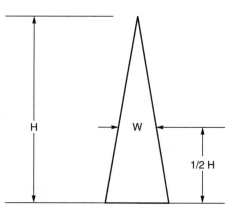

FIGURE 26.20 Triangulation technique by width at half height (W) times height (H) method.

the procedure for integrating an area with a planimeter. A weighted reference point is located at point A so that stylus C can trace the entire perimeter of the chromatographic peak after the base has been enclosed by a straight line. Start stylus C at any convenient point on the peak, but preferably at a corner such as point D. A wheel mechanism located at point B follows the motion at the apex (also point B). A reading is taken from the wheel mechanism before moving the stylus from point D, then the GC peak and straight-line base are traced completely around until point D is again reached. The change in the wheel mechanism readings is calculated, which gives the perimeter distance (area enclosed) in arbitrary units. A conversion factor can be calculated by tracing a square of known area and recording the equivalent arbitrary units from the planimeter. The actual area can then be calculated by multiplying the arbitrary number determined from a GC peak by the conversion factor. In most GC integration work, it is not necessary to convert to actual area since relative ratios between arbitrary numbers comparing standards with unknowns works just as well.

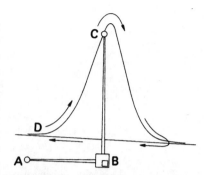

FIGURE 26.21 Planimeter integration calculation. (*Courtesy American Chemical Society, Modern Chemical Technology Series.*)

E. Electronic Integrator/Computers

These devices measure the area under each peak automatically, calculate the relative areas, and print the calculated results with the retention time for each peak to the nearest thousandth of a minute, if desired. This technique is by far the most accurate integration technique, but these integrators cost several thousand dollars. These integrators can be calibrated by internal and external standardization techniques as described below and are used in both GC and LC. A typical electronic integrator display is shown in Fig. 26.22.

QUANTITATIVE ANALYSIS

The various integration techniques listed above simply provide areas or numbers proportional to the area under each GC peak. There are four mathematical procedures for converting area responses into actual concentrations: (1) normalization, (2) external standard, (3) internal standard, and (4) the standard addition method.

A. Normalization

In Fig. 26.23, assume that the total area under all peaks (A, B, and C) represents 100 percent of the sample.

Each peak's area (A, B, and C) represents that compound's fraction of the total composition, and the relative percentage can be calculated using the following equation:

$$\%A = \frac{\text{area of peak A}}{\text{total peak areas}} \times 100$$

The denominator (total peak areas) in this equation is the sum of peak areas of compounds A, B, and C. There are two assumptions with this "simple" normalization method: (1) all compounds were eluted and (2) all compounds have the same detector response. By design, the percent of all components should add up (normalize) to 100 percent.

544 CHAPTER TWENTY-SIX

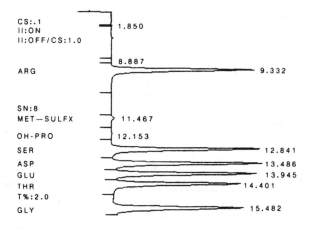

FIGURE 26.22 Typical electronic integrator display. (*Courtesy Varian Associates, Inc.*)

If **response factors** are known for each compound, then each peak area can be corrected before using the normalization formula. Absolute response factors can be determined by plotting a calibration curve with a minimum of three concentrations for each of the compounds expressed in mass and plotted against the corresponding peak areas as shown in Fig. 26.24.

The best straight line is then selected through the data points. The **slope** of this curve (mass/area) is the absolute response factor for that particular compound. The corrected area for each compound must now be used in the normalization formula by multiplying the peak area of each compound by its absolute response factor.

$$\%A = \frac{\text{area of peak A} \times \text{response factor of A}}{\text{total corrected peak areas}} \times 100$$

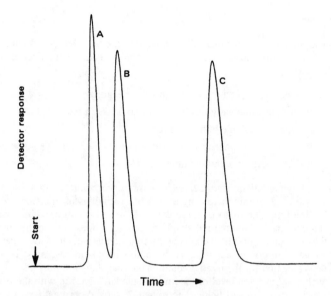

FIGURE 26.23 Three-component mixture separation.

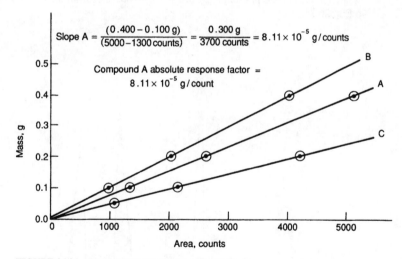

FIGURE 26.24 Absolute response-factor calibration curve.

B. External Standard

This technique uses the exact procedure described above for determining the absolute response factor for each compound of interest. The advantage of this technique is that calibration plots are necessary for only compounds of interest and it is not necessary to resolve or to determine response factors for the remaining peaks. The main disadvantage to this external standard calibration technique is that the sample injection size must be very reproducible.

C. Internal Standard

This method requires that all samples and standards be spiked with a fixed amount of a substance called the **internal standard**. The substance selected as the internal standard must meet the following criteria:

1. It must not be present in the sample to be analyzed.
2. It must be completely resolved from all peaks in the sample.
3. It must be available in high purity.
4. It is added at a concentration level similar to the unknown compound(s).
5. It does not react with the unknown mixture.

This internal standard serves as a reference point for all subsequent peak-area measurements. This procedure minimizes any error from variations in injection size or GC parameters. The calibration procedure is to measure the peak areas for both the compound of interest and the internal standard peak. The internal standard technique does not require that all peaks be measured, only the compounds of interest and the internal standard itself. As was described with external standards above, a response factor for each compound of interest must be determined and an additional determination for the fixed amount of internal standard must also be included. Chromatograms for each concentration level of the known standards are run with the peak areas for both the standard and internal standard being determined. A typical chromatogram with internal standard is shown in Fig. 26.25.

The ratio of peak area for each compound divided by the peak area for the internal standard is calculated. This ratio is plotted against the known concentrations of each compound as is shown in Fig. 26.26.

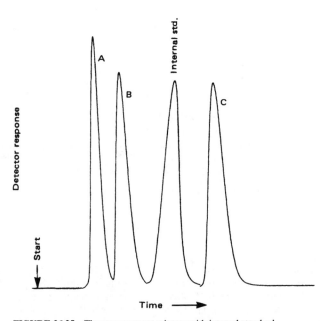

FIGURE 26.25 Three-component mixture with internal standard.

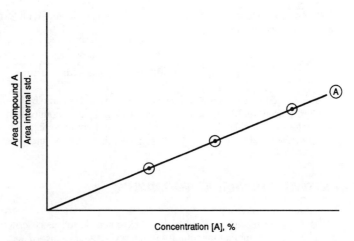

FIGURE 26.26 Calibration curve with ratio of standard to internal-standard response versus concentration.

D. Standard Addition

This method actually combines both external and internal standardization techniques and is used in gas chromatographic work only when the sample matrix has many interferences. A standard amount of one of the compounds of interest is spiked into the sample. This standard-addition technique is performed ideally by adding one-half the original concentration and twice the concentration. The resulting peak areas are measured and plotted versus the concentration added. The unknown's original concentration is then determined graphically by extrapolating the curve back to the x axis (zero peak size) as is shown in Fig. 26.27.

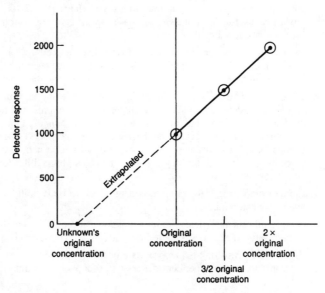

FIGURE 26.27 Standard-addition calibration graph.

A more mathematical calculation simply uses the following ratio formula, where X represents the unknown concentration:

$$\frac{X}{X + \text{addition}} = \frac{\text{area original}}{\text{area with addition}}$$

This procedure as described does not compensate for sample size variations as did the internal-standard procedure. In addition, if the standard addition amounts are kept small with respect to the total volume of sample, then dilution effects should not be a major source of error; otherwise it will be necessary to dilute all samples to the same volume before analysis.

TROUBLESHOOTING GAS CHROMATOGRAPHY

1. Before any testing is done, a set of expected performance criteria must be established. The instrument manufacturer often will supply standard performance benchmarks along with the conditions under which they were measured. Usually a test column and a test mixture are involved. Detector sensitivity, noise, minimum detectable quantity, peak resolution, and other general measurements can be specified. These criteria should be determined by the user when the instrument is installed and periodically thereafter even though the test may not be directly related to the intended application.

2. Instrument manuals usually include a troubleshooting section. This section should be read both as a reference and to learn how to operate the instrument properly.

3. Establish performance standards for the applications for which the instrument was purchased. Then check these performance standards regularly. Perform frequent and regular calibration of detector response factors along with visual assessment of the calibration chromatogram. Control samples interspersed with the application samples will enable the user to detect decreasing detector response, gradually broadening peaks, or drifting retention times.

4. Anticipate trouble. Keep on hand a good supply of nuts, ferrules, and fittings as well as inlet and detector spares such as septa, inlet liners, and flame jets. Extra syringes and autosampler parts are always useful. A set of tools purchased specifically for your instrument is a good investment. Store tools safely in a drawer.

5. Perform the recommended periodic maintenance procedures on schedule. An instrument log is an essential tool for problem diagnosis. This includes everything done with the instrument and auxiliary units. Keep notes on peculiarities that occur—they may warn of upcoming trouble.

6. *General troubleshooting:* If the problem appears to be chromatographic in origin, check the various inlet, detector, and column temperatures, pressures, and flows and compare them to their normal values. If possible, measure some basic chromatographic parameters such as unretained peak times, capacity factors, or theoretical plate numbers. Check for loose column or tubing connections. If the problem appears to be electronic in nature, carefully inspect the instrument (with the power cord unplugged). Are the cables plugged in? Are boards seated firmly in their connectors?

7. *Column problems:* If the column is determined to be at fault, consider the following problems and suggested remedies.

    (a) A duplicate analytical column with known acceptable separation capacity would be extremely useful. If the poor resolution is corrected, then the problem is with the original column.

    (b) If solvent flushing is inappropriate or fails, remove the first two or three column turns of a capillary or the upper centimeter or so of a packed column.

(c) Excessive column heating may cause the stationary film to coalesce into uneven hills and valleys on the inner column wall. Usually a replacement column is needed.

(d) Microcontaminants may gradually deposit at the beginning of the column and interact with solutes, leading to adsorption and possibly to catalytic decomposition or rearrangement. Removing the beginning column section is an effective remedy. A guard column in front of the regular column is another solution.

(e) Oxygen, moisture, or hydrocarbon contamination in the carrier gas can cause excessive column bleed or drift. The remedy is to install a fresh, high-quality-oxygen, molecular sieve, and/or charcoal trap(s).

(f) Loss of resolution due to band-broadening effects may have been caused by large solvent injections displacing the stationary phase. Although bonded phase columns are relatively immune to this problem, back-flushing such columns with solvent is recommended for restoration. (Do not back-flush nonbonded columns.)

(g) If large peaks from previous injections are eluted as broad peaks during a subsequent run, a thinner stationary phase film should be used if higher column temperatures are not possible.

8. *Injector problems:* Septum bleeding, leaking, residues or high-boiling, strongly retentive solutes from a previous run, may raise the baseline. The use of low-quality, high-bleed septa can interfere with sample analyses, especially on very sensitive detectors such as ECD, NPD, and MSD. Try several bake-out temperature cycles with no sample injection and check for a stable baseline. Remember that special, high-temperature septa are commercially available. Snoop® or other leak-detection aids can be used to quickly identify leaks. Great care should be taken that none of this liquid enters and contaminates the chromatographic system.

NOTE: A leaking septum is a leading cause of baseline drift.

Problems with injection techniques:

(a) The volume of the sample is too large, which causes the original bandwidth of the sample "plug" to be excessive. Reduce the sample size; as a rule, the smaller the sample size the better the resolution.

(b) The sample injection is not carried out rapidly enough, thus the width of the sample vapor plug is increased. Increase the temperature on the injector.

(c) The (void) volume between the point of injection and the entrance to the stationary phase is too large, which causes excessive band spreading as a result of diffusion of the sample vapor into the carrier gas. Use inserts inside the injector to minimize the dead volume.

(d) The tubing connecting the column outlet and the detector must be kept to an absolute minimum because diffusion in the carrier gas results in the remixing of the separated components. Addition of makeup gas to the column effluent will speed up the gas velocity and sweep the column effluent through this "dead" or void volume. All couplings should possess zero dead volume.

(e) A new syringe can be used to help isolate or identify the cause of ghost peaks.

9. *Dirty detector:* A dirty or noisy detector, especially FID, can be cleaned in place with special cleaning sprays and/or solvents in an ultrasonic bath.

10. Temperature checks can be made with a simple thermometer or thermocouple device to determine if any temperature controls are defective.

If the problem persists and is electronic, it is time to call the manufacturer's service technician.

LIQUID CHROMATOGRAPHY

Liquid chromatography (LC) is a separation technique which uses two phases in contact with each other; the stationary phase can be an immiscible liquid or solid, but the mobile phase must be a liquid. Technically, ion-exchange chromatography, gel-permeation chromatography, supercritical-fluid chromatography, electrophoresis, thin-layer chromatography, and even paper chromatography can be included as liquid chromatographic systems because they all use liquid mobile phases. The first liquid chromatographic columns were glass tubes packed with calcium carbonate. Reliable liquid pumps were not available; thus gravity was used to cause the mobile phase to flow. A typical liquid chromatography column is shown in Fig. 26.28.

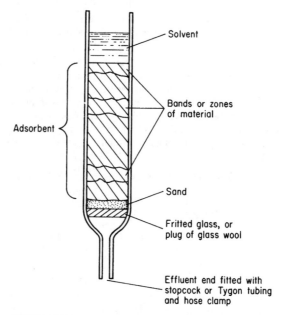

FIGURE 26.28 A column for liquid chromatography.

The fastest-growing field of liquid chromatography is high-performance liquid chromatography (HPLC). Conventional liquid chromatography is a very slow separation technique compared to HPLC, with the faster flow rates allowed by high-pressure pumps. In addition, the smaller packing particles used in HPLC yield higher resolution than found with conventional column chromatographic systems. The number of HPLC instruments sold annually is now greater than that of GC. In GC, liquid and solid stationary phases are used, but the mobile phase is a gas. In both GC and HPLC, the sample is introduced as a liquid into the mobile phase, but for HPLC analysis the sample does not have to be volatile as was necessary in gas chromatography. Therefore, liquid chromatography is very useful for analyzing mixtures of nonvolatile and thermally labile compounds. Liquid chromatography can be conveniently divided into three classifications: liquid/solid chromatography (LSC), liquid/liquid chromatography (LLC), and high-pressure liquid chromatography (HPLC). The basic components of a typical liquid chromatographic system are shown in Fig. 26.29, and a typical liquid chromatogram is shown in Fig. 26.30.

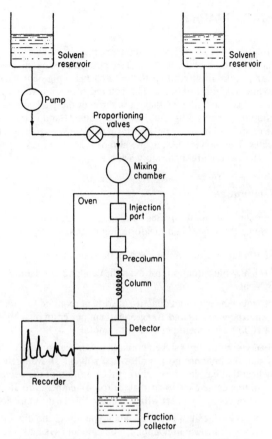

FIGURE 26.29 Basic components of a typical liquid chromatography system. (*Courtesy* American Laboratory, *vol. 8, no. 5, May 1976, p. 14. Reproduced by permission of International Scientific Communications, Inc.*)

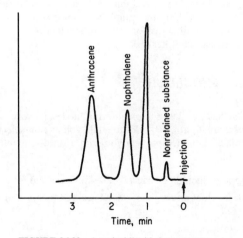

FIGURE 26.30 A typical liquid chromatogram.

COLUMN CHROMATOGRAPHY

The simplest example of LSC is shown in Fig. 26.28 and is represented by column chromatography. This is the type of equipment that Tswett used to separate plant pigments. Tswett coined the word *chromatography* from reading the colored plant-pigment bands through the glass column packed with white calcium carbonate. The colored bands are found not only to separate from each other but to gradually broaden as they pass through the packed column. This phenomena is called **band broadening** or **spreading** and is not desirable. Band broadening cannot be totally eliminated, but can be minimized by using only uniform particle size in the packing material and rapid mobile-phase flow rates. The mobile phase is the **eluant** (or solvent) as it passes through the stationary adsorbent phase. The word **elute** means to wash out.

A. Liquid Column Techniques

Column chromatography is one of the most useful and versatile of the chromatography classifications. Some techniques and tips for using column chromatography follow.

1. Use a column with a minimum length-to-diameter ratio of 20:1.
2. Check any plugs (cotton, for example) under a UV lamp to avoid contamination from an optical brightener.
3. Determine the optimum quantity of sorbent required to effect the separation. Generally adsorption requires between 50 and 100 g of sorbent per gram of sample, and partition requires between 500 and 1000 g of sorbent per gram of sample.
4. When you use a dry-packed column, cover the sorbent with solvent immediately after carefully packing the column; do not allow the solvent to evaporate so that the sorbent dries out. Allow any heat that has developed during the addition of the solvent to the sorbent to dissipate, and let the column come back to normal (or room) temperature. If you desire to shorten the cooling time, the use of cooling jackets will accelerate this heat dissipation.
5. Slurry pack instead of dry pack when you desire the highest resolution in a procedure. Isooctane is well suited for use in nonpolar (nonaqueous) systems. Gently tap or bump the column continuously while you fill it with a slurry of the sorbent in isooctane by pouring in one small portion at a time. This disturbance will free any trapped air bubbles and pack the column more tightly and more uniformly, so that it will yield better results.
6. The way in which the sample is introduced into the system bears directly upon the results. The sample should be introduced uniformly and symmetrically, without disturbing the column sorbent. You can (1) slurry the sample with some sorbent and pour this slurry on top of the column bed, or (2) seat a filter disk or pad of filter paper on top of the column bed and then gently pipette the sample onto the filter disk.

GENERAL OPERATING PROCEDURE

There are almost limitless combinations of modes, solvents, sorbents, and procedures to select from, and yet there are no ironclad rules to guide you in your selection. However, Table 26.10 offers some general guidelines which provide you with a starting point, and the following list* gives hints on LC operations.

1. The septum should be checked daily for leaks and must be changed often.
2. Check the flow rate regularly at a specified pressure to detect buildup of pressure (or decrease of flow).

---

*Used by permission of Gow-Mac Instrument Co.

TABLE 26.10 Suggested Chromatographic Techniques

Categories of samples	Liquid chromatography modes
Positional isomers, moderate-polarity molecules	Liquid-solid
Insect molting steroids	Liquid-solid
Compounds with similar functionality	Liquid-solid or liquid-liquid
Polar and polynuclear aromatics	Liquid-solid
Barbiturates	Liquid-solid
Ionizable species	Ion exchange
Polysulfonated hydroxynaphthalenes	Ion exchange
High-polarity compounds	Liquid-liquid
Metallic chelates	Liquid-liquid
Compounds with differing solubilities	Liquid-liquid
Mixtures of varied sizes of molecules	Gel-permeation
Lubricating oils	Gel-permeation

3. Pressure buildup can be caused by small pieces of septum which become deposited at the head of the column after many injections. To correct this situation, remove a few millimeters of sorbent from the top of the column and repack with new material.
4. Allow sufficient time for the LC system to stabilize after being turned on. Plan ahead.
5. The activity of a solid stationary phase can vary with the purity of the solvents being used and the polarities of the samples. It may be necessary to regenerate the column if it appears to have lost its separating capability.
6. If possible, samples should be dissolved in the liquid mobile phase.
7. Exercise care with flammable and/or toxic solvents.
8. Only high-purity solvents should be used as mobile phases. Some may require distillation prior to use.
9. Try to dissolve samples in the mobile phase or in a less polar solvent than the mobile phase. This technique tends to concentrate the injection on the tip of the column and yields better resolution.
10. When filling the pump, hold the funnel slightly above the opening in the pump; this maneuver allows air to escape from the reservoir.
11. Never remove or loosen the lower ¼-in column fitting; this disturbs the column bed and destroys column efficiency.
12. If the syringe is pushed too far into the column packing, the needle becomes plugged. To clear the needle, hold the syringe with the needle pointed down, allow some solvent to collect around the plunger, and then rapidly remove the plunger, causing a vacuum to form inside the syringe barrel. The vacuum sucks in some of the liquid. If you now replace the plunger, pushing the liquid through the needle, you will force out the plug of packing material.
13. After the standard column has been used for a period of time, its chromatographic properties may change. The column may be restored to its previous activity by pumping through it 50 mL each of ethyl alcohol, acetone, ethyl acetate, chloroform, and hexane. This treatment should leave the column as active as it was when you received it.
14. If you want to change from a hexane mobile phase to water, pump a solvent miscible in both liquids through the system before making the change. This removes all traces of hexane remaining in the system.
15. Stop-flow injections can be made easily by opening the three-way valve, releasing the pressure, and then making the injection and repressurizing the system by turning the three-way valve back to the OPERATE position.
16. Many times it is possible to inject very large samples (100 to 200 mL) in LC when more sensitivity is needed. If the sample is dissolved in a solvent less polar than the mobile phase, even a 1- or 2-L sample is possible with no deleterious effects apparent in the separation.

A. General Precautions

General precautions for solvent compatibility, flow limitations, and general care should be thoroughly read prior to any analysis. Failure to do this can result in misleading analytical information and could terminate the usefulness of the column(s). Careful attention will increase column life and allow you to return to a stored column with a knowledge of the purging steps required prior to analytical use. Always tag a column to indicate the last solvent passed through it. For lengthy storage, refer to the column-maintenance booklet provided with each column.

It is important to filter all solvents that are used in the liquid chromatograph. The liquid chromatograph is a high-precision instrument, and particulate matter (>0.5 μm) could be the source of problems. It is also important to filter all samples before they are injected into the instrument.

B. Sorbents for Column Chromatography

Since the sorbent used in column chromatography is packed in a vertical column, there is no need for any binders such as those used in thin-layer chromatography (TLC). However, the critical parameters for sorbents are particle size and size distribution. These factors directly affect the flow of the solvent, of which the driving force may be either hydrostatic or low pump pressure. A narrow particle-size distribution usually will provide better separation, all other factors remaining the same.

LC STATIONARY PHASES

The simplest liquid chromatographic system is one that contains a stationary phase of a solid adsorbent and a mobile (pure solvent) phase. The most common adsorbents are alumina and silica gel with magnesium silicate, charcoal, calcium carbonate, sucrose, starch, powdered rubber, and powdered cellulose being used less frequently. (See Table 26.11.)

A. Alumina

Aluminum oxide, Al_2O_3, is a very polar adsorbent with a surface area of approximately 150 m²/gram. Alumina normally contains approximately 3 percent water, and the degree of activity can be controlled by its water content. It can be reactivated by dehydrating at 360°C for 5 hours and then allowing the desired moisture content to be readsorbed.

TABLE 26.11 HPLC Column Stationary Phases

Normal-phase packings (polar)	Reverse-phase packings (nonpolar)
Silica Gel, H_2SiO_3	ODS or octadecylsilane
Alumina Al_2O_3	Partisil® ODS-2
Chromegabond®-DIOL	μBondapak®-C18
LiChrosorb®-DIOL	μBondapak®-Phenyl
Nucleosil® NH_2	Spherisorb®-ODS
Zorbax®-NH_2	Hyposil®-SAS
MicroPak®-CN	Nucleosil®-C8
μBondapak®-CN	Supelcosil®-RPLC
Supelcosil® LC-Si	Supelcosil® LC-ABZ

B. Silica Gel

Silicic acid, H_2SiO_3, is a very polar adsorbent with a surface area of approximately 500 m^2/gram, it is less chemically active than alumina and is preferred when dealing with chemically active organic compounds. Silica gel is an acidic compound and very stable in acidic or neutral solvents, but will dissolve in solvent of pH greater than 7.5.

Ordinarily, liquid chromatographic processes use a polar stationary phase such as silica gel or alumina and a nonpolar mobile phase. **Reverse-phase chromatography (RPC)** uses a nonpolar stationary phase and a polar mobile phase. This modified (polar to nonpolar) column requires a polar mobile phase to elute the solute components. Just as in GC, nonpolar compounds are retained by the nonpolar stationary phases and polar compounds are retained by the polar stationary phases. Another popular reverse-phase stationary packing contains phenyl groups, which are more polar than the C18 (see Bonded Stationary Phases below) and have an affinity for double-bonded compounds. The C8 (octyl) and the cyano (—RCN) bonded phases have polarities that are intermediate between C18 and silica gel.

C. Bonded Stationary Phases

The first, commercial liquid chromatographic, nonpolar phases were called C18 or ODS with *o*ctyl *d*ecyl*s*ilane (18 carbons) groups. This stationary phase is prepared by silylation of silica gel to produce a less polar column by reacting the silicic acid stationary phase with chlorodimethylalkylsilane. The very polar sililol (Si—OH) groups on the silicic acid surface are blocked or made sterically inaccessible by large, nonpolar hydrocarbon groups as shown in Fig. 26.31. Normally about 50 percent of the sililol groups remain unreacted in the silica gel and can be further reduced, perhaps to only 20 percent, with additional chemical treatments for specialized applications. Various other nonpolar, bonded columns can be prepared by substituting different alkyl groups: ethyl (C2), hexyl (C6), octyl (C8), and so forth for the octadecane substitution on the monochlorosilane.

$$\text{(Silicic acid)}-Si-OH + Cl-\underset{\underset{CH_3}{|}}{\overset{\overset{CH_3}{|}}{Si}}-(CH_2)_{17}CH_3 \rightarrow \text{(Silicic acid)}-Si-O-\underset{\underset{CH_3}{|}}{\overset{\overset{CH_3}{|}}{Si}}-(CH_2)_{17}CH_3 + HCl$$

FIGURE 26.31 Production of C18 bonding reverse-phase packing using monochlorosilane group to silica gel.

LC MOBILE PHASES

The mobile phase in an LC analysis must be carefully chosen because of the many roles it serves. The mobile-phase liquid not only transports the solute through the stationary phase, but its solvent power is critical for the proper distribution of the solute between the mobile phase and the adsorption sites on the stationary phase. A solvent that readily eludes the solute from the column will not resolve the components of the mixture, and a solvent that eludes the solute too slowly will be too time-consuming. A large number of LC solvents is available and the use of mixed solvents expands the versatility even more.

Solvents should have a viscosity of 0.5 centipoise or less and be "Spectrograde" or "HPLC" grade for best results. The use of only one solvent throughout the entire LC analysis is called **isocratic elution**. If necessary, a series of solvent mixtures called **gradient elution** can be used in LC applications where the polarity of the solvent mixture is gradually changed to effectively resolve components in a mixture. Isocratic and gradient elutions are discussed in more detail under section HPLC Solvent Mixing Systems in this chapter. Gradient systems can be created in an LC system by using two or more pumps to introduce different solvents from separate reservoirs and mixing before introducing the sample. Newer gradient elution systems use a single pump fitted with multiple solvent lines, a controller, solvent proportioning valves, a series of solenoids, and a mixing chamber.

HIGH-PERFORMANCE LIQUID CHROMATOGRAPHY

High-performance liquid chromatography, sometimes called **high-pressure liquid chromatography (HPLC)** is at maximum efficiency at low flow rates because of typically slow diffusion rates between liquid phases. The time required to elute solute through a column can be decreased by using a shorter column, but short columns tend to have less separating ability. Thus, liquid chromatographic systems that use short columns have been developed, but with very small particles of stationary phase to increase the resolution of solute compounds. Particles with diameters of 10 μm and smaller are found to be the most efficient, but tend to be almost impermeable in long columns. Particles of 3-μm diameter are theoretically the optimum size for these columns.

HPLC INSTRUMENT DESIGN AND COMPONENTS

Most HPLC systems have six major components: (1) a solvent delivery system, which includes a reservoir, a proportioning valve, and a pump, (2) a solvent mixing system, (3) a sample injector, (4) a column, (5) a detector, and (6) a recorder/integrator. These components are shown in the schematic given in Fig. 26.32.

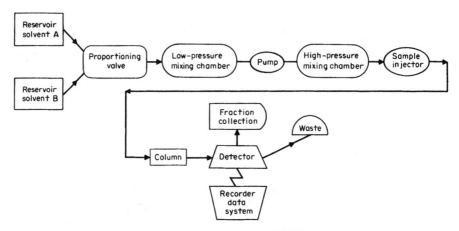

FIGURE 26.32 Schematic showing the components of a basic HPLC system.

More sophisticated HPLC systems might include additional sources of solvent, a gradient pump, proportioning valves, mixing chambers, oven(s), automatic integrator, programmable autosampler, and a computer-controlled workstation. A sophisticated HPLC unit is shown in Fig. 26.33.

FIGURE 26.33 Modern high-performance liquid chromatograph (HPLC) with workstation. (*Model LC-2010 HT, courtesy Shimadzu Scientific Instruments.*)

A. HPLC Solvent Delivery Systems

There are two basic types of high-pressure pumps used in HPLC solvent delivery systems. These pumps should be capable of delivering between 0.5 and 4.0 mL/min of mobile phase at pressures approaching 6000 psi or higher. High-pressure pumps have been perfected within the past few years that can produce an almost pulse-free flow rate of several milliliters per minute. A continuous displacement (syringe-type model) forces solvent from a single reservoir and tends to be very smooth and pulse-free; however, the chromatographic run is limited by the volume of solvent housed in the syringe reservoir or requires frequent shutdown of the system for refilling.

A more complicated delivery system uses the intermittent displacement approach using a single piston. Most of these pumps are of motor-driven, reciprocating, dual-piston design with very small displacement volumes. The pump motor drives a cam that generally operates two pumping chambers as shown in Fig. 26.34a. This type of pump refills itself intermittently, which disrupts the column flow and introduces a fluctuation in the detector baseline. Flow-sensitive detectors such as refractive index or electrochemical detectors may not be suited for this type of pump. One piston is pumping solvent into the system while the other is filling for the next stroke. The two pistons are designed to be 180° out of phase with each other and tend to minimize the pulsation of the solvent flow. This type of pump usually contains a check value on the inlet side and another check value on the outlet. These check valves allow the solvent to flow only in one direction. Figure 26.34b shows a typical double-piston-designed pump.

In addition, some LC manufacturers have further reduced this undesired pulsation by installing a pulse-damping device. A pulse damper can be an air compression baffle or long, narrow-bore coil of tubing between the pump and injector.

As a precautionary step, these HPLC pumps containing potentially corrosive solvents (acidic solvents, aqueous buffer of low pH, halogenated solvents, etc.) should not be allowed to stand for long periods of time. In addition, the entire LC system should be flushed before storage (see Table 26.11) with a nonaqueous solvent such as methanol or isopropanol to minimize bacterial growth.

B. HPLC Solvent Mixing Systems

The mobile phase can be passed through the HPLC system in two very different modes: isocratic and gradient. As described earlier, in isocratic elutions the solvent or solvent mixture remains constant throughout the run, and in gradient elutions the composition of the mobile phase is varied

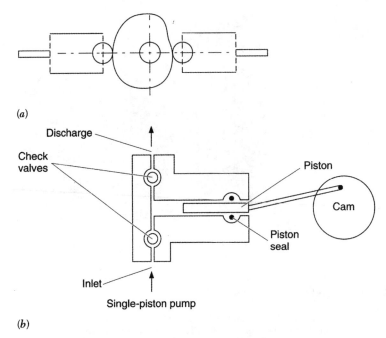

FIGURE 26.34 (*a*) Motor-driven cam with two pumping chambers, (*b*) Intermittent displacement pump. (*Courtesy Thermal Separation Products.*)

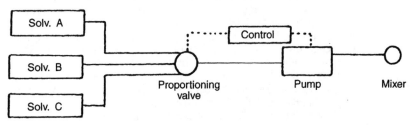

FIGURE 26.35 Low-pressure HPLC mixing system. (*Courtesy Thermal Separation Products.*)

during the course of the chromatographic separation. Varying the solvent composition can greatly improve the selectivity and resolution of sample components, plus it routinely shortens the analysis time. The overall effect of gradient flow in HPLC is similar in effect to that of temperature programming in gas chromatography. Manufacturers produce both low-pressure and high-pressure mixing systems. The pump and the proportioning valve in the mixing system are controlled by a microprocessor. The microprocessor allows the system to provide a very reproducible isocratic or gradient-solvent flow. The low-pressure systems have the mixing chamber located between the solvent reservoir(s) and the pump as shown in Fig. 26.35.

The high-pressure design calls for the mixing chamber to be downstream from the pumps and usually requires a separate pump for each solvent. Most laboratories will use the low-pressure configuration since this mixing system is cheaper, can routinely handle multiple solvents, and can quickly be changed from one solvent to another. The big disadvantage to low-pressure mixing systems is the need to degas solvents on a continuous basis.

C. HPLC Sample Injection

In gas chromatography the sample is introduced into the instrument using a syringe and a self-sealing septum inlet. This technique is not very practical with HPLC because of the high column pressures and the rapid degradation of septum material by the various liquid phases. Most HPLC systems use a fixed-volume **loop injector** system which consists of a rotary valve fitted with several ports and loops. Samples are simply purged through a loop (normally 10 µL to 2 mL capacity) with a liquid syringe, and at injection time the valve is rotated, causing the filled sample loop to be an integral part of the mobile-phase flow as shown by the load and inject positions in Fig. 26.36. The entire loop volume of liquid sample is swept away by the liquid (mobile) phase and introduced onto the column.

All solute components must be soluble in the mobile phase. If the sample needs to be dissolved in a solvent, it is best to use the mobile phase as the solvent. If a different solvent is used, it must be tested (before injection) to demonstrate that the solute does not precipitate upon mixing with the mobile phase. This precipitation or immiscibility problem can block the chromatographic system very quickly.

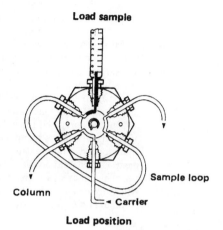

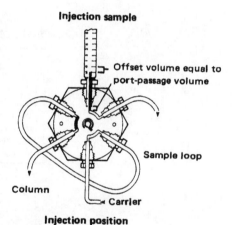

FIGURE 26.36 HPLC sample-injection system shown in load and inject positions. (*Courtesy Varian Associates, Inc.*)

D. HPLC Columns

Liquid chromatographic columns packed with very small beads and/or bonded particles require extremely high pressure to force liquid to flow. Pressures as high as 10,000 psi can be handled with small-diameter (2- to 3-mm) columns, but most analytical LC applications occur at pressures of 4000 psi or less. HPLC columns are made of type-316 stainless steel for higher pressures (less than 10,000 psi) and glass for lower pressures (less than 1500 psi). Titanium metal columns are sometimes recommended for high-pressure HPLC columns where the stainless steel is too reactive with the solvent or sample. Three types of solid supports are found in HPLC columns: solid, pellicular, and porous resin beads. Table 26.11 gives some common normal-phase and reverse-phase column packings.

Pellicular beads are solid glass beads of 20- to 50-µm diameter that are manufactured with a thickness of approximately 1 µm of porous material on the surface. This porous surface is then used to coat a thin layer of liquid phase and to provide a large surface area. Because of their high surface areas, pellicular particles offer very high resolution but have very limited sample-handling capacity due to the thin liquid-phase coating. Most thin films are simply washed away by the mobile phase, thus the chemical bonding of the liquid phase to the solid support has been developed. Bonded columns were discussed earlier and can be exposed directly to flowing solvents without adverse effects. **Porous resin beads** have been developed as a column packing where the stationary support and the surface are composed of the same insoluble material. This type of packing is sometimes called **microporous particles** or **microreticular resins** containing both micro- and macropores.

Analytical LC columns are normally 4.6-mm diameter × 5 to 25 cm long and packed with 5-μm particles with a maximum sample load of 150 μg and flow rates of approximately 1 mL/min. Preparative LC columns are 4.6- to 21-mm diameter × 25 cm long and packed with 40- to 50-μm particles with a maximum sample load of 5 g and flow rates of approximately 30 mL/min.

Most new columns are shipped from the manufacturer with a test chromatogram and test mixture. Column performance should be evaluated upon installation and periodically by analyzing this test sample under the listed instrumental conditions. Manufacturer and actual retention times and operating pressures should be comparable; however, test result differences of as much as 20 percent can be attributed to instrument parameter differences (injector design, detector geometry, connecting tubing, etc.) other than column efficiency.

During storage HPLC columns should be filled with the appropriate solvent and tightly capped. Most manufacturers recommend that the system be flushed with 10 to 20 column volumes of the storage solvent before capping. Columns should always be stored in a vibration-free, cool environment. Care should be taken with certain mobile phases (halogenated hydrocarbons, buffer solutions, etc.) since they can corrode stainless steel components and should not be allowed to stand in the HPLC for extensive periods (over 3 hours) of time without pumping. Table 26.12 provides a general guide to HPLC column storage solvents.

TABLE 26.12 HPLC Column Storage Solvents

HPLC column type	Storage solvent
Reverse phase C1, C2, C3, C6, C8, C18 and phenyl	10% water 90% menthol
Normal phase Silica, alumina, CN, NH$_2$, OH, and diol	100% isopropanol
Normal phase Silica, alumina, CN, NH$_2$, OH, diol, and carbohydrates	100% menthol

Most manufacturers recommend the use of a **precolumn** or a **guard column** to protect the analytical column from strongly retained compounds and/or between the injector and the pump to protect against contaminants in a sample. Cartridge guard columns can be connected directly to the head of an analytical column. These guard columns can be customized to provide effective protection with high-performance 5- to 10-μm packing and built-in 2-μm stainless steel filter elements. Several models of these guard columns are refillable with any chromatographic media and can be connected without tubing or tools. Figure 26.37 shows a typical guard column that connects directly in-line before the analytical column.

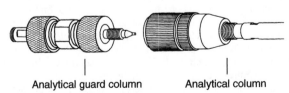

FIGURE 26.37 Refillable analytical guard column. (*Courtesy Alltech Associates, Inc.*)

E. HPLC Detectors

Flow-through detectors with low-dead-volume designs are required for these high-pressure/low-flow-rate systems. There are currently five types of detectors in general use: (1) ultraviolet spectrophotometers, (2) differential refractometers, (3) fluorescence, (4) conductivity, and (5) diode array or multichannel detectors. The **ultraviolet detectors** rely on a mercury lamp and a characteristic wavelength of 254 nm, where most organic compounds with double bonds or aromatic groups tend to absorb. Additional ultraviolet and visible wavelengths of 214, 229, 280, 308, 360, 410, 440, and 550 nm are available. A **chromophore** is defined as a group of atoms that absorb energy in the near-ultraviolet region (400–190 nm). Certain solvents are strong UV absorbers and cannot be used with an ultraviolet detector because they would cut off any detector response. Only solvents with cutoffs below the detector absorption wavelength can be used. Table 26.13 lists some useful HPLC solvents along with their boiling points, refractive indexes, and their respective UV transmittance cutoffs.

TABLE 26.13 Common HPLC Solvents with Boiling Points, Refractive Indexes, and UV Transmittance Cutoffs

Solvent	UV cutoff (max), nm	Refractive index at 20°C	Boiling point, °C
Acetone	330	1.358	56–57
Acetonitrile-UV	190	1.344	81–82
Chloroform*	245	1.445	60–61
Ethyl acetate	256	1.372	77–78
Heptane	200	1.387	98–99
Hexane-UV	195	1.375	68–69
Methanol	205	1.328	64–65
Methylene chloride	233	1.424	40–41
Propanol-2	205	1.376	82–83
Tetrahydrofuran-UV	212	1.407	66–67
Toluene	284	1.496	110–111
2,2,4 Trimethylpentane	215	1.391	99–100

*Contains hydrocarbon preservative
SOURCE: Varian Associates, Inc.

The **differential refractometer** monitors the refractive index difference between pure mobile phase entering the column and the column effluent. For additional discussion of this principle, review the section on refractive index in Chapter 19, "Physical Properties and Determinations." This type of detector is almost universal since all substances have their own unique refractive index, but it is not well suited to gradient liquid chromatography because the baseline (reference) will always be shifting. Refractometer detectors tend not to be very sensitive and are subject to drift with even minor temperature fluctuations.

Some commercially available are HPLCs equipped with detectors based on effluent fluorescence. These are very sensitive, but selective. As the fluorescing solutes elute from the column, a photocell measures the light intensity and generates a proportional electrical signal. HPLC systems utilizing ion-exchange columns and conductivity detectors for measuring the column effluent have recently been perfected. The more expensive HPLC systems are equipped with diode-array or multichannel detectors for increased selectivity and sensitivity.

F. Recorder/Integrator

The same recording/integrating systems described with gas chromatography are applicable with HPLC. In addition, computer-controlled automatic injection systems, multipump gradient controllers, sample fraction collectors, and other devices can be added to increase instrument efficiency.

MOBILE-PHASE PREPARATION AND STORAGE

One of the most common problems in HPLC applications is gas bubbles in the system, especially the detector. There are two main reasons for the gas bubbles: degassing of the mobile phase and leaks in the HPLC system's fittings. The mobile phase can degas as it leaves the relatively high pressure of the column and enters the low-pressure region of the detector. Some manufacturers recommend that back-pressure regulators be used at the detector's waste outlet. These regulators typically control the pressure at between 50 and 75 psi to prevent atmospheric pressure conditions and reduce the possibility of bubble formation. Leaks can actually cause air to be drawn into the mobile phase. Most manufacturers recommend that the mobile-phase solvents be continuously degassed by helium sparging or a vacuum membrane degasser system.

Particulate matter in the mobile phase and/or samples can create another set of problems. Particles in the mobile phase as small as 5×10^{-6} cm in diameter can damage or plug the column or cause excessively high operating pressures. It is strongly recommended that all solvent systems be filtered and degassed before starting an HPLC analysis. In addition, all samples, if possible, should be pre-filtered before injection into an HPLC system.

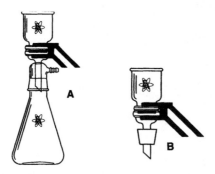

FIGURE 26.38 A typical HPLC vacuum filtration apparatus. (*Courtesy Supelco, Inc.*)

A vacuum apparatus as shown in Fig. 26.38 works well on removing particulate matter and degassing solvents at the same time. An aluminum, spring-action clamp is normally used to hold the filter membrane between the flanges of the funnel top and the vacuum funnel base. Filter membranes composed of Nylon 66®, Teflon® or polytetrafluoroethylene (PTFE), cellulose, and other materials for the 25-, 47-, and 90-mm filter apparatus are commercially available. Most mobile-phase filters have between a 0.20- and 0.45-μm pore size. Nylon 66® is compatible with all solvents commonly used in HPLC. Paper filters should not be used for aqueous solutions. These filter membranes must be wetted first with sample or solvent to filter properly. It is recommended that this type of filtration apparatus be used only with a water aspiration line or controlled vacuum line. Controlled vacuum is necessary to prevent the boiling of various mobile phases.

Because of the clogging and contamination potential of the pump, column packing, column frits, and so forth, solvent filters are highly recommended for all HPLC systems. In general, 2- to 5-μm range in-line filters between the pump and injector and a 10-μm slip-on filter in each solvent reservoir would be required. Water and buffer solution should be passed through a 0.2-μm filter prior to any HPLC use. These filters can be placed on the inlet of the solvent reservoir and/or near the pump inlet (see Fig. 26.39). These slip-on mobile-phase filters are normally made of 316 stainless with a porosity range of 2 to 10 μm and can be quickly connected to tubing ranging in sizes from 1.5 to 3.0 mm. Stainless steel should not be used with halogenated solvents, as corrosion might occur. Mobile-phase filters are also available with porous polypropylene filters and PEEK (polyether ether ketone—thermoplastic material) fittings for metal-free (ion chromatography and biochemical) applications. Titanium filters offer another alternative for IC and biochemical analyses since they do not leach ions into the mobile phase.

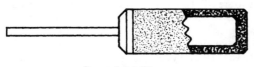

Pump inlet filter

FIGURE 26.39 Cutaway view of typical mobile-phase filter.

Most manufacturers recommend replacing these in-line filters periodically (every 6 months). If cleaning is necessary follow these four steps:

1. First sonicate the contaminated filter for at least 10 minutes in 50 percent nitric acid.
2. Next sonicate the filter for at least 30 minutes with HPLC-grade water.

3. Next sonicate the filter for an additional 30 minutes in fresh deionized water.
4. Finally sonicate for approximately 5 minutes in HPLC-grade methanol.

Degassing is the elimination of oxygen and other gases dissolved in HPLC solvents. This not only prevents the formation of gas bubbles in the system, but oxygen can affect UV absorption in the detector at wavelengths below 230 nm. This oxygen contamination can cause baseline drift and high detector noise levels. Degassing can be accomplished by several techniques:

- Purging the solvent with a steady stream of pure nitrogen or helium.
- Applying a vacuum to the solvent and drawing the liquid through a filter (thin-walled tubular) membrane.
- Heating the solvent under reduced pressure to minimize the quantity of dissolved gases.
- Exposing the untreated solvent to ultrasonification in a vacuum environment.

QUALITATIVE AND QUANTITATIVE HPLC TECHNIQUES

The same separation evaluations, spiking techniques, integrating procedures, and calculation methods described in the Gas Chromatography section in this chapter are used in HPLC.

HPLC PUMP TROUBLESHOOTING

A. No or Low Pump Pressure

Air or Gas Bubbles. The most common problem is air bubbles trapped in the detector cell. This is caused primarily by two factors: leaks in the HPLC system fittings or degassing of the mobile phase as it exits the high pressure of the column and enters the lower pressure (approaching atmospheric pressure) of the detector.

Contaminated Check Valves. Check valves are very sensitive to particulate contamination. One easy step to investigate a check-valve problem is to substitute a good valve for the one in question. It is recommended that at least two discharge and two suction valves be maintained as spares. Check valves can be cleaned by leaving them in place in the pump and carefully passing 30 to 50 percent nitric acid through them for at least 10 minutes. Follow the nitric acid wash with filtered deionized water for at least 30 minutes. The sonic cleaning of check valves equipped with ruby balls is not recommended because of possible damage. One last caution: do not confuse the inlet and outlet check valves (always mark them accordingly), as the pump will not function.

Broken Piston Plungers. Piston plungers are usually made of sapphire or borosilicate glass and can be easily broken by operator error. Applying too much torque to the pump head while tightening check valves or fittings is the most common cause of malfunction. A simple visual inspection should indicate any grooves, dirt, or scratches.

Leaking Seals. The pump pistons operate through high-pressure seals. The seals do not prevent all liquid from passing and some are designed to wet the surface as a lubricant. All seals eventually wear out and must be replaced. Replace the *entire* seal unit: O-ring, bushing, and seal backup ring. Immediately discard old seals and do not attempt to salvage or save any of these parts.

B. High Pump Pressure

In the majority of cases this problem is caused by particulate buildup somewhere in the system. Check guard column, filter samples, and mobile phases, and inspect in-line filters.

AUTOMATED HPLC SYSTEMS

Powerful commercial software is available for fully automating an HPLC system with user-definable parameters for measuring system performance and suitability, including a minimum and maximum relative standard deviation for each parameter. Retries can be prescribed for system failures or noncompliance, and the user preselects the cause of action if the system continues to fail. System suitability can provide extensive high-resolution, color graphics of each peak and can use multiple colors to show peaks at 50, 10, 5 percent, and baseline. The software can automatically calculate resolution, theoretical plates, USP tailing factor, and other measurements to verify system suitability. Some software provides automatic peak purity, graphically displaying the purity value across the entire peak, thus allowing the user to see immediately if the peak is pure or where an impurity is eluting. A peak-purity report shows the calculated peak-purity value. Some units provide automatic library searching, permitting unattended identification of compounds such as polynuclear aromatic hydrocarbons in the air and in crude oils, pesticides in drinking water, antibiotics in foods, drug metabolites, and naturally occurring pharmaceutically active compounds.

ION-EXCHANGE CHROMATOGRAPHY

A. Introduction

Ion-exchange chromatography (IC) is a modern-day example of adsorption chromatography and can be classified as a form of liquid chromatography. As the name implies, this technique deals with the separation of ionic mixtures. Many naturally occurring inorganic substances (clays, zeolites, etc.) have a strong attraction for certain ions in solution. In addition, stationary phases composed of solid polymeric material of styrene cross-linked with divinylbenzene in "bead" form having anion or cation sites on the surface have been developed. An acidic functional group, such as sulfonic acid (SO_3H), hydroxyl (OH), thiol (SH), or carboxylic acid (COOH), is bonded to the polymeric material, and the acidic hydrogen group tends to dissociate, leaving a negative site on the bead's surface to attract other cations. Similarly, basic functional groups such as secondary amines (R_2N), primary amines (NH_2), and other nitrogen-containing groups tend to convert to quaternary ammonium groups with a positive charge on the polymer bead and attract anions. These cation and anion functional groups are classified as strong or weak depending on their tendency to dissociate. Ion exchange resins are ridged polymer beads bonded or cross-linked with divinylbenzene and contain exchange sites as shown in Fig. 26.40. The sites may be strong or weak acid, strong or weak bases, or a combination of both acidic or basic groups. The resins are selected for the ion process to be used. The resin properties to be considered are cross-linkage, particle size, exchange groups, and ion-exchange capacity. The cross-linkage affects a resin's properties by making it more or less dense. The more cross-linkage there is, the tighter the gaps between the exchange sites. Therefore, ions larger in size are excluded, but at the same time more exchange sites are available. With less cross-linkage the opposite is true. Smaller-sized ions will be allowed to flow through the resins, and there will be fewer exchange sites. Particle size (mesh) will affect both the amount of solution and the amount of ions that the resin can handle. The larger the mesh, the faster the solution can flow through the column, and the fewer exchange sites the column can handle. Conversely, the smaller the mesh, the less solution flow the column can handle, but a greater number of sites exist.

Ion exchange takes place when ions in a mobile phase are exchanged with ions of opposite charge in a stationary phase. As previously discussed, there are two types of stationary phases: **anion-exchange**, and **cation-exchange**. In an anion-exchange, cationic exchange sites are attached to the stationary resin. In cation-exchange, anionic exchange sites are attached to the resin. Since different ions each have a different equilibrium quotient, each will stay at the exchange sites for a different length of time. Eventually the differences in equilibrium will cause each of the ionic solutions to elute at different times, resulting in separation.

The **exchange capacity** of an ion-exchange resin can be defined as the total number of replaceable ions per unit of volume expressed in milliequivalents per milliliter. The exchange capacity can be

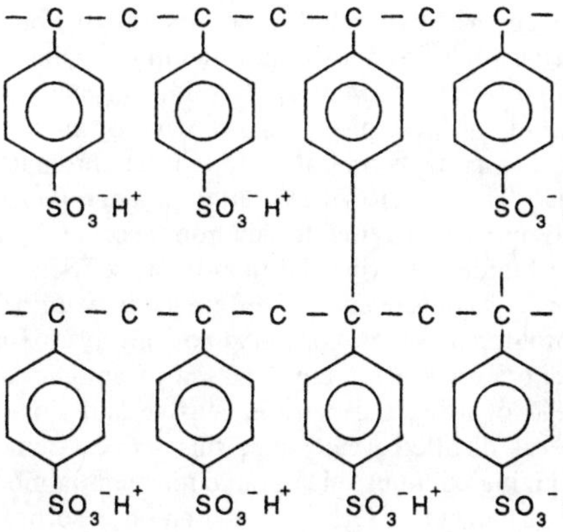

FIGURE 26.40 Typical ion-exchange resins.

determined by: (1) weighing a quantity of the dried resin (R—H) in the H^+ form; (2) adding deionized water to hydrate (swell) the resin; (3) pouring the hydrate resin into a column (note that if dried resin is first added to a column and then water is added, it possibly could break the column); (4) adding a measured concentration of cations (for example, K^+) causing the H^+ ions to be quantitatively displaced; and (5) determining the milliequivalents of H^+ by titration with a known concentration of base. The following equation represents the exchange equilibrium reaction:

$$R\text{—}H + K^+ + H_2O \leftrightarrow R\text{—}K + H_3O^+$$

The used column can now be regenerated by washing the column resin with concentrated HCl, which reverses the equilibrium by flushing off the K^+ ions and restoring the H^+ groups.

Table 26.14 gives the names of some common ion-exchange resins and their chemical properties. Water is normally used with these cationic and anionic exchange resins as the mobile phase, but changes in pH may be required to flush various ions from the beads selectively.

Most ion chromatographic systems consist of two liquid reservoirs, two pumps, a mixer, a pulse damper, a sample introduction system, an analytical column, a suppressor column, and a

TABLE 26.14 Names and Properties of Some Common Ion-Exchange Resins

Resin composition	Type of exchange resin	Resin common name	Exchange capacity
Polystyrene-divinylbenzene with sulfonic acid group	Strong Cation	Amberlite IR-200 Aminex A-7 Dowex 50 Ionac C-242	1.7–2.0 meq/mL
Methacylic acid with carboxylic acid group	Weak Cation	Amberlite IRC-50	3.5–4.0 meq/mL
Polystyrene-divinylbenzene with quaternary ammonium group	Strong Anion	Amberlite IRA-900 Aminex A-28 Dowex 1	1.0–1.4 meq/mL
Polystyrene-divinylbenzene with polyamine	Weak Anion	Amberlite IRA-93 Dowex 3	1.3–2.0

detector system. The two liquid reservoirs, the two pumps, and the mixer are provided in order to be able to change the composition of the mobile phase. Chromatography basically involves separation due to differences in the equilibrium distribution of sample components between two different phases. One of these phases is a moving or mobile phase and the other is a stationary phase. The sample components migrate through the chromatographic system only when they are in the mobile phase. The velocity of migration of a component is a function of the equilibrium distribution. The components having distribution favoring the stationary phase migrate slower than those having distributions favoring the mobile phase. Separation then results from different velocities of migration as a consequence of difference in equilibrium distributions. The mobile phase for all ion chromatographic separations is almost always pH-buffered water solutions.

Many laboratories use a combination of anionic and cationic resins to purify (deionize) tap water instead of using the more expensive and slower distillation process. Deionized water is not as pure as distilled water since this process removes only cation and anion contaminants. These deionizing columns release an equivalent number of H^+ ions for each cation adsorbed and an equivalent number of OH^- for each anion adsorbed. Thus, a molecule of water is released or substituted for each equivalent of cation and anion contaminants retained on the column. Since ion-exchange resins attract only charged particles, few organic contaminants are removed by this process. Figure 26.41 shows a typical cation exchange separation. Figure 26.42 is typical for an anion exchange application.

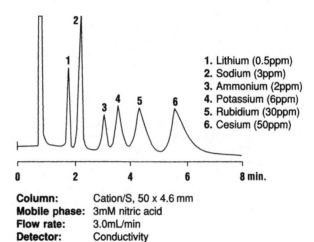

Column: Cation/S, 50 x 4.6 mm
Mobile phase: 3mM nitric acid
Flow rate: 3.0mL/min
Detector: Conductivity

FIGURE 26.41 Typical cation ion-exchange chromatogram. (*Courtesy Alltech Associates, Inc.*)

B. Preparing an Ion-Exchange Column

The following list gives information that might be found on a typical ion-exchange resin bottle label and an interpretation of each item.

Strong-acid resin	Contains acid functional group such as R—SO_3H or R—COOH, that dissociates readily in water.
Sodium form	The exchangeable ion is Na^+. In other words the resin exists as R—Na.
20–60 mesh	The bead spheres are 20 to 60 particles per inch.
10X	The polymer is cross-linked with 10 percent divinylbenzene.
Capacity 2.0 meq/g	The resin has an exchange capacity of ions up to an equivalent of 2.0 milliequivalents per gram of dried resin.

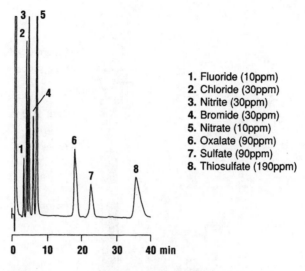

FIGURE 26.42 Typical anion ion-exchange chromatogram. (*Courtesy Alltech Associates, Inc.*)

Fresh ion-exchange resins should be, respectively, washed with 2 *M* HCl, rinsed with water, washed with 2 *M* NaOH, and again rinsed with water until the wash solution is neutral and salt-free. Ion-exchange resins labeled "Analytical Grade" have already been treated.

1. Soak a calculated quantity (based on equivalents of ions to be exchanged) of the freshly washed resin in water for at least 2 hours in a large beaker. Resins with greater *X* values require less time for soaking. More highly cross-linked (for example, 10*X*) resins swell less than those with lower *X* values (for example, 2*X*).
2. Using a conventional liquid chromatographic column, place a glass-wool plug in the bottom, and fill it half full with water.
3. With the aid of a powder funnel, transfer the soaked resin into the column and drain the excess water without allowing any of the resin to "go dry."
4. Back-flush the packed resin column with a stream of water to remove any air bubbles, and then allow the resin to settle.
5. Open the stopcock at the bottom of the column and, using a graduated cylinder, determine the flow rate. Never at any time allow the resin to go dry as "channeling" can occur, drastically reducing the efficiency of the packing.
6. The column is now ready for use. A separatory funnel is a convenient device for adding sample at a controlled rate to the ion-exchange column (see Fig. 26.43).

One common IC packing material is 10-μm particles of styrene-divinylbenzene containing very polar functional groups to selectively remove cations or anions. A second common packing is silica gel that has been modified chemically to produce an ion exchanger. These silica-based packings can be prepared from 5- to 10-μm porous silica spheres; they have mechanical properties similar to

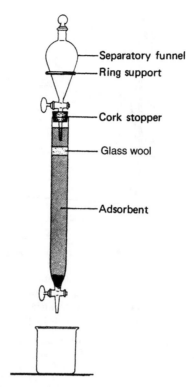

FIGURE 26.43 Conventional ion-exchange column with separatory funnel.

those of the silica liquid chromatographic packings. These two types of packing have advantages and disadvantages. The styrene-divinylbenzene polymer bead exchangers can be used at almost any pH, whereas the silica packings are restricted to pHs between 2 and 8. The silica-based packings are not as prone to dimensional changes with variations in ionic strength and solvent composition as the polymer beads, but they are sensitive to mechanical shock.

A second ion-exchange column sometimes used with IC is called a **suppressor column**. The purpose of the suppressor column is to decrease the concentration of conductive species in the mobile phase. Many ion chromatographic processes utilize solution of high ionic strength as the mobile phase; thus conductivity detectors cannot distinguish the analyte ions. The suppressor column packed with a second ion-exchange resin effectively converts the ions of the mobile phase to a molecular species of limited ionization without affecting the analyte ions. The suppressor column is normally very short in length and is located immediately after the analytical column. Currently, a "mixed bed" of ion-exchange resins is widely used to quantitatively remove all cations and anions while converting them to water molecules. In ion chromatography, the suppressor column removes all mobile-phase ions but does not affect the analyte ions.

C. IC Detectors

The detectors used in IC are the common ones used in most conventional liquid chromatography (UV absorption, diode-array, fluorescence, and refractive index) plus electrical conductivity and amperometric detectors. The typical electrical conductivity detector that is used in ion chromatography consists of a very small tubular flow cell. The conductivity flow cell is usually powered with 3 volts or less and the conductivity of the solution passing through the cell is found by measuring the resulting current. All ions will increase the conductivity of a solution, and the absence of ions, as in deionized water, will not readily conduct a current. The current flow in the detector is directly proportional to the ion strength or concentration. These detectors frequently have circuitry that compensates for small temperature changes. They also have electronic provisions for subtracting the background conductivity of the mobile phase.

SUPERCRITICAL-FLUID CHROMATOGRAPHY

A. Introduction

Supercritical-fluid chromatography (SFC) is a process with similarities to both gas chromatography and high-performance liquid chromatography. However, SFC has some specific differences which in many applications surpass GC and HPLC. In supercritical-fluid chromatography, the mobile phase is neither a gas nor a liquid. At temperatures above its critical temperature (see "Cryogenic Gases and Liquids" section in Chapter 4), a substance called a **supercritical fluid** and can no longer theoretically be referred to as a liquid or a gas. These supercritical fluids have the properties of both liquids and gases. Supercritical fluids are like gases that have been compressed to densities at which they can exhibit liquidlike interactions. If a compound is thermally unstable or has low volatility, GC is inefficient and SFC or LC are alternative choices. Thermally labile and nonvolatile samples can be analyzed using SFC, resulting in high-resolution separations performed at relatively low temperatures.

Viscosities found in supercritical fluids are typically less than those of liquids but considerably higher than gases, and densities are less than liquids but much greater than gases. The improved mass-transfer processes, due to lower fluid viscosities, is an advantage over liquid phases, while increased molecular interactions, due to higher densities, are an advantage over gas phases. The nine most common supercritical fluids are CO_2, N_2O, NH_3, $n\text{-}C_3$, $n\text{-}C_4$, SF_4, Xe, CCl_2F_2, and CHF_3.

Silica gel and alumina have been used as the stationary phase in SFC for nonpolar compounds but are not efficient for polar compounds because of the absorption of polar solutes, particularly when CO_2 is the mobile phase. The most widely used SCF polar stationary phases in open tubular columns are the cyanopropyl polysiloxanes. With CO_2 as the mobile phase, stationary phases have been particularly useful for the analysis of highly polar compounds such as carboxylic acid functional groups. The mobile phase in SFC is the most influential parameter governing solute retention on the column. Unlike in GC, where the mobile phase is relatively inert, SFC mobile phases play an active role in altering the distribution coefficient of the solute between the stationary phase and a compressed carrier fluid phase. SFC has found applications using both packed- and open-tubular-column types. In packed columns, LC packing materials perform well under SFC conditions, since both techniques depend on the ability of the mobile phase to solvate analyte molecules. Particle sizes normally vary from 3 to 10 μm in diameter with pore sizes ranging from 100 to 300 Å. The most popular particle size is 5-μm diameter. A typical packed column (10 m high by 4-mm ID, packed with particles 5 μm in diameter) yields approximately 80,000 plates per meter.

SFC is carried out at temperatures low enough to process most thermally sensitive materials without decomposition while being compatible with sensitive gas-phase detectors. Changing the mobile-phase pressure or adding organic solvents (modifiers) can dramatically alter the solvation properties of supercritical fluids, allowing SFC to separate highly polar as well as nonpolar substances. SFC can be used with both GC and HPLC detectors simultaneously, providing great flexibility in matching the detection scheme to the analyte of interest. Because of this range of options, many complex samples that ordinarily require multiple analyses (e.g., both GC and HPLC) may be amenable to a single SFC analytical solution.

SIZE-EXCLUSION CHROMATOGRAPHY

Size-exclusion chromatography is also called **gel-permeation chromatography (GPC)** or **gel-filtration chromatography (GFC)** and is another example of adsorption chromatography. It is a separation process that employs a gel composed of an insoluble, cross-linked polymer as the stationary phase. These gels swell in aqueous solution and create cavities that can trap molecules of various sizes. The degree of polymeric cross-linking determines the size of the holes in the gel matrix. Because of the sievelike molecular structure of these materials, compounds ranging from molecular weights of one hundred to several million can be concentrated and separated. When the molecules are too large to enter the pore, they will be excluded from the pore (this is size-exclusion chromatography); therefore they must travel with the solvent front.

Size-exclusion chromatography is a simple interaction that involves separation of solute molecules based on size or shape differences. Size-exclusion chromatography has several names: gel-filtration chromatography, gel-permeation chromatography, and steric-exclusion chromatography. The name recommended by ASTM (ANSI/ASTM E 682-79) is **steric-exclusion chromatography**. Size-exclusion chromatography has the following characteristics: (1) The stationary phase consists of a gel in which the solute molecules can penetrate through the entire macroscopic volume of the particles of the packing material. (2) Conditions are chosen so that adsorption (specific or nonspecific) is avoided or reduced. If adsorption does not take place, the elution volume depends on the size of the sample molecules in a way that would be expected from a molecular sieving mechanism, and elution volumes higher than the total bed volume do not occur. (3) Solutes are ideally eluted in the order of the size of the molecules, the larger molecules being eluted first. Size-exclusion chromatography is mainly used for separation and characterization of proteins, polysaccharides, nucleic acids, and other substances of biological interest.

Gel-permeation procedures are well suited for the separation of polymers, copolymers, proteins, natural resins and polymers, cellular components, viruses, steroids, and dispersed high-molecular-weight compounds. The gels prepared should have a narrow or broad range of pore sizes, depending upon the specific need of the technician and the composition of the substance being separated. Gels are available in pore sizes ranging from 25 to 25,000 A and most have excellent temperature stability; the silica gels remain stable up to 500°C. The eluting solvents can be either polar or nonpolar. The volume of solvent that is required for a procedure elution equals the volume of solvent in the column, because there is no affinity of the substances in the sample for the packing. All sample components elute completely with one column volume of solvent. Gel-permeation sorbents may be polymers or copolymers of vinyl acetate, polyethylene glycol dimethylacrylate, polystyrene, phenol-formaldehyde, and others.

PROCESS CHROMATOGRAPHY

Process Chromatography can be defined as a method used to separate a mixture into its individual components on-line during the chemical manufacturing process. Chromatography is perhaps the most often used technique for both qualitative and quantitative analysis of chemicals. Qualitative information is obtained by comparing times of travel (elution) of a standard with times of elution of a test sample. Quantitative information is obtained by comparing the area size of a peak produced by a known concentration to the area of an unknown concentration.

A typical chromatographic system consists of the following components: (a) a sample inlet, (b) separations column, (c) detector, and (d) record or data handling device. Most chromatographic analyses require liquid samples. If the test sample is a solid, it is dissolved in a solvent of known composition. The sample carrier and detection will depend on the type of chromatography being performed. The separation of test samples into individual components can be based on polarity, boiling points, molecular size, and/or charge. Recorders can vary from chart recorders, where deflections or peaks are manually measured, to more sophisticated electronic integrators that will directly report concentrations.

Three major types of process chromatography exist: liquid, ion, and gas. **Liquid chromatography** (LC) utilizes liquid as the sample carrier and light absorption for detection. **Ion chromatography** (IC) also uses liquid as the sample carrier, but uses conductivity for detection. With **gas chromatography** (GC), the sample is vaporized to the gas state, and then inert gases are used as the sample carrier. A variety of detectors can be used in gas chromatographic procedures, with the most common being a thermal conductivity detector. The choice of which chromatographic method to use for an analysis is based on the sample being analyzed. If the test sample is **heat labile** (destroyed by heat), liquid chromatography is the chosen method. Ionic samples are almost always analyzed using ion chromatography. Gas chromatography is more commonly used in the petrochemical industry, whereas high-performance liquid chromatography is used in the pharmaceutical industry.

REFERENCES

1. Guiochon, G., and Guillemin, C. L., "Quantitative Gas Chromatography for Laboratory Analysis and On-Line Process Control," *Journal of Chromatography Library*, Vol. 42, Elsevier Science Publishers (2006).
2. Meyer, V. R., *Practical High-Performance Liquid Chromatography*, 5th edition, John Wiley & Sons, Inc. ISBN 978-0470-68217-3 (2010).
3. Giddings, J. C., *Unified Separation Science*. Wiley-Interscience. ISBN 978-0471-52089-4 (1991).
4. *Chromatographic Theory and Basic Principles*, J. A. Jonsson, editor. CRC Press. ISBN 978-0824-77673-2 (1987).
5. Jork, H., Funk, W., Fischer, W., and Wimmer, H., *Thin-Layer Chromatography: Reagents and Detection Methods*, Vol. 1b. VCH Publishers, Inc. ISBN 978-3527-27834-3 (1994).
6. *High Resolution Gas Chromatography*, 3rd edition, K. J. Hyver and P. Sandra, editors. Hewlett-Packard. ISBN 978-9991847160 (1989).
7. McNair, H. M., and Bonelli, E. J., *Basic Gas Chromatography*, 5th edition. Varian Instrument Division (1969).
8. Perry, J. A., *Introduction to Analytical Gas Chromatography*, Chromatographic Science Series, Vol. 14, Marcel Dekker. ISBN 978-0824-71537-3 (1981).
9. *Ion Exchange Chromatography: Principles and Methods*, 3rd edition. Pharmacia (1991).
10. *High Performance Liquid Chromatography in Biotechnology*, W. S. Hancock, editor, John Wiley & Sons, Inc. ISBN 978-0471-82584-5 (1990).
11. McMaster, M. C., *HPLC: A Practical User's Guide*, Wiley–VCH Publishers, Inc. ISBN 978-0471-18586-4 (1994).
12. Gordon, A. J., and Ford, R. A., "Chromatography Section," *The Chemist's Companion: A Handbook of Practical Data, Techniques, and References*, John Wiley & Sons, Inc. ISBN 0-471-31590-7 (1972).

CHAPTER 27
SPECTROSCOPY

THE ELECTROMAGNETIC SPECTRUM

A. Introduction

Energy can be transmitted by **electromagnetic (em)** waves, and there are many types of these. They are characterized by their frequency f, the number of waves passing a fixed point per second, and their wavelength λ, the distance between the peaks of any two consecutive waves (Fig. 27.1).

Wavelengths of waves in the electromagnetic family or spectrum vary greatly, from fractions of angstroms to kilometers (Fig. 27.2). An examination of this spectrum reveals that as the frequency increases, the wavelength decreases, and as the frequency decreases, the wavelength increases.

The types of radiations are distinguished by the different characteristics exhibited by radiations of different frequencies. For example, visible light waves cannot pass through opaque substances, whereas radio waves, x-rays, and gamma rays can. The human eye cannot detect x-rays or ultraviolet rays, yet they are there. The lines of demarcation between the types of radiations are not sharp and distinct, and ranges may overlap.

In vacuum, the whole family of electromagnetic (em) waves has a constant velocity of 2.998×10^{10} cm/s (usually written as 3×10^{10} cm/s). In any em wave, the product of the wavelength λ and the frequency f equals the velocity of the wave:

$$\text{Wavelength} \times \text{frequency} = \text{velocity}$$

or

$$\lambda \times f = v$$

Types of Electromagnetic Radiation. The values of the wavelength (λ) and frequency (f) are what differentiate one kind of radiation from another within the em radiation family (Fig. 27.3).

Conversion of Units. Wavelengths are often expressed in different units, and this may lead to confusion and error when one attempts to determine equivalent values. Table 27.1 enables you to interconvert wavelengths, regardless of how they are expressed.

Frequency and Wavelength Relationship. Figure 27.4 shows values for wave frequencies in hertz and wavelengths in centimeters for the band-spectrum spread of the various types of em radiations.

B. The Visible Spectrum and Refraction

White light, emitted from the sun or from an incandescent tungsten-filament bulb, produces a continuous spectrum consisting of all the wavelengths visible to the human eye (Fig. 27.5).

When white light is passed through nonparallel surfaces such as hose of a prism, there is a permanent bending of the beam, because different wavelengths (each color has its own wavelength) passing through a material are slowed down to different extents; each one is bent or refracted differently. Generally red is not slowed down as much as blue, and therefore the blue is bent or refracted through a larger angle (Fig. 27.6).

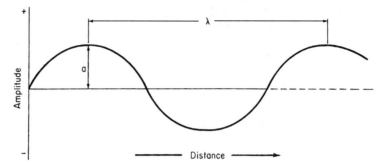

FIGURE 27.1 An electromagnetic wave. λ = wavelength; a = amplitude, a measure of intensity.

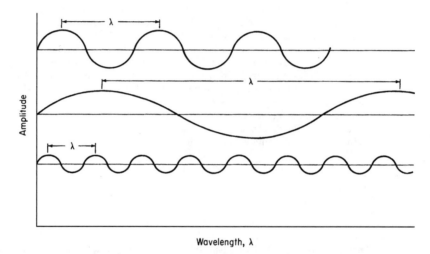

FIGURE 27.2 Variation of wavelength; as wavelength increases, frequency decreases.

C. Diffraction

When waves of em radiation strike an obstacle that does not reflect or refract them, a change occurs in their amplitude or phase; this change is called **diffraction**.

A **diffraction grating** is a flat piece of metal, glass, or plastic that has a great many parallel grooves ruled on it. Nonvisible electromagnetic radiations that are directed onto its surface are separated into their individual wavelength components (Fig. 27.7). In effect, such gratings separate nonvisible radiations just as the prism separates the components of the visible spectrum. Crystals of salt have been used as diffraction gratings for x-rays.

D. Sodium Vapor and Ultraviolet Lamps

Although white light is a combination of all the components of the visible spectrum, not all incandescent bulbs emit "white" light. Some of them emit higher intensities of one component than of the others.

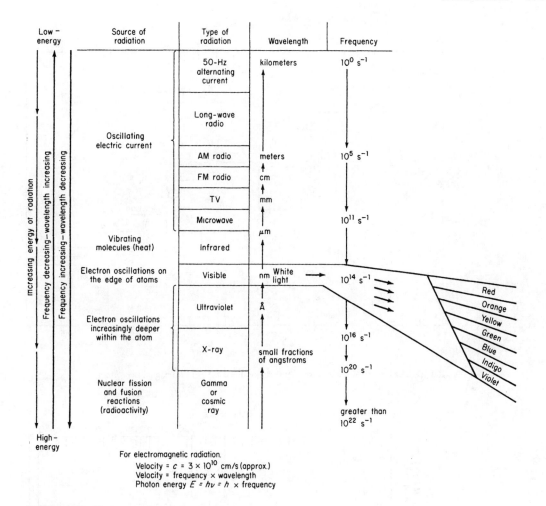

FIGURE 27.3 Electromagnetic radiation spectrum.

TABLE 27.1 Conversion Table

Multiply number of → ↓ To obtain number of	By ↘	Angstroms (Å)	Millimicrons* (nm)	Microns† (μm)	Millimeters (mm)	Centimeters (cm)	Meters (m)
Angstroms (Å)		1	10	10^4	10^7	10^8	10^{10}
Millimicrons (nm)*		10^{-1}	1	10^3	10^4	10^7	10^9
Microns (μm)†		10^{-4}	10^{-3}	1	10^3	10^4	10^6
Millimeters (mm)		10^{-7}	10^{-4}	10^{-3}	1	10	10^3
Centimeters (cm)		10^{-8}	10^{-7}	10^{-4}	10^{-1}	1	10^2
Meters (m)		10^{-10}	10^{-9}	10^{-6}	10^{-3}	10^{-2}	1

*More properly called *nanometers* (therefore nm).
†More properly called *micrometers* (therefore μm).

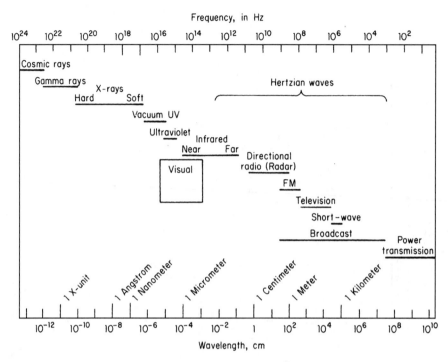

FIGURE 27.4 Frequency and wavelength relationships of the electromagnetic spectrum.

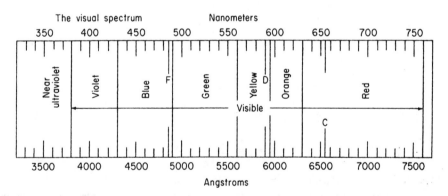

FIGURE 27.5 The visible spectrum.

Some lamps are designed to emit light of a specific wavelength. The sodium vapor lamp emits yellow light (Fig. 27.8), the ultraviolet (UV) lamp provides a concentrated beam of ultraviolet radiation on demand.

E. Use of the Electromagnetic Spectrum in the Study of Substances

The use of the electromagnetic spectrum in the study of substances is divided according to the region or range of radiations used in the procedure. These regions are shown in Table 27.2.

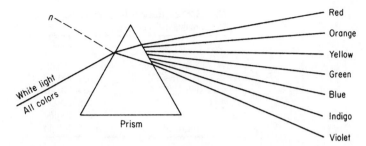

FIGURE 27.6 Light traversing the nonparallel faces of a prism is dispersed according to wavelength.

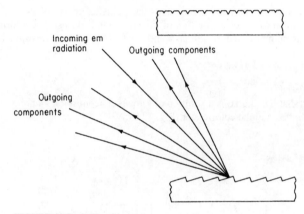

FIGURE 27.7 Diffraction gratings.

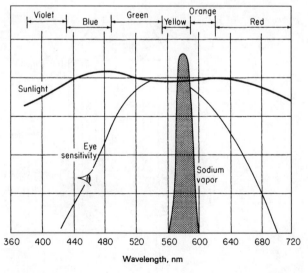

FIGURE 27.8 The sodium vapor lamp spectrum compared to sunlight and the sensitivity of the human eye.

TABLE 27.2 Regions of Study in the Electromagnetic Spectrum

Region	Range
Vacuum ultraviolet	100–180 nm
Ultraviolet	180–400 nm
Visible	400–750 nm
Near-infrared	0.75–2.5 nm
Infrared	2.5–15 µm
Far-infrared	15–300 µm

F. Colorimetry

The solutions of many compounds have characteristic colors. The intensity of such a color is proportional to the concentration of the compound. By using a colorimeter to match samples, one can perform many analyses quickly and easily. One of these instruments is the **Duboscq colorimeter**.

Duboscq Colorimeter.

Materials

All that is required is a solution of known concentration and one of unknown, a pipette, a graduated cylinder, and the colorimeter (Fig. 27.9).

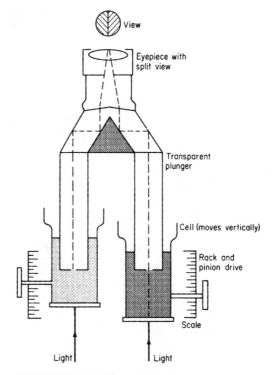

FIGURE 27.9 A Duboscq colorimeter.

Procedure

1. Be sure that the cups and plungers are clean both before and after use. Use a soft cloth or lens tissue to wipe optical glass.
2. Test the zero point of the scale by carefully raising the cups until they touch the plungers. The zero point of the scale is adjusted by screws at the bottom of the cup holders or at the side of the holders. Cups should not be changed from one side to the other.
3. Adjust the instrument for equal light intensity on both sides by filling cups with standard solution (three-fourths full). Set both sides at the same value, and adjust the position of the light until it is equal on both sides. This may be accomplished by shifting the position of the instrument or mirror if daylight or an external light source is employed or by changing the position of the bulb or reflector. If a light source is attached to the instrument, both rods should appear equally bright.
4. Test the light adjustment by filling both cups with the standard solution and setting one cup at a convenient depth; then move the other cup until a balance is obtained. Half of the balance point should be obtained by approaching the point of balance from one direction, and the other half from the opposite direction. Repeat this balancing until 6 to 10 readings have been obtained. The average should agree within 1 or 2 percent with the reading of the stationary cup. If this is not the case, then use one side as a fixed reference as follows: Set the cup just moved (cup 2) at the average reading obtained for it. Leave the standard solution in this cup. Replace the standard solution in the other cup (cup 1) with an unknown. Adjust the unknown cup (cup 1) until a balance is obtained. Then

$$C_{unknown} = C_{standard} = \frac{R_{standard}}{R_{unknown}}$$

where

$R_{standard}$ = reading of standard solution in cup 1
$R_{unknown}$ = reading of unknown in cup 1
$C_{standard}$ = concentration of standard
$C_{unknown}$ = concentration of unknown

5. Be sure there are no air bubbles beneath the plungers when the plungers are inserted beneath the liquid.

G. Nephelometry

Nephelometry is the measurement of the light that is reflected from a finely divided suspension or dispersion of small particles, the **Tyndall effect**. Optically controlled beams of light directed into a suspension of finely divided particles in a fluid will measure the concentration of the particles in a linear relationship. Nephelometric analysis measures the intensity of the Tyndall light that results from the incidence of a controlled light beam of constant intensity upon the suspension. The Tyndall light is detected and measured by photocells mounted at right angles to the incident beam (Fig. 27.10), and their output is measured on the detector, usually a galvanometer or a precisely calibrated potentiometer.

H. Spectroscopy and Spectrophotometry

Spectroscopy is the study of the interaction of electromagnetic radiations with matter. Instruments that measure em **emission** are called **spectroscopes** or **spectrographs**. Those that measure em **absorption** are called **spectrophotometers**. All spectroscopic instruments separate electromagnetic radiation into its component wavelengths to enable one to measure the intensity or strength of the radiation at each wavelength.

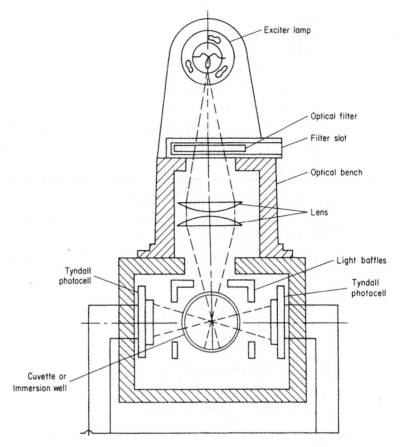

FIGURE 27.10 In the nephelometer, the Tyndall photocells are placed at right angles to the incident beam.

There are three kinds of emission spectra:

1. **Continuous spectra**, which are emitted by incandescent solids (Fig. 27.11)
2. **Line spectra**, which are characteristic of atoms that have been excited and are emitting their excess energy (Fig. 27.12)
3. **Band spectra**, which are emitted by excited molecules

Electrons in an atom are normally in their lowest energy states, known as the **ground state**. When sufficient energy is added, electrically or thermally, one or more electrons may be raised to higher-energy states. When these electrons lose their energy and return to their ground state, they emit

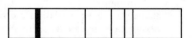

FIGURE 27.11 A continuous spectrum.   **FIGURE 27.12** A line spectrum.

electromagnetic radiations. In returning to their ground state, they may do this in several discrete jumps or energy changes, emitting light of different wavelengths for each jump. When high-energy excitation is used, more lines appear in the spectrum.

Definitions and Symbols. Spectroscopic terminology frequently uses different names and symbols to identify the same property, and there are both modern terms and terms that have been used in the past.

Radiant power (P) The rate at which energy in a beam of radiation arrives at some fixed point. *Intensity (I)* is the same term.

Transmittance (T) The ratio of the radiant power (P) in a beam of radiation after it has passed through a sample to the power of the incident beam (P_0). See Fig. 27.13. It is also referred to as *percent T* or $T \times 100$.

$$T = \frac{P}{P_0}$$

Absorbance (A) Also called the *optical density*, and it is the logarithm (base 10) of the reciprocal of the transmittance (T).

$$A = \log \frac{1}{T} = \log \frac{P_0}{P}$$

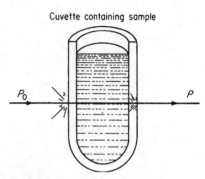

FIGURE 27.13 Transmittance of a sample—the mathematical expression is P/P_0.

where
 P = radiation transmitted by the solution
 P_0 = radiation transmitted by the pure solvent

Molar absorptivity ε (Also known as the *molar extinction coefficient.*) The absorbance of a solution divided by the product of the optical path b in centimeters and the molar concentration c of the absorbing molecules or ions:

$$\varepsilon = \frac{A}{bc}$$

I. Spectrophotometry Concentration Analysis

According to **Beer's law** (also known as the Beer-Lambert Law), the absorbance of a solute in solution is a function of its concentration at a particular wavelength, and therefore absorbance measurements can be used to determine the concentration of solutions. When the optimum wavelength is selected for an analysis, the concentration of the solute can be established.

Absorption of Radiant Energy. When radiation of a specified wavelength is passed through a solution containing only one solute which absorbs that wavelength, the absorbance (which has no units) can be calculated by the formula:

$$A = \log \frac{P_0}{P} = \log \frac{1}{T} = \varepsilon bc$$

where
 P = radiation transmitted by the solution
 P_0 = radiation transmitted by the pure solvent

b = optical length of the solution cell
c = molar concentration of the absorbing solute
ε = the molar extinction coefficient (in liters per mole-centimeter)

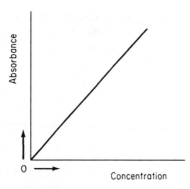

FIGURE 27.14 Plot of absorbance versus concentration.

Practical application of Beer's law is shown in Figs. 27.14 and 27.15.

When a beam of polychromatic radiation is directed into a sample of some substance, the wavelengths of certain components may be absorbed, while others pass through essentially undisturbed. See Fig. 27.16. Components of the radiation are absorbed only if its energy exactly matches that energy which is required to raise molecular or ionic components of the sample from one energy level to another. Those energy transitions may involve vibrational, rotational, or electronic states. After it has been absorbed, that energy may be emitted as fluorescence, utilized to initiate chemical reactions, or actually dissipated as heat energy.

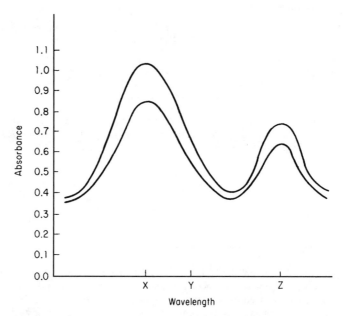

FIGURE 27.15 Absorption spectra for solutions of the same substance at two different concentrations.

Measurement of Absorption Spectra. The spectrophotometer (Fig. 27.17) is used to measure radiation absorption. Its basic components are:

1. A monochromator, a device that attempts to provide a beam of radiation at a single frequency that is to be passed through a sample.
2. A sample holder to contain the solution.
3. A detector to measure the intensity of the transmitted radiation after it has passed through the sample.

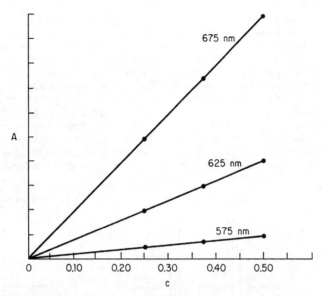

FIGURE 27.16 Effect of varying the wavelength upon the absorbance of a substance. This substance has greatest absorbance at 675 nm.

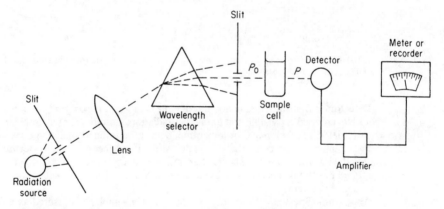

FIGURE 27.17 Schematic diagram showing the basic components of a prism spectrophotometer. The detector screen-scale is viewed by meter, recorder, photographic film, or the human eye.

Visible spectrophotometers can be converted into ultraviolet spectrophotometers by changing the source of radiation, from a tungsten filament—for visible light—to a hydrogen or deuterium discharge lamp (Fig. 27.18)—for ultraviolet.

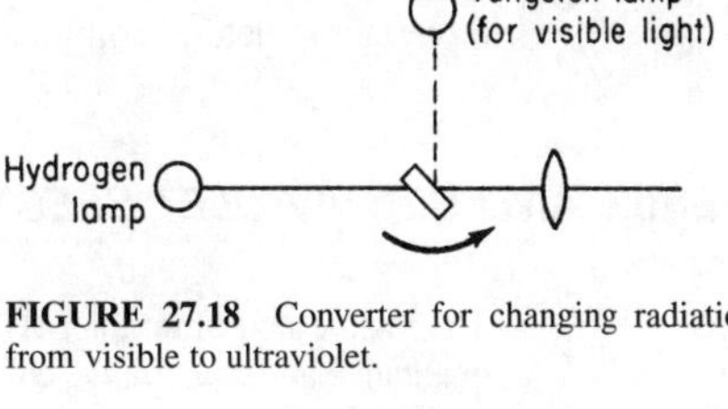

The prisms, lens, and cells must also be changed from glass to quartz or fused silica, because glass absorbs UV radiations. At wavelengths less than 180 nm the system must be operated in a vacuum (vacuum UV), because air will absorb UV radiations that are shorter than 180 nm.

FIGURE 27.18 Converter for changing radiation from visible to ultraviolet.

Monochromator. Radiation sources emit a broad band of wavelengths, which pass through a narrow slit and are focused by lenses and mirrors onto a wavelength selector. The selector is a prism or diffraction grating that disperses the continuous radiation; however, only a narrow band directed toward a slit in the screen is able to pass through the sample. By rotating the prism or grating, it is possible to direct components of different wavelengths through the sample and thus to select desired wavelengths. Those instruments that pass a broad band through the sample are simply called **photometers**.

Sample Cell or Cuvette. A sample solution is contained in an optically transparent cell or *cuvette* (Fig. 27.19) with a known width and optical length. Cells are made of optical glass. Protect them therefore from scratches. Avoid the use of abrasive cleaning agents; clean them well with soft cloths, avoid finger marks and lint or dirt, and handle them only by the top edge when inserting them into the instrument. Because of additional reflection from air-to-glass surfaces, empty cuvettes transmit less radiation than do cells filled with reference standards, such as distilled water.

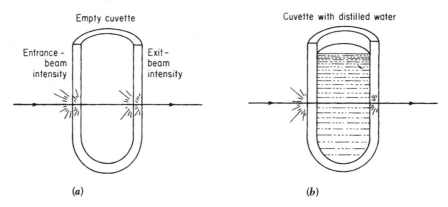

FIGURE 27.19 (*a*) Empty cuvette. The empty cell has radiation loss. (*b*) Cuvette filled with distilled water. Filled cells are used as standards.

Detector. Transmitted radiation through the solution is picked up by a photosensitive detector. Photocells are used for visible and ultraviolet radiations; infrared (IR) instruments use temperature-sensitive detector, a bolometer or a thermopile, because infrared radiation does not have sufficient energy to activate photocells. Radiations that strike the detector generate electric current, which is amplified with an amplifier and is then transmitted to a recorder or a meter. Anything that is sensitive to infrared radiations (heat) can be used; some instruments, however, are more sensitive and useful at certain wavelengths.

Absorption Spectra. When the absorbance of a sample is measured as a function of wavelength, the result is an absorption spectrum. Some instruments vary wavelength manually, while others vary it automatically. The absorption spectrum can be used to identify unknown substances, because particular molecules and ions have specific absorptions at characteristic wavelengths. The comparison of an unknown spectrum with spectra of known substances enables one to identify unknown substances. In the infrared spectrum, particularly, specific wavelengths are characteristic of certain structural components of molecules.

VISIBLE AND ULTRAVIOLET SPECTROSCOPY

The visible and ultraviolet (UV) spectra of ions and molecules were the first to be used to obtain both qualitative and quantitative chemical information. The typical visible-UV spectrum consists of a plot of the **molar absorptivity (ε)** as a function of the wavelength expressed in nanometers as shown in Fig. 27.20. The symbol lambda with subscript "max" (λ_{max}) stands for the **absorption maximum**, the wavelength at which a maximum absorbance occurs. Molar absorptivity can be defined from Beer's law by rearranging the Beer's law equation to

$$\varepsilon = \frac{A}{bc}$$

where
ε = molar absorptivity
A = absorbance = $-\log T$ (where T represents transmittance)
b = cell path length, cm
c = concentration, mol/L

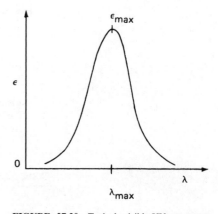

FIGURE 27.20 Typical visible-UV spectrum with molar absorptivity versus wavelength.

The absorption of visible and/or UV radiation is associated with the transition of electrons between different energy levels within the ion group or molecule. Visible and UV radiation is energetic enough to cause not only various electronic transitions, but vibrational and rotational changes thus these absorption bands tend to be very broad. The visible region is considered to be from 800 to 400 nm and the UV region is subdivided into two distinct regions: near-UV (400 to 180 nm) and the far-UV (180 to 100 nm). Figure 27.21 shows the visible and UV spectroscopic regions and the associated energy levels. Some visible-UV spectrophotometers are capable of reaching into the near-infrared region (approximately 2600 nm). The newer "UV-Vis" instruments have **Fourier transform** for very rapid analysis and computer capabilities for data manipulation, storage, and full-spectrum searches.

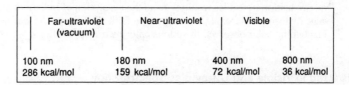

FIGURE 27.21 Visible and UV spectroscopic regions and associated energy levels.

A. Ultraviolet Spectrophotometer Components

The radiation source for a UV spectrophotometer is either a mercury or deuterium (hydrogen) lamp. The optical components (prisms, lens, and cells) are usually composed of quartz or fused silica because glass absorbs ultraviolet radiation. A special UV-sensitive photomultiplier tube is used as a detector. Most routine UV spectroscopy work is limited to the near-UV because the silica optics and atmospheric oxygen absorb an excessive amount of the radiation. **Far-ultraviolet** is sometimes referred to as **vacuum ultraviolet** (vacuum UV) because the instrument is evacuated to remove the atmospheric oxygen in the monochromator.

B. Chromophores

Functional groups that absorb visible and/or UV are called **chromophores**. Chromophores tend to have unsaturated bonds or contain functional groups with multiple bonds as shown in Table 27.3.

TABLE 27.3 Characteristic Ultraviolet Absorption Bands for Some Chromophores

Type	Example	Absorption band, nm
Alkenes	$CH_2\!=\!CH_2$	165–193
Alkynes	$HC\!\equiv\!CH$	195–225
Aldehydes	CH_3CHO	180–290
Ketones	CH_3COCH_3	188–279
Carboxylic acids	CH_3COOH	208–210
Aromatics	C_6H_6	204–254

TABLE 27.4 Absorption of Visible Light and the Corresponding Colors

Wavelength, nm	Color (absorbed)	Color (observed)
400–435	Violet	Yellowish green
435–480	Blue	Yellow
480–490	Greenish blue	Orange
490–500	Bluish green	Red
500–560	Green	Purple
560–580	Yellowish green	Violet
580–595	Yellow	Blue
595–650	Orange	Greenish blue
650–800	Red	Bluish green

When a beam of light is passed through an absorbing solution, the intensity of the incident radiation will be greater than the intensity of the emerging radiation. In general, the excited electrons, atoms, and/or molecules resulting from the absorption of radiation, return very rapidly to the ground state by losing electromagnetic radiation. Table 27.4 shows the wavelength, color absorbed, and corresponding color observed in various colorimetric applications.

C. Solvents Used in Visible and Ultraviolet Spectroscopy

Most compounds with only single bonds tend to absorb radiation in the far-UV (less than 180 nm), but make reasonable solvents for working in the near-UV region. Water, which has an absorption-band maximum at 167 nm, should provide a solvent medium for most polar chromophores that absorb in the visible and/or near-UV regions. Other commonly used polar solvents are 95 percent ethanol and methanol, while nonpolar aliphatic hydrocarbons (like hexane and cyclohexane) can be used if the alkenic and aromatic trace impurities have been removed. Table 27.5 provides a listing of solvents and their wavelength cutoff limits. **Cutoff limits** for the solvents listed below are defined as the wavelength at which the absorbance approaches one in a 10-mm cell.

D. Sample Preparation

All volumetric glassware and UV cells must be clean and dry. Solid samples should be dried to constant weight and stored in a desiccator to minimize moisture and solvent interference. A typical 1-cm cell holds approximately 3 mL of solution; however by preparing samples in greater volume the inherent error that results from weighing small solute quantities can be reduced. Most solutions are analyzed at very low concentrations (10^{-2} to $10^{-6}\,M$) to minimize solvent interactions and shifts in the absorbance maxima. Since both visible and UV spectroscopy are nondestructive techniques, the

TABLE 27.5 Ultraviolet Cutoff Limits for Various Solvents

Solvent	Cutoff point, nm	Boiling point, °C
Acetonitrile	190	81.6
Hexane	195	68.8
Cyclohexane	205	80.8
Ethanol (95%)	204	78.1
Water	205	100.4
Methanol	205	64.7
Diethyl ether	215	34.6
1,4-Dioxane	215	101.4
Carbon tetrachloride	265	76.9
Benzene	280	80.1
Toluene	285	110.8
Acetone	330	56.0

sample can be readily recovered if a volatile solvent is selected. In general, a sample should transmit in the range from 20 to 80 percent (0.7 to 0.1 A) for greatest accuracy in reading the meter. Most UV-Vis meters are normally calibrated in both percent transmittance (%T) and absorbance (A), but %T readings are used because the scale is linear and easier to interpolate. The %T can be converted to its equivalent in absorbance value by the equation:

$$A = 2 - \log \%T$$

Matched Cuvettes. If more than one cuvette is used in an analysis, they should be determined to be "matched." Matched cuvettes have identical path lengths plus similar reflective and refractive properties in the spectrophotometer beam. Sets of matched cuvettes can be purchased, and some are marked with a reference line on the cuvette wall to assure proper and reproducible placement in the beam. Obviously, these cells can become mismatched with use, cleaning procedures, and scratches. A guideline for selecting a set of matched cuvettes is to analyze a 50 percent transmittance solution and use only those cuvettes in a single analysis with a 1 percent deviation or less.

E. Using the Spectrophotometer

Bausch and Lomb Spectronic 20.

Principle

The spectrophotometer measures the intensity of visible light after passage through a sample. (See Figs. 27.22 and 27.23)

Components

1. A monochromatic source of light vibrating at a single frequency (between 375 and 650 nm) (1 nm = 1×10^{-7} cm = 10 Å).
2. A sample holder for supporting the solution in the beam of light.
3. A detector to measure the intensity of light passing through the sample.

Procedure

1. Switch to ON by rotating the ZERO CONTROL knob clockwise until a click is heard. This turns the power on. Allow the unit to warm up for 20 minutes.
2. Adjust the ZERO CONTROL knob until the meter reads 0 percent T (no tube in the sample holder).

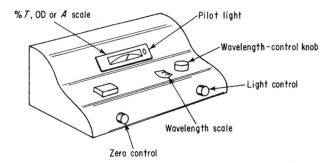

FIGURE 27.22 Bausch and Lomb Spectronic 20. (*Courtesy of Bausch & Lomb*)

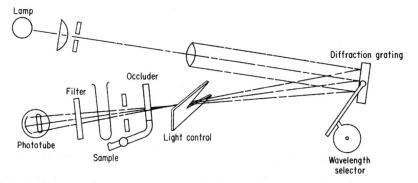

FIGURE 27.23 Schematic optical diagram of the Spectronic 20. (*Courtesy of Bausch & Lomb*)

3. Obtain two Spectronic 20 colorimeter tubes.
   (a) Use one for a solvent blank.
   (b) Use the other as a sample holder.
   (c) *Handle the tubes with extreme care:* Wipe them clean with soft, absorbent, lint-free paper. Never touch the lower part of the tubes with your fingers. Fingerprints absorb and scatter light.
4. Rotate the right-hand LIGHT CONTROL knob counterclockwise as far as it will turn. *Do not force.*
5. Insert a solvent-blank cuvette.
   (a) Zero the meter.
   (b) Set the wavelength.
   (c) Close the cover lid on the sample holder.
   (d) Rotate the LIGHT CONTROL knob clockwise until the meter reads 100 percent T.
6. Remove the blank cuvette holder.
7. Insert the sample cuvette. *Close the cover.*
8. Read the absorbance on the dial.
9. Repeat step 5 each time the wavelength is changed.

CAUTIONS:

1. It may be necessary to rezero the meter because of fluctuations in the electronic components.
2. The meter needle should drop to zero whenever the colorimeter tube is removed from the sample holder and the occluder drops into position.
3. *Always close the sample cover* to exclude stray light when making measurements.

F. Diode-Array Spectro-Photometers

The recent development of the **diode-array spectrophotometer** allows visible, ultraviolet, and fluorescence spectrometers to perform very rapid spectral analyses. These diode-array detectors allow the spectrometers to simultaneously measure all wavelengths. In the classical UV/VIS/NIR spectrophotometers, light from a polychromatic source passes through a monochromator (a prism or a grating followed by a slit) which selects only a narrow band of light. The resulting *almost* monochromatic light passes through a measurement sample. Light transmitted through the sample strikes a photodetector, and the signal output is then used to calculate sample absorption at each wavelength interval. By rotating the prism/grating, the instrument scans through the entire wavelength of interest. Accuracy and repeatability of a classical spectrophotometer are limited by the mechanism used to turn the prism or grating and tune the monochromator output wavelength. Measurements of samples with low absorption, requiring long detector-integration times, takes a long time because the monochromator scans each wavelength interval sequentially. The precise sensitive scan mechanism also reduces the suitability of the spectrophotometer for applications in harsh process-control and quality-assurance environments.

Diode-array spectrometer models with spectral ranges of either 190 to 800 nm or 190 to 1065 nm are currently available. These particular models have silicon photodiode arrays with 512 detector channels and 1024 channels. Each photosensitive element in the array detects a portion of the entire spectrum and outputs a voltage proportional to the integrated intensity at that wavelength. Units with a combination of carefully selected diode arrays, state-of-the-art signal processing electronics and fiber-optic probes, can offer real advantages of speed, sensitivity, and productivity. Once all measurements are completed, the stored spectra are sent through an interface to the host computer for analysis.

G. Computerized Spectra Analytical Systems and Fluorometers

Computerized monochromators not only disperse and select light to allow data acquisition but influence and take an active role in the analysis. Analysis of spectral events in chemistry requires that monochromators can be fixed at a wavelength or scanned at particular rates, scanned over variable ranges, or sequentially scanned with various wavelength steps. Slit widths, grating turrets, entrance/exit ports should all be selectable. This allows the employment of different spectral sources and/or detectors. Modern technology provides complete signal detection and data acquisition systems of great dynamic range featuring 16-, 20-, and 24-bit I/O boards. This allows detector signals to be highly digitized and resolved. The digital data stored in memory and processed by software can be readily manipulated. Collection and display of data in suitable analytical formats is given.

Software drivers exist for a complete range of monochromators and scan control for entire product line of computerized spectroscopic systems. Special attention has been given to data manipulation and drivers for the fluorescence product line. It allows ratioing, differentiating, autozeroing, energy collection, and sorting of luminescence data including blank-signal subtracting, calculation and display of arithmetic operations on spectra, and smoothing. Stored data can subsequently be scale expanded, overlapped for comparisons, and displayed in selected sections over any wavelength interval.

INFRARED SPECTROSCOPY

A. Introduction

Infrared spectroscopy is primarily used as a nondestructive technique to determine the structure, identity, and quantity of organic compounds. As explained in "The Electromagnetic Spectrum," the infrared region extends from approximately 0.75 to 400 micrometers (μm) (1 μm = 1×10^{-6} m); however, the region from 2.5 to 16 μm is most often used by organic chemists for structural and quantitative determinations (Fig. 27.24). There are two other useful infrared regions: 0.75 to 2.5 μm, which is referred to as the **near-infrared** (near the visible region), and the region continuing from 16 to 400 μm, referred to as *the* **far-infrared**. The prefix "infra," which means "inferior to," was added to "red" because this particular radiation is adjacent to but slightly lower in energy than red light in the visible spectrum.

The interpretation of an infrared spectrum is not a simple exercise, and typically an organic compound is ultimately identified by comparing its infrared absorption spectrum to that of a known compound. Most molecules are in constant modes of rotation and/or vibration at temperatures above absolute zero. This rotation, stretching, and bending of molecular bonds happens to occur at frequencies (typically on the order of 100 trillion per second) found in the infrared spectral region. Of course, these vibrational frequencies will vary as the bonded atoms, functional groups, and bond strengths are changed; as a result, each molecule has a distinctive infrared absorption spectrum or "fingerprint." To be more specific, in the **functional-group** region (2.50 to 7.14 μm), it is relatively easy to assign structures to the absorption bands, while in the **fingerprint region** (7.14 to 16.0 μm) it is more difficult to assign structures; nonetheless, this fingerprint region is very useful because it contains many very characteristic and unique absorption bands of the compounds.

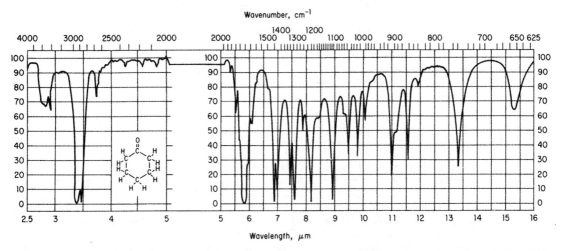

FIGURE 27.24 Typical infrared absorption spectrum of an organic compound (cyclohexanone) with grating change at 2000 cm^{-1}. The ordinate reads in percent transmittance and the abscissa indicates the wavelength expressed in micrometers (μm) or wave numbers (cm^{-1}).

B. Wavelength versus Wave Number

The frequency or energy level at which these characteristic infrared absorptions occur can be expressed in terms of wavelengths (μm) or wave numbers (cm^{-1}). Research has shown that a typical organic molecule requires an average energy of 1.6×10^{-20} joules (J) for vibrational excitation. Given **Planck's equation**, which relates energy and frequency:

$$E = h\nu$$

where

E = energy, in J
h = Planck's constant = 6.6×10^{-34} J · s
v = frequency, in s^{-1} or the equivalent measure hertz (Hz)

Substituting the average energy found necessary to induce vibrational excitation in a typical organic molecule:

$$1.6 \times 10^{-20} \text{J} = (6.6 \times 10^{-34} \text{J} \cdot s) v$$

$$v = \frac{1.6 \times 10^{-20} \text{J}}{6.6 \times 10^{-34} \text{J} \cdot s}$$

$$v = 2.4 \times 10^{13} \ s^{-1}$$

Finally, substituting our calculated frequency into the following **frequency-to-wavelength equation**,

$$\lambda = \frac{c}{v}$$

where

λ = wavelength, in cm
c = velocity of light = 3.0×10^{10} cm/s
v = frequency, in s^{-1}

we find that

$$\lambda = \frac{3.0 \times 10^{10} \frac{\text{cm}}{\text{s}}}{2.4 \times 10^{13} \ s^{-1}}$$

$$\lambda = 1.2 \times 10^{-3} \text{cm}$$

$$\lambda = (1.2 \times 10^{-3} \text{cm}) \left(1 \times 10^{4} \frac{\mu\text{m}}{\text{cm}} \right) = 12 \ \mu\text{m}$$

Note that an average energy of 1.6×10^{-20} J found necessary to induce vibrational adsorption corresponds to a wavelength of 12 μm (infrared radiation). [In the infrared region, the term **micrometer (μm)** is the preferred one, not the older term **micron (μ)** used to describe the same unit.]

The reciprocal of the wavelength is called the **wave number** and represents the number of cycles passing a fixed point per unit of time. The use of a scale expressed in wave numbers (cm^{-1}) in infrared spectroscopy is preferred to a scale of wave-lengths (μm) because wave numbers are linear and proportional to the energy and frequency being absorbed. Wave numbers are sometimes incorrectly referred to as **frequencies**, an error because the wave number is expressed in reciprocal centimeters (cm^{-1}) and frequency is expressed in units of reciprocal time (s^{-1}). Figure 27.25 gives a wave number-to-wavelength conversion scale for the infrared spectral region. Wave numbers can be calculated from wavelengths using the following equation:

$$\text{Wave number (expressed in cm}^{-1}) = \frac{1 \times 10^{4}}{\text{wavelength (expressed in μm)}}$$

For example, if the wavelength is 12 μm, the wave number in cm^{-1} is calculated as follows:

$$\text{Wave number} = \frac{1 \times 10^{4}}{12}$$

$$= 830 \text{ cm}^{-1}$$

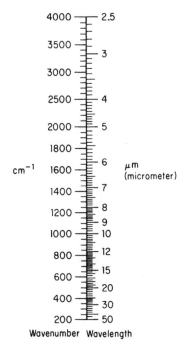

FIGURE 27.25 Wavenumber-to-wavelength conversion scale for infrared data.

Vibrational and rotational absorptions in the infrared region tend to occur over a rather broad range of wavelengths called **bands**. These infrared absorption bands do not usually occur as sharp lines, as might be the case with visible or UV spectral absorptions involving electron excitations.

C. Infrared Instrument Components and Design

A schematic diagram of a simple dispersive-type infrared spectrophotometer is shown in Fig. 27.26. An infrared source of energy is usually provided by a **Nernst glower**, which is composed of rare earth oxides (zirconium, cerium, etc.) formed into a cylinder of 1 to 2 mm diameter and 20 mm long. This filament material is electrically heated to approximately 1500°C, and the resulting infrared radiation is split into a reference and a sample beam by mirrors. The reference beam is passed through air or a beam attenuator, and the sample beam is passed through a sample where selective radiation absorption occurs.

The reference and sample beams are then passed alternately by a **chopper** (a rotating half mirror) to a dispersive device which separates the radiation into specific frequencies (wavelengths). A **prism** is one type of dispersive device that separates radiation of different wavelengths based on differences in refractive indexes at different wavelengths (see the section "The Electromagnetic Spectrum," in this chapter for details on monochromator components). A prism requires the radiation to pass through the material, and since glass will absorb most infrared radiation, sodium chloride or other special materials are required. A **grating** can be used as a dispersive device and consists of a series of closely spaced parallel grooves scribed into a flat surface. A grating simply reflects the diffracted light from its surface and absorbs very little radiation. Many infrared spectrophotometers require special temperature and/or desiccants to prevent moisture damage to the optics in the monochromator.

The radiation reflecting from the grating is directed through a series of narrow **slits**, which eliminate most undesirable frequencies, and onto a **detector**. An infrared detector must sense very small

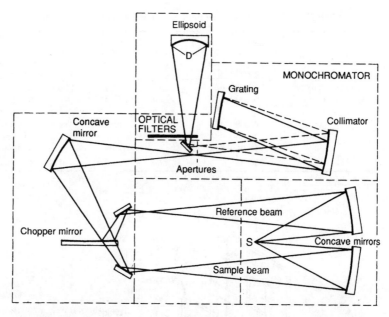

FIGURE 27.26 Schematic diagram of Perkin-Elmer Model 700. A typical double-beam, optical null infrared dispersive-type monochromator. S = source; D = detector. (*Courtesy of Perkin-Elmer Corporation*)

energy changes (typically in the 1×10^{-9} W range). **Thermocouples** are routinely used in infrared detectors and consist of two dissimilar metals fused together at a junction. A potential difference develops as the temperature varies between the thermocouple junctions monitoring the sample beam and the reference beam. These thermocouples can respond to temperature differences in the $1 \times 10^{-6}\,°C$ range. **Golay** or gas thermometers filled with xenon are used as infrared detectors above 50 µm.

Most dispersive-type infrared spectrophotometers are of the double-beam **optical null** type (Fig. 27.26). The radiation passing through the reference beam is reduced or *attenuated* to match the intensity of the sample beam. The **attenuator** is normally a fine-toothed comb which moves in and out of the reference beam's path. This movement is synchronized with a recorder so that its position gives a measure of the relative intensity of the two beams and thus the transmittance of the sample.

D. Infrared Cell Materials

Many different inorganic salts are used in infrared spectroscopy to provide optical components and cells. Table 27.6 lists some of the properties of various window materials.

E. Infrared Liquid Cells

Infrared cells for liquid samples are available in three basic types:

1. **Demountable cells** (Fig. 27.27) are used primarily for "mulls" and highly viscous liquids. These cells have no inlet ports and must be disassembled to clean and fill.
2. **Semipermanent cells** (Fig. 27.28) allow the sample to be introduced through injection ports and can be disassembled for cleaning and varying the cells' path length.
3. **Permanent cells** (Fig. 27.29) are similar to the semipermanent cells except that the windows and spacers are sealed together forming a leakproof cell. This type of cell is recommended for highly

TABLE 27.6 Properties of Some Infrared Window Materials

Material	Transmission range, cm^{-1}	Water solubility 20°C, g/100 g H$_2$O	Comments
KBr	40,000–400	65	Hygroscopic, inexpensive.
NaCl	40,000–625	36	Hygroscopic, easy to polish.
CaF$_2$	67,000–1110	0.0015	Water insoluble; do not use with ammonium salts.
AgCl	20,000–435	0.0015	Water insoluble; darkens with UV.
CsBr	20,000–250	110	Hygroscopic, fogs easily, easily deformed.
ZnS	10,000–715	0.0008	Water insoluble; called Irtran-2® (trademark of Kodak); attacked by oxidizing agent.

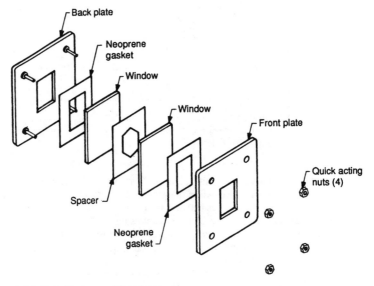

FIGURE 27.27 Demountable liquid cell.

volatile samples and precise quantitative work requiring reproducible pathlengths. These liquid cells are permanently sealed by the manufacturer and are not designed to exchange spacers. This type of cell is sometimes referred to as **amalgamated**, meaning literally "merged into a single body," and mercury is sometimes used for creating this seal.

An economical variation of a liquid cell is the liquid **minicell**. These cells are useful only for qualitative analysis and consist of a threaded, two-piece plastic body and two silver chloride windows. The AgCl window material withstands much abuse and is not as affected by water as most other infrared window materials. However, some darkening will occur upon exposure to sunlight. The AgCl windows each contain a 0.025-mm circular depression. The 0.025-mm circular depression in each window allows the cell's path length to be varied as shown in Fig. 27.30.

Cell path lengths in infrared spectroscopy normally range from 0.01 to 1.0 mm and can be produced by using **spacers** made of lead, copper, Teflon®, etc. Some manufacturers provide a **variable-path-length cell** equipped with a vernier scale in which the path length can be continuously adjusted from

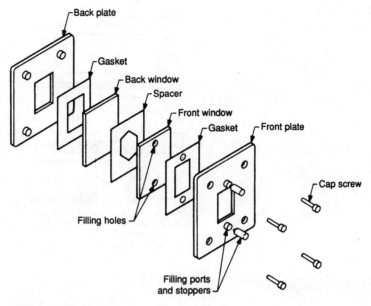

FIGURE 27.28 Semipermanent cell.

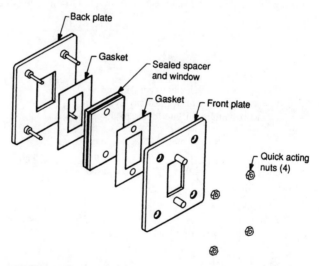

FIGURE 27.29 Permanent cell.

0.005 to 5 mm reproducible. This type of cell can be used for solvent compensation and differential analysis. The cell is filled with solvent and placed in the reference beam, and the path length is adjusted to compensate for the unwanted solvent absorption in the sample beam. Pure liquids are referred to as **neat** liquids in infrared analysis and usually require shorter path lengths (0.01 to 0.025 mm) because of the high concentration.

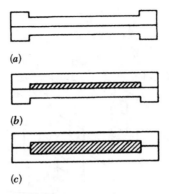

FIGURE 27.30 Path length variations with minicell windows. (*a*) Window positioned for smears. (*b*) Window positioned for 0.025-mm pathlength. (*c*) Window positioned for 0.050-mm pathlength.

F. Calculating Cell Path Length

The path length for a cell (or thickness of a film) can be calculated using the cyclic interference pattern produced by the reflected radiation from the walls of an empty cell (or surface of the film). The following formula can be used to determine the cell path length in centimeters. An infrared spectrum showing the interference fringe pattern from a 0.150-mm empty cell is given in Fig. 27.31.

$$b = \frac{n}{2(W_1 - W_2)}$$

where
 b = path length, cm
 n = number of cycles between W_1 and W_2
 W_1 = wave number at peak of any cycle, cm^{-1}
 W_2 = wave number of later cycle peak, cm^{-1}

$$b = \frac{n}{2(W_1 - W_2)} = \frac{13.5}{2(3000 - 2500 \text{cm}^{-1})} = \frac{13.5}{1000} = 0.0135 \text{ cm}$$

$b = 0.135$ mm

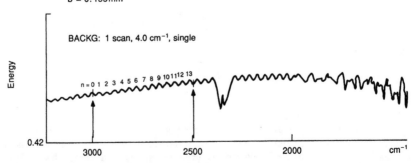

FIGURE 27.31 Technique for calculating the path length in an empty cell by measuring the number of cycles between known wave number values.

G. Solid Samples

There are two primary methods of analyzing solids by infrared spectroscopy. The **mull method** requires that the solid sample be pulverized (mulled) and mixed in a mortar and pestle to an ointment consistency with an oil medium. Only mortars and pestles made of stainless steel or agate are recommended for preparing infrared samples because of possible contamination problems in grinding. The two most common suspending media are **Nujol**® (trademark of Plough, Inc.—a high-molecular-weight hydrocarbon recommended for use from 1370 cm^{-1} to the far-infrared) and **Fluorolube**® (trademark of Hooker Chemical Co.—a high-molecular-weight fluorinated hydrocarbon recommended for use from 4000 to 1370 cm^{-1}). The infrared absorption spectra of pure Nujol® and Fluorolube® are given in Figs. 27.32 and 27.33.

It should be noted that both Nujol® and Fluorolube® have undesirable absorption bands, but they complement each other in that these bands are on opposite ends of the spectrum. A common technique is to run the sample in both mulling agents separately to cover the entire 4000- to 650-cm^{-1} range.

The mull (suspended solids) is then spread between two salt plates and placed in a demountable cell holder. A good mull will have a brown-blue color when held to the light and will spread uniformly between the salt plates. The mull technique is quick, but the resulting spectra are more complicated than those resulting from the KBr pellet technique described below because the Nujol® and Fluorolube® have infrared absorption bands. If there is a large slope to the baseline (greater than 0.5 absorbance units from 4000 to 1500 cm^{-1}), then the mull should be prepared again. Always disassemble the salt plates in a mull by sliding the plates (shearing them) apart. Do not attempt to pull the plates apart. The salt plates should be wiped clean with paper tissue and finally cleaned with chloroform and stored dry.

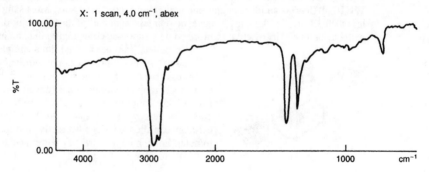

FIGURE 27.32 Infrared spectrum of Nujol®.

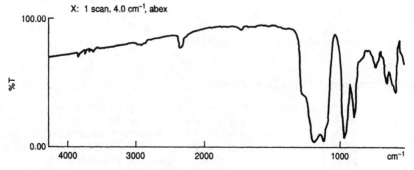

FIGURE 27.33 Infrared spectrum of Fluorolube®.

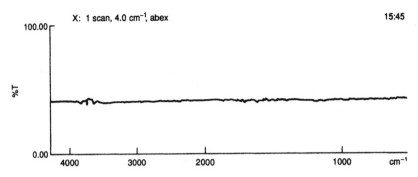

FIGURE 27.34 Infrared spectrum of pure KBr pellet.

A second method of preparing solid samples for infrared analysis is called the **potassium bromide pellet technique**. The solid sample is ground and incorporated into a thin, transparent disc (pellet) of pure KBr. The KBr becomes transparent (Fig. 27.34) when subjected to pressures of 10,000 to 15,000 psi. The KBr must be pure and moisture-free. A major advantage to pressing pellets by the die technique is that the sample pellet can be saved as a reference. This KBr die technique produces a high-quality (usually 13-mm-diameter) pellet, but does require a hydraulic press. Some KBr dies are equipped with a vacuum port to facilitate evacuating the cell and reducing moisture contamination. Prior to pressing, 1 mg or less of finely ground sample is mixed intimately with approximately 100 mg of KBr. The KBr bottle should always be returned to a desiccator when not in use. A grinding mill or **WIG-L-BUG®** is an ideal sample-mixing and -grinding accessory. Most samples can be prepared in less than 10 s using Plexiglas®, agate, or stainless steel ball pestles. A major disadvantage to the KBr pellet technique is its reactivity toward some samples as compared to the common mulling materials.

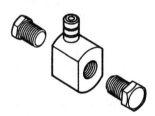

KBr pellets can be formed in a simpler mechanical device called a **minipress**, which does not require hydraulic pressure. The minipress consists of two highly polished bolts which are manually tightened against each other in a rugged steel cylinder as shown in Fig. 27.35. Some minipress cells come equipped with a vacuum outlet for removing moisture. The KBr pellet formed inside the cylinder can then be placed directly into the infrared spectrophotometer. Some manufacturers provide special wrenches and sockets to facilitate pressing the pellet.

FIGURE 27.35 Infrared sample minipress equipped with vacuum takeoff.

CAUTION: Never apply pressure to a die or tighten bolts without sample and KBr present as scoring might occur. One major disadvantage of the minipress is that the pellet must be destroyed in removing it from the cylinder. A newer KBr die technique called a **quick press** is now available which does not require a hydraulic press or wrenches. The quick press has three die sets (1, 3, and 7 mm) and is manually operated (Fig. 27.36).

H. Fourier Transform Infrared Spectroscopy

The Fourier transform spectroscope provides speed and sensitivity in making infrared measurements. The Michelson interferometer is a basic component of the Fourier transform instrument. A fundamental advantage is realized as the interfereometer "scans" the infrared spectrum—in fractions of a second at moderate resolution, a resolution that is constant throughout its optical range. These "scans" can be co-added tens, hundreds, or thousands of times. Co-addition of interferometer signals (called *interferograms*) reduces the background noise of the infrared spectrum dramatically.

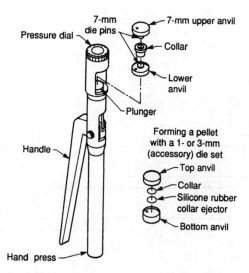

FIGURE 27.36 Infrared sample "quick press" with various dies.

An interferometer has no slits or grating; its energy throughput from the infrared source is high. High efficiency of energy passage through the spectral discriminating device means more energy at the detector, where it is most needed.

The Michelson Interferometer. The Michelson interferometer (Fig. 27.37) consists of two mirrors and a beam splitter. The beam splitter transmits half of all incident radiation from a source to a moving mirror and reflects half to a stationary mirror. Each component reflected by the two mirrors returns to the beam splitter (Fig. 27.38), where the amplitudes of the waves are combined to form an interferogram as seen by the detector. It is the interferogram that is then Fourier-transformed into the frequency spectrum.

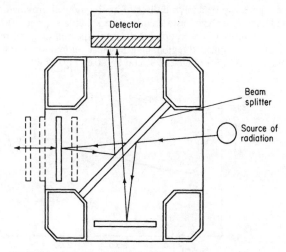

FIGURE 27.37 The Michelson interferometer.

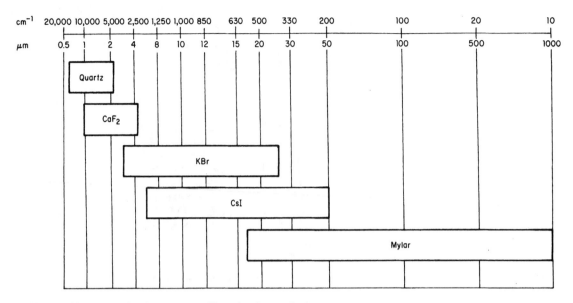

FIGURE 27.38 The wavelength ranges covered by various beam splitters.

The use of the interferometer allows almost all the source energy to be passed through the sample, since no radiation dispersion is required. This total energy availability is known as *Jacquinot's advantage* and increases both the speed of analysis and the sensitivity. The use of computers in Fourier transform spectroscopy has improved the signal-to-noise ratio in interferograms by allowing the co-addition of several scans, known as the *Fellgett's advantage*. The interferometer is coupled to the dedicated microprocessor which quickly yields on-line Fourier transformations. This immediate data processing and display allows the spectrometer parameters and results to be used almost instantaneously.

I. Infrared Absorption Spectra

Pure liquids can be analyzed directly in a liquid cell provided a suitable thickness is available. The liquid can be analyzed as a solution if a suitable solvent is available. A major requirement of the solvent is a lack of absorption bands in the spectral range of the solute. There are no nonabsorbing solvents in the infrared region, and the most commonly used ones are the nonpolar, non-hydrogen-containing CCl_4 (see Fig. 27.39) and CS_2 (see Fig. 27.40).

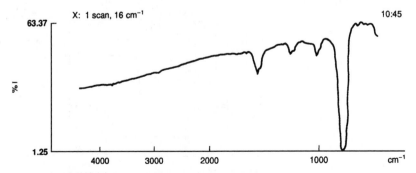

FIGURE 27.39 Infrared spectrum of carbon tetrachloride, CCl_4 (neat).

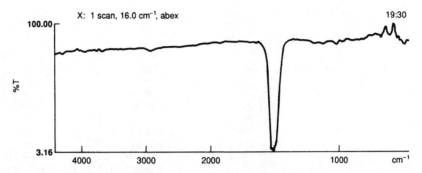

FIGURE 27.40 Infrared spectrum of carbon disulfide, CS_2 (neat).

CAUTION: CCl_4 and CS_2 are excellent solvents for infrared analyses but are potentially hazardous and should only be used with proper ventilation and precautions. A rough guide for analyzing solutions is 2 percent (wt/vol) solution for a cell with a 0.5-mm path length.

It would be impossible to include comprehensive coverage of infrared spectra on all common organic compounds in this *Handbook,* but the characteristic absorption bands for the major organic families (Table 27.7) and a listing of references on interpretation of infrared spectra (see References below) are given.

NEAR INFRARED (NIR) SPECTROMETER

Near infrared (NIR) analysis is a spectroscopic technique that uses the electromagnetic spectrum in the wavelengths between 800 nm (12,500 cm^{-1}) and 2500 nm (4,000 cm^{-1}). Wavelengths expressed in reciprocal centimeters (cm^{-1}) are called wave numbers. NIR is situated between the visible region and the mid-infrared region (see Fig. 27.41). NIR was developed in the 1970s in a United States Department of Agriculture (USDA) laboratory. Improvements in computers (speed, storage capacity, and new software) have contributed greatly to the development of new applications. The use of NIR has increased rapidly in the past few years, for several reasons:

- NIR instruments tend to be easier for nontechnical operators to use than conventional infrared spectroscopy and many other analytical techniques.
- NIR is extremely fast, requiring only seconds as compared to several hours by conventional wet-chemical methods.
- There is little or no preparation time required for an NIR analysis.
- NIR can be used for both liquid and solid samples and requires no chemicals. It is thus cheaper and more environmentally friendly.

As was discussed in Chapter 22, "Inorganic Chemistry Review," all matter is made up of atoms that contain protons, neutrons, and electrons. Neutrons are neutral, protons are positively charged, and electrons are negatively charged. These opposite charges help to hold molecules together by forming bonds. All molecular bonds stretching, bending, and twisting absorb specific wavelengths of light energy. These absorptions occur at specific wavelengths in the mid-infrared region and move through the NIR region as combinations and overtones of the fundamental absorptions.

NIR is an indirect method. Through the use of **chemometric** (chemical measurement) software, the spectral information is correlated to laboratory reference values (wet method results) for measuring constituents of interest. This NIR software uses multivariate regression methods, with the result being a calibration model that is used for sample purposes. The near infrared area has proved to be the ideal spectral area to make natural product and chemical property measurements of liquids, solids, slurries, and even gases.

TABLE 27.7 Characteristic Infrared Absorption Bands for Organic Families

Family	Bond	Types of vibration	Frequency range, cm^{-1}
Alkanes	C—H	Stretching	2800–3000
	C—H	Bending	1370 and 1385
Alkenes	C—H	Stretching	3000–3100
	C=C	Stretching	1620–1680
	cis	Bending	675–730
	trans	Bending	950–975
Alkynes	C—H	Stretching	2100–2260
	C≡C	Stretching	3000
Aromatics	C—H	Stretching	3000–3100
		Monosubstitued	690–710 and 730–770
		o-disubstituted	735–770
		m-disubstituted	680–725 and 750–810
		p-disubstituted	790–840
Alkyl halides	C—F	Stretching	1000–1350
	C—Cl	Stretching	750–850
	C—B	Stretching	500–680
	C—I	Stretching	200–500
Alcohols	O—H	Hydrogen-bonded	3200–3600
	O—H	Not hydrogen-bonded	3610–3640
	C—O	Primary	1050
	C—O	Secondary	1100
	C—O	Tertiary	1150
	C—O	Phenol	1230
Aldehydes	C=O	Stretching	1690–1740
Ketones	C=O	Stretching	1650–1730
Esters	C—O	Stretching	1080–1300
	C=O	Stretching	1735–1750
Ethers	C—O	Stretching	1080–1300
Amines	N—H	Stretching	3200–3500
	N—H	Bending	650–900 and 1560–1650
	C—H	Stretching	1030–1230
Amides	N—H	Stretching	3050–3550
	N—H	Bending	1600–1640 and 1530–1570
	C=O	Stretching	1630–1690
Carboxylic acids	C—O	Stretching	1080–1300
	C=O	Stretching	1690–1760

For example, a typical NIR determination on whole corn for percent moisture, percent oil, percent protein, and percent starch is shown in Table 27.8. The data are recorded on a computer hard drive or disk and ultimately converted on a PC using very specialized calibration software.

The combination and overtone bands in the NIR are much less intense than the primary bands in the mid-IR. NIR has a major advantage over conventional Mid-IR because of the latitude of materials that can be used in sampling and the instrument's optics, such as quartz, glass, and even some plastics. Typical molecular bonds that are analyzed by NIR include C—H, O—H, and N—H. The specific chemical groups are very useful for compounds that include moisture, fat, protein, starch, alcohol, and so on.

A. Continuous Spectra Instruments

Continuous spectra instruments (see Fig. 27.42) allow absorption spectral data at all wavelengths to be collected simultaneously. The instrument shown in Fig. 27.42 uses a broadband light source (usually

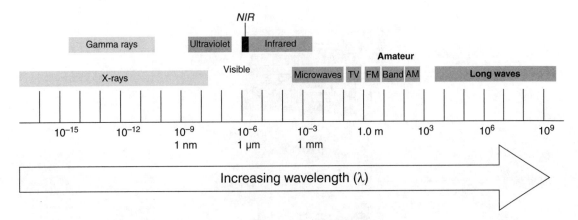

FIGURE 27.41 Electromagnetic spectrum showing the relative sizes of both the infrared and near infrared regions. (*Courtesy Perten Instruments North America, Inc.*)

TABLE 27.8 Typical NIR Concentration Ranges and Specific Absorption Bands

Moisture: 5–45% (O—H)
Oil: 3.29–10.7% (triglycerides, C—H; maybe O—H)
Protein: 5.83–14.79% (N—H, C—H)
Starch: 60.8–74.4% (O—H, C—H)

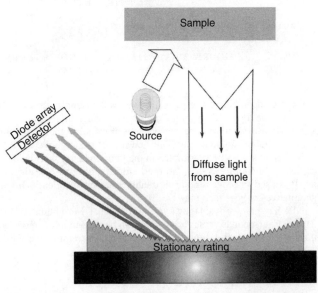

FIGURE 27.42 NIR monochromator using a stationary grating and a diode array detector. (*Courtesy Perten Instruments North America, Inc.*)

FIGURE 27.43 Perten Model DA7200 diode-array NIR analyzer. (*Courtesy Perten Instruments North America, Inc.*)

tungsten-halogen bulb). The light illuminates the sample surface, where it can be both absorbed and reflected. The light is then spread out on the surface of a stationary grating and broken down into its component wavelengths. The sensors in a diode array then detect and measure these various wavelengths of light simultaneously. This newer NIR technology can collect 600 raw spectra per second.

NIR analysis has been accepted in the grain industry worldwide. New applications are now being applied in ethanol processes, the food industry, clinical investigations, and environmental monitoring. It is used in process control, quality assessment, identification of raw material, and chemical quantitative analysis (see Fig. 27.43).

NIR has many advantages over conventional wet chemistry methods: (1) it is nondestructive, (2) very fast (seconds versus hours), (3) can be used in inline monitoring, (4) can handle a wide range of sample forms (solid, liquid, and slurry), and (5) requires very little sample preparation.

B. Process Spectroscopy Analyzers

Spectral information (especially turbidity, refractive index, visible, ultraviolet, and infrared) has been used for many years in process control. Other forms of spectroscopy—for example, X-ray and gamma radiation—have had wide use in measuring physical parameters (film thickness, container fillage, etc.) in process control. The recent development of the diode-array spectrophotometer allows spectrometers to perform very rapid spectral analyses. These diode-array detectors allow the spectrometers to measure all wavelengths simultaneously. In the classical UV/VIS/NIR spectrophotometers, light from a polychromatic source passes through a monochromator (a prism or a grating followed by a slit) that selects only a narrow band of light. The resulting almost monochromatic light passes through a measurement sample. Light transmitted through the sample strikes a photodetector; the signal output is then used to calculate sample absorption at each wavelength interval. By rotating the prism/grating, the instrument scans through all wavelengths of interest. Historically, accuracy and repeatability of a classical spectrophotometer were limited by the mechanism used to turn the prism or grating and tune the monochromator output wavelength. Measurements of samples with low absorption required long detector integration times because the monochromator scans each wavelength interval sequentially. The precise sensitive scan mechanism also reduces the suitability of the spectrophotometer for applications in harsh process control and quality-assurance environments. The new computer-controlled diode-array spectrophotometer found in both laboratory and process instruments has eliminated or minimized most of these problems.

REFERENCES

1. Association of Official Analytical Chemists (AOAC), *Official Methods of Analysis of the Association of Official Chemists*, 14th edition. AOAC (1985).
2. Ewing, G. W., *Instrumental Methods of Chemical Analysis,* 4th edition. McGraw-Hill, Inc. (1975).
3. Willard, H. H., Merritt, L. L., Dean, J. A., and Settle, F. A., *Instrumental Methods of Analysis,* 6th edition. Wadsworth Publishing Co. (1981).
4. Skoog, D. A., *Principles of Instrumental Analysis*, 3rd edition. Saunders College Publishing (1980).
5. Crooks, J. E., *The Spectrum in Chemistry*, Academic Press. ISBN 978-0121-95550-2 (1978).
6. Silverstein, R. M., Bassler, G. C., and Morrill, T. C., *Spectrometric Identification of Organic Compounds,* 3rd edition. John Wiley and Sons Inc. (1974).
7. Gordon, Arnold J., and Ford, Richard A., "Spectroscopy Section," *The Chemist's Companion: A Handbook of Practical Data, Techniques, and References.* John Wiley & Sons, Inc. ISBN 0-471-31590-7 (1972).

CHAPTER 28
ATOMIC ABSORPTION SPECTROSCOPY

INTRODUCTION

When atoms are placed in a flame or other energized (electrical) source, their electrons can be elevated to an excited state. Not all electrons are elevated to this excited state at the same time; millions remain in their unexcited or ground state. These ground-state atoms are capable of absorbing resonance energy, while the excited atoms are capable of emitting resonance energy. Spectroscopists take advantage of these two states by measuring the atoms in the ground state in atomic absorption spectroscopy and atoms in an excited state by emission spectroscopy (see Fig. 28.1).

The word **resonance** describes the energy that is exactly equal to the difference between the ground state and the excited state of a given atom. This energy, in the form of light, has a particular wavelength called the **resonance wavelength**. Since the ground-state electrons may be shifted into one or more permissible energy states, a given element exhibits more than one resonance line. These wavelengths form the emission spectrum of an element. Nickel, when excited, emits radiation at the following wavelengths: 231.1, 232.0, 341.5, 346.2, and 352.4 nm. Conversely, nickel in the ground state will absorb radiation only at these same wavelengths (or resonance lines). Although all of these lines appear in the emission spectrum, some are more intense than others. The lines in an atomic absorption spectrum also vary in intensity. The strongest line is known as the *primary resonance line,* and for nickel it just happens to be 232.0 nm.

Atomic absorption spectroscopy (AA) is a spectrophotometric technique based on the absorption of radiant energy by atoms. To carry out the atomic absorption process, we need a source of monochromatic radiation and a means to vaporize the sample producing ground-state atoms of the element being analyzed. Unlike AA, emission spectroscopy is based on a principle that requires free atoms to be excited by a source of energy (flame), and these excited atoms emit a characteristic radiation as they return to their ground (unexcited) state. If a beam of light is passed through the flame, these ground-state atoms will absorb this energy. In order for absorption to occur, the wavelength of this radiation must be characteristic of atoms (element) present. Table 28.1 shows the characteristic wavelengths for some specific elements.

Approximately 70 elements can be determined by AA in concentrations ranging from perhaps 10 ppm for some of the more difficult rare earths to less than 1 ppb for mercury by the graphite furnace method.

The absorption of this specific radiation follows Beer's law and is directly proportional to the concentration of atoms in the flame. These spectral lines necessary for AA applications (described in The Electromagnetic Spectrum in Chapter 27) are approximately 0.002 nm wide. Therefore a continuous radiation source, such as a tungsten lamp, cannot provide sufficient energy. A source called a **hollow cathode lamp**, as described later, is usually required for each element to be investigated. Since each element usually requires a different source, AA is a poor qualitative tool but is extremely useful for quantitative determinations. Figure 28.2 shows a typical response curve with concentration versus absorption for an element by atomic absorption spectroscopy.

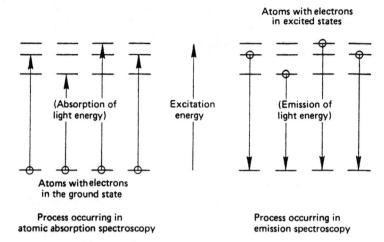

FIGURE 28.1 Electron transitions in atomic absorption and emission spectroscopy.

TABLE 28.1 Characteristic Atomic Absorption Wavelengths for Some Elements

Element	Wavelength, nm	Element	Wavelength, nm
Aluminum	309.3	Lead	217.0
Antimony	217.6	Manganese	279.5
Arsenic	193.7	Nickel	232.0
Barium	553.6	Potassium	766.5
Beryllium	234.9	Silicon	251.6
Cadmium	228.8	Silver	328.1
Chromium	357.9	Sodium	589.0
Copper	324.7	Tin	286.3
Gold	242.8	Vanadium	318.5
Iron	248.3	Zinc	213.9

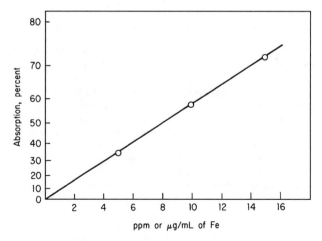

FIGURE 28.2 Typical atomic absorption spectroscopy calibration curve for iron.

ATOMIC ABSORPTION SPECTROPHOTOMETER COMPONENTS

Schematic diagrams of the monochromator for both single- and double-beam atomic absorption spectrophotometers are shown in Fig. 28.3. The double-beam instruments are not as subject to fluctuations in source intensity, require less warm-up time, and give more reproducible results.

The source of radiant energy is provided by a hollow cathode lamp in most conventional AA applications. These hollow cathode lamps usually contain two electrodes: a cathode and an anode, as shown in Fig. 28.4. The cathode is cup-shaped and made of the element that is being investigated. The anode can be made of any suitable material—for example, tungsten. The lamp is filled with neon or argon gas at a low pressure. When an electric current passes through the cathode, metal atoms are vaporized from the cathode. These lamps are sealed and usually contain a noble gas: helium, argon, or neon gas at 4 to 10 torr pressure. These noble gases prevent the vaporized metal from condensing on the surface of the tube. When energized, the metal atoms contained in the lamp's cathode are bombarded by the ionized noble gas, elevated to an excited state, and ultimately emit light energy as they return to their ground state. The window material in the lamp is normally made of quartz for wavelengths shorter than 250 nm, and glass is used for wavelengths longer than 250 nm. These lamps routinely have an operating life of approximately 1000 hours (at 5 to 25 milliamps range), but the lamp can be abused with higher operating currents. The manufacturer supplies specifications for maximum operating conditions above which the cathode may be destroyed. Most hollow cathode lamps require between 5 and 30 minutes to warm up; if a lamp requires more than 30 minutes to warm up, it probably should be replaced. Multielement lamps are also available in which a metallic powder from several highly purified elements is combined into a sintered cathode. These multielement lamps are normally limited to two or three elements because of their more complex spectral emissions.

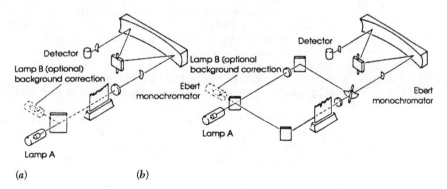

FIGURE 28.3 Schematic diagram of typical AA spectrophotometers: (*a*) single-beam; (*b*) double-beam. (*Courtesy McGraw-Hill, Inc.*, The Chemist's Ready Reference Handbook.)

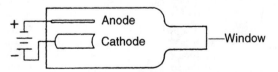

FIGURE 28.4 The hollow cathode tube.

Other types of radiation sources are available for atomic absorption spectroscopy. Vapor discharge lamps are used in AA applications, and these lamps operate by having an electric current pass through a vapor containing the element of interest, such as mercury. This type of lamp is also used in conventional spectrophotometers as a source of ultraviolet radiation. Another radiation source is the **electrodeless discharge lamp (EDL)**. This radiation source does not contain a cathode or an anode. It is constructed of a small-diameter quartz tube which contains the metal or metal salt of interest. These tubes are filled at a very low pressure with argon gas and wrapped with an electric coil. The tube is inductively coupled to an RF field and the metal atoms raised to an excited atomic state. Approximately 20 different elements are available in these EDLs, and they generally produce a much stronger spectrum than conventional hollow cathode lamps. The characteristic line spectrum emitted from a lamp is directed through a flame that contains unexcited atoms of a metal sample. Only if the metal atoms suspended in the flame and the hollow cathode filament atoms are the same will light-energy absorption occur. This absorbance is proportional to the concentration of metal atoms in the flame.

Each double-beam AA monochromator has a **chopper** located in the optical beam between the cathode lamp source and the flame. This chopper is actually a rotating half mirror; the beam is allowed to pass directly through the flame half the time and is reflected to bypass the flame for half the time (called the **reference beam**). This allows the detector to compare the radiation from the source and the flame itself. A **photodetector** measures the light passing through the flame both before and after the sample is introduced into the flame. Most AA detectors are basically the same as those used in ultraviolet and visible spectroscopy, photomultiplier tubes being the most common.

All conventional AA instruments require a burner and a flame. The **burner head** provides a means for suspending and exposing the free metal atoms to the hollow cathode light beam. There are two commonly used gas mixtures: (1) air-acetylene, which burns at approximately 2300°C, and (2) a nitrous oxide-acetylene combination that produces flame temperatures in the 3000°C range. Usually a 10-cm-slot burner is used with air-acetylene mixtures and a smaller 5-cm-slot burner with the hotter nitrous oxide-acetylene mixtures. A water-filled loop in the drain tube prevents air from coming through the drain into the burner where it might cause an explosion. The burner orifice consists of a long, narrow slit, which produces a non-turbulent, quiet flame. The advantages of this type of burner are a long path length for the light beam, stability, and a low-noise (both audible and optical) reference path. The sample and reference beams are recombined and the ratio of the intensity of the two types of pulses is measured electronically. The major advantage of this system is that effects of lamp drift are overcome; consequently the baseline tends to be more stable. Its major disadvantage is that half of the light energy is lost because the beam is split.

Table 28.2 provides a listing of some common elements that may be determined with different AA flames.

Most atomic absorption burners are not of the total consumption type, sometimes referred to as **turbulent-flow**. AA burners tend to be of the premix (**laminar-flow**) type, which aspirate the sample solution through a capillary tube into a mixing chamber where the sample is atomized and mixed with the fuel (acetylene). Some AA instruments use a **nebulizer**, which simply reduces the

TABLE 28.2 Elements Analyzed by Atomic Absorption Spectroscopy and Flame Mixtures Commonly Used

Air-acetylene mixture	Nitrous oxide-acetylene mixture
Antimony	Aluminum
Chromium	Beryllium
Cobalt	Germanium
Copper	Silicon
Lead	Tantalum
Iron	Titanium
Magnesium	Tungsten
Zinc	Zirconium

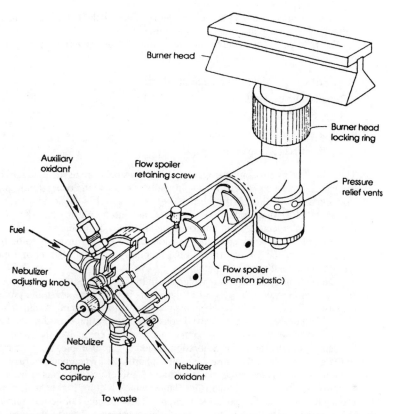

FIGURE 28.5 Typical atomic absorption spectrophotometer nebulizer and burner assembly. (*Courtesy Perkin-Elmer Corporation.*)

sample solution by a pneumatic process to a spray of droplets of various sizes. The larger droplets of the sample settle to the bottom of the mixing chamber because of a series of baffles. A majority (approximately 90%) of the aspirated sample drains from the bottom of the mixing chamber and is discarded; only a small (but reproducible) portion of the sample solution passes through the burner head. A typical nebulizer and slot burner assembly is shown in Fig. 28.5. Periodic cleaning of the nebulizer and burner assembly will help to minimize noise levels. Scraping the burner head to remove carbon deposits and placing the dismantled assembly in an ultrasonic cleaner should be a routine task. Consult the manufacturer's manual for specific details.

FACTORS AND CONDITIONS THAT EFFECT AA SPECTROPHOTOMETER RESPONSE

Atomic absorption spectrophotometry measures only the elemental composition of a sample. Most molecules in the sample are broken down into atoms by the flame. It is the electrons of the individual atoms that absorb radiation. AA is dependent on the emission of characteristic radiation from a hollow cathode tube and the absorption of that radiation by the atoms in a sample. Thus, only one element is measured at a time. For instance, to determine silver, you would use a silver lamp as a source. Even if you used a multielement lamp, you would set the monochro-mator for the wavelength of one element at a time. However, instruments are available that can look at several wavelengths at a time and simultaneously measure more than one element.

The following factors influence the quantity of the material reaching the flame and the overall response:

- Aspiration rate of the material to the flame
- Temperature of the flame
- Viscosity of the material
- Surface tension of the solvent
- Matrix effects and interferences

The aspiration rate is controlled largely by the gas pressure used, although the viscosity of the material can also have an effect. If the solution contains a high concentration of dissolved solids, it will flow more slowly through the burner. The use of organic solvents may increase the absorbance by a factor of 2 to 4. Many industrial procedures are based on using **MIBK (methylisobutyl ketone)** as a solvent. An organic solvent improves the efficiency of the burner since its surface tension is lower than that of water. A lower surface tension allows smaller droplets to be formed. Smaller droplets permit more of the sample to reach the flame in the premix burner and more of the sample to be burned in the total consumption burner.

The percentage of the atoms which are in the ground state is controlled by the temperature of the flame. In most systems nearly all of the atoms in the flame are in the ground state; however, there are cases in which interfering elements prevent the formation of ground-state atoms. This interfering effect of other elements has received considerable attention over the past few years. Sometimes the element to be determined forms compounds or stable complexes that do not dissociate into atoms at the same temperature as the calibration standard. For example, fewer calcium atoms are converted to the ground state in the presence of phosphorus than when phosphorus is absent. Silicon and aluminum have a marked effect on strontium. These effects are caused by the formation of silicates, aluminates, phosphates, etc., in the flame, which require higher than normal temperatures to be broken down to neutral ground-state atoms. These effects can be minimized by adding a variety of substances that can "tie up" these interfering substances. These substances have come to be known as **competing ions**. For example, lanthanum, has been used successfully in a variety of cases.

ATOMIC ABSORPTION SAFETY CONSIDERATIONS

Acetylene is the primary fuel source in AA applications. This particular gas is very flammable. A phenomenon *called* **flashback**, which is a minor explosion inside the burner assembly, can occur with an improper mixture of fuel and air. Most instruments come equipped with safety cables and shields to protect the operator; these safety devices should never be removed. The potential of flashback can be reduced by periodically cleaning the burner nebulizer and head following the manufacturers suggested procedures. Flashbacks can be caused by air being drawn through the drain line on the premix burner. This drain line should always contain a loop filled with solution, and the end of the drain tube should be below the surface of the waste solution as shown in Fig. 28.6. As with any potential source of hazard, follow the directions as provided by the manufacturer for your specific AA instrument.

CAUTION:

- ALWAYS wear safety glasses while operating an AA.
- ALWAYS have a fire extinguisher available.
- ALWAYS turn the AA ventilation system on before starting any work.
- ALWAYS turn the air on first when igniting the AA burner.
- ALWAYS turn the air off last when extinguishing the AA burner.

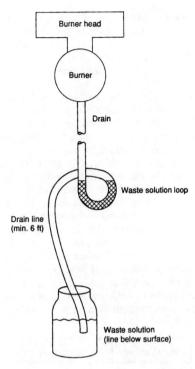

FIGURE 28.6 Proper drain-line procedure for an AA instrument.

Each AA instrument should have a dedicated ventilation system above the burner. The actual operation of the instrument can create a potential hazard with free metal atoms being exhausted into the laboratory atmosphere. Elements such as beryllium and mercury are extremely toxic in the free atomic state. Remember that the instrument could be exhausting potentially toxic metal atoms, even if not analyzing for them specifically.

ATOMIC ABSORPTION CALIBRATION

The performance of an atomic absorption spectrophotometer is generally expressed in terms of **sensitivity** and **detection limit**. Sensitivity is defined as the concentration of an element necessary to absorb 1 percent of the incident light energy. Detection limit is defined as the concentration of an element necessary to cause a reading equal to twice the standard deviation of the background signal. Another, perhaps more useful definition of detection limit is the concentration that gives a signal twice the size of the noise level of the background.

In any spectrophotometer, the amount of radiation absorbed or transmitted can be registered or recorded. As you find with many instrumental techniques, the output for atomic absorption spectrophotometry has not been standardized and varies from one instrument to another. The phototube detector measures the intensity of transmitted radiation. It is set with a blank at 100 percent transmission (or 0 percent absorption). The other end of the meter scale is adjusted to some arbitrary value for a "standard" (known) concentration. Atomic absorption, like any other method of absorption spectrophotometry, relies on the applicability of **Beer's law**:

$$A = -\log T = abc$$

Some instruments indicate absorbance (*A*); others indicate percent transmittance (%*T*); some have both scales on the same meter. In atomic absorption analysis the term **% absorption** is commonly used:

$$\% \text{ absorption} = 100 - (\% \text{ transmittance})$$

Note that although % absorption increases with concentration, the relationship is not a simple, direct proportion. If the meter on your instrument reads % absorption, you must change the reading to absorbance. The relationship between these quantities is shown in Fig. 27.7 along with the types of calibration curves they give. Note that since absorbance (*A*) is proportional to concentration (*c*), a linear calibration curve expresses the relationship. However, **percent transmittance (%T)** versus concentration (*c*) is a logarithmic relationship. Consequently, percent absorption (% absorption) versus concentration (*c*) is also a logarithmic relationship. Any meter scale that is linear in absorbance units can be calibrated to read concentration directly (for a given standard procedure only).

Another calibration term that should be considered is linearity. **Linearity** is defined as the linear relationship between sample concentration and sample absorption or response. Most elements at higher concentrations become nonlinear in their absorption or response on a spectrophotometer. This phenomenon is demonstrated in the graph shown in Fig. 28.8.

Notice how the absorption is linear up to about 50 ppm and then remains almost constant at higher concentrations. This loss of linearity can be caused by several factors, one of which is a **"saturation"** of the detector as the concentration (signal) increases. A calibration curve like this should be determined for each analysis to determine the optimum concentration range. In this example, the optimum concentration range would be from 0 to 50 ppm.

Absorbance	% Transmittance	% Absorption
0	100	0
0.045	90	10
0.097	80	20
0.155	70	30
0.229	60	40
0.301	50	50
0.398	40	60
0.523	30	70
0.699	20	80
1.00	10	90
∞	0	100

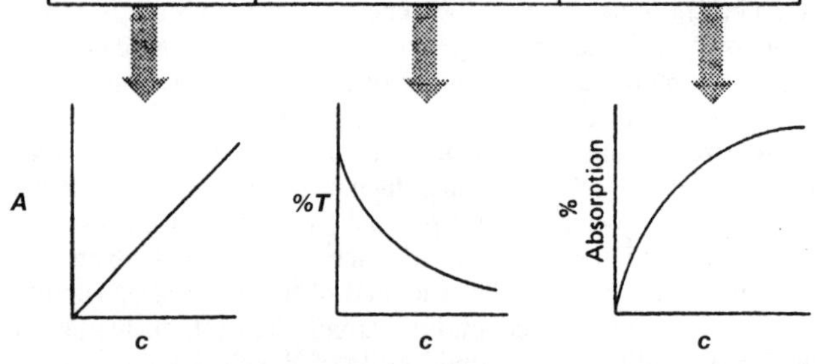

FIGURE 28.7 Relationship between Absorbance, % Transmittance, % Absorption, and Concentration.

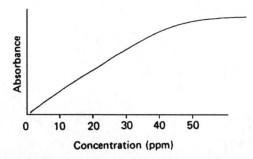

FIGURE 28.8 Effects of concentration on response in AA spectroscopy.

A. Calibration by Standard Additions Method

In atomic absorption, as for all techniques requiring calibration, the calibration standards should be very similar to the material being analyzed. For example, if you wish to analyze samples containing phosphoric acid for sodium, you should make your sodium standards in sodium-free phosphoric acid. Suppose it is impossible to prepare standards of the same composition as the sample. We may not even know what else may be in the sample. To avoid errors caused by not being able to prepare standards identical to the unknowns, we can use a technique known as **standard additions**. Unknown solutions can vary widely in their viscosity and effect the rate of sample aspiration into the flame. The technique of standard additions can reduce viscosity and matrix effect. Known quantities of the element of interest are spiked directly into the unknown solution. The calibration technique of standard additions consists of adding small increments of known concentrations directly to the unknown solution and determining the absorbance after each addition. This technique virtually eliminates errors caused by the sample's complexed **matrix** (background) and viscosity since these variables are constant in both the sample and standard addition samples.

The most accurate technique for using standard additions would be to plot the standard additions concentrations against the absorbance data and extrapolate the graph to zero absorbance. The length of the x-axis (concentration) from the origin to the point of intersection represents the original concentration of the unknown. The standard additions technique should be used only at low concentrations where we expect the calibration curve to be linear. Figure 28.9 shows a typical set of data and standard additions graph.

ATOMIC ABSORPTION INTERFERENCES

Atomic absorption spectroscopy is not subject to spectral interferences because of the narrow bandwidth created by the hollow cathode lamp. A potential source of interference can be certain anions with specific cations. For example, sulfates, phosphates, and silicates are known to interfere with some alkaline earth cations (namely, calcium, strontium, and barium). This particular type of chemical interference can be reduced by the addition of **lanthanum ions** to the sample solution, which compete for the interfering anions.

Background absorption occurs in atomic absorption spectroscopy as molecular absorption and as light scattering by particulate matter (unevaporated droplets and unevaporated salt particles remaining after desolvation of the aerosol). The absorption is usually broad in nature compared to the monochromator band pass and line-source emission widths. Background correction is often required for elements with resonance lines in the far-ultraviolet region and is essential to achieve high accuracy in determining low levels of elements in complex matrices. One commonly used form of background

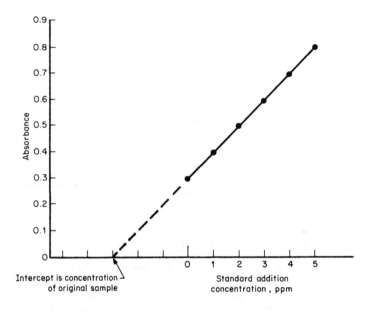

FIGURE 28.9 Standard additions calibration curve for AA.

correction, called the **continuum source**, uses a **deuterium lamp** for elements in the range of 180 to 350 nm and a **tungsten-halide lamp** for elements in the range of 350 to 800 nm. The continuum source is aligned in the optical path of the spectrometer so that light from the continuum source and light from the primary lamp source are transmitted alternately through the flame. The continuum source has a broadband emission profile and is only affected by the background absorption, whereas the primary lamp source is affected by background absorption and absorption resulting from the analyte in the flame. Subtraction of the two signals (electronically or manually) eliminates or at least lessens the effect of background absorption. It is difficult to match the reference and sample beams exactly because different geometries and optical paths exist between the two beams.

COMPUTERIZED ATOMIC ABSORPTION SPECTROPHOTOMETERS

Computerized AA instrumentation modernizes AA analyses. It can offer advanced capabilities with simple operations (multitasking, using PC to develop new methods, permitting simultaneous access to commercially available software) to enhance productivity of the laboratory. Results can be archived into built-in databases, automatically ensuring safe storage of data. Graphics can be stored and reviewed during and after runs. Some have report windows that allow customization of in-run and post-run results, including calibration, signal graphics, all statistics and method parameters, and compatibility with third-party word processors and spreadsheets. They offer fast and accurate correction of background signals, ensure precise and accurate answers for each sample off-line sample preparation system, and are connectable to laboratory PCs. Data can be stored in both height and area in all peak modes, allowing measurement modes to be reviewed later. All raw data, signals, methods, sequences, mean absorbance, mean concentration, replicate readings, and mean background signals can be stored. Error logs are available on-screen with complete fault diagnosis.

GRAPHITE FURNACE TECHNIQUE

Graphite furnace atomic absorption spectrometry (GFAAS) has grown in popularity due in part to its enhanced sensitivity and the ability to determine samples with a relatively complicated matrix. Trace elements in samples such as blood, urine, and environmental samples are routinely measured by GFAAS. This technique is more efficient than conventional AA because most of the sample is atomized and stays in the optical beam. The typical graphite furnace consists of a small cylindrically shaped graphite tube equipped with an injection opening on the top. A cross-sectional drawing of a heated graphite atomizer is shown in Fig. 28.10. A measured amount (usually 5 to 50 µL) of sample is placed directly into the interior of the furnace. The furnace is normally heated electrically in three stages: (1) to a relatively low temperature to drive off the solvent, (2) to a higher temperature to ash the sample, and (3) finally to incandescence (approximately 2500°C) to atomize the sample. The graphite tube must be surrounded by a nitrogen or argon atmosphere to prevent air oxidation of the sample. The detection limits of this technique, of course, depend upon the element and operating conditions, but detection results in the 1×10^{-12} g range can be obtained. For example, mercury is routinely analyzed by this technique, where a detection limit of 3×10^{-10} g/mL is not uncommon. The graphite furnace technique has several advantages over conventional atomic absorption: (1) it requires only a very small sample volume (5 to 50 µL) versus several milliliters, (2) atomization of the sample is nearly 100 percent complete versus perhaps only 0.01 percent by AA, and (3) detection limits are improved for some elements by a factor of 1000 or greater. Matrix interferences and background absorption effects are common disadvantages to the graphite furnace technique.

Automatic GFAAS systems employ advanced software programs that perform a range of tasks, including spectrometer control, furnace temperature control, and data acquisition and reporting. The process is totally automated to virtually eliminate operator input. Ash-atomized software allows the analyst to select a range of ash temperatures and a range of atomization temperatures. A specified

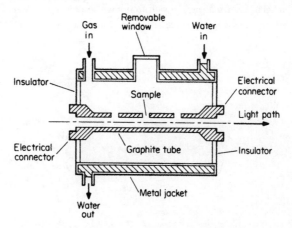

FIGURE 28.10 Cross-sectional view of a heated graphite atomizer. (*Courtesy Perkin-Elmer Corp.*)

sample is then measured at the range of ash. Modern GFAAS instruments provide automatic QC software facilities that should include automatic spike-recovery measurements. Recovery measurements are a very useful indicator of chemical or vapor phase interferences. A typical sample is analyzed to measure the concentration of analyte, and the analysis is then repeated after the sample has been spiked with a specified addition of analyte. The recovery of the addition is automatically calculated and flagged if it is outside specified limits. Recoveries should normally be 100% ± 10%; values outside this range are normally indicative of interference effects that must be corrected.

MERCURY ANALYZERS

Flameless atomic absorption spectroscopy has become the common method for the determination of mercury, using the cold vapor technique. Mercury behaves and performs as a monoatomic gas even at room temperature due to its high vapor pressure. The mercury vapor is transported by a supporting gas into a measuring cuvette, where its concentration is determined according to the Beer-Lambert principle using the selective absorption of the Hg 253-nm line. Analyzers that use fiber optics and nonadjustable optics exhibit exceptional stability. A commercially available automated unit consists of one mercury lamp and two mercury absorption cuvettes installed in a heated case along with the detection system. This system can consist of an interference filter and photomultiplier both optimized for mercury 253.7 nm. The fiber optics transfer the light from the mercury lamp to the cuvette and from the cuvette to the detector system. Stable measuring conditions are achieved by using the double-beam technique (sample and reference cuvette) along with automatic intensity tuning. The computer controls the supporting gas flow rates permitting measuring cycles of around 2 minutes. For the determination of extreme trace levels, the mercury vapor can be amalgamated into a precious-metal gauze and then volatilized to achieve detection limits of 2 ppt or better. Upper limits can be defined for added security against contamination by high mercury concentrations by automatically switching to a flushing (cleaning) solution.

REFERENCES

1. *Applied Atomic Spectroscopy,* vols. 1 and 2, E. L. Grove, editor. Plenum (1978).
2. Slavin, M., *Atomic Absorption Spectroscopy*. John Wiley and Sons (1978).
3. Van Loon, J. C., *Analytical Atomic Absorption Spectroscopy—Selected Methods*. Academic Press (1980).
4. Welz, B., *Atomic Absorption Spectrometry,* 2nd edition. VCH (1985).

APPENDIX A
COMMONLY USED ABBREVIATIONS

Å	angstrom (unit wavelength measure)
A	ampere
abs	absolute
ac	alternating current
ADC	analog to digital converter
AI	analyzer indicator
AIChE	American Institute of Chemical Engineers
amor or amorph	amorphous
anhyd	anhydrous
ANSI	American National Standards Institute
API	American Petroleum Institute
APSS	automatic pump start system
aq	aqueous
AR	analyzer recorder
ASQC	American Society for Quality Control
ASTD	American Society for Training and Development
atm	atmosphere
at no	atomic number
at wt	atomic weight
B&W	Babcock and Wilcox, contruction company
B/D	blow down
B/L	bill of lading; battery limits
B/V	ball valve or block valve
Bé	Baume degree
BFW	boiler feed water
bp	boiling point
Btu	British thermal unit
°C	degree Celsius
C	coulomb, electric charge, A-s
ca	approximately

Cal	large calorie (kilogram calorie or kilocalorie)
cal	small calorie (gram calorie)
cd	candela, luminous intensity
CD-ROM	compact disk-read only memory
cg	centigram
cgs	centimeter-gram-second (system of units)
Ci	curie
cm	centimeter
cm^2	square centimeter
cm^3	cubic centimeter (cc, or milliliter)
CMA	Chemical Manufacturers Association
conc'd	concentrated
cos	cosine
cp	candlepower
cP	centipoise
CSE	confined space entry
cu	cubic
CV	compliance verification documents
D	density diopter
d	day; deci (10^{-1}), one-tenth
DAC	digital to analog converter
dB	decibel
dc	direct current
deg or °	degree
dg	decigram
dil	dilute
DOT	Department of Transportation
dr	dram
dyn	dyne
E	electric tension, electromotive force
e.g.	for example
EHS	Environmental Health and Safety
emf	electromotive force
ESD	emergency shutdown device
esu	electrostatic unit
etc.	and so forth
et. seq.	and the following
eV	electron volt
°F	degree Fahrenheit
F	frictional loss
F	farad, capacitance, C/V
f	frequency

FD	forced draft (when applied to fans)
FI	flow indicator
FIC	flow indicator control
FIR	flow indicator recorder
F/m	farad per meter, permittivity
FQ	flow quanity
FR	flow recorder
FRC	flow recorder chart
FT	flow transmitter
ft	foot
ft^2	square foot
ft^3	cubic foot
ft c	foot-candle
ft lb	foot-pound
g	acceleration due to gravity
g	gram
G	gauss
gal	gallon
g cal	gram-calorie
gr	grain
h	Planck's constant
h or hr	hour
H	henry, inductance, WB/A
HAZMAT(s)	hazardous materials
HAZOP(s)	hazardous operations
HAZWOPER(s)	hazardous waste operations
HCV	hand control valve
HIs, HIHIs	high-level indicators, HIHIHI used for extremely high levels
H/m	henry per meter, permeability
hp	horsepower
HP	high pressure
hph	horsepower-hour
HRU	heat recovery unit or hydrogen recovery unit
hyg	hygroscopic
Hz	hertz (formerly cycles per second, cps)
I	electric current
$_a^z$I	symbol for isotope with atomic number z and atomic mass a
IA	current alarm
ibid.	in the same place
ID	induced draft (when applied to fans)
IDLH	immediate danger to life and health
i.e.	that is

APPENDIX A

in	inch
in²	square inch
in³	cubic inch
insol.	insoluble
ISO	International Standards Organization
iso	isotropic
J	joule; mechanical equivalent of heat
J/K	joule per Kelvin, entropy (heat capacity)
J/(kg-K)	joule per kilogram-Kelvin, specific heat capacity
J/m³	joule per cubic meter, energy density
J/mol	joule per mole, molar entropy, molar energy
J/(mol-K)	joule per mole-Kelvin, molar heat capacity
k	kilo (1000)
K	Kelvin
kc	kilocycle
kcal	kilogram-calorie
kg	kilogram
K/O	knock out
kW	kilowatt
kWh	kilowatt-hour
λ	lambda; wavelength; coefficient of linear expansion
L	liter
l	lumen
l	length
lb	pound
lb/ft³	pound per cubic foot
LCV	level control valve
LED	light-emitting diode
LEL	lower explosion level
LI	level indicator
LIC	level indicator control
LO(s), LOLO(s)	low-level indicators, LOLOLO used for extremely low levels
lm	lumen, luminous flux, cd-sr
ln	natural logarithm
log	logarithm
LTT	lock/tag/try
lx	lux, illuminance, cd-sr/m²
M	molar, as 1 M
M	mega (10^6), one million
m	meter
m²	square meter
m³	cubic meter

μ	micro- (10^{-6}), one millionth
μm	micrometer (micron)
meq	milliequivalent
MeV	million (or mega-) electronvolt
mg	milligram
MHz	million (or mega-) hertz
min	minute
′	minute, 1/60 degree
″	second, 1/60 minute
mks	meter-kilogram-second (system of units)
mL	milliliter
mm	millimeter
mm^2	square millimeter
mm^3	cubic millimeter
mm Hg	millimeter of mercury; torr
mol	mole
MOV	motor-operated valve
mp	melting point
mph	miles per hour
MSA	Mine Safety Appliance (certified as safe in hazardous environments)
MSDS	material safety data sheet
MW	molecular weight; mass
N	newton, force, $kg\text{-}m/s^2$
N	normality, as in 1 N
n	index of refraction; neutron (subatomic particle)
N/A or n/a	not applicable
N-m	meter newton, moment of force
N/m	newton per meter, surface tension
nm	nanometer (10^{-9}), one billionth
Ω	ohm, electric resistance, V/A
oz	ounce
P	poise
Pa	pascal, pressure, stress energy, N/m^2
Pa-s	pascal-second, dynamic viscosity
pH	measure of hydrogen ion concentration of a solution
PM	preventive maintenance
ppb	parts per billion
PPE	personal protective equipment
ppm	parts per million
ppt	precipitate
PRD	pressure relief device
PSC	process safety coordinator

PSI	process safety information
psi	pounds per square inch
psia	pounds per square inch-absolute
psid	pounds per square inch-differential
psig	pounds per square inch-gauge
p. sol.	partly soluble
PSV	pressure safety valve
Q	energy of nuclear reaction
qt	quart
q.v.	which see
R	roentgen (international unit for x-rays)
rad	radian, plane angle (about 57.3°)
RAM	random access memory
Re	Reynolds number
ROM	read only memory
rpm	revolutions per minute
rps	revolutions per second
RTD	resistance temperature detector
s	second
S	siemens, conductance, A/V
sat'd	saturated
sin	sine
SMP	standard maintenance procedure
SOCL	standard operating condition(s) and limit(s)
sol'n	solution
SOP	standard operating procedure
SP	special purpose
SW	switch
sp	specific
sp gr	specific gravity
sp ht	specific heat
sq	square
sr	steradian, solid angle
T	temperature
T	tesla, Wb/m^2
t	time
tan	tangent
TI	temperature indicator
TT	temperature transmitter
UEL	upper explosion limit
UPS	uninterruptable power supply
USCG	United States Coast Guard

V	volt
W	watt, power, J/s
Wb	weber, magnetic flux, volt-s
Wh	watt-hour
W/m^2	watt per square meter
W/(m-K)	watt per meter-Kelvin, thermal conductivity
W/O	work order
W/sr	watt per steradian, radiant intensity
y or yr	year

APPENDIX B
PHYSICAL PROPERTIES OF WATER

VAPOR PRESSURE AND DENSITY OF WATER AT VARIOUS TEMPERATURES

Pure water has a maximum density of 1.00000 g/mL at 4°C and at one atmosphere of pressure. The density of water at all other temperatures is less than 1 g/mL, as is shown by this table. The vapor pressure of water increases as the temperature increases: note that at 100°C, water has a vapor pressure of one atmosphere (760 torr).

Temperature, °C	Density, g/mL	Vapor pressure, torr
0	0.99987	4.6
1	0.99993	4.9
2	0.99997	5.3
3	0.99999	5.7
4	**1.00000**	6.1
5	0.99999	6.5
6	0.99997	7.0
7	0.99993	7.5
8	0.99988	8.0
9	0.99981	8.6
10	0.99973	9.2
11	0.99963	9.8
12	0.99953	10.5
13	0.99941	11.2
14	0.99927	12.0
15	0.99913	12.8
16	0.99897	13.6
17	0.99880	14.5
18	0.99863	15.5
19	0.99844	16.5
20	0.99823	17.5
21	0.99802	18.7
22	0.99780	19.8
23	0.99757	21.1
24	0.99733	22.4
25	0.99708	23.8
26	0.99681	25.2
27	0.99654	26.7
28	0.99626	28.3
29	0.99598	30.0
30	0.99568	31.8

(*Continued*)

Temperature, °C	Density, g/mL	Vapor pressure, torr
31	0.99537	33.7
32	0.99506	35.7
33	0.99473	37.7
34	0.99440	39.9
35	0.99406	42.2
36	0.99372	44.5
37	0.99336	47.1
38	0.99300	49.7
39	0.99263	52.4
40	0.99225	55.3
41	0.99186	58.3
42	0.99147	61.5
43	0.99107	64.8
44	0.99066	68.3
45	0.99024	71.9
46	0.98982	75.7
47	0.98939	79.6
48	0.98896	83.7
49	0.98852	88.0
50	0.98807	92.5
51	0.98762	97.2
52	0.98715	102.1
53	0.98669	107.2
54	0.98621	112.5
55	0.98573	118.0
60	0.98323	149.4
61	0.98270	156.4
62	0.98217	163.8
63	0.98164	171.4
64	0.98111	179.3
65	0.98058	187.5
70	0.97779	233.7
71	0.97721	243.9
72	0.97663	254.6
73	0.97605	265.7
74	0.97547	277.2
75	0.97487	289.1
80	0.97182	355.9
85	0.96864	433.6
90	0.96534	525.8
95	0.96192	633.9
100	0.95838	**760.0**
105	*	906.1

*Water is a gas at this temperature and one atmosphere of pressure.

APPENDIX C
GLOSSARY OF CHEMICAL PROCESS TERMS

Abscissa The horizontal axis in a graph, usually symbolized by x.

Absolute A chemical substance that is not mixed with any other substances. An absolute ethanol solution would be without any other substances, for example, water. Material safety data sheet (MSDS) term for a single dose of or exposure to a substance.

Absolute zero The lowest temperature theoretically possible, called zero degrees Kelvin (−273.15°C or −459.67°F).

Absorber Equipment used specifically to remove components from a gas stream by contact with other gas or liquid.

Absorption (1) in the human body, the direct passage of chemicals through the outer tissue barrier of the body and into the bloodstream; (2) in chemical processes, the removal of components in a gas by contact or exposure to another gas or liquid phase.

Accumulation The difference (ΔP) between the initial lift pressure and full-lift pressure on a safety valve.

Accuracy The degree to which a given answer agrees with the true value. See Chapter 7, "Standard Operating Procedures."

Activity The effective concentration of ions in solution.

Activity coefficient The ratio of activity to actual concentration.

Acute effects Rapidly developing and severe effects, as opposed to chronic effects.

Address The number or label identifying the memory location where a unit of information is stored.

Adsorber Any device (usually filled with porous solids) used to remove gases and/or liquid from a process stream.

Aerator Any device used to add air to a process system.

Aftercooler Heat exchanger located on discharge side of a compressor to remove excess heat.

Aliquot A measured volume of a liquid that is a known fractional part of a larger volume.

Alkali metals Group IA elements on the periodic table: lithium, sodium, potassium, rubidium, and cesium. These elements tend to lose one electron in a binary reaction.

Alkaline earth metals Group IIA elements on the periodic table: beryllium, magnesium, calcium, strontium, and barium. These elements tend to lose two electrons in a binary reaction.

Alkylation The process of exposing small molecules to catalyst and converting them into larger molecules, usually hydrocarbons.

Allotropic Two or more forms of an element that exist in the same physical state (example: coal and diamond).

Alphanumeric Set of computer characters that contains both letters and numbers.

Alpha particle A helium nucleus (two neutrons and two protons) emitted from a nuclear decay.

Alumel® An aluminum nickel alloy used in the negative element of a K-type thermocouple. Trademark of Hoskins Manufacturing Company. See Chapter 17, "Process Analyzers."

Ambient temperature The average or mean temperature of the surrounding air that comes in contact with the equipment under test.

Amphoteric compound Compound with the ability to act as an acid or base.

Analog signal A directly measurable, continuous quantity or signal (voltage, temperature, resistance, pressure, rotations, etc.).

Analog to digital converter (A/D) A converter that changes the typical "analog" signal from a sensor (for example, GC millivolt signal) to equivalent "digital values" that the computer can process.

Anion A negatively charged atom or group of atoms.

Anode The electrode at which oxidation occurs in an electrochemical cell.

Anodize To coat a metal surface with a metal-oxide by anodic oxidation.

Antiseize compound Any lubricant, sealant, or substance used on exposed pipe and valve threads to prevent galling or seizing.

Approach to tower The temperature difference (ΔT) between water vapor leaving a cooling tower and the wet-bulb temperature of entering tower air.

Arch The narrow connection between the convection section and the stack or chimney.

ASCII Abbreviation for American Standard Code for Information Interchange. ASCII is an industry-wide standard used to represent alphanumeric characters and to interface all computers and peripheral devices.

Assembler A program that translates assembly language instructions into machine language.

Assembly language A machine-oriented language in which mnemonics (a technique used to improve memory) are used to represent each machine language instruction.

Atmosphere A unit of pressure equivalent to the normal pressure of our atmosphere experienced at sea level, 760 torr, 14.7 pounds per square inch. See Chapter 4, "Handling Compressed Gases."

Atom The smallest particle of an element retaining the properties of that element.

Atomic mass or atomic weight The relative weight of an atom with the lightest isotope of carbon being arbitrarily set at 12.0000 atomic mass units. See Chapter 22, "Inorganic Chemistry Review."

Avogadro's hypothesis Equal volumes of gases under the same conditions contain equal numbers of molecules. See Chapter 22, "Inorganic Chemistry Review."

Avogadro's number The number (6.023×10^{23}) of atoms, molecules, particles, etc., found in exactly one mole of that substance. See Chapters 22, "Inorganic Chemistry Review," and 24, "Chemical Calculations and Concentration Expressions."

Background radiation Radiation extraneous to an experiment; usually the low-level natural radiation from cosmic rays and trace radioactive substances present in our environment.

Backup A file, device, or system used to store data in the event of a malfunction or loss of data.

Baffles Plates or partitions mounted inside of equipment (such as tube heat exchangers or standpipes) to increase the turbulent flow, thus reducing hot spots or undesirable vortexing action.

Bandwidth A symmetrical region around the set point at which proportional control occurs.

Batch processing A system primarily used by mainframe (large) processors in which data are collected and stored on an input medium such as magnetic tape. The information is scheduled to run as time permits, with the output stored on a peripheral device. Batch processing tends to optimize use of the central processing unit (CPU).

Baud A unit of data transmission speed equal to the number of bits per second. For example, 400 baud equals 400 bits per second.

Bellows pump A pump designed to move fluids through a reciprocating bellow (expanding and contracting chamber) that is coupled to a drive rod.

Beta particle (beta radiation, beta ray) An electron that has been emitted by an atomic nucleus.

BIOS Acronym for basic input/output system. The commands used to direct a CPU in how to communicate with the rest of the computer.

BIT Binary digit, a single digit (1 or 0) in the binary system. The binary system is used primarily in digital electronics and computers. A binary sequence can be used to represent any number or letter in the alphabet. For example, the binary code of just five digits is used to represent the following numbers: 00000 = 0; 00001 = 1; 00010 = 2; 00011 = 3; 00100 = 4; 00101 = 5; and so forth. Each of the five binary code digits represents an exponent of 2, as follows: $2^4\ 2^3\ 2^2\ 2^1\ 2^0$. A 1 in the digit position farthest left represents the value 16, whereas a 1 in the bit position farthest right represents the value 1. Combinations of ones and zeros can equate to any number between 0 and 31 in a five-bit binary code.

Block valve A valve used to isolate or totally block flow.

Boiler A type of fired, industrial furnace used to produce steam by boiling water under pressure.

Boiler load The total demand for steam from a plant.

Bonnet The bell-shaped dome that covers and protects the body of a valve.

Boot up To turn the computer on so that the operating system takes over and provides a useful work environment.

Bottoms Residue from processing operations consists of heavy liquids and/or solids (coke).

Buffer (chemical) A solution of a substance or combination of substances that is resistant to quick changes in the pH level; usually a combination of a weak acid and the salt of that weak acid, or a weak base and the salt of that weak base.

Buffer (computer) A storage area [normally found in an input/output (I/O) operation] for data that is used to compensate for speed differences when transferring data from one device to another. For example, most printers require a buffer because they cannot process data as quickly as the computer can generate data.

Bus The transmission line or path that connects the components (functions) of a computer. Examples include a data bus, control bus, address bus, and video bus.

Byte A grouping or sequence of bits that constitutes a discrete piece of information. Bytes are usually no more than eight bits long. A series of bytes arranged in a sequence represents a piece of data or instruction and is called a *word*.

C See *Ceiling*. The maximum allowable exposure to an airborne substance for humans; this limit is never to be exceeded, even momentarily.

Capillary A tube having a very small inside diameter.

Carbanion An organic ion carrying a negative charge on a carbon atom.

Carbonium ion An organic ion carrying a positive charge on a carbon atom.

Carcinogen A substance capable of causing or producing cancer in mammals. A carcinogen is any substance that meets one of the following criteria:
   (a) It is regulated by the Occupational Safety and Health Administration (OSHA) as a carcinogen; or
   (b) it is listed under the category "known to be carcinogens" in the *Annual Report on Carcinogens* published by the National Toxicology Program; or
   (c) it is listed under Group 1 ("carcinogenic to humans") by the International Agency for Research on Cancer Monographs.

CAS # (Chemical Abstract Service number) Number assigned by the American Chemical Society (ACS) to substances. Over a 100-year period, ACS has routinely produced a chemical abstract for each common substance and most new substances and assigns each substance a unique CAS number.

Catalyst A substance that changes the rate of a reaction, but is not itself consumed.

Cathode The electrode at which reduction occurs in electrochemical cells.

Cathode-ray tube (CRT) The screen or monitor that provides visual output for the computer.

Cation A positively charged ion.

Cavation The process of gas pockets forming and collapsing around the impellers during pump operation.

Ceiling The maximum allowable exposure to an airborne substance for humans; this limit is never to be exceeded, even momentarily. See also *C* or *TLV-C*.

Central processing unit (CPU) The combination of a computer's control unit (CU) and the arithmetic logic unit.

Chatter The rapid cycling on and off of a control relay.

Check valve A valve designed to allow fluids to flow only in one direction. A check valve is primarily used in inlet lines and discharge lines to prevent reverse flow in reciprocating pumps.

Chemical hygiene officer An employee, designated by the employer and qualified by training or experience, who is assigned to provide technical guidance in the development and implementation of the provisions of the Chemical Hygiene Plan (CHP).

Chemical Hygiene Plan (CPH) A written program, developed and implemented by the employer, that sets forth procedures, equipment, personal protective equipment, and work practices that are capable of protecting employees from the health hazards presented by dangerous chemicals used in that particular workplace.

CHEMTREC Chemical Transportation Emergency Center, a national center established to provide pertinent information concerning specific chemicals on request. Its toll-free, 24-hour number is 1-800-424-9300.

Chromel® A chromium-nickel alloy that makes up the positive element of a Type K or Type E thermocouple. Tradename of the Hoskins Manufacturing Company.

Cis- Prefix used to indicate that groups are located on the same side of a bond about which rotation is restricted. See Chapter 23, "Organic Chemistry Review."

Classifier Any device used to classify or separate good pellets from other pellets that are too large, too small, or misshaped.

Coefficient of expansion The ratio of the change in length or volume of a body to the original length or volume for a unit change in temperature.

Coking The buildup of carbon and other solid deposits inside furnace tubes.

Combustible Classification of liquid substances that will burn on the basis of flash points. Specifically, combustible liquid means any liquid having a flash point at or above 100°F (37.8°C) but below 200°F (93.3°C), except any mixture having components with flash points of 200°F (93.3°C) or higher, the total volume of which make up 99 percent or more of the total volume of the mixture.

Compensating alloy An alloy used to connect a thermocouple to electronics. These alloys have similar thermal electric properties as the thermocouple alloys themselves.

Compiler A program that translates a high-level language, such as BASIC or COBOL, into machine language.

Compressed gas (a) A gas or mixture of gases having, in a container, an absolute pressure exceeding 40 psi at 70°F (21.1°C); (b) a gas or mixture of gases having, in a container, an absolute pressure exceeding 104 psi at 130°F (54.4°C) regardless of the pressure at 70°F (21.1°C); or (c) a liquid having a vapor pressure exceeding 40 psi at 100°F (37.8°C) as determined by ASTM D-323-72.

Compression packings Various materials made of yarns, ribbons, asbestos, and so forth, braided together in different shapes to improve lubrication retention around bearings and other moving parts.

Condensate Moisture or other liquids that form when hot gases are allowed to cool and condense (undergo gas-to-liquid phase change).

Conjugated Two double bonds separated by one single bond, for example (—C≡C—C≡C—).

Continuous duty device Any device designed to operate continuously.

Control air header In-house network of piping that provides dry, clean air to instruments and process units.

Control loop A system of process instrumentation (chromatographs, beta ray detectors, etc.) networked together to control a process automatically.

Control point The temperature or other selected parameter at which a system is to be maintained.

Control unit (CU) The unit responsible for coordinating the operation of the entire computer system. The CU receives information from the input device, retrieves instructions and data from the memory, and transmits instructions to the arithmetic unit.

Corrosion The slow conversion of a metal to an oxidized form. Example: iron to rust. Most acids are corrosive to active metals.

Corrosive chemical A chemical that causes burns to the skins or is corrosive to other substances (see *Corrosion*). Examples include ammonia, phenol, nitric oxide, oxidizing agents, reducing agents, sulfuric acid, etc.

Coulometry The quantitative application of Faraday's law to the analysis of materials. The current and the time are the usual variables measured.

Counterflow Heat exchanger term used to describe flow through tube-side and shell-side but in opposite directions.

Curie (Ci) The basic unit used to describe the intensity of radioactivity in a sample of material. One curie equals 37 billion disintegrations per second or approximately the amount of radioactivity given off by 1 g of radium.

Curie point The temperature at which a magnetic material becomes nonmagnetic.

Cyclone A separations device used to remove or reduce solids entrained in a gas stream.

Damper A butterfly-type valve used to control air flow in a stack or chimney.

Daughter A nuclide formed by the radioactive decay of a different (parent) nuclide.

Dead band The minimum change of input necessary to cause a deflection of the pen position on a chart recorder.

Dead head The ability of a pump to continue to run without damage when the discharge line is closed off. Normally, only centrifugal pumps are recommended for this duty.

Debugging The process of locating and correcting mistakes in a computer program.

Decay (radioactive) The change of one radioactive nuclide into a different nuclide by the spontaneous emission of alpha, beta, or gamma rays.

Default The options or values that a program automatically assumes unless otherwise instructed.

Demineralizer A purification-type process (filtration and/or ion exchange) used to remove dissolved solids from fluids. Water put through this process is sometimes referred to as deionized water.

Demister Any device (usually a cyclone) used to remove moisture from a gas.

Denaturation A process pertaining to a change in structure of a protein from a regular to irregular arrangement of the polypeptide chains.

Denatured Commercial term used to describe ethanol that has been rendered unfit for human consumption because of the addition of harmful ingredients to make its sale tax-exempt.

Dermal toxicity Adverse health effects resulting from skin exposure to a substance. See Chapter 2, "Chemical Plant and Laboratory Safety."

Designated area An area that may be used for work with carcinogens, reproductive toxins, or substances that have a high degree of acute toxicity. A designated area may be the entire laboratory, an area of a laboratory, or a device such as a laboratory hood.

Detoxification The body's method of converting toxic substances into nontoxic substances.

Deuterium An isotope of hydrogen whose atoms are twice as massive as ordinary hydrogen; deuterium atoms contain both a proton and a neutron in the nucleus.

Diaphragm pump A pump that utilizes one or more flexible diaphragms to displace fluids. These pumps are well suited for pumping liquids that contain solids. They can usually be run dry without damage and are self-priming.

Digital to analog (D/A) converter An operational amplifier can summarize the input currents from laboratory instruments and convert them to a scaled output voltage. D/A converters are used to operate any analog input signal and interface it to a digital source such as a computer.

DIN Deutsche Industrial Norms, a German agency that sets dimensional and engineering standards.

Disk operating system (DOS) A program used as the operating system for many types computers, such as MS DOS (in PCs) or Mac OS-8 (in Macintosh). See *Operating system software*.

Distilland The material in a distillation apparatus that is to be distilled. See Chapter 21, "Distillation and Evaporation."

Distillate The material in a distillation apparatus that is collected in the receiver. See Chapter 21, "Distillation and Evaporation."

Dosimeter A small, calibrated electroscope worn by laboratory personnel and designed to detect and measure incident ionizing radiation or chemical exposure.

DOT The Department of Transportation. See Chapters 1, "Chemical Process Industry Worker and Governmental Regulations," and 2, "Chemical Plant and Laboratory Safety."

DNA (deoxyribonucleic acid) The molecule that contains hereditary information or code (genes) for an organism to duplicate themselves or which determine the organism's characteristics.

Dose The amount of chemical that the body absorbs.

Doublet Two peaks or bands of about equal intensity appearing close together on a spectrogram.

Downcomers Tube-shaped weirs used in distillation towers that allow liquid to drop from higher-level to lower-level trays.

DP number The "degree" of polymerization, more specifically, the average number of monomer units per polymer unit.

Economizer A furnace design feature that preheats feed water before it enters the main boiler system.

Electrolytic cell An electrochemical cell in which chemical reactions are forced to occur by the application of an outside source of electrical energy.

Electrophile Positively charged or electron deficient.

Electrophoresis A technique for separation of ions by rate and direction of migration in an electric field.

Electroplating The deposition of a metal onto the surface of a material by an electrical current.

Eluant or eluent The solvent used in the process of elution.

Eluate The liquid obtained from a chromatographic column; same as *effluent*.

Enantiomer One of the two mirror image forms of an optically active molecule.

Energy balance Thermodynamic calculation that shows heat (energy) added to a process equals the heat (energy) removed from the process.

Engineer's console A dedicated workstation used by authorized plant personnel to change control schemes and modify control systems.

Enzyme Any chemical manufactured by the body that can moderate the chemical reactions that take place in the body. Enzymes are catalysts.

Equilibrium A state in which two opposing processes are occurring at the same rate in a closed system.

Essential oil An extract of a plant that has a pleasant odor or flavor.

Eutectic temperature The lowest possible melting point of an alloy mixture.

Evaporation rate The rate at which a particular substance will vaporize (evaporate) when compared to the rate of a known substance such as ethyl ether. This term is especially useful for health and fire hazard considerations.

Explosion-proof motor (XPRF) A motor that is totally enclosed to prevent sparks or arcs within its housing from igniting surrounding vapor or gas. This designation is also given to motors designed to withstand minor explosions within the motor/pump system itself.

Explosive A chemical that causes a sudden, almost instantaneous release of pressure, gas, and heat when subjected to sudden shock, pressure, or high temperature.

Explosive limits The range of concentrations over which a flammable vapor mixed with proper ratios of air will ignite or explode if a source of ignition is provided.

Extrapolate To estimate the value of a result outside the range of a series of known values. It is a technique used in the standard additions calibration procedure.

Extruder A process device used to convert solid plastics into products. These extruders typically consist of a pellet hopper, heaters, pressurization drivers, and heated dies.

Film badge A small patch of photographic film worn on clothing to detect and measure accumulated incident ionizing radiation.

Flammable A liquid as defined by the National Fire Protection Association (NFPA) and the Department of Transportation (DOT) as having a flash point below 100°F (37.9°C).

Flare A safety device used to burn excess process vapors, usually during a process upset.

Flash Point The temperature at which a liquid will yield enough flammable vapor to ignite. There are various recognized industrial testing methods; therefore, the method used should also be stated.

Flooded suction The gravitational flow of liquids into a pump inlet from an elevated source needed for priming. It is not necessary on self-priming pumps.

Foot valve A valve used at point of fluid intake to retain fluids in the system and prevent loss of prime. A foot valve basically contains a check valve with a built-in strainer.

Fouling Buildup of solids (usually mineral deposits) on surface of heat exchangers, ultimately resulting in reduced flow or plugging.

Free radical A highly reactive chemical species carrying no charge and having a single unpaired electron in an orbital.

Fuel cell A voltaic cell that converts the chemical energy of a fuel and an oxidizing agent directly into electrical energy on a continuous basis.

Functional group A group of atoms having the same properties when appearing in various molecules, especially organic compounds.

Gain The amount of amplification used in an electrical circuit, such as a recorder.

Gamma ray A highly penetrating type of nuclear radiation similar to x-ray radiation, except that it comes from within the nucleus of an atom and has higher energy. Energy-wise, a gamma ray is very similar to a cosmic ray, except that cosmic rays originate from outer space.

Galvanizing Placing a thin layer of zinc on a ferrous material to protect the underlying surface from corrosion.

Galvanometer An instrument used to measure small electrical current flows by means of deflecting magnetic coils.

Geiger counter A gas-filled tube that discharges electrically when ionizing radiation passes through it.

Governor Any device used to regulate the flow, speed, or output of a process.

Half-life (a) In nuclear physics, the time in which half the atoms of a particular radioactive nuclide disintegrate; (b) in toxicology, the time it takes for the initial concentration of a chemical in the body to be reduced by half.

Halogens Group VIIA elements on the periodic table: fluorine, chlorine, bromine, and iodine. These elements tend to gain one electron in a binary reaction.

Handshake A computer interface procedure that is based on status/data signals that ensure orderly data transfer.

Hand wheel Handle attached to a valve stem used to control flow. Most hand wheels are turned clockwise to close and counterclockwise to open. Remember: "righty tighty, lefty loosey."

Hardcopy A microprocessor output in a permanent form such as a printout, not on a disk or display terminal.

Hazard The property of a substance to cause an adverse effect.

Hazardous chemicals Chemicals for which there is statistically significant evidence, based on at least one study conducted in accordance with established scientific principles, that acute or chronic health effects may occur to exposed employees. The term "health hazard" includes chemicals that are carcinogens, toxic or highly toxic agents, reproductive toxins, irritants, corrosives, sensitizers, hepatotoxins, nephrotoxins, neurotoxins, agents that act on the hematopoietic systems, and agents that Appendices A and B of the Hazard Communication Standard (29 CFR 1910.1200). This standard provides further guidance in defining the scope of health hazards and determining whether or not a chemical is to be considered hazardous for purposes of this standard.

Heat exchanger Process device used to either heat or cool fluids without the two systems coming into direct contact. The tube-side system has an inlet and outlet circulating one fluid system whereas the shell side has an inlet and outlet circulating a second fluid system.

Heat of combustion The quantity of heat released when one gram or mole of a substance is oxidized.

Heat of fusion The quantity of heat required to melt 1 gram of a solid substance. For example, the heat of fusion of ice is 80 calories per gram at 0°C.

Heat of vaporization The quantity of heat required to vaporize one gram of a liquid substance. For example, the heat of vaporization is 540 calories per gram for water at 100°C.

Heavy metals Those metals that have ions that form an insoluble precipitate with sulfide ion.

Heavy water Water formed of oxygen and deuterium; deuterium oxide. Its mass is similar to that of water, except that approximately it is 1.1 times heavier than that of ordinary water.

Hertz (Hz) A unit of frequency equal to one cycle per second. Normally, the higher the hertz rating, the faster the computer.

Heterocyclic compound A compound containing a ring of atoms of which at least one atom is not carbon.

HETP Height Equivalent to a Theoretical Plate. It is calculated by dividing the column length by the number of equilibrium steps in the column or tower.

Hexadecimal A base 16 number system using the characters 0 through 9 and A through F to represent the values. Many machine language programs are written in hexadecimal notation.

Host The primary or controlling computer in a multiple part system.

Hydrogen bond A loose bond between two molecules, one of which contains an active hydrogen atom; the other contains primarily a very electronegative atom of O, N, or F.

Hydrolysis A reaction in which water molecules react with another species with the release of H^+ or OH^- ions from the water molecules.

Icon A graphic or pictorial symbol display. Icons are used to simplify or represent commands or functions to be performed by the computer.

Immunosuppression The reduction in the ability of the immune system to fight disease.

Impedance The total opposition to electrical flow in a circuit.

Incompatible materials Materials that could cause dangerous reactions from direct contact with one or more other substances.

Inert gases Group VIIIA elements and several other elements (nitrogen) that are extremely stable and tend not to react with other elements. Helium, neon, argon, krypton, and xenon are referred to inert gases because of no or very low reactivity.

In situ At the original location.

Instrumentation (process) Any device used to measure or monitor parameters (flow, temperature, pH, level, etc.) of the plant processes.

Interface A junction between the laboratory instrument/process and the computer where signal matching or adjusting is accomplished to make the two systems compatible.

Interpolate To find or insert an unknown point between two known points in a series, like a graph.

In vitro In an artificial environment, such as a test tube or glass.

Ionizing radiation Radiation that is capable of producing ions either directly or indirectly.

IPTS International Practical Temperature Scale of 1948 or 1968.

Irradiate To expose to some form of radiation.

Irritant A substance of sufficient concentration and period of exposure that can cause an inflammatory response or reaction of the eye, skin, or respiratory system.

Isoelectric point The pH at which there is no migration in an electric field of dipolar ions.

Isothermal system System or process in which the temperature is held constant.

Isotopes Atoms that have the same number of protons but a different number of neutrons in their nuclei.

Kinetic molecular theory The theory that states that all matter is composed of particles that are in constant motion.

Laboratory scale Way of working with substances in which the containers used for reactions, transfers, and other handling of substances are designed to be easily and safely manipulated by one person. Laboratory scale excludes those workplaces whose function is to produce commercial quantities of materials.

Laboratory-type hood A device located in a laboratory, enclosed on five sides with a movable sash or fixed partial enclosure on the remaining side; it is constructed and maintained to draw air from the laboratory and to prevent or minimize the escape of contaminated air.

LD_{50} A dose that is lethal to 50 percent of a population.

LEL or LFL Lower Explosive Limit or Lower Flammable Limit, the lowest concentration of vapor that will produce a fire or flash when an ignition source is available.

Ligands (coordination groups) The molecules or anions attached to the central metal ion.

Linearity The ability of an instrument to produce a straight line with multiple responses.

Lipid A nonpolar solvent-soluble product occurring in nature, usually limited to oils, fats, waxes, and steroids.

Louver Vents or openings located on cooling towers and other process units to release heat.

Lyophilization A process whereby the material is frozen, a vacuum applied, and the water and low boiling compounds removed by sublimation.

MAC Maximum allowable concentration of a toxic substance under prescribed conditions in an atmosphere to be breathed by humans.

MSDS Material safety data sheet; such documents are required by current OSHA regulations on all chemicals as to their possible health, fire, and other hazards.

Manometer A device for measuring the pressure of gases in a system. See Chapter 4, "Chemical Handling and Hazard Communication."

Manostat A device for maintaining a constant pressure. See Chapter 4, "Chemical Handling and Hazard Communication."

Mantissa The decimal part of a logarithm. In the logarithm 3.6222, the characteristic is 3 and the .6222 is called the mantissa. See Chapter 6, "Mathematics Review and Conversion Tables."

Mass number Approximately the sum of the number of protons and neutrons found in the nucleus of an atom. See Chapter 22, "Inorganic Chemistry Review."

Material balance Calculations to show that the mass of products (output) equals the mass of reactants (input), not overall loss of materials. See Chapter 24, "Chemical Calculations and Concentration Expressions."

Medical consultation A consultation that takes place between an employee and a licensed physician for the purpose of determining what medical examinations or procedures, if any, are appropriate in cases where a significant exposure to a hazardous chemical may have taken place.

Melting point The temperature at which a pure substance changes from the solid state to the liquid state. An impure substance will have a melting range and tend to be lower than the pure substance. *Melting point* and *freezing point* are identical terms.

Meniscus The curved upper surface of a liquid column. See Chapter 25, "Volumetric Analysis."

Metallic bond The type of bond found in metals as the result of attraction between positively charged ions and mobile electrons.

Metalloid An element that exhibits properties intermediate between metals and nonmetals.

Metric system A system of measurement in which the units are related by powers of 10. The preferred reference is System International (SI).

Metric ton Unit of mass in the metric system equal to 1000 kilograms. Also called a *megagram*.

Modum Modulator/demodulator. A device that transforms digital signals into audio tones for transmission over telephone lines and does the reverse process for reception.

Molecular distillation The distillation of viscous materials under high vacuum so that the mean free path is longer than the distance from the material to the condenser.

Molecular weight The sum of the atomic weights of all atoms in a molecule. The relative weight of a molecule based on the standard carbon-12 = 12.0000 atomic mass units.

Molecule The smallest particle of a compound that retains the properties of that substance.

Monosaccharide A small carbohydrate molecule that is the monomeric unit from which the polymeric carbohydrates are composed.

Motherboard The part of a microcomputer that contains the microprocessor and connections for all necessary expansion boards.

Mutagen A chemical substance that causes irreparable damage to DNA, thus resulting in a mutation.

Mutarotation The interconversion of two forms of a sugar molecule.

Neat liquid Pure liquids as opposed to solutions. Another definition means without water present.

Net positive suction head (NPSH) For best efficiency, fluids cannot be sucked or pulled into the impeller of a centrifugal pump; fluid must be pushed.

Network A group of computers that are connected to each other to share information and resources.

Noble gases Group VIIIA elements that are extremely stable and tend not to react with other elements. Helium, neon, argon, krypton, and xenon are referred to as inert gases because of no or very low reactivity.

Noise Unwanted electrical interference, especially on the signal wires.

Nonpolar molecule A molecule in which the electrical charges are symmetrically distributed around the center.

Nucleon A constituent of the nucleus; a proton or a neutron.

Nuclide Any species of atom that exists for a measurable length of time, distinguished by its atomic weight, atomic number, and energy state.

Off-line computer A computer not dedicated to a particular instrument or application.

Oil separator A device used to remove oil from compressed gases.

On-line computer A computer dedicated to a specific laboratory instrument and/or process. Such a computer allows technicians to operate much more sophisticated instrumentation like gas chromatograph-mass spectrometers, Fourier transform infrared spectrophotometers, etc.

Operating system software The machine language software that controls the very basic functions of the computer. It provides an interface between key stokes and mouse movement to machine language commands that the hardware understands.

Optical activity The property of a molecule involving rotation of plane polarized light.

Ordinate The vertical axis on a graph, normally symbolized by y.

Organic peroxide An organic compound that contains the bivalent –O–O– structure and which may be considered to be a structural derivative of hydrogen peroxide where one or both of the hydrogen atoms have been replaced by an organic radical.

Oxidation A reaction with oxygen or oxygen compounds; a loss of electrons; a positive change in valence.

Oxidation number An arbitrary number assigned to represent the number of electrons lost or gained by an atom in a compound or ion.

Oxidizer A chemical other than a blasting agent or explosive, as defined in 1910.109(a), that initiates or promotes combustion in other materials, thereby causing fire either of itself or through the release of oxygen or other gases.

Oxidizing agent A substance that oxidizes another substance while it is reduced.

Packing gland Any mechanical device used to compress packing material, used primarily around valve stems to prevent leaking.

Parent A radionuclide that decays to another nuclide that may be either radioactive or stable.

Parts per billion (ppb) A unit of concentration expression primarily used for trace impurities. Parts of gas, liquid, or solid per billion parts of sample.

Parts per hundred (pph) Parts per hundred parts, usually expressed as percent.

Parts per million (ppm) An expression for concentration; one part of the substance per million total parts.

PEL Permissible Exposure Limit. An OSHA-established exposure limit that may be time-weighted average (TWA) limit or maximum concentration (C) exposure limit.

Peptide Two or more amino acids joined by peptide linkages between the terminal carboxylic acid group of one amino acid and an amine group located on another molecule.

Peripheral Any computer-related device external to the central processing unit and main memory. Printers, scanners, modems, and so forth are all considered peripherals.

Permissible Exposure Level (PEL) An exposure limit established by OSHA. It can be a maximum concentration exposure limit or a time-weighted average (TWA).

Permissive A special type of interlock that controls a set of prerequisite (required) conditions before a piece of equipment can be started.

Peroxide A molecule containing -O–O- bonds.

Physical hazard A chemical for which there is scientifically valid evidence that it is a combustible liquid, a compressed gas, explosive, flammable, an organic peroxide, an oxidizer, pyrophoric, unstable (reactive), or water-reactive.

Pixel Picture element. Locations on a computer monitor that are used to form the ultimate images on the screen. The greater the number of pixels, the higher the screen resolution.

Plasticizer A liquid of low volatility (high boiling point) that is added to soften polymers.

Polarity In electricity, the quality of having two oppositely charged poles, one positive and one negative, like a battery.

Polar molecule A molecule in which the electrical charges are not symmetrically distributed; the center of positive charge is separated from the center of negative charge.

Potentiostat An instrument designed to maintain a constant potential between two electrodes in a solution.

ppb Parts per billion. A unit of concentration with the mass or volume of a specific substance divided by the mass or volume of the total sample multiplied by a factor of one billion.

ppm Parts per million. A unit of concentration with the mass or volume of a specific substance divided by the mass or volume of the total sample multiplied by a factor of one million.

Precision A measure of the reproducibility of results.

Prime A charge of liquid required to beginning pumping action when the liquid source level is lower than the pump. The priming charge can be manually added to the pump's intake line or held by a foot valve. In many pumps, the priming liquid provides a seal and lubricant for the pumping mechanism.

Progressive cavity pump A pump that uses a screw-type rotor turning inside an elastomeric (rubber-like) stator. Fluids are trapped between the tightly fitting rotor and stator and are forced through the pump by the screw-type rotation. It is an excellent pump for high-viscosity, abrasive, and/or particulate-containing fluids. It is sometimes referred to as a screw-type rotary or Moyno® pump.

Protocol A formal definition or procedure that describes how data are to be handled and exchanged.

PROM Programmable Read Only Memory. Computer memory medium that allows the user to modify the read-only memory (ROM) established by the manufacturer. It is sometimes referred to as Erasable Program Read Only Memory (EPROM). EPROM chips have a small window through which ultraviolet light can be passed to erase the contents of the memory locations prior to reprogramming.

Proton The nucleus of a hydrogen atom having a mass of approximately one amu and carrying one unit of positive charge.

Psi Pound per square inch. See Chapter 5, "Pressure and Vacuum."

Puking Upset phenomenon that occurs in distillation when vapor pressure becomes excessive and forces liquid up and out the overhead line.

Pyrolysis The breakdown of a material by heating, usually in the absence of oxygen.

Pyrophoric material A substance which can heat up substantially or ignite spontaneously upon exposure to air.

Pyrophoric substance Any substance that will spontaneously ignite in air at temperatures of less than 130°F (54.4°C).

Racemate A mixture of having no optical activity which consists of equal amounts of enantiomers.

Radial bearing A bearing that prevents the shaft on any motor, pump, compressor, and so forth from moving side to side.

Radioactive Any material that give off harmful radiation. This radiation can be in the form of particles such as alpha rays (helium nuclei) or beta rays (electrons) or energy such as gamma rays (electromagnetic radiation).

Radioactive date A technique for estimating the age of an object by measuring the amounts of various radioisotopes compared to stable isotopes. Carbon-13 dating is an example of this technique.

RAM Random access memory. Data information located in the CPU that the user can change. This type of memory is described as volatile because the instructions contained in it are lost when the power is switched off.

Reactive A substance that will react violently or even self-explode when mixed with another substance under certain temperatures, pressures, or shock conditions.

Reactivity The tendency of a substance to undergo a chemical reaction, especially with the rapid release of energy. Other properties such as rapid pressure changes and formation of toxic or corrosive products are also included in reactivity descriptions.

Real-time processing Computer processing in which the computer is directly connected and usually dedicated to one or more laboratory instruments/processes. Data from ongoing analysis are collected by the computer system, processed, and made available to the operator immediately. The computer can also be used to modify instrument conditions in "real time" (as changes are being made).

Reboiler A heat exchange used to maintain the heat balance on a distillation tower.

Reboot To boot up the computer again, often used as a kind of reset button.

Reducing agent A substance that reduces another while it is being oxidized.

Reduction The removal of oxygen or the addition of hydrogen; a gain of electrons; a negative change in valence.

Reflux ratio In distillation, the ratio of the amount of material returned to a column compared to the amount collected per unit of time.

Relief valve An adjustable, spring-loaded valve used to release pressure if it is above the preset level. Such a valve is used primarily on positive-displacement–type pumps to protect the pump and motor.

Reproductive toxins Chemicals that affect reproductive capabilities, including chromosomal damage (mutations) and effects on fetuses (teratogenesis).

Resolution The smallest detectable increment of measurement.

Resonance The stability shown by a molecule, ion, or radical to which two or more structures differing only in the distribution of electrons can be assigned.

Response An effect observed in toxicology when an organism reacts to the intake or application of a chemical substance.

Rf value Ratio of the distance a compound moved to the distance the solvent moved on a paper or thin-layer chromatogram.

Risk The probability of an adverse effect following the intake of or application of a chemical substance.

Robotics An automated system used to perform repetitious, manual laboratory procedures. The system is usually a microprocessor-controlled robotics arm specifically designed for sample preparation or handling.

ROM Read-only memory. Information located in the CPU that has been fixed by the manufacturer and cannot be changed by the user. This type of memory is described as "nonvolatile" because the instructions contained in it are not lost when the power is switched off. Nonvolatile system instructions are often referred to as firmware.

Scale Dissolved solids that adhere to the internal surfaces of equipment in the form of deposits.

Scaler An electronic instrument for counting radiation-induced pulses from radiation detectors such as a Geiger-Muller tube.

Scintillation counter An instrument that detects and measures gamma radiation by counting the light flashes (scintillations) induced by the radiation.

Scroll To move rapidly from one screen location to another.

Seal Materials used to prevent a pump from leaking. There are three types of seals: (1) Packing seal consists of multiple flexible rings mounted around the pump shaft and compressed together by tightening gland nuts. Some leakage can occur with this type of seal as it serves to lubricate the rotating shaft. (2) Lip seal consists of a flexible rubber ring with the inner edge held closely against the rotating shaft by a spring. (3) Mechanical seal consists of a rotating part and a stationary part with highly polished, touching surfaces. Mechanical seals can be damaged by dirt and grit in the pumping liquids.

Sealless (magnetic drive) A type of pump that does not contain the conventional drive shaft and thus requires no seals. Magnetic force is used to transmit power from the motor to the pump.

Self-priming A type of pump designed to draw liquids up from below the pump inlet (suction lift). Pumps that require priming are described as requiring flooded suction.

Shell side See *Heat exchanger*.

Side stream plumbing Plumbing that removes process components from the side of a tower, not directly from the top or bottom.

Slip The amount of fluids allowed to leak (slip) past the internal clearances of any pump, usually expressed in percentage/time.

Solubility in water The solubility of any substance in water, usually in percentage of material (by weight) that will dissolve in a stated quantity of water at ambient temperature. Solubility data are useful in determining spill cleanup and fire-extinguishing procedures.

Specific gravity The ratio of the density of a substance to the density of a standard substance. For solids and liquids, water is the standard substance; for gases, air is the standard substance. Specific gravity is unitless because of this definition.

Specific heat The number of calories required to raise the temperature of one gram of a substance by 1°C. The specific heat of water is 1 calorie per gram per degree Celsius.

Spuds Bundle or group of gas-filled tubes in a natural gas burner.

Stability The tendency of a substance to remain unchanged with time, storage, or exposure to other chemicals.

Standard conditions In dealing with gases, the set of conditions adopted for standard pressure and temperature (STP), which is one atmosphere and 0°C.

Standard electrode potential The potential compared to a hydrogen electrode, which exists when the electrode is immersed in a solution of its ions at unit activity.

Steam trap Device used to remove condensate from live steam.

Stereoisomers Isomers having different arrangement of the atoms in space but the same number of atoms, kinds of atoms, and sequence of bonding of the atoms.

String A sequence of characters.

Superheated steam Water vapor that has been stripped of liquid water and heated under pressure to above water's normal boiling point.

Syntax The rules governing the structure of a programming language.

Temperature gradient The natural change of temperatures (ΔT) from the bottom to top of a process unit, primarily due to design, loss by radiation, and poor insulation.

Teratogens Chemical substances that can cause fetal malformations or birth defects.

Ternary mixture A three-component mixture.

Threshold The lowest dose of a chemical that will cause an effect in an animal, human, or other living organism.

Thrust bearing A bearing that prevents axial (not radial) movement in a motor or pump shaft.

Threshold Limit Values (TLV) Concentration expression for airborne substances that have no adverse effects on most people who are in daily contact with the substances.

Threshold Limit Values (TLV)–Time Weighted Average (TWA) The short-term exposure limit or maximum concentration for continuous 15-minute exposure period. The exposures episodes must not exceed four per day and must allow at least 60 minutes between exposure periods.

TLV-C The ceiling exposure limit, which is a concentration that should never be exceeded.

Toxicology The study of the adverse effects of chemical substances on living things.

Tracer A small amount of radioactive isotope introduced into a system in order to follow the behavior of some component of that system.

Trans- Prefix used to indicate that groups are located on the opposite sides of a bond about which rotation is restricted.

Transducer Any device that converts a parameter being measured into a proportional output signal. For example, thermocouples transform heat into a millivolt output response or pneumatic signal to electrical.

Transmitter A device used to measure process variables and produce a signal that goes to a process controller.

Tube side See *Heat exchanger*.

UEL Upper Explosive Limit. Concentration of a flammable substance will not support combustion because the vapor is too rich in fuel for the amount of oxygen present.

Unstable chemical A chemical that, when in the pure state or as produced or transported, vigorously polymerizes, decomposes, condenses, or becomes self-reactive under conditions of shock, pressure, or temperature.

Vapor density The mass of gas vapor compared to the mass of an equal volume of air at the same temperature. A gas with a vapor density less than 1.0 tends to rise and dissipate, even though some mixing will occur. A gas with a vapor density of greater than 1.0 tends to sink in air and concentrate in low places.

Vapor pressure The pressure exerted by a saturated vapor in equilibrium with the pure liquid phase in a closed container. The vapor pressure of a substance is directly related to the temperature; as the temperature increases, the vapor pressure increases. By convention, most substances have their vapor pressures expressed in psi and reported at 100°F. However, MSDS vapor pressures tend to be expressed in torr and at 20°C. (See Chapter 5, "Pressure and Vacuum.")

Voltaic cell An electrochemical cell in which an electrical current is generated by a chemical reaction.

Water-reactive substance Any substance that reacts with water to release a flammable gas or toxic substance. For example, all alkali metals react with water to produce hydrogen gas.

Wheatstone bridge A circuit containing four resistances, a power source, and a galvanometer connected in such a way that when the four resistances are matched, the galvanometer shows no deflection or has a null reading. The thermal conductivity detector in a gas chromatograph is an example.

Zooming In computer graphics, to make an object larger or smaller on the monitor.

Zwitterion An ion that contains both a negative and a positive charge; same as a dipolar ion. Some neutral amino acids are examples of zwitterions.

INDEX

Abbe refractometer, 398
Abbreviations, 619–625
Abderhalden drying apparatus, 289
Abscissa, 126
Absolute error, 143
Absolute pressure, 105, 106
Absolute temperature scale, 374
Absolute value, 143
Absorbance (A):
   versus concentration, 581, 582
   definition, 581, 614
   equation for, 581
Absorbance curves, 614
Absorption:
   versus concentration curve, 614
   defined, 579
   of gases, 221
   of radiant, 585
Absorption bands, atomic absorption, 608
Absorption bands, infrared, 602
Absorption bands, near infrared, 603
Absorption bands, ultraviolet light, 586
Absorption bands, visible light, 586
Absorption maximum, 584
Absorption spectra, 582
AC/DC, 202
Accuracy, 142, 268
Acetic acid, 486
Acetone, drying glassware, 503
Acetylene gas, 80, 363, 610
Acid-base, common commercial concentrations, 494
Acid-base, common laboratory concentrations, 324
Acid-base indicators, 172
Acid-base primary standards, 314, 514
Acid-base titration equation, 514
Acid gases, 81
Acid into water rule, 324
Acid solution spills, 70
Acids:
   carboxylic acids, 477–478
   commercial, concentrations of, 494
   defined, 171

Acids (*Cont.*):
   laboratory, concentrations of, 324
   naming, 463, 477
   relative strengths of, 172
   strong, 170, 172
   weak, 170, 172
Activation of TLC plates, 526
Activity, analyte, 193
Actual yield, 488
Acute toxicity, 23
Adapters, glassware, 153, 249
Addition reactions, 480
Adhesion, 396
Adsorbed water, 285
Adsorbing (drying) agents, 290
Adsorption chromatography, 521
Aerial lift safety, 40
Affected employee, 44
AFFF, aqueous film forming foam, 72
Ag/AgCl reference electrode, 338
Agate, 270, 283
Aging and digestion, 251
Air (oxygen-deficient), 30
Air-acetylene flame mixtures, 610
Air changes per hour (ACH), 23
"al" suffix, 476
Alcoholic potassium hydroxide cleaning solution, 502
Alcohols, 473–474
Alcohols, common names, 474
Alconox, 499
Aldehydes, 476–477
Algorithms (*see* Glossary, 629–643)
Alkali metals, 455
Alkali solution spills, 71
Alkaline earth metals, 455
Alkaline gases, 81
Alkanes, 467
Alkanes, simple, names, 468
Alkenes, 469
Alkoxy prefix, 476
Alkyl, simple, names, 468
Alkyl groups, 468
Alkyl halide, common names, 473

Alkyl halides, 472
Alkylation, 482
Alkynes, 469
Alloys, solder, 358
Alternating currents (AC), 202
Alumina, 99, 264, 554, 569
Aluminum oxide, 523
Amalgamated cells, infrared, 594
Amalgamates, 72
American Board of Industrial Hygienists (ABIH), 14
American Chemical Society (ACS), 1, 14, 314
American Conference of Governmental Industrial Hygienists (ACGIH), 14, 22, 72
American Industrial Hygiene Association (AIHA), 14
American National Standards Institute (ANSI), 14, 25, 40, 51, 53, 59, 67, 364
American Organization of Analytical Chemists International (AOAC), 14
American Petroleum Institute (API), 14, 389
American Red Cross, 14
American Society for Mechanical Engineers (ASME), 66, 67
American Society for Quality Control (ASQC), 141, 142
American Society for Testing and Materials, 15, 75, 130, 143, 146, 335, 337, 374, 396, 570
American Society of Safety Engineers (ASSE), 15
American Standards Association (ASA), 356
American Welding Society (AWS), 356
American Wire Gauge (AWG), 201
Amides, 478
Amines, 475–476
Amino acids, 479
Amino prefix, 475
Ammeter, 208
Ammonia, 81, 170, 340, 475
Ammonia, leak test, 81
Ampere, 199
Amplitude, 574
Analog computer (*see* Glossary, 629–643)
Analytical balances, 271
Analyzer, Nicol prism, 400
Analyzing plant materials, 5
Anchoring, fall protection, 41
"ane" suffix, 467
Angle lift check valve, 368
Angle of rotation, 403
Angstrom, 575, 576
Anion-exchange, 564
Anions, 458
Anions, common, 459
ANSI thermocouple color codes, 336
Antifreeze, 474
Apothecaries, conversion, 129–130
Apparel, protective, 28
Aqua regia, 294
Aqua regia cleaning solution, 508
Aqueous extraction solvents, 411
Aqueous Film Forming Foam (AFFF), 72

Archimedes' principle, 386
Area equivalent tables, 127
Area measurement conversions, 127
Arenes, 471
Aromatic, nomenclature, 471–472
Aromatic compounds, 471
Arrhenius, 169
Asbestos (insulating) gloves, 28
Asbestos mats, 249
Ascarite®, 296, 297, 304
"ase" suffix, 490
Ashing, filter paper, 236, 253
Aspirator, water, 108, 109, 110, 240, 246, 446
Association of Official Analytical Chemists (AOAC), 143
ASTM (*see* American Society for Testing and Materials)
"ate" suffix, 463
ATEX, ATmospheres EXplosibles, 64
ATEX directive (EU), 65
Atmosphere, pressure unit, 105
Atmospheric pressure, 106
Atmospheric pressure unit conversions and equivalents, 106
Atom components, 451–452
Atomic absorption calibration curve, 608
Atomic absorption interferences, 612
Atomic absorption spectroscopy, 607–618
Atomic absorption wavelengths, 608
Atomic mass of elements, table of, 452–453
Atomic mass units, 451
Atomic number, 451
Attenuator, 533, 593
Augers, 225
Authorized employees, 44
Auto-ignition temperature, 384
Automatic pipette and tube cleaners, 501
Average deviation, 144
Avogadro's law, 107
Avogadro's number, 486
Avoirdupois, conversions, 129
AWG, American Wire Gage, 201–202
Azeotropic distillation, 423

Back titration, 497, 515
Backstreaming, vacuum pumps, 115
Balances, 269–279
Balancing equations, 484, 493
Ball and socket joints, 149
Ball check valve, 368
Ball mill, 230
Ball valves, 361
Ballast, gas, 113
Band broadening, chromatography, 552
Band spectra, 580
Bands, 592
Bar coding, 61

Bases:
    definition, 171
    laboratory, concentration of, 324
    relative strengths, 172
    strong, 170, 171
    weak, 170, 171
Basic, 173
Batch:
    decolorization, 264
    production and control records, 139
    records, guidelines, 140
Baumé degrees, 389
Baumé degrees to specific gravity, 390
Bausch and Lomb Spectronic 20, 587
Bearing assemblies, 370
Bearing lubrication, 369
Bearings, 369
Beer's law, 581, 613, 618
Bellows pressure element, 118
Belt guards, 114
Belts, 210
Bending glass tubing, 166
Bending metal tubing, 348
Benzene, 472
Beral pipettes, 508
Berl saddles, packing, 427, 443
Bernoulli's principle, 108, 360
Bidentate, 517
Bimetallic thermometers, 334
Binary compounds, naming, 460
Bit sizes for drills, 136
Blank sample readings, 402
Blanking of process fluid lines, 44
Bleeding, septum, 527
Bleeds, column, 530
BLEVE, 82
Body belt, safety, 40
Body harnesses, 40
Boiling points determinations, 381–382
Bomb samplers, 225
Bonded stationary phases, 555
Bonding and grounding, 64
Bonds:
    covalent, 457
    definition, 455
    ionic, 455
Borax, 317
Bound moisture, 463
Bourdon gauges, 118, 338
Boyle's law, 105
Braid over braid packing, 371
Branched groups, organic, 467
Breaking emulsions, 413
British thermal unit (Btu), 131
Brix degrees, 389
Brookfield® viscometer, 391
Brushes, cleaning, 500
Bruun column, 427
Btu, 131
Bubble cap column, 155

Bubble-cap fractionating column, 428, 442
Bubble flowmeter, 538
Büchner funnel, 110, 246, 247, 250, 259
Buffer solutions, 181, 189–192, 339
Bulb, pipette, 506
Bulk containers, 61
Bulk modulus, 405
Buoyancy, 269
Burette tips, removal of grease from, 501
Burettes, 509–510
Burettes, accuracy tolerances, 509
Burner head, atomic absorption, 610

Calcium chloride, 94
Calcium sulfate, 290
Calculations:
    acid-base titrations, 513
    gravimetric, 301–302, 488–495
    oxidation/reduction titration, 515
    volumetric, 183–189, 487–495, 513–517
Calculators, scientific, 121
Calomel reference electrode, 179, 192, 193, 339
Calorie, 131
Canadian Centre for Occupational Health and
    Safety (CCOHS), 16
Canadian Standards Association, 16
Cans, safety, 61
Capacitor plates, moisture measurements, 343
Capacity, 268
Capillary, 396
Capillary tube, 166, 377–382
Capillary air-inlet tube, vacuum distillation, 433
Capillary columns, GC, 528
Capillary-tube method:
    for boiling point, 382
    filling of, 377
    for melting point, 377
Carbon dioxide, absorbed, 302–305
Carbon dioxide fire extinguishers, 38
Carbon dioxide venting, extractions, 409
Carbon disulfide, solvent hazards, 601
Carbon tetrachloride, solvent hazards, 85, 601
Carbonate-free NaOH solution, 324
Carbonyl group, 476
Carboxylic acids, 477–478
Carboys, 61
Cargo tanks, 64
Carrier gas, gas chromatography, 527
Cartridge gas mask, 34
Catalytic cracking, 481
Cathode lamp, 607
Cation, common, 459
Cation-exchange, 564
Cations, 458
Cavitation, 364, 502
CE marking (CE), 65
Ceiling (C), 22
Celite, 234, 264
Cell path length, calculations, 596
Celsius temperature, 373

Celsius temperature scale conversions, 375–376
Centerline, 147
Centers for Disease Control (CDC), 16
Centigrade, 374
Centipoise, 390
Centistokes, 394
Central processing unit (CPU) (*see* Glossary, 629–643)
Centrifugal pumps, 364
Chainomatic® balances, 270
Characteristic, logarithm, 127
Charcoal, carrier gas, 527
Charcoal, decolorizing, 264
Charles' law, 105
Check valves, 89, 91, 363, 368
Chelates, 517
Chelating agents, 518
Chemical Abstracts Service (CAS), 16, 61
Chemical bonding, 455–457
Chemical calculations:
 dilutions, 493
 mass percent, 489
 molarity, 491
 normality, 492
 percent yield, 488
 pH, 170
 titration, 183
 volume percent, 489
 weight percent, 489
Chemical change, 373, 374
Chemical formula, 458
Chemical hygiene plan (CHP), 12
Chemical laboratory technicians, 1, 6
 critical job functions, 6
 professional references and resources, 14
Chemical Manufacturers Association (CMA), 316
Chemical nomenclature, inorganic, 458
Chemical nomenclature, organic, 467
Chemical plant operators, 1
Chemical process industry (CPI), 1
Chemical process operators, 1
 core competencies, 1
 critical job function, 1
 professional references and resources, 14
Chemical spills:
 acids, 70
 alkali, 71
 mercury, 72
 solids, 70
 volatile/flammable solvents, 71
Chemical symbols, 451
Chemical Transportation Emergency Center (CHEMTEC), 16, 69
Chemical waste disposal, 72
Chemicals:
 chemically pure (CP), 313
 chromatographic grade, 314
 commercial grade, 313
 grades of purity of, 313
 hazardous, common, 314

Chemicals (*Cont.*):
 incompatible, common, 314
 practical grade, 313
 primary standard grade, 314
 reagent analyzed, 314
 reagent grade, 314
 solution preparation of, 324, 331–335
 spectroscopic grade, 314
 storage of, 314
 United States Pharmaceutical (USP) grade, 313
 warning labels for, 315
Chemicals Hazard Information and Packaging for Supply (CHIP), 48
Chemistry:
 inorganics, fundamentals of, 451–465
 organics, fundamentals of, 467–482
Chemometric, 601
CHEMTEC, 16, 69
Chemtotes, 63
CHEMTREC, 69
Chlorine gas, 80
Chopper, 592, 610
CHP (chemical hygiene plan), 12
Chromatographic grade chemicals, 314
Chromatographic techniques, 553
Chromatography, 521–571
 adsorption, 521
 column, 521
 definition, 521
 gas (GC), 526
 high performance liquid (HPLC), 556
 high pressure liquid (HPLC), 556
 ion-exchange (IC), 564
 liquid (LC), 550
 mobile phase, 521
 partitioning, 521
 stationary phase, 521
 thin-layer (TLC), 522
Chromel, 336
Chromophores, 561, 585
Chromosorbs®, 535
Circuit breakers, 205
Cis/trans isomers, 471
Claisen columns, laboratory, 155
Claisen flask, 434
Claisen fractionating column, 434
Clamps, laboratory, 161
Classifications of fire, 37
Classifying particle size, 228
Clean Air Act, 14
Cleaning solution:
 alcoholic sodium hydroxide, 502
 Alconox®, 499
 aqua regia cleaning solution, 502
 dichromate-sulfuric acid, 501
 dilute nitric acid, 502
 metal decontamination of glassware, 502
 Nochromix®, 502
 trisodium phosphate, 502

INDEX **649**

Closed-cup flash point apparatus, 383
Closed-end manometer, 117
Coast Guard, U.S., 16, 69
Code of Federal Regulations (CFR), 10
Coefficients, balancing equations, 484
Coefficients, exponential notation, 121
Cold finger sublimation, 449
Cold trap, dry ice, 222
Coliwasas samplers, 224
Collecting and measuring gases, 94
Colloidal dispersions, 321
Colorimetric determination, indicators, 176
Colorimetry, 578
Column chromatography, 521, 552
Column oven, 529
Columns, fractionating, 155
Columns efficiency, GC, 536
Combined gas law, 107
Combustion furnaces, 297
Combustion of organic materials, 303
Combustion-tube methods, 296
Commercial grade chemicals, 313
Common gases, characteristics, 79
Common logarithm, 126
Complex ions, solubilities, 320
Compounds:
    definitions of, 458
    trivial names for, 464
Compressed air, 80
Compressed Gas Association (CGA), 16, 78
Compressed gases, 75–104
    collecting gases, 94
    compressed air, 80
    cylinder markings, 76
    dispersers, 95
    drying, 94
    flowmeter for, 92
    gas-leak detectors, 93
    gas-washing bottles, 93
    general precautions chart for, 77
    handling cautions on, 75
    humidifying, 94
    installation, regulators, 85
    lecture bottle for, 78
    liquefied, 82
    OSHA violations, 76
    physical properties, some common gases, 79
    pressure-reduction stages for, 85
    quick couplers, 91
    regulators for, 84
    safety devices for, 91
    securing a cylinder, 75, 76
    troubleshooting for, 87
    valves, gas, 88
Compression fittings, 351
Compression packing, 370
Compressors, 99
Computerized balances, 277
Computers, terminology (*see* Glossary, 629–643)

Concentration:
    definition, 489
    dilutions, 493
    expressions, 489
    molarity, 491
    normality, 492
    parts per billion, 490
    parts per million, 490
    percent, by mass, 489
    percent, by volume, 489
    percent by weight/volume, 489
Concentration of commercial acids and bases, 494
Concentration of stock acids and bases, 324
Condensate, 419
Condensation, freeze-out, 221
Condensation, reaction, 480
Condensers, laboratory, 154, 441
Conductance, specific, of water, 319
Conductivity, 201, 336, 337
Conductivity of various aqueous solutions, 337
Conductivity of various metals, 202
Conductivity of water, 319
Conductivity probes, 336
Conductors, 201
Confidence intervals, 144, 145
Confidence limits, 144, 145
Confined space entry permit, 43
Confined spaces, 43
Conformance marking (CE), 65, 66
Coning and quartering samples, 226
Conservation of mass law, 484
Constant mass, 254
Constantan, 336
Continuity meter, 208
Continuous duty pump (*see* Glossary, 629–643)
Continuous liquid-liquid extractions, 415
Continuous spectra, 580, 602
Continuous spectra instruments, 602
Continuum source, 616
Control charts, 146
Control of Substances Hazardous to Health (CSHH), 19, 48
Control records, 139
Control room operators, 137
Control samples, 146
Controlling spills, 69
Conversion factors, SI and engineering units, 130–133
Conversion factors, SI and USCS, 130–133
Conversion table, millimeters to decimal inches, 134
Conversion table, pressure, 106
Conversion table, units of area, 127
Conversion table, units of length, 127
Conversion table, units of liquid measure, 128
Conversion table, units of mass, 129–130
Conversion table, units of temperature, 376
Conversion table, units of volume, 128
Conversion table, wave number to wavelength 591–592

650 INDEX

Conversion table, wavelengths, 575
Conversion tables, 127–133
Copolymerization, 480
Copper tube fittings, 358
Core competencies for chemical technicians, 6
Core competencies for process workers, 1
Coriolis effect, 341
Corrosive and hazardous chemicals, 314
Coulometry (see Glossary, 629–643)
Covalent bonds, 457
Covalent (binary) compounds, naming, 460
CP (chemically pure) chemicals, 313
Cracking, 481
Cradle to grave, EPA, 15, 73
Crane safety, 67, 68
Crane safety lights, 68
Crawl, gas regulators, 87
Crawler cranes, 67
Creeping, precipitates, 244
Critical density, 83
Critical job functions, laboratory technicians, 6
Critical job functions, process operators, 1
Critical pressure, 83
Critical temperature, 83
Crucibles, 248
Crude oil, 467
Crushed stone, distillation, 443
Crushing samples, 228
Cryogenic gases and liquids, 82, 83
Cryogenic handling precautions, 84
Cryogenic properties of some common gases, 83
Crystalline membrane, 194
Crystallization, 257
Cubic centimeter, 497
Current good manufacturing practice (CGMP), 137, 138, 140
Cut and weigh integration technique, 541
Cutoff limits, solvents, 561, 586
Cutting glass tubing, 163
Cuvettes, 584, 587
Cyclic hydrocarbons, 470
Cyclo prefix, 470
Cycloalkanes, 470
Cylinder marking, gas, 78
Cylinder sizes, gas, 78

D line, sodium lamp, 396, 401
D-optical rotation, 400
Dalton's law, 107
Dangerous Substances and Explosive Atmospheres Regulations (DSEAR), 65
Dead head, 364
Decantation, 240, 241
Decimal inch to millimeter conversion table, 133–135
Decolorization, 263
Decomposition point, 377
Degassing, HPLC, 562
Deionized water, 318, 566
Demineralized water, 319
Demister, 116

Demountable cells, infrared, 593
Density determination, 385–388
  by density-bottle method, 386
  by Dumas method, 387
  by float method, 387
  of gases, 387
  by hydrometer method, 388
  of irregularly shaped solids, 386
  of liquids, 386
  by pyknometer method, 388
  of regularly shaped solids, 385
  by water-displacement method, 386
Density of water at various temperatures, 627–628
Department of Agriculture (USDA), 601
Department of Education, 1
Department of Labor (DOL), 1
Department of Transportation (DOT), 13, 16, 51, 52, 75, 78
  DOT gas cylinder markings, 76
  DOT hazard class or division number for materials, 52
  DOT hazard labels color code, 52
  DOT signs, placards, and labels, 51
Desiccants, 287
Desiccators, 254, 287
Detection limit, atomic absorption, 613
Detectors, GC, 530
Determinate errors, 143
Deuterium lamp, 560, 582, 616
Deviation, 143
Dew point, 342
Dewar flask, 83, 222, 292
Dextrorotation, 400
Dial-O-Gram® balances, 271
Dialysis, 323
Diaphragm, regulator, 85
Diatomaceous earth, 535
Diatomic elements, 483
Dichromate-sulfuric acid cleaning solution, 501
Diethyl ether, peroxides in, 411
Diethyl ether in extractions, 411
Differential pressure (DP), 341, 342
Differential refractometer, 561
Diffraction, 574
Diffraction grating, 574, 577
Diffusing stones, gas, 96
Diffusion pump oil, 111
Diffusion pumps, 111
Digestion and aging, 251
Dilute nitric acid cleaning solution, 502
Dilution to the mark, 504
Dilutions, calculations, 493
Dimensional analysis, 127
Dimethyldichlorosilane, 522
Diode-array spectrophotometer, 568, 589, 603
Dimers, 479
Diols, 474
Dip tube, 81
Dipper, 225
Direct current (DC), 203

INDEX 651

Direction valve, gas, 89
Dispersed phase, 321
Dispersers, gas, 95
Dispersing phase, 321
Dispersions, colloidal, 321
Displacement pumps, 366, 558
Dissociation constant, k, 173
Dissolutions, 320
Distillate, 419
Distillation:
    assemblies, 157–158
    azeotropes, 423
    columns, laboratory, 154–160
    defined, 419
    fractional, 426
    maximum boiling azeotropes, 423
    minimum boiling azeotropes, 434
    plate, 538
    process unit, 442
    pure liquids, 421
    rate of, 423
    receivers, 160, 435
    simple, 419
    solutions, 422
    steam, 435
    superheated steam, 439
    two liquids, 422
    vacuum, 430
    vapor pressure, 419
Distilled water, 318
Distribution coefficient, 407
Dopes, pipe sealants, 354
DOT, 13, 16, 51, 52, 75, 78
Double beam, 609
Double bond, 469
Downcomer, 442
Drierite®, 94
Drill bits, size of, 136
Drum cradle, 63
Drum transport, 63
Drums, 61, 63
Dry ashing procedure, 299
Dry-bulb temperature, 99
Dry chemical fire extinguishers, 38
Dry ice cooling, 222, 292
Dry-ice vapor trap, 222, 432
Dryer tubes, 99
Drying agents, 290
Drying gases, 94, 95
Drying glassware, 503
Drying organic solvents, 290
Drying pistol, 289
Drying samples, 286–293
Drying tubes, 94, 99
DSEAR, 65
Duboscq colorimeter, 578
Dumas determination of nitrogen, 298, 308–309
Dumas method for density of gases, 387
Dust and particulate respirators, 31

Ear muffs, 27
Ear plugs, 27
Eductor tube, 81
Effective, analyte, 193
Elastic limits, 404
Elastic modulus, 404
Elasticity, 408
Elastomeric stator, pump, 367
Electric circuit elements, 217
Electric circuit symbols, 217
Electric connections, 210–213
    soldering, 212
    splicing, 211–212
    taping, 213
Electric current, 197
Electric devices, servicing of, 207
Electric melting point apparatus, 380
Electric motors, 208
Electric power, 206
Electric resistivity of metals, 202
Electrical conductivity, 201, 315, 317, 336, 337
Electrical connections, 210
Electrical grounding, determination of, 215
Electrical power, 206
Electrical safety practices, 197
Electrical testing instruments, 207
Electricity, 197–217
    circuit breakers, 205
    compared to fluids, 199
    connections, 210
    current symbols for, 217
    distance, safety, 197
    fuses, 204
    grounding for, 213
    insulating shell (wire nut), 211
    parallel, 206
    personal protective equipment, 197
    physiological effects, 198
    power, 206
    safety, guidelines and practices, 198
    soldering, 212
    splicing wires, 211–212
    testing meters, 207
    wire conductors, 211
Electrification, static electricity, 267
Electrode, calomel, 179, 192
Electrode, glass, 179
Electrode, pH maintenance, 182
Electrodeless discharge lamp (EDL), 610
Electrolyte, 192
Electromagnetic radiation 573, 575
Electromagnetic (em) spectrum, 573, 575
Electromagnetic (em) waves, 573
Electron, 451
Electron flow, 197
Electronic balances, 275
Electronic integrators, chromatography, 543
Elements:
    definition of, 451
    symbols for, 451

Eluant, 552
Elute, 552
Emergency hood, gases, 96
Emergency response plan (ERP), 69, 70
Emission spectrographs, 579
Emulsions, 413
End point, 497
"ene" suffix, 469
Engineering notation, 122
Engineering to SI conversion factors, 133–135
Environmental Protection Agency (EPA), 14, 15, 25, 61, 73, 146
Enzyme, 482
Enzyme substrate membrane, 195
Eppendorf micropipettes®, 509
Equal arm balances, 269
Equilibrium balances, 269
Equipment cleaning and use log, 139
Equipment specific, SOP, 137
Equivalence point, 497
Equivalent mass in acid/base reactions, 492, 516–517
Equivalent mass in redox reactions, 492, 516–517
Equivalent masses, 492, 516–517
Errors:
   absolute, 143
   control samples, 146
   determinate, 143
   indeterminate, 143
   method of least squares, 145
   random error, 143
   relative, 143
   standard deviation, 144
Essential water, 285
Esters, 478–479
Ether (*see* Diethyl ether)
Ethers, 475–476
Ethers, caution use, 476
Ethylenediaminetetraacetic acid (EDTA), 518
Evaporating dishes, 444
Evaporating dishes, direct heating of, 444
Evaporation:
   of liquids, 443
   reduced pressure, 445
   under vacuum, 446
Evaporator, rotary, 447
Exchange capacity, 564
Exhaust hoods, 23
Explosion limits, 384
Explosion-proof motor (XPRF), 364
Exponential notation, 121
External standard method, 545
Extinguishers, fire:
   carbon dioxide, 38
   dry chemical, 38
   halogenated, 38
   Met-L-X, 38
   water, 37

Extraction, 407–418
   using dilute aqueous acid solution, 410
   using dilute aqueous basic solution, 410
   using disposable phase separators, 414
   using selective technique, 410
   using Soxhlet extractor, 407
   using water, 410
Extractions, continuous liquid-liquid, 415
Extractions, solvent choices, 410
Extrapolation, 126, 547
Eye protection, 19, 25
Eyewash fountains, 39

Face shield, 26
Factor-label method, 127
Factors, gravimetric, 301, 302, 331
Factory Mutual (FM), 16, 61
Fahrenheit temperature scale conversions, 375–376
Fall protection, 39
Falling-ball viscometer, 392
Falling-piston viscometer, 391
Families, organic, 481
Family, periodic table, 455
Far infrared region, 590
Far ultraviolet region, 585
Federal Insecticide Fungicide and Rodenticide Act (FIFRA), 138
Feedback loop (*see* Glossary, 629–643)
Fellgett's advantage, 600
Fermentation, 482
Ferrules, 351
FID, flame ionization detector, 531
Filter aids and media, 238
Filter media, 233
Filter membrane, 236
Filter nail, 250
Filter paper, 233, 235, 243
Filter paper, fluted, 244
Filter pump setup, 246
Filter pumps, 240
Filter respirators, 31
Filter supports, 238
Filtrate, 234
Filtration, laboratory, 233–254
Fingerprint region, infrared, 590
Fire, hazards, NFPA, 53
Fire classifications, 37
Fire extinguishers, 37–38
Fire polishing, glassware, 20, 163
Fire prevention from welding activities, 358
Fire safety, terminology, 37
Fire triangle, 37
Fire watchers, 358
First-derivative titration curve, 513
Fisher burner, 254
Fisher degrees, 389
Fit checks, respirators, 33
Fittings, tubing, 351
Fittings and valves, GC, 535
Fixed notation, 123

INDEX

Flame arrestor, 62
Flame ionization detector (FID), 531
Flame test, elements, 283–284
Flameless atomic absorption spectroscopy, 618
Flammability hazards, NFPA, 53
Flammable liquids, OSHA and NFPA classification, 383
Flanges, valve, 362
Flap check valve, 368
Flared tubing, 350
Flareless tubing connections, 351
Flaring copper tubing, 350
Flaring tools, 350
Flash arrestor, 363
Flash point, 383, 384
Flash point determinations, 382–384
Flashback, 612
Flask, suction adapter, 246, 251, 562
Flasks, laboratory, 154, 434
Flexible vane pumps, 366
Float method for density, 387
Flooded suction, 366
Flow control valves, 91
Flow rate measurements, 340
Flowmeters, 92, 538
Fluid, 359
Fluid flow, 359
Fluid mechanics, 359
Fluid surface tension, 395–396
Fluidity, 390
Fluidized bath method, 502
Fluorescence detector, 568
Fluorine, liquid, 83
Fluorolube®, infrared, 597
Fluorometers, 589
Fluted filter paper, 244
Fluxes, 294, 296
Food and Drug Administration (FDA), 16, 61, 137, 138
Foot valve, 366
Forklift operations, 66
Formaldehyde, caution, 477
Formalin, caution, 477
Formulas, 456
"Foundations for Excellence in the Chemical Process Industries", 1
Fourier transform, 585, 598
Fraction conversion chart, 134
Fractional crystallization, 265
Fractional distillation, laboratory, 157–158, 426
Fractionating column heating, 428
Fractionating column packings, 427
Fractionating columns, laboratory, 155
Frangible disc, 90, 363
Free-moisture, 463
Freeze drying, 292
Freeze-out, 221
Frequencies, 591
Frequencies in electromagnetic spectrum, 573
Frequency to wavelength equation, 591

Frequency-wavelength relationship, 573, 576
Fritted disk, 95
Fritted glass, gas dispersers, 95
Fritted glassware, grades, 237
Fritted glassware cleaning, 237
Frostbite from cryogenic liquids, 84
Fuel, 37
Fulcrum, forklift, 67
Fume cupboards, 23
Fume hoods, laboratory, 23
Fuming nitric acid, 299
Functional group region, infrared, 590
Functional groups, organics, 481
Funnel, wire liners, 248
Funnels, 243
Funnels, heating, 259
Fuse-holding blocks, 204
Fuse replacement, 204
Fused silica, UV, 585
Fuses, 203
Fusible plugs, 90, 363
Fusion, 295, 299

Galvanized pipe fittings, 355–358
Gas, valves, 88
Gas ballast, 113
Gas chromatography, 526–549
    automatic sampler, 527
    capillary columns, 528
    carrier gas, 527
    column efficiency, 528
    column fittings, 351, 535
    column oven, 529
    columns, 528
    detectors, 530
    flame ionization detector, 531
    flowmeters, 538
    HETP, 538
    injection port, 527
    integration techniques, 541
    liquid phases, 539
    microsyringes, 534
    operational procedures, 532
    packed columns, 528
    qualitative analysis, 539
    quantitative analysis, 543
    recorder/integrator, 530
    resolution, 538
    rotameters, 539
    sampling valve, 527
    septa (septum), 527
    solid supports, 532
    solute classification, 537
    support-coated open tubular (SCOT) columns, 528
    temperature programming, 529
    thermal conductivity detector, 529
    troubleshooting, 548
Gas collecting and measuring, 220
Gas control in a reaction, 96

Gas cylinder markings, 75, 78
Gas cylinders, handling precautions, 75
Gas density determinations, 387
Gas diffusing stone, 96
Gas diffusion stones, 95
Gas dispersers, 95
Gas displacement collector, 223
Gas flow controllers, 88–91
Gas gauges (*see* Gauges)
Gas humidifying and drying, 99
Gas laws:
    Avogadro's, 107
    Boyle's, 105
    Charles', 105
    combined, 107
    Dalton's, 107
    Gay-Lussac's, 107
    ideal, 107
    Pascal's, 108
Gas-leak detectors, 93
Gas-liquid filtration, 233
Gas-masks, 31
Gas oil, 472
Gas-permeable membrane, 194
Gas regulators, 84, 85
Gas safety devices, traps, check valves, 91
Gas thermometer, 593
Gas-washing bottles, scrubbers, 93, 94, 441
Gases (*see* Compressed gases)
Gate valves, 361
Gauge, pressure:
    Bourdon-type, 118
    closed-end manometer, 117
    McLeod, 91, 119
    open-end manometer, 116
    Pirani, 119
    thermocouple, 120
    U-tube manometer, 116
Gauge, vacuum:
    closed-end manometer, 117
    McLeod gauge, 91, 119
    open-end manometer, 116
Gauge pressure, 91, 105
Gaussian distribution, 145
Gaussian shaped, 145
Gay-Lussac's law, 107
GC (*see* Gas chromatography)
Gear pumps, 366
Gel filtration chromatography (GFC), 570
Gel permeation chromatography (GPC), 570
Generators, hazardous waste, 73
Glass beads, 427
Glass electrode, 179, 339
Glass helices, packings, 427
Glass tubing, 163–167
Glass tubing safety, 163
Glassblowing, 163
Glasses, safety, 25
Glassware, 149–168
Glassware, cleaning, 500

Glassware, heating caution, 503
Glassware storage, 158
GLC (*see* Gas chromatography)
Globally harmonized system (GHS), 47, 51, 60
Globe valve, 361
Glossary, 629–643
Gloves, 28
Glycine (glycerin), 20, 152
Glycols, 474
Goggles, safety, 25, 26
Golay thermometer, 593
Gooch crucibles, 233, 248, 331
Good laboratory practice (GLP), 138, 520
Gooseneck educator tube, 81
Governmental and other information resources, 14–16
Governmental regulations, 10–14
Grab samples, 222
Gradient elution, 555
Grain alcohol, 474
Gram, 129, 385
Gram-equivalent weight, 491–492
Graphite furnace, 617
Graphite furnace technique, 617
Graphs, 125
Grating, 577, 592
Gravimetric analysis, 281
Gravimetric calculations, 301, 331
Gravimetric factor, 301, 302, 331
Gravity filtration, 233, 240, 242, 259, 286
Greek prefixes, naming, 468
Grindability, 228
Grinding equipment, 230
Grinding samples, 230
Gross sample, 219
Grote method, 298
Ground glass equipment, 150
Ground fault circuit interrupters (GFCI), 216
Ground state, electrons, 580, 608
Grounding and bonding, 64
Grounding electrical equipment, 213
Grounding wire, 214
Grounds, 214
Groups, periodic table, 455
Guard columns, 560
Guide numbers, DOT, 53
Gypsum, 285

Half-width method, 542
Haloalkanes, 473
Halogen, determinations, 309
Halogenated hydrocarbon fire extinguishers, 38
Halogens, 309, 455
Hammer mill, 230
Hand protection, 28
Handling chemicals, 3, 19
Handling compressed gases, 75
Handling cryogenic liquids, 82
Hard hat, 30
Hard water, 317
Hardness, 228, 406

Hardness scale (Mohs scale), 406
Harness, safety, 40
Hasp, multiple lock, 44
Hazard Communication Standard, 19
Hazard statements, 48
Hazardous materials, warning labels, 315
Hazardous materials identification system, 59
Hazardous materials information system (HMIS), 51, 53
Hazardous materials table, 13
Hazardous waste disposal, 72
HAZWOPER, 69
Head protection, 30
Health hazards, NFPA, 53
Hearing protection, 27
Hearing protective devices, HPD, 27
Height equivalent to a theoretical plate (HETP), 538
Helium, carrier gas, 530
Helmets, 30
HEPA, high efficiency particulate air, 33
Heterogenous material, 219
Hierarchy of control principles, 24
High efficiency particulate air (HETP), 33, 70
High performance liquid chromatography (HPLC), 556–564
   analytical guard column, 560
   automated HPLC systems, 564
   column storage, 560
   columns, 559
   design and components, 556
   detectors, 561, 568
   pump inlet filters, 562
   mobile phases, 562
   recorder/integrator, 561
   sample injection systems, 559
   solvent delivery systems, 557
   solvent mixing systems, 557
   troubleshooting, HPLC pumps, 563
High pressure liquid chromatography, 556–564
High volume sampler (Hi Vol), 220
Higher-density solvent extraction, 415
Highway tankers, 64
Hirsch funnels, 250
Histogram, 146
Hollow cathode lamp, 607
Homogeneity, 320
Homogeneous materials, 219
Homogeneous solution, 320, 413
Hood, fume, 23
Hooke's law, 404
Hot work, 357
Hot work permits, 357
HPLC (*see* High performance liquid chromatography)
HPLC grade solvents, 314, 555
HPLC pump troubleshooting, 563
Humidifying gases, 94
Humidity, relative, 99
Hydrates, 285, 463
Hydraulic swaging unit (HSU), 353
"Hydro" prefix, 463

Hydrocarbon grease, 152
Hydrocarbons, 467
Hydrochloric acid, 293
Hydrofluoric acid, 294
Hydrogen, carrier gas, 530
Hydrogen gas, 81
Hydrogen ion, 171
Hydrogen sulfide, safety, 43
Hydrolyze, 482
Hydrometers, 388
   Baumé conversion table, 390
Hydronium ion, 169, 173
Hydrophobic, 522
Hydroxide ion (hydroxyl ion), 171
Hydroxyl group, alcohols, 473
Hygiene plan, chemical (CHP), 12
Hygroscopic substances, 286

"ic" suffix, 463
"ic/ate" suffix, esters, 478
ID numbers, DOT, 52
"ide" suffix, 458, 460
Ideal gas constant, 107
Ideal gas law, 107
Ignition, filtration, 252
Immediately Dangerous to Life or Health (IDLH), 22
Immiscible solvents, 259, 407, 488
Impellers, 365
Implosion, 223, 431
Incompatible chemicals, 314, 316
Indeterminate errors, 143
Index of refraction, 399
Indicator solutions, preparation of, 176
Indicators, acid/base, 175, 176
Inducing crystallization, 263
Inert (noble) gases, 79
Information, governmental and other sources, 14–17
Infrared absorption spectra, 600
Infrared liquid cells, 593
Infrared spectroscopy (IR), 590–601
Infrared window materials, 593
Injector splitters, GC, 528
Inorganic chemical solubility rules, 320
Inorganic chemistry review, 451–466
Inorganic compound nomenclature, 458
Inorganic compound solubilities, 322
Inorganic sample dissolving, 320
Inserting glass tubing, caution, 22
Instability hazard, NFPA, 53, 55
Insulating shell (wire nut), 211
Insulators, 199
Integration techniques, 541
Integrator, electronic, 531
Intensity, 581
Interferences, atomic absorption, 615
Interferogram, infrared, 599
Internal standard method, 546
International Air Transportation Association (IATA), 51, 60

International Standards Organization (ISO), 16, 51, 53, 54, 141
International system of units (*see* System International, SI system)
International system of units (SI), conversion factors, 128–133
International Union of Pure and Applied Chemistry (IUPAC), 467
Internet, MSDS, 51
Interpolation, 126
Inventory and tracking of chemicals, 60
Ion, defined, 456
Ion exchange chromatography (IC), 564
Ion exchange chromatography, detectors, 568
Ion exchange columns, 566
Ion exchange resins, 310, 317, 565
Iron pipe fittings, 355
Ion product constant for water (K_w), 173
Ion selective electrode (ISE), 192–195
Ionic bonds, 456
Ions, common anions, 459, 462, 463
Ions, common cations, 459, 460, 462
ISO, 141, 147
ISO 9000 GLP common deficiencies, 142
Isocratic elution, 555
Isomers, 471
Isothermal, 529
"ite" suffix, 463
IUPAC, nomenclature, 467

Jacked Büchner funnel, 261
Jacquinot's advantage, 600
Jaw crusher, 230
Job specific, SOP, 137
Job task analysis for chemical laboratory technicians, 6
Job task analysis for chemical operators, 1
Joint numbers, glassware, 149
Joints and clamps, glassware, 150
Joule, 590

Karl Fischer, 342, 519
Kelvin temperature scale, 374
Kerosene, 472
Ketones, 477
Kevlar®, 28
Kinematic viscosity, 394
Kinetic pumps, 364
Kjeldahl method, 299, 305–307
Knock-out (KO) drums, 99, 116
Kuderna Danish sample extractors, 417
K_w, water ion-product constant, 173

L-optical rotation, 400
Labeling, chemicals and materials, 51, 314, 315
    ANSI, 51, 53, 59
    DOT, 51
    GHS, 51, 55
    HMIS, 51, 53, 59
    IATA, 51, 60

Labeling, chemicals and materials (*Cont.*):
    ISO, 51, 54
    NFPA, 51, 53
Labeling gas cylinders, 75
Labels, hazardous chemicals, 315
Labile, heat, 571
Laboratory and chemical plant safety, 19–45
Laboratory chemicals, handling techniques, 47–73
Laboratory information management system (LIMS), 60
Laboratory notebook, 140
Laboratory solutions, preparation of, 313–331
Laboratory technicians:
    job tasks and skills, 6
    professional sources of information, 14
    role, 6
Laboratory Standard (OSHA), 19
Ladders, safety, 40
Laminar flow, 359
Laminar flow burner, 610
Lanthanum, 615
Lanyard, 39
Large quantity generator (LQG), 73
Law of conservation of mass, 484
LD_{50}, 21
Leaking gas cylinder, 75–78
Leaks, vacuum systems, 113
Least squares method, 146
Lecture bottles, 78
Length equivalent tables, 128
Lessing rings, distillation, 443
Lethal concentration (LC), 21
Lethal dose (LD), 21
Level measurements, 342
Levorotation, 400
Lifelines, 39
Lifting safety, 42
Ligands, 517
Likes dissolve likes, 258
LIMS (laboratory information management system), 60
Line spectra, 580
Linear, 126
Linear regression, 146
Linearity, 614
Lip seal, 371
Liquefied gas, 82
Liquefied petroleum gas (LP), 81
Liquid chromatography (LC), 550–556
Liquid evaporation, 443
Liquid extractors, 410
Liquid-filled remote thermometers, 334
Liquid fluorine, 83
Liquid-liquid membrane, 194
Liquid-liquid separation, 233
Liquid measures, equivalent tables, 128
Liquid nitrogen, 83
Liquid nitrogen cooling, 83, 222
Liquid oxygen (LOX), 83

INDEX **657**

Liquid percent by volume, 489
Liquid phases, GC, 535, 539
Liquid sampling, 223
Liquid transfer pumps, 69
Liquid volume equivalent tables, 128
Liter, 123, 130, 497
Litmus paper, 81
Load capacity, forklift, 67
Load center, forklift, 67
Local ventilation systems, 23
Lockout/Tagout, 44, 137, 362
Lockup, gas regulator, 84
Log book, guidelines, 140
Logarithms, 126, 174
Loop injector, 559
Loops, sample, 559
Lower action limit, 146
Lower density solvent extraction, 415
Lower explosion limit (LEL), 80, 384
Lower warning limit, 146
LOX, liquid oxygen, 82
Lubrication, ground glass, 151
Lyophilization, 292

Magnesium perchlorate, 304
Magnetic-driver centrifugal pumps, 365
Magnetic starter, motor, 209
Maintenance, duties, 4
Manometers:
   closed-end, 117
   filling of, 118
   open-end, 116
   U-tube, 91, 116
Mantissa, 127
MAPP, methyl acetylene propadiene, 82
Mass:
   definition of, 268
   determinations of, 267–279
   by difference technique, 301
   errors in determining, 269
   measurement of, 278
   sources of error, 269, 278
   terminology, 123
Mass equivalent tables, 129, 130, 133
Mass measurement, factors, 129, 130, 133
Mass number, 451
Mass percent by weight, 489
Mass relationships from equations, 485
Massing bottles, 300
Master production and control records, 139
Material balance, 483
Material Safety Data Sheets (MSDS), 3, 13, 21, 47, 48, 49, 75
Mathematical signs and symbols, 127
Mathematics review and conversion tables, 121–138
Matrix effect, 615
Matter, 451
Maximum boiling azeotropes, 423
Maximum folk height, 67
Maximum storage capacities, solvents, 62

McLeod gauge, 119
Mean, statistical, 143, 147
Measurement, 267
Measuring pipettes, 504
Mechanical pump, vacuum, 109, 446
Mechanical seal, 371
Median, statistical, 143
Meker burner, 254
Melting point determinations, 377–380
   capillary-tube method, 377
   electric apparatus for, 380
   hot-oil bath for, 378
   Thiele tube for, 378
Membrane:
   crystalline, 194
   enzyme substrate, 195
   filters, 236
   gas permeable, 194
   liquid-liquid, 194
   solid state, 194
   tube filter, 418
Meniscus, 497, 498
Mercury, clean-up procedure, 71
Mercury, danger of, 72
Mercury analyzers, 607, 618
Mercury spills, 72
Mercury use, 119
Mesh sizes, 229, 535
Met-L-X fire extinguishers, 38
Metal-arc welding (MIG), 356
Metal decontamination of glassware, 502
Metal fumes, 31
Metal resistivity, electrical, 202
Metal spiral, packing, 427
Metallochromic indicators, 518
Metalloids, 455
Metals, 455
Meter, 123, 132, 575
Metering pumps, 369
Metering valves, 89, 362
Method of least squares, 146
Methyl acetylene propadiene (MAPP), 82
Metric prefixes, 123
Metric (SI) system, 122, 123, 575
Metric ton, 135
   (*See also* Glossary, 629–643)
Mho, 201, 336
MIBK, methyl isobutyl ketone, 612
Michelson interferometer, infrared, 599
Micro, prefix, 123, 337
Microbes, 237
Microdetermination of carbon and hydrogen, 302
Micrometer, 575, 591
Microohm, 318
Microparticles, 237
Micropipets, tolerances, 509
Microporous particles, 559
Microreticular resins, 559
Microsiemens, 318
Microliter syringe, 534

658 INDEX

Micrometers, 575
Micrometers versus nanometers, 576
Micron, 575, 591
Microsiemens, 318
Microwave, level measuring devices, 342
Milli, prefix, 337
Milliequivalents, 492, 516–517
Milliliter, 497
Millimeter conversion to decimal inches, 134
Millipore® membrane, 234, 236
Mine Safety and Health Administration (MSHA), 16
Minicells, infrared, 594
Minimum boiling azeotropes, 424
Minipress, infrared, 598
Miscible, 257–259, 408, 488
Mistakes and errors in log books, 140, 279
Mists, 31
Mixed ethers, 475
Mixtures, 219
Mobile phases, 521, 561
Mode, statistical, 143
Mohr method, 310
Mohs scale, 406
Moisture in samples, 269, 284
Moisture measurements, 342
Molar absorptivity, 581, 584
Molar extinction coefficient, 581
Molar solutions, 491, 516
Molarity, 491, 516
Mole, definition of, 491
Mole relationships from equations, 486
Molecular formulas, 457
Molecular masses, 485
Molecular relationships from equations, 483
Molecular sieves, 99
Molecular weights, 457
Molecule, 451
Monatomic ions, 458
Monitoring processes, 3
Monochromator, 584, 593, 609
Monodentate, 517
Monomers, 479
Monroe crucible, 234, 248
Mortar and pestle, 283
Motor starting, 208
Motors:
  capacitor, 210
  dual-voltage, 210
  grounding, 213
  magnetic starters, 209
  repulsion-induction, 210
  split-phase, 210
  thermal overload, 210
  three-phase, 210
  types of, 210
Moyno® pumps, 367
MPT pipe tread, 355
MSDS (see Material safety data sheets)
MSDS on the internet, 51
MT, gas cylinders, 76

Mull method, infrared, 597
Mullers, 230
Multimeter, 208
Multiple alkyl branches, 468

"N" amine prefix, 475
"NA" drum identification, 63
Naming:
  acids, 463
  binary compounds, 460
  cations, common, 459
  common names, 464
  compounds containing polyatomic ions, 461
  covalent compounds, 461
  hydrates, 463
  inorganic compounds, 458–464
  monatomic anions, 459
  multiple charged metals, 460
  organic compounds, 467–482
  polyatomic ions, 461, 462
  prefixes, 461
  trivial names, 464
Naming organic compounds, 467–482
Nanometers, 575, 576
NaOH, carbonate-free, 324
Naphtha, 472
Naphthenic compounds, 470
National Bureau of Standards (NBS), 268
National Fire Protection Association (NFPA), 16, 51, 53, 54, 64
National Fire Protection Association (NFPA), numerical codes, 53, 55–58
National Fire Protection Association classifications, 53
National Institute for Occupational Safety and Health (NIOSH), 24, 75
National Institute of Standards and Technology (NIST), 16, 374
National Paint and Coating Association (NPCA), 53
National Response Center (NRC), 16, 69
National Safety Council (NSC), 16
Natural logarithms, 126
Near infrared absorption bands, 603
Near infrared monochromator, 592
Near infrared spectroscopy (NIR), 590, 601–604
Neat, liquid, 595
Nebulizer, 610
Needle valves, 88
Negative ions (anion), table of, 459
Negative pressure fit check, 34
Nephelometry, 579
Nernst equation, 193
Nernst glower, 592
Net positive suction head (NPSH), 364
Neutral, 173
Neutralization, pH, 494
Neutron, 451
Newton, pressure unit, 105
Nichrome wire, 283
Nichrome wire heater, 428

Nicol prism, 400
NIOSH/OSHA guide to chemical hazards, 24, 75
Nitric acid, 293
Nitric acid cleaning solution, 502
Nitrogen analyzers, 307
Nitrogen liquid, 83
Nitrous oxide-acetylene mixture, 610
Noble gases, 455
Nochromix® cleaning solution, 502
Nomenclature (*see* Naming)
Nominal, pipe diameter, 356
Nomograph, vacuum distillation, 430
Nonmetals, 455
Nonpolar, 257, 535
Norit, 263
Normal solutions, 492
Normality, 492, 516
Normalization, 543
Notebook (*see* Log book)
NPSH, net positive suction head, 364
NRC (National Response Center), 16
Nuclear Regulatory Commission (NRC), 17
Nujol®, 597

"oate" suffix, 478
Occluder, spectrophotometer, 588
Occupational exposure limits (OELs), 22
Occupational Safety and Health Administration, 11, 17, 21, 25, 30, 47, 61, 75, 99, 114, 137, 356
Octane, 482
Ohm's law, 200
Ohm, 200
"oic" suffix, 478
Oiling out, 266
"ol" suffix, 473
"one" suffix, 477
Open cup flash point apparatus, 383
Open-end manometer, 116
Optical null, 593
Optical rotation, 400
Ordinate, 126
Organic chemistry, 460
Organic chemistry review, 467–482
Organic extraction solvents, 411
Organic families, functional groups, 481
Organic reactions, 481
Orifice plates, 341
Orsat analysis, 223
OSHA, 11, 17, 21, 25, 30, 47, 61, 75, 99, 114, 137, 356
OSHA HOTLINE, 17
Ostwald viscometer method, 393
"ous/ic" suffix, 460
Overtones, infrared, 590
Oxidation, 493
Oxidizing agent, 493
Oxygen-deficient atmosphere, 22
Oxygen gas, 82
Oxygen liquid, 83

Packed columns, GC, 528
Packing glands, 370
Packing lubrication, 370
Packing materials, distillation columns, 443
Packing seal, 371
Pan crusher, 230
Paraffins, 467
Parallax error, 498
Parallel circuits, 205
Parent, naming, 467
Pareto charts, 146
Parr bomb method, 309, 311
Partial-immersion thermometers, 374
Particle size distribution, 228
Partition chromatography, 521
Partition rings, distillation, 443
Parts per million (ppm), 490
Pascal, pressure unit, 105
Pascal's law, 108
Pasteur pipettes, 508
Path length, infrared cells, 596
Peak height, chromatography, 542
PEEK, 562
PEL, 25 (*see also* Permissible exposure limits)
Pellicular beads, 559
Percent, by mass, 489
Percent, by volume, 489
Percent, definition, 489
Percent, yield, 488
Percent absorption, 614
Percent relative humidity (RH), 99
Percent transmittance, 614
Percentage, calculation of, 498
Perchloric acid, 294
Periodic table, of the elements, 451–455
Periods, 455
Peristaltic pump, 369
Permanent cells, infrared, 593
Permanent connection, grounding, 213
Permanent threshold shift (PTS), 27
Permanently hard water, 317
Permissible exposure limit (PEL), 25, 356
Permit-required confined space, 43
Peroxide bomb method, 299
Peroxide fusion, 299
Peroxides in ethers, caution, 412
Personal protective equipment (PPE), 19, 24, 25, 30, 36
Pestle, and mortar, 283
pH analyzers, process, 338
pH measurements:
  acid and bases, 169
  Arrhenius theory, 169
  calculations, 170
  defined, 170
  electrode, 180
  indicators, 176
  scale, 170
  theory, 171
  titration, 183

pH meters, 179, 338
　buffer standardization of, 181, 189
　ion-selective electrodes, 192
　procedure with, 176
　troubleshooting, 340
pH test paper, 178
pH values, common acids, 172, 176
pH values, common bases, 172, 176
Phase separators, extraction, 414
Phases, liquid, 535
Phenol, 472, 473
Phenolphthalein, indicator, 175
Phenyl group, 472
Photocells, 592
Photodetector, 610
Photometer, 584
Photomultiplier tube, 585
Physical change, 373
Physical properties, 373–406
　boiling points, 381
　bulk modulus, 405
　compressibility, 405
　density, 385
　flash point, 382
　hardness, 406
　measuring, 8
　melting point, 377
　optical rotation, 399
　refractive index, 396
　specific gravity, 388
　surface tension, 395
　viscosity, 390
Physical properties of water
　(see Appendix B, 627–628)
Pipe fittings, 355, 358
Pipe joints, 355
Pipe sealants, 354
Pipe thread size, 355
Pipe thread taper, MPT, 355
Pipes, 355–356
Pipette, volume tolerances, 505
Pipette squeeze bulbs, 506
Pipettes, 22, 504–509
　to contain (TC), 504
　to deliver (TD), 504
Pipettes, automatic cleaner, 501
Pipettes, electronic, 520
Pirani gauges, 119
Pistol, drying, 289
Piston, pump, 368
Pixel (see Glossary, 629–643)
Placards, ANSI, 59
Placards, CE, 65
Placards, DOT, 51, 63
Placards, hazardous materials, 315
Placards, NFPA, 54
Planar liquid chromatography, 522
Planck's equation, 590
Planimeter, 542
Plates, distillation, 429, 442

Plates, theoretical, 538
Platinum crucible, 249
Plumbing, 345
Plunger, pump, 368
Pneumatic, 368
pOH, 171, 174
Poise, 390
Poisons, 21
Polar, 257, 535
Polarimeter, 401
Polarimeter tubes, 402
Polarimetry, 399–404
Polarity:
　liquid phases, 535
　solute, in gas chromatography, 537
　solvent strength, 258
Polarity, solvent, 258
Polarized light, 400
Polarizer (Nicol prism), 400
Policeman, rubber, 242
Polyatomic ions, 461
Polychromatic radiation, 573–578
Polymers, 479–480
Polyprotic acids, 463
Porapak®, 535
Porous resin beads, 559
Portable tanks, 63
Positive displacement pump, 367
Positive ions (cation), table of, 459, 460
Positive pressure fit check, 34
Potassium bromide pellet technique, infrared, 598
Potassium chloride, 184, 340
Power, electric, 206
Power, exponential numbers, 121
Powered industrial truck, 66
PPE, personal protective equipment, 19, 24, 25, 30, 36
Practical grade chemicals, 313
Pre-column, HPLC, 560
Precautionary statements, 48
Precision, 142, 268
Prefixes, alkyl IUPAC, 468
Prefixes, Greek, naming, 468
Pregl method, 298, 309
Preparation of standard laboratory solutions, 324–331
Pressure and vacuum, 105–120
Pressure equivalent tables, 106, 131–133
Pressure gauge (see Gauges, pressure)
Pressure measurement, factors, 106, 131–133
Pressure measurements, 338
Preventative maintenance, 4
Primary, nomenclature, 473, 475
Primary alcohol, 474
Primary amine, 475
Primary resonance line, 607
Primary standard, 314

Prime, pumps, 364
Prism, 577, 592
Process analyzers, 333–343
　chromatographic, 570
　conductivity, 336
　flow rate measurements, 340
　level measurements, 342
　moisture measurements, 342
　pH, 338
　pressure, 338
　spectroscopy, 604
　temperature, 333
Process distillation, 442
Process operator, 1
Process plumbing, 345–371
Process pumps, 364–371
Process sensors, 333
Process valves, 88, 361
Products, 483
Professional sources of information, 15–17
Programming temperature, 529
Progressive cavity pump, 367
Proportioning valve, HPLC, 558
Protective apparel, 28
Proton, 452
Prudent Practices for Disposal of Chemicals from Laboratories, 12
Prudent Practices for Handling Hazardous Chemicals in Laboratories, 12
Psychrometer, 99
Pulleys, 210
Pump maintenance guidelines, 369
Pumps:
　centrifugal, 364
　flexible vane, 366
　gear, 366
　general maintenance guidelines, 369
　HPLC, 557
　kinetic, 364
　maintenance, 369
　mechanical, 109
　metering, 369
　Moyno®, 367
　peristaltic, 369
　pneumatic, 368
　positive displacement, 367
　progressive cavity , 367
　reciprocating, 364, 367
　rotary, 366
　screw-type, 367
　sliding vane, 366
　slip, 364
　vacuum, 109
　vapor-diffusion, 111
　variable-frequency drive, 369
Purity of chemicals, 313
Pyknometer, 388
Pyrolysis, 519
Pyrometer, 335

Quadratic equation, 125
Qualitative analysis, chromatography, 540, 563
Qualitative fit check, 33
Quality control charts, 146
Quality control statistics, 143
Quartering samples, 226
Quartz, 585
Quaternary amines, 475
Quick couplers, gases, 91
Quick press, infrared, 598

R phrases (R), 48
Radar, level measuring devices, 342
Radiant power, 581
Radiation and toxicity warning labels, 315
Rail tank car, 64
Rain water, 317
Rankine temperature scale, 373
Raschig rings, packing, 427, 443
REACH, 48
Reactants, 483
Reaction assemblies, 156–158
Reactivity hazards, NFPA, 53, 55–58
Readability, 268
Reagent analyzed, 314
Reagent grade chemicals, 314
Reagent-grade water, 319
Reciprocating pump, 367, 558
Recorder-integrator, chromatography, 531
Recrystallization, 257–266
Recrystallization of a solid, 259
Redox reactions, 493
Reducing agent, 493
Reduction/oxidation reactions, 493
Reduction ratio, 228
Reference beam, 610
Reference electrode, Ag/AgCl, 338
Reflux, total, 429
Reflux condensers, laboratory, 155, 159
Refluxing, 439
Refluxing ratio, 429
Refraction of light, 573
Refractive index, 396–399
Refractive index detector, 568
Refractometers, 397
Registrar accreditation board (RAB), 142
Regulating valves, 89
Regulations, governmental, 10–17
Regulator, gases, 84
Regulator, troubleshooting, 87
Regulator safety, 84–87
Regulators, compressed gases, 84
Relative deviation, 144
Relative error, 143
Relative humidity, 99
Relative standard deviation, 145
Relief valves, 90, 91, 99, 363
Removal of interfering substances, 303

Replusion-induction motors, 210
Representative sample, 219
Reproducibility, 142
Residue, 419
Residuum, 472
Resistance, 200
Resistance temperature detectors (RTD), 336
Resistivity, 336
Resistivity, aqueous solutions, 337
Resistivity, electric metals, 202
Resistor, 202
Resistors, color code, 202, 203
Resolution, gas chromatography, 536
Resonance, 607
Resonance wavelength, 607
Resource Conservation and Recovery Act (RCRA), 15, 72
Respiratory protection, 30
Respiratory protective equipment (RPE), 30
Respiratory safety:
    care and maintenance of, 35
    chemical cartridge respirators, 31
    color coding, 32
    dust masks, 31
    filtering face pieces, 31
    fit checks, 33
    self-contained breathing apparatus (SCBA), 34
    supplied air respirators, 34
Response factors, chromatography, 544
Retention time, 540
Reverse osmosis, 318, 319
Reverse phase chromatography (RPC), 555
Reynolds numbers, 360
Rhe, 390
Rheology, 404–406
Risk assessment, 23
Riffle, 225
"Right to Know" law, 12
Ring, support, 160
Risk and safety statements, 48
Risk assessment, 23
Rivers and Harbors Act, 14
Role of chemical laboratory technician, 6
Role of chemical process industry worker, 1
Rolling and quartering samples, 227
Roman numerals, naming, 460
Rotameters, 340, 539
Rotary evaporator, 447
Rotary pumps, 366
Rotating-concentric cylinder viscometer, 391
Round braid packing, 371
Rounding rule, mathematics, 124
Rubbing alcohol, 474
Run charts, 146

S phrases (S), 48
Safety bottle, 91
Safety cans, 61
    Type I, 62
    Type II, 62
Safety, chemical plant and laboratory, 19–45
Safety, health, and environmental standards (SHE), 2
Safety data sheets (SDS), 47
Safety equipment, laboratory:
    gloves and protective garments, 28
    safety bottle trap, 246
    safety eyewash fountain, 39
    safety glasses, 25, 26, 431
    safety goggles, 26
    safety helmets, 30
    safety shields, 26, 431
    safety showers, 39
Safety release valves, 90, 91, 99, 363
Safety rules, plant and laboratory, 19
Safety traps, 246, 432
Salting out, 413
Sample cells, 584, 587
Sample injection port, GC, 526
Sample valve and loop, GC, 527
Sampling, 219–231
    coning of, 226
    crushing of, 282
    dissolving and decomposing, 293
    drying, 286
    of gases, 220
    grinding of, 282
    handling of, 7, 281
    homogenous solids, 225
    of liquids, 223
    of metals, 227
    moisture in, 284
    nonhomogenous solids, 226
    notebook, logbook, 281
    pretreatment of, 281
    quartering of, 226
    rolling of, 226
    rules for, 220
    screening of, 228
Saturated hydrocarbons, 467
Saturated solutions, 320
Saturation, atomic absorption, 614
Saybolt universal viscosity (SSU), 394
Saybolt viscometer method, 391
Scheduled number, pipe, 356
Schellbach burette, 510
Schöninger oxidation, 298, 309, 311
Scientific calculators, 121
Scientific notation display, 122
Scoops, sampling devices, 20
Scott air pack, 35
Screen analysis, 228
Sea water, 319
Sealants, pipe, 354
Sealless (magnetic) pumps, 371
Seals, HPLC leaking, 563
Seals, pumps, 370
Secondary alcohol, 473
Secondary amines, 485
Seebeck voltage, 333
Seeding crystals, 263

Segregation Table for Hazardous Materials, DOT, 14
Selective ion calibration, 193
Self-contained breathing apparatus (SCBA), 34
Self-priming pumps, 366
Semipermanent cells, infrared, 593
Sensitivity, atomic absorption, 613
Sensitivity, balances, 269
Separatory funnels for extraction, 408
Septa, 527
Septum, 527
Series circuits, 205
Shewhart, 146
Shields, safety, 26
Shipping and handling chemical materials, 47–73
Shoes, safety, 28
Short-term exposure limits (STEL), 22, 72
Showers, safety, 39
SI base units, 123
SI conversions to engineering units, 131–133
SI prefixes, 123
SI system, metric, 122, 575
Siemens, 201, 319, 336
Sieve shaker, 229
Sieves:
    Alternate, 228
    Dimensions, 229
    Standard, 228
    Tyler, 228
Significant figures, choosing correct balance, 278
Significant figures, definition of, 123
Signs, safety, 42, 44, 51, 54, 59, 60, 63, 65, 68, 71
Signs and symbols, common mathematical, 127
Silanized, 535
Silencers, compressors, 99
Silica gel (silicic acid), 99, 522, 555, 569
Silicone grease, 152
Silver chloride windows, infrared, 594
Simple distillation, 419
Sintered-glassware, 233, 237
Size exclusion chromatography, 570
Sizes, glass tubing, 164
Sliding vane pumps, 366
Slip, pumps, 364
Slits, 583, 592
Slope, graphing, 126, 146, 544
Slurries, 367
Slurry, thin-layer chromatography, 523
Small-bore tube viscometer method, 391
Small-quantity generators (SQG), 73
Snyder columns, 417
Sodium hydrogen carbonate, 409
Sodium hydroxide, preparation of carbonate-free, 324
Sodium peroxide fusion, 295
Sodium vapor lamp, D line, 396, 401, 574, 577
Soft glass, heating caution, 20
Softening hard water, 317
Solder, alloys, 358
Soldering, electrical, 212
Soldering, plumbing, 358

Solderless connectors, 211
Solid-gas filtration, 233
Solid-liquid filtration, 233
Solid phase extractors (SPE), 414
Solid samples, infrared, 597
Solid state membrane, 194
Solid supports, 535
Solubility:
    complex ions, 320
    curve, typical, 266
    discussion of, 320
    gases in water, 93
    inorganic compounds, 320–322
    organic compounds, 257–258
    rate of solution, 321
    rules, 320
Solute, 320, 488
Solute classification in gas chromatography, 537
Solution rate, increase of, 321
Solutions:
    concentration expressions, 489
    discussion, 320, 488
    molarity, 491
    normality, 492
    parts per billion, 490
    parts per million, 490
    percent, by mass, 489
    percent, by volume, 489
    percent by weight/volume, 489
    preparation of, 320–331
    primary standards, 314
    saturated, 320
    standard laboratory, 324–331
    supersaturated, 320
Solvency, 257
Solvent, 320
Solvent delivery system, HPLC, 557
Solvent mixing system, HPLC, 557
Solvent pairs, 259
Solvent polarity chart, 258
Solvent pumps (*see* Pumps)
Solvent requirements for recrystallization, 257–259
Solvents, defined, 324, 488
Solvents, extraction, 410
Solvents, ultraviolet and visible, 586
Sorbed water, 285
Sorbents, chromatographic, 554
Sources of information, 15–17
Soxhlet extraction, 407
Spacers, infrared, 594
Spatulas, scoops, sampling devices, 20
Special hazards, NFPA, 53
Specific conductance, 317, 319
Specific gravity, 388–389
Specific gravity and viscosity table, 395
Specific gravity conversions, 389
Specific heat (*see* Glossary, 629–643)
Specific refraction, 397
Specific resistance, 319
Specific rotation, 401

664 INDEX

Spectra, band, 580
Spectra, continuous, 580
Spectra line, 580
Spectrograde solvents, 314, 561
Spectrographs, 579
Spectronic 20® spectrophotometer, 587
Spectrophotometers, 579, 583
   conversion from visible to ultraviolet, 583–584
Spectrophotometric analysis, 581–613
Spectrophotometry, 587
Spectroscopic grade chemicals, 314, 555
Spectroscopy, 573–605
   infrared (IR), 590
   near infrared (NIR), 601
   theory, 573
   ultraviolet (UV), 575, 583
   visible, 573, 583
Spiking, chromatography, 540
Spills, containment, 69
   acids, 70
   action, 23
   alkali, 71
   mercury, 72
   pillows, 70
   reporting, 69
   solids, 70
   types, 21
   volatile/flammable solvents, 71
Spiral rings, distillation, 443
Splash goggles, 26
Splash shields, 26
Splicing wires, 211, 212
Split field image, 403
Split-phase motors, 210
Splitting a drop, titration, 513
Spontaneous combustion, 384
Spotting in thin-layer chromatography, 526
Spraying reagents, TLC, 525
Square braid packing, 371
Stability triangle, forklift, 67
Standard, primary, 314
Standard additions method, 547, 615
Standard atmospheric pressure, 105
Standard conditions, gases, 105
Standard deviation, 144, 268
Standard hydrogen electrode (SHE), 192
Standard laboratory solutions, preparation of, 324–331
Standard operating conditions and limits (SOCL), 137
Standard operating procedure (SOP), 3, 124, 137–148
Standard solution, 497
Standard solutions, acid and bases, 324
Standard temperature and pressure (STP), 107
Starter, magnetic, electric motor, 210
Starter, manual, electric motor, 208
Static electricity, balance use, 269
Stationary phases, 521, 554

Statistical process control (SPC), 147
Statistics, 142
Steam distillation, 435–439
Steam generator, 438
Stedman column, 155
Step ladder, safety, 40
Stereoisomers, 471
Steric exclusion chromatography, 570
Stirrer, electric, 441
Stoichiometric equation, 516
Stoichiometric relationship of compounds, 517
Stoichiometry, 483, 516
Stoke, viscosity, 394
Stopcocks:
   frozen, loosening of, 167
   holding, 510
   lubrication of, 151
   removal of, 168
   use of, 223, 409, 435
Storage of liquids, limits, 61–64
Storing chemical materials, 61
Straight-chained hydrocarbons, 467
Straight run gasoline, 472
Strong acid/strong base titration, 184
Strong acid/weak base titration, 189
Strong electrolytes, 336
Structural formulas, 457
Subatomic particles, 452
Sublimation, 292, 447
Suction flask adapter, 153, 249
Suction head, 364
Suction lift, 364
Sugar solutions, specific rotation, 401
Sulfur analysis, 310
Sulfuric acid, 20, 94, 294
Sulfuric acid and water, safety, 20
Supercritical fluid, 569
Supercritical fluid chromatography (SFC), 569
Superfund, EPA, RCRA, Hazardous Waste Hotline, 16
Superfund amendments and reauthorization act (SARA), 11
Superheated steam, 439, 440
Supernatant liquid, 244
Supersaturated solutions, 320
Supplied-air respirators, 34
Support bases, 162
Support coated open tubular (SCOT) columns, 528
Support rings, 160
Support stands, laboratory, 162
Suppressor column, 568
Surface active materials, 396
Surface tension, 395–396
Suspensions and colloids, 321
Swagelok® compression fittings, 351
Symbols, commonly used mathematical, 127
Symbols, electrical, 217
Symbols, elements, 452–453
Symmetrical ethers, 475

INDEX **665**

Synthesizing compounds, 10
Syringe-handling techniques, 534
System International (SI), metric system, 122, 123, 575

Tagout/Lockout, 46
Tailing effects, GC, 535
Tapered pipe tread, 355
Tapers, glass joints, 149
Taping, electrical, 213
Taping pipe, Teflon®, 354
Target, 147
TC, to contain, pipette, 504
TD, to deliver, pipette, 504
Technical grade chemicals, 313
Teflon®, synthesis, 480
Teflon® tubing, 346
Telemetry, thermistor, 334
Temperature measurements, 373
Temperature measuring, 373
Temperature programming, 529
Temperature scale conversions, 374, 375
Temperature scales, 375
Temperature sensors, 333
Temporary fume hood, 97
Temporary hard water, 317
Tertiary alcohols, 473
Tertiary amines, 475
Tesla coil, 113
Tetramers, 479
Theoretical plates, 538
Theoretical yield, 488
Thermal conductivities of gases, 530
Thermal conductivity detector (TCD), 530
Thermal cracking, 481
Thermal overload protection devices, 208
Thermistors, 333
Thermocouple, ANSI, color codes, 335
Thermocouple gauges, 120
Thermocouples, 333, 335, 593
Thermocouples, types and materials, 335–336
Thermometer corrections, 377
Thermometer wells, 154
Thermometers, 375
  bimetallic, 344
  correction, 377
  filled, 333
  liquid filled, 334
  partial-immersion, 374
  total immersion, 374
Thermoplastic, 480
Thermosetting, 480
Thief, probe sampler, 225
Thiele tube, melting point, 378
Thimble, extraction, 408
Thin-layer chromatography (TLC), 522
  developing tank, 524
  edge effects in, 526
  plate activation, 526
  plate preparation, 522

Thin-layer chromatography (*Cont.*):
  sandwich chamber in, 525
  spotting in, 526
  spraying reagents, 525
Three-phase motors, 208
Threshold limit values (TLVs), 22
  ceiling (C), 22
  short-term exposure limit (STEL), 22
  time-weighted average (TWA, PEL, WEL, REL, MAK, etc.), 22
Time-weighted average (TWA), 22, 72
Titrants, splitting drops of, 513
Titration error, 497
Titrations, 494, 497, 510
  acid-base, curves, 183–184
  complexometric, 517
  curves, 513
  electronic, automatic, 520
  Karl Fischer, 519
  oxidation-reduction, 515
  pH, 183–189
  splitting a drop, 513
  strong acid/strong base, 184
  strong acid/weak base, 189
  weak acid/strong base, 186
TLC, thin layer chromatography, 522
TMU, ton multi unit, 64
Toluene, 472
Top-loading balance, 277
Torr, pressure, 106, 111
Torsion balance, 277
Total immersion thermometer, 374
Total oxidizable carbon (TOC) analyzer, 304
Total quality management (TQM), 1
Total reflux distillation, 429
Totes (*see* Chemtotes), 63
Toxic chemicals, 21
Toxic exposure, 22
Toxic Substances Control Act (TSCA), 138
Tracing of pipe systems, 356
Tracking and inventory, chemicals, 60
Trans/cis isomers, 471
Transfer, precipitate, 240, 242, 252
Transfer pipette, 504
Transferring liquids:
  carboys, 61
  chemtotes, 63
  drums, 63
  highway tankers, 64
  pipettes, 504
  pumps, 69
  rail tank cars, 64
  safety cans, 61
Transition, pipe flow, 359
Transition elements, 455
Transmittance (T), 581
Transporting chemical materials, 61–68
Trap bottle with water aspirator, 110, 246
Traps, vacuum flask, 246, 251

Traps, water, 91, 246
Traps and check valves, 91
Trays, distillation, 442
Triangulation, chromatography, 542
Trimers, 479
Triols, 474
Triple-beam balances, 271
Triple bonds, 470
Trisodium phosphate, 317, 502
Trivial names, 464
Troubleshooting, GC, 548
Troubleshooting, HPLC, 563
Troy, conversions, 129, 133
Tswett, Mikhail, 521
Tube cutters, 346
Tube fittings, 351
Tube trailers, 64
Tubes, capillary, 377
Tubing:
    bending of, 348
    bur removal from, 348
    compression fittings, 351
    connecting of, 350
    copper, 346
    cutting, 163, 346
    flaring, 350
    glass, size chart for, 164
    maximum temperatures, 347
    metal cutting of, 346
    polyvinyl chloride (PVC), 345
    selecting, guidelines, 345
    Tygon®, 345
    uncoiling of metal, 347
Tubing connection, flareless, 350
Tungsten halide lamp, 616
Tungsten inert gas welding (TIG), 356
Turbogrid®, 442
Turbulent flow, 359
Turbulent flow burner, 610
Twisted wire gauge column packing, 427
Two-stage gas regulators, 86
Tygon® tubing, 345
Tyndall effect, 323, 579
Tyvek™, 28

U-tube manometer, 91
Ultrafiltration, 416
Ultrasonic, level measuring devices, 342
Ultrasonic cleaning, 502
Ultraviolet detectors, HPLC, 561
Ultraviolet lamps, 561, 585
Ultraviolet light, 525, 552, 584
Ultraviolet (UV) spectroscopy, 584
"UN" drum identification, 63
Unbalanced equation, 484
Underwriters Laboratories, Inc. (UL), 17, 61
Unit-conversions method, 127
Unit conversions tables (*see* Conversion tables)
Unit-factor method, 127

Unit operations, 483
United States Department of Agriculture (DOA), 60
Universal detector, 530
Universal motors, 210
Unsaturated hydrocarbons, 469
Unterzaucher method, 298
Upper action limit, 146
Upper explosion limit (UEL), 80, 384
Upper warning limit, 146
USP chemicals, 313

Vacuum, 108
Vacuum, laboratory sources, 108
Vacuum desiccators, 288
Vacuum distillation:
    capillary air inlet in, 433
    dry-ice trap in, 432
    nomograph in, 430
    principle, 430
    safety precautions, 431–433
Vacuum filtration, 246, 259, 562
Vacuum filtration apparatus, 246, 251, 562
Vacuum fractionation, 434, 437
Vacuum leaks, Tesla coil, 113
Vacuum measuring devices, 116–120
Vacuum pumps:
    connections to, 113
    diffusion, 111
    maintenance of, 112
    mechanical, 109
    oil for, 111
    troubleshooting, 114
Vacuum rotary evaporator, 447
Vacuum safety, implosions, 223, 431
Vacuum safety traps, 246, 432
Vacuum symbols, 109
Vacuum ultraviolet, 583, 585
Valence electrons, 455
Valves, process, 361–363
    ball check valve, 368
    ball valve, 361–362
    check valve, 89, 363
    direction valve, 89, 363
    flap check valve, 368
    gate valve, 361
    globe valve, 361
    metering valve, 89, 362
    needle valve, 88
    regulating valve, 89
    relief valve, 90, 363
    safety relief valve, 363
Vapor-diffusion pump, 111
Vapor pressure, 419, 420
Vapor pressure of water at various temperatures, 627–628
Vapor trap, dry ice, 222, 432
Variable area flow meter, 340
Variable frequency drive pumps (VFD), 369
Variable-path length cells, infrared, 594

INDEX

Variable transformer, 428
Velocity, 573
Ventilation, 23
Ventilation rates, 23
Venture tubes, 341
Verifying step, 61
Vernier, 272
Very small quantity generator (VSQG), 73
Vigreux column, 155, 427
Viscosity, 390
Viscosity and specific gravity table, 395
Viscosity determinations, 390–395
Viscosity standards, 390
Viscosity units and conversions, 390
Visible spectroscopy, 573, 5 79, 584–589
Visible spectrum, 573, 575, 576
Volatile and flammable solvent spills, 71
Volatility, 258, 419
Volhard method, 310
Volt, 200
Voltage, 199
Voltmeter, 207
Volume, units, 497
Volume displacement pumps, 367
Volume equivalent tables, 128
Volume percent of solutions, 325, 497
Volume relationships from equations, 487
Volumetric analysis:
    back titration, 497, 515
    burettes, 500, 509
    calculations, 513
    chelates and complexometric titrations, 517
    cleaning glassware, 500
    electronic pipettes, 520
    end point in, 497
    equivalence point in, 497
    flasks, 504
    Karl Fischer titration, 519
    meniscus, 497, 498
    micropipettes, 508
    oxidation-reduction titrations, 507
    pipettes, 500, 504
    primary standards in, 514
    standard solution, 497
    titration, 497, 510
    titration curves, 513
    titration errors, 497
    volumetric flasks in, 504
Volumetric calculations, 498–520
Volumetric concentrates, dilution techniques with, 493
Volumetric flasks, 6, 504
Volumetric glassware, cleaning procedure for, 500
Voluntary Industry Skill Standards, 1

Wall-coated open tubular (WCOT) columns, 528
Warning labels, chemicals, 314–315
Wash bottles, 239
Washing, precipitates of, 240, 241, 263

Washing glassware:
    alcoholic sodium hydroxide, 502
    Alconox®, 499
    aqua regia cleaning solution 502
    dichromate-sulfuric acid, 501
    dilute nitric acid, 502
    metal decontamination of glassware, 502
    Nochromix®, 502
    trisodium phosphate, 502
Waste chemical disposal, 72
Water:
    absolute pure, 318
    adsorbed, 285
    of constitution, 285
    deionized, 318
    demineralized, 319
    density (see Appendix B, 627–628)
    distilled, 318
    essential, 285
    ion exchange, 317
    nonessential, 285
    occluded, 285
    permanently hard, 317
    reagent-grade, 319
    by reverse osmosis, 318
    sorbed, 285
    temporary hard, 317
    vapor pressure (see Appendix B, 627–628)
Water aspirator, 108, 109, 110, 240, 246, 446
Water purity specification, 317
Water-softening substances, 317
Water trap, steam, 439
Water vapor pressure at various temperatures, 627–628
Watts, 200
Wave number to wavelength conversions, 592
Wave numbers, 590, 591, 592
Wavelength, 573
Wavelength conversion table, 575
Wavelength/frequency relationship, 576
Wavelength in electromagnetic spectrum, 575
Wavelength versus wave number, 591, 592
Weak acids, 170–172
Weak bases, 170–172
Weak electrolytes, 336
Weighing bottles, 300
Weight, definition, 268
Welding filter lenses, 27, 356
Welding safety, 356
Welding safety rules, checklist, 357
Western Electric rules, 147
Wet-ashing procedures, 295, 299
Wet-bulb temperature, 99
Wet, surface tension, 395
Wheatstone bridge, 119, 336, 530
White light, 573, 574
Wig-L-Bug®, 598

Wind socks, 23
Window materials, infrared, 594
Wing-tip burners, 166
Wire conductors, 201
Wire gauze funnel liner, 248
Wire nuts, 211
Wire sizes, 202
Wires, splicing of, 211–212
Witt plate, funnel, 250
Wood alcohol, 474
Working fluids, 359
Workplace Hazardous Materials Information System (WHMIS), 48
Workstation, chromatographic, 533

X-ray radiation, 575
Xylene, 472

Y connector, 117
Yeast, 482
Yield percent, 488
"yl" suffix, 467
"yne" suffix, 469
Young's modulus, 404
Young's modulus and elastic modulus, 404–405

Zero, balances, 274
Zero control, 587